T0342806

ROGER SWAINSTON is one of the foremost marine-life artists in the world, with more than 35 years of artistic and scientific endeavour. His paintings are renowned around the world for their accuracy and vibrancy.

After training as a zoologist at the University of Western Australia, Swainston spent several years in the Department of Ichthyology at the Western Australian Museum, working on the taxonomy of fishes and the re-cataloguing of the museum's extensive collection of specimens. In the ensuing years he has taken part in scientific surveys of a wide range of environments: from the deep waters off the North West Shelf to the inshore waters of the southern coast of Australia and the coral reefs of Papua New Guinea. As expedition artist with a team of international scientists, he spent two months surveying the fish life of the remote Clipperton Atoll in the far eastern Pacific. Later he worked with the Natural History Museum of Paris and Pro-Natura International, portraying the biodiversity of the marine environments and freshwater streams of Vanuatu. More recently he has accompanied scientists surveying the biodiversity of northern Mozambique and Madagascar.

Swainston has dived, snorkelled and painted fishes right around the coast of Australia and in many other locations worldwide, including the Mediterranean, Red Sea, Caribbean, and areas throughout the Indian and Pacific oceans. For many years he has closely monitored a number of sites on Ningaloo Reef, off Western Australia, spending countless hours under water drawing the reefs and observing and photographing fishes to produce panoramic scenes of the environments. He has also worked extensively with professional fishers around Australia, studying their catches and fishing methods.

His work has been included in many well-known field guides to fishes, including *Sea Fishes of Southern Australia*, *Marine Fishes of North-western Australia*, *Reef Fishes of New Guinea*, *Freshwater Fishes of New Guinea* and *Fishes of the Eastern Pacific*. He has also illustrated major scientific works, such as *Sharks and Rays of Australia* and guides to the Emperors and Snappers of the world, as well as contributing illustrations to numerous other publications. His artworks have been used by numerous scientific institutions, including the Smithsonian Institute, the Food and Agriculture Organization of the United Nations (FAO), Australia's Commonwealth Scientific and Industrial Research Organisation (CSIRO), and the Natural History Museum of Paris, as well as many government environmental and fisheries agencies both in Australia and abroad.

Swainston has held solo exhibitions in Sydney and Fremantle, and in France. His paintings are held by private collectors around the world, and by institutions including the Ministry of the Environment and the Musée des Arts et Traditions Populaire in France, the Oceanographic Institute of Monaco, the National Archives of Australia, and the Western Australian Museum. Several documentaries about his underwater work in the Red Sea and on Ningaloo Reef have been widely broadcast internationally.

Roger Swainston presently lives and works in Fremantle, Western Australia, with his partner Catherine and their three children.

For more information visit:
rogerswainston.com
anima.net.au

SWAINSTON'S FISHES *of* AUSTRALIA

Roger Swainston

VIKING
an imprint of
PENGUIN BOOKS

MERCI CATHERINE

CONTENTS

Contents

Contents

FOREWORD

Over many years I have had the opportunity to see and handle an enormous variety of fishes; always looking at them with one eye as a scientist, the other as an artist. At times I have illustrated other subjects, but I always return to fishes and remain captivated by their astounding diversity and beauty.

The illustrations in this book are selected from nearly 6000 paintings of fishes I have created over the past 25 years. All are executed in acrylics on watercolour paper. Each is a synthesis of all the information I have been able to find, visual or otherwise, on a particular fish. Ideally this means photographing a fish while it is still alive, with all its vibrant natural colour; keeping a specimen for a detailed, measured drawing; and completing the painting in the studio. This process is of course not always possible, and some paintings have had to be conjured from a few poor photographs, a faded preserved specimen, or simply a detailed description.

In my travels over the years, finding the local fish market or the port where boats land their catch has always been a priority. I have seized any chance to look at fishes, photograph a specimen, or get a feel for regional differences in the fauna – always with the anticipation of discovering something new or previously unseen. I have dived and snorkelled and accompanied scientists and fishers, working in every type of marine and freshwater habitat, from the ocean depths to still jungle pools.

What drives my fascination with fishes? A Longtom shimmering in crystal-clear tropical waters could be a pure distillation of light, water and colour; a perfectly camouflaged Stonefish could have spontaneously generated from the lumpy, algae-covered rocks themselves. Fishes are form and function and habitat made flesh; they display the most extraordinary responses to the constraints of the underwater world. Two species can be almost identical in form, swimming side by side in the freezing darkness thousands of metres below the surface, and one is jet black, the other completely transparent. There are fish that can fly, others that walk on land at the edge of the sea, and even some that move mysteriously through the total darkness of underground streams in fissures beneath the earth. Still others excavate and live in burrows; cultivate algae farms; work in groups to herd rushing shoals of prey; or lie motionless for weeks, awaiting the moment to explode from the sand and engulf a passing meal.

The diversity of fishes is endlessly intriguing, and every colour imaginable can be found in their beauty. Yet the more I study fishes, the more I come to realise how little we really know about them. Apart from a few commercially important and much-studied species, the vast majority of fishes are out of sight and thus out of mind, their lives and habits unknown to us.

Fishes are now under intense pressure around the world. Fishing has become industrialised and technologically advanced, and many species have been exploited to the verge of total population collapse. Although overfishing is occurring in Australia, we are fortunate perhaps that it has not yet progressed to such a sad extent as in other parts of the world – and we still have a chance to ensure that it will not do so. However, there is a lot of work that needs to be done and some difficult decisions to be made before the fishes of Australian waters can be considered safe from overexploitation. There is room for a great deal of improvement in the way we target and exploit fishes, and our fishing methods and strategies must become much more refined. This is not so that we can take even more from a diminishing resource, but so that we can build a much larger resource and take a smaller, sustainable percentage of it.

During the past three decades I have seen the numbers of many fish species gradually decrease in Australian waters. We can now only imagine what must once have been: the multitudes that teemed in our oceans, before a century or more of ever-increasing fishing effort. We must turn and begin on the path back towards that natural state. Across society the realisation is steadily growing that strong measures must be taken to restore balance in our waters and rebuild the diversity and abundance of fishes.

Fishes are fascinating and beautiful creatures, and they have sustained humankind since prehistoric times. They deserve our respect and protection. We must ensure that our children and grandchildren can experience the same extraordinary diversity and abundance of fishes as our grandparents. We must learn more, take less, and cherish and protect our natural environment; everyone has a part to play in this.

There is so much we have yet to discover about these remarkable animals. Great cycles, migrations and movements are constantly in play beneath the surface of the waters. Imagine this: every evening around the globe, as the sun sets, countless billions of small luminous fishes, many of them still unknown to us, rise from the ocean depths in a glowing constellation towards the surface to feed. As the world turns and the sun rises they sink into the darkness again.

I wish I could see it. Perhaps there is a way . . .

INTRODUCTION

ABOUT THE BOOK

Australia is an island continent, blessed with an extraordinary array of aquatic environments. Our waters stretch from the warmth of near-equatorial tropical shallows to the freezing currents around sub-Antarctic islands. In between lie coral reefs, deep ocean trenches and seamounts, flooding monsoonal rivers, isolated desert springs and vast, hypersaline coastal bays. In all of these environments, even the most extreme, there are fishes – displaying an astounding variety of anatomical and behavioural adaptations.

There are more than 4500 fish species presently known to occur in Australian waters and the purpose of this book is to provide, for the first time, a complete overview of this remarkable fauna. Artworks take pride of place here, portraying the extraordinary diversity and range of adaptations fishes have developed to survive in even the most hostile environments. The wealth of visual material will hopefully allow the reader to place any fish into its family, and the accompanying text will offer some understanding of how it lives and where it fits in the broad spectrum of the fauna.

In this book, families are grouped into their orders. A brief introduction to each order gives, wherever possible, the characteristics that unite the families within the order. These characteristics are not always obvious, or even consistent across the order; some are complex internal structures, and they are not always included here. Each family entry (Fig. 1)

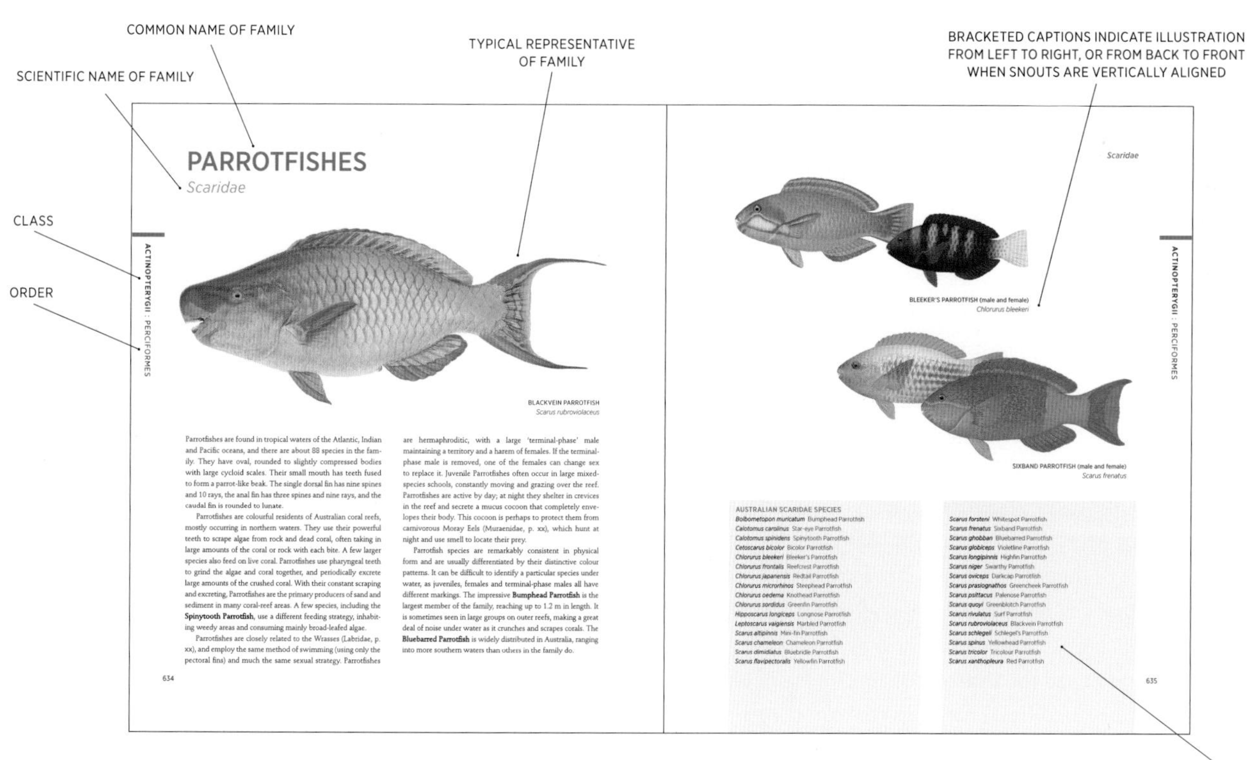

Figure 1 : A typical family entry

shows the group's scientific and common name, and the text provides a brief outline of the defining physical characteristics of the family, its distribution, and as much information on its diversity and biology as is feasible. Sometimes the amount of information given is constrained by space and sometimes by availability. There are many families, particularly of deepwater fishes, about which very little is known other than their physical characteristics and whatever inferences can be drawn from them. Other families are well known and so diverse that it is impossible to impart, in the available space, all of the adaptations and life strategies they have developed. A typical representative of each family is pictured, followed by illustrations of other members of the family. These have been chosen, where possible, to show the variety and range of adaptations apparent within the family. Though the additional illustrations are not exactly to scale, they give an indication of the relative differences in size between fishes within each family.

The exact number of fish species and families is in a constant state of flux. New species are discovered, relationships between groups are revised, and families are merged or separated as new information is uncovered. There are also differing opinions amongst scientists on how species and families are classified. In this book, the sequence and grouping of families into orders is (with a few minor differences) taken from J. S. Nelson's *Fishes of the World* (2006), an indispensable and rigorous overview of all the fish families of the world.

The language here has been kept simple and easy for a layperson to understand, while at the same time introducing those scientific terms that are necessary in order to keep the text concise, precise and unambiguous. A glossary at the back of the book (p. 787) explains all such terms.

For each family, there is a list of all species known to occur in Australian waters. This information is adapted from *Standard Names of Australian Fishes* (Yearsley, Last and Hoese 2006), a publication that provides a complete set of common and scientific names for every Australian fish. Many readers may disagree with some of these (as indeed I do). However, a set of consistent names that can be applied across all states and territories is long overdue. Common names such as Queenfish and Sweetlip are currently used for a variety of species (sometimes from different families, in different regions of Australia), while a single fish may be known by a dozen different names around the country. While this continues, confusion remains.

EVOLUTION & GEOLOGICAL HISTORY

Fishes are the most ancient, diverse and numerous of all the vertebrates. Modern fishes can be divided into three main groups: the jawless fishes or Agnathans, including the Hagfishes (class Myxini, p. 25) and Lampreys (class Petromyzontida, p. 31); the cartilaginous fishes or Chondrichthyans (class Chondrichthyes, p. 37), comprising the sharks, rays and chimaeras; and the bony fishes or Osteichthyans (classes Actinopterygii, p. 141 and Sarcopterygii, p. 783), containing all other forms.

The earliest known fish fossils date back about 500 million years. At that time all fishes were jawless but had already developed bone, and the different forms that gave rise to modern jawed and jawless fishes were already apparent.

About 400 million years ago, the landmasses of Australia, New Zealand, Africa, South America, Antarctica and India were joined together to form the ancient southern super-continent of Gondwana. Much of present-day Australia was at that time covered by shallow lakes and seas, with vast fringing reefs of calcareous algae and ancestral forms of coral. For the next 50 million years, during the Devonian era (sometimes known as the Age of Fishes), there was an explosion in the evolution and radiation of fishes, followed at the end of the period by mass extinctions. Abundant fish fossils from these Devonian reefs have been found in Australia and on other continents.

Many different fish groups evolved in the shallow seas around Gondwana. The ancestors of modern Chondrichthyans diverged into two main lineages: the Holocephali (chimaeras), and the Elasmobranchii (sharks and rays). The Elasmobranchii further diverged into the Selachii (sharks) and the Batoidea (rays). Ancestral bony fishes also diverged into two main lineages: the Actinopterygii (ray-finned fishes), and Sarcopterygii (Coelacanths and Lungfishes). The ancestors of all other four-limbed vertebrates also evolved from this latter group.

The most successful group of all the modern fishes, the Teleosts or spiny-rayed fishes, developed from the Actinopterygii about 250 million years ago and has since evolved into myriad forms and spread to every aquatic habitat where fishes are found.

Gondwana began to fragment and drift apart about 150 million years ago, with Australia, New Zealand and Antarctica breaking away as a single block. New Zealand split off about 80 million years ago. Australia began to slowly rift and separate from Antarctica about 65 million years ago, but for a long time the mainland remained attached via a land bridge to Tasmania.

The Australian continental plate continued imperceptibly drifting northwards, taking with it the ancestors of modern fishes, and Australia has been completely isolated from all these other southern landmasses for more than 45 million years. This has been ample time for Australian waters to develop a large number of unique species. Only a very few species, such as the Australian Lungfish (p. 786), have persisted, virtually unchanged, since the fragmentation of Gondwana. Fossils very similar in form to today's Australian Lungfish have been found in both South America and Australia, clearly illustrating that it originated on this ancient supercontinent.

By about 30 million years ago, Australia had drifted into tropical waters, allowing coral reefs to develop around the northern coast. These reefs have existed – expanding, contracting and evolving with changing sea levels – ever since. Continuing on its inexorable journey northwards, Australia collided with the South-East Asian continental plate about 20 million years ago, and the fish faunas of these two regions became mixed. To the north of Australia, the Indo-Australian archipelago (formed by this collision) is an area of great and complex ongoing tectonic activity, surrounded by three major bodies of water: the North Pacific, South Pacific and Indian oceans. Changing sea levels and tectonic movements have continuously isolated and merged fish populations in this region, resulting in immense diversity. The epicentre of the diversity now appears to be located around the western tip of Papua New Guinea. About 1500 species of reef fish may be found in this area, and northern Australian waters share some of this wealth of marine species.

Moving outward from the centre of diversity into the Indian and Pacific oceans, and down the west and east coasts of Australia, fish diversity gradually decreases, while the percentage of endemic fish species increases. In tropical Australian waters as a whole, only about 13 per cent of fish species are endemic, reflecting the constant mixing of fish populations in the region. However, along the south coast of Australia – isolated for many millions of years, geologically and by converging currents – the number of endemic species reaches a peak, with approximately 85 per cent of all coastal fishes occurring there found nowhere else in the world.

THE MOVEMENT OF THE WATERS

Water movement is a major factor influencing present-day fish distribution and diversity. Currents transport nutrients, carry the pelagic larvae of some fishes far from their origins, and control the water temperature, salinity and other conditions of all marine habitats. In the Pacific and Indian oceans, the surface waters rotate ponderously in an anticlockwise direction (Fig. 2, p. 6). Where the southern edges of these oceans touch the circumglobal Southern Ocean, they mesh with its great eastward flow of water.

Oceanic water flows move not only in a horizontal plane; there are also great vertical circulations driving water flow around the world's oceans (Fig. 3, p. 7). A number of factors influence these circulations. Salt water is denser and heavier than fresh water. Warm water is less dense and lighter than cold. In the tropics, high rainfall and heat from the sun create a less dense layer of low-salinity warm water at the surface, which flows outward from the equator. Around Antarctica in winter, the surface waters are cold, highly saline (due to lack of rain, and to ice formation removing fresh water) and therefore heavy. These waters sink down the continental slope and flow outward, creating a uniform layer covering vast areas in the deep oceans. In summer, as Antarctic ice melts, another layer of slightly less cold, less saline water flows outward above the dense saline layer. Because there is no surface turbulence to mix these two layers, they flow down the continental slope and spread out horizontally over the abyssal plains – the two cold, dense layers beneath the warmer, lighter water above. These distinct layers can be clearly defined in terms of depth by their different temperatures and salinity. The continuity of these discrete layers means most deepwater fish species inhabiting them are widely dispersed around the world in what could be thought of as globally continuous habitat.

These cold bottom waters have an important role in bringing nutrients into coastal waters. They collect nutrients in their slow passage across the abyssal plains, and when they encounter the Australian continental plate they are forced upward to mix with and displace coastal waters of widely differing temperature, salinity and nutrient content. Prevailing winds play a part in the cycle by driving masses of water at the surface: for example, wind can push warmer, nutrient-poor water at the surface away from an area, allowing cold, nutrient-rich water to well upwards from beneath.

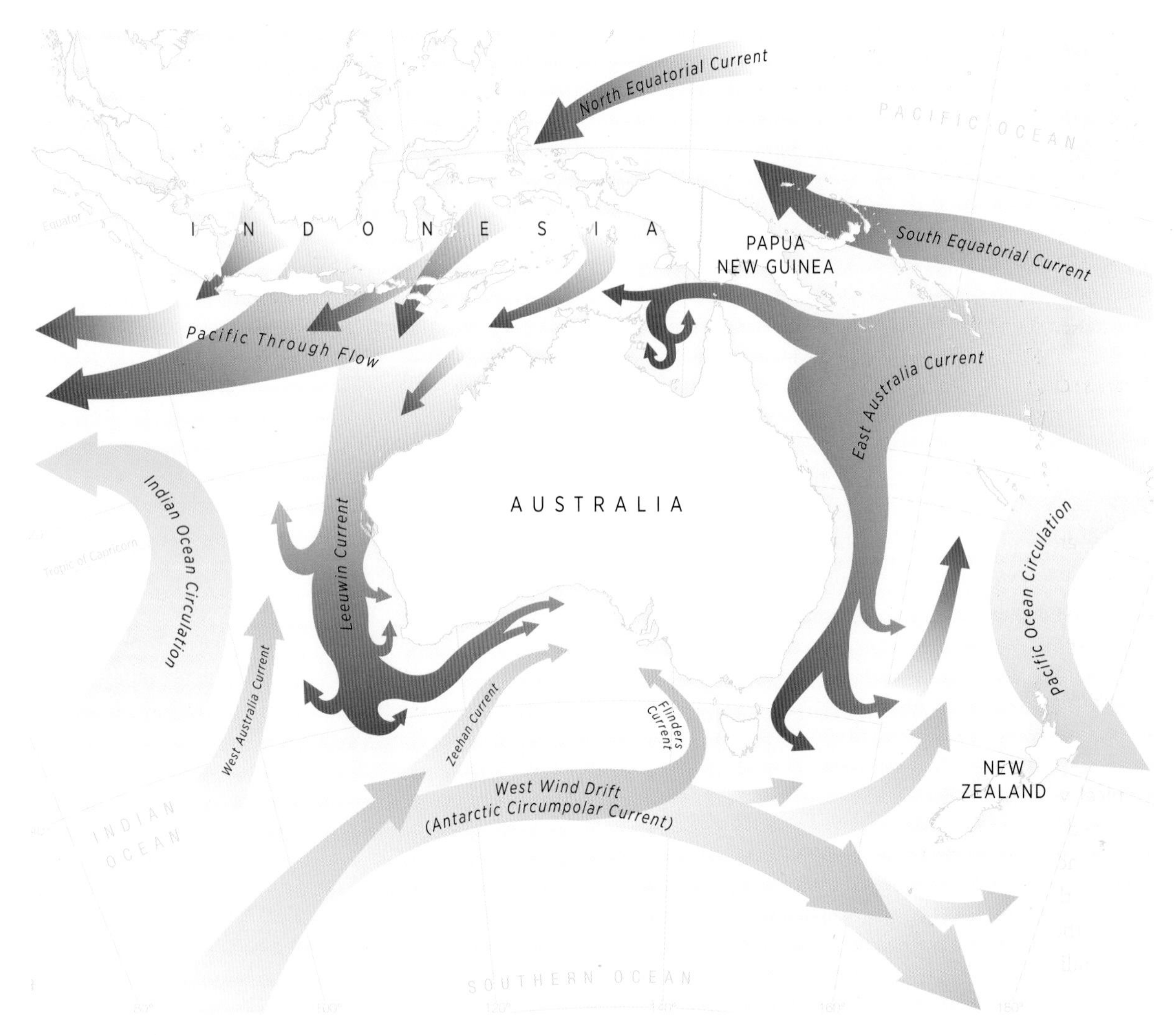

Figure 2 : Ocean currents around Australia : cold currents are shown in blue and warm currents in red.

To the north-west of Australia, water movements are complex. Strong currents flow from the North Pacific through the Indonesian archipelago carrying enormous amounts of water into this area (calculated to be up to 10 million tonnes per second), and monsoon winds temporarily reverse surface flows. Combined with the Indian Ocean circulation, these forces create upwellings and downwellings, and currents of differing salinity and density, which carry water away from the area in several different directions.

Pushed by the flow of water through Indonesia, the Leeuwin Current originates in the triangle of ocean between Indonesia and north-western Australia. It flows southward, carrying warm, nutrient-poor water down the west coast during autumn and winter, moving offshore as it progresses. It turns eastward at Cape Leeuwin and finally dissipates as it moves across the Great Australian Bight. At times the effects of the Leeuwin Current are felt as far east as Tasmania.

Further offshore on the west coast, a weak, cold, nutrient-rich current flows northward, moving closer inshore when the Leeuwin Current flows less strongly. The interaction between these two currents forms huge eddies, at times bringing warm water close to the coast and at others moving it hundreds of kilometres out to sea. The warm Leeuwin Current, flowing southward down the west coast of the continent, is unique.

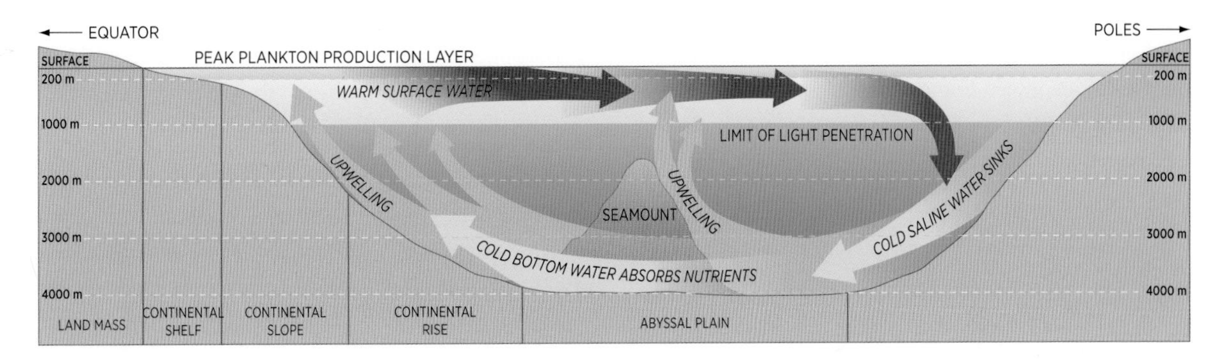

Figure 3 : Cross-section of the ocean showing vertical water circulation

South America and South Africa both have cold, nutrient-rich currents flowing northward up their west coasts and as a consequence a far greater biomass of fish in their western waters. The Leeuwin Current maintains warmer, nutrient-poor conditions on the west coast of Australia, causing slower growth rates and a lower biomass of fishes.

On the east coast, the equatorial westward flow of the Pacific Ocean circulation meets the Australian continental plate and is forced southward, forming the East Australian Current. This carries tropical water from the Coral Sea down the coast of Queensland and New South Wales. The flow peaks in summer and autumn, and periodically reaches as far south as Tasmania. Moving further offshore at times, it pushes colder, deeper water close to the south-east coast. At its southernmost reach, the East Australian Current meets cold water flowing northward and is forced out into the Tasman Sea. Some of its water flows back northward and some meanders out into the Tasman Sea towards New Zealand. Occasionally, great masses of warm water are pinched off from the East Australian Current, forming spinning eddies up to 200 km in diameter that can veer and wander independently off south-eastern Australia for up to two years.

Further to the south of Australia, the West Wind Drift – a cold current that circles around Antarctica – flows parallel to the southern coastline. This current, combined with winds from the Southern Ocean and the tail of the Leeuwin Current, creates the weaker Zeehan Current that flows eastward along the continental shelf of the Great Australian Bight.

The Flinders Current branches off from the West Wind Drift and flows northward up the west coast of Tasmania carrying cold, nutrient-rich water. It turns along the south coast and circles back to the south off South Australia. A combination of these currents and prevailing winds causes permanent, seasonal upwellings of nutrient-rich cold water at several locations along the southern coast.

Seasonal and annual variability of all of these currents influences the complex coastal environments and fish distributions around the long coastline of Australia.

PELAGIC FISHES

Pelagic fishes are those that spend their lives in open water, not associated with the bottom; most depend directly or indirectly on the production of plankton. The upper 200 m of the ocean is where primary production by phytoplankton (which consists of microscopic single-celled algae) takes place. The level of production is determined by the availability of nutrients such as phosphates and nitrates, and sunlight.

The nutrients that phytoplankton require for growth accumulate in deep water from the breakdown of organic matter by bacteria – mainly faeces of zooplankton and other animals that falls to the sea floor. Zones of high plankton production occur where these deep, nutrient-rich waters upwell into sunlit conditions. It has been estimated that every year the growth of phytoplankton in surface waters uses 15–18 billion tonnes of carbon, which flows on to every level of the food chain throughout the oceans. The waters around Australia, however, with few upwellings, generally have low levels of plankton.

Areas of upwelling make up about 1 per cent of the area of the oceans, yet provide nearly 50 per cent of the total world fish catch. They usually occur near continental landmasses – further out to sea there is less vertical mixing between water layers, and nutrient levels are lower. In the centre of the great rotating masses of surface water in the Indian and Pacific oceans there is very little plankton at all.

Phytoplankton provide food for herbivorous zooplankton, which in turn support carnivorous zooplankton, invertebrates and fishes. A few pelagic fishes, such as certain Anchovies (Engraulidae, p. 181), strain phytoplankton directly from the water using very fine gillrakers. The majority of planktivorous fishes, though, feed on zooplankton, filtering it from the water as they swim or selectively picking individual animals from the water column.

Zooplankton is composed of minute animals, predominantly crustaceans such as copepods, ostracods, isopods and mysid shrimps. Many other invertebrate groups, including hydrozoans, jellyfishes, ctenophores and chaetognaths, make up part of the zooplankton. Most of these microscopic animals carry out their entire life cycle in the pelagic zone. About 70 per cent of all marine animals have eggs or larvae that spend some time in the zooplankton and these also form an important component.

Phytoplankton and zooplankton grow seasonally and are moved by surface currents across large areas, so pelagic fishes have evolved to be mobile. They either drift or swim constantly in their search for food. Some larger pelagic fishes, such as the Whale Shark (Rhincodontidae, p. 60) and Manta Rays (Myliobatidae, p. 136), feed by swimming steadily and filtering the largest zooplankton from the water. They are often seasonal visitors in waters with high plankton production and may travel very large distances.

Pelagic predators higher up the food chain, such as Tunas (Scombridae, p. 717), follow currents rich in zooplankton and the small fishes, like Flyingfishes (Exocoetidae, p. 332) and Sardines (Clupeidae, p. 184), that feed on it. They are preyed on in turn by larger fishes such as Marlins (Istiophoridae, p. 723) and Mackerel Sharks (Lamnidae, p. 69). These fast-swimming pelagic predators follow long, seasonal migratory routes in open waters. They have the fusiform body and strong, forked tail of constant swimmers. Some of them are among the most highly evolved of all the fishes, having developed circulatory heat-exchange mechanisms to warm their bodies. This allows them to follow their prey into colder, deeper waters than they would otherwise be able to enter, and thus to utilise the rich food resource of deepwater pelagic fishes such as Lanternfishes (Myctophidae, p. 251), which rise from the depths every night to feed in the lower reaches of the plankton-rich layer.

In open waters of the central oceans, further from zones of high plankton production, small amounts of zooplankton remain in the water. Gelatinous invertebrates such as ctenophores and jellyfishes are found here. They expend little energy and are efficient at harvesting the meagre pickings available. The Sunfishes (Molidae, p. 780) and many of the Lampridiforms (p. 255) are solitary pelagic travellers in these far expanses, feeding on the gelatinous prey, and are widely distributed around the world. Other wide-ranging pelagic species such as Driftfishes (Nomeidae, p. 729) use the larger gelatinous invertebrates as protection, living within the stinging tentacles of jellyfishes and siphonophores.

Drifting flotsam such as sargassum weed creates another moving habitat in this open ocean environment, sheltering juveniles of many pelagic fish species. The Sargassum Fish (p. 294), spends its entire life in floating clumps of sargassum and consequently is widespread around the world. Large floating rafts of weed and other debris can travel enormous distances and attract a variety of pelagic fishes, including Triggerfishes (Balistidae, p. 758) and Mahi Mahi (Coryphaenidae, p. 494).

The pelagic world merges with coastal waters, and rare open-ocean fishes such as the Louvar (Luvaridae, p. 701) may occasionally approach the shallows or be washed ashore in storms. Many large, solitary pelagic fishes are known only from such strandings.

DEEPWATER FISHES

More than 70 per cent of the surface of the Earth is covered by a layer of water with an average depth of nearly 4 km. At its deepest point, near the Mariana Islands in the Pacific, the sea floor drops to more than 11 km in depth. Almost all of the energy supply to the deepwater environment comes from the upper 200 m of the ocean, where there is enough light for photosynthesis to take place and where vast amounts of phytoplankton are produced, driving the food chain that supplies energy to the depths.

From the surface of the oceans, temperature and light levels drop rapidly until, at about 1000 m, all is darkness and the water temperature is 5°C or less. Below 1000 m, temperatures continue to fall, though more gradually, reaching around 2°C at about 4000 m on the abyssal plains. This is the realm of deepwater fishes, and the volume of their deepwater environment (below 200 m) constitutes by far the most extensive habitat on the planet. However, deepwater fishes are the least well known of all the fishes, because sampling such volumes, at such depths, is difficult and expensive.

Most of the ocean, then, is cold, dark, immense in volume and low in energy and food. Compared to shallow water, the conditions in the depths are extreme, uniform and very slow to change, enabling the development of extraordinary specialisations and adaptations in fishes that live there. Many of the most highly specialised deepwater fishes are primitive forms that have developed entirely in very deep water over extremely long periods of time. Less specialised are the fishes that have evolved in shallower water and have moved far more recently into the upper layers of the deepwater realm. Although these latter generally show less extreme adaptations to deepwater conditions, they have evolved many of the same modifications as the more primitive fishes, such as reduced skeletons and light-producing organs.

Many deepwater fishes from both these groups produce bioluminescence, either through symbiotic bacteria or by chemical means. This ability is most highly developed in fishes living in the upper 1000 m of the water column: up to 90 per cent of all animals in this region are bioluminescent. Bioluminescence is used by deepwater fishes to find mates, lure prey, signal schooling neighbours, and camouflage predators and prey alike. Many adaptations to bioluminescence have evolved in deepwater fishes, such as tubular eyes with increased light sensitivity and depth perception, and black stomach linings to mask glowing prey within. A fish living in the upper 1000 m of the ocean experiences a steady uniform glow from above and infinite darkness below. Predators such as Pearleyes (Scopelarchidae, p. 241) and planktivores such as Barreleyes (Opisthoproctidae, p. 203) have developed upward-pointing, highly sensitive eyes to distinguish the silhouettes of their prey in these conditions. On the other hand, many fishes produce downwardly directed light to camouflage themselves in the steady feeble glow from above.

The majority of fishes in the upper 1000 m of the ocean undertake a nightly vertical migration, rising towards the productive upper layers of the ocean to feed, then returning to the depths to avoid surface predators during the day. Vast numbers of small planktivorous fishes, such as Lanternfishes (Myctophidae, p. 251), Hatchetfishes (Sternoptychidae, p. 220) and Bristlemouths (Gonostomatidae, p. 218), make this daily migration, and they are among the most abundant of all vertebrates. Other families, such as the Dragonfishes (Stomiidae, p. 223), are predators of these schooling fishes and have developed enormous fangs and expandable stomachs to swallow large prey whole. Some follow the nightly migration of their prey, while others remain stationed at certain depths to intercept the migrators, and still others move upward from greater depths to meet the migrators at their lowest point. Lanternfishes and Dragonfishes comprise nearly half the number of fish species in this upper midwater environment (from about 200 to 1000 m).

Below 1000 m, in deep midwaters, there is no light or turbulence and far less food and energy. Many fishes here have formed passive drifting habits to conserve energy, and have reduced skeletal development (building bones requires large amounts of nutrients). Most are opportunistic predators, taking any prey they encounter, and many are capable of swallowing animals as large as themselves. Whalefishes (Rondeletiidae, p. 347; Barbourisiidae, p. 348; Cetomimidae, p. 349), Gulper Eels (Saccopharyngidae, p. 177) and the extremely diverse and specialised Deepsea Anglerfishes (Lophiiformes, p. 291) are the dominant groups.

Deepsea Anglerfishes are fascinating fishes that entice prey towards them with a luminous lure and then engulf it in their enormous, fanged jaws. A mate can be difficult to find in the enormous volume of deep midwater, and some Deepsea Anglerfishes have evolved a unique system of sexual parasitism. The males, which are far smaller than the females, locate a mate, usually by following pheromones, then attach to the female with special teeth. In some cases the body of the male fuses with that of the female, the male then surviving on the blood supply of the host.

Down on the great plains of the deep ocean floor, the stillness and darkness are near absolute. The only energy input is the rain of dead and decaying organic material from the layers of water above, and the only movement the slow drift of great masses of freezing, salty water. Seasonal changes, though, can still be felt. For example, algal blooms at the surface are quite rapidly translated to the bottom as they die off and sink. Benthic fishes are few and far between in very deep water, as food is scarce and energy levels very low. Bottom-dwelling fishes feed mainly on the invertebrates associated with the bacteria and decaying organic matter on the substrate, and on each other.

Because the waters here are so slow moving, some species, like the Deepsea Tripodfishes (Ipnopidae, p. 239), have developed extremely sensitive vibration detection systems, in the form of long sensory fin rays, to detect the least movement. Most fishes in this environment move very little to conserve energy, and those that move more have developed eel-like forms, as seen in the Spiny Eels (Notocanthidae, p. 152), Whiptails (Macrouridae, p. 268) and Halosaurs (Halosauridae, p. 151). These elongate fishes feed on benthic invertebrates and move sinuously and slowly, leaving virtually no turbulence to signal their passage. Deepsea Lizardfishes (Bathysauridae, p. 232) are predators in this domain; they may rest immobile on the bottom for weeks or even months waiting for prey to pass by.

Occasionally, large dead animals such as whales drop to the bottom, creating instant islands of huge productivity. A whole community – bacteria and invertebrates, and the fishes that feed on them – has evolved to rapidly colonise these islands which, in the case of a large whale carcass, may supply sustenance for a decade or more. It is thought that fishes such as Aphyonids (Aphyonidae, p. 284) may have diversified on the sea floor beneath the ancient migratory routes of whales.

On the abyssal plains of the deepsea floor, seamounts and the edges of continental plates create barriers to the movement of deepwater fishes. Seamounts (the remains of ancient underwater volcanoes) force currents towards the surface, mix waters and provide a far greater range of habitats than the abyssal plains. Harder substrates, scoured of sediment by water currents, allow an extraordinary diversity of invertebrates – such as sponges, sea anemones, corals and sea fans – to develop, and these support a great diversity of fishes. Many seamounts, such as those found to the south-east of Tasmania, have only recently been sampled, and numerous new species have been found there. Snailfishes (Liparidae, p. 436), for example, were virtually unknown in Australian waters until deepwater surveys revealed about 30 new species, some seemingly restricted to a single seamount.

Families such as the Whiptails (Macrouridae, p. 268) and Slickheads (Alepocephalidae, p. 206), as well as many deepwater sharks and rays, predominate on the continental slopes above the gentle rise from the abyssal plains. The slopes gradually steepen as they approach the continental shelf; often they are gouged by deep canyons such as the Rottnest Trench, off the west Australian coast, bringing deepwater fishes closer to the coast. Occasionally catastrophic 'waterslides' occur, where masses of water suddenly slump down the continental slope like an avalanche. These may flow far out over the abyssal plain and can cause enormous fish mortality as they drastically alter temperature and salinity in deep water.

Further up the continental slope, habitats change more rapidly and fish diversity increases as the seafloor rises towards the sunlit upper layers of the ocean, where surface currents and more turbulent conditions impose their influence on the fish fauna.

COASTAL FISHES

In this book, coastal fishes are considered to be those that occur over the continental shelf (Fig. 3, p. 7) from the high-tide line down to about 200 m. Australia has an immense coastline and the vast area of continental shelf surrounding the continent (Fig. 4) contains a tremendous diversity of environments, from steaming tropical mudflats to freezing southern kelp forests.

The far northern coast of Australia extends as a broad shelf beneath the Timor Sea and the Arafura Sea, nowhere deeper than about 200 m. It is heavily affected by the numerous monsoonal rivers that drain into it. The rivers carry large amounts of sediment and organic material, creating muddy, nutrient-rich inshore waters that are rich in invertebrate life such as prawns and other crustaceans. Benthic, invertebrate-feeding fishes such as Stingrays (Dasyatidae, p. 131), Threadfins (Polynemidae, p. 541) and Flatfishes (Pleuronectiformes, p. 735) are common in these northern waters.

Extensive mudflats here create favourable conditions for mangroves to thrive. These tangled, muddy forests provide a rich environment for fishes, but with harsh conditions: high water temperatures, and periodic flooding and drying. The remarkably adapted Mudskippers (Gobiidae, p. 678) thrive here and are able to survive for lengthy periods out of the water, even climbing the mangrove trees. Small invertebrate feeders such as Ponyfishes (Leiognathidae, p. 507) and Gobies (Gobiidae, p. 678) are abundant. Mangroves are vitally important nursery areas and juveniles of many different species, including Breams

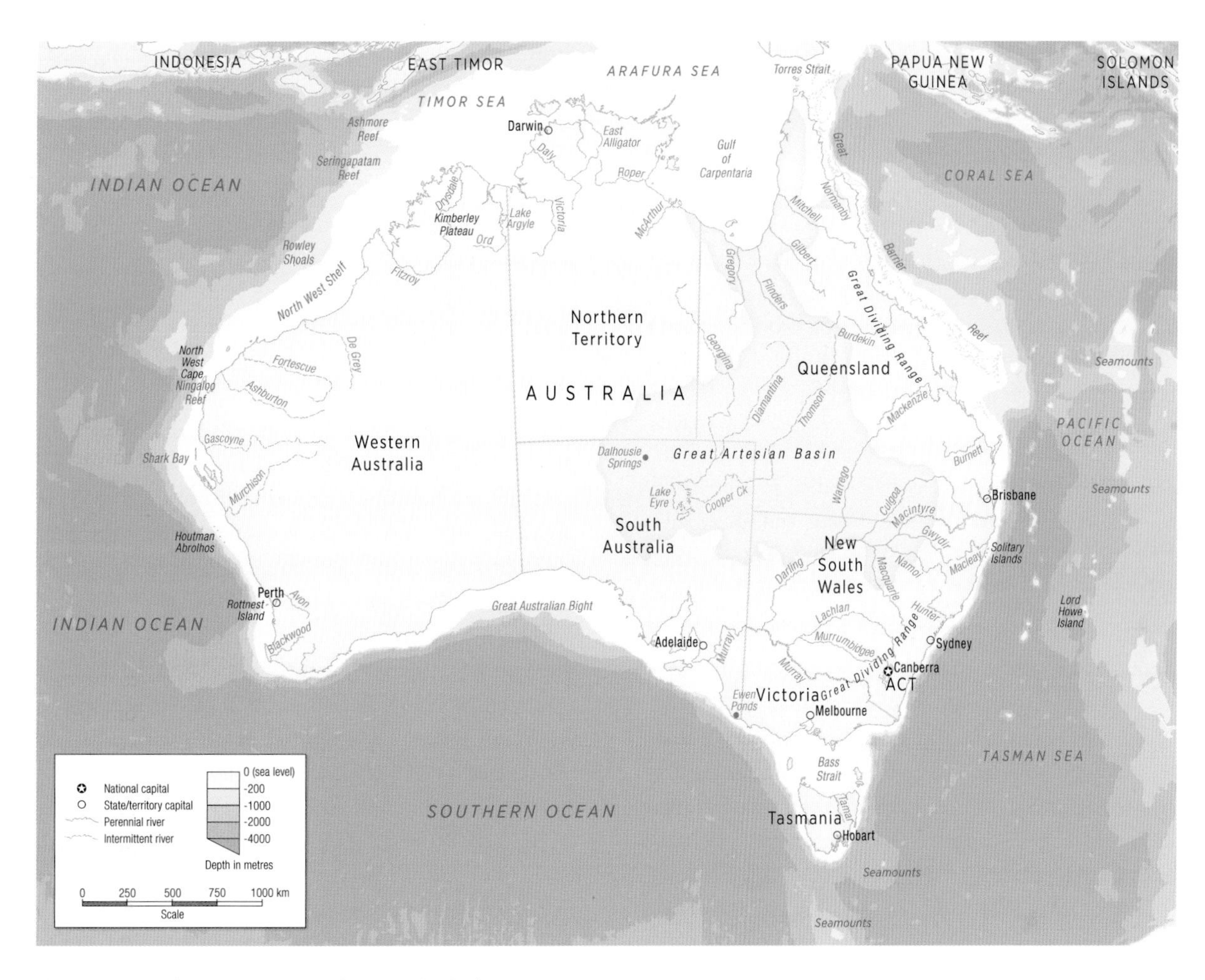

Figure 4 : Water depths around Australia and major freshwater systems

(Sparidae, p. 538), Snappers (Lutjanidae, p. 514) and Mullets (Mugilidae, p. 318), form the largest part of the fish populations in mangroves. More than 200 species of fish have been recorded from mangroves in northern Australia; some are permanent residents, but many move in and out with the tides or use the mangroves as nurseries.

Large tides ebb and flow along the north-west coast, moving in and out of numerous coastal creeks, mangroves, wetlands and rivers. The fish fauna includes many species adapted to turbid waters, such as Catfishes (Siluriformes, p. 195), and others that are capable of surviving in widely varying salinities, such as Archerfishes (Toxotidae, p. 557). The Barramundi (p. 442), for example, lives and breeds in shallow coastal waters, but requires the brackish conditions of tidal creeks and estuaries for its larvae to develop, and fresh water for the juveniles to mature.

In clear, warm waters further offshore in the north-west, the Leeuwin Current begins its journey southward. Coral atolls here, close to the centre of marine biodiversity in Indonesia, are home to a vast number of fish species. Transported by the great flow of water through the Indonesian archipelago, the fish fauna here is diverse and contains all the major coral-reef families, including the Parrotfishes (Scaridae, p. 634), Butterflyfishes (Chaetodontidae, p. 570), Damselfishes (Pomacentridae, p. 606), Blennies (Blenniidae, p. 658) and others too numerous to list here.

To the north-east of Australia, similar conditions exist in the Coral Sea, where clear, warm waters flow over the offshore atolls of the Great Barrier Reef, heading southward down the east coast. Corals proliferate closer inshore down the east and west coasts, and create a multitude of different habitats for fishes. Although the nutrients in these waters are mostly used up by the growth of phytoplankton, there is still enough zooplankton to support numerous midwater-feeding fishes, such as Fusiliers (Caesionidae, p. 521) and Fairy Basslets (Serranidae, p. 451). The convoluted reef topography also shelters enormous amounts and myriad forms of bacteria – on the surface and within the porous reef itself – that in turn feed micro-invertebrates, then larger invertebrates and the fishes that feed on them.

Coral reefs create a wide range of habitats for fishes, including sand flats, rubble areas, seagrass meadows, silty lagoons, stands of coral, algae-covered platforms and wave-battered coral barriers. Often these habitats are in close proximity to each other and provide a large variety of ecological niches.

Diversity of fishes, and consequently competition for food and space, is very high in these complex reef ecosystems, and many species show great specialisation, both in feeding strategies and habitat preference. Some fishes, such as the Cardinalfishes (Apogonidae, p. 479), may only be found amongst the branches of a single species of coral; others, like some Angelfishes (Pomacanthidae, p. 579), may feed only on a particular species of algae or sponge. On the other hand, there are species, including some Wrasses (Labridae, p. 617) and predatory Whaler Sharks (Carcharhinidae, p. 79), that move between many habitats.

On the east coast, the Great Barrier Reef, with its rich coral growth, extends south from the northernmost tip of Australia to around the Solitary Islands, off New South Wales, where southern and northern fishes coexist. On the west coast, tropical coral reefs extend as far south as the islands of the Houtman Abrolhos, near Geraldton. Here too, tropical and temperate species mingle, and southern seaweeds and northern corals grow side by side. These areas mark, very broadly, the distribution boundaries of those fishes described in this book as being found in northern or southern Australian waters.

Pelagic larvae of many tropical fishes are transported to the limits of the warm, southward-flowing currents and settle to the substrate to develop where local conditions allow. Juveniles of tropical species are thus often seen far to the south of their normal habitat, though they generally do not survive harsh southern winter conditions to establish ongoing populations there. At Rottnest Island, off the south-west coast of Australia, the Leeuwin Current brings warm water and a seasonal influx of larvae, thus maintaining a small southern hotspot of tropical reef fishes. Up to 100 tropical species have been recorded there. Norfolk Island, off the east coast, is a similar outpost for tropical species.

Waters are very nutrient poor further down the west coast and much of the energy cycle sustaining fishes here revolves around the growth and decay of seaweed and seagrass.

Seagrass, which carpets shallow bays and estuaries, is a vitally important habitat for fishes, occurring around the entire circumference of Australia and covering more than 50 000 km^2 of seabed. It forms meadows in flat sandy bays and estuaries around the south coast, scattered beds in lagoons along the Great Barrier Reef, and vast fields around the Torres Strait. This habitat in Australia is the most diverse of its type anywhere in the world. Seagrass is like the rainforest of the sea, covered in epiphytes and providing abundant food and shelter for marine organisms. It harbours immense numbers of juveniles of many fish species, as well as a great diversity of mature fishes adapted to this habitat. Numerous species of Weed Whiting (Odacidae, p. 632) and Shore Eel (Gobiesocidae, p. 668) are found here, along with juvenile Breams (Sparidae, p. 538), Snappers (Lutjanidae, p. 514) and countless others.

As the warm currents move south their influence gradually decreases and waters become progressively cooler. Rocky limestone reefs – with abundant kelp and other algae in shallow inshore waters, and rich invertebrate growth in deeper waters – are found along the lower part of both east and west coasts. Herbivorous fishes such as Drummers (Kyphosidae, p. 560) are common, and the coastline of long, sandy beaches between rocky headlands provides expansive stretches of water for Whitings (Sillaginidae, p. 487), Tailor (Pomatomidae, p. 493) and Mullets (Mugilidae, p. 318) to move along the shore.

Offshore in the south-east, cooler, nutrient-rich waters flowing from the south have more influence, and large schools of Dories (Zeidae, p. 374) and other deepwater species such as Gemfishes (Gempylidae, p. 713) and Lings (Ophidiidae, p. 280) occur. Closer to the southern coast, granite outcrops and headlands punctuate the sandy beaches, and pounding waves and storms from the Southern Ocean create often harsh and turbulent inshore conditions.

Australia is unique in having an enormous stretch of southern coastline facing directly onto the cold, nutrient-rich waters of the Southern Ocean. Isolated for millions of years by geology and ocean currents, this coast has developed an immensely rich marine fauna and flora, the majority of which is found nowhere else in the world. More than 600 species of fish have been recorded from this region. Fish populations here have diversified due to the wide range of water temperatures that occur. Areas to the west are influenced by the warm Leeuwin Current, while areas towards the centre of the coast and around Tasmania are impacted more by cold upwellings and storms from the Southern Ocean. Water temperatures can vary from as low as 10°C to as high as 25°C across the southern coast.

Over southern inshore reefs, forests of kelp and sargassum weed cling tightly to the rocky substrate, providing shelter for numerous species of fish. Many have adaptations that allow them to hold their position in the face of violent water movement. Clingfishes (Gobiesocidae, p. 668), which possess a ventral sucking disc, are diverse along this coast, while Kelpfishes (Chironemidae, p. 597) and Marblefishes (Aplodactylidae, p. 598) have strong pectoral fins to hold them in position in surging water. Several other families, such as the Leatherjackets (Monocanthidae, p. 762), many Wrasses (Labridae, p. 617) and small Weedfishes (Clinidae, p. 665) – all closely associated with rich algal growth – are very diverse in these southern coastal waters.

Deeper, offshore southern reefs are difficult to access and are perhaps the least well-known, though some of the richest, environments in Australian waters. Upwellings, and the plankton and nutrients they carry, support an astounding array of sessile invertebrates such as sponges, corals and ascidians. Anglerfishes (Lophiiformes, p. 291), which are closely linked to the substrate, are also diverse on these invertebrate-rich reefs. Planktivorous fishes such as the Butterfly Perch (p. 451) and large schools of commercially important species such as Blue Grenadier (p. 275) and Dories (Zeidae, p. 374) congregate around the reefs, and Boarfishes (Pentacerotidae, p. 586) also patrol them.

The southernmost waters of Australia, around Tasmania, are rich in fish life. Influenced on occasion by both the furthest reaches of the East Australian Current and the Leeuwin Current, they are churned and mixed by storms and upwellings from the Southern Ocean. Fish diversity is lower here than in other areas around the country, but abundance increases off Tasmania's western coast, where vast schools of species such as the Common Jack Mackerel (p. 498) and Barracouta (p. 713) occur. Inshore on the west coast, however, the constant pounding from storms creates a very harsh environment with few fish species.

On Tasmania's southern coast, upwellings and currents bring species usually found in deep waters, such as Skates (Rajidae, p. 121) and Gurnards (Triglidae, p. 425), into much shallower waters. The east coast of Tasmania, on the other hand, appears to be increasingly influenced by the warm East Australian Current. Habitats are changing rapidly here as the current more frequently penetrates further south. Kelp forests are declining as nutrient-poor warm water displaces nutrient-rich cold water, and southern species such as Trumpeters (Latridae, p. 602) are being replaced by more northern fishes like the Herring Cale (p. 632).

FRESHWATER FISHES

There are about 300 species of fish found in Australia's freshwater habitats. This number is small compared to other areas of the world: South America, for example, is home to more than 2000 freshwater species. The low number of freshwater fishes is not surprising, however, when one considers that, after Antarctica, Australia is by far the driest continent on earth. Most of the interior is arid and major environmental challenges face freshwater fishes in inland Australia.

The ephemeral and highly variable nature of many Australian freshwater habitats means large areas of the country are completely devoid of freshwater fishes. The Australian freshwater fauna, though, is remarkable for its hardiness and adaptation to this harsh environment. Fishes such as the Spangled Perch (p. 588) cope with very high temperatures and fluctuating salinity and water quality, and this hardy survivor is found across most of the interior of Australia.

With the exception of the ancient Australian Lungfish (p. 786) and the Saratogas (Osteoglossidae, p. 144), Australian freshwater fishes have all evolved relatively recently from marine ancestors. Many species retain links to these marine origins through their larvae, which are flushed from rivers to be dispersed by ocean currents. Others, such as Lampreys

(Petromyzontida, p. 33), still spend part of their adult life in the sea, and some marine fishes, such as Mullets (Mugilidae, p. 318), have juveniles that venture into the lower reaches of freshwater streams and rivers.

Freshwater environments vary widely across the continent. In south-eastern Australia and Tasmania, conditions are cooler and there is relatively high rainfall. The freshwater environments here have been isolated for many millions of years and, although overall diversity is low, several ancient fish families with high levels of endemism occur. These include the Galaxids (Galaxiidae p. 211) and Temperate Perches (Percichthyidae, p. 445), both of which have ancestral lines leading back to the time of Gondwana.

The major feature affecting freshwater fish distributions in the eastern part of Australia is the Great Dividing Range, which runs from western Victoria to northern Queensland. The Murray-Darling river system, which is Australia's largest drainage, carries water from the western side of the Great Dividing Range southward to the river mouth near Adelaide. The Murray-Darling river system is home to Australia's largest freshwater fish, the Murray Cod (p. 445). More than 30 other fish species are also found there, including several popular angling and food fishes, such as the Golden Perch (p. 446) and Silver Perch (p. 588).

The Murray-Darling was once a thriving ecosystem, with a cycle of regular flooding supporting numerous wetlands and freshwater billabongs. It has now been degraded by agriculture, excessive water use for irrigation, artificial barriers, and restrictions to its normal flow and flood cycles to such an extent that the collapse of large parts of the ecosystem seems imminent. Many endemic species are now threatened by this habitat degradation.

Further north, the western side of the Great Dividing Range drains inland, mainly into Lake Eyre. River flows in this area are highly irregular and many years may pass before water runs in the dry river beds and massive floods spread out over the plains. Found here are species that reproduce quickly and can disperse rapidly to take advantage of rare flooding events and the accompanying explosion of food and habitat. The Golden Perch (p. 446), some Hardyheads (Atherinidae, p. 328) and the Bony Bream (p. 184) are typical examples, able to survive in the isolated billabongs and waterholes that remain when floodwaters recede.

Water also flows to this area underground. The seas have invaded the continent of Australia several times in the ancient past, and the sediment deposited then forms the vast arid plains of the interior and the porous limestone rocks holding the waters of the Great Artesian Basin. Water in this reservoir, deep underground, may have taken 10 000 years to arrive, having fallen as rain on the Great Dividing Range and then slowly crept through porous rocks. Forced to the surface in a few isolated springs, the ancient water maintains unique and hardy fishes such as Desert Gobies (Gobiidae, p. 678) and Hardyheads (Atherinidae, p. 328), endemic species with very restricted distributions in the arid desert environment. Artesian bores tapped into the underground reservoir have been left to run uncapped for decades, forming small permanent creeks. These have largely been colonised by the Mosquitofish (p. 343), a species introduced into Australia.

On the eastern side of the Great Dividing Range there is a greater diversity of freshwater habitats: from alpine streams in the south to jungle pools in the north, via coastal swamps and meandering rivers. In the north-east there is high rainfall, and freshwater fish diversity is high in the numerous streams, lagoons and rivers. Many species found here, such as the Rainbowfishes (Melanotaeniidae, p. 323) and Gudgeons (Eleotridae, p. 674) also occur in Papua New Guinea, reflecting the relatively recent connection that existed between the two countries.

Across the north of Australia are many seasonally flooding monsoonal rivers, which spread out over surrounding coastal plains in the wet season and then retreat, leaving isolated waterholes. High tides move up and down numerous tidal creeks, creating brackish-water habitats. Freshwater fishes from predominantly marine fish families are diverse in this area, such as the Freshwater Thryssa (p. 181) and many species of Catfish (Siluriformes, p. 195). Swamps, wetlands and lagoons are common, harbouring ancient Saratogas (Osteoglossidae, p. 144) and highly evolved Swamp Eels (Synbranchidae, p. 398).

Further west, seasonal rivers run through deep gorges, replenishing isolated pools in the rocky Kimberley plateau. These pools are home to many endemic species of Grunter (Terapontidae, p. 588), Rainbowfish (Melanotaeniidae, p. 323) and Gudgeon (Eleotridae, p. 674). Floods cannot disperse fishes over wide plains here and some of these species are restricted to a single river system.

Further along the west coast, freshwater habitats become very scarce and rivers flow for only a few short weeks during the wet season. North West Cape, near the mid-west coast, is home to two unique subterranean fish species: the Cave Gudgeon (p. 674) and the Blind Cave Eel (p. 398), which survive in the permanent darkness of an extensive underground aquifer. Only a dozen or so species of fish are found in the mid-west region; about half of them are Grunters (Terapontidae, p. 588) and Gudgeons (Eleotridae, p. 674).

The south-west of Australia has more regular rainfall and many small coastal streams. However, fish diversity remains low, reflecting once more the long isolation of southern Australia. About 80 per cent of the species found here, mainly Galaxids (Galaxiidae, p. 211), are endemic. Surviving in ephemeral pools on the sandy coastal plain is one of the most interesting freshwater fishes in Australia, the Salamanderfish (p. 211). It has adapted to the transient nature of its habitat by digging into the soil as its pool dries up and then waiting in a cool, damp burrow for water to return. The Salamanderfish is an extremely ancient species with ancestry that can be traced back to the time of Gondwana and it is found nowhere else in the world except on this hot, sandy plain. This tiny fish embodies many of the features that make the Australian freshwater fish fauna unique.

FISHERIES AND CONSERVATION

In general, the waters surrounding Australia are nutrient poor due to a lack of major upwellings of cold water, such as those that support the immense quantities of Anchovies (Engraulidae, p. 181) off the coast of South America. Consequently, the Australian fish fauna, though diverse, is low in biomass and cannot sustain such large-scale fisheries. However, Australia does have a vast, territorially exclusive fishing zone, extending 200 km from the coast around the continent and including waters around offshore islands such as Macquarie, Heard and McDonald in the Southern Ocean, and Christmas, Lord Howe and the Cocos further north. This is the third-largest fishing zone in the world, but ranks only about 50th in terms of fish production.

Subsistence fishing in inshore and fresh waters has been carried out by Indigenous Australians for tens of thousands of years. Fishing with discretion and respect for the environment, using hooks and lines, traps, nets and spears, they seem to have had little long-term impact on fish populations. For thousands of years, fishers have travelled seasonally from South-East Asia to fish in northern Australian waters. Fishing also played a vital role in sustaining the first European settlers as they gained a foothold on the coast, providing a reliable source of food as agriculture struggled in its infancy. Small-scale inshore fishing with nets, and hook and line, has continued ever since.

Commercial fishing on a larger scale has been carried out in Australian waters since the early part of the last century. The first steam trawlers arrived in about 1915, but until the 1950s most fishing was still carried out from small boats in coastal waters. In the following 40 years, commercial fisheries expanded rapidly and the Australian commercial catch rose 470 per cent. Much of this was due to continuing discoveries of new fishing grounds and greater fishing effort. As technology improved and inshore fisheries were depleted, larger vessels began to fish new grounds in deeper waters further offshore.

The recreational fishing sector has shown similar growth and recent times have seen an explosion in the number and size of boats. In some cases recreational fishing now puts more pressure on fish populations than the commercial sector. The increasing availability of electronic aids such as GPS navigation systems and echo sounders has meant that productive offshore reefs can be precisely targeted by large numbers of recreational anglers. Most reefs near human population centres have now been severely depleted of fishes and recreational fishing pressure is expanding further offshore into deeper waters.

More than 200 different fish species are currently caught and sold commercially in Australia on a regular basis. They come from a very wide range of habitats, including mangroves and coral reefs, seagrass beds and deep continental slopes. Countless more fish species and other animals are directly affected by commercial fishing. Many non-target species are taken as bycatch, and many more suffer from the impact fishing has on their ecosystem. In northern prawn fisheries, the bycatch (of which two-thirds is fish) can be up to 10 times the amount of target species taken, and almost all of it is discarded. In areas heavily fished by prawn trawlers, the diversity and composition of the fish fauna have been completely altered.

Responsibility for commercial and recreational fisheries and the effects they have on fish populations is divided between the state and federal governments, and is overseen by the

Australian Fisheries Management Authority (AFMA). Most inshore fisheries are managed by the individual states, though some are jointly managed, and the AFMA is generally responsible for offshore and large-scale fisheries. State and federal authorities produce annual assessments of the status of targeted fish populations (see Bibliography, p. 789). These provide scarce details of population trends and clearly reflect the limits of our knowledge of these populations. In the 2007 federal report, the status of 52 of the 96 fish populations assessed was classified as 'uncertain'.

In order to manage fishing pressures and protect fish populations, a great deal of information must be gathered on the fishes themselves. Fisheries scientists are unable to keep up with the amount of work that needs to be done in this area. Even accurately measuring the size of fish populations is difficult and time consuming, let alone understanding their fluctuations over time or their links and importance within an ecosystem.

An enormous amount of work has been done in this area by fisheries workers and scientists around the country to try and ensure the protection of Australian fish populations. All too often their recommendations are ignored or buried in political or bureaucratic debate and lobbying, and the delay can have an enormous impact. The Orange Roughy (p. 356) catch in south-eastern waters rose from about 1000 tonnes in 1985 to 40 000 tonnes in just four years, then rapidly declined as populations were overfished. By the time management practices caught up and maximum catch quotas were put in place, the Orange Roughy population had already decreased to such an extent that the quotas were never reached. Nearly 25 years on, the slow-growing Orange Roughy is still classed as 'overfished', with only a small increase in its population since it was first assessed in 1993. Fisheries for other species such as Gemfishes have followed a similar pattern.

Not all fisheries have such dramatic impacts on fish populations. One of the oldest fisheries in Australia is on the south coast and uses beach seine nets to take Australian Salmon. Started in the early 1930s, extensive information has been gathered on this easily accessible fishery. The long-standing fishery seems to be sustainable, but long-term population variations caused by factors such as the Leeuwin Current (see The Movement of the Waters, p. 5) are still not fully understood. Fine-tuning of catch limits and regulation is still being carried out to protect Australian Salmon populations.

Aquaculture is a rapidly growing industry in Australia and its success is vital in order to ease pressure on wild fish populations. However, farming of fish has its own environmental impacts, such as the high levels of waste material that falls to the sea floor, leading to algal blooms and eutrophication of benthic habitats. Disease escaping into wild fish populations is also an issue. Imported fishes used as bait and food for Bluefin Tunas (p. 718) grown in sea cages are thought to have passed on a virus that decimated wild Australian Pilchard (p. 184) populations along the south coast.

Fish Habitats

About 80 per cent of Australians live within 50 km of the coast. It is largely due to this fact that many fish habitats around Australia are seriously degraded. Coastal development, industry and inappropriate land use have had an enormous impact, particularly on shallow-water nursery habitats.

Studies carried out in New South Wales have shown that 60 per cent of the commercial fish species caught there rely on seagrass habitats for juvenile development. Seagrass beds have been decimated in some areas. In Cockburn Sound, an important seagrass habitat in south-western Australia, an estimated 97 per cent of the seagrass has been lost. A major portion of the West Australian population of Snapper (Sparidae, p. 538) spawns and grows as juveniles there, and although the effect of this habitat loss is impossible to quantify, it can only be detrimental to the Snapper population. Around the Australian coast the picture is little different. In northern waters natural climatic effects such as cyclones have also caused the loss of large areas of seagrass.

Estuaries are similarly important nursery areas for fishes. They are severely impacted by pollution from urban runoff, as well as by agricultural fertilisers, which cause algal blooms. Estuaries are fragile environments, affected by both marine and freshwater inputs, and vulnerable to environmental disturbances. Hardy fish and invertebrate species that have been introduced from overseas tend to establish themselves and thrive in estuaries, outcompeting local species. About 70 per cent of all exotic marine fish species in Australian waters are found in estuaries.

Coastal mangroves are another vital fish habitat under increasing pressure, and the area they cover is decreasing near centres of coastal development. Mangroves are increasing slightly in some undisturbed areas, but unfortunately such areas are becoming more and more rare.

Some freshwater habitats are even more severely degraded. The great Murray-Darling river system has been put under such stress by agriculture and other pressures that some downstream habitats have completely disappeared. Oxbow lakes (formed when meanders of the river are cut off by silt) were previously replenished by regular flooding, but have now dried up, and the acid sulphate soils subsequently exposed to the air have turned once-thriving habitats into sterile, sulphuric-acid sludge.

Few rivers in Australia have escaped damage. South-western rivers are becoming more saline as the catchments that feed them deteriorate. Water use for agriculture has dried up many small coastal streams. The lowering of the water table in some areas of the Great Artesian Basin is putting at risk artesian springs in desert areas and the unique fishes that inhabit them.

Deepwater habitats, which are extremely fragile and slow-growing, are being devastated by the passage of bottom trawls. These environments contain extraordinary diversity, much of it new or as yet unknown to science, yet destructive fishing practices continue. In the North Atlantic an area of deepwater reefs about the size of Western Europe has already been completely destroyed. The Australian government has moved to protect areas of remarkably diverse deepwater habitat in the Huon Commonwealth Marine Reserve. However, only 370 km^2 of this reserve has been closed to bottom trawling. In comparison, New Zealand has now closed 33 per cent of all its deepwater habitats to bottom trawling – an area of more than 1.1 million km^2.

Climatic changes are also affecting Australian fish habitats. The east coast of Tasmania appears to be increasingly influenced by the warm East Australian Current, while corals on the Great Barrier Reef and atolls off north-western Australia have suffered from higher sea-surface temperatures. It seems clear that habitats are being transformed by changing temperatures, as they certainly have been in the past. The processes driving these changes are vast and immensely complex, and operating on a global scale, perhaps with a geological time frame. The accelerating rate of change highlights how imperative it is for us to protect the diversity and health of fish habitats, so that they can maintain a full complement of species, and thus the resilience to adapt.

Conservation and Protection

Commercial and recreational fishing make a large contribution to the Australian economy and employ many people. However, for these activities to continue, fish populations and the marine environment as a whole must be protected and conserved. The realisation of this necessity is growing and some trends are encouraging.

Unfortunately, management of marine habitats and fish populations often operates from a starting point that is already far removed from an undisturbed state. Numerous Australian fish species have already disappeared completely from previous parts of their range, and many are now perceived as never having occurred in those areas. Blue Gropers (Labridae, p. 617) and the West Australian Dhufish (p. 551) were once common on inshore shallow reefs, but are now thought of as deepwater reef dwellers. Due to fishing pressure, some species no longer reach the age or size they once did: Barramundi can grow to nearly 2 m in length and weigh over 60 kg, but it is many years since specimens old enough to reach that size have been seen.

The pressures on fishes are manifold: habitat degradation, pollution, coastal development, and fishing. Realisation of the importance of closed areas – where fish populations and ecosystems can regain health, maintain balance and replenish the fishes continually being removed from other areas – is beginning to permeate through the general population. Most people now understand that a healthy percentage of fishes must be allowed to grow to full size in an undisturbed environment in order to maintain diversity and population sizes. Fully mature large female fishes usually breed more often and produce many more eggs than smaller females, and their eggs are larger and more likely to survive.

Studies have shown that in areas closed to fishing, previously targeted fish populations increase rapidly and spill outward to increase numbers in surrounding areas. Closed areas are vital buffers to environmental change and insurance against errors in management of fish populations and the environment.

Large Marine Protected Areas have been established in some locations, mostly in northern waters such as the Great Barrier Reef. However, very few of these areas prohibit taking fishes, and the unique and fragile southern-coast marine environment is barely protected. The actual area of Australian marine environments where fishing is completely prohibited for conservation purposes is minuscule. Substantial areas have recently been earmarked for intensive investigation – a preliminary step to the establishment of marine reserves. The process of selecting the location, type and size of these zones is fraught with difficulties and subject to intense pressure by many stakeholders. It can only be hoped that the legislative process necessary for the creation of reserves can proceed rapidly and that sufficient area is set aside to provide effective protection. Meanwhile, coastal development continues almost unabated around Australia, often in the face of community and scientific opposition. Mining and industry continue to discharge waste into fish nurseries, and marinas and canals continue to be built in estuaries.

It is imperative to maintain healthy, balanced marine ecosystems and high biodiversity in all Australian waters so that both marine and freshwater fishes and their habitats can absorb pressures and adapt to changes, however they may present themselves in the future. For this to occur, the status of fishes and our perception of them must change. Fishes must become far more than a simple economic resource; wealth to be mined from the sea for profit or pleasure. They can no longer be disregarded by coastal developers or perceived as rubbish to be discarded from trawlers. Fishes must be recognised as a precious natural heritage, a national treasure, and their protection considered a national and international responsibility.

ANATOMY

The anatomy of fishes (Fig. 5) follows basically the same plan as it does in the higher vertebrates, and most aspects of human anatomy can be related to similar structures in fishes. However, variations and modifications to this plan are extraordinarily diverse in fishes, and far too numerous to detail here. The following is a basic outline of the anatomy of fishes, with some examples included to show the diversity of their adaptations.

Bone

The skeleton of fishes is composed of two different types of bone. Dermal bone, which develops in the skin layers, forms structures such as scales and teeth, and the skin denticles of sharks. Cartilage replacement bone, which develops deep within the body from cartilage in the embryo, forms the internal skeleton. Chondrichthyans (Chondrichthyes, p. 37) grow and strengthen this embryonic cartilage with calcification, but it does not develop into true bone as it does in the Osteichthyans (Actinopterygii, p. 141). However, Chondrichthyans do develop distinct vertebrae, distinguishing them from the Lampreys (Petromyzontida, p. 31) and Hagfishes (Myxini, p. 25), which have only a rod of cartilage serving as a backbone.

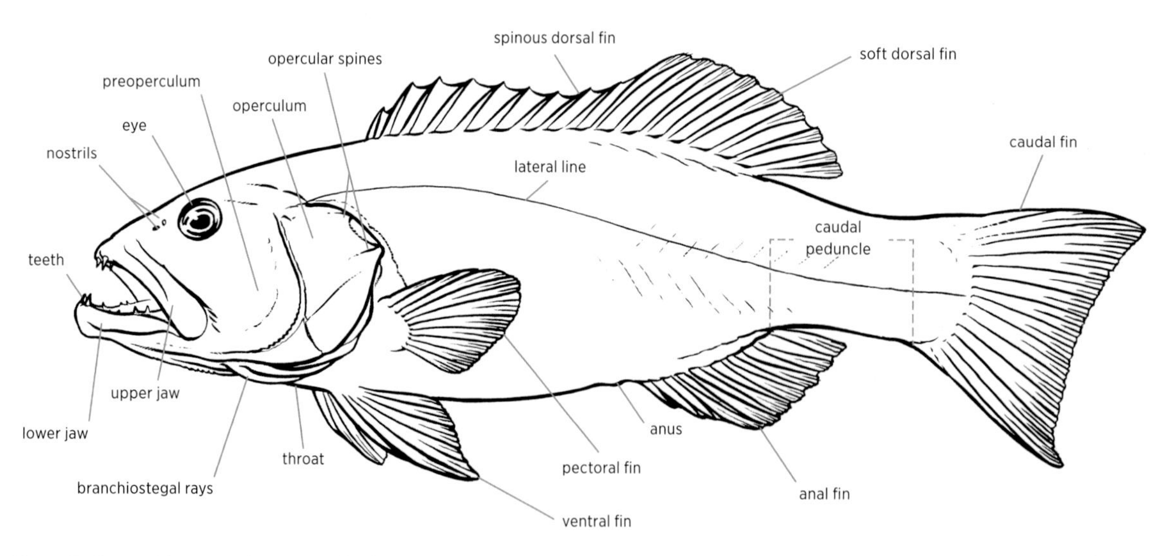

Figure 5 : External features of a typical fish of the order Perciformes: the Coral Trout (*Plectropomus sp.*)

Scales, Spines and Teeth

There are several different types of fish scales. The placoid scales or dermal denticles of sharks have a hard outer layer of enamel-like material and a rearward-pointing spine on a base plate that is embedded in the skin. This structure is similar to that of teeth found in most vertebrates.

The Teleosts, encompassing the vast majority of fishes, have cycloid or ctenoid scales. These thin, overlapping plates of dermal bone lack an outer layer of enamel-like material and each scale displays concentric growth rings. The age of a Teleost fish can often be ascertained from these rings in the same way that the age of a tree can be determined from the growth rings in its trunk. Ctenoid scales have a row of small tooth-like projections on the exposed edge, while cycloid scales have smooth edges (Fig. 6). Teleost fishes are able to replace lost scales by producing new ones; these scales grow rapidly to fill the empty space, without forming concentric rings.

Scales along a fish's sensory lateral line have small holes to allow water passage to the sensory cells beneath. Other scales may be modified to form scutes, such as those found in Trevallies (Carangidae, p. 497), or interlocking plates like those seen in Boxfishes (Ostraciidae, p. 768). Many fishes have greatly reduced scales or have lost their scales altogether. Catfishes (Siluriformes, p. 195), for example, have no scales, although many retain some bony dermal plates. Eels (Anguilliformes, p. 153) have very small scales that are deeply embedded in the skin.

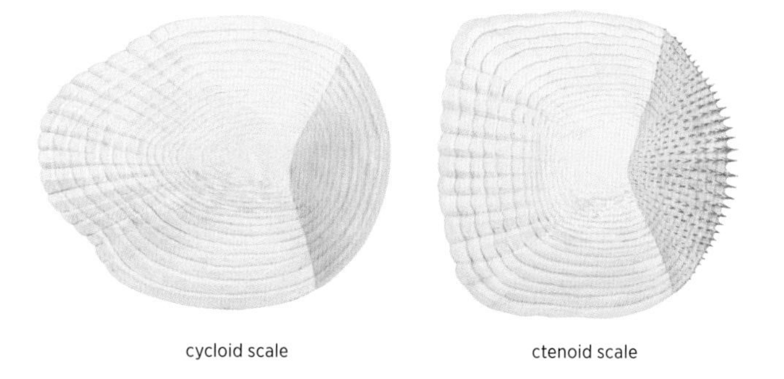

Figure 6 : Typical examples of the two main scale types of fishes

Fin spines and fin rays have evolved from scales. Spines are generally solid, hard, pointed and not divided into segments. Rays are flexible, usually branched, divided into segments, and composed of left and right halves fused together. Primitive fishes generally have only soft rays. Fin spines are present in more advanced bony fishes and in some cases, such as in the Remoras (Echeneidae, p. 496), they have become extremely modified. Remoras have broad lamellae developed from the spines, which they use to attach (using suction) to other animals. Chondrichthyans have horny cartilaginous rays that support the fins, and these are supported in turn by rods of cartilage, known as radials, that run from the base of the fins toward the vertebral column. The spines anterior to the fins of some sharks are modified and fused radials. In Teleosts each spine or ray is supported by an internal bone known as a pterygiophore, to which muscles are attached for control of the fins.

Fish teeth come in an enormous variety of shapes and sizes, with the teeth of each species corresponding very closely to the fish's dietary habits. Serrated, blade-like teeth such as those of Whaler Sharks (Carcharhinidae, p. 79) are used for cutting and biting chunks of flesh. Long, thin fangs, as seen in the Fangtooth (p. 352), are for seizing slippery prey; while the flat, plate-like teeth found in many rays (Rajiformes, p. 115 and Myliobatiformes, p. 125) are for grinding hard-shelled prey such as molluscs. Herbivores such as Nibblers (Girellidae, p. 562) and some Grunters (Terapontidae, p. 588) have flattened incisors for scraping and cutting algae. Wrasses (Labridae, p. 617) have combined dentition: canines and conical teeth anteriorly, for picking up and holding; and molariform teeth posteriorly, for crushing. Wrasses, along with a number of other groups, have developed pharyngeal teeth (flat grinding teeth in the pharynx or throat), which are used to grip or further crush hard food items.

Feeding habits of fishes also correspond to the development of the gillrakers (bony projections on the inner surface of the gill arches). Water passes over the gillrakers during respiration and feeding. Fishes that feed on plankton have numerous long, thin gillrakers to trap tiny prey items as water flows through the gills, while fast-swimming predators of larger food items have far fewer and less-developed gillrakers.

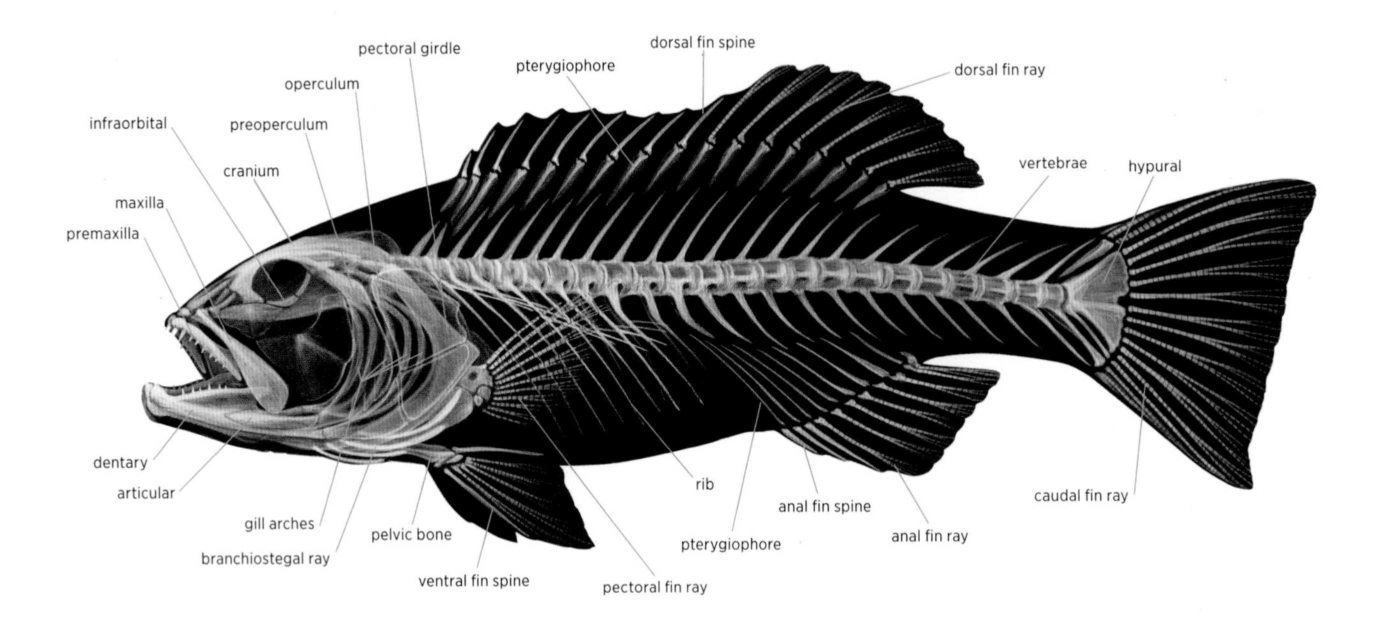

Figure 7 : Skeleton of a typical fish of the order Perciformes: Coral Trout (*Plectropomus sp.*) with preopercular bones rendered transparent to show gill arches and pectoral girdle beneath

Skeleton

The form and structure of the skeleton of fishes (Fig. 7), particularly the development of the skull and jaw apparatus, are major elements used to trace evolutionary relationships.

In fishes the pelvic structure is generally not attached to the vertebral column. In Chondrichthyans (Fig. 8) and primitive Osteichthyans it is attached only to muscles. In the more evolved Teleosts the ventral fins and associated pelvic bones have moved forward and are attached to the pectoral girdle, either directly or by ligaments. The pelvic structures have become greatly modified in some fishes. Eels (Anguilliformes, p. 153), for example, have lost the ventral fins altogether, while fishes such as the Australian Lungfish (p. 786) have strongly developed fin structures that almost resemble limbs.

The pectoral girdle of fishes is also not attached to the vertebral column. In Chondrichthyans it is effectively unattached and floats freely in the musculature of the body wall, while in Osteichthyans it is suspended from the rear of the skull. In Chondrichthyans the upper jaw is either movable and suspended from the skull by ligaments (as in the sharks and rays), or immobile and fused to the skull (as in chimaeras).

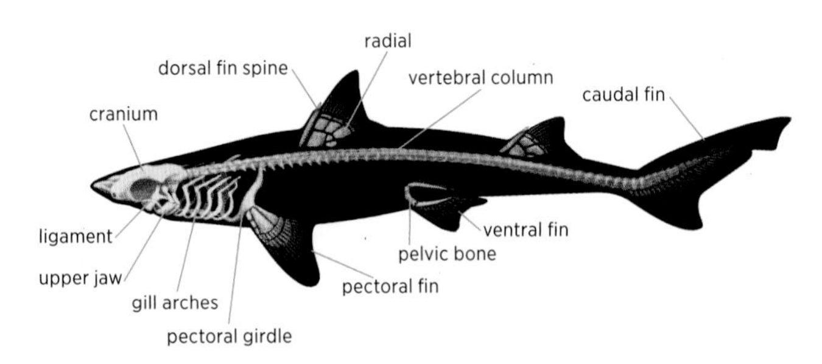

Figure 8 : Typical shark skeleton: the Dogfish (*Squalus sp.*)

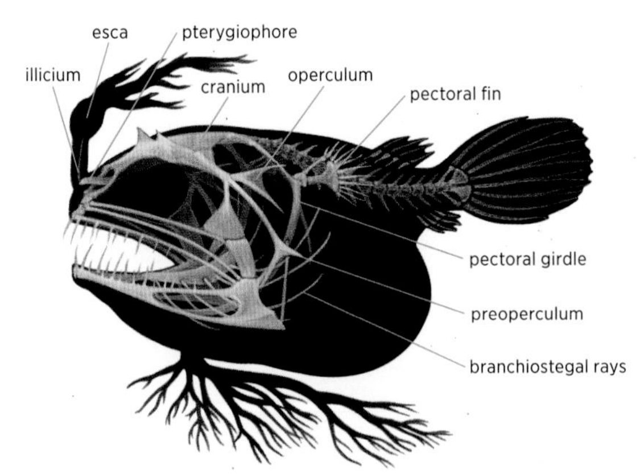

Figure 9 : Deepwater Anglerfish: showing the greatly reduced skeleton

The evolutionary trend in Osteichthyans has generally been towards the fusing together and loss of bones in the skull and skeleton. More primitive fishes such as Giant Herrings (Elopidae, p. 146) have a greater number of bones in the skeleton, while highly evolved fishes such as Puffers (Tetraodontidae, p. 773) have far fewer. Some deepwater fishes, despite having primitive lineages, have evolved greatly modified and reduced skeletons in order to conserve energy. The deepwater Anglerfishes (Lophiiformes, p. 291), for example, have lost many of the bones normally found in fishes, such as ribs, gill arches, the operculum and the pelvic structure (Fig. 9).

PHYSIOLOGY

Respiration and Circulation

Fish breathe by passing water over their gill filaments. They use various methods to ensure a constant flow of water over the gill filaments, which are richly supplied with fine capillaries to absorb oxygen and excrete carbon dioxide. In Agnathans (see Evolution & Geological History, p. 4) water can be pumped in and out through the gill openings using muscular gill pouches. In other fishes water is either taken in through the mouth, or through spiracles (as in some sharks and rays).

Many bottom-dwelling sharks and rays have their mouth, which is located on the ventral surface of the body, permanently on the substrate, preventing them from taking in water through the mouth. Instead, water flows into the body through spiracles on the dorsal surface, then over the gill filaments and out through the gill openings on the ventral surface.

Some fishes, such as Tunas (Scombridae, p. 717), have evolved to swim constantly, their forward movement forcing water over the gills, and are unable to breathe if held still. Fast-swimming fishes such as these usually have large gill openings to allow high water flow over the gills. For fishes with a more sedentary lifestyle, such as Anglerfishes (Lophiiformes, p. 291), high oxygen uptake is not a priority, and they often have greatly reduced gill openings, sometimes little more than a small pore.

The circulatory system of fishes is similar to that of higher vertebrates, with one major difference. In fishes the blood passes from the heart to the gills, where it absorbs oxygen, and is then transported directly to the rest of the body. In higher vertebrates the blood is loaded with oxygen from the lungs and then returns to the heart to be given another push around the rest of the body. This means that fish have lower blood pressure and much slower circulation than higher vertebrates. They have developed very large central veins with low friction to allow easy passage of the blood from the body back to the heart. This process is assisted by a system of peripheral veins that pass between the muscle segments along the body, so that muscular action involved in swimming helps pump blood back to the heart.

Fish absorb oxygen from the water into the blood via a network of fine capillaries in the gill filaments. This process is much slower than absorbing oxygen from the air and means that the blood spends a relatively long time very close to the water. The temperature of the blood as it leaves the gills therefore approximates that of the fish's surroundings. In most species this cool, oxygenated blood travels through arteries in the core of the fish and keeps the body at the same temperature as the environment. However, some highly evolved fishes, such as Tunas (Scombridae, p. 717), have developed a limited ability to warm their bodies by passing cool, oxygenated blood from the gills past a network of fine capillaries that have been warmed by muscular activity. The blood is then slightly warmer when it arrives in the core of the body (up to 7°C warmer in some Tunas) and this allows a higher metabolic rate and greater muscle performance.

Digestion and Excretion

The simplest digestive system of fishes is found in Agnathans such as the Lampreys (Petromyzontida, p. 31). Their mouth cavity branches into a respiratory tube leading to the gill pouches and oesophagus, where a simple valve leads directly into the intestine. There is no stomach. Instead the intestine has longitudinal folds to increase the area for absorption.

Chondrichthyans (Chondrichthyes, p. 37) have a well developed stomach that empties into a short intestine, the interior of which has a spiral form that greatly increases its surface area. The intestine empties into the cloaca, a chamber where all waste, as well as reproductive material (eggs and sperm), gathers before being expelled from the body. Chondrichthyans also have a very large liver that serves a dual purpose: energy conversion and energy storage. Energy is stored in the form of low-density oil called squalene, which also helps to maintain neutral buoyancy.

Osteichthyans (Actinopterygii, p. 141 and Sarcopterygii, p. 783) have a somewhat similar digestive tract to the Chondrichthyans, but the stomach is often divided into two chambers. The smaller, posterior chamber usually has numerous finger-like extensions to increase the absorption area. Mullets (Mugilidae p. 318) have an extremely muscular stomach that is used to grind plant material mixed with sharp, hard shell fragments or sand. Wrasses (Labridae, p. 617) and Parrotfishes (Scaridae, p. 634) on the other hand, lack a stomach altogether and thus cannot produce the acids and enzymes necessary to break down shells. They crush shells using pharyngeal teeth and then simply excrete the fragments.

In Osteichthyans the stomach empties into the intestine, which is variably developed according to the diet of the particular fish. Herbivores tend to have a long, voluminous intestine to deal with hard-to-digest plant material, while carnivores have a much shorter intestine. After passing through the intestine, waste is excreted through the anus. Osteichthyans do not have a cloaca, and the urinary and reproductive systems are vented to the exterior through a separate opening, usually a small pore just behind the anus.

Another important modification in Osteichthyans is their air-filled swim bladder, which has evolved from an outgrowth of the digestive tract and provides buoyancy. This has been lost in some forms and is highly developed in others. Many Lizardfishes (Synodontidae, p. 232) always rest on the substrate and have no need for a swim bladder. Some fishes, such as the Australian Lungfish (p. 786), have a large swim bladder with a rich blood supply, which functions as a primitive lung.

In general, the more primitive groups of fishes still have a connection between the swim bladder and the digestive tract, and can regulate buoyancy by swallowing air. Higher Teleosts have developed a completely separate swim bladder that must be regulated by gas-producing glands and a complex network of blood vessels. To combat the challenges of pressure changes, some deepwater fishes such as Lanternfishes (Myctophidae, p. 251) have developed swim bladders filled with low-density fat, which provides buoyancy in a similar manner to the oil in the liver of Chondrichthyans.

The swim bladder also serves to improve sensitivity to sound vibrations and in some groups, such as the Clupeiformes (p. 179) and Cypriniformes (p. 191), it is linked to the inner ear. The swim bladder of Whitings (Sillaginidae, p. 486) has numerous thin outgrowths, which probably increase sensitivity to vibrations. Many groups display other adaptations to the swim bladder. Jewfishes (Sciaenidae, p. 543), for example, can use their muscular swim bladder to produce sound.

Senses

Fishes live in a liquid medium that transmits vibration (and hence sound) more effectively than air. They detect vibration by two means: auditory receptors in chambers in the inner ear, and the lateral line system along the body. In some fishes sound is magnified by connections between the inner ear and swim bladder. The inner ear is also responsible for balance, as it is in humans.

The lateral line senses vibrations not only from sound but also from the movements of other animals, and from pressure waves from schooling neighbours and obstacles in the water. There is a series of sensory hairs, or neuromasts, in canals or chambers along the lateral line, and the neuromasts are usually protected by scales, with a small hole to allow water passage. Neuromasts may also be free-standing on the skin, particularly in deepwater fishes, where turbulence is minimal. Many fishes have developed a 'sixth sense': a sensitivity to the weak electric fields produced by all living animals. Sharks have electro-receptors in many small pores around the snout, giving them extremely good electrical sensitivity.

The sense of smell is well developed in most fishes. Some groups, such as the Salmons (Salmonidae, p. 214), have an extremely powerful sense of smell that the adults use to find their way back to the streams in which they were born. Most fishes have nostrils in pairs: an anterior nostril and a posterior nostril. Water enters through the anterior nostril, passes over sensory cells in the nasal organ and exits the posterior nostril. Agnathans, though, have a single nostril, and Chondrichthyans have incompletely separated nostrils. Fishes use their sense of smell to locate prey. In some cases it is also used to trace pheromones and locate mates – this is particularly evident in males of some deepwater Anglerfish species (Lophiiformes, p. 291), which have greatly expanded nasal organs.

The eyes of fishes are very similar to those of higher vertebrates. However, only the Chondrichthyans can adjust the iris and hence the size of the pupil. Osteichthyans have a rigid iris and many possess thin plates of bone behind the iris to protect the fragile interior of the eye. There are two types of light receptor in the retina of fishes: rods, which are sensitive to low light levels, and cones, which are sensitive to brighter light. Shallow-water fishes have more cones and are sensitive to the entire colour spectrum. Deepwater fishes have more rods and are more sensitive to colours that penetrate furthest in water: blue, violet and ultraviolet (most photophores found in deepwater fishes emit blue light). Nocturnal fishes also have far more rods than cones, for better night vision. Some deepwater fishes have developed tube-like eyes for better light sensitivity and depth perception, while others have accessory lenses and reflective layers within the eye to rebound light onto the retina and increase sensitivity. Conversely, many fishes living in permanent darkness, such as subterranean Gudgeons (Eleotridae, p. 674), have lost the use of their eyes entirely and rely on the other senses to navigate and function in their environment.

MYXINI

HAGFISHES

MYXINIFORMES

Myxiniforms are the most primitive of the modern fishes. Along with other jawless fishes, they are thought to belong to a sister group of all other fishes and vertebrates, diverging from a common ancestor up to 500 million years ago. Myxiniforms have an eel-like body with no scales, a cartilaginous skeleton with no vertebrae, and no true bone. They do not have a jaw structure, although a horny biting apparatus is present. (In other fishes, the jaws are derived from the gill arches.) There are no ventral, pectoral, dorsal or anal fins, but a ventral skin fold resembling an anal fin may be present. The eye lenses and eye muscles are absent, and there are 1–16 pairs of gill openings along the sides of the body. There is a single family in the order and it is present in Australian waters.

HAGFISHES

Myxinidae

BROADGILLED HAGFISH
Eptatretus cirrhatus

There are about 70 species of Hagfish and they are found in temperate waters worldwide. The head bears three pairs of barbels, a single large nostril at the tip of the snout, and rudimentary eyes buried under the skin. Although they lack true jaws, Hagfishes have rows of sharp, horny, biting plates in the mouth, and sharp, horny teeth on the tongue and palate. They have rows of pores along the lateral surface of the body that secrete large amounts of slimy mucus when the fish is threatened or handled.

The two Australian species, both from the genus *Eptatretus,* each have 6–7 pairs of gill openings on the ventral surface of the body. They inhabit muddy bottoms in deep water from about 50 to 700 m. Hagfishes are opportunistic scavengers and feed by boring into the bodies of dead or dying animals. In deeper waters, much of their time is spent in a state of torpor – they lie buried in the muddy bottom awaiting foodfalls (when dead or injured animals sink to the bottom from upper layers of the ocean), which they detect using their keen sense of smell. They are occasionally taken in traps set for other species, which they enter to feed on the fish or crustaceans caught inside. Hagfishes grow to a maximum length of about 1 m.

AUSTRALIAN MYXINIDAE SPECIES

Eptatretus cirrhatus Broadgilled Hagfish
Eptatretus longipinnis Longfin Hagfish

PETROMYZONTIDA

LAMPREYS

PETROMYZONTIFORMES

The order Petromyzontiformes, along with the Myxiniformes (p. 27) and other, extinct, forms of jawless fishes, is thought to be a sister group to all other fishes and vertebrates. Petromyzontiforms have an eel-like body with no scales and no true bone, supported by cartilage and fibrous tissue. There is no jaw structure and the mouth is in the form of a sucking disc that bears sharp, horny teeth – these teeth are also present on the tongue. (In other fishes, the jaws are derived from the gill arches.) They have no pectoral or ventral fins, 1–2 dorsal fins, and seven pairs of gill openings along the sides of the body. A single, prominent nostril is on the dorsal surface of the head. There are three families in the order, two of which are found in Australia.

SHORTHEAD LAMPREYS

Mordaciidae

SHORTHEAD LAMPREY
Mordacia mordax

Shorthead Lampreys are found only in the southern hemisphere, in south-eastern Australia, Tasmania and Chile. There are three species in the family. They have an eel-like body that is typical of Lampreys, but their eyes are positioned on the dorsal surface of the head, rather than laterally. Most adult Shorthead Lampreys, like other Lampreys, are parasitic, and feed by attaching themselves to prey with their oral disc and rasping the flesh with their teeth and tongue to consume blood and tissue. They reach a maximum length of about 50 cm.

The **Shorthead Lamprey** lives in the sea or in the lower reaches of rivers in southern Australia; adults migrate to the upper reaches of freshwater streams to spawn before dying. The eggs hatch into larvae called ammocoetes, which live in sediment in the upper reaches of rivers for about three and a half years, feeding on microscopic algae filtered from the water, before metamorphosing into adults and migrating downstream to the sea. The **Non-parasitic Lamprey**, which is found only in two rivers in New South Wales, does not feed at all during its brief adult life, instead relying on stored fat for energy. Populations of both Australian species are threatened by the general decline in river conditions, as well as by man-made barriers to upstream migration.

AUSTRALIAN MORDACIIDAE SPECIES

Mordacia mordax Shorthead Lamprey
Mordacia praecox Non-parasitic Lamprey

POUCH LAMPREY

Geotriidae

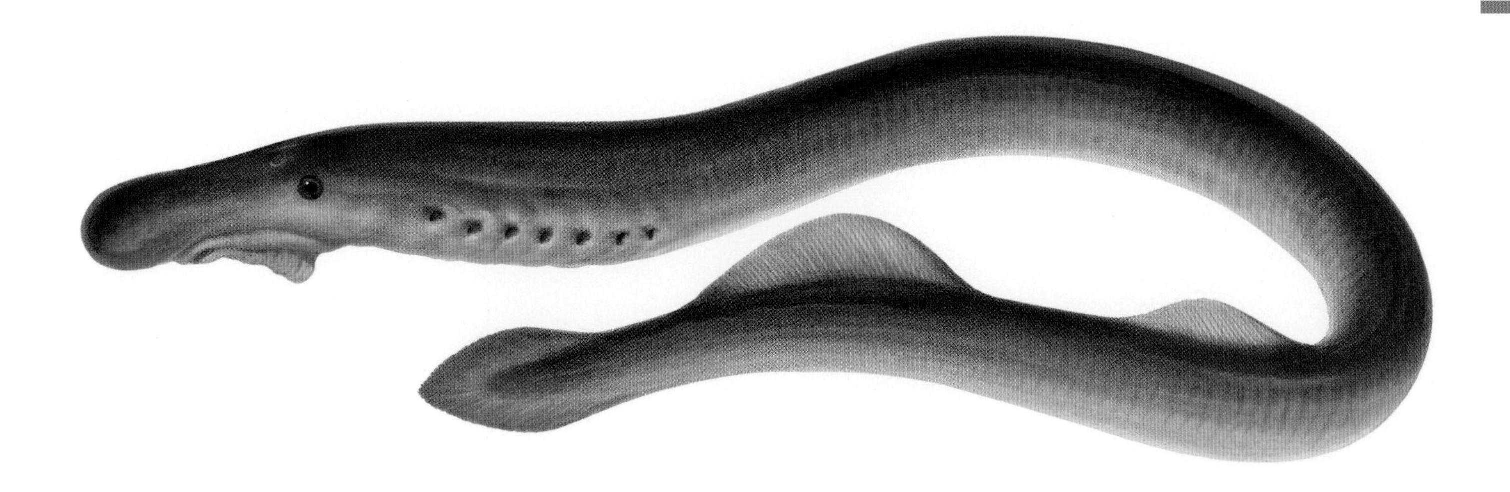

POUCH LAMPREY
Geotria australis

The **Pouch Lamprey** is the only member of this family. It has a wide distribution in the southern hemisphere and is found in waters of southern Australia, New Zealand, South America and islands in the Southern Ocean. It is anatomically very similar to the Shorthead Lampreys (Mordaciidae, p.34), but has finely branched flaps of tissue around the mouth, and eyes positioned laterally instead of on the dorsal surface of the head. Mature males also have a loose pouch beneath the head and an enlarged snout.

The biology of the species is not well known, but its life cycle closely resembles that of the Shorthead Lampreys. The worm-like larvae, or ammocoetes, burrow into muddy substrates in the upper reaches of freshwater streams and filter microscopic organisms and detritus from the water. While metamorphosing into an intermediate phase they move downstream into the sea, where as adults they adopt a parasitic mode of feeding, attaching to fish with their sucking oral disc and rasping the flesh with their teeth and tongue to consume blood and tissue. After several years in the sea, adults return to fresh water to spawn. The Pouch Lamprey grows to a maximum length of about 60 cm. It is highly esteemed as a food fish in some areas and is caught in traps during the migration upstream.

AUSTRALIAN GEOTRIIDAE SPECIES
Geotria australis Pouch Lamprey

CHONDRICHTHYES

SHARKS & RAYS

CHIMAERIFORMES

Chimaeriforms have a cartilaginous skeleton, but differ from all other sharks and rays in that the upper jaw is completely fused to the cranium (rather than being suspended from the cranium by connecting elements of cartilage and ligaments). This characteristic places them in their own subclass of the Chondrichthyans or cartilaginous fishes; the Holocephali. All other sharks and rays form another subclass: the Elasmobranchii.

Chimaeriforms have a fleshy flap covering the gills, leaving a single opening on each side. They have no spiracles, and smooth skin with no dermal denticles. There are two dorsal fins: the first is short-based, tall and movable with a spine at the leading edge; the second is long-based, low and fixed with no spines. The males of all shark and ray species possess claspers (sexual organs used for internal fertilisation), but Chimaeriforms have additional clasping organs on the head and also anterior to the ventral fins to aid in copulation. The teeth are fused into plates, with two pairs in the upper jaw and one pair in the lower jaw. Chimaeriforms lay large, spindle-shaped egg cases on the substrate. There are three families in the order, all of which occur in Australian waters.

ELEPHANT FISHES

Callorhinchidae

ELEPHANT FISH
Callorhinchus milii

The three species in this family are found in temperate waters of the southern hemisphere. They venture into shallow waters and estuaries to breed, but are generally found at depths from 200 to 1000 m. Also known as Plownose Chimaeras, Elephant Fishes have a snout that is formed into a flattened hook-like structure, which they use to fossick in soft sediment for the molluscs and other benthic invertebrates on which they feed. Their teeth are formed into crushing plates, to cope with their hard-shelled prey. Like the Chimaeridae (p. 43), they have a spine on the first dorsal fin, accessory clasping organs, and sensory pores and canals on the head.

The **Elephant Fish** grows to about 1.2 m in length and is common in Victorian and Tasmanian waters, where it is often taken as bycatch. It is an excellent food fish.

AUSTRALIAN CALLORHINCHIDAE SPECIES

Callorhinchus milii Elephant Fish

SPOOKFISHES

Rhinochimaeridae

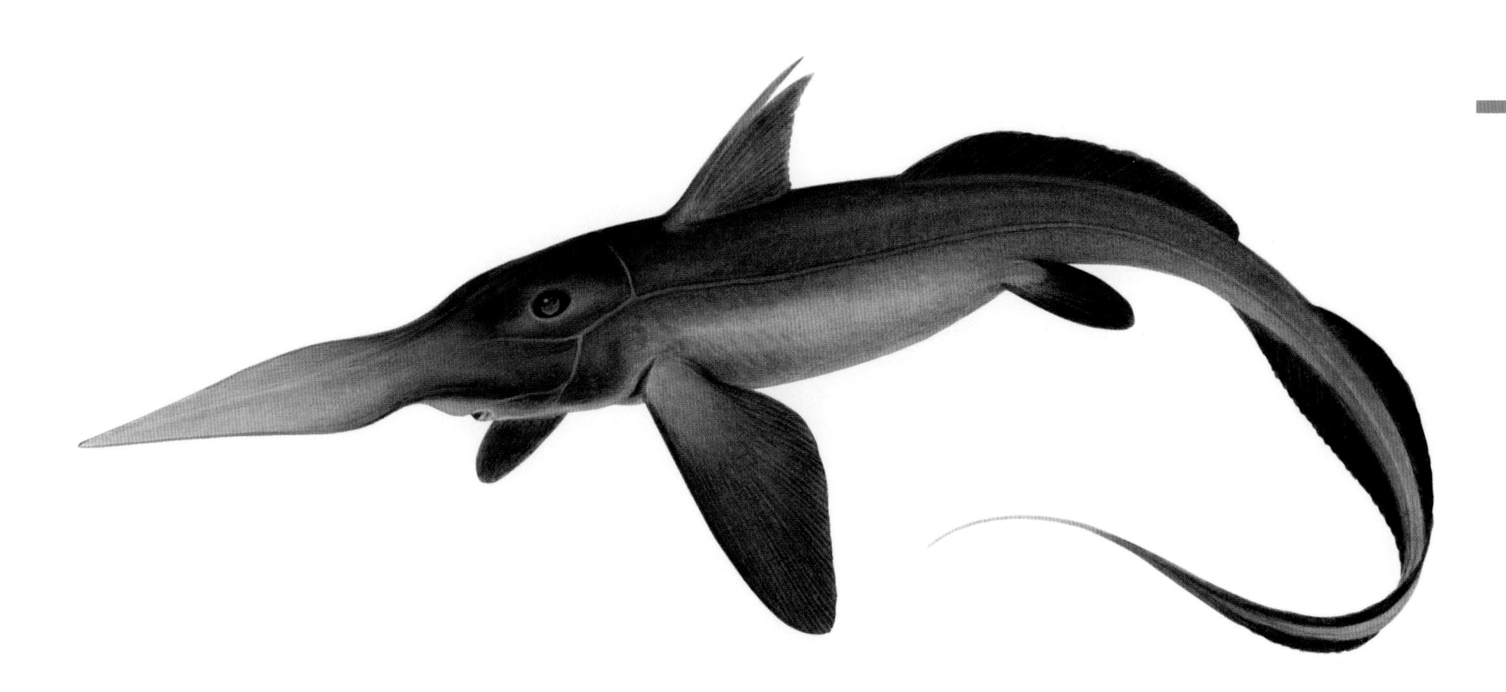

PACIFIC SPOOKFISH
Rhinochimaera pacifica

Spookfishes (also known as Longnosed Chimaeras) are found in deep waters of the Atlantic, Indian and Pacific oceans. There are about eight species in the family. They are very similar to the Chimaeridae (p. 43), but possess a long, pointed snout instead of a blunt, rounded one, and usually lack an anal fin. They have a system of mucus-filled sensory canals on the head that extend onto the ventral surface of the snout. It is likely that the sensitive, enlarged snout is used for detecting invertebrates in the muddy substrates of their deepwater habitat.

Found in scattered locations in tropical and temperate waters around the world, Spookfishes inhabit continental slopes and ocean waters to depths of at least 2600 m. They are rarely encountered and information about their biology is scarce. They are known to lay horny egg capsules on the substrate and to feed on crustaceans and other benthic invertebrates. Spookfishes reach a maximum length of about 1.2 m.

SMALLSPINE SPOOKFISH
Harriotta haeckeli

PADDLENOSE SPOOKFISH
Rhinochimaera africana

AUSTRALIAN RHINOCHIMAERIDAE SPECIES

Harriotta haeckeli Smallspine Spookfish
Harriotta raleighana Bigspine Spookfish
Rhinochimaera africana Paddlenose Spookfish
Rhinochimaera pacifica Pacific Spookfish

SHORTNOSE CHIMAERAS

Chimaeridae

MARBLED GHOSTSHARK
Hydrolagus marmoratus

Shortnose Chimaeras are found around the world in tropical and temperate seas. There are about 22 species in the family. They are mostly restricted to deep water, from about 200 m to at least 2000 m. These strange fishes have teeth fused into plates that in some species form a parrot-like beak, and in others develop as grinding plates further back in the mouth. They have no scales, and smooth, often very fragile skin. The gills are covered by a fleshy flap, leaving only one opening on each side of the body. Water for respiration is taken in principally through the nostrils rather than the mouth. Shortnose Chimaeras have a long, sharp spine at the front of the first dorsal fin, and *Hydrolagus* species usually lack an anal fin. The tapering tail ends in a long filament, which is often damaged during capture, and the smooth, rounded snout is covered with a network of sensory pores and mucus-filled canals.

Slow-swimming bottom dwellers, Shortnose Chimaeras feed mainly on benthic invertebrates and small fishes. They reach a maximum length of about 1.3 m in the **Giant Chimaera**. In Australia several species, such as the **Southern Chimaera**, **Black Ghostshark** and **Ogilby's Ghostshark**, are often taken as bycatch by deepwater trawlers; they are reported to be good food fishes.

GIANT CHIMAERA
Chimaera lignaria

OGILBY'S GHOSTSHARK
Hydrolagus ogilbyi

ABYSSAL GHOSTSHARK
Hydrolagus trolli

AUSTRALIAN CHIMAERIDAE SPECIES

Chimaera argiloba Whitefin Chimaera
Chimaera fulva Southern Chimaera
Chimaera lignaria Giant Chimaera
Chimaera macrospina Longspine Chimaera
Chimaera obscura Shortspine Chimaera
Hydrolagus homonycterus Black Ghostshark
Hydrolagus lemures Blackfin Ghostshark
Hydrolagus marmoratus Marbled Ghostshark
Hydrolagus ogilbyi Ogilby's Ghostshark
Hydrolagus trolli Abyssal Ghostshark

HETERODONTIFORMES

Heterodontiforms are small sharks that have two dorsal fins, both with spines at the base of the leading edge, and an anal fin – a combination of characteristics that distinguishes them from all other sharks. They have five pairs of gill slits, and a small spiracle behind each eye. There are bony crests over the eyes and a prominent groove connecting the mouth and nostrils. Heterodontiforms are oviparous and females lay distinctive spiral-shaped egg cases. There is a single family in the order and it occurs in Australian waters.

HORNSHARKS

Heterodontidae

CRESTED HORNSHARK
Heterodontus galeatus

Hornsharks are found in tropical and warm temperate waters of the Indian and Pacific oceans. There are about eight species in the family. They have strong spines at the base of both dorsal fins, broad pectoral fins and an anal fin. Hornsharks are nocturnal bottom dwellers and are found in coastal waters down to about 200 m. They feed on molluscs, sea urchins, crustaceans and other invertebrates, which they grasp with their small, sharp anterior teeth and crush with their flattened molar teeth.

Growing to about 1.5 m in length, these placid, slow-swimming sharks lay distinctive spiral-shaped egg cases, which they wedge amongst weeds and rocks on the sea floor. Empty cases are often found washed ashore on southern Australian beaches. The **Port Jackson Shark** is frequently caught by anglers off the southern half of Australia, but is not generally considered to be a food fish.

PORT JACKSON SHARK
Heterodontus portusjacksoni

ZEBRA HORNSHARK
Heterodontus zebra

AUSTRALIAN HETERODONTIDAE SPECIES

Heterodontus galeatus Crested Hornshark
Heterodontus portusjacksoni Port Jackson Shark
Heterodontus zebra Zebra Hornshark

ORECTOLOBIFORMES

Orectolobiforms have two dorsal fins without spines, and an anal fin. All species except the Whale Shark have a long, almost horizontal caudal fin with a well-developed lower lobe. They have five pairs of small gill slits, with the fourth gill slit overlapping the fifth in all species except the Whale Shark, which has large, non-overlapping gill slits. A spiracle is present below or behind each eye and there are prominent grooves connecting the mouth to the nostrils. The transverse mouth is entirely anterior to the eyes. Orectolobiforms are either oviparous or ovoviviparous. There are seven families in the order, all of which occur in Australian waters.

COLLARED CARPETSHARKS

Parascyllidae

COLLAR CARPETSHARK
Parascyllium collare

Collared Carpetsharks inhabit tropical and warm temperate waters of the western Pacific and around southern Australia. There are about eight species in the family. They are small, slender sharks, with two equal-sized dorsal fins, a short, rounded snout and a very small mouth that is connected by shallow grooves to the nostrils. They have a tiny spiracle below each slit-like eye, and small nasal tentacles.

In Australia Collared Carpetsharks are found in coastal waters from the shallows down to about 180 m, mainly around the southern half of the continent. The five species of Collared Carpetshark found in Australia are all endemic, but very little is known of their biology. The **Elongate Carpetshark** is known only from a single specimen that was found in the stomach of a School Shark (Triakidae, p. 76). Growing to a maximum length of about 90 cm, Collared Carpetsharks are nocturnal and feed on a range of invertebrates and small fishes. Females lay flattened egg cases with tendrils that anchor them to the substrate.

ELONGATE CARPETSHARK
Parascyllium elongatum

RUSTY CARPETSHARK
Parascyllium ferrugineum

VARIED CARPETSHARK
Parascyllium variolatum

AUSTRALIAN PARASCYLLIDAE SPECIES

Parascyllium collare Collar Carpetshark
Parascyllium elongatum Elongate Carpetshark
Parascyllium ferrugineum Rusty Carpetshark
Parascyllium sparsimaculatum Ginger Carpetshark
Parascyllium variolatum Varied Carpetshark

BLIND SHARKS

Brachaeluridae

COLCLOUGH'S SHARK
Brachaelurus colcloughi

The two small sharks in this family are endemic to Australia. They are found in tropical and temperate waters of New South Wales and Queensland, usually on shallow rocky or coral reefs. Blind Sharks are stout-bodied and sluggish, with two equal-sized dorsal fins. They have sharp, tricuspid teeth, a small mouth that is connected to the nostrils by shallow grooves, and nasal barbels. Blind Sharks also have a large spiracle situated below and close to each eye. They are viviparous and give birth to litters of 6–8 pups. Growing to a maximum size of just over 1 m, they feed on small fishes and a wide range of reef invertebrates.

Blind Sharks are not blind (they have normal vision), but are so named because of their habit of closing their eyes when removed from the water. These sharks are able to survive long periods out of the water; this is probably an adaptation to frequent strandings in the intertidal zones in which they are often found. They are often caught by anglers fishing from the shore along the east coast of Australia.

BLIND SHARK
Brachaelurus waddi

AUSTRALIAN BRACHAELURIDAE SPECIES

Brachaelurus colcloughi Colclough's Shark
Brachaelurus waddi Blind Shark

WOBBEGONGS

Orectolobidae

ORNATE WOBBEGONG
Orectolobus ornatus

Wobbegongs are found in coastal waters of the Indo-West Pacific. There are about 11 species in the family, 10 of which occur in Australian waters. They have a wide, flattened head with the eyes on the dorsal surface, short, broad pectoral fins and two large, almost equal-sized dorsal fins. There is a small anal fin and a long, almost horizontal caudal fin without a ventral lobe. Wobbegongs have a very large spiracle behind each eye to allow continuous water flow for respiration. Most have skin flaps visible around the edges of the head. They grow to a maximum length of about 3 m.

Wobbegongs are found in most coastal waters around Australia, from the shallows down to about 100 m. Although the biology of many species is not well known, they are all thought to be viviparous and to give birth to quite large litters. A litter of 47 pups has been recorded for the **Gulf Wobbegong**. Wobbegongs rest immobile on the sea floor in caves or beneath ledges during the day and feed nocturnally on a wide variety of invertebrates and fishes. They have powerful jaws with long, dagger-like teeth in the front of the mouth. Wobbegongs have been responsible for numerous attacks on humans, often as the result of accidental contact by divers – they have superb camouflage – but swimmers have also been bitten well above the sea floor. Despite their innocuous and somewhat sleepy appearance, Wobbegongs should always be treated with caution and never provoked.

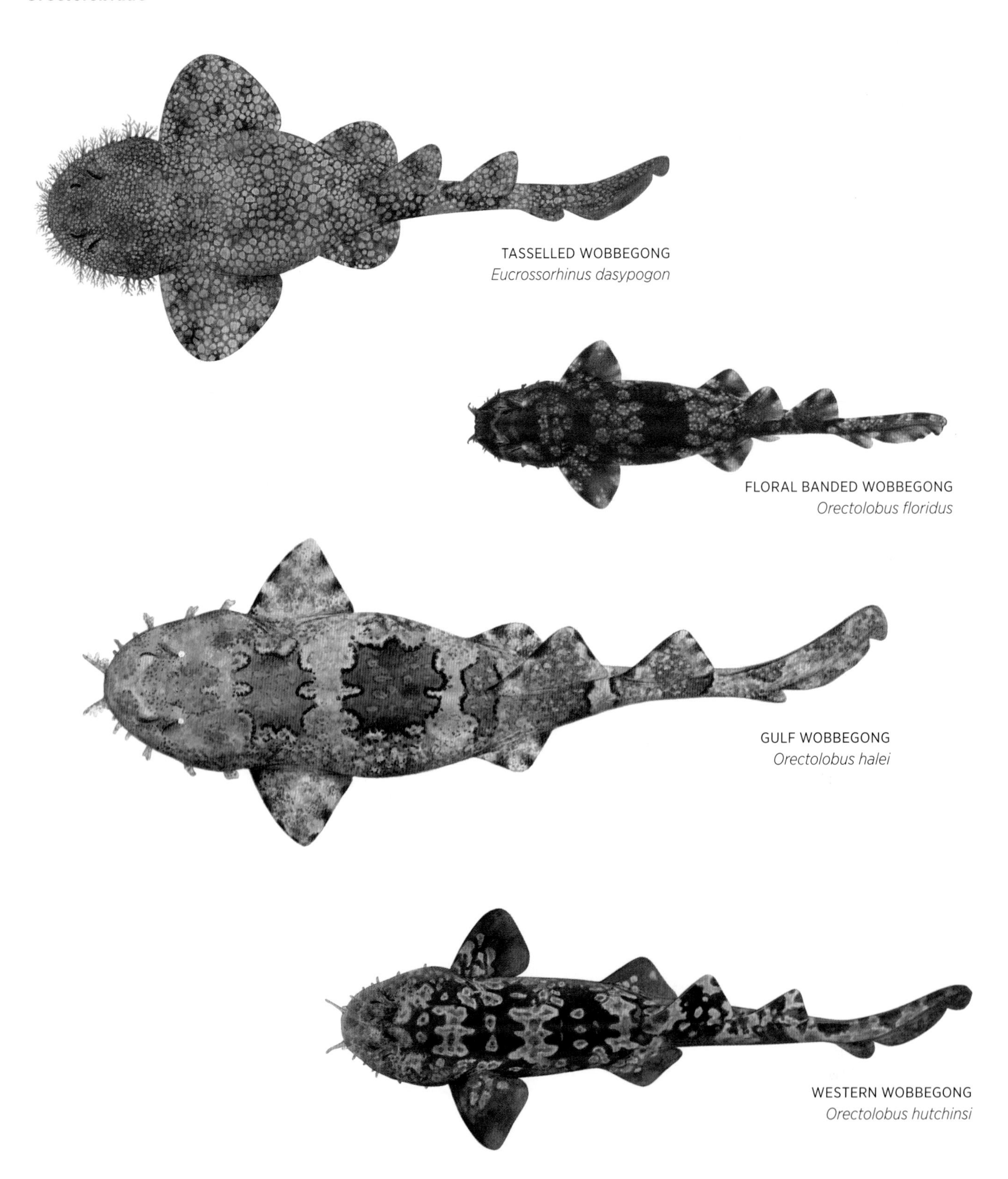

TASSELLED WOBBEGONG
Eucrossorhinus dasypogon

FLORAL BANDED WOBBEGONG
Orectolobus floridus

GULF WOBBEGONG
Orectolobus halei

WESTERN WOBBEGONG
Orectolobus hutchinsi

SPOTTED WOBBEGONG
Orectolobus maculatus

DWARF SPOTTED WOBBEGONG
Orectolobus parvimaculatus

COBBLER WOBBEGONG
Sutorectus tentaculatus

AUSTRALIAN ORECTOLOBIDAE SPECIES

Eucrossorhinus dasypogon Tasselled Wobbegong
Orectolobus floridus Floral Banded Wobbegong
Orectolobus halei Gulf Wobbegong
Orectolobus hutchinsi Western Wobbegong
Orectolobus maculatus Spotted Wobbegong
Orectolobus ornatus Ornate Wobbegong
Orectolobus parvimaculatus Dwarf Spotted Wobbegong
Orectolobus reticulatus Network Wobbegong
Orectolobus wardi Northern Wobbegong
Sutorectus tentaculatus Cobbler Wobbegong

LONGTAIL CARPETSHARKS

Hemiscylliidae

EPAULETTE SHARK
Hemiscyllium ocellatum

There are about 12 species of Longtail Carpetshark and they are distributed from Madagascar in the western Indian Ocean to the western Pacific. Similar to the Blind Sharks (Brachaeluridae, p. 52), they possess small nasal barbels, grooves connecting the nostrils and the mouth, and a large spiracle below and close to each eye. They have a longer tail than Blind Sharks and a small anal fin that is almost continuous with the caudal fin.

These small, harmless sharks grow to about 1 m in length and inhabit tropical waters of Australia's north. They are reasonably common in intertidal zones and are often encountered around shallow reefs. They use their short, muscular fins almost like legs to manoeuvre themselves in very shallow water and restricted spaces. They are also able to survive long periods out of water, an adaptation to the frequent strandings they endure in this habitat. Longtail Carpetsharks forage nocturnally for invertebrates and small fishes. They are oviparous and lay roughly retangular or oval flattened eggs on the substrate.

GREY CARPETSHARK
Chiloscyllium punctatum

SPECKLED CARPETSHARK
Hemiscyllium trispeculare

AUSTRALIAN HEMISCYLLIIDAE SPECIES

Chiloscyllium punctatum Grey Carpetshark
Hemiscyllium ocellatum Epaulette Shark
Hemiscyllium trispeculare Speckled Carpetshark

ZEBRA SHARK

Stegostomatidae

ZEBRA SHARK (adult and juvenile)
Stegostoma fasciatum

The **Zebra Shark** is the only member of this family and is an unmistakeable inhabitant of tropical and subtropical inshore waters of the Indo-West Pacific. It is named for the vivid black and yellow stripes of the juveniles – this colouration gradually changes to the dark, spotted pattern seen in the adults. Several strong ridges of skin run the length of the rounded body, and the caudal fin is extremely broad and elongate. Reaching about 2.5 m in length, this docile shark is found in northern Australian waters. It is often found resting on sandy bottoms during the day in areas of high water current, facing into the flow with its mouth open to aid respiration. The Zebra Shark appears to be nocturnal, feeding mainly on molluscs, although other invertebrates and small fishes are also taken. It is oviparous and females lay flattened egg cases with bunches of hair-like fibres to anchor the eggs to the substrate.

AUSTRALIAN STEGOSTOMATIDAE SPECIES
Stegostoma fasciatum Zebra Shark

NURSE SHARKS

Ginglymostomatidae

TAWNY SHARK
Nebrius ferrugineus

The three species of Nurse Shark inhabit inshore tropical and subtropical waters around the world. The **Tawny Shark** is the only member of the family found in Australian waters. It has small, fan-shaped serrated teeth, nasal barbels, grooves connecting the nostrils to the mouth, and a small spiracle behind and below each eye. The Tawny Shark inhabits shallow tropical waters and forages at night over sand and reef areas, often resting in small groups in caves during the day. It feeds on a range of small fishes, as well as invertebrates such as crustaceans, cephalopods and sea urchins, and can capture small prey by powerfully sucking water into its mouth. It grows to about 3 m in length but is quite docile and is harmless to humans. The Tawny Shark is viviparous, bearing litters of around 20–30 pups.

AUSTRALIAN GINGLYMOSTOMATIDAE SPECIES

Nebrius ferrugineus Tawny Shark

WHALE SHARK

Rhincodontidae

WHALE SHARK
Rhincodon typus

The well-known and spectacular **Whale Shark** is the only member of this family. It is the largest of all the fishes and grows to at least 12 m in length. The Whale Shark migrates over large distances and is found worldwide in tropical and subtropical waters. Details of its migrations are largely unknown but research is being carried out to track its movements. Usually found near the surface, it is known to dive to depths of 1000 m or more, although the amount of time it spends at such depths is unknown.

The broad, flattened mouth of the Whale Shark contains more than 300 rows of tiny, hooked teeth. These sharks are gentle giants, feeding mostly on small planktonic prey such as euphausiid shrimps, as well as on small schooling fishes and squid. Occasionally they will take larger fishes, and even small Tunas have been found in their stomachs. The Whale Shark's normal mode of feeding is to swim slowly, mouth open, through waters rich in plankton. However, unlike other passive filter feeders such as the Basking Shark (Cetorhinidae, p. 68), the Whale Shark can also suck water and prey into its mouth using muscular action, and has been observed feeding suspended vertically in the water with mouth near the surface. I have also watched a 9 m Whale Shark approach surface schools of euphausiid shrimps at high speed from below, with its mouth wide open, powering through the schools and surging out of the water as far as the first dorsal fin. Each autumn, Whale Sharks aggregate off Ningaloo Reef on the West Australian coast and snorkellers often swim with them there. The Whale Shark is viviparous and produces litters of up to 300 pups.

AUSTRALIAN RHINCODONTIDAE SPECIES

Rhincodon typus Whale Shark

LAMNIFORMES

Lamniforms are large sharks with two dorsal fins, both without spines, and an anal fin. They have five pairs of gill slits; in some species all five gill slits are anterior to the pectoral-fin base, while in others the last two may be positioned above the pectoral fin. A small spiracle is usually present behind each eye, and the eyes are without a nictitating membrane. The large jaws extend behind the eyes and most species have well-developed teeth for a carnivorous diet, though several are filter feeders. Lamniforms are ovoviviparous. There are seven families in the order, all of which occur in Australian waters.

GREY NURSE SHARKS

Odontaspidae

GREY NURSE SHARK
Carcharias taurus

Grey Nurse Sharks are found in tropical and temperate waters of the Atlantic, Indian and Pacific oceans, and there are three species in the family. They are large, deep-bodied sharks with two dorsal fins of almost equal size. They grow to over 3.5 m in length and occur from the shallows to depths of 800 m or more. They have long, curved, pointed teeth and feed mainly on fishes.

The **Grey Nurse Shark** is found right around Australia but is more common in temperate waters. Once greatly feared as a maneater and indiscriminately killed, this shark is now protected in much of its range. Despite this, numbers are still critically low in many areas. Occasionally encountered in small groups by divers, it is usually slow-moving and rarely aggressive. Its graceful movement, as well as its fearsome-looking teeth, make it a popular choice for large public aquariums. The rarely encountered **Sand Tiger Shark** is very similar, but has a slightly smaller second dorsal fin and is usually found in much deeper waters. Uterine cannibalism (where one embryo consumes all others in the uterus) occurs in this family, so litters contain no more than two pups, one from each uterus.

SAND TIGER SHARK
Odontaspis ferox

AUSTRALIAN ODONTASPIDAE SPECIES

Carcharias taurus Grey Nurse Shark
Odontaspis ferox Sand Tiger Shark

GOBLIN SHARK

Mitsukurinidae

GOBLIN SHARK
Mitsukurina owstonii

The strange **Goblin Shark**, the only member of this family, has been found in scattered localities around the world, including off south-east Australia. It lives near the bottom in the deeper waters of continental shelves, to 1300 m and more. It is rarely encountered and little is known of its biology. Growing to nearly 4 m in length, it has a soft, flabby body, small fins and eyes, and a long, flattened, blade-like snout that is covered with sensory cells. The jaws are highly protrusible and bear very long, slender teeth. The Goblin Shark probably feeds by moving slowly along the bottom, using the electoreceptors on its snout to detect prey in the darkness, and snapping up fishes, crustaceans and cephalopods with its protrusible jaws.

AUSTRALIAN MITSUKURINIDAE SPECIES
Mitsukurina owstonii Goblin Shark

CROCODILE SHARK

Pseudocarchariidae

CROCODILE SHARK
Pseudocarcharias kamoharai

The **Crocodile Shark** is the only member of this family. It is widespread in tropical oceanic waters around the world and is found from the surface to depths of about 600 m. This small, stout-bodied shark grows to about 1 m in length and has large gill slits, quite low dorsal fins, and fleshy keels on the caudal peduncle. Although little is known about the behaviour and preferred habitat of the Crocodile Shark, its very large eyes indicate that it is nocturnal or occupies deep waters. It has protrusible jaws with slender, curved teeth, like those typically found in predators of fishes, squid and crustaceans. It appears to undertake nightly migrations towards the surface and is often caught by offshore longline fisheries. The Crocodile Shark is ovoviviparous, with litters of 2–4 pups. The embryos feed on unfertilised eggs while still in the uterus. Little else is known of its biology. The species gets its name from its habit of snapping furiously when removed from the water.

AUSTRALIAN PSEUDOCARCHARIIDAE SPECIES

Pseudocarcharias kamoharai Crocodile Shark

MEGAMOUTH SHARK

Megachasmidae

MEGAMOUTH SHARK
Megachasma pelagios

The only member of this family, the **Megamouth Shark** was discovered in 1976, when a large specimen became entangled in a ship's anchor-line off Hawaii. Since then, about 40 more specimens have been collected from various localities around the world. Growing to over 5 m in length, the Megamouth Shark has a long, flabby body and small gill slits. Its enormous, wide mouth has a reflective silver interior, which perhaps helps to camouflage it when approaching prey. The highly protrusible jaws bear numerous tiny, hooked teeth.

This shark is slow-swimming and inhabits mid to deep waters. It makes nightly migrations towards the surface, filter feeding on planktonic animals such as small fishes, euphausiid shrimps, copepods and jellyfishes. I was lucky enough to observe a large specimen that washed ashore near Mandurah in Western Australia, in excellent condition and still alive (though barely). Attempts to revive it failed; it was subsequently preserved and for many years was on display at the Western Australian Museum.

AUSTRALIAN MEGACHASMIDAE SPECIES
Megachasma pelagios Megamouth Shark

THRESHER SHARKS

Alopiidae

THRESHER SHARK
Alopias vulpinus

The three species in this family occur worldwide in tropical and temperate waters. They are mainly found offshore, from the surface to depths of 500 m or more. The upper lobe of the caudal fin is enormously elongated (almost as long as the body), while the anal and second dorsal fins are minute. The long tail is used to round up schooling fishes and squid, and also to stun them using a whip-like action. Thresher Sharks are often caught by the tail on baited longline hooks. They have quite a small mouth and small teeth, and feed mainly on pelagic fishes and squid, though some bottom-dwelling species are also taken.

The **Thresher Shark** grows to more than 5 m in length. The extraordinary **Bigeye Thresher Shark** has enormous eyes that extend onto the dorsal surface of the head. It is usually found in deeper waters than the other two species in this family.

Like the closely related Mako Sharks (Lamnidae, p. 69), Thresher Sharks have a modified circulatory system that acts as a heat exchanger, enabling them to maintain a body temperature higher than the temperature of the surrounding water – this allows for the greater muscle activity needed for their fast-swimming habits.

PELAGIC THRESHER SHARK
Alopias pelagicus

BIGEYE THRESHER SHARK
Alopias superciliosus

AUSTRALIAN ALOPIIDAE SPECIES

Alopias pelagicus Pelagic Thresher Shark
Alopias superciliosus Bigeye Thresher Shark
Alopias vulpinus Thresher Shark

BASKING SHARK

Cetorhinidae

BASKING SHARK
Cetorhinus maximus

The **Basking Shark** is the only member of this family. It is found seasonally in coastal temperate waters around the world, its appearances corresponding with high concentrations of plankton. Second only to the Whale Shark (Rhincodontidae, p. 60) in terms of size, the Basking Shark can reach more than 10 m in length. A passive filter feeder, it swims slowly with mouth wide open, snaring plankton on its hair-like gill-rakers. These gillrakers are periodically shed and it is thought the Basking Shark descends to feed in deep water during the winter months.

The Basking Shark is only occasionally seen in southern Australian waters, but elsewhere in the world it has been hunted for its meat and enormous oil-rich liver. Slow-moving and easily approached and harpooned, populations of these sharks are quickly reduced and they are now protected in many areas.

AUSTRALIAN CETORHINIDAE SPECIES

Cetorhinus maximus Basking Shark

MACKEREL SHARKS

Lamnidae

GREAT WHITE SHARK
Carcharodon carcharias

The family of Mackerel Sharks includes five species of large, highly active and potentially dangerous sharks. They are found worldwide, in temperate and tropical zones, from very shallow coastal waters to depths of more than 1200 m for the **Great White Shark**. All have a muscular, spindle-shaped body, two dorsal fins (the second very small), a pointed conical snout, a fleshy keel on the caudal peduncle, and the stiff, crescent-shaped tail characteristic of powerful fast-swimming fishes. Their circulatory system has a built-in heat exchanger, which enables them to maintain a body temperature higher than the water temperature and thus remain active and fast-swimming even in cold waters. All Mackerel Sharks are ovoviviparous, and embryos feed on unfertilised eggs within the uterus.

The **Shortfin Mako** is widespread in oceanic waters, where it preys on Tunas and other pelagic fishes. It is capable of great speed and tremendous leaps out of the water, making it a popular target for sportfishers. The **Porbeagle** prefers cooler waters and is occasionally found over the continental shelf off the southern coast of Australia. All members of this family are capable of taking large prey and feed mainly on fishes and other sharks. Seals and other marine mammals form an important part of the diet of large Great White Sharks, which can reach over 6 m in length. The large, serrated triangular teeth of the Great White, along with its enormous size, make it arguably the most dangerous of all sharks and it has been responsible for numerous fatal attacks on humans. The species is capable of remarkable migrations; during a nine-month period, one tagged individual travelled from South Africa to Australia and back again. Listed internationally as an endangered species, the Great White is now protected in Australian waters and the number of large individuals appears to be increasing.

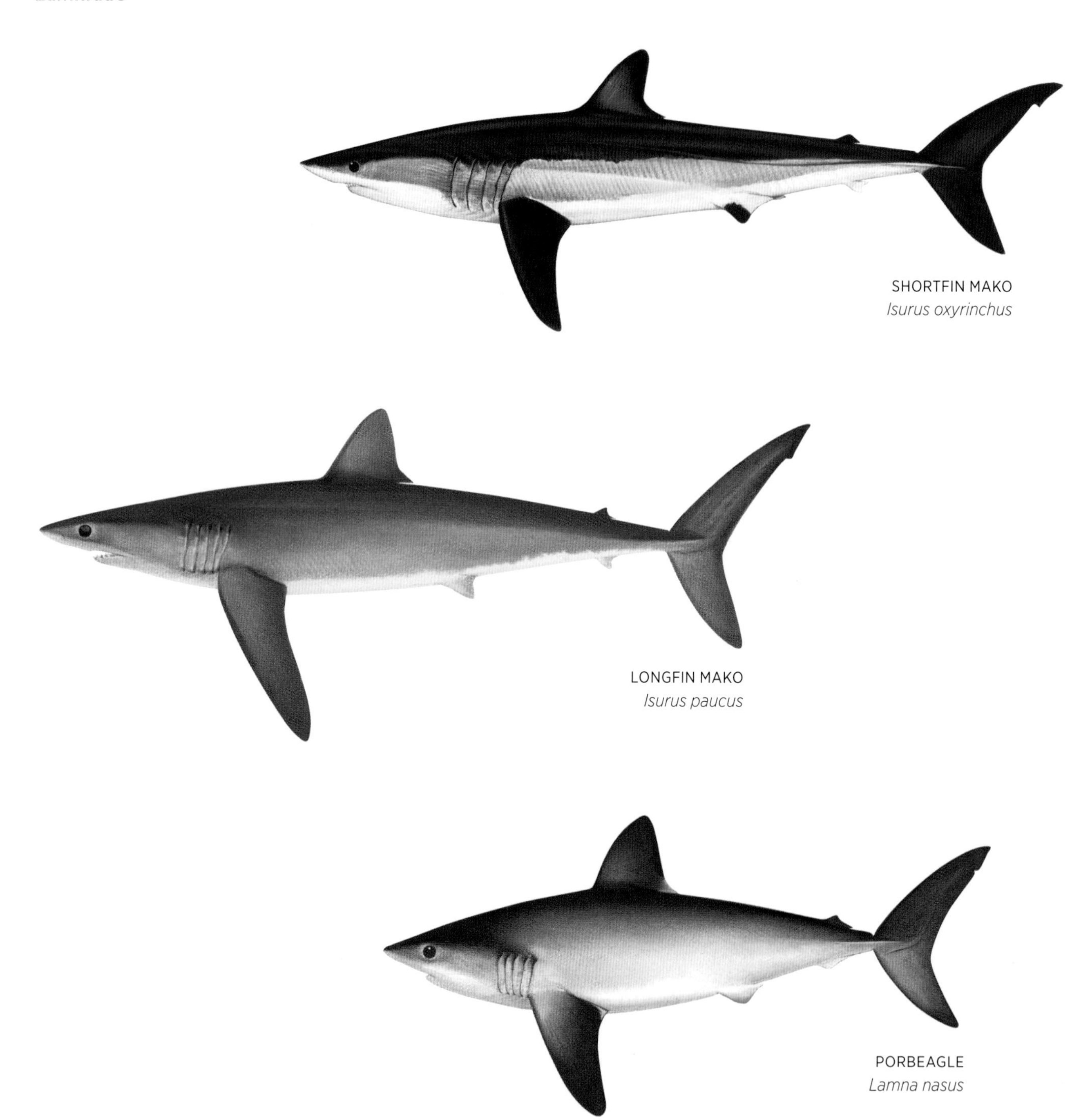

SHORTFIN MAKO
Isurus oxyrinchus

LONGFIN MAKO
Isurus paucus

PORBEAGLE
Lamna nasus

AUSTRALIAN LAMNIDAE SPECIES

Carcharodon carcharias Great White Shark
Isurus oxyrinchus Shortfin Mako
Isurus paucus Longfin Mako
Lamna nasus Porbeagle

CARCHARHINIFORMES

Carcharhiniforms have two dorsal fins without spines, an anal fin, and five pairs of gill slits, the last 1–3 pairs of which are positioned over the pectoral-fin base. Spiracles are usually small or absent altogether. They have large jaws that always extend behind the eyes, well-developed teeth, and nictitating eyelids developed to varying degrees. They may be ovoviviparous, viviparous or oviparous. There are eight families in the order, six of which are found in Australian waters.

CATSHARKS

Scyliorhinidae

DRAUGHTBOARD SHARK
Cephaloscyllium laticeps

This is the largest family of sharks, with more than 160 species found worldwide. Catsharks occur from shallow inshore waters to ocean depths of more than 2000 m. They have moderately large spiracles, two small dorsal fins without spines, and oval or slit-like eyes with rudimentary nictitating membranes. They have quite a large, curved mouth, with small, pointed teeth. Their bodies are rounded or slightly flattened and they reach a maximum length of about 1 m.

Catsharks are usually found close to the bottom and their diet includes a wide variety of invertebrates and fishes. They are either oviparous or ovoviviparous. Most are sluggish and not powerful swimmers, with some having quite restricted distribution. Many new deepwater species have been found in recent years; no doubt more remain to be discovered.

The Swellsharks are so named because when threatened they can inflate their bodies by swallowing water, and the Sawtail Catsharks because they have a line of enlarged denticles on the upper edge of the caudal fin. The **Draughtboard Shark**, named for its blotched colour pattern, is common along the south coast of Australia and often enters rock-lobster pots. Most shallow-water species are rarely encountered due to their nocturnal habits, though I have several times found the **Blackspotted Catshark** sheltering under large plate corals during the day.

SMOOTHBELLY CATSHARK
Apristurus longicephalus

ORANGE SPOTTED CATSHARK
Asymbolus rubiginosus

GULF CATSHARK
Asymbolus vincenti

EASTERN BANDED CATSHARK
Atelomycterus marnkalha

DUSKY CATSHARK
Bythaelurus incanus

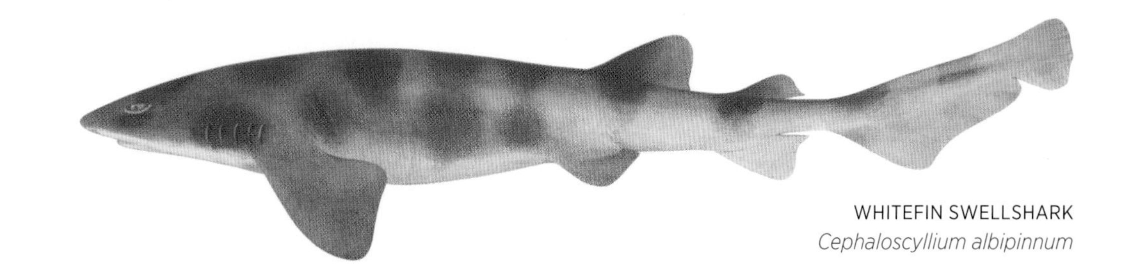

WHITEFIN SWELLSHARK
Cephaloscyllium albipinnum

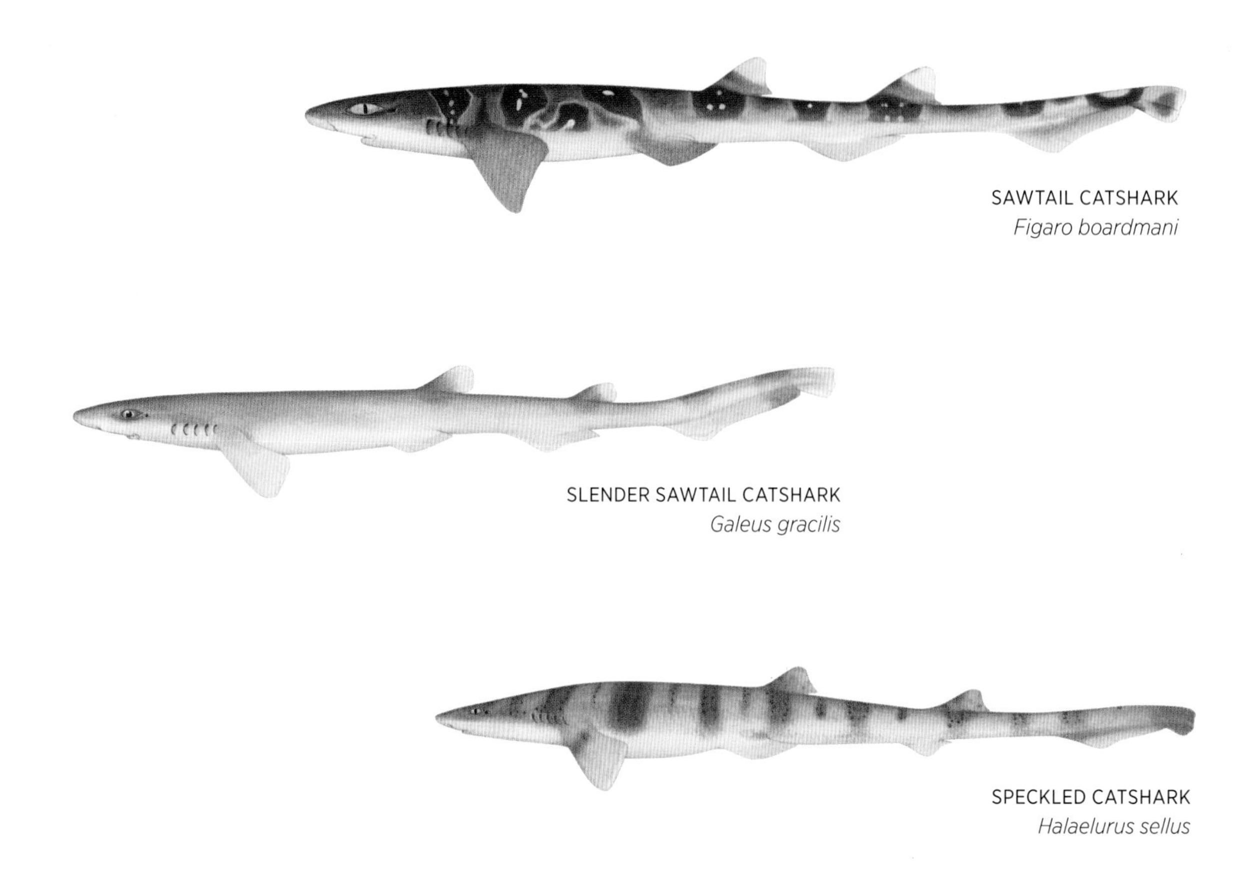

SAWTAIL CATSHARK
Figaro boardmani

SLENDER SAWTAIL CATSHARK
Galeus gracilis

SPECKLED CATSHARK
Halaelurus sellus

AUSTRALIAN SCYLIORHINIDAE SPECIES

Apristurus ampliceps Roughskin Catshark
Apristurus australis Pinocchio Catshark
Apristurus bucephalus Bighead Catshark
Apristurus longicephalus Smoothbelly Catshark
Apristurus melanoasper Fleshynose Catshark
Apristurus pinguis Bulldog Catshark
Apristurus platyrhynchus Bigfin Catshark
Apristurus sinensis Freckled Catshark
Asymbolus analis Grey Spotted Catshark
Asymbolus funebris Blotched Catshark
Asymbolus occiduus Western Spotted Catshark
Asymbolus pallidus Pale Spotted Catshark
Asymbolus parvus Dwarf Catshark
Asymbolus rubiginosus Orange Spotted Catshark
Asymbolus submaculatus Variegated Catshark
Asymbolus vincenti Gulf Catshark
Atelomycterus fasciatus Banded Catshark
Atelomycterus macleay Marbled Catshark
Atelomycterus marnkalha Eastern Banded Catshark
Aulohalaelurus labiosus Blackspotted Catshark
Bythaelurus incanus Dusky Catshark
Cephaloscyllium albipinnum Whitefin Swellshark
Cephaloscyllium cooki Cook's Swellshark
Cephaloscyllium hiscosellum Reticulate Swellshark
Cephaloscyllium laticeps Draughtboard Shark
Cephaloscyllium signourum Flagtail Swellshark
Cephaloscyllium speccum Speckled Swellshark
Cephaloscyllium variegatum Saddled Swellshark
Cephaloscyllium zebrum Narrowbar Swellshark
Figaro boardmani Sawtail Catshark
Figaro striatus Northern Sawtail Catshark
Galeus gracilis Slender Sawtail Catshark
Halaelurus sellus Speckled Catshark
Parmaturus bigus Short-tail Catshark

FALSE CATSHARKS

Pseudotriakidae

FALSE CATSHARK
Pseudotriakis microdon

There are about five species of False Catshark and they are found in scattered locations around the world, in deepwater habitats to about 2000 m. The only species found in Australia, the **False Catshark** is a large, heavy-bodied shark that grows to about 3 m in length. It can be recognised by its long, low first dorsal fin, slit-like eyes, large spiracles, and wide, angular mouth, which contains more than 200 rows of small, sharp teeth.

The false Catshark is rarely encountered and only occasionally taken by deepwater longline and trawl fisheries. Very little is known of its biology, though the watery, soft body indicates a sedentary lifestyle. The large mouth would enable it to take quite large prey and its diet probably includes fishes, other sharks and cephalopods – automatic deepwater cameras have captured it taking fish bait. The False Catshark is ovoviviparous. Embryos are nourished by an attached placental yolk sac, but will also consume unfertilised eggs while in the uterus.

AUSTRALIAN PSEUDOTRIAKIDAE SPECIES
Pseudotriakis microdon False Catshark

HOUNDSHARKS

Triakidae

GUMMY SHARK
Mustelus antarcticus

The family of Houndsharks contains about 43 species and they are found worldwide in tropical and temperate waters. Houndsharks have slender, fusiform bodies and small spiracles. They have a moderately large mouth with small to medium-sized sharp teeth or, as in Gummy Sharks, molariform crushing teeth. Most are strong swimmers and live close to the bottom, feeding on fishes, cephalopods, crustaceans, molluscs and other invertebrates, though some take midwater prey. Several species are known to congregate in large schools and undertake quite long migrations.

The **Gummy Shark** can grow to about 1.8 m in length, but most species in the family grow to a little over 1 m. Highly prized as food fishes, the **Whiskery Shark**, **School Shark** and Gummy Shark are important commercial species fished off southern Australia. All of them suffer from overfishing, but populations appear to be recovering following the introduction of management controls such as quotas. Gummy and School Sharks are occasionally taken by anglers from the shore, particularly at night. Similar species, such as the **Grey Gummy Shark**, **Western Spotted Gummy Shark** and **Eastern Spotted Gummy Shark**, are less common and occur around northern Australia. Houndsharks are viviparous, and embryos are nourished either by an attached yolk sac or with the female supplying nutrients to the embryos in the uterus via a placenta.

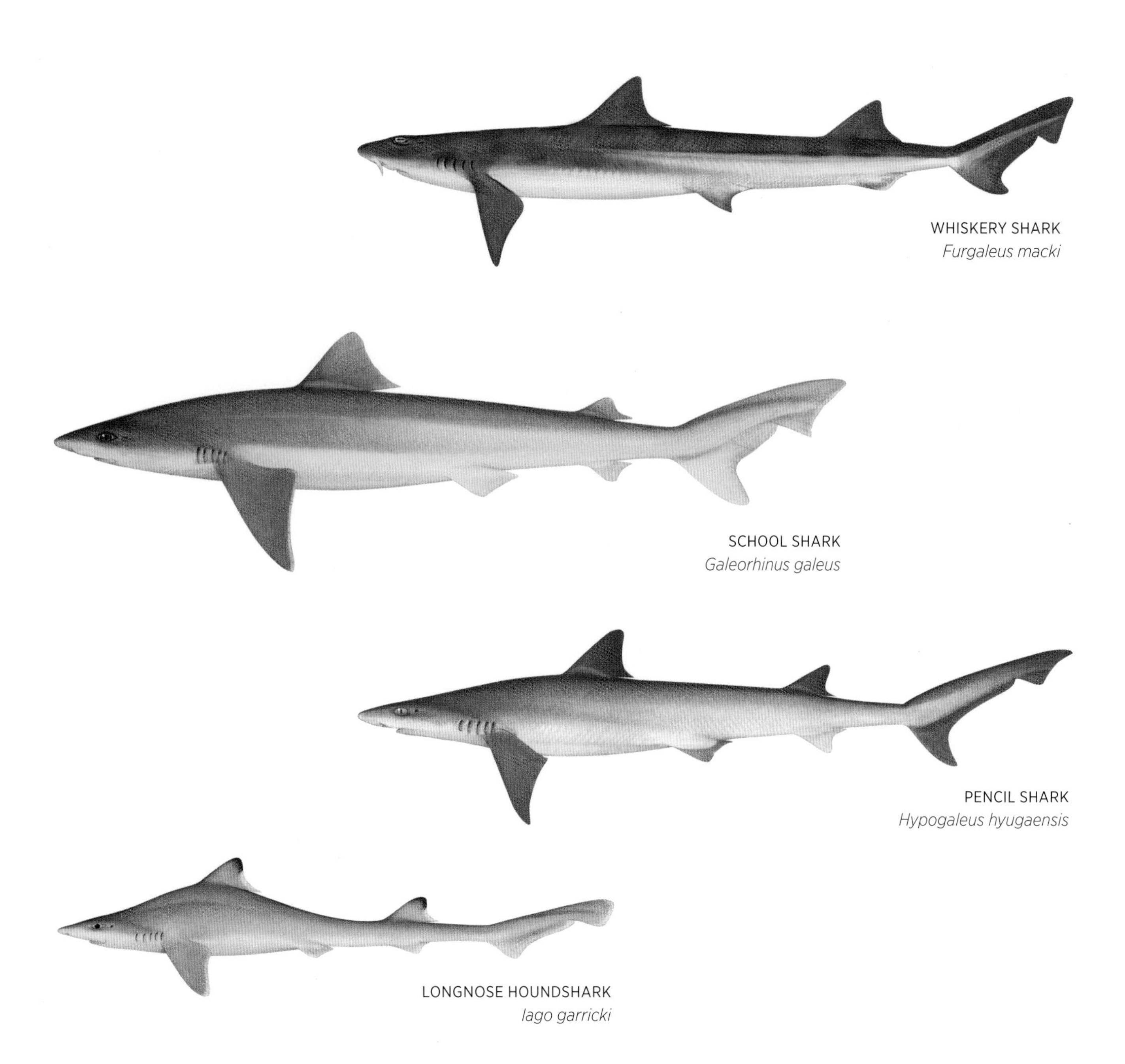

WHISKERY SHARK
Furgaleus macki

SCHOOL SHARK
Galeorhinus galeus

PENCIL SHARK
Hypogaleus hyugaensis

LONGNOSE HOUNDSHARK
Iago garricki

AUSTRALIAN TRIAKIDAE SPECIES

Furgaleus macki Whiskery Shark
Galeorhinus galeus School Shark
Hemitriakis abdita Darksnout Houndshark
Hemitriakis falcata Sicklefin Houndshark
Hypogaleus hyugaensis Pencil Shark
Iago garricki Longnose Houndshark
Mustelus antarcticus Gummy Shark
Mustelus ravidus Grey Gummy Shark
Mustelus stevensi Western Spotted Gummy Shark
Mustelus walkeri Eastern Spotted Gummy Shark

WEASEL SHARKS

Hemigaleidae

WEASEL SHARK
Hemigaleus microstoma

Weasel Sharks are found in coastal waters of the tropical Eastern Atlantic and Indo-Pacific. The family contains about eight species. These sharks are physically similar to the Whaler Sharks (Carcharhinidae, p.79), but have very small spiracles and more curved, tapering fins. The two species in Australia are found mainly in northern waters, from the shallows down to around 170 m. The **Weasel Shark** grows to 1.5 m and the **Fossil Shark** to 2.5 m. The Weasel Shark has angled, serrated teeth and feeds predominantly on cephalopods. The Fossil Shark (also known as the Snaggletooth Shark) has long, hooked teeth that protrude from the lower jaw, and preys on fishes and crustaceans as well as cephalopods. Members of this family are viviparous and the embryos are nourished by a placental yolk sac.

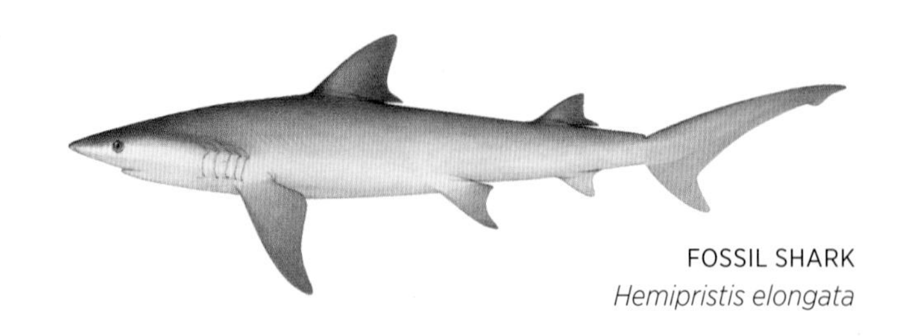

FOSSIL SHARK
Hemipristis elongata

AUSTRALIAN HEMIGALEIDAE SPECIES

Hemigaleus microstoma Weasel Shark
Hemipristis elongata Fossil Shark

WHALER SHARKS

Carcharhinidae

TIGER SHARK
Galeocerdo cuvier

About 54 species of Whaler Shark are found throughout the world's tropical and temperate seas. They have a rounded, muscular body, notches dorsally and ventrally on the caudal peduncle, and a notch before the tip of the caudal fin. They have no spiracles (except for the **Tiger Shark**), eyes with a nictitating membrane, and sharp, serrated, blade-like teeth. They are viviparous, with embryos of all species (except the Tiger Shark) nourished by a placental yolk sac.

Whaler Sharks are common in tropical Australian waters, particularly around reefs and offshore islands. The **Bronze Whaler** and **Dusky Whaler** are found right around the southern coast, often following migrating schools of Australian Salmon (Arripidae, p. 558). Some species, such as the **Blue Shark**, **Silky Shark** and **Oceanic Whitetip Shark**, are wide ranging in the open ocean, while others, such as the **Speartooth Shark** and **Bull Shark**, may move quite long distances up freshwater rivers. The **Whitetip Reef Shark** and **Blacktip Reef Shark** – small species that reach a maximum length of about 1.7 m – are very common around northern coral reefs.

Whaler Sharks are fast-swimming, opportunistic predators that feed on invertebrates and a wide range of fishes, including other sharks. The Tiger Shark, in particular, is known for its varied diet, taking fishes, invertebrates, mammals, turtles, birds and even floating rubbish. Whaler Sharks range in size: Sharpnose Sharks grow to only about 70 cm in length, while the Tiger Shark can reach 7 m.

Whaler Sharks are often portrayed in the media as ferocious maneaters, and many of the larger members of this family are indeed dangerous. The Bull Shark, which frequents murky inshore waters, estuaries and even fresh water, is thought to be responsible for more attacks on humans worldwide than any other shark. The Oceanic Whitetip Shark has made many attacks on humans and the Tiger Shark is responsible for numerous fatalities in Australian waters. All large Whaler Sharks should be regarded as potentially dangerous and treated with caution. However, they are not the bloodthirsty maneaters they are often portrayed to be. Whaler sharks are usually inquisitive and will pass divers for a closer look, but they are actually quite timid and difficult to approach. Divers should always heed warning signs, though – such as a shark arching its back and lowering the pectoral fins, or making rapid, nervous swimming movements – and move away or get out of the water as quickly as possible.

Whaler Sharks are commercially fished all over the world and populations in many areas have been decimated. The trade in shark fins, which are considered a delicacy in some Asian countries, is one of the main reasons for their capture. As peak predators, Whaler Sharks play an important role in the ecosystem and a decline in their numbers may cause serious ecological imbalances. Many scientists and environmentalists are calling on governments and fishing industries around the world to impose stronger measures to protect Whaler Sharks.

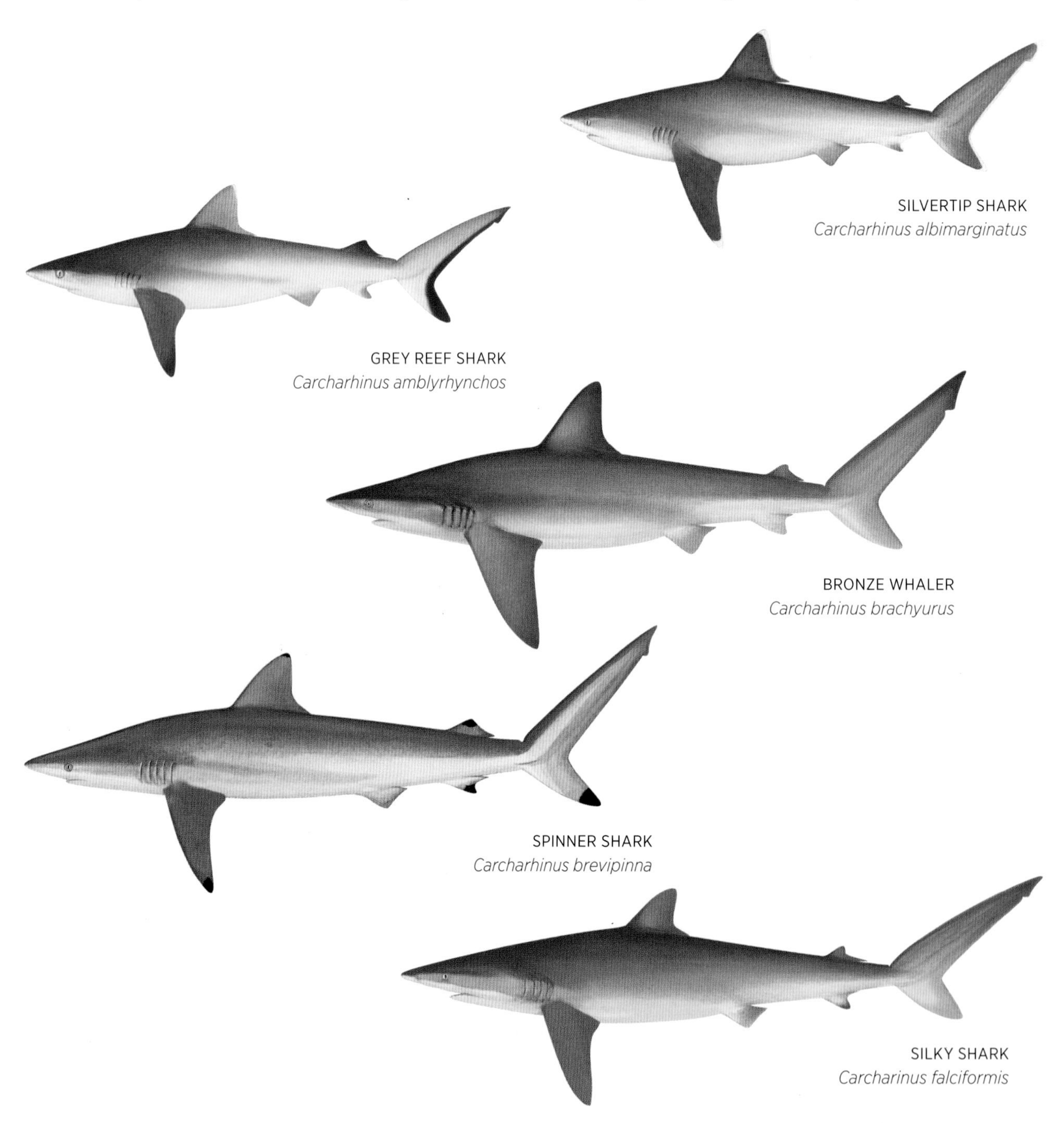

SILVERTIP SHARK
Carcharhinus albimarginatus

GREY REEF SHARK
Carcharhinus amblyrhynchos

BRONZE WHALER
Carcharhinus brachyurus

SPINNER SHARK
Carcharhinus brevipinna

SILKY SHARK
Carcharinus falciformis

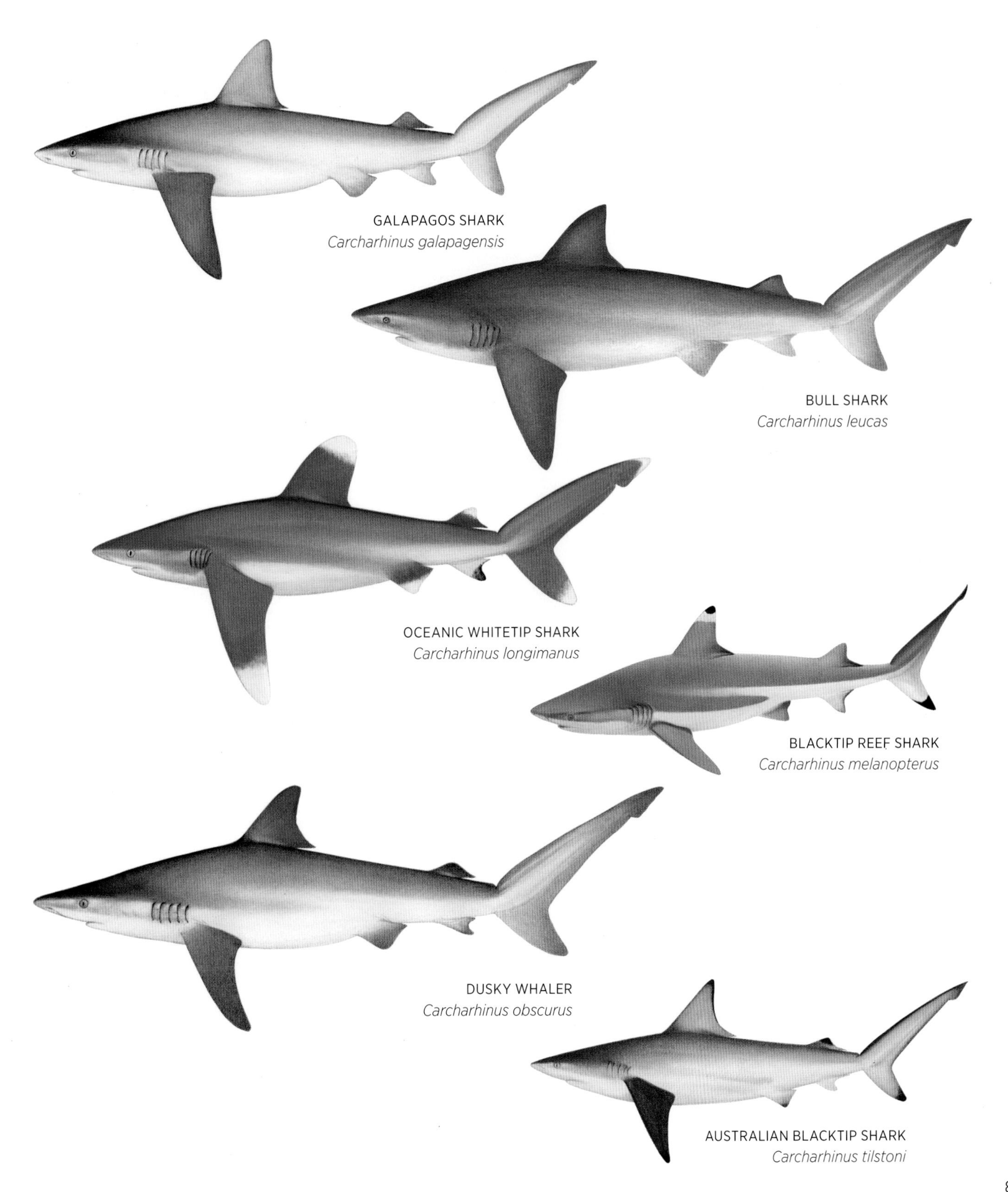

GALAPAGOS SHARK
Carcharhinus galapagensis

BULL SHARK
Carcharhinus leucas

OCEANIC WHITETIP SHARK
Carcharhinus longimanus

BLACKTIP REEF SHARK
Carcharhinus melanopterus

DUSKY WHALER
Carcharhinus obscurus

AUSTRALIAN BLACKTIP SHARK
Carcharhinus tilstoni

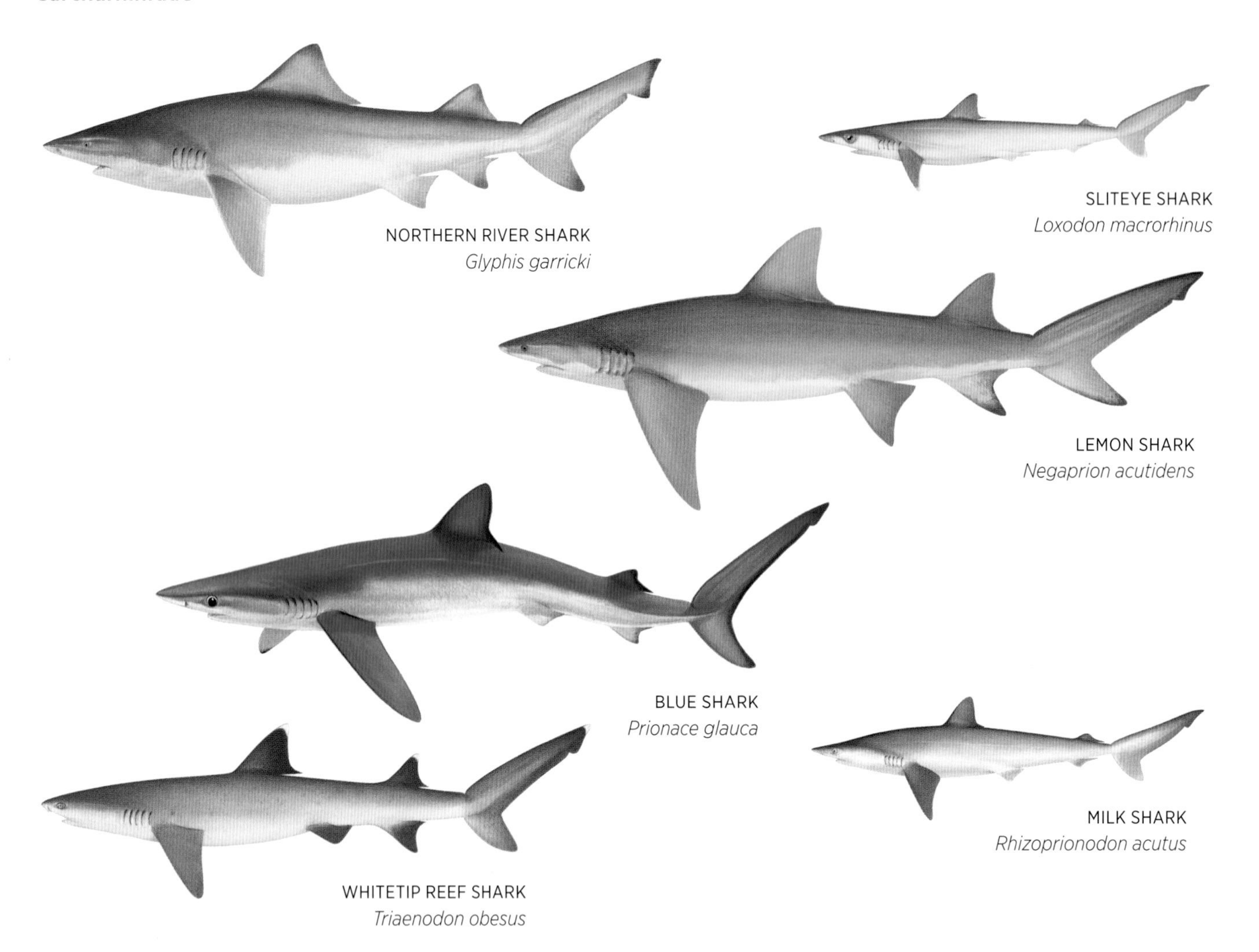

NORTHERN RIVER SHARK
Glyphis garricki

SLITEYE SHARK
Loxodon macrorhinus

LEMON SHARK
Negaprion acutidens

BLUE SHARK
Prionace glauca

MILK SHARK
Rhizoprionodon acutus

WHITETIP REEF SHARK
Triaenodon obesus

AUSTRALIAN CARCHARHINIDAE SPECIES

Carcharhinus albimarginatus Silvertip Shark
Carcharhinus altimus Bignose Shark
Carcharhinus amblyrhynchoides Graceful Shark
Carcharhinus amblyrhynchos Grey Reef Shark
Carcharhinus amboinensis Pigeye Shark
Carcharhinus brachyurus Bronze Whaler
Carcharhinus brevipinna Spinner Shark
Carcharhinus cautus Nervous Shark
Carcharhinus dussumieri Whitecheek Shark
Carcharhinus falciformis Silky Shark
Carcharhinus fitzroyensis Creek Whaler
Carcharhinus galapagensis Galapagos Shark
Carcharhinus leucas Bull Shark
Carcharhinus limbatus Common Blacktip Shark
Carcharhinus longimanus Oceanic Whitetip Shark
Carcharhinus macloti Hardnose Shark
Carcharhinus melanopterus Blacktip Reef Shark
Carcharhinus obscurus Dusky Whaler
Carcharhinus plumbeus Sandbar Shark
Carcharhinus sorrah Spot-tail Shark
Carcharhinus tilstoni Australian Blacktip Shark
Galeocerdo cuvier Tiger Shark
Glyphis garricki Northern River Shark
Glyphis glyphis Speartooth Shark
Loxodon macrorhinus Sliteye Shark
Negaprion acutidens Lemon Shark
Prionace glauca Blue Shark
Rhizoprionodon acutus Milk Shark
Rhizoprionodon oligolinx Grey Sharpnose Shark
Rhizoprionodon taylori Australian Sharpnose Shark
Triaenodon obesus Whitetip Reef Shark

HAMMERHEAD SHARKS

Sphyrnidae

SCALLOPED HAMMERHEAD
Sphyrna lewini

Hammerhead Sharks are found worldwide in tropical and warm temperate coastal waters. There are about eight species in the family, and their unique appearance makes them difficult to confuse with any other shark. In profile, the body of Hammerheads resembles the Whaler Sharks (Carcharhinidae, p. 79), but the head is flattened and expanded laterally to form a wide hammer shape, with the eyes and nostrils positioned on the outside edges, and electroreceptors on the underside. This 'cephalofoil' has several advantages: it acts as a hydroplane to increase manoeuvrability, it provides a wide surface area for sensory receptors, and it gives improved binocular vision. The **Winghead Shark**, which grows to about 1.8 m, has developed the cephalofoil to an extraordinary extent; the width of its head can equal half the length of its body.

In Australia Hammerheads occur mainly in northern waters. They usually feed on fishes and squid, but also take other sharks, rays and crustaceans. Hammerheads are viviparous, with the embryos nourished by an attached placental yolk sac.

The **Great Hammerhead** can reach 6 m in length and is generally solitary. The **Scalloped Hammerhead** and **Smooth Hammerhead** are smaller, reaching about 3.5 m in length, and sometimes form large schools. Large Hammerheads have a reputation for being dangerous and there have been reports of attacks on humans (although not in Australian waters). While not usually aggressive towards humans, Hammerheads should be treated with caution. Like the Whaler Sharks, Hammerheads are caught and killed for their fins, and populations have declined drastically in many areas of the world.

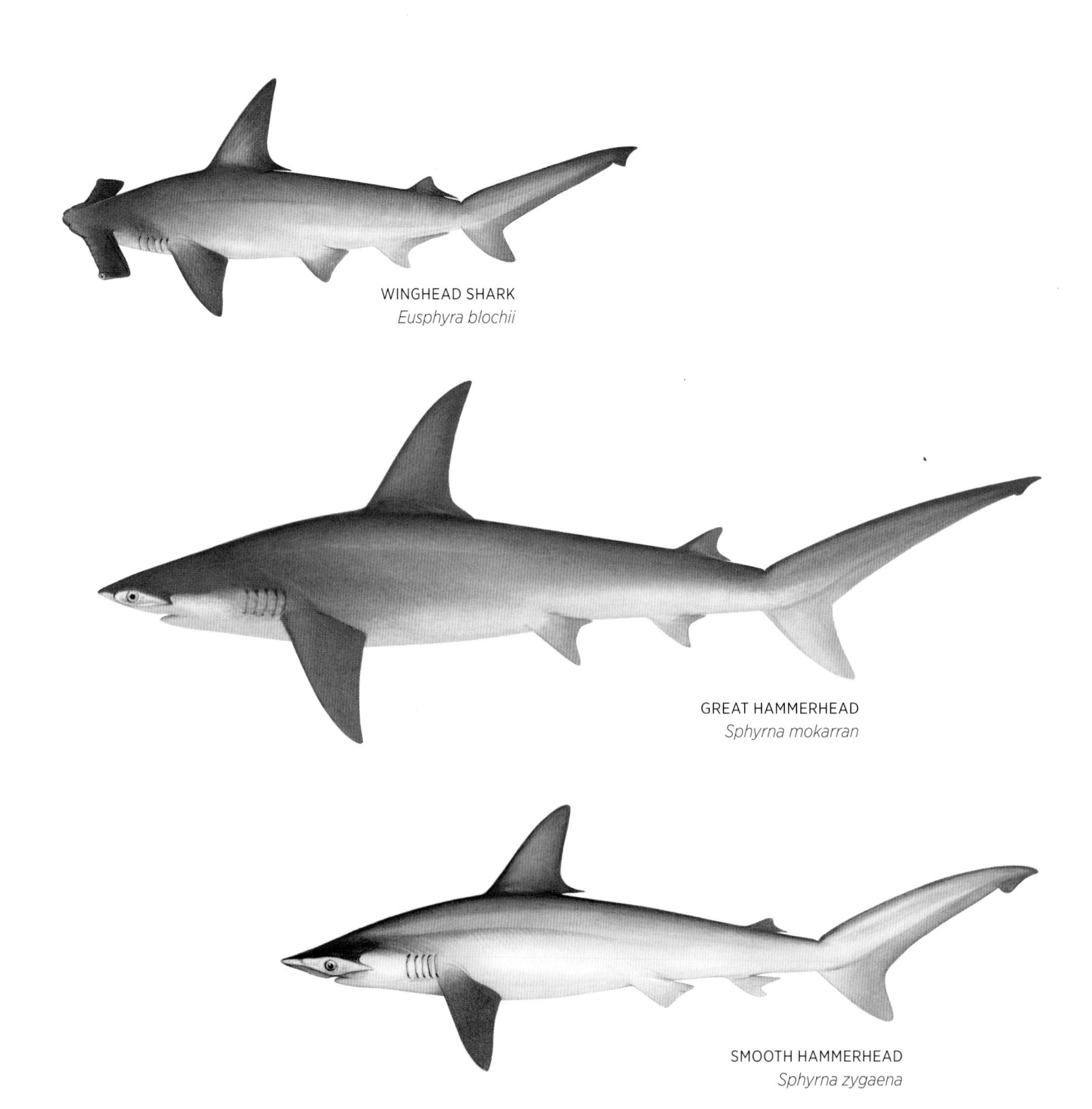

WINGHEAD SHARK
Eusphyra blochii

GREAT HAMMERHEAD
Sphyrna mokarran

SMOOTH HAMMERHEAD
Sphyrna zygaena

AUSTRALIAN SPHYRNIDAE SPECIES

Eusphyra blochii Winghead Shark
Sphyrna lewini Scalloped Hammerhead
Sphyrna mokarran Great Hammerhead
Sphyrna zygaena Smooth Hammerhead

HEXANCHIFORMES

Hexanchiforms have one dorsal fin, with no spines, and an anal fin. They are distinct in that they have six or seven pairs of gill slits, while all other sharks have five pairs. Small spiracles are present and they do not have a nictitating membrane over the eyes. Hexanchiforms are ovoviviparous. There are two families in the order, both of which occur in Australian waters.

FRILL SHARK

Chlamydoselachidae

FRILL SHARK
Chlamydoselachus anguinensis

The **Frill Shark** is the only member of this family and is found worldwide in deep water, to 1200 m or more. The Frill Shark has a very long, slender body and grows to a maximum length of about 2 m. It has a single dorsal fin located towards the rear of the body and six pairs of gill slits with frilled margins, the first pair connected underneath the head. It has small spiracles and unusual spiky, three-cusped teeth. The Frill Shark is ovoviviparous, with litters of 8–15 young. The gestation period is thought to be as long as three and a half years, which is by far the longest of any vertebrate. This shark feeds mainly on squid, but also takes fishes and other small sharks.

AUSTRALIAN CHLAMYDOSELACHIDAE SPECIES

Chlamydoselachus anguinensis Frill Shark

COWSHARKS

Hexanchidae

BROADNOSE SEVENGILL SHARK
Notorynchus cepedianus

Cowsharks occur in all the oceans of the world and there are four species in the family. They inhabit deep water, to more than 2000 m, but are occasionally encountered in shallow coastal areas. All have a single dorsal fin, set well towards the rear of the body, and six or seven pairs of gill slits. Unlike the closely related Frill Shark (Chlamydoselachidae, p. 86), the first pair of gill slits don't meet underneath the head. In the lower jaw they have long-based, deeply serrated teeth, which are very large in relation to the size of the shark.

Cowsharks range in size from 1.5 m up to 4.8 m for the **Bluntnose Sixgill Shark**. They feed on a wide range of fishes and invertebrates and, although generally heavy-bodied and sluggish, are also known to take large prey such as seals and Tunas (Scombridae, p. 717). Cowsharks are ovoviviparous and may have litters of up to 100 young. Although not known to attack humans, their size and formidable teeth make them potentially dangerous. The **Broadnose Sevengill Shark** commonly enters shallower waters and is regularly seen by divers in some Tasmanian and Victorian waters.

BIGEYE SIXGILL SHARK
Hexanchus nakamurai

AUSTRALIAN HEXANCHIDAE SPECIES

Heptranchias perlo Sharpnose Sevengill Shark
Hexanchus griseus Bluntnose Sixgill Shark
Hexanchus nakamurai Bigeye Sixgill Shark
Notorynchus cepedianus Broadnose Sevengill Shark

ECHINORHINIFORMES

Echinorhiniforms are large, heavy-bodied sharks. The order is closely related to the Squaliformes (p. 91) and was previously classified as belonging to that group. They have two small dorsal fins without spines, located well towards the rear of the body, no anal fin, five pairs of gill slits, and very small spiracles. Their skin is covered with enlarged denticles in the form of sharp, broad-based spines, and the lateral line is in the form of an open groove. Echinorhiniforms are viviparous. There is a single family in the order and it is present in Australian waters.

BRAMBLE SHARKS

Echinorhinidae

PRICKLY SHARK
Echinorhinus cookei

The two species in this family are deepwater sharks, occurring in temperate and tropical waters around the world but rarely encountered. They have a large, heavy body and can reach up to 4 m in length. The denticles on the skin are greatly enlarged, creating a prickly surface over the body. In the **Bramble Shark**, the base of each denticle can measure up to 1.5 cm across, and groups of denticles may be fused together to form bony plates.

Members of this family are sluggish and found close to the bottom, where they feed on fishes and invertebrates such as crustaceans. Both species have been recorded off southern Australia, at depths from 400 to 900 m.

BRAMBLE SHARK
Echinorhinus brucus

AUSTRALIAN ECHINORHINIDAE SPECIES

Echinorhinus brucus Bramble Shark
Echinorhinus cookei Prickly Shark

SQUALIFORMES

Squaliforms are very diverse, the order including some of the smallest and some of the largest sharks in the world. All families have two dorsal fins (both usually with spines) and no anal fin. The first dorsal fin is positioned well towards the anterior of the body, originating before the ventral fin. They have a large spiracle behind each eye and five pairs of gill slits. Squaliforms are oviparous. There are six families in the order, all of which occur in Australian waters.

DOGFISHES

Squalidae

WHITESPOTTED SPURDOG
Squalus acanthias

Dogfishes are widely distributed throughout the world in cool temperate to tropical waters. There are about 23 species in the family. They have a large spiracle behind each eye, no anal fin, and strong spines without grooves on both dorsal fins. There are fleshy keels on each side of the caudal peduncle and no notch before the tip of the caudal fin.

Most Dogfishes grow to no more than 1 m in length. They mainly inhabit colder waters to depths of about 800 m – though in southern Australia the **Whitespotted Spurdog** is often found in shallow inshore waters. Dogfishes have small, blade-like teeth and feed on small fishes, squid, crustaceans and molluscs. They are slow-growing and reported to live for up to 70 years. Dogfishes are viviparous, with a gestation period of 18–24 months. Most species form small to large schools, making them a prime target for fisheries around the world. However, because they grow and reproduce so slowly they are easily overfished.

The Whitespotted Spurdog is perhaps the most abundant, and heavily fished, of all sharks: tens of thousands of tonnes are taken each year around the world. In some areas, populations have come close to being eradicated. Although this species is not highly valued as a food fish in Australia, it forms a significant part of the bycatch for trawl fisheries off south-eastern Australia and populations have been much reduced. Several other Dogfish species, including the **Eastern Longnose Spurdog** and **Piked Spurdog**, are also often taken as bycatch. The **Mandarin Shark** closely resembles the Spurdogs but can be differentiated by its very long nasal barbels.

MANDARIN SHARK
Cirrhigaleus australis

EDMUND'S SPURDOG
Squalus edmundsi

PIKED SPURDOG
Squalus megalops

AUSTRALIAN SQUALIDAE SPECIES

Cirrhigaleus australis Mandarin Shark
Squalus acanthias Whitespotted Spurdog
Squalus albifrons Eastern Highfin Spurdog
Squalus altipinnis Western Highfin Spurdog
Squalus chloroculus Greeneye Spurdog
Squalus crassispinus Fatspine Spurdog
Squalus edmundsi Edmund's Spurdog
Squalus grahami Eastern Longnose Spurdog
Squalus megalops Piked Spurdog
Squalus montalbani Philippine Spurdog
Squalus nasutus Western Longnose Spurdog
Squalus notocaudatus Bartail Spurdog

GULPER SHARKS

Centrophoridae

WESTERN GULPER SHARK
Centrophorus westraliensis

Gulper Sharks are found worldwide in temperate and tropical waters. There are about 17 species in the family. They have two dorsal fins with strong, grooved spines at the base, no anal fin and a notch before the tip of the caudal fin. Gulper Sharks have large spiracles and large, luminous green eyes. Their blade-like teeth are larger in the lower jaw.

These medium-sized sharks grow to lengths of 1–1.6 m, and are often abundant in deeper waters, from 100 m down to more than 2000 m. Gulper Sharks feed on benthic and mid-water fishes, squid and crustaceans. Their large, oil-rich liver makes them an important catch for deepwater trawl fisheries in other parts of the world – insufficient quantities are taken in Australian waters to support a processing industry. Gulper Sharks are slow-growing and viviparous, bearing litters of just 1–2 pups, making them particularly vulnerable to overfishing.

The **Brier Shark** and **Longsnout Dogfish** occur in high concentrations in the oceans off southern Australia. These two species have a long, flattened snout, long, low dorsal fins and very fragile skin. Along with several other species of Gulper Shark, including the **Southern Dogfish** and **Endeavour Dogfish**, they have suffered serious population declines in southern Australian waters as a result of deepwater trawl fishing.

HARRISSON'S DOGFISH
Centrophorus harrissoni

TAIWAN GULPER SHARK
Centrophorus niaukang

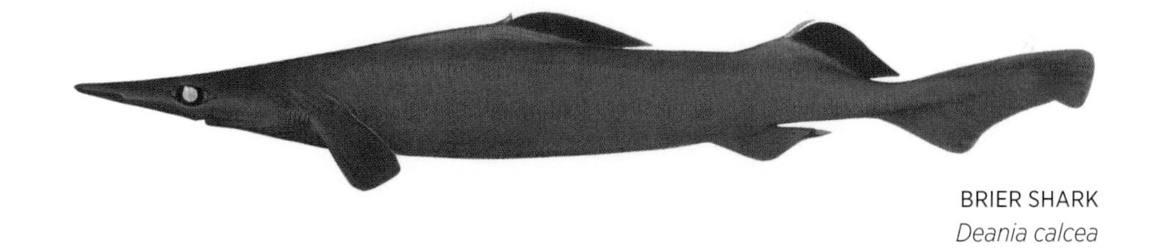

BRIER SHARK
Deania calcea

AUSTRALIAN CENTROPHORIDAE SPECIES

Centrophorus acus Gulper Shark
Centrophorus harrissoni Harrisson's Dogfish
Centrophorus moluccensis Endeavour Dogfish
Centrophorus niaukang Taiwan Gulper Shark
Centrophorus squamosus Leafscale Gulper Shark
Centrophorus westraliensis Western Gulper Shark
Centrophorus zeehaani Southern Dogfish
Deania calcea Brier Shark
Deania quadrispinosa Longsnout Dogfish

LANTERNSHARKS

Etmopteridae

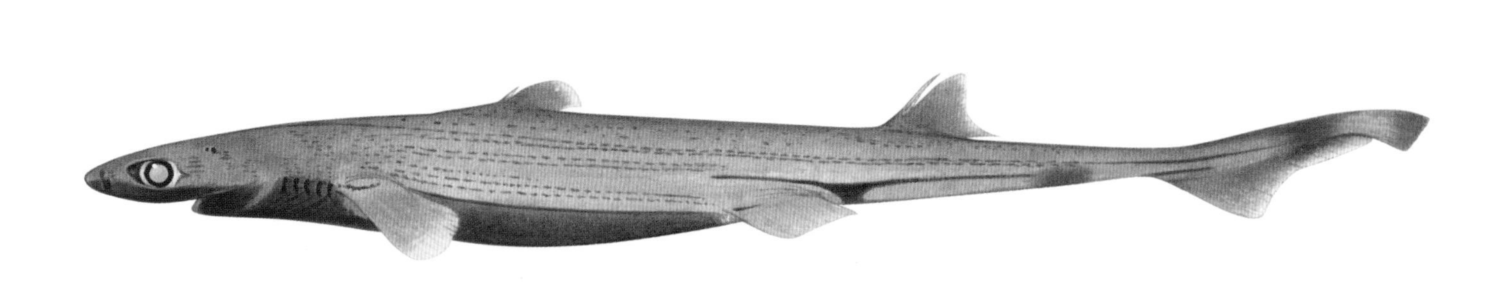

LINED LANTERNSHARK
Etmopterus dislineatus

Lanternsharks are found in the Atlantic, Indian and Pacific oceans in waters down to 2000 m or more. The family contains about 40 species. They take their common name from the numerous bioluminescent photophores on their ventral surfaces. They have grooved spines at the base of both dorsal fins, no anal fin, and a notch before the tip of the caudal fin. Many have distinctive black markings on the body. Lanternsharks feed on fishes and squid, and grow to about 80–90 cm in length. They are viviparous, with litters of 2–20 pups. Many species are known from only a few recorded catches by deep-water trawls and their biology is not well known.

BARESKIN DOGFISH
Centroscyllium kamoharai

SOUTHERN LANTERNSHARK
Etmopterus baxteri

BLACKBELLY LANTERNSHARK
Etmopterus lucifer

AUSTRALIAN ETMOPTERIDAE SPECIES

Centroscyllium kamoharai Bareskin Dogfish
Etmopterus baxteri Southern Lanternshark
Etmopterus bigelowi Slender Lanternshark
Etmopterus brachyurus Short-tail Lanternshark
Etmopterus dianthus Pink Lanternshark
Etmopterus dislineatus Lined Lanternshark
Etmopterus evansi Blackmouth Lanternshark
Etmopterus fusus Pygmy Lanternshark
Etmopterus lucifer Blackbelly Lanternshark
Etmopterus molleri Moller's Lanternshark
Etmopterus pusillus Smooth Lanternshark
Etmopterus unicolor Bristled Lanternshark

SLEEPER SHARKS

Somniosidae

SOUTHERN SLEEPER SHARK
Somniosus antarcticus

Sleeper Sharks are widely distributed in deep waters from the Arctic to the Antarctic. The family contains about 18 species. Most have no spines, or very small spines, on both the dorsal fins, and bioluminescent organs (photophores) on the surface of the body.

The **Southern Sleeper Shark** is a giant of deep water, reportedly reaching lengths of up to 6 m. It can be found anywhere between the surface and depths of at least 2000 m. It preys on a wide range of fishes, crustaceans and cephalopods (including Giant Squid), as well as marine mammals such as seals. The Dogfishes in this family are much smaller, with maximum lengths of 1.2–1.7 m, and are found in deep waters from 250 to 1500 m or more. They feed on fishes and cephalopods, and several species form large schools.

Dogfishes have large livers that contain high concentrations of squalene oil, which provides buoyancy. This oil is highly desirable for use in cosmetics, health products and high-grade industrial lubricants, making Dogfishes a valuable fisheries resource. Being slow-growing and long-lived, they are vulnerable to overfishing, and strict quotas and trawl depth limits now apply to all Dogfish species. Members of the family are viviparous, with females often bearing large litters – up to 59 pups have been recorded for the **Whitetail Dogfish**.

GOLDEN DOGFISH
Centroscymnus crepidater

WHITETAIL DOGFISH
Scymnodalatias albicauda

AUSTRALIAN SOMNIOSIDAE SPECIES

Centroscymnus coelolepis Portuguese Dogfish
Centroscymnus crepidater Golden Dogfish
Centroscymnus owstonii Owston's Dogfish
Centroscymnus plunketi Plunket's Dogfish
Scymnodalatias albicauda Whitetail Dogfish
Scymnodalatias sherwoodi Sherwood's Dogfish
Scymnodon squamulosus Velvet Dogfish
Somniosus antarcticus Southern Sleeper Shark
Zameus squamulosus Velvet Dogfish

PRICKLY DOGFISHES

Oxynotidae

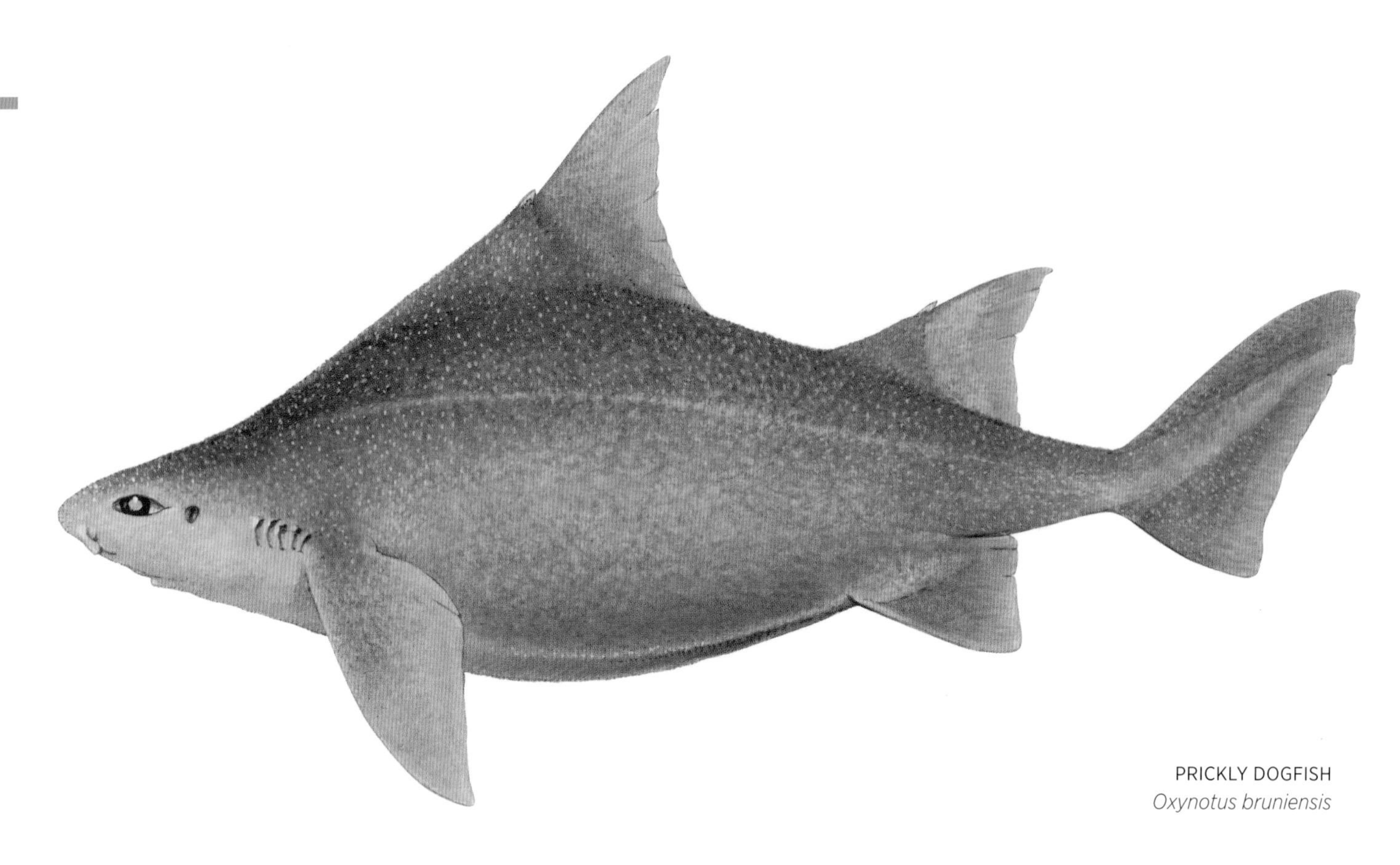

PRICKLY DOGFISH
Oxynotus bruniensis

The five species of Prickly Dogfish are found in the Atlantic and West Pacific. They were previously included in the Squalidae (p. 92) and share many features with that family. Their 'humpback' and two tall, pointed dorsal fins (both with small spines at the base), make them distinctive.

The only Australian species, the **Prickly Dogfish**, is found off southern Australia and New Zealand. It has enlarged denticles on the skin that give it a prickly texture. A prominent ridge runs along the lateral surface of the body between the pectoral and ventral fins. The body is almost triangular in cross-section, with a wide ventral surface and high, ridged dorsal surface. Living near the bottom, at depths of 50–650 m, it grows to about 70 cm in length. It has small, blade-like teeth and probably feeds on benthic invertebrates and small fishes. The Prickly Dogfish is ovoviviparous but very little else is known of its biology.

AUSTRALIAN OXYNOTIDAE SPECIES
Oxynotus bruniensis Prickly Dogfish

KITEFIN SHARKS

Dalatiidae

BLACK SHARK
Dalatias licha

This family of about 10 species is found worldwide. It contains what is possibly the world's smallest shark – the 20 cm long **Smalleye Pygmy Shark**. Most species in the family live in deep water, with some migrating from depths as great as 2000 m towards the surface to feed each night. They have a cigar-shaped body with a conical snout, and bioluminescent organs (photophores), predominantly on the ventral surface of the body. They have no spines on the dorsal fins, except the Smalleye Pygmy Shark, which has a small spine on the first dorsal fin.

Cookiecutter Sharks, which have been found at depths of more than 3500 m, reach a maximum length of about 50 cm. They feed on other sharks, large fishes and marine mammals, using specialised lips to suction onto their prey, then rotating their body to cut a round plug of flesh with their razor-sharp, saw-like teeth. This leaves a characteristic round scar on the prey. I have seen a large Mahi Mahi (Coryphaenidae, p. 494), in very poor condition, with six or seven such marks, several completely healed and others freshly made. The **Black Shark**, which can reach up to 1.6 m in length, may also occasionally feed in this way, but it normally preys on small fishes, cephalopods and crustaceans.

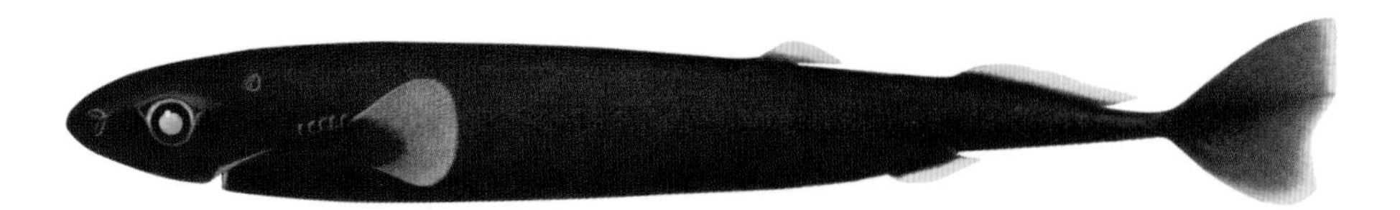

PYGMY SHARK
Euprotomicrus bispinatus

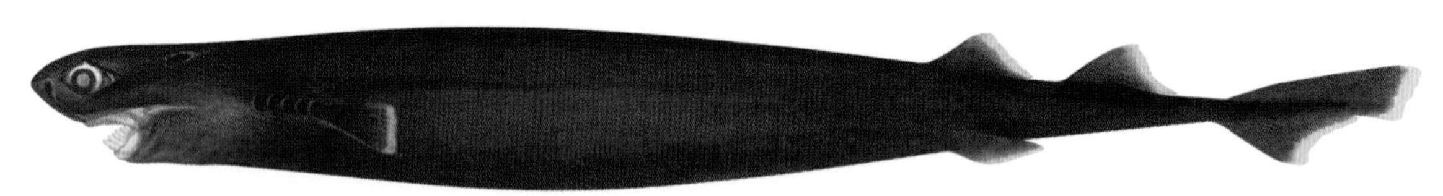

LARGETOOTH COOKIECUTTER SHARK
Isistius plutodus

AUSTRALIAN DALATIIDAE SPECIES

Dalatias licha Black Shark
Euprotomicrus bispinatus Pygmy Shark
Isistius brasiliensis Smalltooth Cookiecutter Shark
Isistius plutodus Largetooth Cookiecutter Shark
Squaliolus aliae Smalleye Pygmy Shark

SQUATINIFORMES

Squatiniforms are somewhat similar in appearance to rays: they have a dorsoventrally flattened head and body, and expanded pectoral fins. However, unlike in rays, the pectoral fins are not fused to the sides of the head. The large spiracles (for intake of water for respiration while the fish is lying on the bottom) are on the dorsal surface, as are the eyes. The five pairs of gill slits are on the ventral surface of the head. They have two dorsal fins without spines, and no anal fin. The mouth is at the tip of the snout. Squatiniforms are ovoviviparous. The order contains only one family and it occurs in Australian waters.

ANGELSHARKS

Squatinidae

ORNATE ANGELSHARK
Squatina tergocellata

Angelsharks are found worldwide, inhabiting warm temperate and tropical waters from inshore areas to depths of more than 1300 m. There are about 18 species in the family. Their mouth, nostrils and branched nasal barbels are at the tip of the snout. The broad pectoral fins extend along both sides of the head as unattached lobes, and the lower lobe of the caudal fin is longer than the upper. The **Australian Angelshark**, common in shallow waters off southern Australia, grows to 1.5 m in length; the other Australian species are less than 1 m long.

Angelsharks are principally ambush predators: they lie camouflaged and immobile on the bottom, sometimes for weeks at a time, then surge forth and snap up passing prey. Their protrusible, trap-like jaws bear long, sharp teeth. They are also known to swim and feed off the bottom, and their diet includes small fishes, crustaceans, squid and other invertebrates. The Australian Angelshark and **Ornate Angelshark** are good food fishes and are frequent bycatches of southern trawl fisheries – populations of both have declined due to overfishing. Anglesharks are viviparous, with litters of up to 20 pups.

AUSTRALIAN SQUATINIDAE SPECIES

Squatina australis Australian Angelshark
Squatina tergocellata Ornate Angelshark
Squatina albipunctata Eastern Angelshark
Squatina pseudocellata Western Angelshark

PRISTIOPHORIFORMES

Pristiophoriforms have a long, slender, slightly flattened body, two dorsal fins without spines, and no anal fin. The snout is formed into a long, tapering blade, with alternating long and short tooth-like denticles weakly embedded along its edges, and a pair of long barbels on its ventral surface. They have five pairs of gill slits on the sides of the head (except for one South African species that has six pairs) and a large spiracle behind each eye to take in water for respiration while the fish is lying on the bottom. Pristiophoriforms are ovoviviparous. The order contains one family and it is present in Australian waters.

SAWSHARKS

Pristiophoridae

TROPICAL SAWSHARK
Pristiophorus delicatus

There are about five species of Sawshark, and they are found in tropical and temperate waters of the western Atlantic, southwest Indian, and western Pacific oceans. Sawsharks are superficially similar to the Sawfishes (Pristidae, p. 112): both have an elongated, flattened, blade-like snout with rows of sharp tooth-like spines (actually enlarged denticles) along the edges. However, Sawsharks have alternating large and small spines (rather than spines of uniform size), which continue along the sides of the head to the level of the eyes. The snout of Sawsharks is quite delicate and bears sensory cells on the underside. Their gill slits are on the side of the head, not on the ventral surface as in Sawfishes. A pair of long barbels is present about midway along the undersurface of the snout. The barbels help them to detect prey in mud or sand, while the saw-like snout is used to chase out and catch the invertebrates on which they feed. Growing to about 1.5 m in length, Sawsharks are usually found in deep water, from about 100 to 600 m. The **Southern Sawshark** and **Common Sawshark** are frequently caught as bycatch in the southern shark fishery in Bass Strait and the Great Australian Bight, where they often aggregate in large numbers.

COMMON SAWSHARK
Pristiophorus cirratus

AUSTRALIAN PRISTIOPHORIDAE SPECIES

Pristiophorus cirratus Common Sawshark
Pristiophorus delicatus Tropical Sawshark
Pristiophorus nudipinnis Southern Sawshark

TORPEDINIFORMES

Torpediniforms have a soft, flabby body, with the head and pectoral fins forming a flattened, rounded disc. The gill slits are on the ventral surface, and spiracles on the dorsal surface take in water for respiration while the fish is lying on the bottom. All Australian members of the order have two dorsal fins without spines, and a well-developed caudal fin (the dorsal fins may be lacking in species not found in Australia). Torpediniforms have electrogenic organs at the base of the pectoral fins and some are capable of giving a powerful electric shock. They are ovoviviparous. There are two families in the order, both of which are found in Australian waters.

TORPEDO RAYS

Torpedinidae

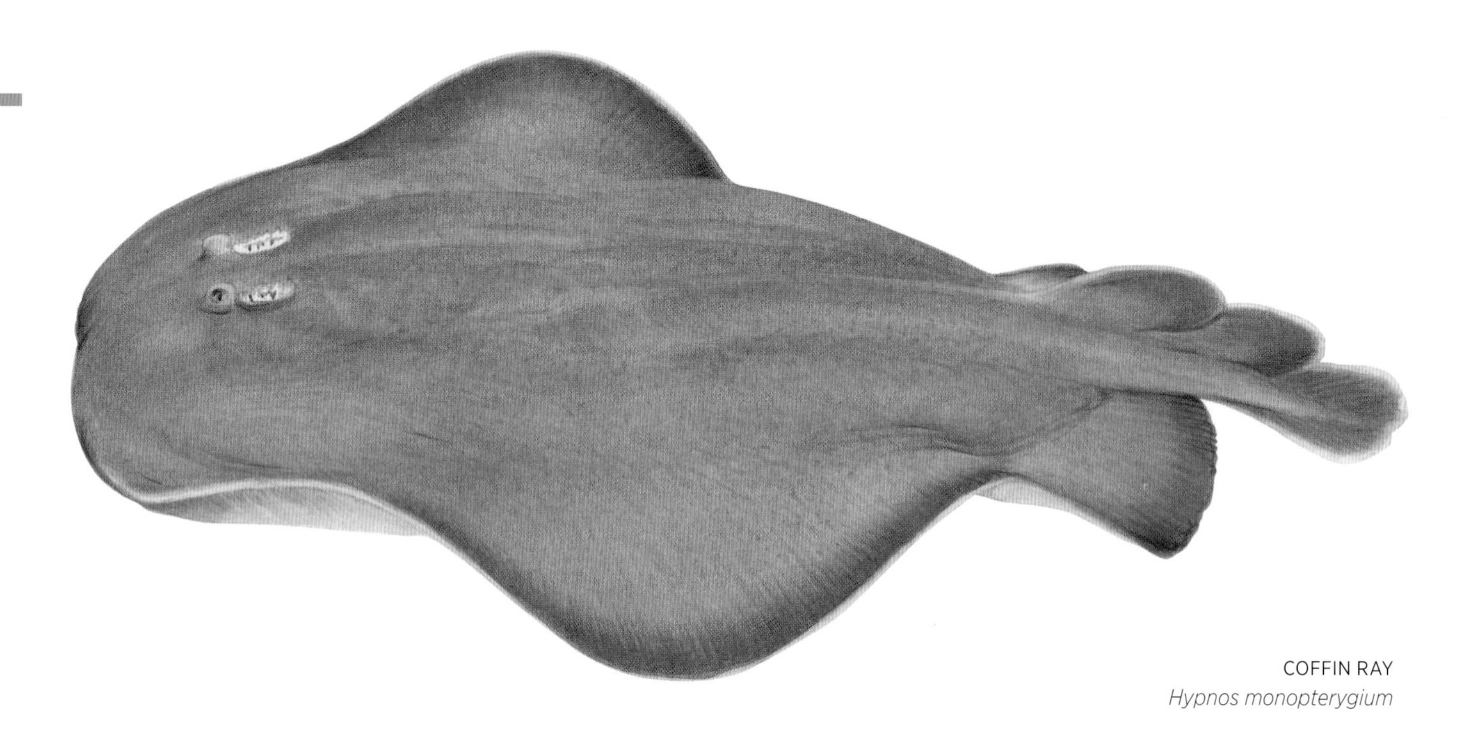

COFFIN RAY
Hypnos monopterygium

Torpedo Rays are found in tropical and temperate waters of the Atlantic, Indian and Pacific oceans. There are about 22 species in the family. Their body and pectoral fins form a wide, rounded disc that is covered with flabby skin. They have a short tail with two small dorsal fins and a well-developed caudal fin. The eyes and spiracles are very small. They have electrogenic organs in the pectoral fins on either side of the head, which are composed of tightly packed, column-shaped hexagonal cells derived from branchial muscle. Torpedo Rays can give powerful electric shocks.

The **Coffin Ray** is found only in Australian waters and is named for the strange swollen shape it assumes when washed ashore. It prefers sheltered environments with sandy or muddy bottoms in warm temperate and tropical waters, and is found from the shallows to depths of more than 250 m. It spends much of its time buried in the sand or lying camouflaged on the bottom. The Coffin Ray feeds on fishes and benthic invertebrates and is capable of taking large prey such as Flounders. Divers and fishers who have accidentally made contact with this small ray will attest to the powerful shock it can produce.

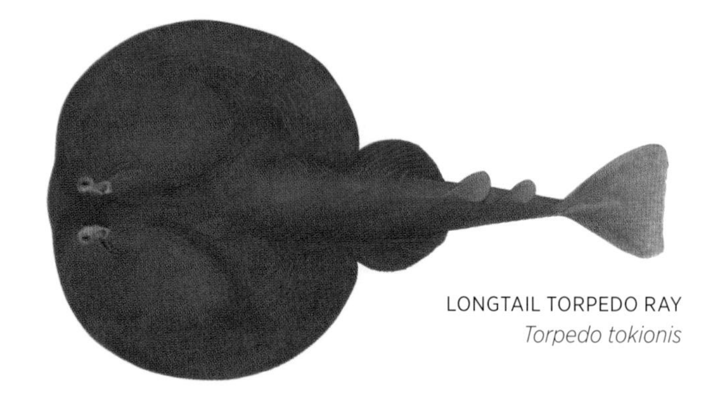

LONGTAIL TORPEDO RAY
Torpedo tokionis

AUSTRALIAN TORPEDINIDAE SPECIES

Hypnos monopterygium Coffin Ray
Torpedo macneilli Short-tail Torpedo Ray
Torpedo tokionis Longtail Torpedo Ray

NUMBFISHES

Narcinidae

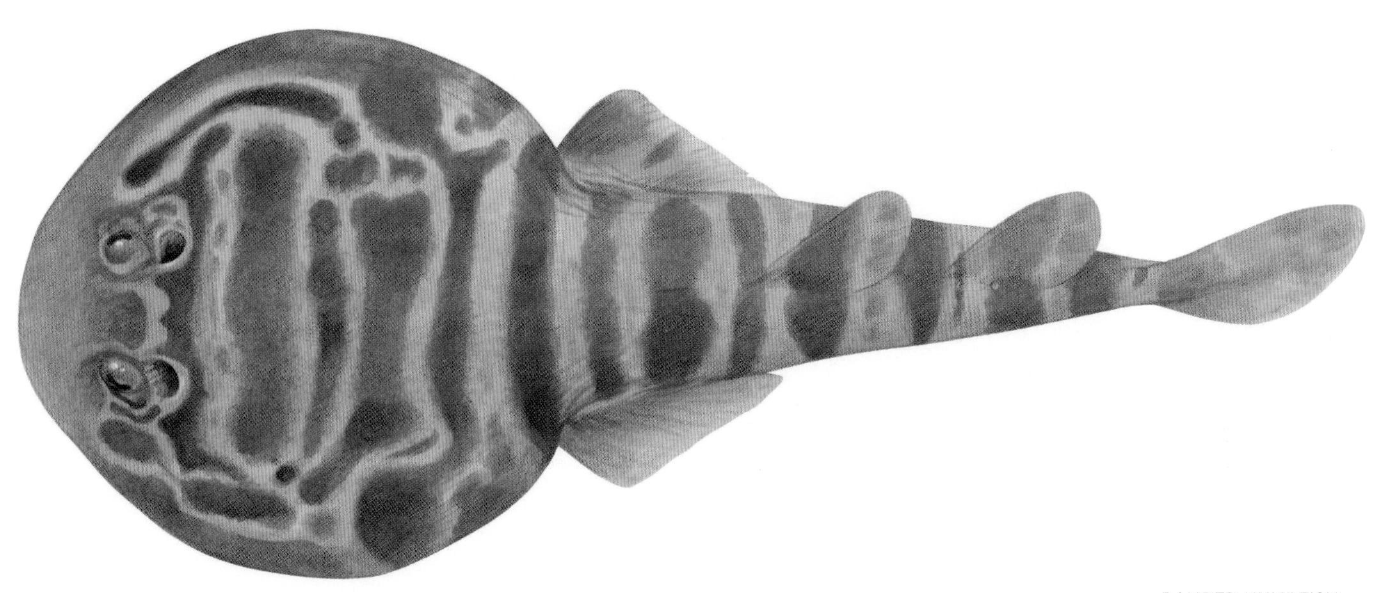

BANDED NUMBFISH
Narcine westraliensis

The family of Numbfishes contains about 37 species. These small rays are found in tropical and temperate waters in the Atlantic, Indian and Pacific oceans. They have a rounded to wedge-shaped head with small eyes and spiracles on the dorsal surface. The body has soft, loose skin, two moderate-sized dorsal fins and a well-developed caudal fin. Numbfishes have kidney-shaped electrogenic organs in the pectoral fins on either side of the head that can produce an electric shock of 20–50 volts. This charge serves to deter predators and is probably also used to stun prey such as small fishes.

The Australian species are found right around the coastline, on muddy and sandy bottoms, from shallow nearshore waters to depths of over 600 m. They grow to a maximum of about 50 cm in length and feed on a range of small fishes and benthic invertebrates, which they snap up from the substrate with their highly protrusible jaws. Numbfishes are viviparous, with small litters of up to about 10 pups.

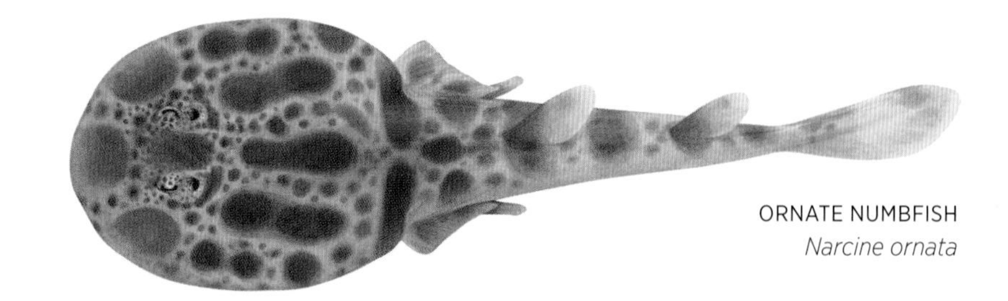

ORNATE NUMBFISH
Narcine ornata

TASMANIAN NUMBFISH
Narcine tasmaniensis

AUSTRALIAN NARCINIDAE SPECIES

Narcine lasti Western Numbfish
Narcine nelsoni Eastern Numbfish
Narcine ornata Ornate Numbfish
Narcine tasmaniensis Tasmanian Numbfish
Narcine westraliensis Banded Numbfish

PRISTIFORMES

Pristiforms have an elongated, flattened body, two dorsal fins without spines, and no anal fin. The snout is formed into a long, saw-like blade with large, firmly embedded teeth along both edges, and no barbels on the ventral surface. The large spiracles (to take in water for respiration while the fish is lying on the bottom) are on the dorsal surface, as are the eyes. The five pairs of gill slits are on the ventral surface of the head. The pectoral fins are expanded and attached to the head. Pristiforms are ovoviviparous. There is a single family in the order and it is found in Australian waters.

SAWFISHES

Pristidae

GREEN SAWFISH
Pristis zijsron

Sawfishes, of which there are about seven species, occur worldwide in tropical waters and are easily confused with the Sawsharks (Pristiophoridae, p. 106). Both have a flattened, bony, blade-like snout with rows of sharp tooth-like spines along both edges. However, Sawfishes have widely spaced, strong, firmly attached bony spines, all of about equal size (Sawsharks have alternating large and small spines that are weakly attached). Sawfishes have no barbels on the undersurface of the snout, and the gill slits are beneath the head, not on the sides as in Sawsharks. The tail of Sawfishes is shark-like and they have two large dorsal fins, large spiracles and no anal fin.

Sawfishes use their saw in a sideways slashing motion to disable and capture schooling fishes, and to dig crustaceans and invertebrates from muddy bottoms. The magnificent **Green Sawfish** and **Freshwater Sawfish**, which are each reported to reach more than 7 m in length, are becoming extremely rare in many parts of their range. The Freshwater Sawfish, found in the rivers and estuaries of northern Australia, Indonesia and Papua New Guinea, as well as occasionally in nearby coastal waters, is particularly vulnerable to overfishing and habitat degradation. Like all the Sawfishes, its saw easily becomes entangled in gillnets. Sawfishes can be very dangerous when caught, as they will thrash their formidable saw violently from side to side; it can be a risky business manoeuvring them back into the water.

NARROW SAWFISH
Anoxypristis cuspidata

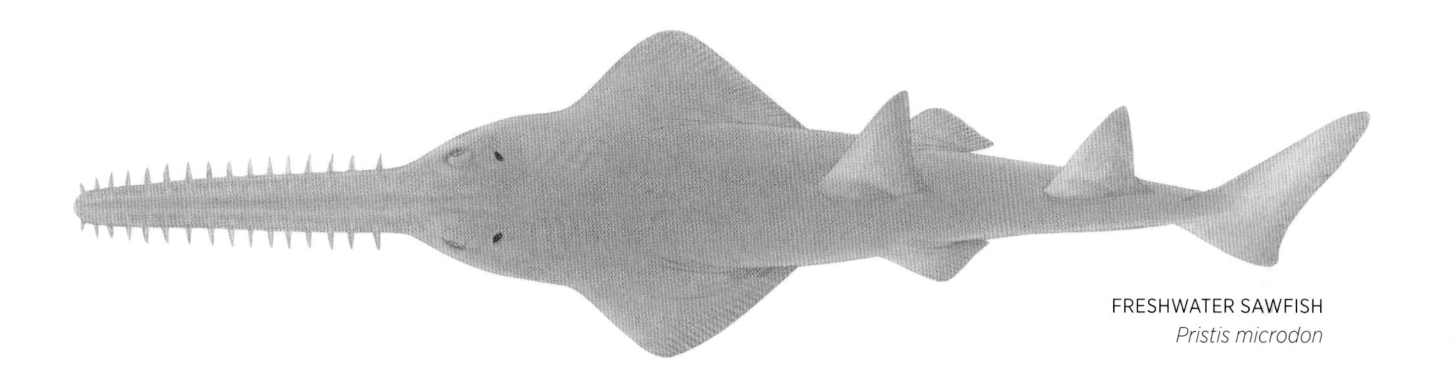

FRESHWATER SAWFISH
Pristis microdon

AUSTRALIAN PRISTIDAE SPECIES

Anoxypristis cuspidata Narrow Sawfish
Pristis clavata Dwarf Sawfish
Pristis microdon Freshwater Sawfish
Pristis zijsron Green Sawfish

RAJIFORMES

In Rajiforms, the head and pectoral fins form a flattened disc, which varies in shape from almost circular to diamond-shaped. The tail is long and slender, usually with two dorsal fins (some species of the Rajidae family, p. 121, lack dorsal fins), and the caudal fin is moderate-sized, reduced or absent. They have large spiracles on the dorsal surface to take in water for respiration while the fish is lying on the bottom, and five pairs of gill slits on the ventral surface. Most species have spines on the skin of the dorsal surface, especially along the midline. Rajiforms are oviparous. There are four families in the order, all of which are found in Australian waters.

SHARK RAY

Rhinidae

SHARK RAY
Rhina ancylostoma

The only member of this family is the distinctive **Shark Ray**, which is, as its name suggests, somewhere between a shark and a ray. It has a deep body with shark-like fins, and a flattened, rounded head that is separate from the pectoral fins. There are distinctive ridges of large, thorny denticles behind the eyes and over the back, particularly along the dorsal midline. It has ridged, crushing teeth and feeds on benthic crustaceans and molluscs. Reaching a maximum length of about 2.7 m, the Shark Ray is found in coastal waters throughout the Indo-West Pacific, from the shallows to depths of about 90 m. It is occasionally taken by trawlers off the northern half of Australia, and is a highly prized catch in South-East Asia, where it is targeted for the shark-fin trade.

AUSTRALIAN RHINIDAE SPECIES

Rhina ancylostoma Shark Ray

GUITARFISHES

Rhynchobatidae

WHITESPOTTED GUITARFISH
Rhynchobatus australiae

The four species in this family are found off the west coast of Africa and in the Indo-West Pacific. They have a shark-like body with two dorsal fins, the first originating over or anterior to the ventral fins, and a well-developed bilobed caudal fin. They are very similar to the Shovelnose Rays (Rhinobatidae, p. 119), but Shovelnoses have the first dorsal fin originating posterior to the ventral fins and do not have a distinctly bilobed caudal fin. Occurring in inshore waters and estuaries, Guitarfishes can grow to over 3 m in length. They have small, blunt teeth and feed mainly on crustaceans, molluscs, cephalopods and small fishes.

The three Australian species are difficult to distinguish from each other and until recently were all thought to be the same species. Divers have encountered a very large black Guitarfish in Shark Bay, Western Australia, which is thought to be a melanistic form of the **Whitespotted Guitarfish**.

SMOOTHNOSE WEDGEFISH
Rhynchobatus laevis

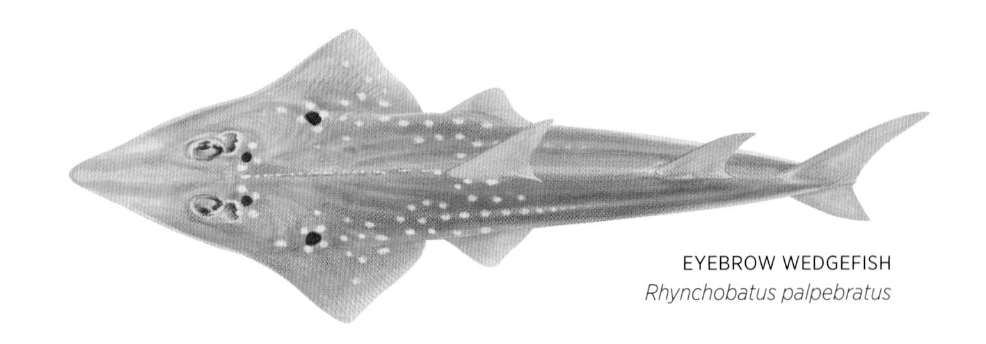

EYEBROW WEDGEFISH
Rhynchobatus palpebratus

AUSTRALIAN RHYNCHOBATIDAE SPECIES

Rhynchobatus australiae Whitespotted Guitarfish
Rhynchobatus laevis Smoothnose Wedgefish
Rhynchobatus palpebratus Eyebrow Wedgefish

SHOVELNOSE RAYS AND FIDDLER RAYS

Rhinobatidae

WESTERN SHOVELNOSE RAY
Aptychotrema vincentiana

Shovelnose Rays and Fiddler Rays are found around the world in tropical and temperate waters. There are about 40 species in the family. They have a flattened head, broad pectoral fins and a pointed or rounded snout, which combine to form a wedge-shaped disc. Members of this family have two quite large dorsal fins and large spiracles on the dorsal surface. Shovelnose Rays have a triangular, pointed snout, whereas Fiddler Rays have a more oval-shaped snout. Both have a series of enlarged spiny denticles along the midline of the dorsal surface.

Fiddler Rays, which can grow to about 1.2 m in length, are found in shallow coastal waters around southern Australia, preferring bottoms with sand or seagrass. Shovelnose Rays can attain lengths of 2.7 m and have a wider distribution. They are found right around the Australian coast, from very shallow beaches, mudflats and mangroves to depths of several hundred metres, usually in areas with sandy or muddy bottoms. At times, large groups of juvenile Shovelnose Rays aggregate along Ningaloo Reef, Western Australia, in water little more than ankle deep. Shovelnose Rays and Fiddler Rays are opportunistic bottom feeders, preying on crustaceans, molluscs and other invertebrates, and sometimes small fishes.

GIANT SHOVELNOSE RAY
Glaucostegus typus

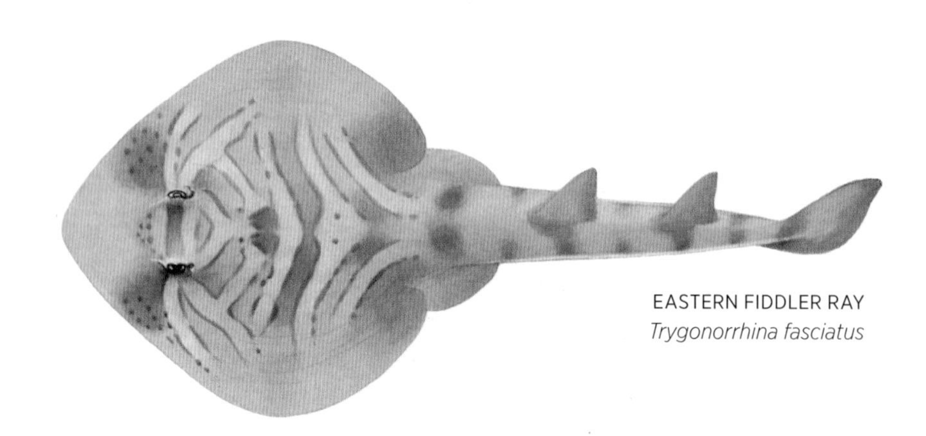

EASTERN FIDDLER RAY
Trygonorrhina fasciatus

AUSTRALIAN RHINOBATIDAE SPECIES

Aptychotrema rostrata Eastern Shovelnose Ray
Aptychotrema timorensis Spotted Shovelnose Ray
Aptychotrema vincentiana Western Shovelnose Ray
Glaucostegus typus Giant Shovelnose Ray
Rhinobatos sainsburyi Goldeneye Shovelnose Ray
Trygonorrhina dumerilii Southern Fiddler Ray
Trygonorrhina fasciatus Eastern Fiddler Ray
Trygonorrhina melaleuca Magpie Fiddler Ray

SKATES

Rajidae

QUEENSLAND DEEPWATER SKATE
Dipturus queenslandicus

The family of Skates is very large, with over 200 species found worldwide, from the Antarctic to the Arctic and from the shallows down to 3000 m or more. They have a diamond- or heart-shaped disc, a long, slender tail, usually with two small dorsal fins towards the tip, and a very small caudal fin. Male Skates have large claspers that project beyond the rear edge of the disc. Most species have patches of thorny denticles on the pectoral fins, dorsal surface and tail.

Skates are divided into three main groups, and these are sometimes classed as distinct families. The Hardnose Skates and Softnose Skates are distinguished based on the presence or absence of strong rostral cartilage running from between the eyes to the snout tip. The Leg Skates have deeply divided ventral fins, the forepart of each resembling a small leg – hence their name. Leg Skates live in deep waters off the north-east and north-west coasts of Australia, at depths from 400 to 1200 m. They are rarely encountered and very little is known of their biology.

Many Skates seem to have quite restricted distribution and one, the **Maugean Skate**, has been found only in brackish water in two estuaries in southern Tasmania. Many species are small, reaching only 30–50 cm in length – although the **Melbourne Skate,** from southern Australian waters, can reach 1.7 m or more. Skates have flattened crushing teeth and feed primarily on benthic invertebrates such as crustaceans and molluscs. Females lay roughly rectangular-shaped, flattened egg cases with tendrils that attach them to the bottom.

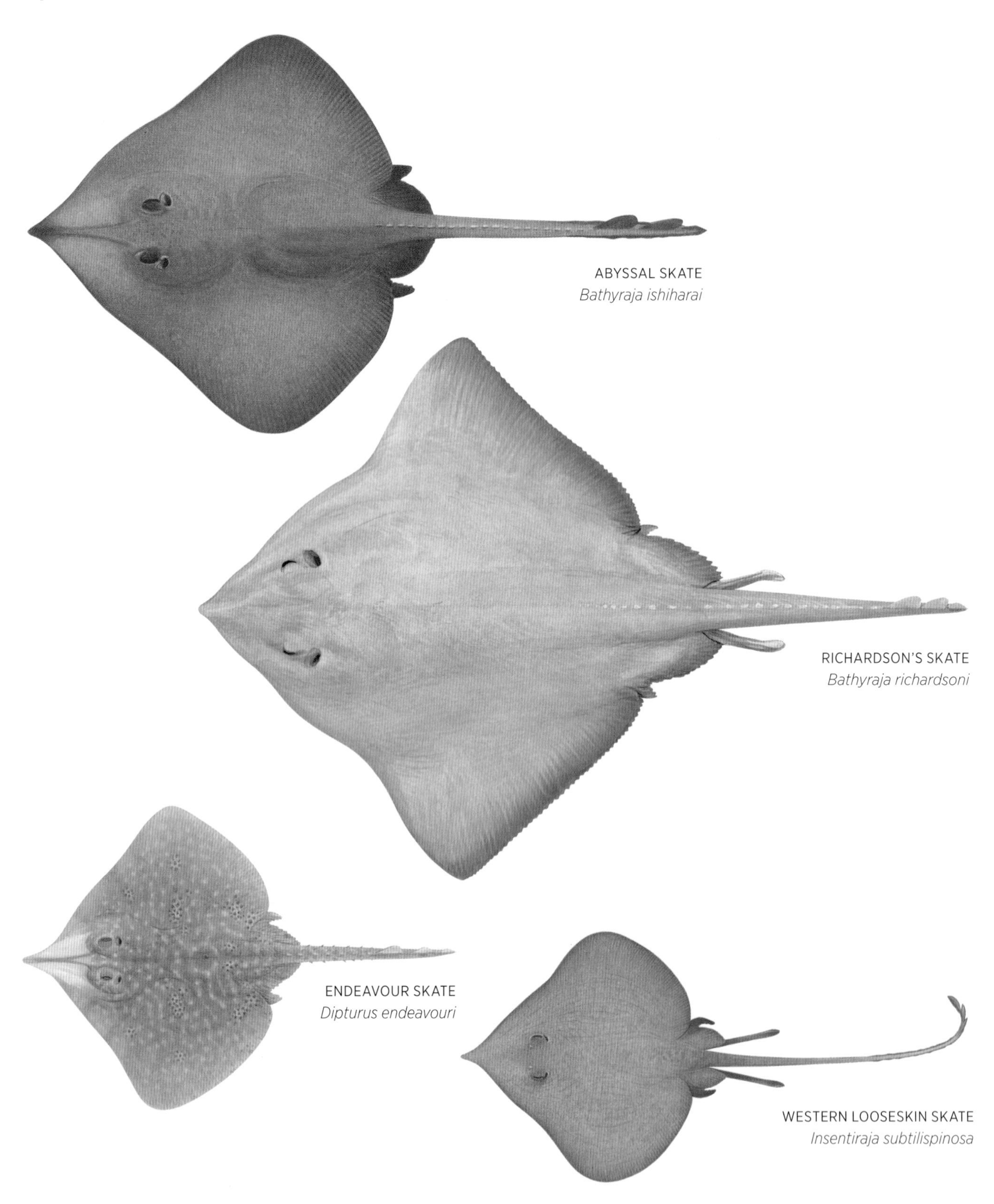

ABYSSAL SKATE
Bathyraja ishiharai

RICHARDSON'S SKATE
Bathyraja richardsoni

ENDEAVOUR SKATE
Dipturus endeavouri

WESTERN LOOSESKIN SKATE
Insentiraja subtilispinosa

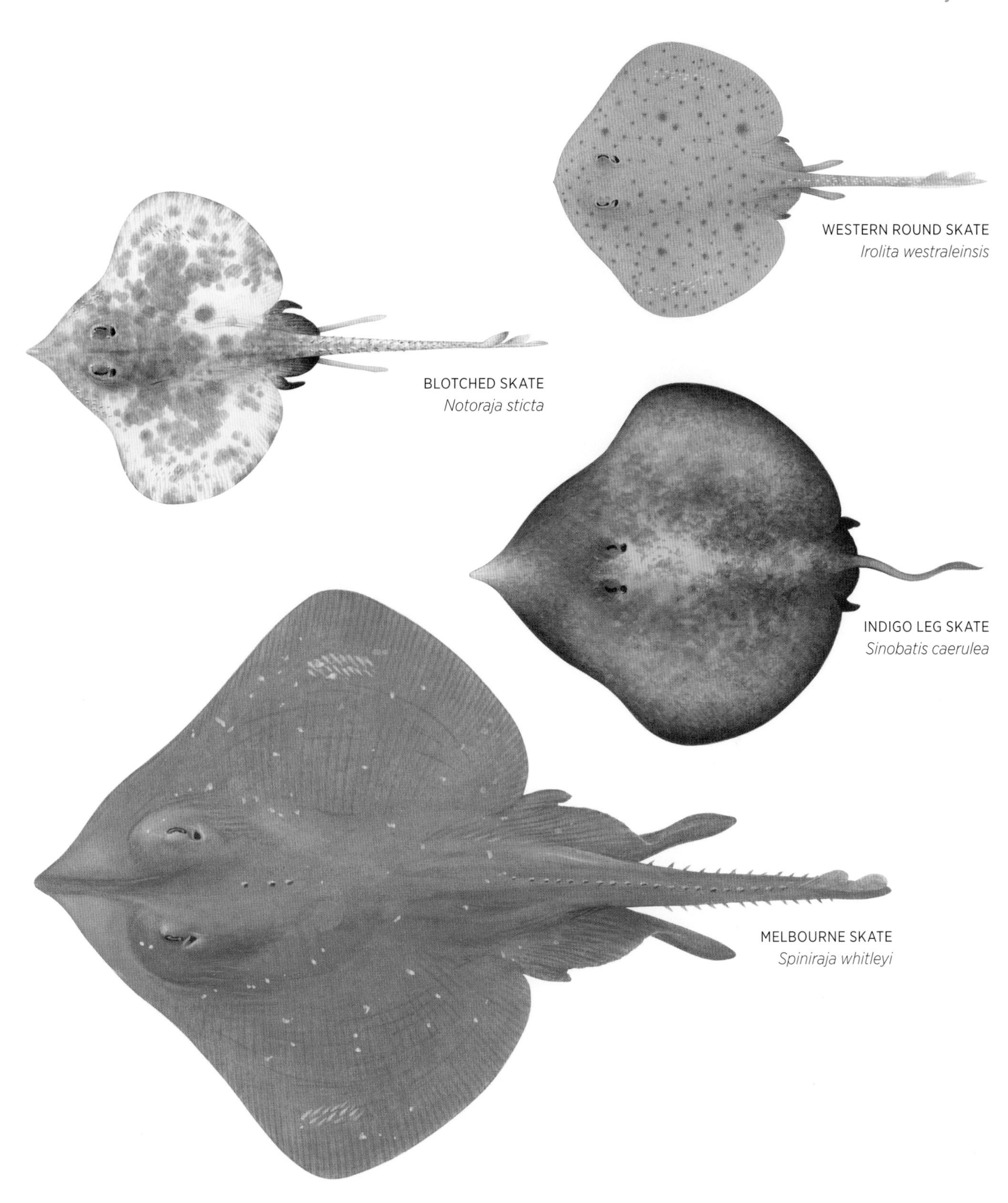

WESTERN ROUND SKATE
Irolita westraleinsis

BLOTCHED SKATE
Notoraja sticta

INDIGO LEG SKATE
Sinobatis caerulea

MELBOURNE SKATE
Spiniraja whitleyi

MAUGEAN SKATE
Zearaja maugeana

AUSTRALIAN RAJIDAE SPECIES

Amblyraja hyperborea Boreal Skate
Bathyraja ishiharai Abyssal Skate
Bathyraja richardsoni Richardson's Skate
Dentiraja flindersi Pygmy Thornback Skate
Dentiraja lemprieri Thornback Skate
Dipturus acrobelus Deepwater Skate
Dipturus apricus Pale Tropical Skate
Dipturus australis Sydney Skate
Dipturus canutus Grey Skate
Dipturus cerva Whitespotted Skate
Dipturus confusus Longnose Skate
Dipturus endeavouri Endeavour Skate
Dipturus falloargus False Argus Skate
Dipturus grahami Graham's Skate
Dipturus gudgeri Bight Skate
Dipturus healdi Heald's Skate
Dipturus melanospilus Blacktip Skate
Dipturus oculus Ocellate Skate
Dipturus polyommata Argus Skate
Dipturus queenslandicus Queensland Deepwater Skate
Dipturus wengi Weng's Skate
Insentiraja laxipella Eastern Looseskin Skate
Insentiraja subtilispinosa Western Looseskin Skate
Irolita waitii Southern Round Skate
Irolita westraliensis Western Round Skate
Leucoraja pristispina Sawback Skate
Notoraja azurea Blue Skate
Notoraja hirticauda Ghost Skate
Notoraja ochroderma Pale Skate
Notoraja sticta Blotched Skate
Okamejei arafurensis Arafura Skate
Okamejei leptoura Thintail Skate
Pavoraja alleni Allen's Skate
Pavoraja arenaria Sandy Skate
Pavoraja mosaica Mosaic Skate
Pavoraja nitida Peacock Skate
Pavoraja pseudonitida False Peacock Skate
Pavoraja umbrosa Dusky Skate
Rajella challengeri Challenger Skate
Sinobatis bulbicauda Western Leg Skate
Sinobatis caerulea Indigo Leg Skate
Sinobatis filicauda Eastern Leg Skate
Spiniraja whitleyi Melbourne Skate
Zearaja maugeana Maugean Skate

MYLIOBATIFORMES

In Myliobatiforms, the flattened head, body and pectoral fins form a disc that varies from circular to diamond-shaped depending on the species. In bottom-dwelling species the gills are on the ventral surface and water for respiration is taken in through spiracles on the dorsal surface. Myliobatiforms have a long, narrow tail, with or without a caudal fin, and occasionally a small dorsal fin. The tail often extends into a whip-like filament and most have one or more strong, serrated, stinging spines on the dorsal surface of the tail. Venom produced in the core of the spine is stored in tissue around the spine and introduced into the wound it makes, causing the sting to be extremely painful. Myliobatiforms are viviparous. There are 10 families in the order, six of which occur in Australian waters.

SIXGILL STINGRAYS

Hexatrygonidae

SIXGILL STINGRAY
Hexatrygon bickelli

The family of Sixgill Stingrays is a unique and recently discovered group of rays. There are possibly five species found in the Indo-West Pacific, but as very little is known about the group it is difficult to determine how many species are actually represented by the few specimens that have been collected. Sixgill Stingrays have six pairs of gill slits, instead of the five pairs that sharks and rays usually have. The tail has a well-developed caudal fin and a strong spine. They have been found on the continental slopes, from depths of about 800 to 1100 m. The body is very flabby, with fragile skin and a large, fluid-filled, triangular-shaped snout that is covered with sensory pores. Sixgill Stingrays can manipulate this flexible snout independently, and it is presumably used for detecting prey and extricating it from the muddy bottoms of their deepwater habitat. These Stingrays have small, blunt teeth and probably feed on a range of benthic invertebrates.

AUSTRALIAN HEXATRYGONIDAE SPECIES

Hexatrygon bickelli Sixgill Stingray

GIANT STINGAREE

Plesiobatidae

GIANT STINGAREE
Plesiobatis daviesi

The **Giant Stingaree** is the only species in this family and is found throughout the tropical Indo-Pacific region from depths of about 275 to 700 m. The Giant Stingaree is, as its name suggests, enormous and can grow to 2.7 m in length. It was formerly included in the family Urolophidae (p. 128) with all other Stingarees, but differs from them in having small denticles all over the dorsal surface, as well as a much longer, pointed snout. The Giant Stingaree, like the Stingrays (Dasyatidae, p. 131), has a long, sharp, venomous spine on the dorsal surface of the tail, but differs from them in having a well-developed caudal fin. It has been taken by deepwater trawl off the north-east and north-west coasts of Australia. It is found over muddy bottoms, where it feeds on small fishes, crustaceans, cephalopods and other invertebrates.

AUSTRALIAN PLESIOBATIDAE SPECIES

Plesiobatis daviesi Giant Stingaree

STINGAREES

Urolophidae

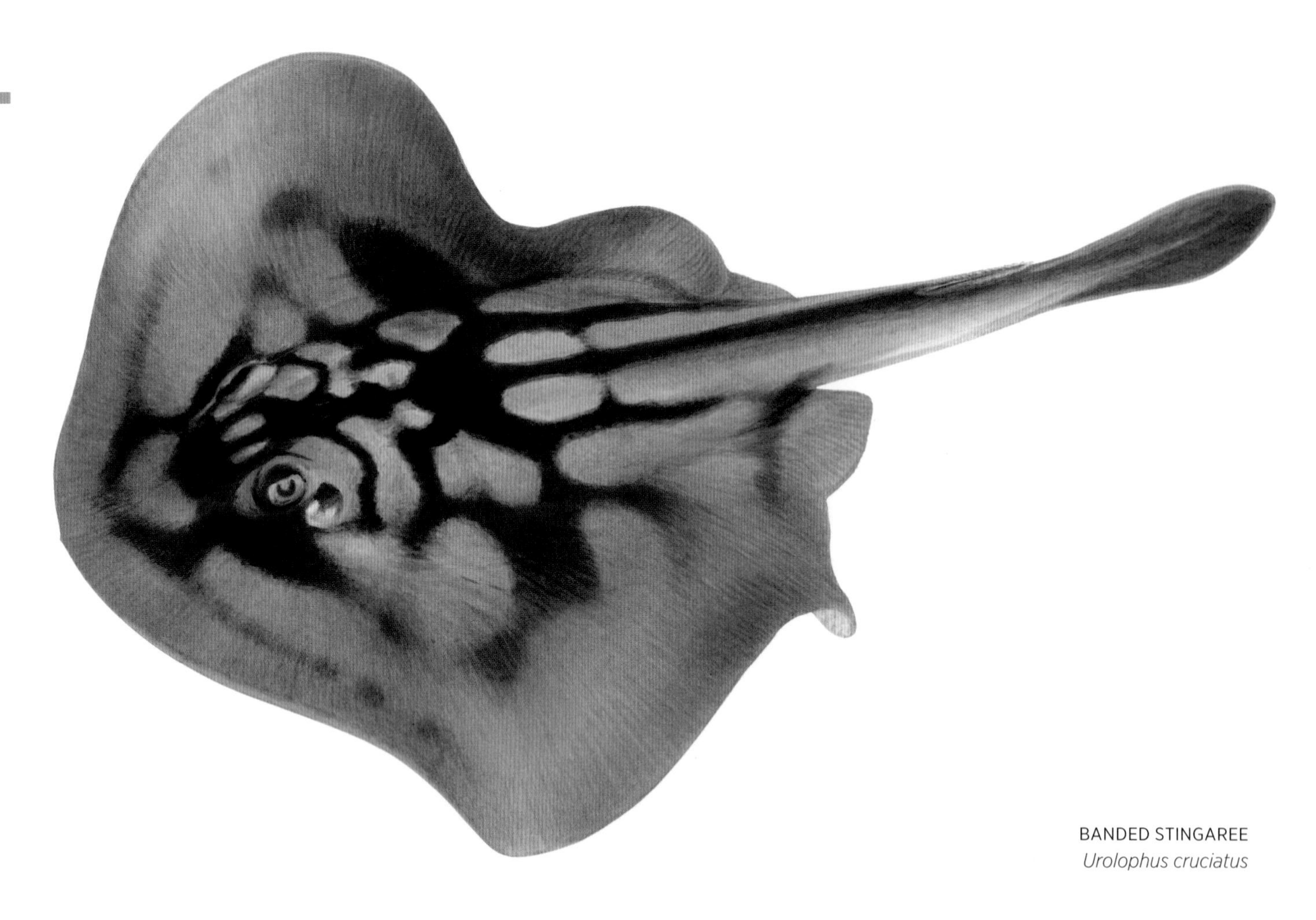

BANDED STINGAREE
Urolophus cruciatus

Stingarees are found in temperate and tropical waters of the Atlantic, Indian and Pacific oceans, and there are about 28 species in the family. They are particularly well represented in Australian waters and several species are endemic. The flattened head, body and pectoral fins form a disc, which varies between species from almost round to diamond-shaped. Unlike other rays, they have smooth skin without small spines on the dorsal surface. They are closely related to the Stingrays (Dasyatidae, p. 131) and possess a barbed, venomous spine on the tail. However, they have a shorter, more muscular tail than Stingrays, with a well-developed caudal fin. The tail is particularly flexible and can arch right over the back.

Stingarees can be found in areas ranging from shallow coastal waters, over reefs, sandflats and weedbeds and in estuaries, to the continental slopes and depths over 350 m. They have numerous small teeth and feed mainly on benthic invertebrates, though little is known about the habits of many members of this family. Most Stingarees are 40–60 cm in length, with the largest Australian species, the **Sandyback Stingaree**, measuring up to about 90 cm. Stingarees are often encountered by divers in the shallow bays and estuaries of south-west Australia, and the family is particularly diverse in this region.

STRIPED STINGAREE
Trygonoptera ovalis

CIRCULAR STINGAREE
Urolophus circularis

PATCHWORK STINGAREE
Urolophus flavomosaicus

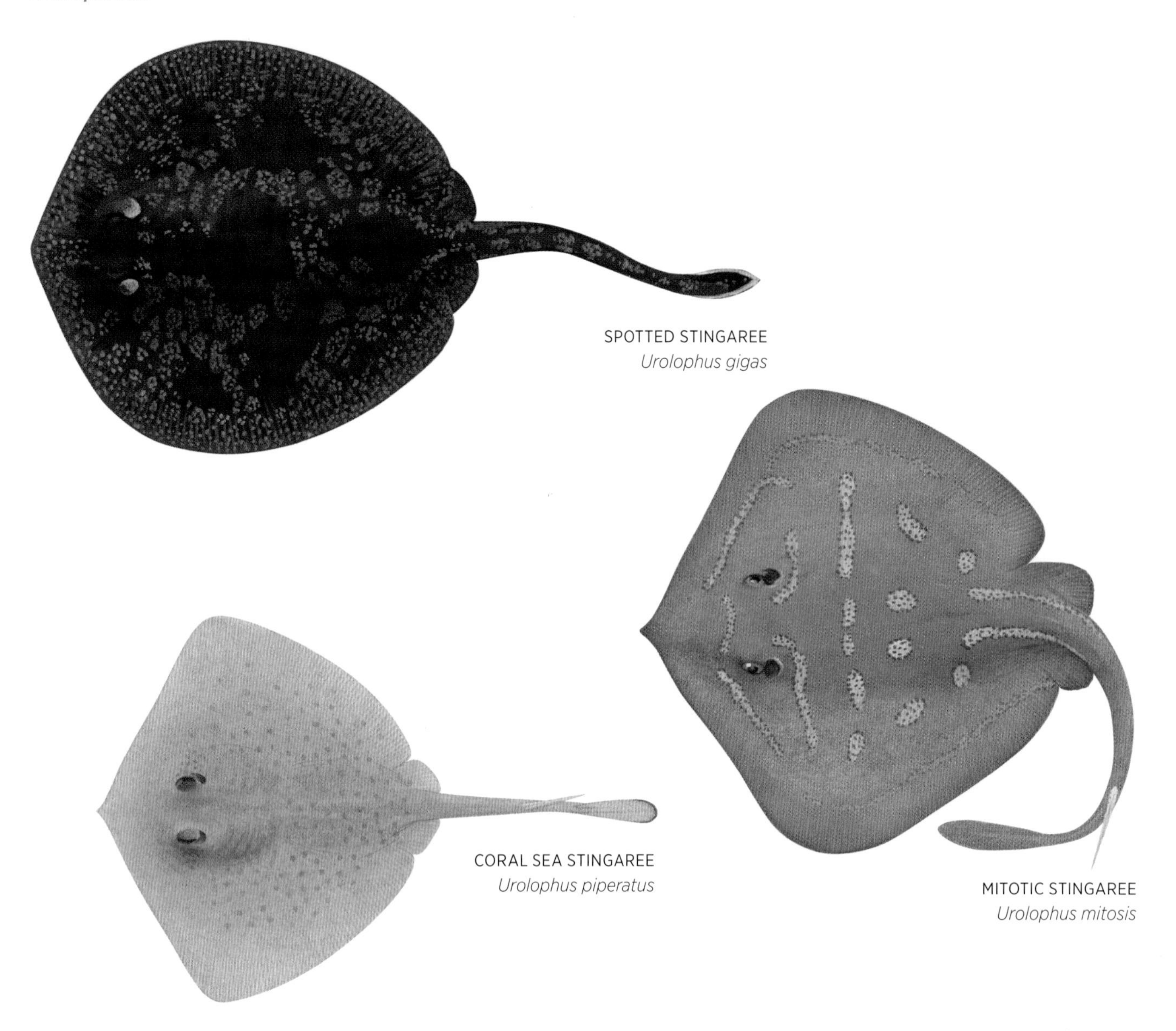

SPOTTED STINGAREE
Urolophus gigas

CORAL SEA STINGAREE
Urolophus piperatus

MITOTIC STINGAREE
Urolophus mitosis

AUSTRALIAN UROLOPHIDAE SPECIES

Trygonoptera galba Yellow Shovelnose Stingaree
Trygonoptera imitata Eastern Shovelnose Stingaree
Trygonoptera mucosa Western Shovelnose Stingaree
Trygonoptera ovalis Striped Stingaree
Trygonoptera personata Masked Stingaree
Trygonoptera testacea Common Stingaree
Urolophus bucculentus Sandyback Stingaree
Urolophus circularis Circular Stingaree
Urolophus cruciatus Banded Stingaree
Urolophus expansus Wide Stingaree
Urolophus flavomosaicus Patchwork Stingaree
Urolophus gigas Spotted Stingaree
Urolophus kapalensis Kapala Stingaree
Urolophus lobatus Lobed Stingaree
Urolophus mitosis Mitotic Stingaree
Urolophus orarius Coastal Stingaree
Urolophus paucimaculatus Sparsely-spotted Stingaree
Urolophus piperatus Coral Sea Stingaree
Urolophus sufflavus Yellowback Stingaree
Urolophus viridis Greenback Stingaree
Urolophus westraliensis Brown Stingaree

STINGRAYS

Dasyatidae

COWTAIL RAY
Pastinachus atrus

Stingrays are found worldwide in marine, brackish and freshwater habitats. There are about 80 species in this large family. The flattened head, body and pectoral fins form a disc, which ranges between species from circular or oval to diamond-shaped. They have a large spiracle behind each eye. In most species the tail is long and thin, with no dorsal, anal or caudal fins, though there may be skin folds that resemble fins, such as in the **Cowtail Ray**.

The tail is usually armed with one or two very sharp, barbed, venomous spines, which can inflict extremely painful wounds and have caused human fatalities. Most stings occur when humans step on Stingrays in shallow waters – Stingrays can arch their tail up over their back and stab the spine into a foot or leg with considerable force. A friend of mine who was stung as he ran into the water said it felt as though someone had struck him on the ankle with an axe. He suffered extensive bleeding, severe pain and swelling.

Australian Stingrays range in size from about 25 cm across for the **Plain Maskray**, up to more than 2 m in width and 350 kg for the massive **Smooth Stingray**, which is found around the southern half of Australia. Stingrays usually occur in coastal waters in areas with muddy or sandy bottoms; a few species, such as the **Bluespotted Fantail Ray**, inhabit coral reefs. The unique **Pelagic Stingray** is found in the open ocean and the **Freshwater Whipray** occurs in most large rivers in northern Australia. Stingrays have flattened crushing teeth and feed on invertebrates such as crustaceans and molluscs, and small fishes.

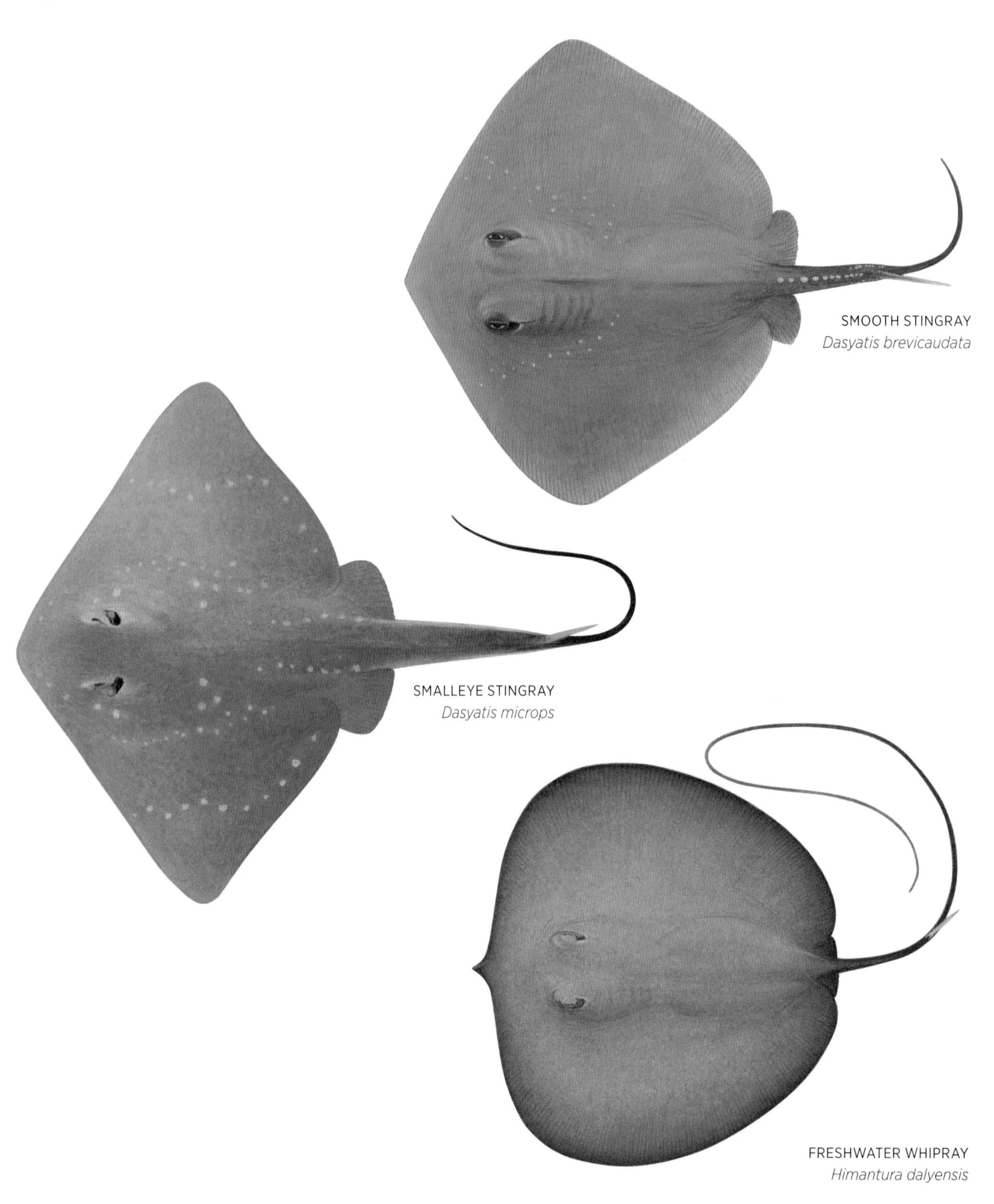

SMOOTH STINGRAY
Dasyatis brevicaudata

SMALLEYE STINGRAY
Dasyatis microps

FRESHWATER WHIPRAY
Himantura dalyensis

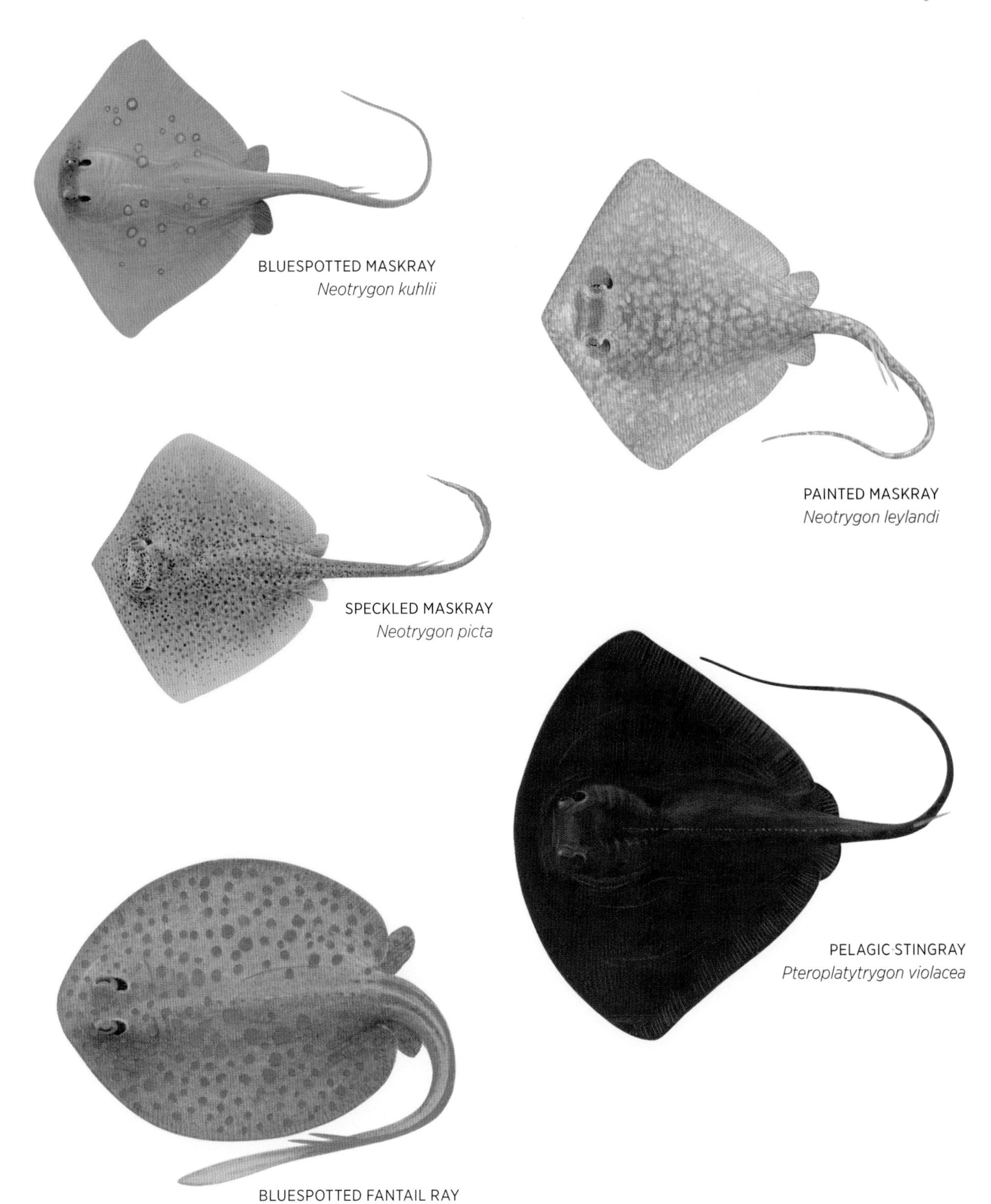

BLUESPOTTED MASKRAY
Neotrygon kuhlii

PAINTED MASKRAY
Neotrygon leylandi

SPECKLED MASKRAY
Neotrygon picta

PELAGIC STINGRAY
Pteroplatytrygon violacea

BLUESPOTTED FANTAIL RAY
Taeniura lymma

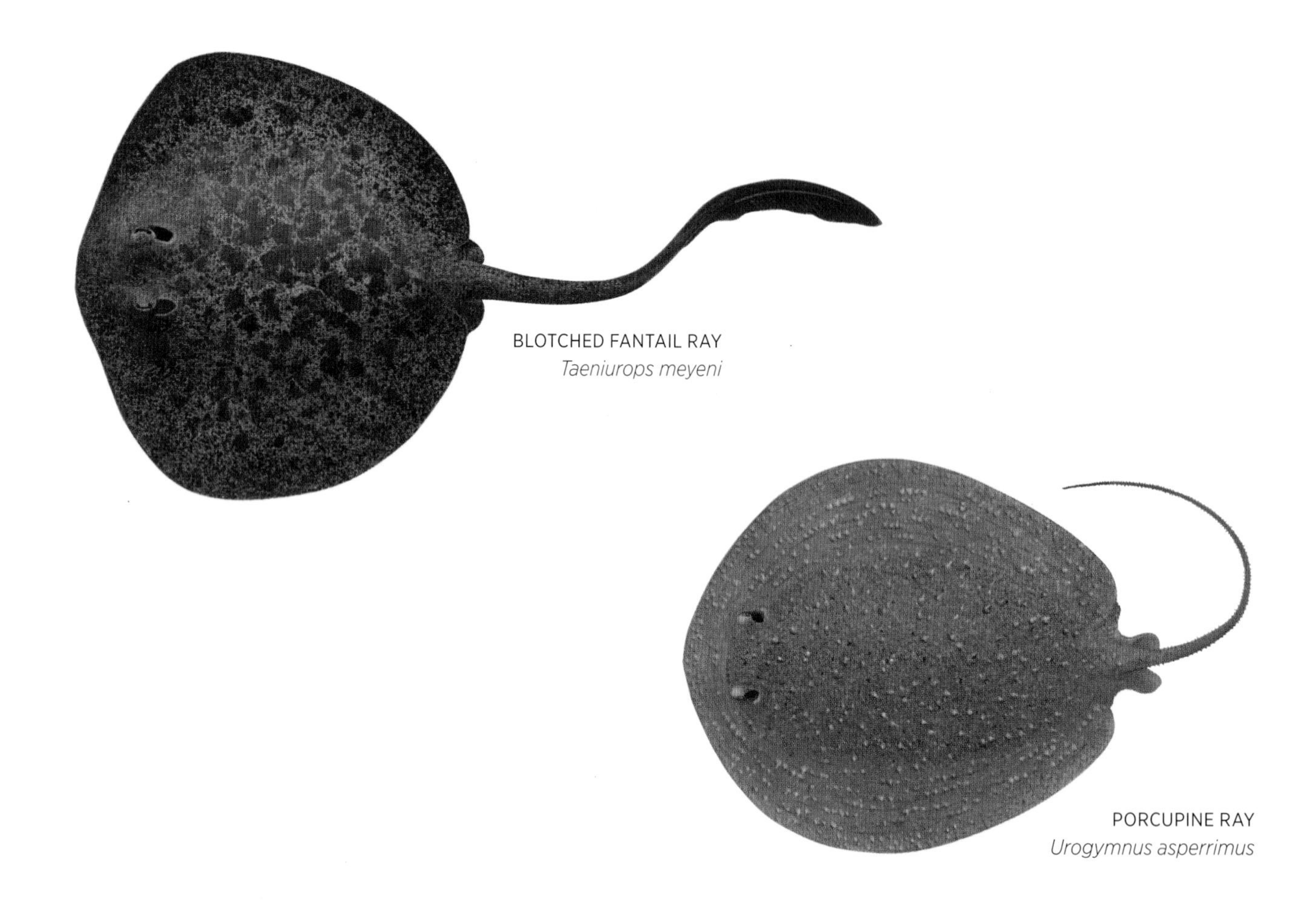

BLOTCHED FANTAIL RAY
Taeniurops meyeni

PORCUPINE RAY
Urogymnus asperrimus

AUSTRALIAN DASYATIDAE SPECIES

Dasyatis brevicaudata Smooth Stingray
Dasyatis fluviorum Estuary Stingray
Dasyatis microps Smalleye Stingray
Dasyatis parvonigra Dwarf Black Stingray
Dasyatis thetidis Black Stingray
Himantura astra Blackspotted Whipray
Himantura dalyensis Freshwater Whipray
Himantura fai Pink Whipray
Himantura granulata Mangrove Whipray
Himantura jenkinsii Jenkins' Whipray
Himantura leoparda Leopard Whipray
Himantura toshi Brown Whipray
Himantura uarnak Reticulate Whipray
Neotrygon annotata Plain Maskray
Neotrygon kuhlii Bluespotted Maskray
Neotrygon leylandi Painted Maskray
Neotrygon picta Speckled Maskray
Pastinachus atrus Cowtail Ray
Pteroplatytrygon violacea Pelagic Stingray
Taeniura lymma Bluespotted Fantail Ray
Taeniurops meyeni Blotched Fantail Ray
Urogymnus asperrimus Porcupine Ray

BUTTERFLY RAYS

Gymnuridae

AUSTRALIAN BUTTERFLY RAY
Gymnura australis

Butterfly Rays are found worldwide in tropical and warm temperate waters. There are about 12 species in the family and they are immediately recognisable by their extremely broad disc, which is much wider than it is long. They also differ from the Stingrays (Dasyatidae, p. 131) in having a dorsal fin and no tail spine.

The **Australian Butterfly Ray** grows to a width of about 70 cm. It inhabits coastal waters of the northern half of Australia to depths of about 50 m. Little is known of its biology, but it is known to be a bottom dweller that spends much of its time immobile and covered with a layer of sand. It feeds on crustaceans, cephalopods and small fishes, and is often captured by prawn trawlers.

AUSTRALIAN GYMNURIDAE SPECIES
Gymnura australis Australian Butterfly Ray

EAGLE RAYS AND MANTA RAYS

Myliobatidae

MANTA RAY
Manta birostris

There are about 37 species in this family and they are distributed worldwide in tropical and temperate waters. They differ from other rays in having the eyes and spiracles on the sides of the head, which rather than being flattened is elevated above the disc. They have a long, thin tail with a small dorsal fin and most possess a venomous spine at the base of the tail. The family is diverse in its habits and is sometimes divided into three subfamilies: the Mobulinae, comprising Devilrays and Manta Rays; the Myliobatinae or Eagle Rays; and the Rhinopterinae or Cownose Rays.

MOBULINAE Manta Rays can reach over 6 m in width and, like the smaller Devilrays, cruise tropical waters, filter feeding on plankton. In members of this group the wide mouth is at the front of the head and there are two large lobes on either side of it that are extensions of the pectoral fins. These cephalic lobes can be rolled up to reduce drag during swimming or extended to direct a greater flow of water into the mouth. Devilrays and Manta Rays are generally placid and often easily approached by divers. When they encounter waters with high concentrations of plankton, they will swim in a continuous vertical loop in order to remain within the area.

MYLIOBATINAE and RHINOPTERINAE These rays are smaller than the Mobulinae, with most less than 1.6 m in width. They have a rounded or bilobed snout with the mouth positioned ventrally, well back from the front of the head. The teeth are fused into crushing plates to deal with hard-shelled prey such as molluscs. The **Whitespotted Eagle Ray**, found in tropical coastal areas and around offshore islands, swims constantly, usually near the surface, and occasionally leaps high out of the water. It will sometimes swim closer to the bottom when feeding on crustaceans, molluscs, other invertebrates and small fishes. The **Southern Eagle Ray** is commonly encountered in inshore coastal waters off the southern half of Australia, where it forages over weed beds and sand for crabs and molluscs. It is often taken by anglers using Pilchard baits to fish for Tailor or Australian Salmon. Cownose Rays are pelagic in coastal and offshore waters of northern Australia and sometimes occur in large schools.

WHITESPOTTED EAGLE RAY
Aetobatus narinari

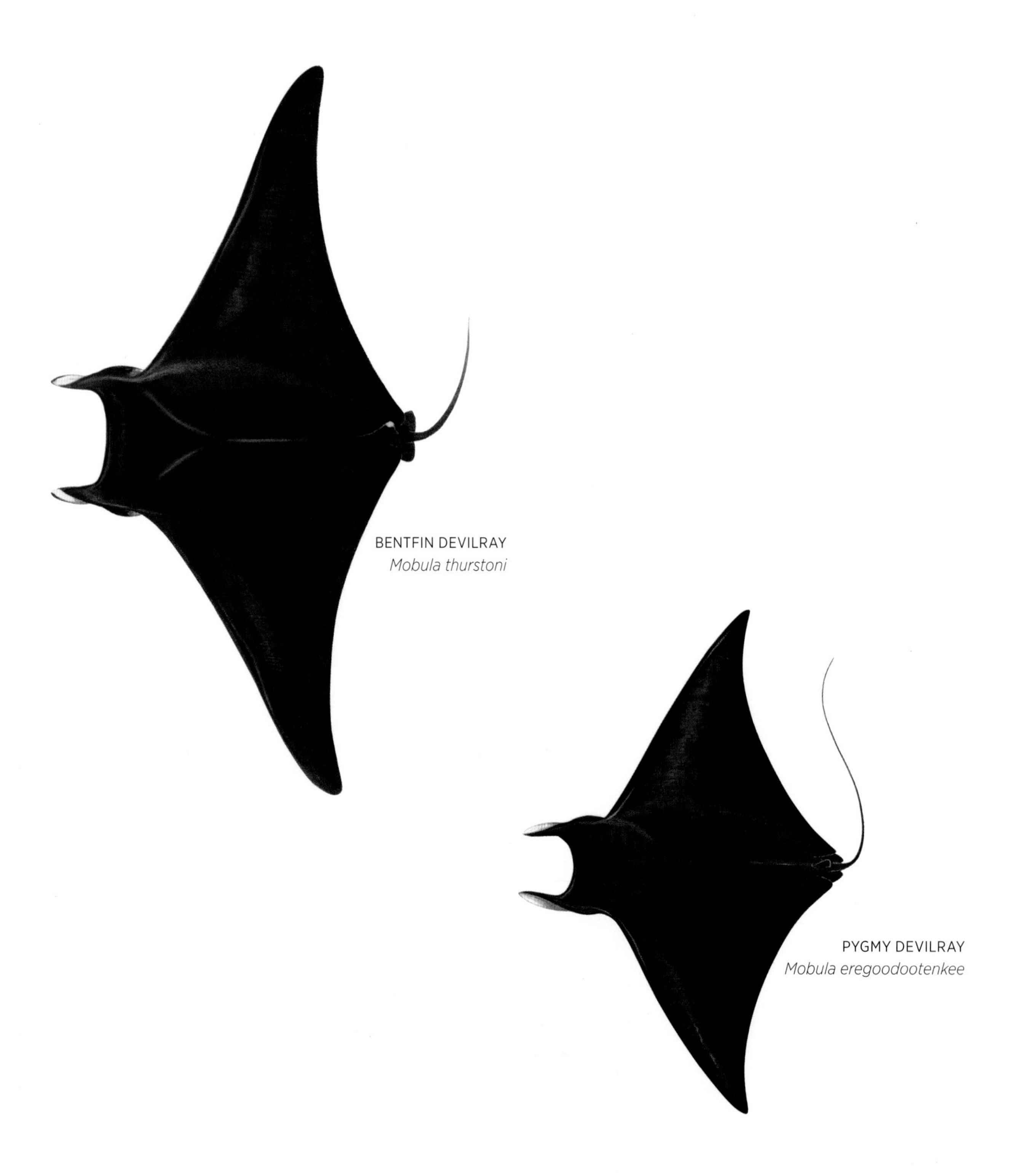

BENTFIN DEVILRAY
Mobula thurstoni

PYGMY DEVILRAY
Mobula eregoodootenkee

SOUTHERN EAGLE RAY
Myliobatis australis

AUSTRALIAN COWNOSE RAY
Rhinoptera neglecta

AUSTRALIAN MYLIOBATIDAE SPECIES

Aetobatus narinari Whitespotted Eagle Ray
Aetomylaeus nichofii Banded Eagle Ray
Aetomylaeus vespertilio Ornate Eagle Ray
Manta birostris Manta Ray
Mobula eregoodootenkee Pygmy Devilray
Mobula japanica Japanese Devilray
Mobula thurstoni Bentfin Devilray
Myliobatis australis Southern Eagle Ray
Myliobatis hamlyni Purple Eagle Ray
Rhinoptera javanica Javanese Cownose Ray
Rhinoptera neglecta Australian Cownose Ray

ACTINOPTERYGII

BONY FISHES

OSTEOGLOSSIFORMES

The Osteoglossiformes is considered to be the most ancient and primitive order of all the living Teleost fishes. Anatomy varies considerably between the different families, but most have well-developed teeth on a bony tongue that is used to bite against toothed bones on the roof of the mouth. They have a single dorsal fin and fewer bones supporting the caudal fin than most other fishes. The intestine passes to the left of the oesophagus, as opposed to the right as in other fishes. The order includes perhaps the world's largest freshwater fish, the Arapaima of Brazil, which can reportedly reach 4.5 m in length. All living Osteoglossiforms inhabit fresh water and have evolved and diversified there. They are mainly found in South America, Africa and South-East Asia. There are seven families in the order, one of which occurs in Australia.

SARATOGAS

Osteoglossidae

NORTHERN SARATOGA
Scleropages jardinii

This family, which contains about eight species, is found in South-East Asia, Africa and South America. Two species are found in the tropical north of Australia. It is a fascinating ancient family, with fossil remains dating back at least 135 million years. Saratogas use their bony tongue rather than their lower jaw to bite against sharp bones in the roof of their mouth. They have a lung-like swim bladder and can gulp air at the surface to assist respiration. The large lower jaw has a pair of fleshy barbels at the tip.

The two very similar Australian species of Saratoga inhabit still waters containing plenty of aquatic vegetation, such as billabongs and lagoons, and attain a maximum length of about 1 m. They are voracious surface and midwater feeders, and take a wide range of prey including insects, small fishes and frogs, along with some plant material. Both Australian species are mouthbrooders. The female holds 100 or more eggs in its mouth for several weeks until they hatch and then protects the young for several more weeks. Saratogas are sought-after by sportfishers for their aggressive nature and ability to perform spectacular leaps when hooked.

AUSTRALIAN OSTEOGLOSSIDAE SPECIES
Scleropages jardinii Northern Saratoga
Scleropages leichhardti Southern Saratoga

ELOPIFORMES

This is a primitive group of fishes with an unusual transparent, leaf-shaped larva known as a leptocephalus. The leptocephali have a forked tail, differentiating them from those of the true eels (Anguilliformes, p. 153), which have a pointed tail. Elopiforms have a gular plate (a bony support in the throat) and 23–25 branchiostegal rays. They have a single dorsal fin, a deeply forked caudal fin, and a long, thin axillary scale at the base of each pectoral and ventral fin – there are no spines in any of the fins. The tip of the snout does not overhang the mouth. There are two families in the order, both of which occur in Australian waters.

GIANT HERRINGS

Elopidae

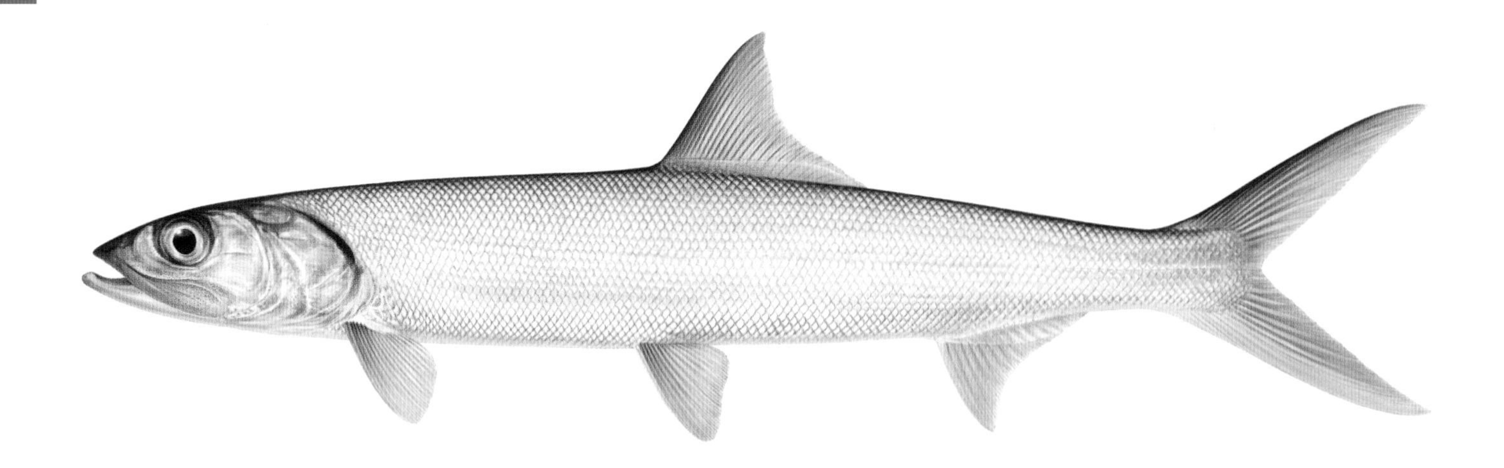

AUSTRALIAN GIANT HERRING
Elops machnata

Giant Herrings are distributed worldwide in warm temperate and tropical seas, and there are about eight species in the family. The two species found in Australian waters are very similar: both have a long, rounded body, with small, very fragile scales, a single dorsal fin and a large, powerful tail. They have large eyes with partial adipose eyelid and are fast-swimming predators of small fishes and invertebrates.

The Australian species mainly occur in northern coastal waters, often in small schools. They readily enter brackish estuaries and as juveniles occasionally venture into fresh water. Giant Herrings are sought-after as sportfishes for their tremendous speed, tenacity and acrobatic leaps when hooked. They are often carried south by tropical currents and surprise anglers in southern estuaries. Giant Herrings reach a maximum length of about 1.2 m.

AUSTRALIAN ELOPIDAE SPECIES

Elops hawaiensis Hawaiian Giant Herring
Elops machnata Australian Giant Herring

TARPONS

Megalopidae

OXEYE HERRING
Megalops cyprinoides

There are only two species of Tarpon: one occurs in the Atlantic Ocean and the other in the Indo-Pacific region. Tarpons are closely related to the Giant Herrings (Elopidae, p. 146), but have larger scales and a deeper body, and the last ray of the dorsal fin is elongated.

The **Oxeye Herring** is found throughout the Indo-Pacific region in tropical inshore waters, mangroves, estuaries and occasionally in the lower reaches of freshwater rivers. It preys on small fishes and grows to a maximum length of 1.5 m, though most of those encountered are about 50–60 cm. Tarpons can tolerate waters with very low oxygen concentration, as their modified swim bladder allows them to gulp air at the surface.

AUSTRALIAN MEGALOPIDAE SPECIES
Megalops cyprinoides Oxeye Herring

ALBULIFORMES

This is a widely varied group, but all members have a leptocephalus larval stage similar to that seen in members of the order Elopiformes (p.145). Adult body shapes range from eel-like with no caudal fin, to fusiform with a deeply forked caudal fin. The tail shape of the leptocephalus larvae corresponds to that of the adults, being either forked or pointed. In all adults the tip of the snout overhangs the mouth. The order contains three families, all of which occur in Australian waters.

BONEFISHES

Albulidae

PACIFIC BONEFISH
Albula forsteri

The Bonefishes are found worldwide in warm temperate and tropical waters. The family was originally thought to contain only one species in the genus *Albula*, but the genus is now separated into five or six different species. It is uncertain exactly how many species occur in Australian waters as all are very similar and difficult to tell apart. The Bonefishes are closely related to the Giant Herrings (Elopidae, p. 146) and Tarpons (Megalopidae, p. 147), but have smaller eyes and the snout projects forward over the mouth. They have a muscular, almost cylindrical, fusiform body and partial adipose eyelids.

Bonefishes are shy, shallow-water residents of northern Australian waters. They feed over sand flats – often in water so shallow that their tails project above the surface – preying on small crustaceans, molluscs and worms. They move into offshore waters to spawn and are occasionally caught by anglers bottom-fishing in deep waters. Bonefishes are highly sought-after by sportfishers, particularly flyfishers, as their large forked tail and powerful body give them great speed and stamina. They are generally regarded as poor food fishes as they have numerous fine bones between muscle segments.

AUSTRALIAN ALBULIDAE SPECIES

Albula forsteri Pacific Bonefish

HALOSAURS

Halosauridae

OVEN'S HALOSAUR
Halosaurus ovenii

About 15 species of Halosaur are found worldwide. They occur on deep continental slopes and abyssal plains, to depths of 5000 m or more. They have an elongated, eel-like body with quite large scales, a single short dorsal fin with no spines, and a long, low anal fin that reaches to the tip of the tail. There is a backward-projecting spine on the rear of the upper jaw. The head and flattened, projecting snout are covered with mucus-filled sensory canals and pores that are used to detect prey in the darkness of very deep water. The sensory lateral line, which runs along the entire length of the body, is also very well developed.

Halosaurs are found right around Australia, and feed on pelagic and benthic invertebrates (stomach contents show that they often capture the latter by ingesting mouthfuls of sediment). They are occasionally encountered in very deep trawl catches.

AUSTRALIAN HALOSAURIDAE SPECIES

Aldrovandia affinis Allied Halosaur
Aldrovandia phalacra Baldhead Halosaur
Halosauropsis macrochir Black Halosaur
Halosaurus ovenii Oven's Halosaur
Halosaurus pectoralis Australian Halosaur

SPINY EELS

Notocanthidae

COSMOPOLITAN SPINEBACK
Notocanthus chemnitzii

About 10 species of Spiny Eel are found worldwide, usually in very deep water, to more than 3500 m. Like their close relatives the Halosaurs (Halosauridae, p. 151), they have a long, low anal fin and no caudal fin. However, their scales are very small and in *Notacanthus* species the dorsal fin is composed of a series of individual short spines. The inferior mouth is used to take benthic invertebrates such as copepods and polychaete worms, although these eels also feed on pelagic crustaceans.

The **Cosmopolitan Spineback** grows to about 1.2 m in length and has specialised knife-like teeth that it uses to bite off the tentacles of anemones, corals, bryozoans and hydrozoans. The **Spiny Sucker Eel** uses its sucker-like mouth to ingest organic material from bottom sediment. The Australian species are rarely encountered, only occasionally appearing in the deepest trawl catches taken off the southern coast.

AUSTRALIAN NOTOCANTHIDAE SPECIES
Lipogenys gillii Spiny Sucker Eel
Notacanthus chemnitzii Cosmopolitan Spineback
Notacanthus sexspinis Southern Spineback

ANGUILLIFORMES

Anguilliforms are the true eels. All families in the order have a leaf-shaped, transparent leptocephalus larval stage, which has a pointed or rounded tail. Adults have a snake-like body that either is scale-less or has very small, embedded scales. They have long, low dorsal and anal fins and usually small pectoral fins, and the ventral fins and supporting bones of the pelvic girdle are absent. There are no spines in any of the fins, and the gill openings are reduced to small slits or holes. In many Anguilliforms the leptocephali remain pelagic for a long time, resulting in many species having wide distribution. Anguilliforms are found in a broad range of environments, from fresh water to the ocean depths. There are 15 families in the order, 13 of which occur in Australian waters.

FRESHWATER EELS

Anguillidae

SOUTHERN SHORTFIN EEL
Anguilla australis

Freshwater Eels are found worldwide in tropical and temperate waters. There are about 15 species in the family, all in the genus *Anguilla*. They have pectoral fins but no ventral fins, and the dorsal and anal fins are continuous with the caudal fin. They have minute scales embedded in thick skin. All have prominent tubular nostrils and distinctive fleshy flanges on the upper and lower lips.

Freshwater Eels spend most of their life in rivers, streams and lakes, and migrate to the sea to breed in deep oceanic waters. Their spawning locations were for a long time a complete mystery. It is now known that European Freshwater Eels spawn in the depths of the North Atlantic, but the spawning areas of Australian eels are still largely unknown. Adult eels die after spawning and the newly hatched leptocephalus larvae drift back towards land on ocean currents. On the way, they metamorpose into slim, transparent juvenile eels, known as elvers. The elvers return to freshwater rivers en masse, where they grow and mature.

The **Indonesian Shortfin Eel** and **Pacific Shortfin Eel** are found in northern Australia. The **Southern Shortfin Eel** and **Longfin Eel** occur in south- and east-coast drainages and Tasmania – both species support small commercial fisheries. These eels feed nocturnally on crustaceans, molluscs, worms, small fishes and other vertebrates such as frogs. The Longfin Eel is the largest Australian member of the family; it can reach lengths of 1.5 m and weigh over 20 kg.

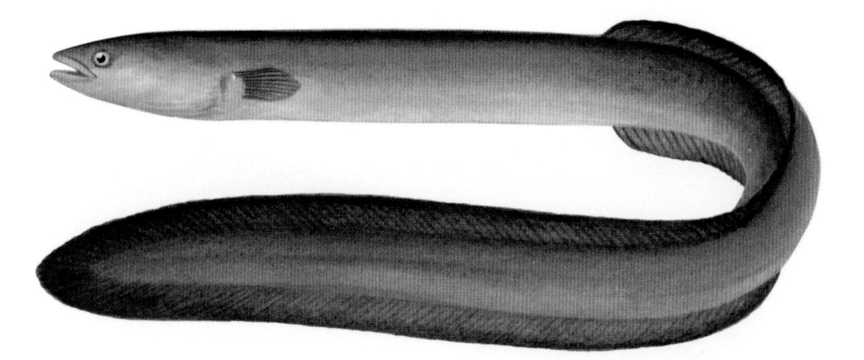

PACIFIC SHORTFIN EEL
Anguilla obscura

LONGFIN EEL
Anguilla reinhardtii

AUSTRALIAN ANGUILLIDAE SPECIES

Anguilla australis Southern Shortfin Eel
Anguilla bicolor Indonesian Shortfin Eel
Anguilla obscura Pacific Shortfin Eel
Anguilla reinhardtii Longfin Eel

WORM EELS

Moringuidae

COMMON WORM EEL
Moringua microchir

Worm Eels are distributed in tropical waters of the Indo-Pacific and Western Atlantic. The family contains about six species. They have an extremely long, thin body, without scales, and small, low dorsal and anal fins originating well back on the body. The pectoral fins are small and weak, or completely absent, and the lower jaw projects beyond the upper jaw. In juveniles the eyes are much reduced and covered with skin; as the eels mature they develop functioning eyes and larger fins. Worm Eels grow to a maximum length of about 1.2 m, with the females much larger than the males. They live in shallow waters, often near coral reefs.

Although quite common in Australia, Worm Eels are rarely seen due to their habit of burrowing into sandy or muddy bottoms. The **Common Worm Eel** is found in the lower reaches of rivers and estuaries in northern Australia, where it feeds on small fishes and invertebrates (mainly crustaceans). Adults move into open water to breed. Like other eels, the leptocephalus larvae find their way back to shallow coastal and estuarine waters, where they mature.

AUSTRALIAN MORINGUIDAE SPECIES
Moringua abbreviata Short Worm Eel
Moringua ferruginea Rusty Worm Eel
Moringua javanica Javan Worm Eel
Moringua microchir Common Worm Eel
Neoconger tuberculatus Swollengut Worm Eel

FALSE MORAY EELS

Chlopsidae

PLAIN FALSE MORAY
Kaupichthys hyoproroides

False Moray Eels are found throughout the warm temperate and tropical waters of the Indian, Pacific and Atlantic oceans. There are about 18 species in the family. Unlike the true Moray Eels (Muraenidae, p. 158), the Australian species of False Morays all have pectoral fins. They have no scales and there is a very small, round gill opening on each side of the body. They have long dorsal and anal fins that are continuous with the caudal fin, and sensory pores on the head, though not along the lateral line. There are two rows of teeth on the roof of the mouth, but no large fangs like in true Morays. False Moray Eels grow to about 30 cm in length and are very rarely seen, living secretively in the deep crevices of coral reefs and rubble.

AUSTRALIAN CHLOPSIDAE SPECIES

Kaupichthys atronasus Blacknose False Moray
Kaupichthys brachychirus Shortfin False Moray
Kaupichthys hyoproroides Plain False Moray

MORAY EELS

Muraenidae

Y-PATTERNED MORAY
Gymnothorax berndti

This large, diverse family contains about 185 species and they are found worldwide in tropical and temperate waters. Most have a long, muscular, scaleless body with tough skin. The dorsal and anal fins are continuous with the caudal fin, and there are no pectoral fins. They have a single small, rounded gill opening on each side of the head, sensory pores and tubular nostrils on the head, and usually large jaws reaching behind the eyes.

Moray Eels typically live in shallow waters on reefs, hiding in crevices by day and hunting at night for cephalopods, fishes and crustaceans. Most Australian reefs contain large numbers of individuals, though typically they remain hidden from view. The beautifully patterned **Starry Moray**, a common resident of Australian reef flats and rocky coastal areas, has blunt crushing teeth for dealing with its usual prey of crustaceans.

Gymnothorax species are mainly fish eaters, with large fang-like teeth. The largest of the genus, the **Giant Moray**, can reach over 2 m in length and is popular with divers on the Great Barrier Reef. Although their reputation for being vicious is perhaps undeserved, large specimens certainly shouldn't be provoked or hand-fed, as they are extremely powerful and their sharp teeth can inflict serious wounds. The **Green Moray** is a common species on rocky reefs, and is one of the few members of the family found right around the southern coast of Australia – most species prefer tropical waters.

Enchelycore species can be recognised by their long, hooked jaws, which cannot close due to their shape. The Snake Morays, of the genus *Uropterygius*, have greatly reduced dorsal and anal fins, with skin-covered ridges at the tail tip being all that remain. The spectacular **Ribbon Eel** is occasionally encountered in the open, but is more commonly seen with just the head and forepart of the body projecting from the sand or rubble. The Ribbon Eel is one of the species of Moray Eel known to be able to change sex from male to female, but it is unclear if this ability is universal within the family.

All Morays have pelagic leptocephalus larvae that remain in open water for a particularly long time. As a result, many species have a very wide distribution. The **Y-Patterned Moray**, for example, is found in waters from Japan to New Zealand. Although it has not been officially recorded in Australian waters, it probably occurs in deep waters off the east coast.

STARRY MORAY
Echidna nebulosa

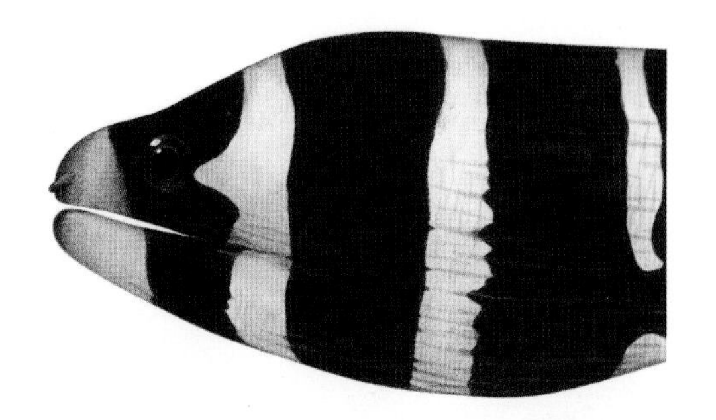

GIRDLED MORAY
Echidna polyzona

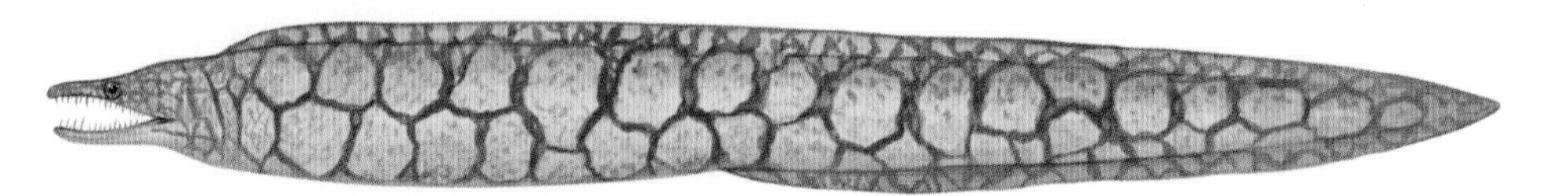

MOSAIC MORAY
Enchelycore ramosa

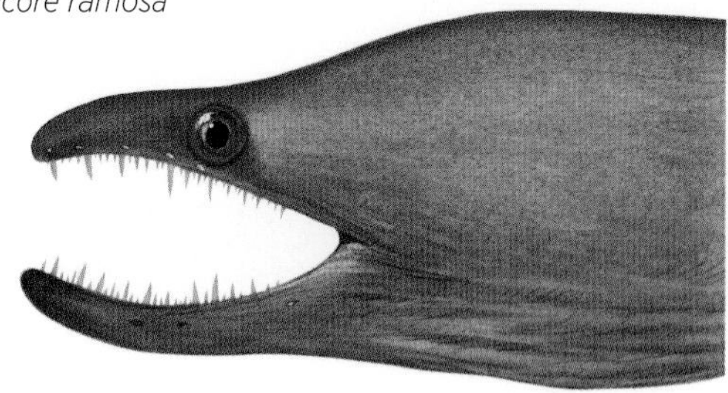

HOOKJAW MORAY
Enchelycore bayeri

LATTICETAIL MORAY
Gymnothorax buroensis

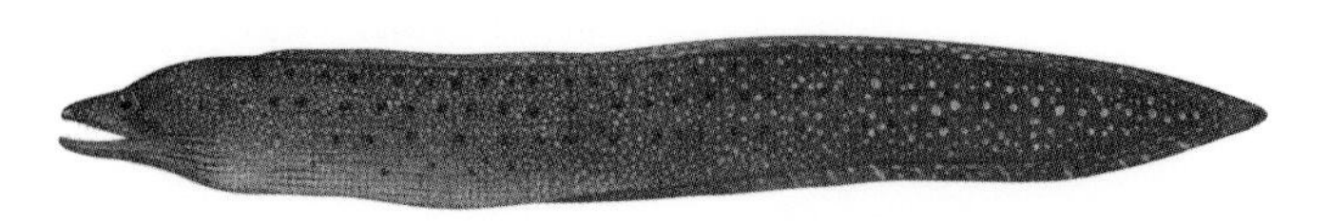

STOUT MORAY
Gymnothorax eurostus

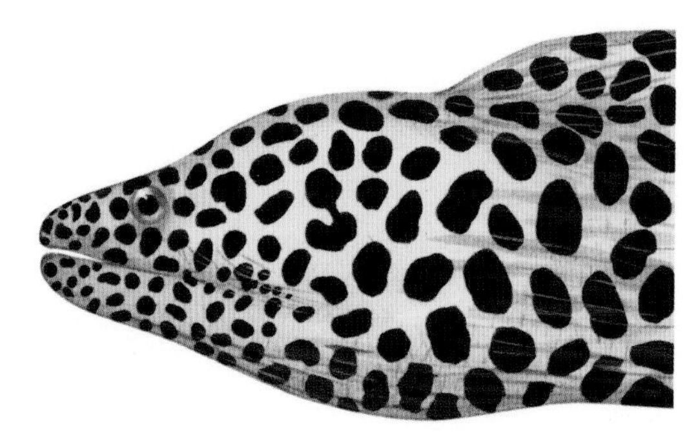

TESSELLATE MORAY
Gymnothorax favagineus

FIMBRIATE MORAY
Gymnothorax fimbriatus

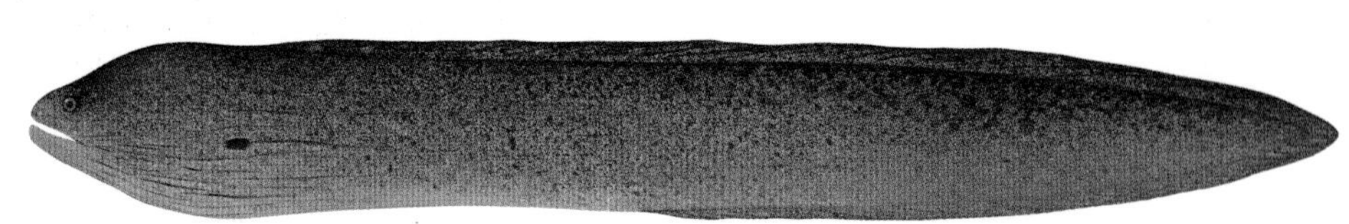

YELLOWMARGIN MORAY
Gymnothorax flavimarginatus

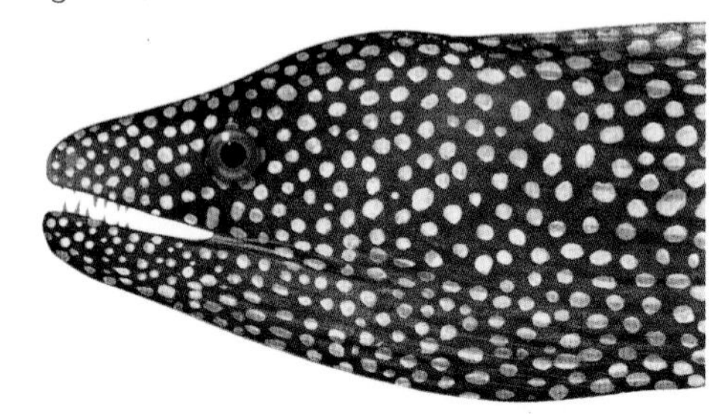

WHITEMOUTH MORAY
Gymnothorax meleagris

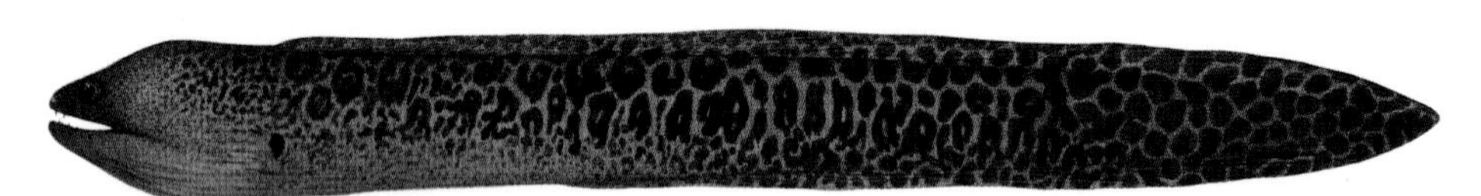

GIANT MORAY
Gymnothorax javanicus

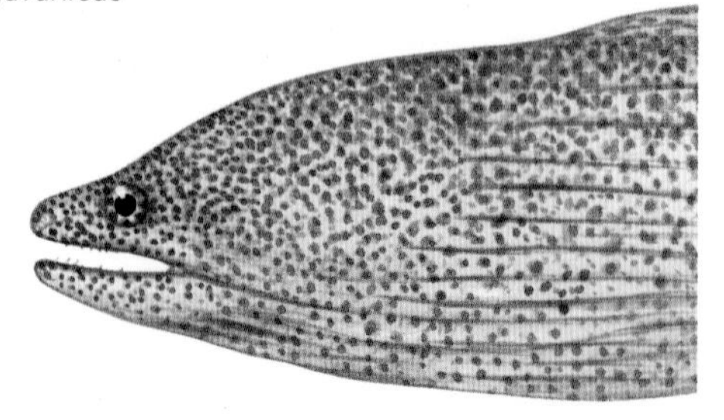

PAINTED MORAY
Gymnothorax pictus

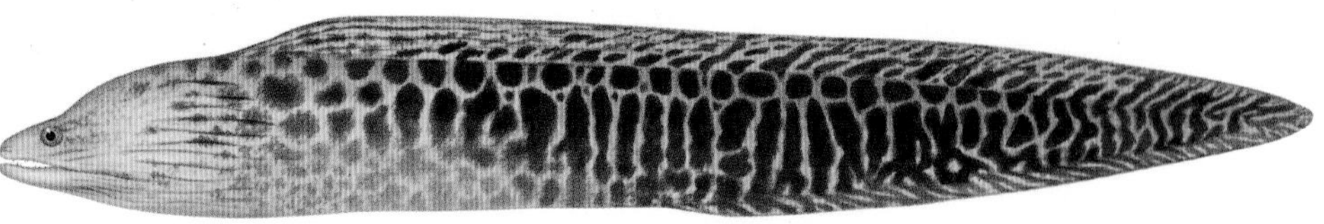

UNDULATE MORAY
Gymnothorax undulatus

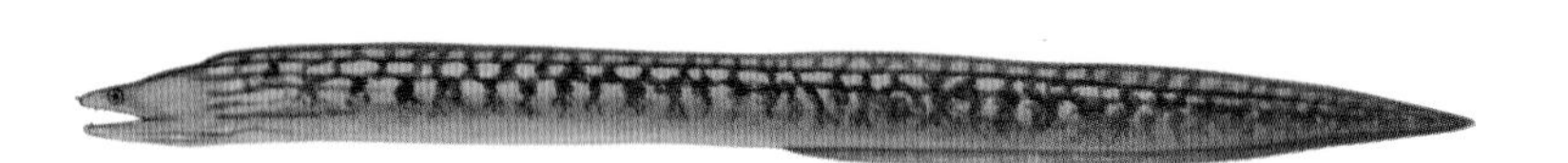

WOODWARD'S MORAY
Gymnothorax woodwardi

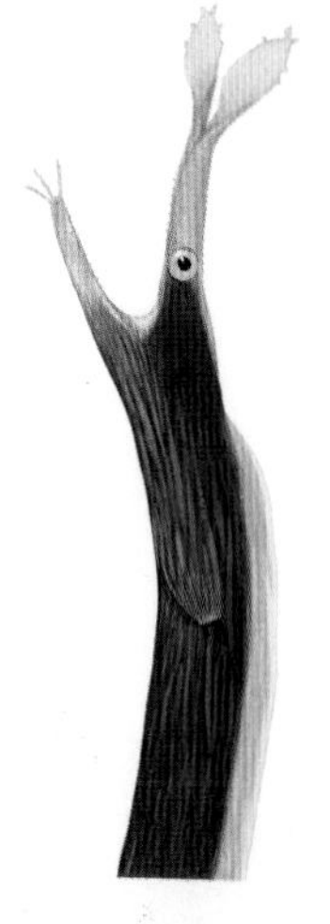

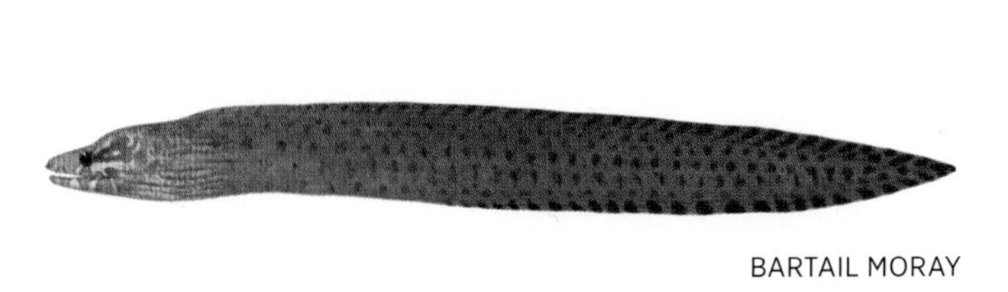

BARTAIL MORAY
Gymnothorax zonipectis

RIBBON EEL
Rhinomuraena quaesita

AUSTRALIAN MURAENIDAE SPECIES

Anarchias allardicei Allardice's Moray
Anarchias cantonensis Canton Moray
Anarchias leucurus Finespot Moray
Anarchias seychellensis Seychelles Moray
Echidna nebulosa Starry Moray
Echidna polyzona Girdled Moray
Echidna unicolor Pale Moray
Enchelycore bayeri Hookjaw Moray
Enchelycore ramosa Mosaic Moray
Enchelynassa canina Longfang Moray
Gymnomuraena zebra Zebra Moray
Gymnothorax annasona Lord Howe Moray
Gymnothorax atolli Atoll Moray
Gymnothorax austrinus Southern Moray
Gymnothorax berndti Y-patterned Moray
Gymnothorax buroensis Latticetail Moray
Gymnothorax castlei Castle's Moray
Gymnothorax cephalospilus Headspot Moray
Gymnothorax chilospilus Lipspot Moray
Gymnothorax cribroris Sieve Moray
Gymnothorax enigmaticus Tiger Moray
Gymnothorax eurostus Stout Moray
Gymnothorax favagineus Tessellate Moray
Gymnothorax fimbriatus Fimbriate Moray
Gymnothorax flavimarginatus Yellowmargin Moray
Gymnothorax fuscomaculatus Brownspotted Moray
Gymnothorax gracilicaudus Slendertail Moray
Gymnothorax intesi Whitetip Moray
Gymnothorax javanicus Giant Moray
Gymnothorax kidako Kidako Moray
Gymnothorax longinquus Yellowgill Moray
Gymnothorax margaritophorus Pearly Moray
Gymnothorax mccoskeri Manyband Moray
Gymnothorax melatremus Dwarf Moray
Gymnothorax meleagris Whitemouth Moray
Gymnothorax microstictus Smallspot Moray
Gymnothorax minor Lesser Moray
Gymnothorax monochrous Monotone Moray
Gymnothorax nubilus Grey Moray
Gymnothorax nudivomer Yellowmouth Moray
Gymnothorax obesus Speckled Moray
Gymnothorax pictus Painted Moray
Gymnothorax pindae Pinda Moray
Gymnothorax polyuranodon Freshwater Moray
Gymnothorax porphyreus Lowfin Moray
Gymnothorax prasinus Green Moray
Gymnothorax prionodon Sawtooth Moray
Gymnothorax pseudoherrei False Brown Moray
Gymnothorax pseudothyrsoideus Highfin Moray
Gymnothorax robinsi Pygmy Moray
Gymnothorax rueppellii Banded Moray
Gymnothorax thyrsoideus Greyface Moray
Gymnothorax undulatus Undulate Moray
Gymnothorax woodwardi Woodward's Moray
Gymnothorax zonipectis Bartail Moray
Rhinomuraena quaesita Ribbon Eel
Strophidon sathete Longtail Moray
Uropterygius concolor Unicolor Snake Moray
Uropterygius fuscoguttatus Brownspotted Snake Moray
Uropterygius kamar Moon Snake Moray
Uropterygius marmoratus Marbled Snake Moray
Uropterygius micropterus Shortfin Snake Moray
Uropterygius nagoensis Nago Snake Moray

CUTTHROAT EELS

Synaphobranchidae

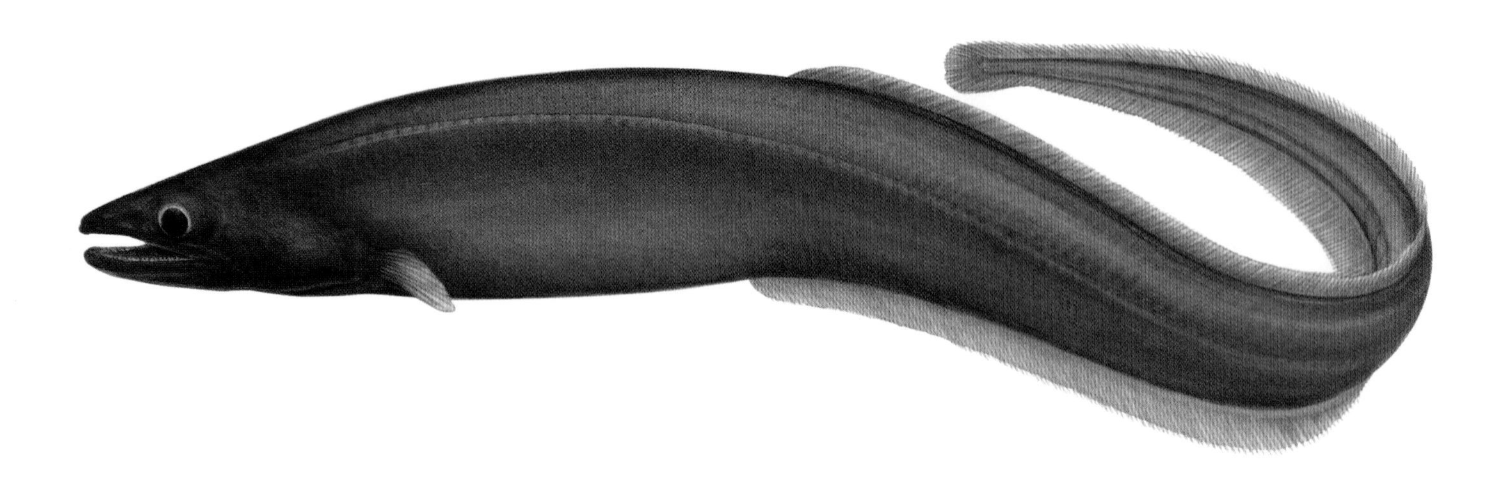

GREY CUTTHROAT EEL
Synaphobranchus affinis

Cutthroat Eels are found worldwide in deep water from 100 to 5000 m. The family contains about 32 species. They take their name from the appearance of their gill openings, which are on the ventral surface and positioned close together or joined, giving the impression of a cut throat. They have a large mouth with small teeth and the body is covered with very small, embedded scales arranged in a cross-hatch or irregular pattern. They have pectoral fins, and the dorsal and anal fins are continuous with the caudal fin. Their leptocephalus larvae have unique telescopic eyes with unknown function. The larvae live closer to the surface than the adults and consequently more larval forms have been identified than adults.

The **Black Cutthroat Eel** grows to about 1.4 m in length and has been found at depths to 5440 m. Like most other species in the family it feeds on cephalopods, crustaceans and fishes. The **Snubnose Eel** is an exception – it is a scavenger and feeds mainly on dead fishes.

AUSTRALIAN SYNAPHOBRANCHIDAE SPECIES

Diastobranchus capensis Basketwork Eel
Histiobranchus australis Southern Cutthroat Eel
Histiobranchus bathybius Black Cutthroat Eel
Histiobranchus bruuni Bruun's Cutthroat Eel
Ilyophis brunneus Muddy Arrowtooth Eel
Simenchelys parasitica Snubnose Eel
Synaphobranchus affinis Grey Cutthroat Eel
Synaphobranchus brevidorsalis Shortfin Cutthroat Eel
Synaphobranchus kaupii Kaup's Cutthroat Eel

SNAKE EELS AND WORM EELS

Ophichthidae

SERPENT EEL
Ophisurus serpens

This very large family comprises about 290 species and is found worldwide in warm temperate and tropical seas. Snake Eels and Worm Eels have a long, rounded body without scales, and downward-pointing tubular nostrils beneath the overhanging snout. There is a basket-like structure of branchiostegal rays under the throat, with a small gill opening at its rear edge. Snake Eels have no caudal fin and the tail is formed into a hard spike that is used for burrowing rearwards into sand. Many Snake Eels are beautifully coloured and patterned, but most species are sand dwellers and are rarely seen. They are most active at night, emerging to feed on small fishes and crustaceans. Occasionally they can be found moving across the bottom in the open and when disturbed they 'swim' into the sand, either head or tail first, with surprising speed. The widely distributed **Serpent Eel** is the largest member of this family and can exceed 2.5 m in length.

Worm Eels are smaller than Snake Eels, reaching about 30 cm in length, and generally plain in colour. Their dorsal and anal fins are continuous with the caudal fin, and the tail is not formed into a spike. Like Snake Eels, Worm Eels are sand or mud dwellers and are rarely seen.

Ophichthidae

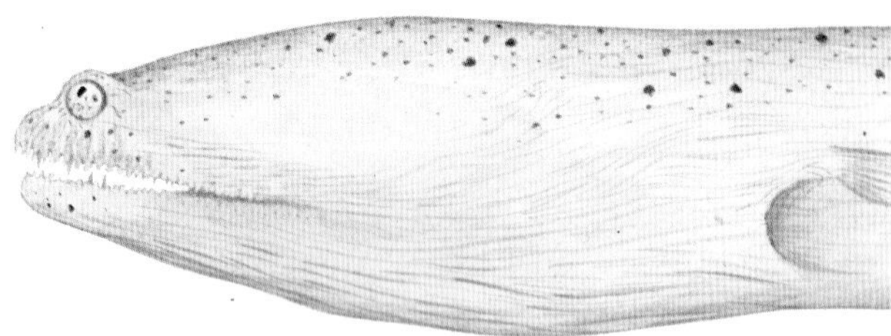

CROCODILE SNAKE EEL
Brachysomophis crocodilinus

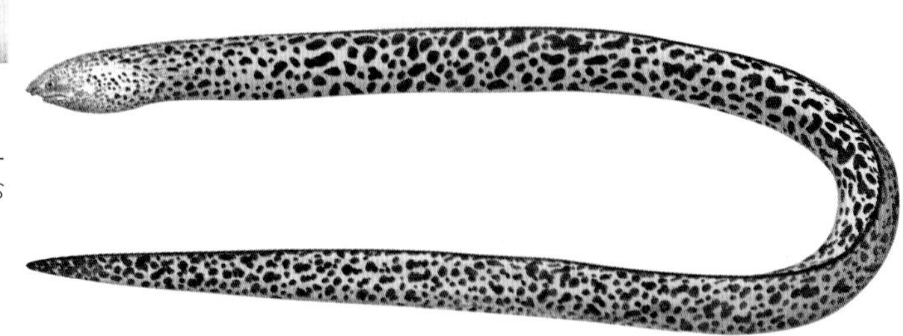

MARBLED SNAKE EEL
Callechelys marmorata

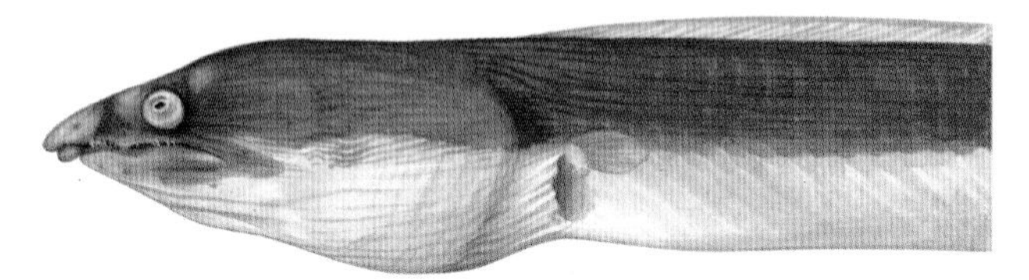

FRINGELIP SNAKE EEL
Cirrhimuraena calamus

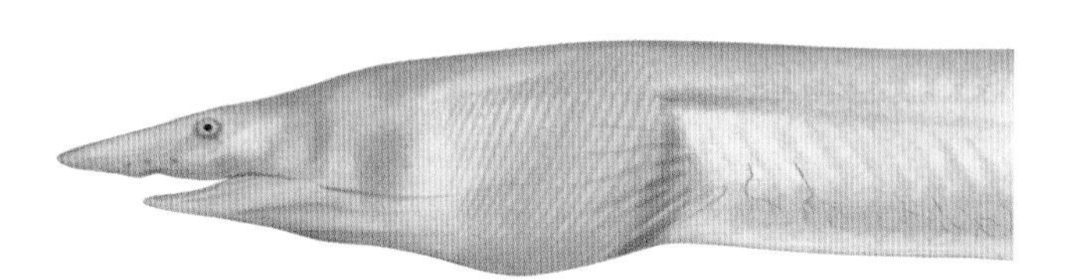

VULTURE EEL
Ichthyapus vulturis

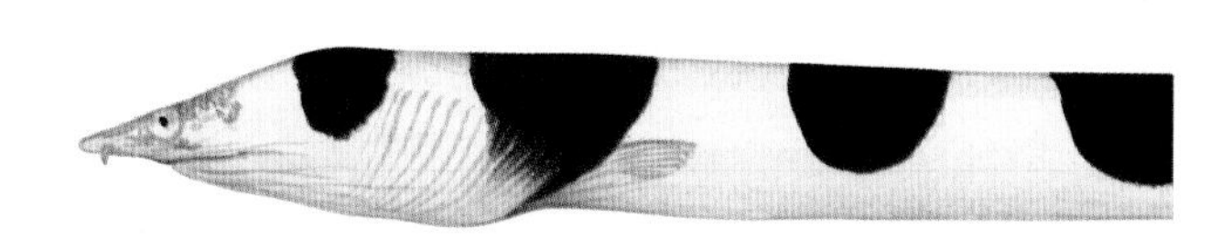

SADDLED SNAKE EEL
Leiuranus semicinctus

HARLEQUIN SNAKE EEL
Myrichthys colubrinus

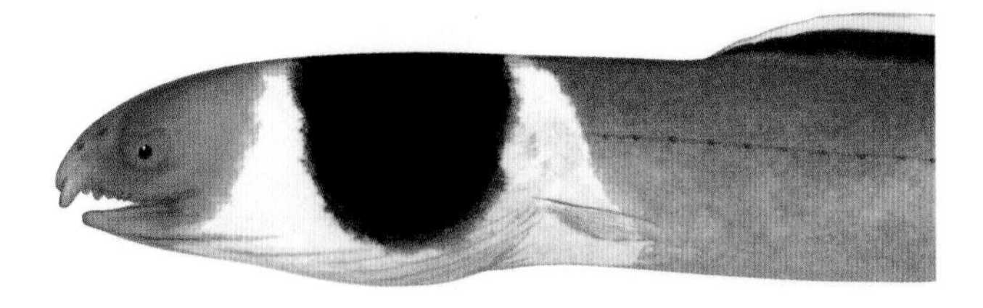

BLACKSADDLE SNAKE EEL
Ophichthus cephalozona

FLAPPY SNAKE EEL
Phyllophichthus xenodontus

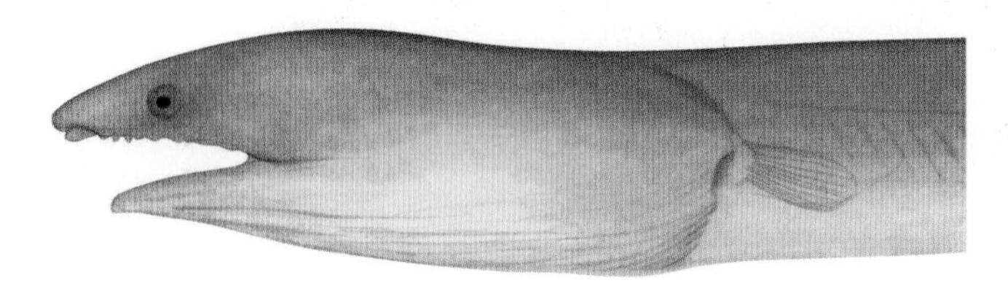

ESTUARY SNAKE EEL
Pisodonophis boro

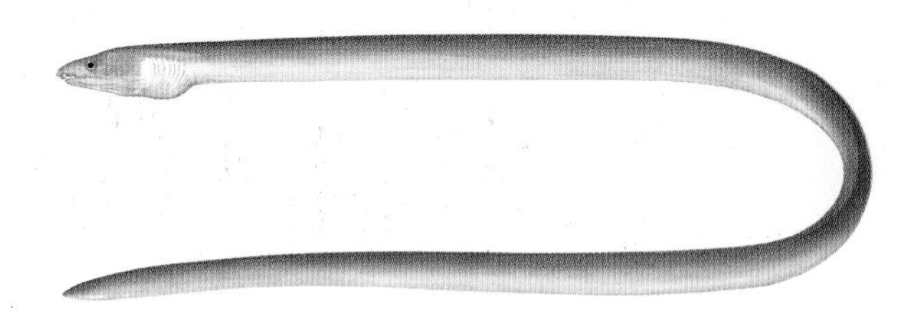

SLENDER WORM EEL
Scolecenchelys gymnota

AUSTRALIAN OPHICHTHIDAE SPECIES

Allips concolor Smalleye Snake Eel
Apterichtus flavicaudus Sharpnose Snake Eel
Apterichtus klazingai Sharpsnout Snake Eel
Bascanichthys sibogae Siboga Snake Eel
Brachysomophis cirrocheilos Stargazer Snake Eel
Brachysomophis crocodilinus Crocodile Snake Eel
Brachysomophis henshawi Henshaw's Snake Eel
Callechelys catostoma Blackstriped Snake Eel
Callechelys marmorata Marbled Snake Eel
Cirrhimuraena calamus Fringelip Snake Eel
Echelus uropterus Finned Snake Eel
Elapsopis cyclorhinus Banded Snake Eel
Ichthyapus vulturis Vulture Eel
Leiuranus semicinctus Saddled Snake Eel
Leiuranus versicolor Convict Snake Eel
Malvoliophis pinguis Halfband Snake Eel
Muraenichthys thompsoni Thompson's Snake Eel
Myrichthys colubrinus Harlequin Snake Eel
Myrichthys maculosus Ocellate Snake Eel
Myrophis microchir Ordinary Snake Eel
Neenchelys retropinna Gaper Snake Eel
Ophichthus altipennis Blackfin Snake Eel
Ophichthus bonaparti Purplebanded Snake Eel
Ophichthus cephalozona Blacksaddle Snake Eel
Ophichthus evermanni Evermann's Snake Eel
Ophichthus rutidodermatoides Olive Snake Eel
Ophichthus urolophus Manetail Snake Eel
Ophisurus serpens Serpent Eel
Phyllophichthus macrurus Leafynose Snake Eel
Phyllophichthus xenodontus Flappy Snake Eel
Pisodonophis boro Estuary Snake Eel
Pisodonophis cancrivorus Burrowing Snake Eel
Schismorhynchus labialis Groovejaw Worm Eel
Schultzidia johnstonensis Johnston's Snake Eel
Scolecenchelys australis Shortfin Worm Eel
Scolecenchelys breviceps Shorthead Worm Eel
Scolecenchelys godeffroyi Godeffroy Worm Eel
Scolecenchelys gymnota Slender Worm Eel
Scolecenchelys iredalei Iredale's Worm Eel
Scolecenchelys laticaudata Redfin Worm Eel
Scolecenchelys macroptera Narrow Worm Eel
Scolecenchelys nicholsae Nichols' Worm Eel
Scolecenchelys tasmaniensis Tasmanian Worm Eel
Skythrenchelys zabra Angry Worm Eel
Xyrias revulsus Strict Snake Eel
Yirrkala misolensis Misol Snake Eel
Yirrkala chaselingi Chingilt

SHORT-TAIL EELS

Colocongridae

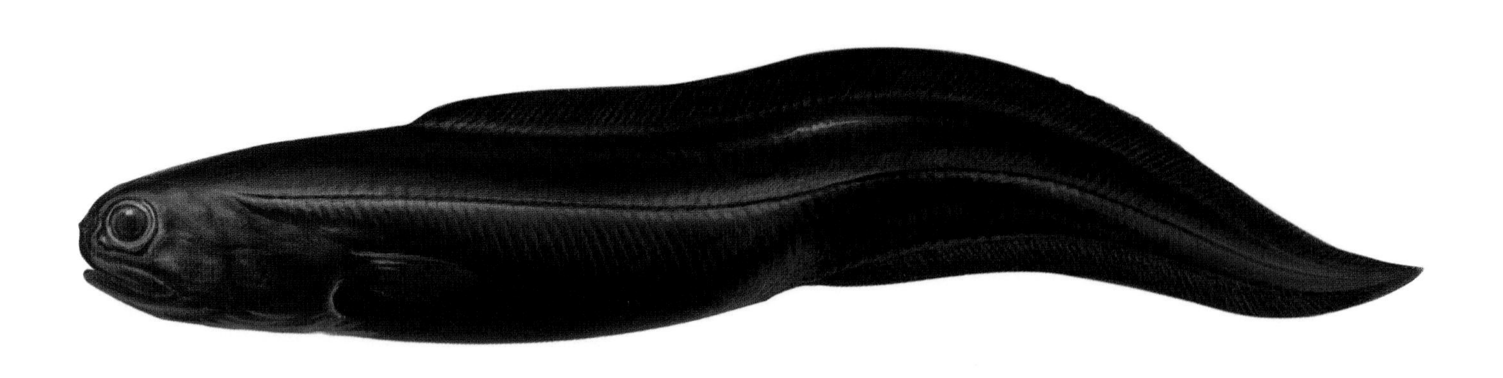

SHORT-TAIL CONGER
Coloconger scholesi

Short-tail Eels are found in the Atlantic, Indian and western Pacific oceans. There are about five species in the family. The **Short-tail Conger** is the only species found in Australian waters. It has a rounded head with a short, blunt snout and large eyes, a flabby body, well-developed pectoral fins, a short tail and the least elongate form of all the eels. The dorsal and anal fins are continuous with the caudal fin, and the jaws have minute conical teeth. It lives on muddy bottoms in deep water to depths of at least 800 m and reaches about 50 cm in length.

AUSTRALIAN COLOCONGRIDAE SPECIES

Coloconger scholesi Short-tail Conger

NECK EELS

Derichthyidae

INGOLF DUCKBILL EEL
Nessorhamphus ingolfianus

The family of Neck Eels is found worldwide and contains only three species, all of which occur in Australian waters. They are pelagic and found in deep water down to at least 1800 m. Neck Eels have a long, thin 'neck' between the gill openings and the pectoral fin. There are parallel striations on the head that are part of the sensory system. The dorsal and anal fins are continuous with the caudal fin, and there are no scales on the body. Growing to about 60 cm in length, they have many fine teeth in the jaws and feed mainly on planktonic crustaceans.

AUSTRALIAN DERICHTHYIDAE SPECIES

Derichthys serpentinus Deepwater Neck Eel
Nessorhamphus danae Dana Duckbill Eel
Nessorhamphus ingolfianus Ingolf Duckbill Eel

PIKE EELS

Muraenesocidae

COMMON PIKE EEL
Muraenesox bagio

Pike Eels are found in freshwater, estuarine and marine environments in tropical and warm temperate waters around the world. There are about eight species in the family. They have a cylindrical body with no scales and a very distinct lateral line, a compressed tail and well-developed pectoral fins. Pike Eels have a long snout, a very large mouth that extends behind the eyes, and the eyes are covered with transparent skin. They have many large canine teeth in the jaws and on the roof of the mouth.

The widely distributed **Common Pike Eel** is an important food fish throughout South-East Asia and is farmed in Japan. It is found right around Australia, except along the southern coast, ranging from shallow estuaries to depths of several hundred metres or more. It grows to a maximum length of about 2.2 m and large specimens can be aggressive and dangerous to handle. The **Darkfin Pike Eel** only reaches about 80 cm in length and is restricted to more northern waters. Pike Eels feed nocturnally on fishes, crustaceans and other invertebrates.

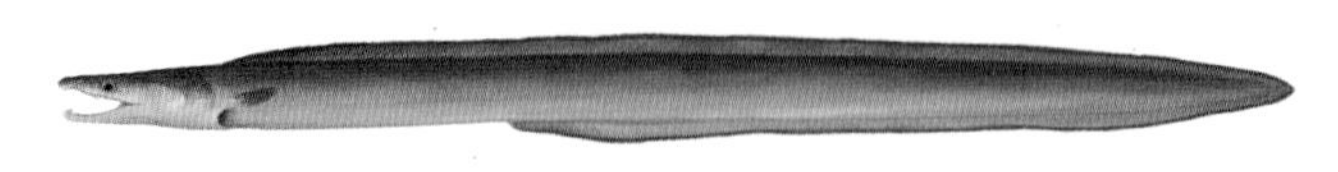

DARKFIN PIKE EEL
Muraenesox cinereus

AUSTRALIAN MURAENESOCIDAE SPECIES
Muraenesox bagio Common Pike Eel
Muraenesox cinereus Darkfin Pike Eel
Oxyconger leptognathus Bigeye Pike Eel

SNIPE EELS

Nemichthyidae

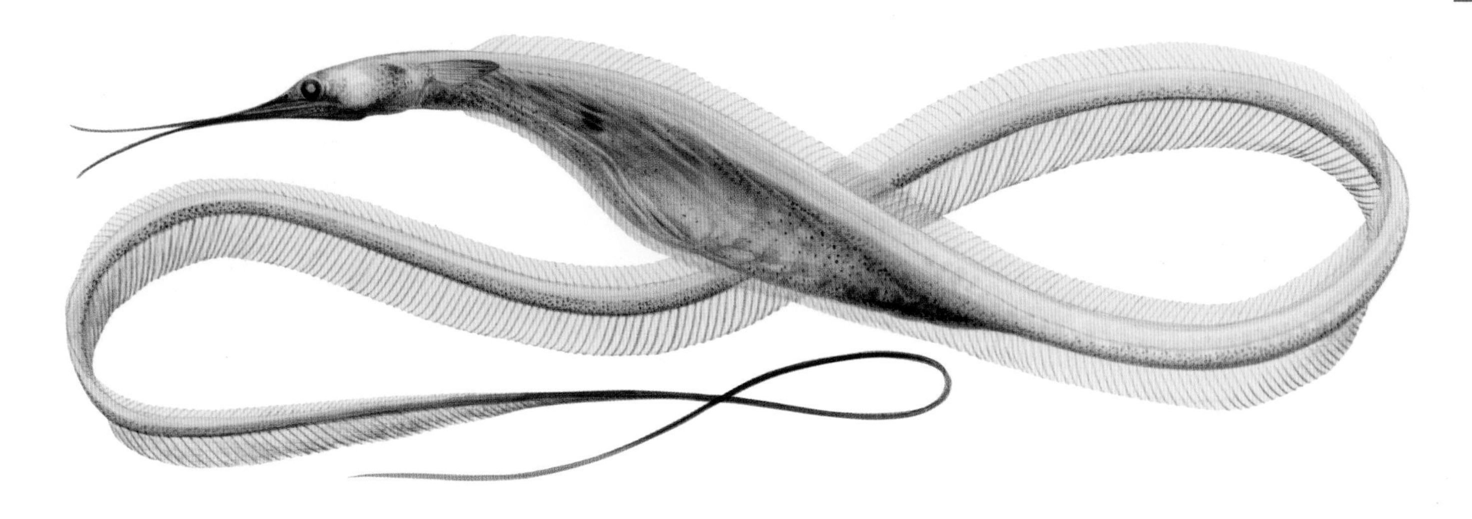

SLENDER SNIPE EEL
Nemichthys scolopaceus

Snipe Eels are found worldwide and there are about nine species in the family. They are pelagic and found in mid to deep water, to about 4500 m. They have a very long, thin body with no scales, and large eyes. Pectoral fins are present and the dorsal and anal fins are continuous with the caudal fin, which extends into a long, thin filament. Their extremely elongate, curved jaws are lined with minute inward-pointing teeth. Mature males lose these elongate jaws and develop large nostrils, presumably to detect female pheromones.

Snipe Eels feed on planktonic shrimps and other crustaceans, and grow to about 1.2 m in length. The Snipe Eel illustrated here was taken by a deepwater trawl off the northwest coast of Australia. A large prawn it had swallowed could clearly be seen through the thin tissue of its body. How these fragile animals are able to capture and ingest such large, spiny and relatively powerful prey is a mystery – although it is possible that this particular specimen swallowed a dead prawn while in the trawl net.

AUSTRALIAN NEMICHTHYIDAE SPECIES

Avocettina acuticeps Southern Snipe Eel
Avocettina infans Avocet Snipe Eel
Labichthys yanoi Yano's Snipe Eel
Nemichthys curvirostris Boxer Snipe Eel
Nemichthys scolopaceus Slender Snipe Eel

CONGER EELS AND GARDEN EELS

Congridae

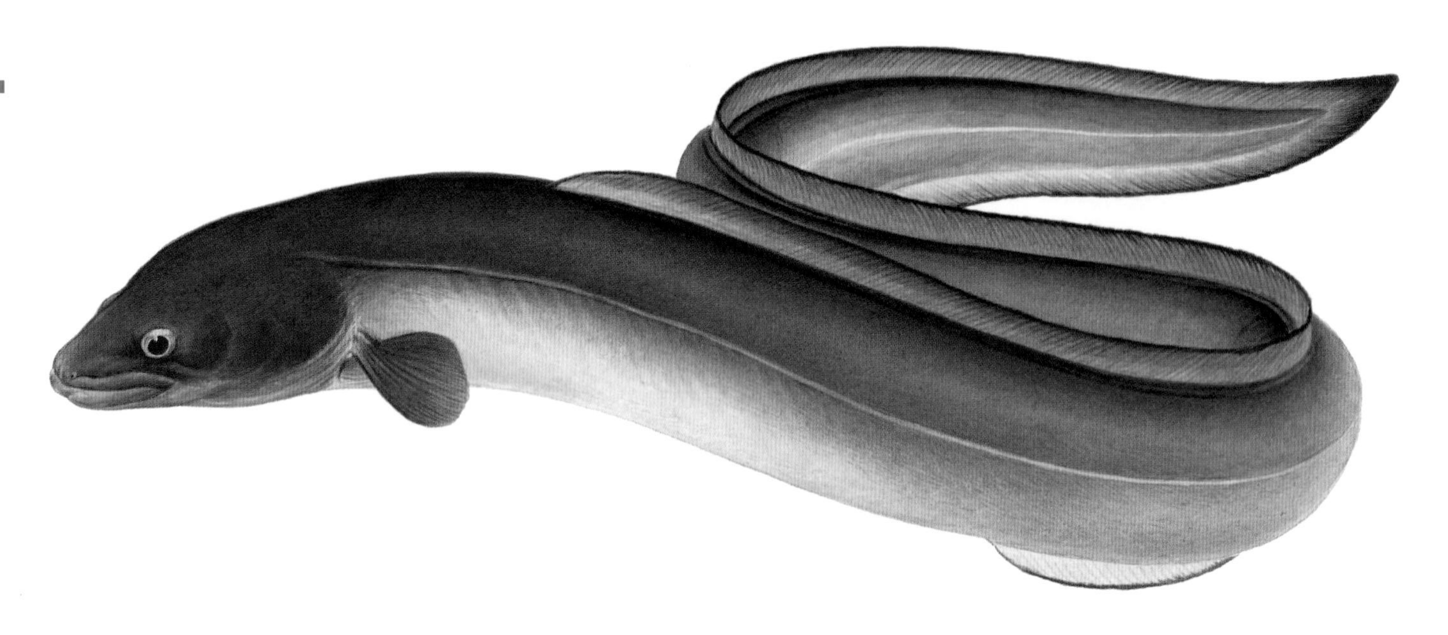

EASTERN CONGER
Conger wilsoni

This is a large and diverse family, with about 160 species found worldwide in temperate and tropical waters. The family is generally divided into three subfamilies:

CONGRINAE Conger Eels have a scaleless body and well-developed pectoral fins, and the dorsal and anal fins have segmented rays and are continuous with the caudal fin. The lips usually have prominent fleshy flanges and the teeth are quite small. They are found right around the Australian coast, on reefs or over muddy bottoms, and some species burrow into soft substrates. The **Swollen-headed Conger** occurs down to 1000 m or more, and is one of several Conger species found in deep water. The **Southern Conger** is typical of the family in that it feeds nocturnally on fishes and crustaceans, and spends the daytime in holes and crevices in reefs. It can grow to 2 m in length and weigh 20 kg, but most Congers measure around 1 m or less. Congers are important commercial fishes in some areas of the world, but are not often used for food in Australia.

BATHYMYRINAE This subfamily is similar to the Congrinae. It comprises Congers of the genus *Ariosoma*, which have unsegmented dorsal and anal fin rays, and well-developed pectoral fins.

HETEROCONGRINAE Garden Eels are found in tropical waters and form colonies in sandy patches on the sea floor with high current flow. They have an extremely elongate, snake-like body, long dorsal and anal fins with unsegmented rays, usually no pectoral fins, and an upturned mouth. Garden Eels grow to a maximum length of about 60 cm. They live in burrows in the sand and, with tails anchored in the substrate, extend their bodies into the current to feed on plankton. The large colonies really do resemble a garden of strange, sinuous reeds gently undulating in the current. These eels are difficult to approach closely, as they retreat into the sand at any hint of danger.

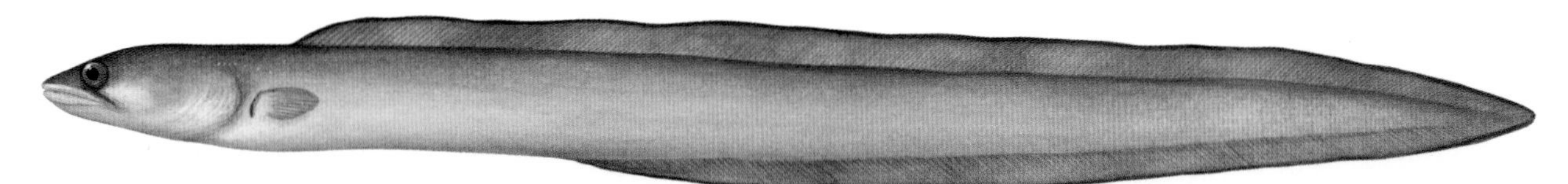

BLACKLIP CONGER
Conger cinereus

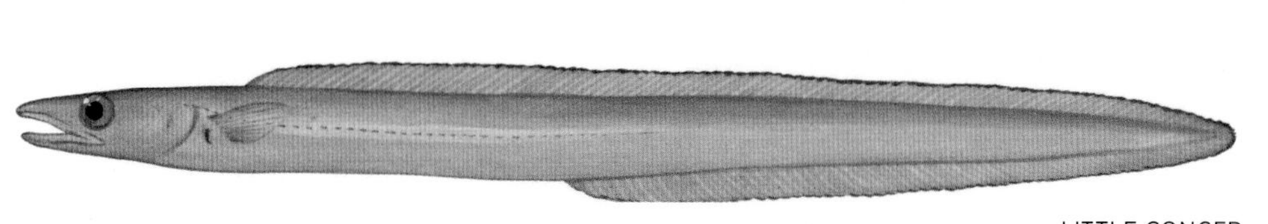

LITTLE CONGER
Gnathophis longicaudus

SPOTTED GARDEN EEL
Heteroconger hassi

AUSTRALIAN CONGRIDAE SPECIES

Ariosoma anago Darkfin Conger
Ariosoma anagoides Sea Conger
Ariosoma mauritianum Blunt-tooth Conger
Ariosoma scheelei Tropical Conger
Bassanago bulbiceps Swollen-headed Conger
Bassanago hirsutus Deepsea Conger
Bathycongrus guttulatus Lined Conger
Bathycongrus odontostomus Toothy Conger
Bathycongrus retrotinctus Blackedged Conger
Bathyuroconger vicinus Largetooth Conger
Blachea xenobranchialis Frillgill Conger
Castleichthys auritus Eared Conger
Conger cinereus Blacklip Conger
Conger verreauxi Southern Conger
Conger wilsoni Eastern Conger
Diploconger polystigmatus Headband Conger
Gavialiceps javanicus Duckbill Conger
Gnathophis castlei Castle's Conger
Gnathophis grahami Graham's Conger
Gnathophis longicaudus Little Conger
Gnathophis macroporis Largepore Conger
Gnathophis melanocoelus Blackgut Conger
Gnathophis microps Smalleye Conger
Gnathophis nasutus Bignose Conger
Gnathophis umbrellabius Umbrella Conger
Gorgasia galzini Speckled Garden Eel
Gorgasia preclara Splendid Garden Eel
Heteroconger hassi Spotted Garden Eel
Lumiconger arafura Luminous Conger
Macrocephenchelys brevirostris Rubbernose Conger
Poeciloconger kapala Mottled Conger
Rhynchoconger ectenurus Longnose Conger
Scalanago lateralis Ladder Eel
Uroconger lepturus Slender Conger

DUCKBILL EELS

Nettastomatidae

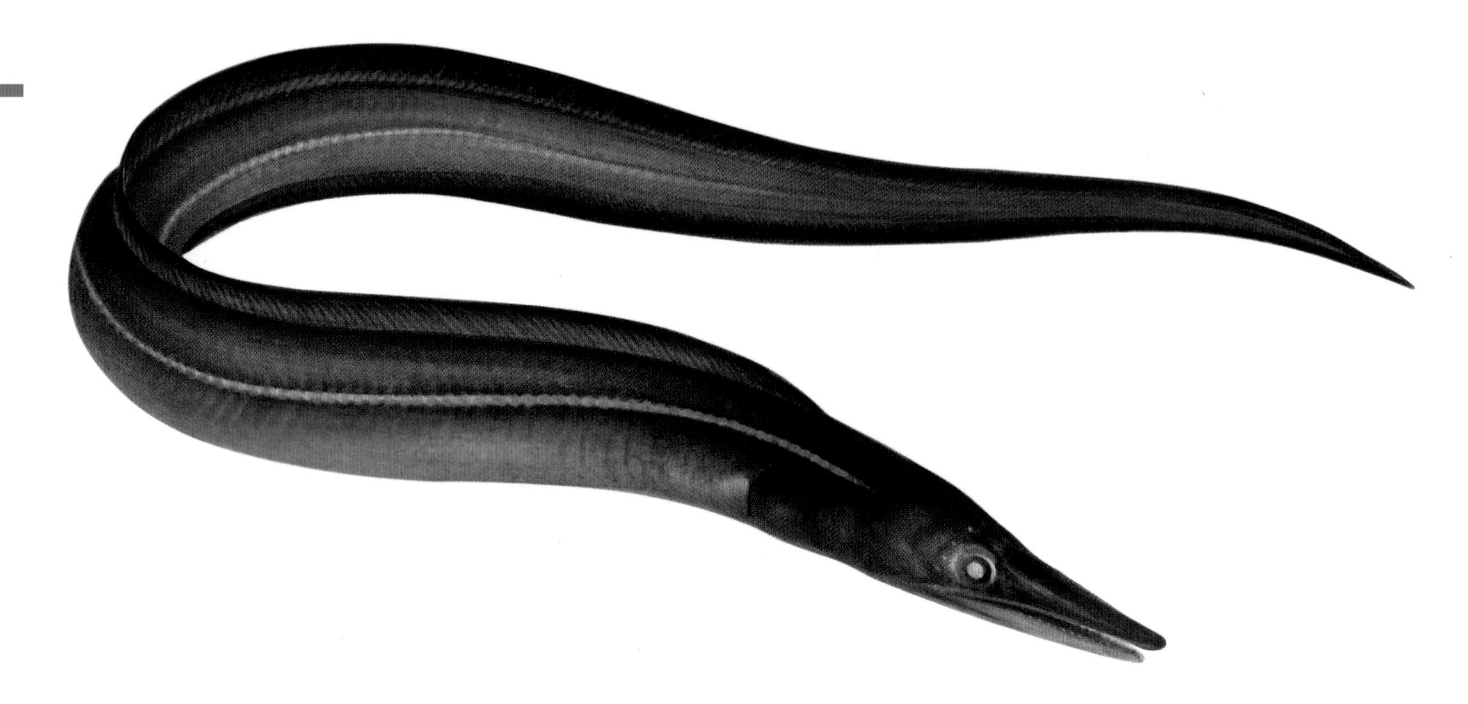

BLACKFIN SORCERER
Nettastoma melanurum

Duckbill Eels are found in tropical and warm temperate zones around the world, in deep water from 100 to 2000 m. There are about 38 species in the family. They have a long, flattened snout, large jaws with numerous small teeth, and a long, compressed body with no scales. The species found in Australia all lack pectoral fins. Some Australian species, such as the **Blackfin Sorcerer**, live in burrows in muddy bottoms on the continental slope, while others, such as the **Pillar Wire Eel**, are pelagic near the bottom in deep water. Duckbill Eels reach a maximum length of about 1 m, feed on small fishes and crustaceans, and are rarely encountered.

AUSTRALIAN NETTASTOMATIDAE SPECIES

Nettastoma melanurum Blackfin Sorcerer
Nettastoma parviceps Smallhead Duckbill Eel
Nettastoma solitarium Solitary Duckbill Eel
Nettenchelys gephyra Bridge Duckbill Eel
Saurenchelys finitimus Whitsunday Wire Eel
Saurenchelys stylura Pillar Wire Eel

SAWTOOTH EELS

Serrivomeridae

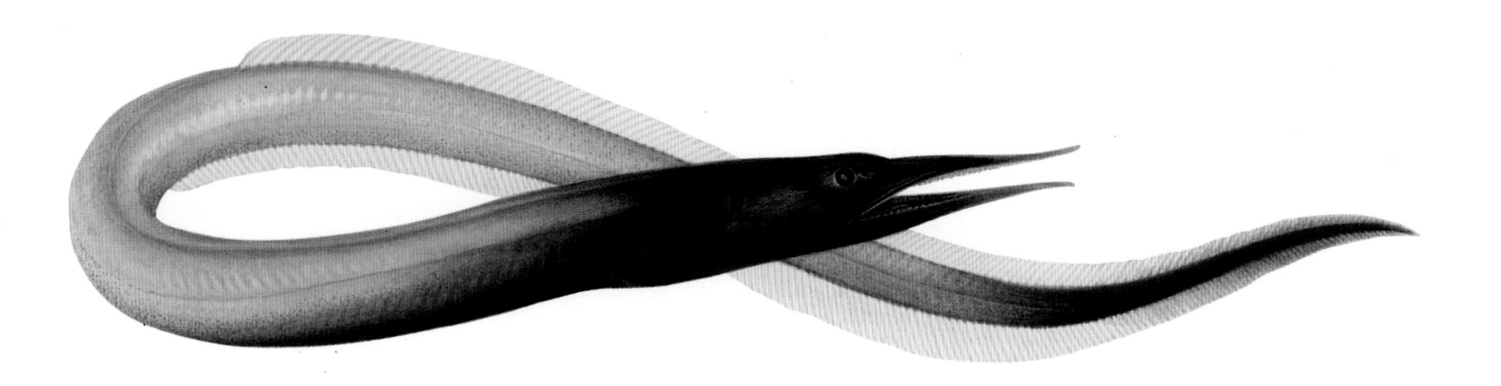

BEAN'S SAWTOOTH EEL
Serrivomer beanii

The family of Sawtooth Eels contains about 10 species and they are found worldwide in deep waters. They have a slender body with no scales, the dorsal and anal fins are continuous with the caudal fin, and they have a rudimentary pectoral fin. The elongate, beak-like jaws bear rows of minute teeth, and there is a row of large, blade-like teeth on the roof of the mouth. They are generally pelagic in midwater at depths of about 500–1000 m, though they have been recorded as deep as 6000 m. Sawtooth Eels make vertical migrations towards the surface at night to feed on small fishes and crustaceans. Growing to about 80 cm in length, they are occasionally taken by deepwater trawls.

AUSTRALIAN SERRIVOMERIDAE SPECIES

Serrivomer beanii Bean's Sawtooth Eel
Serrivomer bertini Thread Sawtooth Eel

SACCOPHARYNGIFORMES

Saccopharyngiforms are extremely specialised eel-like deepwater fishes, with a leptocephalus larval stage and extraordinary modifications to their anatomy. They have a long, thin body with no scales, and a greatly reduced skeleton that lacks many of the elements common to most bony fishes. Their jaws are long and fragile, and some species have an enormously expandable pharynx that enables them to swallow very large prey. The gill openings are reduced to small holes on the ventral surface of the body, the dorsal and anal fins are long and low, and the caudal fin is absent or rudimentary. There are four families in the order, three of which occur in Australian waters.

ARROW EELS

Cyematidae

ARROW EEL
Cyema atrum

The family of Arrow Eels contains only two species; one is found worldwide, including in Australian waters, while the other is known only from a few specimens found off South Africa. They are pelagic, inhabiting deep water from 330 to 5000 m. Their short, scaleless body is quite compressed, and there are dermal papillae on the head and body. They have pectoral fins, and the posterior dorsal and anal fin rays are elongated to form a blunt, arrow-shaped tail. The eyes are vestigial and the large mouth is similar to that of the Snipe Eels (Nemichthyidae, p. 169), with long, thin, curved jaws and numerous minute granular teeth. Arrow Eels grow to about 15 cm long. They are rarely encountered and very little is known of their biology.

AUSTRALIAN CYEMATIDAE SPECIES
Cyema atrum Arrow Eel

GULPER EELS

Saccopharyngidae

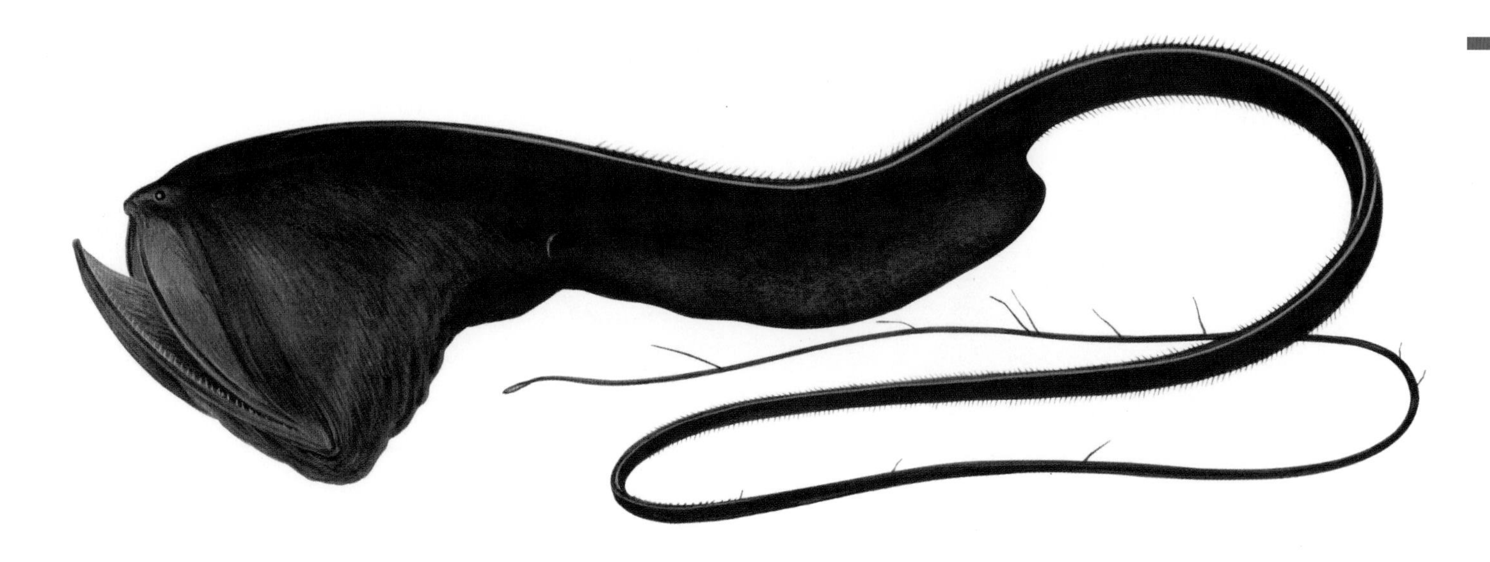

BLACK GULPER
Saccopharynx schmidti

Gulper Eels are found worldwide, in deep water from 1000 to 3000 m. There are about 10 species in the family. They have a snake-like body with a very long, slender tail, and a leptocephalus larval stage. These remarkable animals, like their close relation the Pelican Eel (Eurypharyngidae, p. 178), lack many of the features common to most other bony fishes: they have no rib bones, opercular bones, branchiostegal rays, ventral fins or swim bladders.

Gulper Eels feed mainly on other fishes and their extremely long jaws, sharp curved teeth, and highly distensible stomach enable them to swallow large prey. They are pelagic near the bottom, and are probably ambush predators. They have a light-emitting organ at the tip of the tail that may be used as a lure to entice prey within striking distance or perhaps to distract predators. They reach a maximum length of about 2 m. The **Black Gulper** has been found in New Zealand waters and almost certainly occurs in Australia, though it has not been officially recorded here.

AUSTRALIAN SACCOPHARYNGIDAE SPECIES
Saccopharynx schmidti Black Gulper

PELICAN EEL

Eurypharyngidae

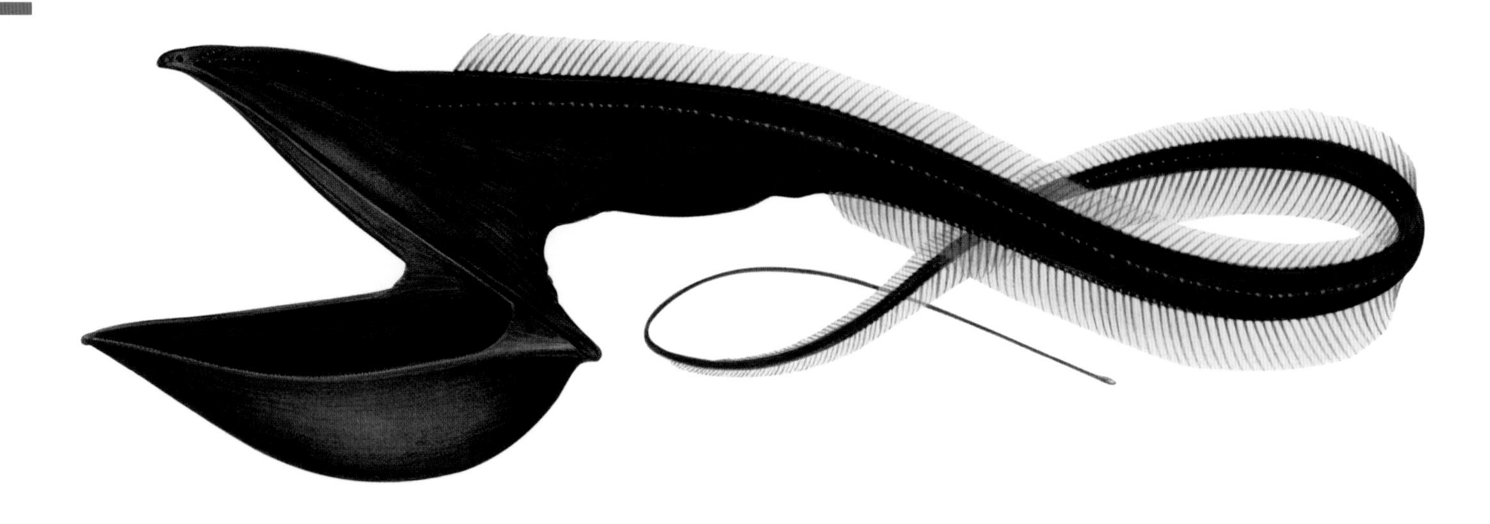

PELICAN EEL
Eurypharynx pelecanoides

The single species in this family is found worldwide, in deep water, from 500 m to as deep as 7000 m. It is one of the strangest of all fishes, with hugely elongated, fragile jaws containing numerous minute teeth. Like the closely related Gulper Eels (Saccopharyngidae, p. 177), it has many skeletal modifications: it has no rib bones, opercular bones, branchiostegal rays, ventral fins or swim bladder. The **Pelican Eel** has tiny eyes positioned very close to the tip of the snout. A luminous organ on the tip of the tail may be used to attract prey or perhaps to distract predators.

The Pelican Eel is believed to use its enormous mouth as a fishing net to capture small pelagic fishes, cephalopods, and crustaceans such as shrimps and copepods. The skin of the mouth is extremely flexible and individuals are thought to engulf a large amount of water containing prey, then slowly expel the water. The eel grows to about 75 cm in length. Although it is widespread and reasonably common, very little is known of its biology.

AUSTRALIAN EURYPHARYNGIDAE SPECIES

Eurypharynx pelecanoides Pelican Eel

CLUPEIFORMES

Members of this primitive group of fishes are characterised by having a connection between the swim bladder and the inner ear. The anterior of the swim bladder is forked and the two branches form sacs in bony swellings within the skull. This adaptation improves sensitivity to low-frequency sounds, such as the tail beats of other fishes, and helps with schooling behaviour and detecting the approach of predators. Many species form vast schools and are important commercially. Clupeiforms have a compressed body with well-developed cycloid scales and most lack a sensory lateral line. They have a single dorsal fin, pectoral fins set low on the body, ventral fins at about mid-body, and no spines in any of the fins. There are five families in the order, four of which occur in Australian waters.

ILISHAS

Pristigasteridae

ELONGATE ILISHA
Ilisha elongata

Ilishas are found worldwide in coastal tropical waters, and in South American freshwater systems. There are about 34 species in the family. Until recently, Ilishas were included in the Clupeidae (p. 184) and they share many characteristics with members of that family, including a row of scutes along the belly. However, Ilishas are generally larger and have a much longer anal fin, and the structure of the tail bones is such that there is no gap between the upper and lower central caudal fin rays, as there is in the Clupeidae. The **Ditchelee** grows to about 16 cm in length, and is widely distributed throughout South-East Asian waters and the Indian Ocean, in marine, brackish and freshwater environments. Ilishas feed mainly on small planktonic crustaceans, although some also take small fishes and cephalopods.

DITCHELEE
Pellona ditchela

AUSTRALIAN PRISTIGASTERIDAE SPECIES

Ilisha elongata Elongate Ilisha
Ilisha lunula Longtail Ilisha
Ilisha striatula Banded Ilisha
Pellona ditchela Ditchelee

ANCHOVIES

Engraulidae

AUSTRALIAN ANCHOVY
Engraulis australis

Anchovies are found worldwide, in shallow coastal waters in temperate and tropical regions, with a few species occurring in fresh water. The family contains about 139 species. They are easily recognised by the elongate mouth that extends well behind the eyes, and the overhanging, rounded snout. They have a single dorsal fin with no spines, and small ventral fins, all situated at about mid-body. Their scales are smooth and easily shed, and they have a silvery stripe running along the centre of the lateral surface. They have scutes along the belly, except in *Engraulis* species.

Anchovies form large schools and are extremely important commercially in many areas of the world. The Peruvian Anchovetta (*Engraulis ringens*) is the most heavily exploited fish species in history – in 1970, before the fishery collapsed due to overfishing, 13 million tonnes were landed in a single year. The **Australian Anchovy**, which occurs in the coastal waters around Australia's southern half, is commercially fished on a much smaller scale; it is used for bait, as well as for human consumption.

Anchovies are also an important part of the marine food chain. Each year large schools migrate northward along the coast of Western Australia, providing a food source for a multitude of larger fishes. Feeding frenzies occur as the schools are encircled and attacked by Tunas, Mackerels and sharks, until only scattered handfuls of Anchovies remain. The only freshwater Anchovy found in Australia is the carnivorous **Freshwater Thryssa**, which occurs in northern rivers draining into the Gulf of Carpentaria. This species is the largest in the family, growing to about 40 cm in length. Most Anchovies are plankton feeders, either selectively taking larger zooplankton or using long gillrakers for passive filter feeding.

Engraulidae

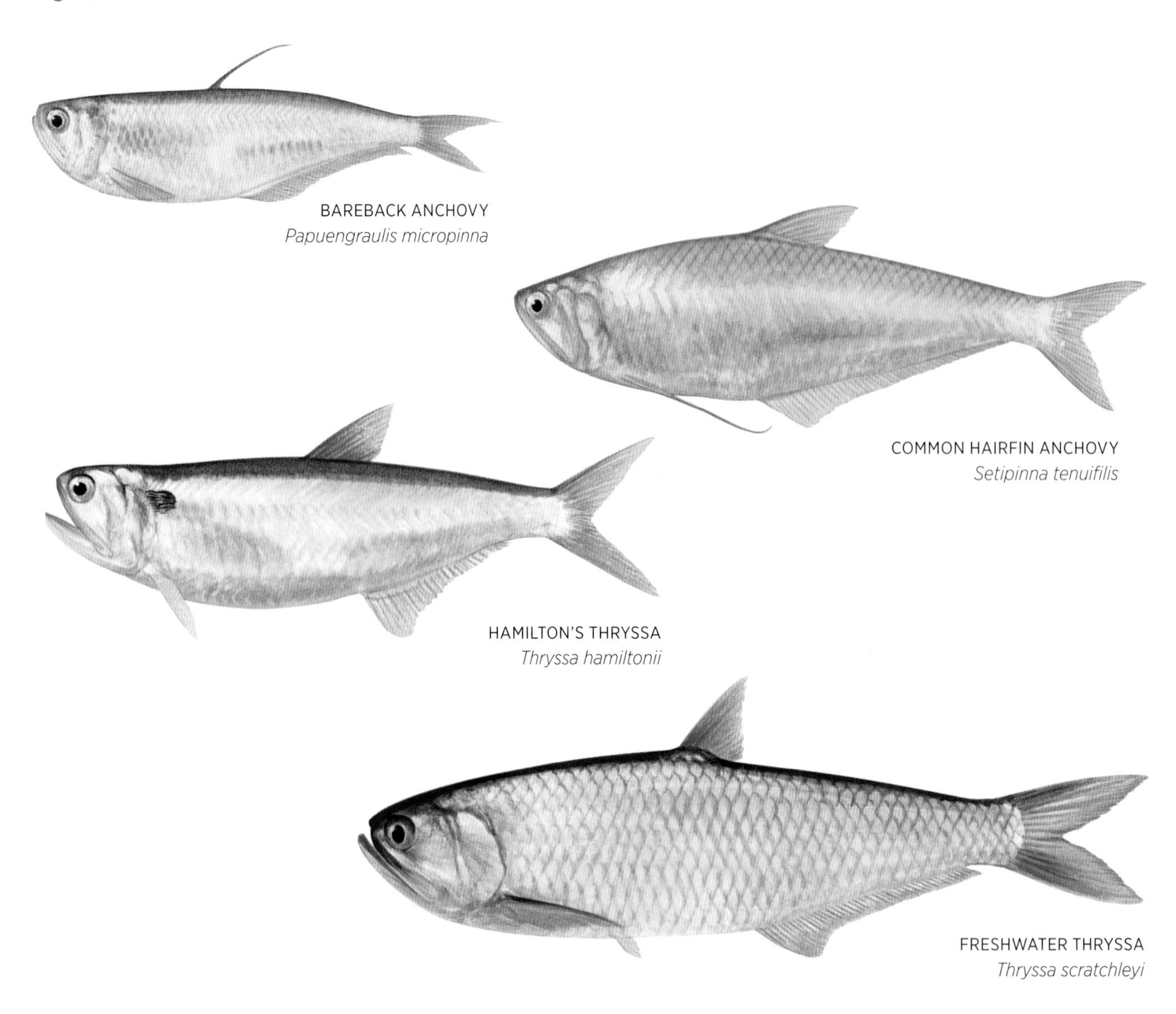

BAREBACK ANCHOVY
Papuengraulis micropinna

COMMON HAIRFIN ANCHOVY
Setipinna tenuifilis

HAMILTON'S THRYSSA
Thryssa hamiltonii

FRESHWATER THRYSSA
Thryssa scratchleyi

AUSTRALIAN ENGRAULIDAE SPECIES

Encrasicholina devisi Devis' Anchovy
Encrasicholina heteroloba Shorthead Anchovy
Encrasicholina punctifer Buccaneer Anchovy
Engraulis australis Australian Anchovy
Papuengraulis micropinna Bareback Anchovy
Setipinna paxtoni Humpback Hairfin Anchovy
Setipinna tenuifilis Common Hairfin Anchovy
Stolephorus advenus False Indian Anchovy
Stolephorus andhraensis Andhra Anchovy
Stolephorus brachycephalus Broadhead Anchovy
Stolephorus carpentariae Carpentaria Anchovy
Stolephorus commersonii Commerson's Anchovy
Stolephorus indicus Indian Anchovy
Stolephorus nelsoni Nelson's Anchovy
Stolephorus waitei Spottyface Anchovy
Thryssa aestuaria Snubnose Thryssa
Thryssa baelama Little Priest
Thryssa brevicauda Short-tail Thryssa
Thryssa encrasicholoides False Baelama Anchovy
Thryssa hamiltonii Hamilton's Thryssa
Thryssa marasriae Marasri's Thryssa
Thryssa scratchleyi Freshwater Thryssa
Thryssa setirostris Longjaw Thryssa

WOLF HERRINGS

Chirocentridae

DORAB WOLF HERRING
Chirocentrus dorab

The two species of Wolf Herring are found in warmer coastal waters of the Indian Ocean and western Pacific Ocean. They have a very elongate, compressed body, with small scales that are easily shed. They are easily recognised by their upturned mouth, which has large fang-like teeth projecting forwards in the upper jaw.

Both of these fast-swimming, voracious predators occur in small schools in northern Australian waters. They feed on small fishes and grow to about 1 m in length. The two species are almost identical and may in fact be the same species – the only apparent differences are that the **Dorab Wolf Herring** has shorter pectoral fins than the **Whitefin Wolf Herring**, and also has a black marking on the dorsal fin tip that is absent in the Whitefin.

AUSTRALIAN CHIROCENTRIDAE SPECIES
Chirocentrus dorab Dorab Wolf Herring
Chirocentrus nudus Whitefin Wolf Herring

HERRINGS, SARDINES AND SHADS

Clupeidae

AUSTRALIAN PILCHARD
Sardinops sagax

Herrings, Sardines and Shads are found worldwide in marine and freshwater environments. There are about 188 species in the family. They are among the most important of all commercially exploited fishes. Morphologically and ecologically diverse, they have small ventral fins and a single dorsal fin with no spines, all located at about mid-body. There are no scales on the head, weakly attached cycloid scales on the body, and usually a row of scutes along the belly.

The majority of species in this family are pelagic and grow to 10–20 cm in length. They feed mainly on planktonic crustaceans and sometimes occur in enormous schools. In tropical waters they have a tendency to form multi-species schools, while in temperate waters – where the bulk of commercial catches are made – they are more often found in single-species schools. As well as being commercially important, they are a vital food resource for larger predators.

The **Australian Pilchard**, which occurs in southern coastal waters, is the principal species caught in commercial quantities in Australia. It is used for human consumption as well as for bait, pet food and fish meal. Other species commercially exploited on a smaller scale include the **Sandy Sprat**, **Blue Sprat**, **Scaly Mackerel** and **Perth Herring**. Various other members of the family are caught incidentally. The **Bony Bream** is a freshwater species that is widely distributed in northern and eastern Australian waters, and it is also taken commercially.

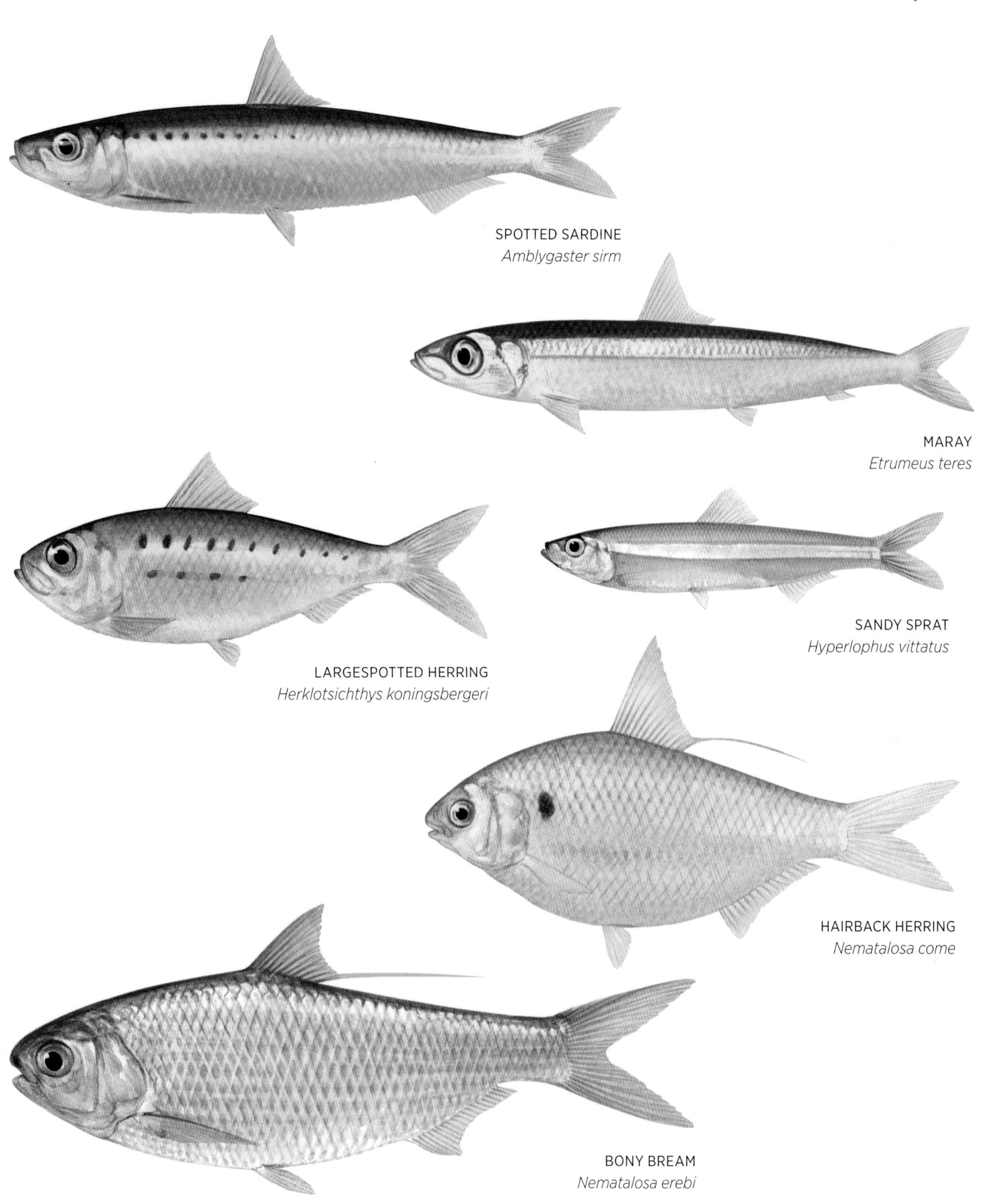

SPOTTED SARDINE
Amblygaster sirm

MARAY
Etrumeus teres

LARGESPOTTED HERRING
Herklotsichthys koningsbergeri

SANDY SPRAT
Hyperlophus vittatus

HAIRBACK HERRING
Nematalosa come

BONY BREAM
Nematalosa erebi

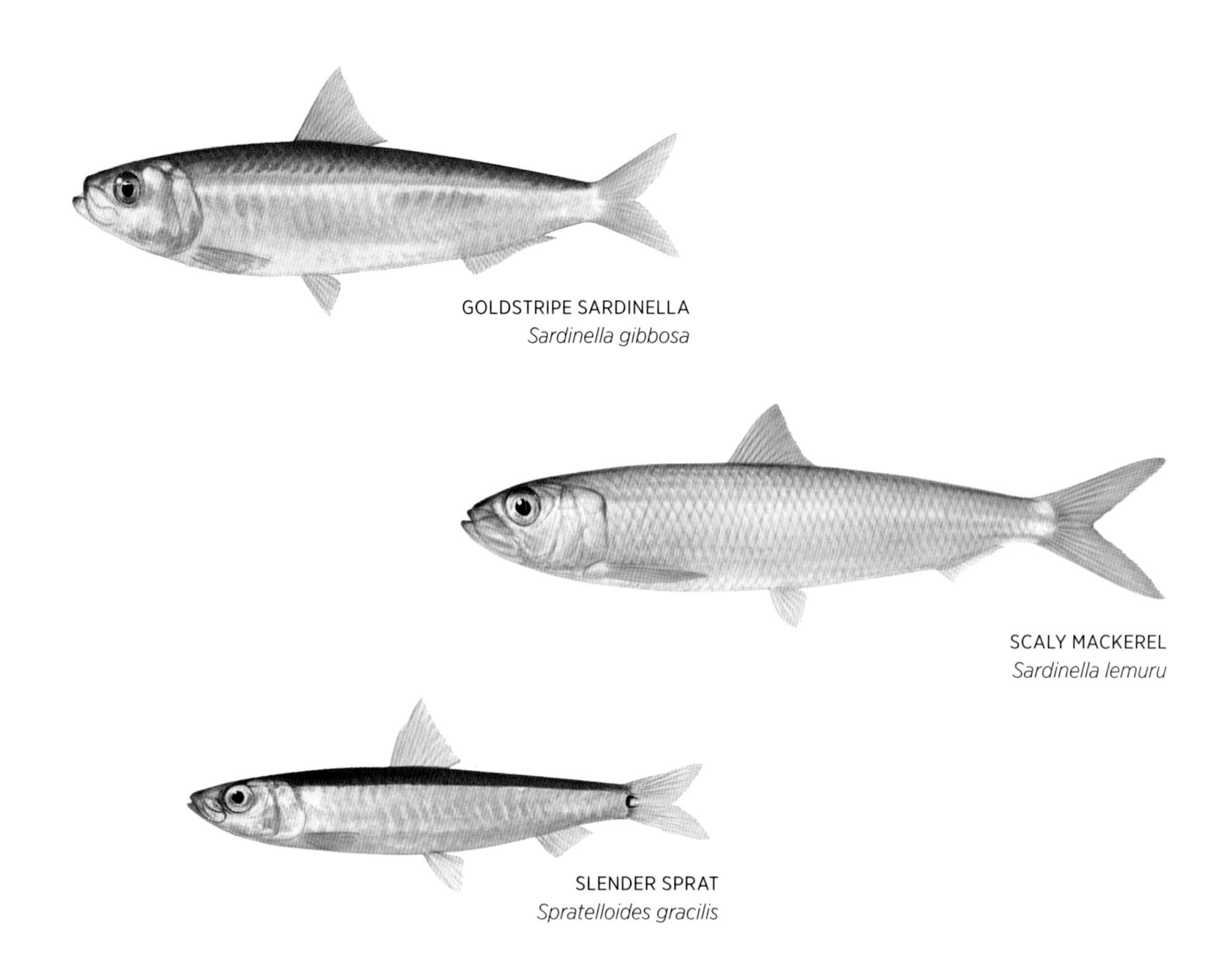

GOLDSTRIPE SARDINELLA
Sardinella gibbosa

SCALY MACKEREL
Sardinella lemuru

SLENDER SPRAT
Spratelloides gracilis

AUSTRALIAN CLUPEIDAE SPECIES

Amblygaster leiogaster Smoothbelly Sardine
Amblygaster sirm Spotted Sardine
Anodontostoma chacunda Gizzard Shad
Dussumieria elopsoides Slender Sardine
Escualosa thoracata White Sardine
Etrumeus teres Maray
Herklotsichthys blackburni Blackburn's Herring
Herklotsichthys castelnaui Southern Herring
Herklotsichthys collettei Collette's Herring
Herklotsichthys gotoi Darwin Herring
Herklotsichthys koningsbergeri Largespotted Herring
Herklotsichthys lippa Smallspotted Herring
Herklotsichthys quadrimaculatus Goldspot Herring
Hyperlophus translucidus Glassy Sprat
Hyperlophus vittatus Sandy Sprat
Nematalosa come Hairback Herring
Nematalosa erebi Bony Bream
Nematalosa vlaminghi Perth Herring
Potamalosa richmondia Freshwater Herring
Sardinella albella White Sardinella
Sardinella brachysoma Deepbody Sardinella
Sardinella gibbosa Goldstripe Sardinella
Sardinella lemuru Scaly Mackerel
Sardinella melanura Blacktip Sardinella
Sardinops sagax Australian Pilchard
Spratelloides delicatulus Blueback Sprat
Spratelloides gracilis Slender Sprat
Spratelloides robustus Blue Sprat
Sprattus novaehollandiae Australian Sprat

GONORYNCHIFORMES

Gonorynchiforms have a rounded body with small scales and a small, toothless mouth. They have a single dorsal fin, and ventral fins at about mid-body, and both the pectoral and ventral fins have a scaly axillary process at their base – there are no spines in any of the fins. The eyes are covered by an adipose eyelid. The order is closely related to the major freshwater fish groups containing Catfishes (Siluriformes, p. 195), Carps (Cyprinidae, p. 192) and Characids (not found in Australian waters). All have the ability to produce and detect alarm pheromones when one of their number is injured. Gonorynchiforms, however, lack the Weberian ossicles (see p. 191) found in the other groups. There are four families in this order, two of which occur in Australian waters.

MILKFISH

Chanidae

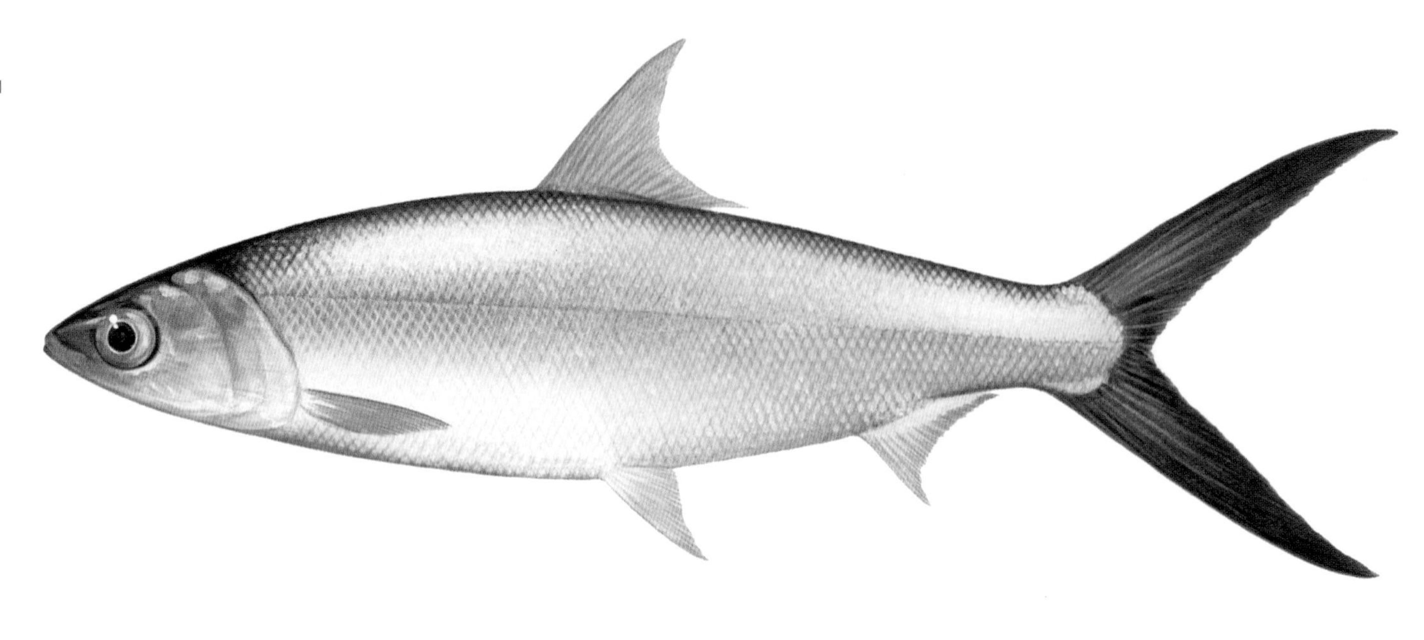

MILKFISH
Chanos chanos

The **Milkfish** is the only member of this family. It is found in shallow tropical waters in the Indian and Pacific oceans and occasionally enters estuaries and the lower reaches of freshwater rivers. The Milkfish has a small, terminal mouth and a single dorsal fin with no spines. It has small, cycloid scales on the body but not on the head, and a large, deeply forked tail.

The Milkfish has pelagic eggs, and the leptocephalus larvae that emerge from them migrate inshore to estuaries and mangroves, and occasionally into fresh water. Juveniles return to the sea to mature, where they feed on phytoplankton. Adults feed mostly on benthic algae and invertebrates, with larger specimens sometimes taking small fishes. Adults can reach a maximum length of about 1.8 m, though most encountered are no more than 1 m.

The species can tolerate high water temperatures (up to 32°C), and is extensively cultivated for food in shallow ponds in South-East Asia. Although quite common in the northern half of Australia, it is not generally eaten here. However, it is a popular target for sport-flyfishers, being difficult to catch and a tremendously powerful fighter once hooked.

AUSTRALIAN CHANIDAE SPECIES
Chanos chanos Milkfish

BEAKED SALMONS

Gonorynchidae

BEAKED SALMON
Gonorynchus greyi

Beaked Salmons are found in the Indian and Pacific oceans, from shallow coastal waters and estuaries to depths of 700 m or more. The five known species are small, bottom-dwelling fishes. They have an elongate, cylindrical body with small, smooth scales extending onto the head, and a single spineless dorsal fin set well back on the body. The pointed snout bears a small barbel at its tip and projects forward over the protrusible, tube-like mouth.

Beaked Salmons prefer sandy substrates in which they forage with their pointed snout, feeding on small benthic invertebrates. They may also dive head first into the sand when threatened. The Australian **Beaked Salmon** grows to about 60 cm in length and is found in southern Australian coastal waters to depths of at least 160 m. It often enters estuaries and is occasionally found well offshore under floating debris. The specimen illustrated here is a dark form from an estuary in the south-west of Australia, but many are a light sandy colour.

AUSTRALIAN GONORYNCHIDAE SPECIES

Gonorynchus greyi Beaked Salmon

CYPRINIFORMES

This is one of the largest orders of freshwater fishes, and members have a diverse range of body forms and lifestyles. All Cypriniforms have a single dorsal fin, no fin spines (though some have hardened fin rays), and a protrusible mouth with no teeth. They have variably shaped pharyngeal teeth, used for grinding or filtering food items. Like members of the Siluriformes (p. 195) and Characiformes (not found in Australian waters), Cypriniform fishes have modifications to the first 4–5 vertebrae, and attached bones and ligaments known as the Weberian ossicles that serve to conduct and amplify vibrations from the swim bladder to the inner ear. This gives them an acute sense of hearing and is a major factor behind their success in populating turbid, low-visibility freshwater environments. Cypriniforms are found mainly in the northern hemisphere, with the greatest diversity occuring in South-East Asia. There are six families (and over 3000 species) in the order – none occurs naturally in Australia, but species from two families have been introduced.

CARPS

Cyprinidae

EUROPEAN CARP
Cyprinus carpio

The Cyprinidae is the largest family of fishes in the world, containing more than 2420 species. The family is native to North America, Africa and Eurasia, where it is easily the dominant group in fresh water. The species found in Australia have all been introduced, either intentionally for freshwater anglers or accidentally through escapes from aquariums or ponds. Members of the family are characterised by having a protrusible mouth with no teeth, a single dorsal fin, no fin spines, and pharyngeal teeth.

The **European Carp** is now found throughout southern Australia and is well established in the Murray-Darling river system. Its ability to withstand high water temperatures and low oxygen levels means that it thrives in poor river conditions – the species actually helps to create such conditions by uprooting vegetation and stirring muddy bottoms in search of food. It is now considered a noxious pest in Australia. The **Rosy Barb** is a popular aquarium fish that has been released into Queensland waters and the **Goldfish** is now also well distributed in the south-west and south-east of the country. **Roach** and **Tench** are found in several locations in south-east Australia and Tasmania. All feed on aquatic invertebrates and some include vegetation in their diet.

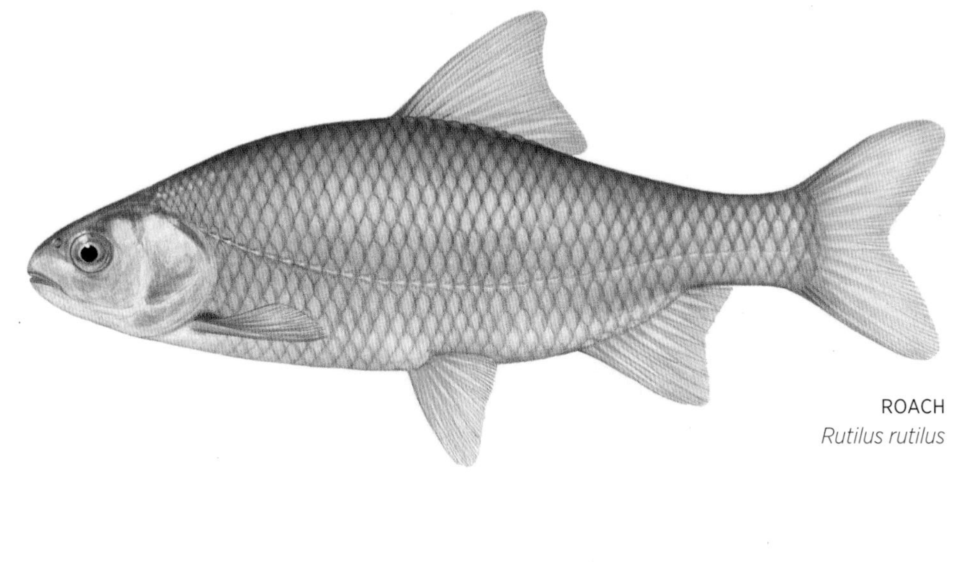

ROACH
Rutilus rutilus

TENCH
Tinca tinca

AUSTRALIAN CYPRINIDAE SPECIES

Carassius auratus Goldfish
Cyprinus carpio European Carp
Puntius conchonius Rosy Barb
Rutilus rutilus Roach
Tanichthys albonubes Mountain Minnow
Tinca tinca Tench

LOACHES

Cobitidae

ORIENTAL WEATHERLOACH
Misgurnus anguillicaudatus

Loaches are native to freshwater systems in Eurasia and North Africa. There are about 180 species in the family, most occurring in South-East Asia. They have a rounded, elongate body, a small, movable defensive spine below the eye, and 3–6 pairs of barbels around the mouth. Loaches are bottom dwellers in rivers, lakes and streams, and use their barbels to help locate the insect larvae and other invertebrates that form their diet.

The **Oriental Weatherloach**, which grows to about 20 cm in length, is a popular aquarium fish and has been introduced into many freshwater systems around the world. It has established populations in a number of watercourses in south-eastern Australia. It can tolerate a wide range of temperatures and can gulp air at the surface to supplement respiration in oxygen-poor environments. It prefers slow-flowing waters with muddy or detritus-covered bottoms, and will often bury itself in the substrate, leaving only the head protruding.

AUSTRALIAN COBITIDAE SPECIES

Misgurnus anguillicaudatus Oriental Weatherloach

SILURIFORMES

Siluriforms are characterised by a scaleless body and usually a number of sensory barbels around the mouth. They have a single strong spine at the origin of the dorsal fin and each pectoral fin. There are patches of teeth on the roof of the mouth and the different shapes of these patches can be used to differentiate the various species. They have modifications to the first 4–5 vertebrae, and attached bones and ligaments known as the Weberian ossicles, which conduct and amplify vibrations from the swim bladder to the inner ear. This gives them an acute sense of hearing – an important advantage in the turbid environments they frequently inhabit. There are about 35 families and more than 2800 species in the order, but only two families are found in Australian waters.

EELTAIL CATFISHES

Plotosidae

HYRTL'S CATFISH
Neosilurus hyrtlii

Eeltail Catfishes are found in the Indo-West Pacific region, mostly in fresh water, though some occur in marine environments. There are about 35 species in the family. They have venomous spines at the origin of the dorsal and pectoral fins, which can inflict an extremely painful sting. There are four pairs of barbels around the mouth, and the dorsal and anal fins are continuous with the caudal fin. Eeltail Catfishes are opportunistic predators, feeding on a wide range of aquatic invertebrates and small fishes. Most are quite small, from 10 to 50 cm in length, though the **White-lipped Catfish** can reach over 1.3 m.

The family includes well-known species such as the **Freshwater Catfish**, which is widely distributed in north-eastern Australia. This species builds nests up to 2 m wide from sand and gravel. Males guard the eggs until hatching and protect the young for several weeks afterwards. The **Estuary Cobbler**, a popular food fish, occurs in coastal waters and estuaries across southern Australia, but has unfortunately disappeared completely from many areas where it was formerly abundant. The **Striped Catfish** is commonly encountered by divers on northern coral reefs, where it forms small, dense schools that are often seen sheltering in caves. Schools of juveniles may be seen moving across the bottom in rolling 'waves': as some individuals slow down to feed briefly on the bottom, the remainder swim over them and take their place at the front of the school, then these take their turn to slow down and feed, and so on. A densely packed school moving in this way at first glance resembles a large Stingray. The **Sailfin Catfish**, on the other hand, is solitary, and a common resident of coral reefs in north-western Australia. *Neosilurus* species are widespread in rivers and creeks in central and northern Australia – **Hyrtl's Catfish** has a particularly wide distribution.

ESTUARY COBBLER
Cnidoglanis macrocephalus

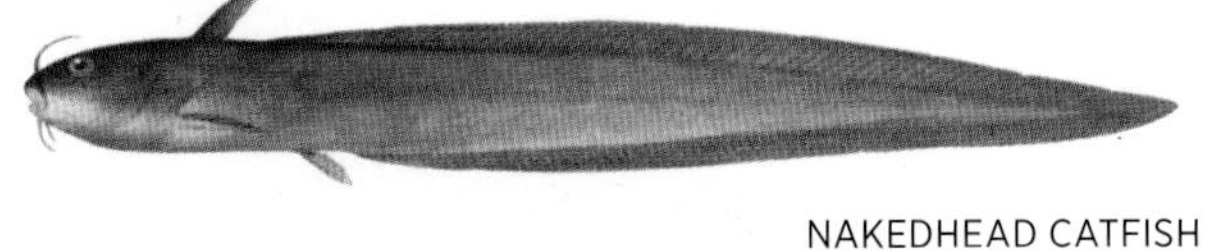

NAKEDHEAD CATFISH
Euristhmus nudiceps

SAILFIN CATFISH
Paraplotosus butleri

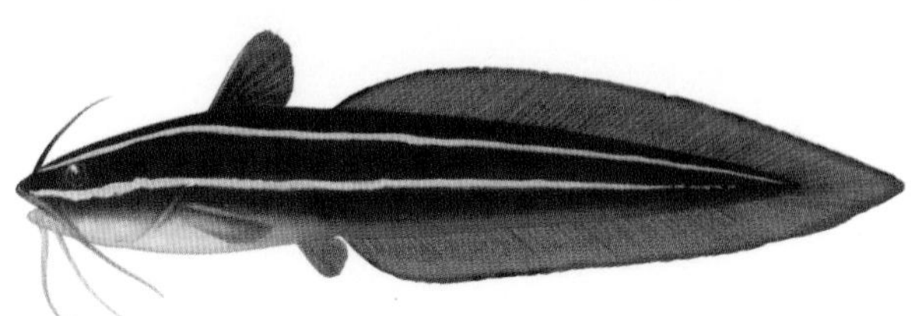

STRIPED CATFISH
Plotosus lineatus

FRESHWATER COBBLER
Tandanus bostocki

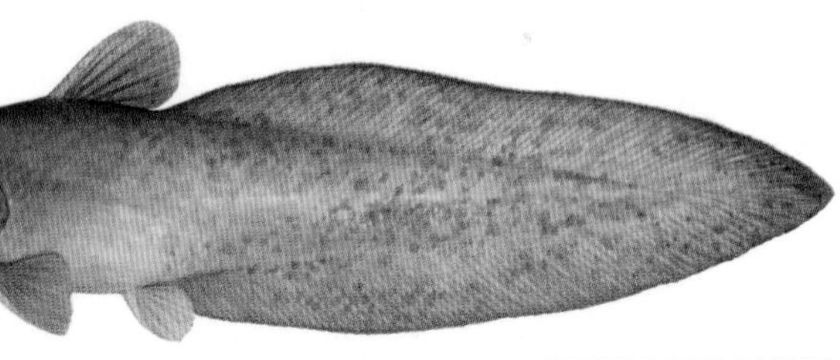

FRESHWATER CATFISH
Tandanus tandanus

AUSTRALIAN PLOTOSIDAE SPECIES

Anodontiglanis dahli Toothless Catfish
Cnidoglanis macrocephalus Estuary Cobbler
Euristhmus lepturus Longtail Catfish
Euristhmus microceps Smallhead Catfish
Euristhmus nudiceps Nakedhead Catfish
Neosiluroides cooperensis Cooper Creek Catfish
Neosilurus argenteus Silver Catfish
Neosilurus ater Black Catfish
Neosilurus brevidorsalis Shortfin Catfish
Neosilurus gloveri Desert Catfish
Neosilurus hyrtlii Hyrtl's Catfish
Neosilurus mollespiculum Softspine Catfish
Neosilurus pseudospinosus Falsespine Catfish
Paraplotosus albilabris White-lipped Catfish
Paraplotosus butleri Sailfin Catfish
Paraplotosus muelleri Kimberley Catfish
Plotosus canius Eel Catfish
Plotosus lineatus Striped Catfish
Porochilus obbesi Obbes' Catfish
Porochilus rendahli Rendahl's Catfish
Tandanus bostocki Freshwater Cobbler
Tandanus tandanus Freshwater Catfish

FORKTAIL CATFISHES

Ariidae

GIANT SEA CATFISH
Netuma thalassina

The family of Forktail Catfishes contains about 150 species distributed worldwide in tropical and warm temperate seas. Many enter brackish or fresh waters, with some living exclusively in fresh water. They have a small adipose fin and usually three pairs of barbels around the mouth. The head is covered by a bony shield, and the first dorsal fin and the pectoral fins have strong spines that can inflict an extremely painful wound. Forktail Catfishes are mouthbrooders, the male carrying the large eggs in its mouth for 4–5 weeks until hatching, then allowing the newly hatched juveniles to take refuge there.

Several species are fished commercially in Australia, including the **Silver Cobbler**. Found in Lake Argyle in Western Australia, this species can grow to 1.4 m in length and weigh 28 kg. The **Giant Sea Catfish** occurs in northern coastal waters down to depths of about 150 m and can reach 1.8 m in length, while the **Blue Catfish** is widely distributed in rivers across northern Australia and grows to about 60 cm. Most species are omnivorous, feeding on a wide range of invertebrates, small fishes and plant material; the **Smallmouth Catfish** is an exception, feeding exclusively on small molluscs. The various species are difficult to differentiate, and the size and shape of tooth patches on the roof of the mouth are the main identifying characteristics.

BLUE CATFISH
Neoarius graeffei

SILVER COBBLER
Neoarius midgleyi

AUSTRALIAN ARIIDAE SPECIES

Arius argyropleuron Longsnout Catfish
Arius armiger Copper Catfish
Arius berneyi Highfin Catfish
Arius bilineatus Two-line Sea Catfish
Arius dioctes Warrior Catfish
Arius hainesi Ridged Catfish
Arius insidiator Flat Catfish
Arius leptaspis Boofhead Catfish
Arius mastersi Masters' Catfish
Arius nella Shieldhead Catfish
Arius paucus Shovelnose Catfish
Arius pectoralis Sawspine Catfish
Arius proximus Arafura Catfish
Cinetodus froggatti Smallmouth Catfish
Neoarius graeffei Blue Catfish
Neoarius midgleyi Silver Cobbler
Netuma thalassina Giant Sea Catfish

ARGENTINIFORMES

The order Agentiniformes is closely related to the Osmeriformes (p. 209) and Salmoniformes (p. 213), and all three groups were formerly placed in the same order. Agentiniforms are found in deep waters. They have a single dorsal fin (some also have a small adipose fin), ventral fins at about mid-body, a forked caudal fin and no fin spines. There are usually no teeth on the upper jaw. They possess a crumenal organ in the throat – a pouch-like structure of cartilage and interlaced gillrakers, which aids in the filtering and breaking down of food. There are five families in the order, all of which occur in Australian waters.

HERRING SMELTS

Argentinidae

SILVERSIDE
Argentina australiae

Herring Smelts are small, silvery schooling fishes. There are about 23 species in the family and they are found worldwide in temperate and tropical waters. They have smooth scales that shed easily and a silver stripe along the side of the body. They are similar in form to the Deepsea Smelts (Microstomatidae, p. 204), but have a longer snout and much smaller mouth, with jaws that don't reach back to the front of the eye. They have large eyes, the first dorsal fin at about mid-body (it has no spines), and a small adipose fin above the anal fin. Herring Smelts feed principally on plankton.

The **Silverside** is endemic to southern Australia, grows to a maximum length of about 19 cm and occurs in large schools from depths of 50 to 400 m. The **Saddled Herring Smelt** is known from only one specimen, which was taken in tropical waters off Western Australia. Herring Smelts are an important food source for predatory fishes. They are fished elsewhere in the world, but in Australia are rarely taken in trawl nets due to their small size.

AUSTRALIAN ARGENTINIDAE SPECIES
Argentina australiae Silverside
Glossanodon australis Southern Herring Smelt
Glossanodon pseudolineatus Saddled Herring Smelt

BARRELEYES

Opisthoproctidae

GRIMALDI'S BARRELEYE
Opisthoproctus grimaldii

These strange pelagic, deepwater fishes are found in the Atlantic, Pacific and Indian oceans, at depths from 100 to 2700 m. There are about 11 species in the family. They have a small mouth with very fine teeth and some have a small adipose fin. Their elongated, barrel-shaped eyes usually point upward, but in some species can be rotated forward.

The **Binocular Fish** has forward-pointing, rather than upward-pointing, tubular eyes. **Grimaldi's Barreleye** holds luminescent bacteria in a special rectal organ; these can be spread over a flattened reflecting surface or 'sole' under the abdomen to direct a diffuse light downwards. The more elongate **Glasshead Barreleye** lacks the reflective sole and instead has scattered photophores on the black ventral surface.

Barreleyes feed principally on gelatinous planktonic organisms such as siphonophores. The upward-directed eyes and downward-directed luminosity may be adaptations to feeding at extreme depths, where light can only just penetrate – its sensitive eyes distinguish the faint silhouettes of prey above, while the luminosity camouflages its own silhouette from predators below.

AUSTRALIAN OPISTHOPROCTIDAE SPECIES

Opisthoproctus grimaldii Grimaldi's Barreleye
Rhynchohyalus natalensis Glasshead Barreleye
Winteria telescopa Binocular Fish

DEEPSEA SMELTS

Microstomatidae

SLENDER SMALLMOUTH
Microstoma microstoma

Deepsea Smelts are found worldwide and there are about 38 species in the family. These small, elongate, fragile fishes are pelagic in deep water from 100 to 1700 m. Similar in form to the Herring Smelts (Argentinidae, p. 202), they have a shorter snout and larger mouth, with the jaws reaching back to the front edge of the eye. The lateral line extends onto the tail and the first dorsal fin is well behind the midpoint of the body. Some also have a small dorsal adipose fin above the anal fin. Deepsea Smelts make nightly vertical migrations to within 100 m or so of the surface. They feed mostly on gelatinous plankton, and have very fine teeth and a long intestine to aid in the digestion of such a diet. They are an important food source for deepwater predatory fishes.

BIGSCALE DEEPSEA SMELT
Melanolagus bericoides

AUSTRALIAN MICROSTOMATIDAE SPECIES

Bathylagichthys australis Southern Deepsea Smelt
Bathylagichthys greyae Grey's Deepsea Smelt
Bathylagoides argyrogaster Silver Deepsea Smelt
Bathylagus antarcticus Antarctic Deepsea Smelt
Dolicholagus longirostris Longsnout Deepsea Smelt
Lipolagus ochotensis Fatty Deepsea Smelt
Melanolagus bericoides Bigscale Deepsea Smelt
Microstoma microstoma Slender Smallmouth
Nansenia ardesiaca Robust Smallmouth

TUBESHOULDERS

Platytroctidae

SMALLSCALE TUBESHOULDER
Maulisia microlepis

There are about 37 species of Tubeshoulder and they are found worldwide from depths of about 200 to 2000 m. They take their name from a tube-like opening on the side of the body, located behind the pectoral fin and just below the lateral line, and connected to a gland that produces blue–green luminescent fluid. This fluid is perhaps used as a defensive mechanism to disorientate predators. It is this feature that differentiates them from the closely related Slickheads (Alepocephalidae, p. 206). Tubeshoulders have sensory canals beneath the skin on the head and body, small teeth and quite large gill openings.

The **Spangled Tubeshoulder**, like many in the family, has multiple light-producing organs on the ventral surface. It is widely distributed around the southern hemisphere, grows to about 15 cm in length, and is one of the more common members of this family to be taken in deep trawl nets – most other species are rarely encountered. Tubeshoulders grow to a maximum length of about 30 cm. They are pelagic near the bottom or in midwater and are known to make nightly migrations towards the surface. Being rare, their biology is not well known, but their dietary habits are probably similar to those of the Slickheads.

SPANGLED TUBESHOULDER
Persparsia kopua

AUSTRALIAN PLATYTROCTIDAE SPECIES

Holtbyrnia laticauda Tusked Tubeshoulder
Maulisia acuticeps Sharpsnout Tubeshoulder
Maulisia mauli Maul's Tubeshoulder
Maulisia microlepis Smallscale Tubeshoulder
Persparsia kopua Spangled Tubeshoulder
Platytroctes apus Legless Tubeshoulder

SLICKHEADS

Alepocephalidae

SOFTSKIN SLICKHEAD
Rouleina attrita

This large family contains about 90 species. They are found worldwide, in tropical and temperate waters as deep as 6000 m. Most are drab in colour – brown or black – and many have small photophores on the head and body. They have no scales on the head, which is instead covered with smooth, slippery black skin. Some species are entirely scaleless. This is one of the dominant families of bottom-dwelling deepwater fishes in terms of abundance and number of species, and it occurs right around Australia.

Most Slickheads feed on gelatinous planktonic animals such as jellyfishes, ctenophores and salps, and have fine teeth and a large coiled intestine to digest this low-protein diet. They have a specialised crumenal organ at the rear of the gill arches that helps break down the poisonous, stinging cells of their prey, and a thick, tough lining in the mouth and pharynx for protection. Their large eyes are very sensitive to the low levels of light produced by their often bioluminescent prey. Some Slickheads have a mixed diet of pelagic and benthic crustaceans and fishes, while some others are thought to feed mainly on fishes. A *Rouleina* species that feeds on organic matter and microplankton has been observed at great depth with sheets of mucus hanging from the jaws and body – this mucus is perhaps used to trap small organic particles, which can then be consumed along with the mucus. Slickheads do not have a swim bladder and have a very high proportion of water in their body tissues, indicating a slow-moving or mostly sedentary lifestyle. Many species have been observed at depth, drifting passively, waiting for prey to cross their path.

SMALLSCALE SLICKHEAD
Alepocephalus australis

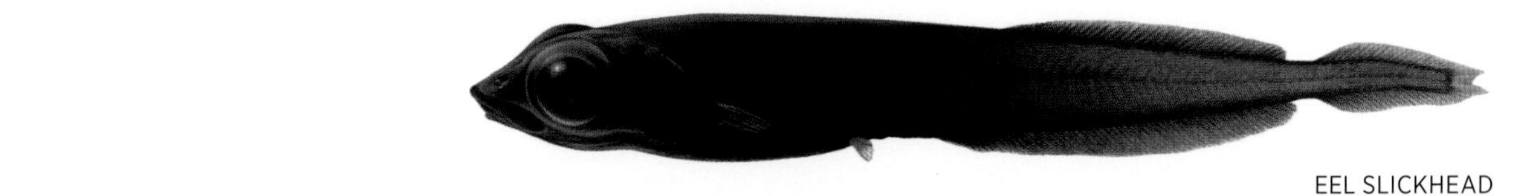

EEL SLICKHEAD
Leptoderma affinis

LONGTAIL SLICKHEAD
Talismania longifilis

AUSTRALIAN ALEPOCEPHALIDAE SPECIES

Alepocephalus antipodianus Antipodean Slickhead
Alepocephalus australis Smallscale Slickhead
Alepocephalus bicolor Bicolor Slickhead
Alepocephalus longiceps Longfin Slickhead
Alepocephalus longirostris Longsnout Slickhead
Alepocephalus owstoni Owston's Slickhead
Alepocephalus productus Smalleye Slickhead
Alepocephalus triangularis Triangulate Slickhead
Asquamiceps caeruleus Blue Slickhead
Asquamiceps hjorti Barethroat Slickhead
Aulastomatomorpha phospherops Luminous Slickhead
Bajacalifornia arcylepis Network Slickhead
Bajacalifornia calcarata Brown Slickhead
Bajacalifornia megalops Bigeye Slickhead
Bathyprion danae Fangtooth Slickhead
Bathytroctes squamosus Deepscale Slickhead
Bathytroctes zugmayeri Zugmayer's Slickhead
Conocara kreffti Wrinkled Slickhead
Conocara microlepis Elongate Slickhead
Conocara murrayi Murray's Slickhead
Conocara nigrum Flathead Slickhead
Einara macrolepis Loosescale Slickhead
Herwigia kreffti Toothless Slickhead
Leptochilichthys microlepis Smallscale Smoothhead
Leptochilichthys pinguis Vaillant's Smoothhead
Leptoderma affinis Eel Slickhead
Microphotolepis schmidti Schmidt's Slickhead
Mirognathus normani Beaked Slickhead
Narcetes lloydi Lloyd's Slickhead
Narcetes stomias Blackhead Slickhead
Photostylus pycnopterus Starry Slickhead
Rouleina attrita Softskin Slickhead
Rouleina eucla Eucla Slickhead
Rouleina guentheri Bordello Slickhead
Rouleina maderensis Madeiran Slickhead
Rouleina squamilatera Sparkling Slickhead
Talismania antillarum Antillean Slickhead
Talismania longifilis Longtail Slickhead
Talismania mekistonema Threadfin Slickhead
Xenodermichthys copei Bluntsnout Slickhead

OSMERIFORMES

The order Osmeriformes is closely related to the Argentiniformes (p. 201) and Salmoniformes (p. 213), and all three groups were formerly placed in the same order. The Osmeriformes is still sometimes categorised as a sub-order of the Salmoniformes. These small fishes have an elongate body, either scaleless or with cycloid scales, and a single dorsal fin (some also have a small adipose fin). They have ventral fins about midway along the body and a caudal fin that can be forked or truncate. There are no spines in the fins. Osmeriforms spawn in fresh water and many spend part of their life in the ocean. There are three families in the order, two of which occur in Australian waters.

SOUTHERN SMELTS

Retropinnidae

AUSTRALIAN SMELT
Retropinna semoni

Southern Smelts are found only in southern Australia and New Zealand, including the Chatham Islands. There are about five species in the family. They are closely related to the Galaxids (Galaxiidae, p. 211) but differ in having scales on the body, an adipose fin and no lateral line. Like Galaxids, they are essentially freshwater fishes. They spawn in fresh water, and the larvae are washed downstream to the sea, where they spend part of their life. (An exception is the **Australian Smelt,** which has some landlocked populations in the Murray-Darling and Cooper Creek river systems.)

Southern Smelts are mostly small fishes, reaching about 7–10 cm in length. They feed on small crustaceans, insects and algae. The **Australian Grayling** is larger, reaching a maximum length of about 33 cm. Algae makes up a large proportion of its diet and it has a long intestine to help it digest this material. Once a popular target for anglers, its numbers have dwindled and it is now found only in a few south-eastern drainages – it is currently a protected species. When freshly caught, Southern Smelts smell distinctly like cucumber.

AUSTRALIAN GRAYLING
Prototroctes maraena

AUSTRALIAN RETROPINNIDAE SPECIES

Prototroctes maraena Australian Grayling
Retropinna semoni Australian Smelt
Retropinna tasmanica Tasmanian Smelt

GALAXIDS

Galaxiidae

TROUT GALAXIAS
Galaxias truttaceus

The 51 species of Galaxid are found in fresh water throughout the southern hemisphere. The family is most abundant and diverse in southern Australia and Tasmania, and 11 of the 17 species found there are endemic. They are small, elongate fishes, most measuring less than 20 cm in length. They have no scales (except the **Salamanderfish**) and no adipose fin (except the **Tasmanian Whitebait**). They all have a single spineless dorsal fin set well back on the body, ventral fins about midway along the body and a lateral line. Some also have a line of sensory cells on the upper body, running from the head to the dorsal fin. Most feed on insects, crustaceans and other small aquatic invertebrates, while some also consume algae.

Galaxids are the dominant freshwater fishes in southern Australia. Some have marine larval stages, but many are confined to fresh water and have limited distribution. The four *Paragalaxias* species, for example, are found only in a few lakes in central Tasmania. Several such Galaxids are listed as critically endangered, due to their extremely small ranges and predation by introduced species like Trout and Redfin Perch. The **Pedder Galaxias** is on the verge of extinction, with only a small number occurring in two streams in Tasmania. Other, more widespread species have larvae that are washed downstream after hatching, and spend part of their life cycle in the sea.

The **Common Galaxias** has an interesting distribution, with separate, very similar populations found at isolated locations right around the southern hemisphere – this is probably due to its long pelagic larval stage. The Tasmanian Whitebait spends most of its short life in the sea, re-entering rivers to spawn when one year old, then dying. It once formed part of a small seasonal fishery, which netted schooling adults as they moved into fresh water en masse. The fishery closed due to dwindling numbers. The **Climbing Galaxias**, which inhabits rocky streams in south-eastern Australia and Tasmania, is able to scale damp vertical rock faces as high as 10 m as it moves upstream – it wriggles upwards, gripping with its downward-facing pectoral and ventral fins.

The fascinating and unique Salamanderfish is sometimes classified as belonging in its own family. It is found only in acidic, ephemeral pools in the sandy coastal plains of south-western Australia. In order to survive when the pools dry up, this much-studied fish burrows into damp sand, where it secretes a mucus cocoon to prevent water loss. It is kept moist by groundwater and is capable of absorbing oxygen through the skin. It reappears within minutes of the pool being refilled. It is also able to move its head from side to side, an odd ability made possible by the presence of a wider-than-normal space between the skull and first few vertebrae. This flexibility serves to compensate for a lack of muscles for controlling eye movement. The Salamanderfish, which grows to a maximum length of only 7 cm, is thought to be the only remaining member of an evolutionary line dating back at least 90 million years.

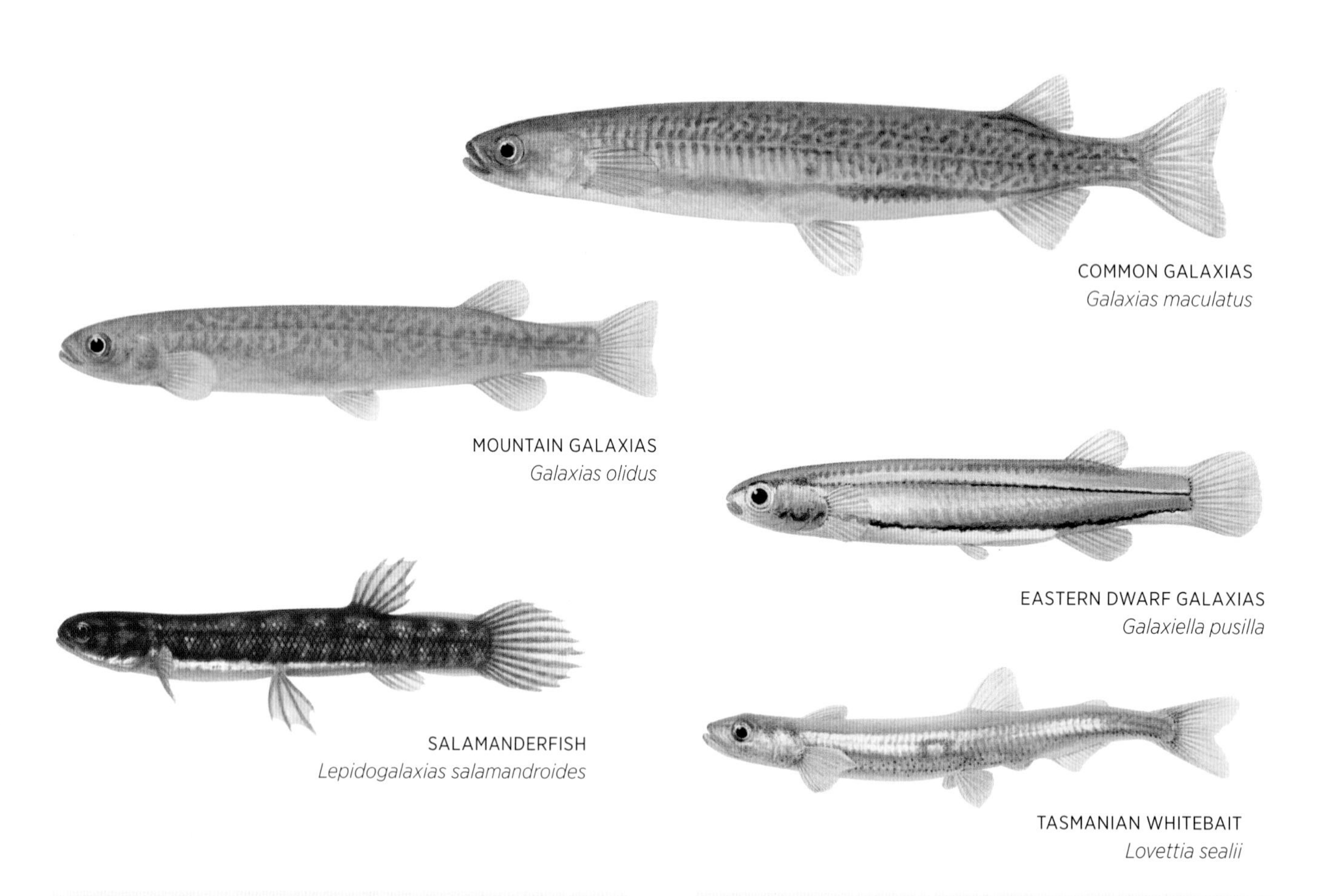

COMMON GALAXIAS
Galaxias maculatus

MOUNTAIN GALAXIAS
Galaxias olidus

EASTERN DWARF GALAXIAS
Galaxiella pusilla

SALAMANDERFISH
Lepidogalaxias salamandroides

TASMANIAN WHITEBAIT
Lovettia sealii

AUSTRALIAN GALAXIIDAE SPECIES

Galaxias auratus Golden Galaxias
Galaxias brevipinnis Climbing Galaxias
Galaxias fontanus Swan Galaxias
Galaxias fuscus Barred Galaxias
Galaxias johnstoni Clarence Galaxias
Galaxias maculatus Common Galaxias
Galaxias niger Black Galaxias
Galaxias occidentalis Western Galaxias
Galaxias olidus Mountain Galaxias
Galaxias parvus Swamp Galaxias
Galaxias pedderensis Pedder Galaxias
Galaxias rostratus Flathead Galaxias
Galaxias tanycephalus Saddled Galaxias
Galaxias truttaceus Trout Galaxias
Galaxiella munda Western Dwarf Galaxias
Galaxiella nigrostriata Blackstriped Dwarf Galaxias
Galaxiella pusilla Eastern Dwarf Galaxias
Lepidogalaxias salamandroides Salamanderfish
Lovettia sealii Tasmanian Whitebait
Neochanna cleaveri Tasmanian Mudfish
Paragalaxias dissimilis Shannon Galaxias
Paragalaxias eleotroides Great Lake Galaxias
Paragalaxias julianus Julian Galaxias
Paragalaxias mesotes Arthurs Galaxias

SALMONIFORMES

The order Salmoniformes is closely related to the Argentiniformes (p. 201) and Osmeriformes (p. 209), and all three groups were formerly placed in the same order. Salmoniforms have an elongate body with small cycloid scales, a single dorsal fin and small adipose fin, ventral fins (with a prominent axillary process) about midway along the body, pectoral fins low on the body, and a forked caudal fin. There are no spines in any of the fins. The single family in the order contains the Trouts and Salmons, which are predominantly northern hemisphere fishes. They spawn in fresh water but many spend large parts of their lives in the sea. Salmoniforms have been widely introduced into freshwater habitats around the world, including in Australia.

SALMONS AND TROUTS

Salmonidae

BROWN TROUT
Salmo trutta

There about 66 species in this family of freshwater fishes and they have a circum-Arctic distribution. Most hatch and develop in fresh water as juveniles, then move into the sea where they spend between one and four years before returning to fresh water as adults to spawn. Salmons and Trouts have been introduced into many countries around the world, including Australia, both for recreational fishing and for aquaculture.

Although the **Brown Trout** and **Rainbow Trout** are widely distributed in southern Australia, there are only a few populations that follow the normal life cycle; for the most part stocks are replenished with juveniles that have been reared in hatcheries. The **Brook Trout** is self-sustaining in a few freshwater locations in Tasmania and New South Wales. **Atlantic Salmon** and Rainbow Trout are bred in large quantities in commercial sea-cage aquaculture operations in Tasmania. Brown Trout are also farmed commercially on a small scale in many areas of southern Australia.

Despite the popularity of recreational Trout fishing, their introduction has had a disastrous effect on native fish populations in some areas as Trouts prey on native species and compete with them for food. In Tasmania, where Trouts thrive, this has contributed to several species of native Galaxids (Galaxiidae, p. 211) becoming critically endangered.

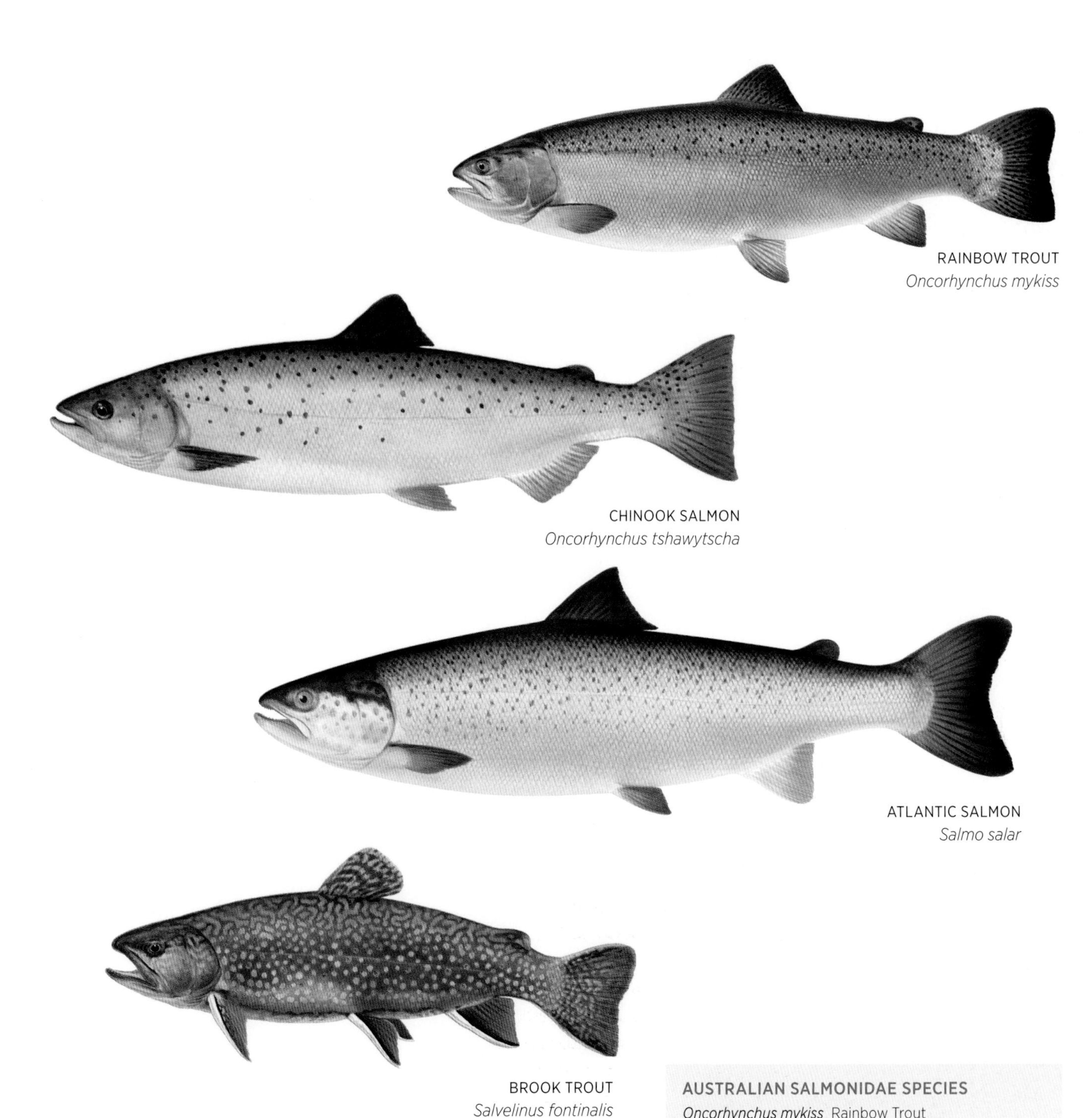

RAINBOW TROUT
Oncorhynchus mykiss

CHINOOK SALMON
Oncorhynchus tshawytscha

ATLANTIC SALMON
Salmo salar

BROOK TROUT
Salvelinus fontinalis

AUSTRALIAN SALMONIDAE SPECIES

Oncorhynchus mykiss Rainbow Trout
Oncorhynchus tshawytscha Chinook Salmon
Salmo salar Atlantic Salmon
Salmo trutta Brown Trout
Salvelinus fontinalis Brook Trout

STOMIIFORMES

Stomiiforms possess photophores that create light through biochemical reactions, rather than from luminous symbiotic bacteria as in most other bioluminescent fishes. These deepwater fishes display a great range of body forms and there are few anatomical features common to all species. Most have a large mouth that extends past the eyes, teeth in both jaws, and a single dorsal fin. Some have small ventral and dorsal adipose fins. Stomiiforms are usually black or silvery, and either have no scales or fragile cycloid scales. There are four families in the order, all of which occur in Australian waters.

BRISTLEMOUTHS

Gonostomatidae

SMALLTOOTH BRISTLEMOUTH
Cyclothone microdon

Bristlemouths are abundant in all the oceans of the world, including the Southern Ocean. There are about 20 species in the family. Most are pelagic in deep water from 200 to 2000 m but some have been found at depths to 5000 m. Bristlemouths have large jaws that extend well behind the eyes, an elongate body with thin skin, and weak scales that are easily lost in nets. They have photophores in rows along the ventral surface, a long anal fin, and a single dorsal fin at about mid-body, and some have a small adipose fin. Bristlemouths feed predominantly on small pelagic crustaceans and have the numerous fine teeth and gillrakers typical of zooplanktivores. Most are only 6–7 cm in length but their abundance means they are major consumers of zooplankton in deeper waters. They are also an important food source for many predatory deepwater fishes.

Larger Bristlemouth species are generally found in deeper waters than smaller species, and adults in deeper waters than juveniles. Some species are always found close to the bottom, but most make nightly migrations towards the surface to feed. They may travel 700 m each way on these migrations, moving between widely differing temperatures and pressures. **Rebains' Portholefish** is one of the larger Bristlemouths, measuring about 25 cm in length. It is widely distributed in southern oceans and occurs at depths of at least 2000 m.

REBAINS' PORTHOLEFISH
Diplophos rebainsi

DEEPSEA FANGJAW
Sigmops bathyphilus

AUSTRALIAN GONOSTOMATIDAE SPECIES

Bonapartia pedaliota Longray Fangjaw
Cyclothone acclinidens Bent-tooth Bristlemouth
Cyclothone alba Pale Bristlemouth
Cyclothone braueri Brauer's Bristlemouth
Cyclothone kobayashii Kobayashi's Bristlemouth
Cyclothone microdon Smalltooth Bristlemouth
Cyclothone obscura Hidden Bristlemouth
Cyclothone pallida Tanned Bristlemouth
Cyclothone parapallida Shadow Bristlemouth
Cyclothone pseudopallida Slender Bristlemouth
Diplophos rebainsi Rebains' Portholefish
Diplophos taenia Pacific Portholefish
Gonostoma atlanticum Atlantic Fangjaw
Gonostoma elongatum Elongate Fangjaw
Margrethia obtusirostra Bighead Portholefish
Sigmops bathyphilus Deepsea Fangjaw

HATCHETFISHES AND PEARLSIDES

Sternoptychidae

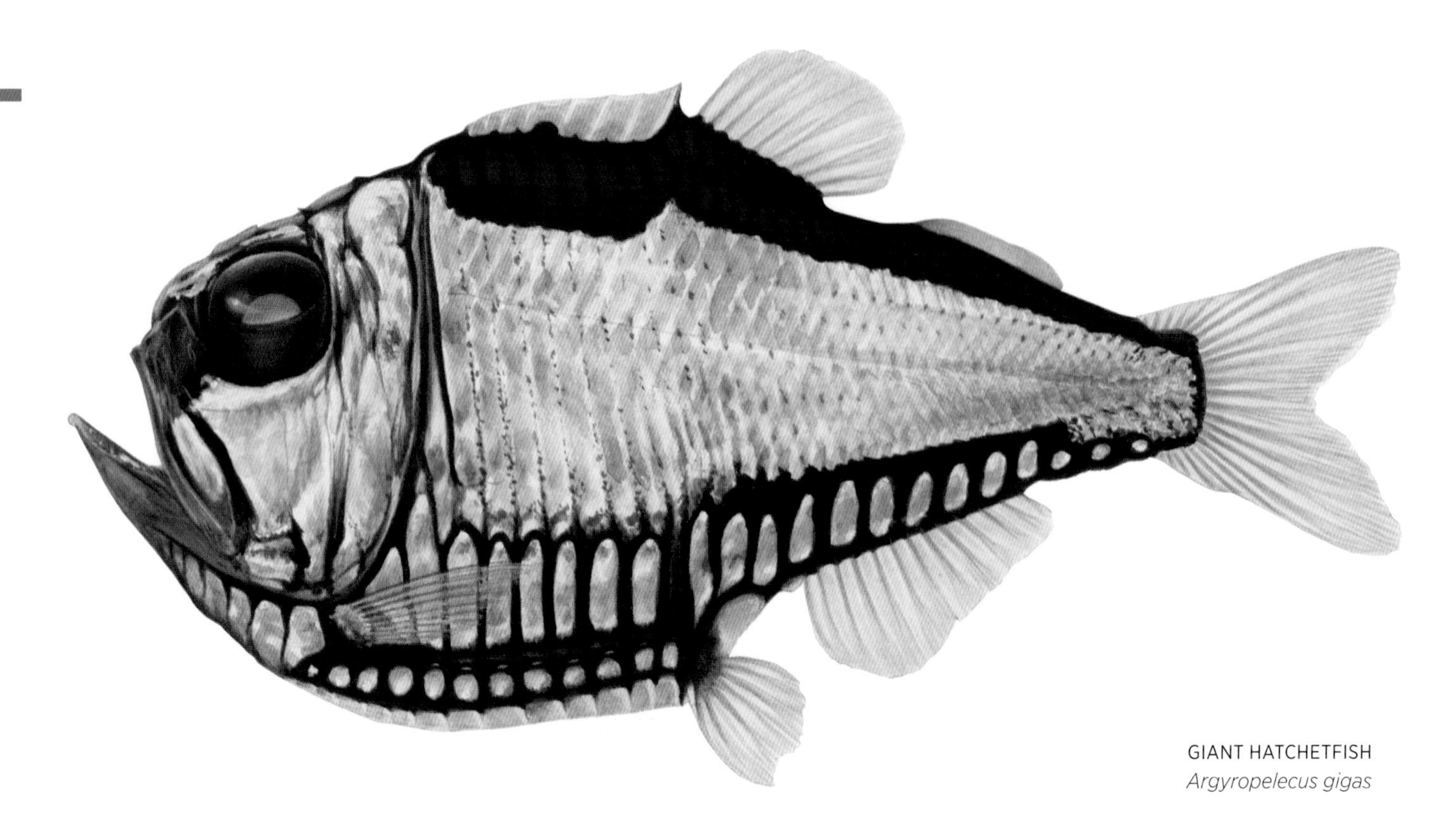

GIANT HATCHETFISH
Argyropelecus gigas

Hatchetfishes and Pearlsides are distributed in all the world's oceans. They are pelagic in midwater or close to the bottom and are found from surface waters to depths of more than 4000 m. There are about 67 species in the family. They grow to a maximum length of about 12 cm and have large photophores characteristically arranged in series or small groups. They have a silvery, reflective body with thin, weak scales, large eyes, and a large mouth with small, sharp teeth. Some have an adipose dorsal fin, which is usually attached along its full length. They have a gas-filled swim bladder, and the well-developed gillrakers typical of zooplanktivores. The family is divided into two subfamilies:

STERNOPTYCHINAE The Hatchetfishes have a deep, compressed body, a keel-like structure on the abdomen, and a transparent blade or spine in front of the first dorsal fin. Some Hatchetfishes have tubular eyes that are directed upward.

MAUROLICINAE Pearlsides are more elongate than Hatchetfishes and do not have a dorsal blade or abdominal keel.

Hatchetfishes and Pearlsides are zooplanktivores and many species make nightly migrations towards the surface to feed. They are abundant, and are one of the major groups of fishes responsible for transporting nutrients from productive surface waters, where phytoplankton and zooplankton abound, to ocean depths, where productivity is very low. However, some species remain at greater depths, where they feed almost constantly. Hatchetfishes and Pearlsides are an important food source for larger predatory fishes.

PENNANT PEARLSIDE
Maurolicus australis

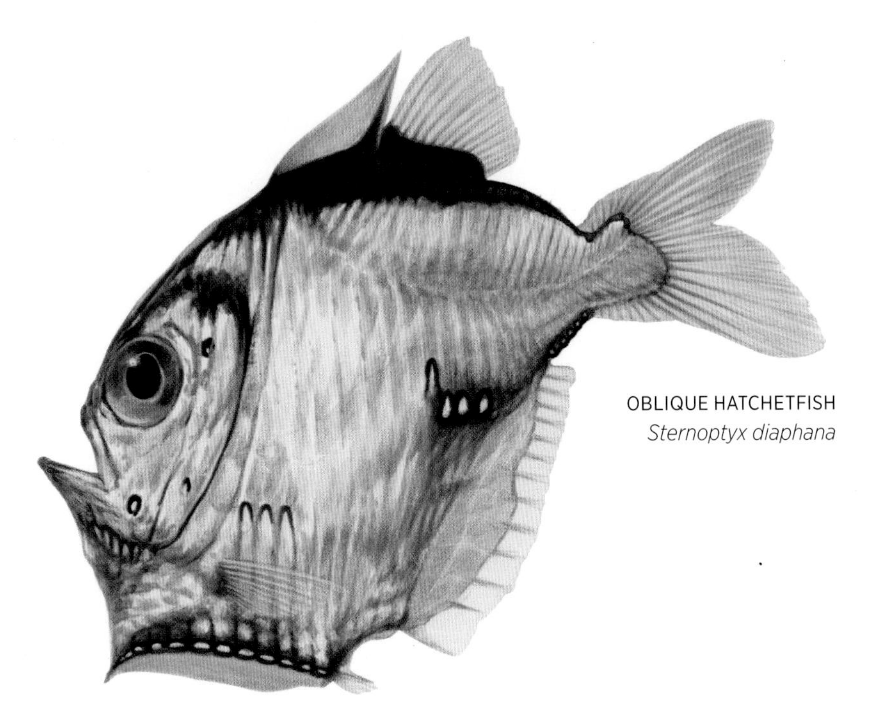

OBLIQUE HATCHETFISH
Sternoptyx diaphana

AUSTRALIAN STERNOPTYCHIDAE SPECIES

Argyripnus ephippiatus Saddled Pearlside
Argyripnus iridescens Brilliant Pearlside
Argyropelecus aculeatus Lovely Hatchetfish
Argyropelecus gigas Giant Hatchetfish
Argyropelecus hemigymnus Halfnaked Hatchetfish
Argyropelecus sladeni Lowcrest Hatchetfish
Maurolicus australis Pennant Pearlside
Maurolicus javanicus Javan Pearlside
Polyipnus aquavitus Aquavit Hatchetfish
Polyipnus elongatus Elongate Hatchetfish
Polyipnus kiwiensis Kiwi Hatchetfish
Polyipnus latirastrus Combside Hatchetfish
Polyipnus paxtoni Paxton's Hatchetfish
Polyipnus ruggeri Rugby Hatchetfish
Polyipnus soelae Soela Hatchetfish
Polyipnus spinosus Spiny Hatchetfish
Polyipnus tridentifer Threespine Hatchetfish
Polyipnus triphanos Threelight Hatchetfish
Sternoptyx diaphana Oblique Hatchetfish
Sternoptyx pseudobscura Highlight Hatchetfish
Sternoptyx pseudodiaphana False Oblique Hatchetfish
Valenciennellus tripunctulatus Constellationfish

LIGHTFISHES

Phosichthyidae

OVATE LIGHTFISH
Ichthyococcus ovatus

Lightfishes are found worldwide and there are about 20 species in the family. They are pelagic from near the surface to depths of 2000 m, but are most commonly found from 200 to 600 m. They are closely related to the Bristlemouths (Gonostomatidae, p. 218) and are similar in body shape, with the dorsal fin at about mid-body, above the origin of the anal fin. Lightfishes have two rows of photophores on the ventral surface running from the head to the anal fin, with a single row continuing to the tail. They have thin scales that are easily lost, an adipose dorsal fin and no chin barbel.

Lightfishes feed mainly on pelagic crustaceans such as copepods, and have the numerous small teeth and gillrakers typical of zooplanktivores. Some species make nightly migrations towards the surface to feed. Lightfishes are extremely plentiful, with the genus *Vinciguerria* estimated to be one of the most abundant of all vertebrate groups. Samples taken from the south-east Pacific have shown that Lightfishes (mainly *Vinciguerria* species) make up 85 per cent of the midwater pelagic fish biomass.

POWER'S LIGHTFISH
Vinciguerria poweriae

AUSTRALIAN PHOSICHTHYIDAE SPECIES

Ichthyococcus australis Southern Lightfish
Ichthyococcus intermedius Intermediate Lightfish
Ichthyococcus ovatus Ovate Lightfish
Phosichthys argenteus Silver Lightfish
Polymetme corythaeola Rendezvous Fish
Polymetme illustris Brilliant Lightfish
Polymetme surugaensis Suruga Lightfish
Vinciguerria attenuata Slender Lightfish
Vinciguerria nimbaria Narooma Lightfish
Vinciguerria poweriae Power's Lightfish
Woodsia meyerwaardeni Austral Lightfish
Woodsia nonsuchae Bigeye Lightfish

DRAGONFISHES

Stomiidae

SLOANE'S VIPERFISH
Chauliodus sloani

The family of Dragonfishes is very diverse and more than 270 species are found worldwide. They occur from the surface (at night) to depths as great as 5000 m. All have photophores in rows along the ventral surface of the body and larger photophores on the head. Most Dragonfishes have large jaws with long, fine teeth, and a barbel tipped with a light-producing organ on the chin. Some have a hexagonal pattern of pigment along the body, which resembles scales. Many members of the family have evolved to prey on the vast numbers of small planktivorous fishes that migrate from the depths towards the surface at night. Dragonfishes are usually ambush predators and are thought to use the light on their barbel as a lure to attract prey. They are usually separated into six subfamilies:

ASTRONESTHINAE Snaggletooths or Stareaters have small photophores over most of the body, as well as two prominent rows along the ventral surface. They have an adipose dorsal fin and no hexagonal scale pattern, and many have a small ventral adipose fin before the anal fin. Some have irregular patches of luminescent tissue on the body. They reach about 30 cm in length and most are pelagic in midwater at depths of 200–1500 m. Snaggletooths are active foragers and have more muscular bodies than other Dragonfishes.They feed on crustaceans and small fishes – some species migrate towards the surface at night, while others remain close to the bottom.

STOMIINAE Scaly Dragonfishes have an elongate, compressed body with 5–6 rows of small photophores, each of which is surrounded by a hexagonal pigmented area, giving a scale pattern. The dorsal fin is at the rear of the body, above the anal fin, and there is no adipose fin. The body is covered with a luminous gelatinous membrane, which is usually rubbed off in trawl-caught specimens. Scaly Dragonfishes have rigid, fang-like teeth on the upper and lower jaws, as well as numerous small, fine teeth on the rear upper jaw. They occur mainly at depths from 300 to 500 m, with some following the nightly migrations of their prey towards the surface. They grow to 20–40 cm in length and feed on pelagic crustaceans and small fishes.

MELANOSTOMIINAE Scaleless Dragonfishes differ from the Scaly Dragonfishes in that they lack a scale pattern and possess a small adipose fin. The chin barbel may be very long. They have large, barbed fangs in the front of the lower and upper jaws that hinge inward to allow consumption of large prey, and smaller teeth towards the rear of the jaws. There are two rows of large photophores on the ventral surface, as well as minute photophores covering the body (usually arranged in vertical rows). Mainly preying on other fishes, they inhabit similar depths to Scaly Dragonfishes and attain about 30 cm in length.

IDIACANTHINAE Black Dragonfishes have an eel-like body, no scale pattern and a very long dorsal fin. This small group shows strong sexual dimorphism: males lack teeth, a functioning gut, the chin barbel and pectoral fin, and only grow to about 15 per cent of the female size (females are 30–50 cm in length). The males retain some of the characteristics of larvae, cannot feed after metamorphosing into adults, and apparently exist only to serve in reproduction. Females have numerous barbed fangs of varying size that can hinge inward, as well as a small, sharp spur at the base of each dorsal and anal fin. Black Dragonfishes have been found down to 2000 m and feed on pelagic crustaceans and fishes – females make nightly migrations towards the surface, while males remain in deep water.

CHAULIODONTINAE Viperfishes grow to about 30 cm in length. They have a large head, and jaws with very long, rigid, fang-like teeth. The dorsal fin is just behind the head and has a greatly elongated first fin ray with a light organ at its tip. The chin barbel is present in juveniles but is very small or absent in adults. Dorsal and ventral adipose fins are present. Viperfishes are similar to Scaly Dragonfishes, with 5–6 rows of photophores, the photophores in the lower two rows being larger. Each photophore is surrounded by a hexagonal pigmented area, giving a scale pattern. The body is covered with a luminous gelatinous membrane, but this is usually lost in trawl-caught specimens. Viperfishes occur in waters from the surface (at night) to depths greater than 2500 m. They feed on crustaceans and fishes, and their jaws can open extremely wide to accommodate large prey.

MALACOSTIINAE Loosejaws are perhaps the most remarkable group of Dragonfishes. These small fishes grow to only about 25 cm in length, but their head can hinge vertically on the first few vertebrae, enabling them to open their enormous jaws to an angle of almost 180 degrees. The jaws bear large, rigid fangs interspersed with smaller teeth, and the two lower jaw bones do not have a membrane between them, so there is no floor to the mouth. This suggests that they consume only large prey items, but the stomach contents of specimens have been found to include small zooplankton. Loosejaws do not make nightly migrations towards the surface to feed, instead remaining at depths of 500–2000 m. Most bioluminescent fishes produce and are sensitive to blue light, which travels furthest through water, but the Loosejaws have a large red-light-producing organ. This perhaps allows them to see and approach prey that cannot detect red light.

IJIMA'S SNAGGLETOOTH
Astronesthes ijimai

COMMON BLACK DRAGONFISH
Idiacanthus atlanticus

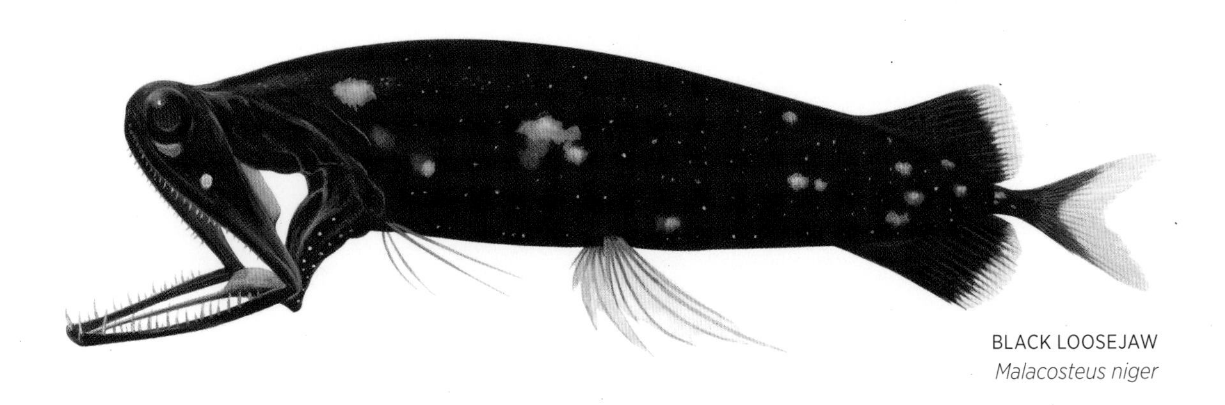

BLACK LOOSEJAW
Malacosteus niger

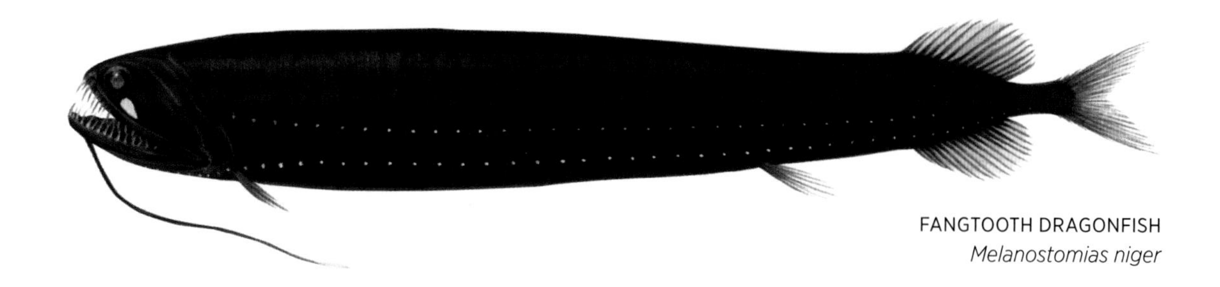

FANGTOOTH DRAGONFISH
Melanostomias niger

AUSTRALIAN STOMIIDAE SPECIES

Astronesthes bilobatus Twinlobe Snaggletooth
Astronesthes boulengeri Boulenger's Snaggletooth
Astronesthes exsul Exile Snaggletooth
Astronesthes ijimai Ijima's Snaggletooth
Astronesthes indicus Black Snaggletooth
Astronesthes indopacificus Indo-Pacific Snaggletooth
Astronesthes kreffti Krefft's Snaggletooth
Astronesthes lucifer Pacific Snaggletooth
Astronesthes lupina Little Wolf
Astronesthes martensii Martens' Snaggletooth
Astronesthes psychrolutes Temperate Snaggletooth
Astronesthes splendidus Splendid Snaggletooth
Astronesthes trifibulatus Triplethread Snaggletooth
Bathophilus abarbatus Barbless Dragonfish
Bathophilus ater Winged Dragonfish
Bathophilus indicus Black Dragonfish
Bathophilus longipinnis Longfin Dragonfish
Bathophilus nigerrimus Scaleless Dragonfish
Bathophilus pawneei Pawnee Dragonfish
Borostomias antarcticus Antarctic Snaggletooth
Borostomias mononema Sickle Snaggletooth
Chauliodus sloani Sloane's Viperfish
Eupogonesthes xenicus Exotic Snaggletooth
Echiostoma barbatum Threadfin Dragonfish
Eustomias achirus Proud Dragonfish
Eustomias australensis Australian Dragonfish
Eustomias bifilis Twinthread Dragonfish
Eustomias bulbornatus Grapevine Dragonfish
Eustomias cryptobulbus Hiddenbulb Dragonfish
Eustomias enbarbatus Barbate Dragonfish
Eustomias macronema Bigbarb Dragonfish
Eustomias macrurus Yellowstem Dragonfish
Eustomias multifilis Multi-thread Dragonfish
Eustomias parini Parin's Dragonfish
Eustomias satterleei Twinray Dragonfish
Eustomias schmidti Schmidt's Dragonfish
Eustomias trewavasae Deepsea Dragonfish
Eustomias vitiazi Vitiaz Dragonfish
Eustomias vulgaris Common Dragonfish
Flagellostomias boureei Longbarb Dragonfish
Heterophotus ophistoma Wingfin Snaggletooth
Idiacanthus atlanticus Common Black Dragonfish
Idiacanthus fasciola Serpent Black Dragonfish
Malacosteus australis Southern Stoplight Loosejaw
Malacosteus niger Black Loosejaw
Melanostomias globulifer Brightchin Dragonfish
Melanostomias niger Fangtooth Dragonfish
Melanostomias paucilaternatus Spothead Dragonfish
Melanostomias pauciradius Three-ray Dragonfish
Melanostomias tentaculatus Tentacle Dragonfish
Melanostomias valdiviae Valdivia Dragonfish
Neonesthes capensis Cape Snaggletooth
Neonesthes microcephalus Smallhead Snaggletooth
Opostomias micripnus Obese Dragonfish
Pachystomias microdon Smalltooth Dragonfish
Photonectes albipennis Whitepen Dragonfish
Photonectes braueri Brauer's Dragonfish
Photonectes caerulescens Bulbless Dragonfish
Photonectes gracilis Graceful Dragonfish
Photonectes mirabilis Blueband Dragonfish
Photonectes parvimanus Fleshyfin Dragonfish
Stomias affinis Honeycomb Scaly Dragonfish
Stomias boa Boa Scaly Dragonfish
Stomias longibarbatus Longbarb Scaly Dragonfish
Thysanactis dentex Broomfin Dragonfish
Trigonolampa miriceps Threelight Dragonfish

ATELEOPODIFORMES

This order contains only a single family, the Ateleopodidae. The family was formerly included in the order Lampridiformes (p. 255) and seems to be closely related to the Stomiiformes (p. 217), but the relationship between this family and other fishes is uncertain. Ateleopodiforms have an eel-like body with a gelatinous consistency, and a mostly cartilaginous skeleton. The order is found in Australian waters.

JELLYNOSES

Ateleopodidae

PACIFIC JELLYNOSE
Ateleopodus japonicus

The family of Jellynoses contains about 12 bottom-dwelling species. They inhabit areas of the Caribbean, eastern Atlantic, Indo-West Pacific and East Pacific, to depths of at least 700 m. Growing to a maximum length of about 2 m, they have an eel-like body with a gelatinous consistency, and a partially calcified, mostly cartilaginous skeleton. They have a single dorsal fin just behind the head, a long anal fin that is continuous with the caudal fin, and small villiform teeth. Their bulbous, jelly-like snout and fragile body are usually damaged when taken in trawl nets. Jellynoses feed on benthic invertebrates. Their biology and relationship to other families are not well known.

AUSTRALIAN ATELEOPODIDAE SPECIES
Ateleopus japonicus Pacific Jellynose
Ijimaia sp. Stumpfin Jellynose

AULOPIFORMES

Members of this large and diverse group of fishes are found mainly in deep water, though several families occur in shallow coastal waters. Aulopiforms have an elongate body with cycloid scales, a single dorsal fin, ventral fins at about mid-body, and no spines in the fins. Some also have a small adipose fin. The bones of the pelvic girdle, which supports the ventral fins, are fused together in the centre. Aulopiforms do not have a swim bladder and many rely on strong ventral fins to support their body on the substrate. Most have a large mouth with well-developed teeth in both jaws and are carnivorous. Many species are hermaphroditic. There are 15 families in the order, 14 of which occur in Australian waters.

CUCUMBERFISHES

Paraulopidae

BLACKTIP CUCUMBERFISH
Paraulopus nigripinnis

Cucumberfishes are found in the Indo-West Pacific region and there are about 15 species in the family. They are bottom-dwelling and usually found in deep waters. Several species are endemic to Australia and there are a number of undescribed species in southern waters. Cucumberfishes reach a maximum length of about 35 cm and have a slender, cylindrical body with smooth, fragile scales. They have a tall first dorsal fin, a small adipose fin and large eyes. They possess numerous villiform teeth, which extend onto the outside of the anterior upper and lower jaws. The infraorbital bones form a cup shape that supports the eyeball.

The **Blacktip Cucumberfish** is abundant in the Great Australian Bight and around New Zealand, and occurs down to 600 m. Cucumberfishes are superficially similar to the Southern Smelts (Retropinnidae, p. 210), but very little is known of their biology. It is unclear how Cucumberfishes got their name – whether, like Southern Smelts, they give off a smell of cucumber (some observers report no smell at all), or if the name simply refers to the physical resemblance.

AUSTRALIAN PARAULOPIDAE SPECIES

Paraulopus brevirostris Shortsnout Cucumberfish
Paraulopus melanogrammus Blackline Cucumberfish
Paraulopus nigripinnis Blacktip Cucumberfish
Paraulopus okamurai Piedtip Cucumberfish

THREADSAILS

Aulopidae

SERGEANT BAKER
Hime purpurissatus

Threadsails are found in tropical and temperate waters of the Atlantic, Indian and Pacific oceans. There are about 10 species in the family. They have an elongate, cylindrical body, covered with cycloid or spiny scales that extend onto the beginning of the caudal fin. There is a tall first dorsal fin and a small adipose fin, and the ventral fins are broad and have thickened outer rays. They have large jaws with numerous fine teeth in several rows.

The endemic **Sergeant Baker** is the best-known Australian example and is common on deeper rocky reefs and adjacent sandy areas along the southern coast, in waters down to about 250 m. It grows to around 70 cm in length and males have an elongated second dorsal fin ray. The Sergeant Baker is often seen in the open, resting on its strong ventral fins and tail with the head well elevated. It feeds on molluscs, crustaceans and other fishes. The **Shortsnout Threadsail** is also endemic and is found over sandy or muddy bottoms off north-eastern Australia, at depths of 135–180 m. The **Japanese Threadsail** is found in the same area and also throughout the western Pacific, from depths of 85 to 500 m.

AUSTRALIAN AULOPIDAE SPECIES
Hime curtirostris Shortsnout Threadsail
Hime japonica Japanese Threadsail
Hime purpurissatus Sergeant Baker

LIZARDFISHES AND SAURIES

Synodontidae

LARGESCALE SAURY
Saurida undosquamis

Lizardfishes and Sauries are found in tropical and temperate waters of the Atlantic, Indian and Pacific oceans. The family contains about 57 species and is well represented in Australian waters. They are small, bottom-dwelling fishes, most measuring 20–50 cm in length. They have a cylindrical body, no spines in the fins, a small adipose fin, and large jaws reaching well past the eyes. The jaws have multiple rows of inward-slanting, needle-like teeth, and similar teeth are present on the tongue. Lizardfishes do not have a swim bladder and spend most of their time resting on the substrate. The inner rays of their strong ventral fins are much longer than the outer ones, and are used to support the body while resting on the bottom.

Lizardfishes and Sauries are ambush predators, mainly targeting other fishes. Many have mottled patterns over the body for camouflage, while others bury themselves in the sediment to avoid detection. They are capable of making lightning-fast lunges from their positions of concealment to take passing prey. Members of this family are commonly encountered in sand and rubble areas near reefs in northern Australian waters – although some, like the **Largescale Saury**, are found in deeper waters of the continental shelf down to several hundred metres.

The **Variegated Lizardfish** is a common resident of coral reefs in northern Australian waters, and can often be seen perched on coral outcrops. The **Painted Grinner**, on the other hand, inhabits sandy areas from the shallows down to more than 400 m, and is almost always buried in the substrate with only the eyes exposed. The **Glassy Bombay Duck** occurs in coastal waters and estuaries across the northern half of Australia, and grows to about 70 cm in length. It has a flabby translucent body, with no scales except for a series running along the lateral line and a small area on the tail. It is partly pelagic, spending some of its time hunting other fishes in midwater. The strange name is thought to come from an Indian recipe that uses this fish.

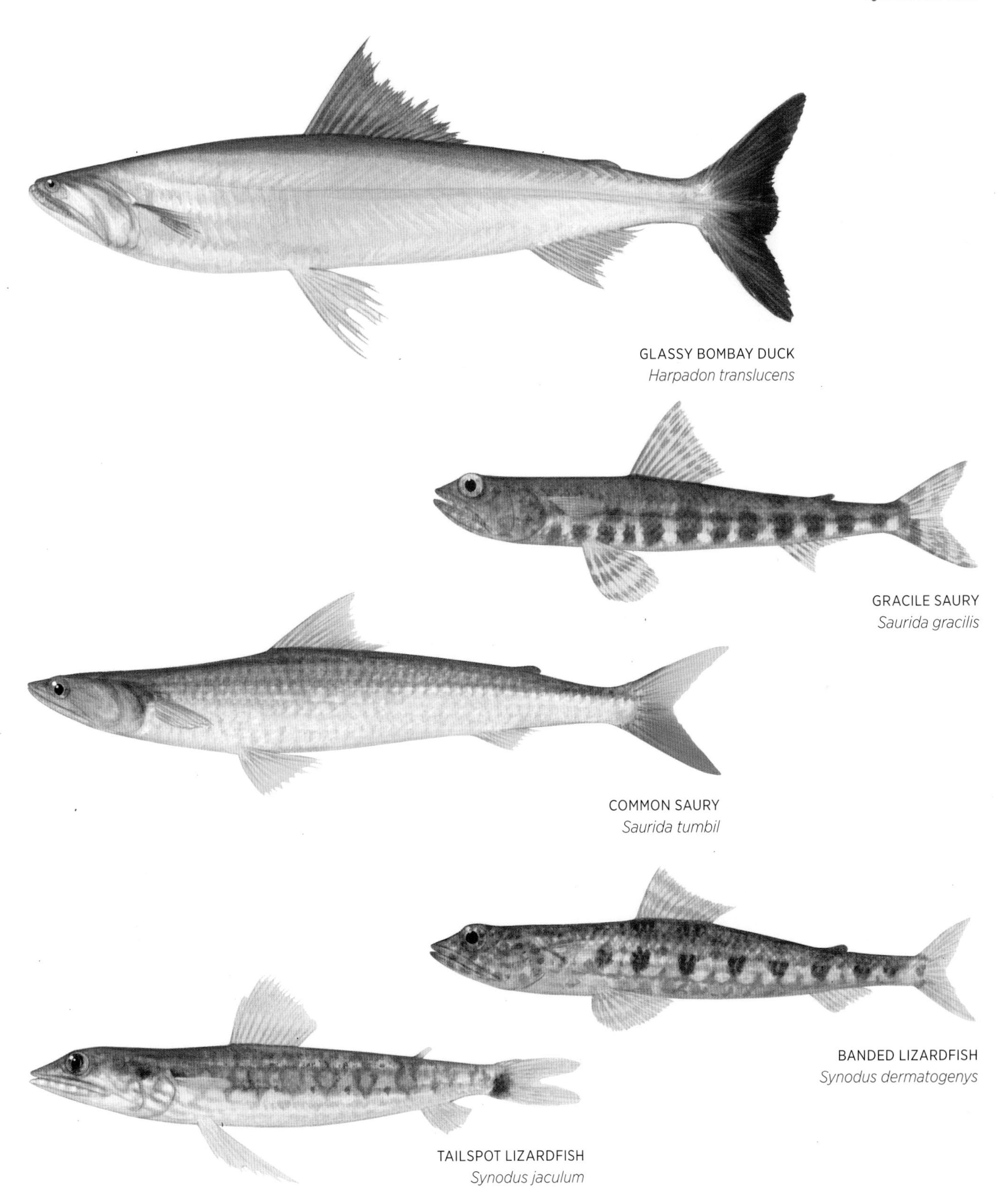

GLASSY BOMBAY DUCK
Harpadon translucens

GRACILE SAURY
Saurida gracilis

COMMON SAURY
Saurida tumbil

BANDED LIZARDFISH
Synodus dermatogenys

TAILSPOT LIZARDFISH
Synodus jaculum

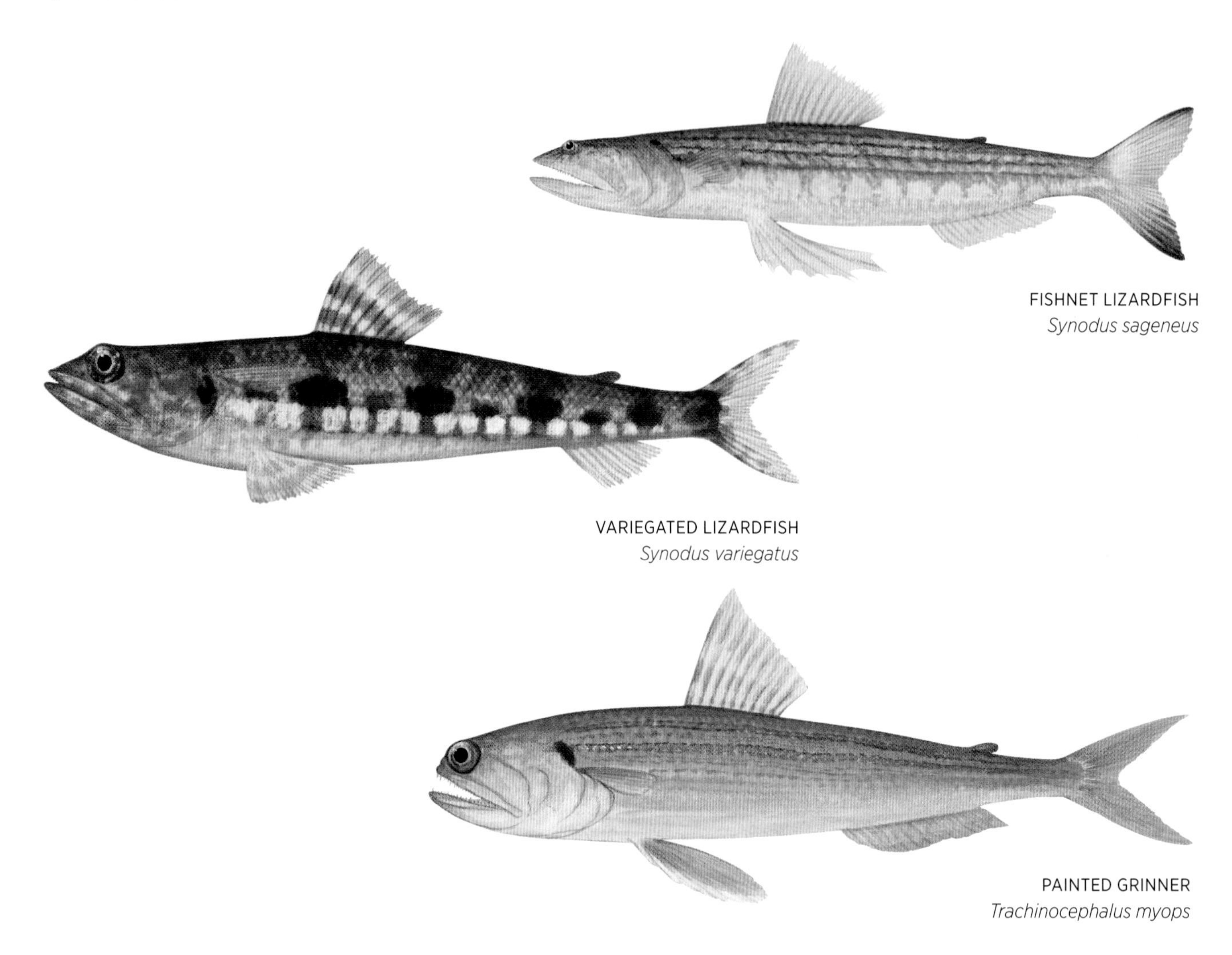

FISHNET LIZARDFISH
Synodus sageneus

VARIEGATED LIZARDFISH
Synodus variegatus

PAINTED GRINNER
Trachinocephalus myops

AUSTRALIAN SYNODONTIDAE SPECIES

Harpadon translucens Glassy Bombay Duck
Saurida argentea Shortfin Saury
Saurida elongata Slender Saury
Saurida filamentosa Threadfin Saury
Saurida gracilis Gracile Saury
Saurida grandisquamis Grey Saury
Saurida longimanus Longfin Saury
Saurida nebulosa Clouded Saury
Saurida tumbil Common Saury
Saurida undosquamis Largescale Saury
Saurida wanieso Wanieso Saury
Synodus binotatus Twospot Lizardfish
Synodus dermatogenys Banded Lizardfish
Synodus doaki Arrowtooth Lizardfish
Synodus hoshinonis Blackshoulder Lizardfish
Synodus houlti Hoult's Lizardfish
Synodus indicus Indian Lizardfish
Synodus jaculum Tailspot Lizardfish
Synodus kaianus Black Lizardfish
Synodus macrops Triplecross Lizardfish
Synodus oculeus Large-eye Lizardfish
Synodus rubromarmoratus Redmarbled Lizardfish
Synodus sageneus Fishnet Lizardfish
Synodus similis Streaky Lizardfish
Synodus tectus Tectus Lizardfish
Synodus variegatus Variegated Lizardfish
Trachinocephalus myops Painted Grinner

PALE DEEPSEA LIZARDFISH

Bathysauroididae

PALE DEEPSEA LIZARDFISH
Bathysauroides gigas

The **Pale Deepsea Lizardfish**, the single species in this family, is found in the western Pacific, in deep waters on the continental slope. It grows to about 30 cm in length and is a bottom dweller. Virtually nothing is known of its biology, and its relationship to the various Deepsea Lizardfishes and other Aulopiforms is not well understood.

AUSTRALIAN BATHYSAUROIDIDAE SPECIES
Bathysauroides gigas Pale Deepsea Lizardfish

GREENEYES

Chlorophthalmidae

SHORTNOSE GREENEYE
Chlorophthalmus agassizi

Greeneyes are found in the Atlantic, Indian and Pacific oceans, and there are about 20 species in the family. They are bottom dwelling and found from 50 to 1000 m. These cylindrical fishes grow to about 40 cm in length and have large eyes, which are green in fresh specimens, with a distinctive teardrop-shaped pupil. They have a small adipose fin, large pectoral fins and variable blotched colour patterns. The two Australian species are found only in northern waters. The **Shortnose Greeneye** occurs in schools over muddy or sandy substrates, where it feeds on benthic and pelagic invertebrates.

AUSTRALIAN CHLOROPHTHALMIDAE SPECIES
Chlorophthalmus agassizi Shortnose Greeneye
Chlorophthalmus nigromarginatus Blackedge Greeneye

BLACK DEEPSEA LIZARDFISHES

Bathysauropsidae

BLACK DEEPSEA LIZARDFISH
Bathysauropsis gracilis

The **Black Deepsea Lizardfish** is one of only two species in this family and is found in subtropical southern hemisphere waters from depths of 1000 to 2800 m. It is very similar to the other Deepsea Lizardfishes (Bathysauroididae, p. 235 and Bathysauridae, p. 247), but the first dorsal fin has a shorter base, there is a small adipose fin, and the teeth do not extend onto the outside of the jaws. The Black Deepsea Lizardfish is bottom dwelling, preys on other fishes, and grows to at least 32 cm in length.

AUSTRALIAN BATHYSAUROPSIDAE SPECIES

Bathysauropsis gracilis Black Deepsea Lizardfish

WARYFISHES

Notosudidae

SMALLSCALE WARYFISH
Scopelosaurus hamiltoni

Waryfishes are found worldwide and are pelagic in deep waters. There are about 20 species in the family. They have elongate, cylindrical or slightly compressed bodies and measure up to 50 cm in length. They possess large cycloid scales, a short-based dorsal fin at about mid-body, and a small adipose fin. The eyes are large and the snout is flattened and rounded at the tip. The jaws are long, extending to underneath or behind the eyes, and bear several rows of small, pointed teeth. Waryfishes occur in midwater or near the bottom at depths of 300–2000 m. They migrate into open oceanic waters to spawn. Adult Waryfishes are hermaphroditic. They lose all their teeth and gillrakers as they approach sexual maturity and presumably die after spawning. They feed on zooplankton, pelagic crustaceans and small fishes.

AUSTRALIAN NOTOSUDIDAE SPECIES

Luciosudis normani Norman's Waryfish
Scopelosaurus ahlstromi Ahlstrom's Waryfish
Scopelosaurus hamiltoni Smallscale Waryfish
Scopelosaurus hoedti Hoedt's Waryfish
Scopelosaurus mauli Maul's Waryfish
Scopelosaurus meadi Blackring Waryfish

DEEPSEA TRIPODFISHES

Ipnopidae

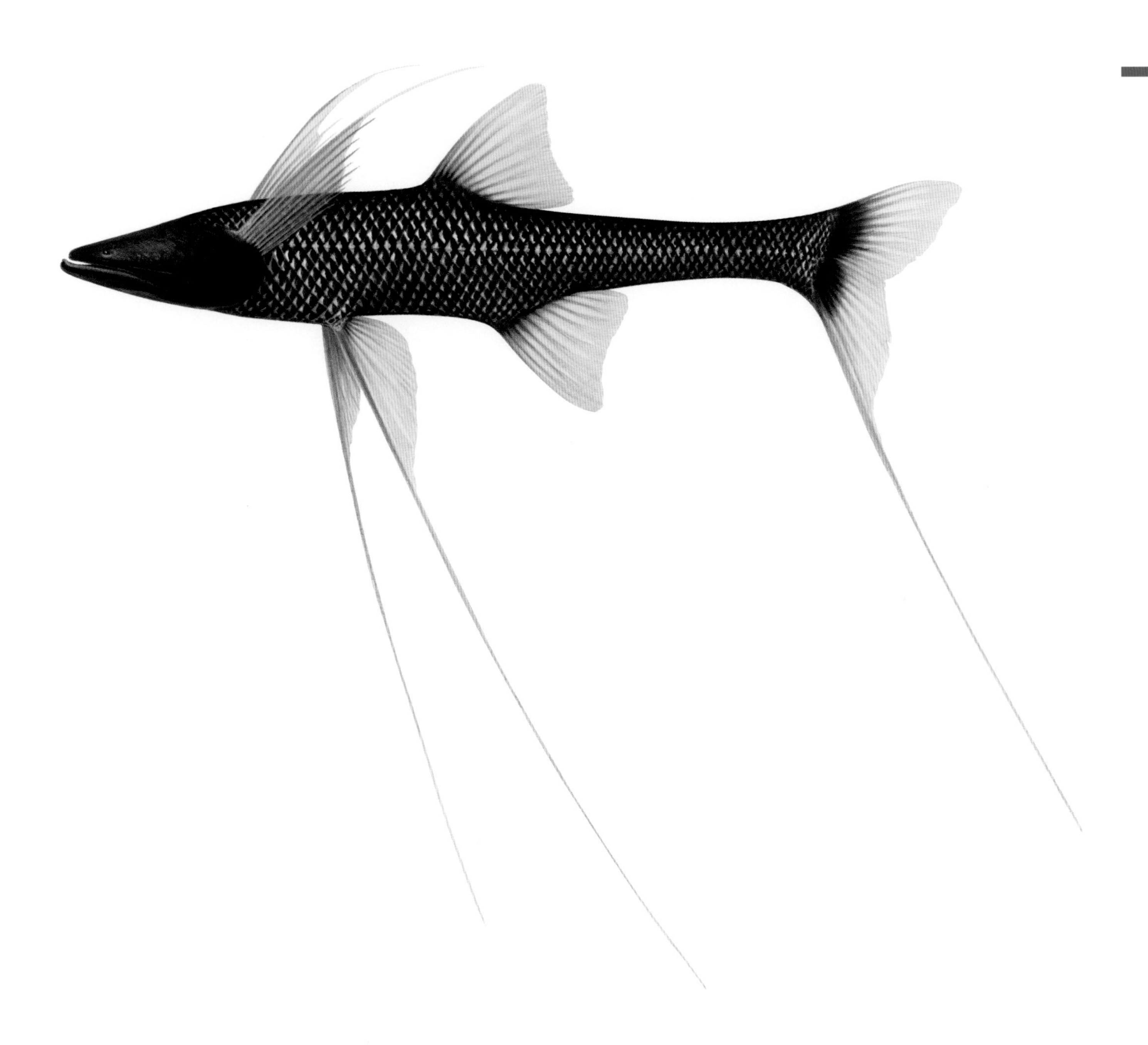

TRIPOD SPIDERFISH
Bathypterois grallator

Ipnopidae

Tripodfishes are found in deep water to more than 5000 m, in the Atlantic, Indian and Pacific oceans. There are about 30 species in the family. They have an elongate, cylindrical body that is covered with fragile scales, a dorsal fin at about mid-body, and a small adipose fin. Their large jaws reach well past the eyes, and they have greatly elongated rays in some of their fins.

These remarkable fishes have minute eyes that in some species are reduced to small plate-like organs that lack lenses. At great depth there is very little water movement and most deep-water fishes have evolved body forms that can move through the water smoothly, causing minimum turbulence, to avoid detection. The Deepsea Tripodfishes have sacrificed vision for the extremely sensitive motion-detection system provided by their elongate fin rays.

In the **Tripod Spiderfish** the long ventral and caudal fin rays are used as a tripod to support the fish on the substrate, and the greatly elongated pectoral fin rays are extended to sense the least vibration in the water. Perched on their tripod of fin rays, they wait for passing prey such as pelagic crustaceans, squid and small fishes. The fin rays appear to be flexible during swimming and rigid when used as supports. Photographs of tracks in the substrate indicate that Tripodfishes can 'walk' on their fin rays. Deepsea Tripodfishes reach a maximum length of about 35 cm. They are rarely encountered due to the great depths at which they live.

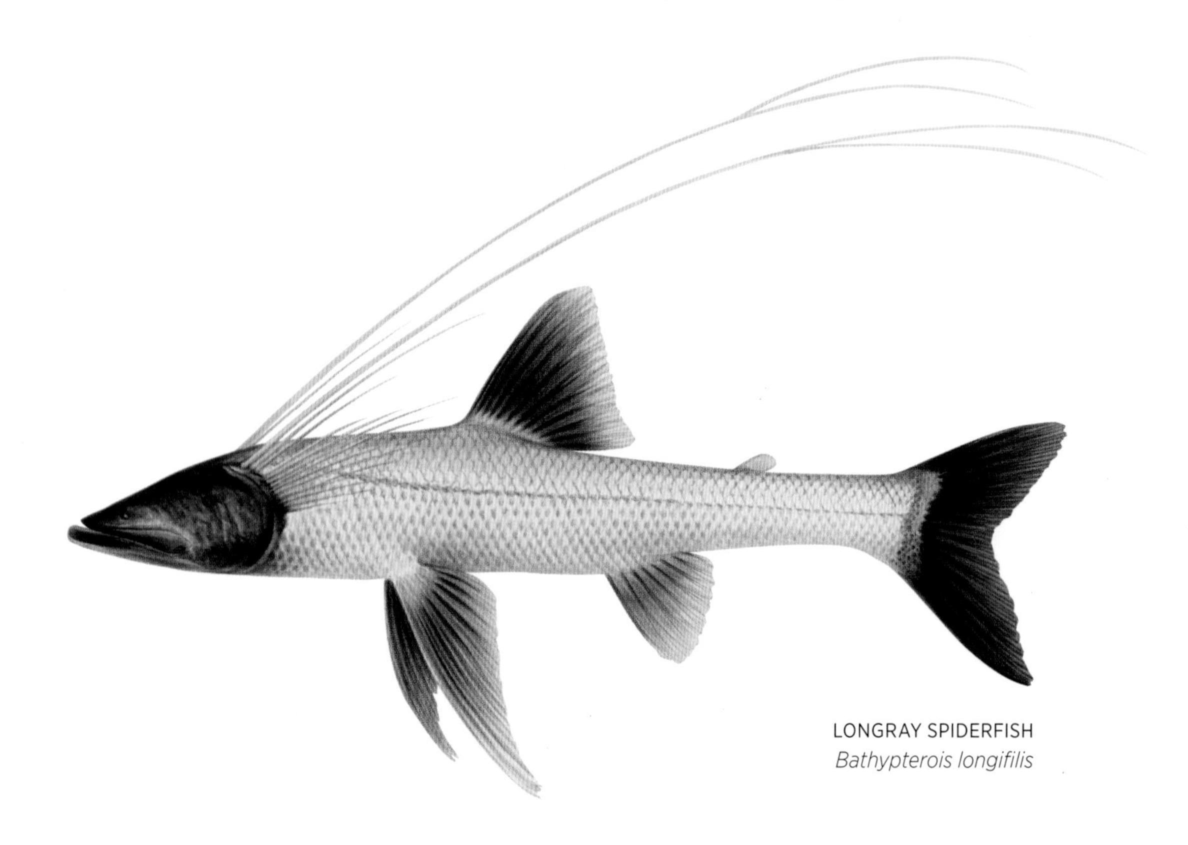

LONGRAY SPIDERFISH
Bathypterois longifilis

AUSTRALIAN IPNOPIDAE SPECIES

Bathypterois grallator Tripod Spiderfish
Bathypterois guentheri Tribute Spiderfish
Bathypterois longifilis Longray Spiderfish
Bathypterois longipes Abyssal Spiderfish
Bathypterois ventralis Ventrad Spiderfish
Bathytyphlops marionae Marion's Spiderfish
Ipnops agassizii Grideye Spiderfish

PEARLEYES

Scopelarchidae

SHORTFIN PEARLEYE
Scopelarchus analis

Pearleyes are found worldwide in all oceans except the Arctic, and there are about 17 species in the family. They grow to 15–23 cm in length and have a compressed body, and large, tubular eyes that are directed forward and upward. Each eye has a supplementary lens structure in the side of the tube, hence the name Pearleye. This structure may serve to increase their field of vision, as tubular eyes give increased sensitivity but a narrow angle of vision. Some species also have a yellow-tinted lens in the upper part of the eye, which may improve their ability to detect the bioluminescent colours produced by their prey.

Pearleyes have smooth, fragile scales, a short-based spineless dorsal fin, and a small adipose fin. They have remarkable dentition, with multiple rows of teeth on the jaws – including curved, barbed canines that can hinge inward, and shorter rearward-sloping teeth – as well as strong hooked teeth on the tongue. Adult Pearleyes occur at depths from 500 to 1000 m, where they feed on planktivorous fishes such as Bristlemouths (Gonostomatidae, p. 218) and Lanternfishes (Myctophidae, p. 251). Juvenile Pearleyes are pelagic and found in the upper 100 m of the water column, where they are thought to feed on zooplankton.

CHILDISH PEARLEYE
Benthalbella infans

AUSTRALIAN SCOPELARCHIDAE SPECIES

Benthalbella infans Childish Pearleye
Rosenblattichthys alatus Winged Pearleye
Scopelarchoides danae Dana Pearleye
Scopelarchus analis Shortfin Pearleye
Scopelarchus guentheri Staring Pearleye

SABRETOOTHS

Evermannellidae

BALBO SABRETOOTH
Evermannella balbo

Sabretooths are found in the Atlantic, Indian and Pacific oceans, in tropical and subtropical waters. The family contains about seven species. They are pelagic in midwater, usually from 500 to 1000 m, and reach a maximum length of about 18 cm. Similar to the Pearleyes (Scopelarchidae, p. 241) in general shape and habits, their body is more compressed and the tail is almost ribbon-like, with a prominent adipose fin. Some have tubular eyes (though they are not as pronounced as in the Pearleyes) and an adipose eyelid. Sabretooths have a blunt snout, a pair of enormous fangs in the front of the roof of the mouth, and more fangs interspersed with smaller teeth along the jaws (but not on the tongue). They have no scales on the body and the musculature of the tail is divided into three distinct bands. Sabretooths feed mainly on other fishes, but some also prey on cephalopods. They are strong swimmers and their large jaws and expandable stomachs enable them to swallow large prey.

AUSTRALIAN EVERMANNELLIDAE SPECIES
Coccorella atlantica Atlantic Sabretooth
Evermannella balbo Balbo Sabretooth
Evermannella indica Indian Sabretooth

LANCETFISHES AND HAMMERJAW

Alepisauridae

HAMMERJAW
Omosudis lowii

The three known species in this family are found in the Atlantic, Indian and Pacific oceans. They are pelagic, from surface waters down to 1000 m and occasionally as deep as 1800 m. They have an elongate, very compressed body with no scales, a fleshy keel on the midline of the caudal peduncle, large eyes that are not tubular, and formidable fangs on the roof of the mouth and in the jaws.

Lancetfishes are among the largest pelagic predators in deep water – the largest member, the **Longnose Lancetfish**, can reach up to 2 m in length. They have a long, sail-like dorsal fin and are voracious predators of fishes, cephalopods and crustaceans. They are frequently caught on baited Tuna longlines set in deep water. They occasionally approach inshore waters and have even been taken by anglers trolling lures.

The **Hammerjaw** is much smaller, growing to about 20 cm in length, and is found worldwide. It feeds mainly on squid and is capable of swallowing prey larger than itself. The lower jaw is massive and very deep, and both the jaws and the roof of the mouth are armed with long fangs of varying sizes. Large adult Hammerjaws are rarely taken in trawls, perhaps because they are fast enough swimmers to evade nets. Like many predators of midwater fishes, their belly cavity has a black lining – an adaptation to conceal the bioluminescence of swallowed prey.

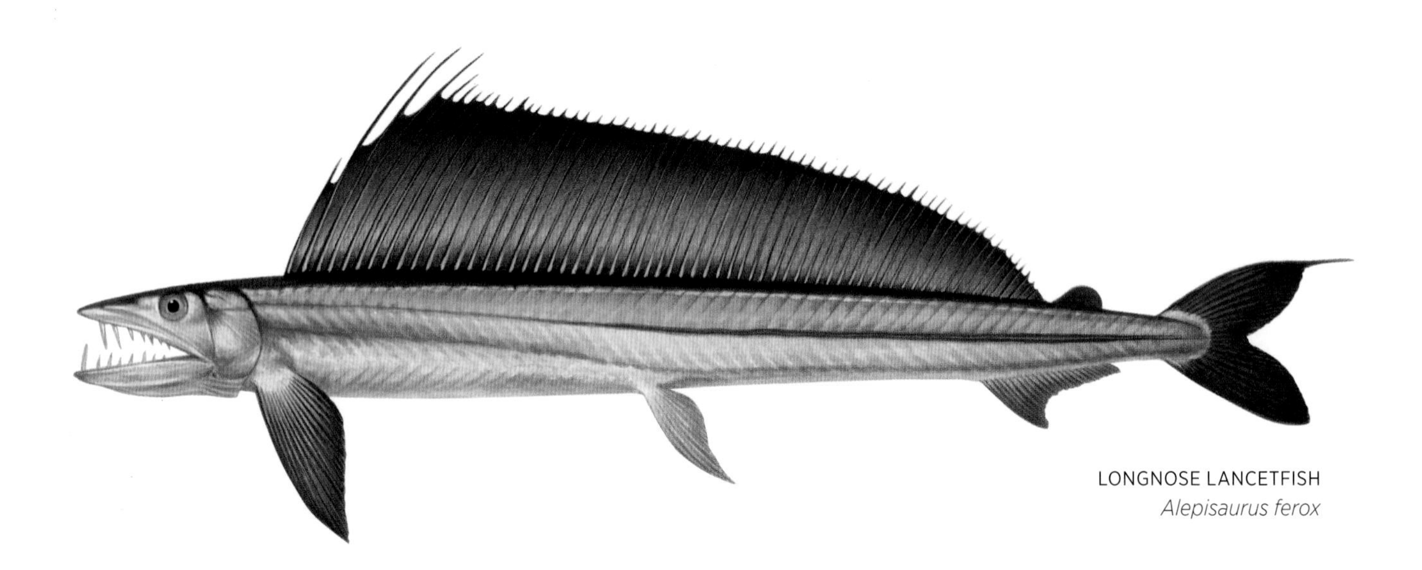

LONGNOSE LANCETFISH
Alepisaurus ferox

AUSTRALIAN ALEPISAURIDAE SPECIES

Omosudis lowii Hammerjaw
Alepisaurus brevirostris Shortnose Lancetfish
Alepisaurus ferox Longnose Lancetfish

BARRACUDINAS

Paralepididae

SOUTHERN DAGGERTOOTH
Anotopterus vorax

Barracudinas are abundant in all the oceans of the world, including in Arctic and Antarctic waters. The 50 or more species in the family are pelagic in midwater, occurring from near the surface to depths of about 2000 m. Despite bearing a superficial resemblance to the Barracudas (Sphyraenidae, p. 711), they are not closely related. Growing to a maximum length of 20–50 cm, they have a very elongate, compressed, translucent body. Scales are usually absent except for a series of modified scales along the lateral line, but a few species do have silvery, fragile scales. They have large eyes, a long pointed snout and a slightly projecting lower jaw with large, fang-like teeth, some of which can hinge inward and others of which are rigid. The dorsal fin is usually at mid-body, and a dorsal adipose fin is positioned towards the rear. Some species also possess a ventral adipose fin anterior to the anal fin.

Barracudinas prey on small fishes and in turn provide an important food source for large pelagic predators such as Tunas (Scombridae, p. 717). The **Southern Daggertooth** grows up to 1 m long. It has a very large head that can make up a quarter of the body length, with enormous fangs in the jaws. It has a fleshy keel on the rear part of the body, no first dorsal fin, and a well-developed adipose fin. Adult Southern Daggertooths inhabit the Southern Ocean, mainly preying on other Barracudinas. Prior to spawning they shed their teeth and reduce the size of their digestive tract. They travel northward to spawn in tropical and subtropical waters, and die after spawning.

ATLANTIC BARRACUDINA
Lestidium atlanticum

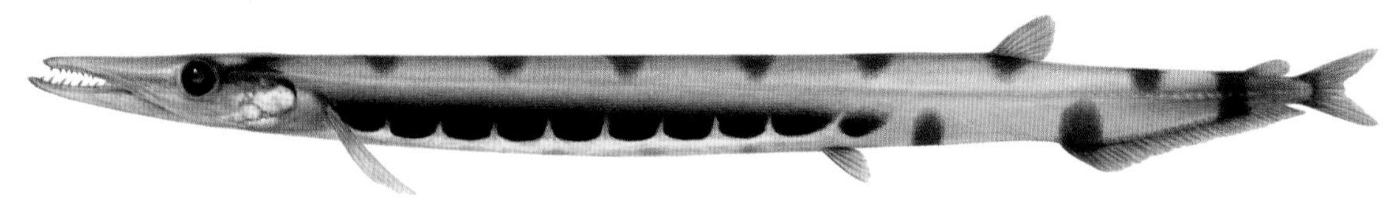

ROTHSCHILD'S BARRACUDINA
Stemonosudis rothschildi

AUSTRALIAN PARALEPIDIDAE SPECIES

Anotopterus vorax Southern Daggertooth
Arctozenus risso Ribbon Barracudina
Lestidiops indopacifica Indo-Pacific Barracudina
Lestidiops jayakari Pacific Barracudina
Lestidium atlanticum Atlantic Barracudina
Lestidium nudum Naked Barracudina
Lestrolepis japonica Japanese Barracudina
Macroparalepis macrogeneion Longfin Barracudina
Magnisudis atlantica Duckbill Barracudina
Stemonosudis elegans Tailspot Barracudina
Stemonosudis macrura Sharpchin Barracudina
Stemonosudis rothschildi Rothschild's Barracudina

DEEPSEA LIZARDFISHES

Bathysauridae

DEEPSEA LIZARDFISH
Bathysaurus ferox

The two species of Deepsea Lizardfish are found worldwide, in deep water from 600 to 5000 m. They have a somewhat flattened head and very large jaws extending well behind the small eyes, with the lower jaw projecting beyond the upper. There are numerous curved teeth on the outside of the jaws and the anterior roof of the mouth, and these can hinge inward. They have a long-based dorsal fin, no adipose fin, and scales on the body that extend onto the caudal fin.

Deepsea Lizardfishes grow to about 80 cm in length. They are bottom-dwelling and prey on fishes and crustaceans. Like their shallow-water relatives (Synodontidae, p. 232), they probably rest immobile on the substrate and ambush passing prey. The single Australian species is rarely seen but has occasionally been trawled at great depth off the southern coast.

AUSTRALIAN BATHYSAURIDAE SPECIES
Bathysaurus ferox Deepsea Lizardfish

TELESCOPEFISHES

Giganturidae

CHUN'S TELESCOPEFISH
Gigantura chuni

The two known species of Telescopefish are pelagic in mid-water from around 500 to 1000 m, and although quite rare, they are distributed worldwide. They have large, tubular eyes that are directed forward. Reaching a maximum length of about 23 cm, they have a slender body with no scales and loose silvery skin. An enormous mouth reaches to below the pectoral fin, and the jaws contain numerous long, curved teeth that can hinge inward. The pectoral fins are unusual in that they are horizontal, positioned high on the body, and quite large and wing-like. The caudal fin often has a greatly elongated lower lobe. Adults lose many of the features present in juveniles, such as the ventral fins and their supporting bones, the dorsal adipose fin, the branchiostegal rays and the gillrakers. The differences are so remarkable that juveniles were originally thought to constitute a different family. Adults feed on fishes and squid, and have a greatly distensible stomach that allows them to ingest prey items that are larger than themselves.

AUSTRALIAN GIGANTURIDAE SPECIES

Gigantura chuni Chun's Telescopefish
Gigantura indica Indian Telescopefish

MYCTOPHIFORMES

Myctophiforms are small deepwater fishes with an elongate, compressed body, and fragile scales and skin. They have a single dorsal fin and an adipose fin, ventral fins at about mid-body (usually with eight rays), and a large mouth and eyes. Almost all have photophores arranged in a series of lines and small groups on the body. They are extremely abundant, and of the orders found in the upper layers of the open ocean, the Myctophiformes contains more species than any other. There are two families in the order, both of which occur in Australian waters.

BLACKCHINS

Neoscopelidae

LARGESCALE NEOSCOPELID
Neoscopelus macrolepidotus

The six species of Blackchin are found in mid to deep waters of the Atlantic, Indian and Pacific oceans. They have a moderately compressed body with fragile, smooth scales, a spineless dorsal fin directly above the ventral fins at about mid-body, and a small adipose fin. The mouth reaches to the rear edge of the eyes and the jaws are lined with an outer series of small villiform teeth and an inner series of larger teeth that can hinge inward.

Some species of Blackchin lack photophores and a swim bladder, but the Australian species all have a swim bladder, and have photophores arranged in series along the body and inside the mouth on the edge of the tongue, though not on the head. Blackchins are found at depths from 700 to 2000 m. They are usually pelagic in bottom waters on the continental slope, feeding on crustaceans such as euphausiid shrimps and other zooplankton. They grow to a maximum length of about 30 cm and do not appear to carry out nightly vertical migrations to feed near the surface.

AUSTRALIAN NEOSCOPELIDAE SPECIES

Neoscopelus macrolepidotus Largescale Neoscopelid
Neoscopelus microchir Shortfin Neoscopelid
Neoscopelus porosus Spangleside Neoscopelid

LANTERNFISHES

Myctophidae

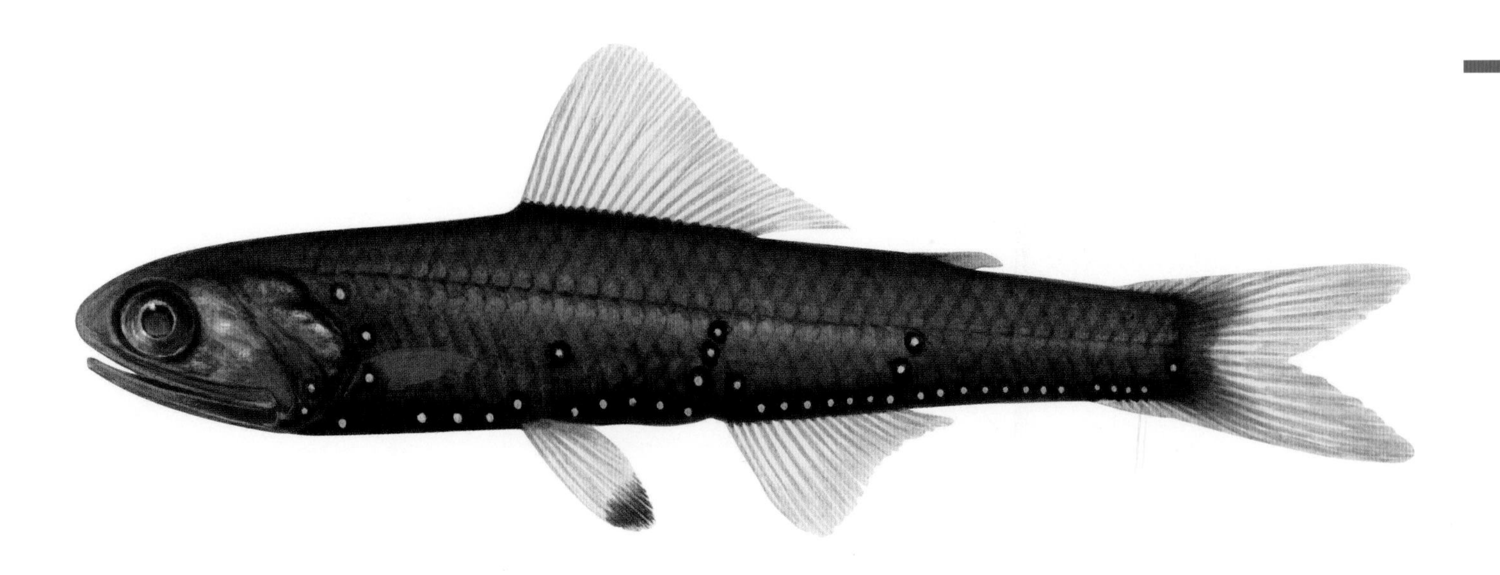

SOUTHERN BLACKTIP LANTERNFISH
Gymnoscopelus piabilis

This is one of the largest families of fishes, with 33 genera and some 250 species found worldwide. They are extremely abundant and make up a major component of the mid- to deepwater pelagic fauna. Most species are only around 10 cm in length, with the largest reaching about 30 cm. They have a moderately compressed body with smooth, fragile scales, large eyes, and a short, rounded snout that overhangs the mouth in many species. The large mouth reaches well behind the eyes, and there are numerous small teeth on the jaws and roof of the mouth. They have a spineless dorsal fin positioned over the ventral fins, a small adipose fin and a swim bladder. Photophores on the body and head are in series and small groups, and usually differ slightly between males and females. The arrangement of photophores is the main characteristic used to differentiate the numerous species. Many also have patches of luminous tissue and some luminous scales.

Lanternfishes are usually found at depths of 300–1200 m by day and most, though not all, make nightly migrations towards the surface to feed on plankton in the upper layers of the ocean. Most species do not venture into depths less than 100 m, but some reach surface waters if conditions are favourable. Such migrations may take up to an hour and cover as much as 1000 m in each direction, exposing the fishes to a wide range of temperatures and pressure gradients. It is not fully understood how Lanternfishes, with gas-filled swim bladders, are capable of withstanding these dramatic changes in pressure. Some species have developed a swim bladder filled with low-density fat to counteract the problem of expansion and contraction, while others have developed extremely efficient gas production and venting capabilities.

Lanternfishes are an important part of the food chain in the open ocean, feeding on a variety of zooplankton in the nutrient-rich upper layers of the water, then transporting this energy into deeper waters as they descend at night. They are preyed on by a wide range of pelagic fishes, as well as mammals such as whales and seals, and constitute the largest proportion, by biomass, of deepwater pelagic fishes.

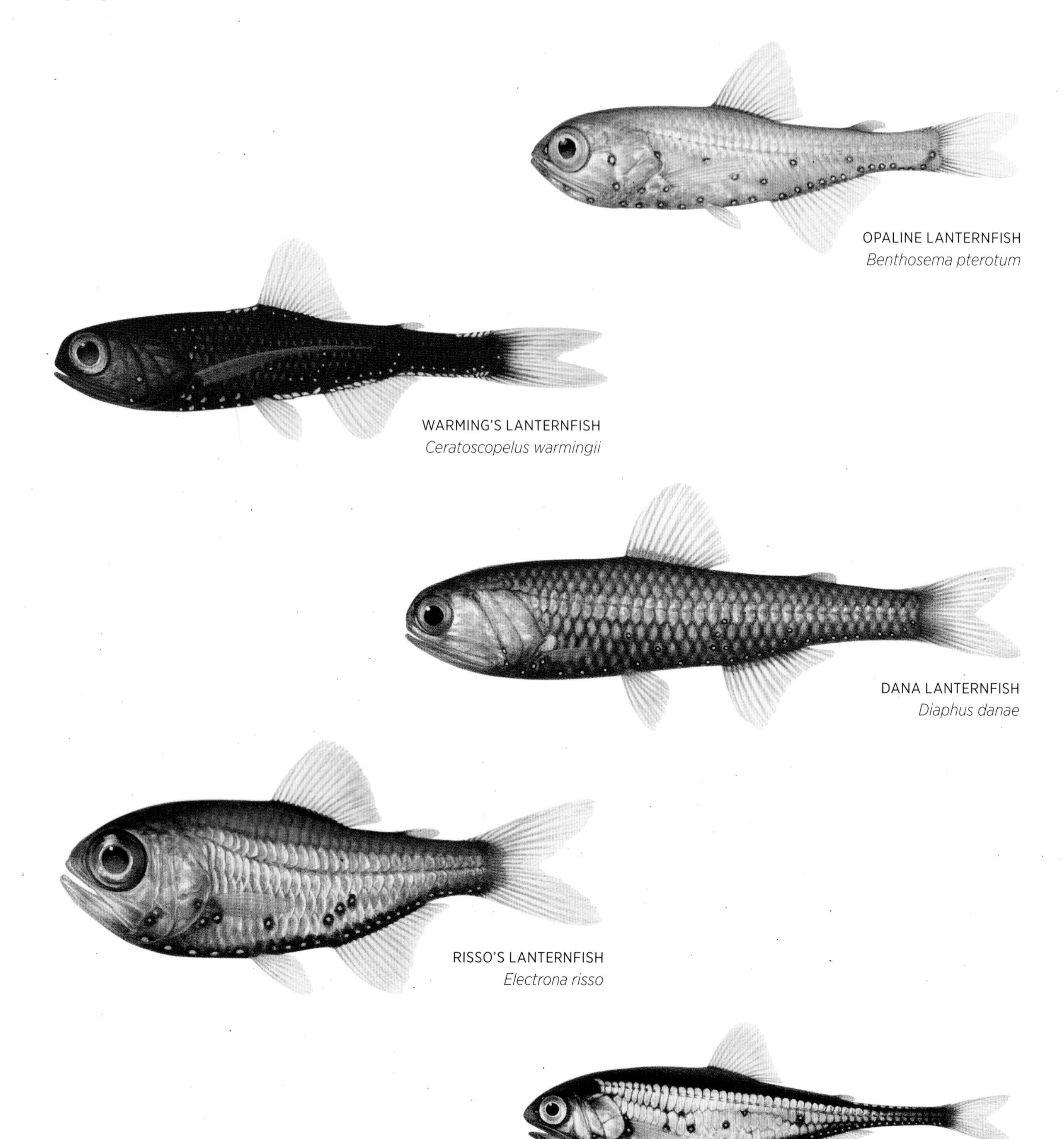

OPALINE LANTERNFISH
Benthosema pterotum

WARMING'S LANTERNFISH
Ceratoscopelus warmingii

DANA LANTERNFISH
Diaphus danae

RISSO'S LANTERNFISH
Electrona risso

BARNES' LANTERNFISH
Gonichthys barnesi

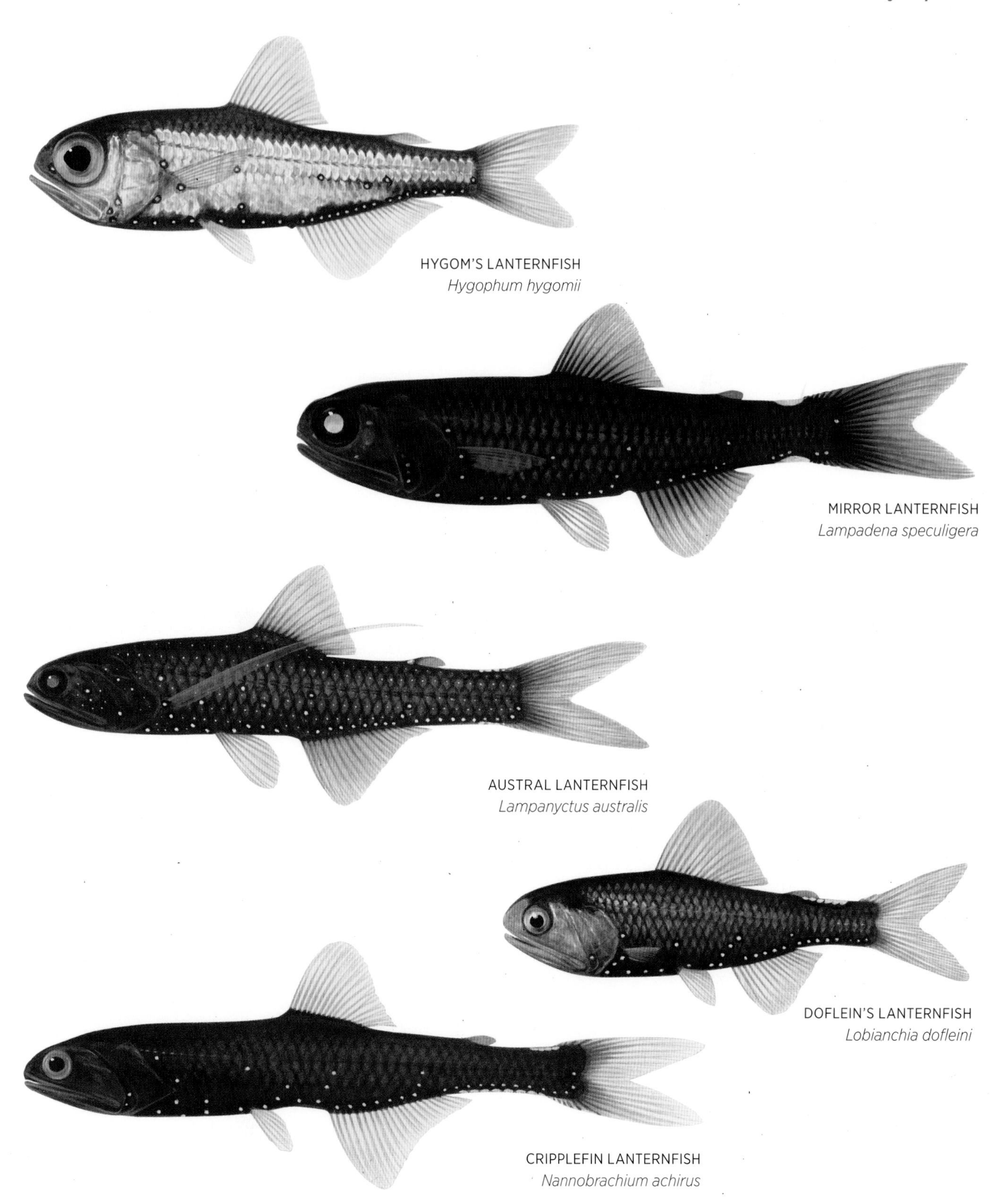

HYGOM'S LANTERNFISH
Hygophum hygomii

MIRROR LANTERNFISH
Lampadena speculigera

AUSTRAL LANTERNFISH
Lampanyctus australis

DOFLEIN'S LANTERNFISH
Lobianchia dofleini

CRIPPLEFIN LANTERNFISH
Nannobrachium achirus

AUSTRALIAN MYCTOPHIDAE SPECIES

Benthosema fibulatum Spinycheek Lanternfish
Benthosema pterotum Opaline Lanternfish
Benthosema suborbitale Dimple Lanternfish
Bolinichthys indicus Smoothcheek Lanternfish
Bolinichthys longipes Popeye Lanternfish
Bolinichthys nikolayi Nikolay's Lanternfish
Bolinichthys photothorax Spurcheek Lanternfish
Bolinichthys pyrsobolus Fiery Lanternfish
Bolinichthys supralateralis Stubby Lanternfish
Centrobranchus andreae Andrea's Lanternfish
Centrobranchus nigroocellatus Roundnose Lanternfish
Ceratoscopelus warmingii Warming's Lanternfish
Diaphus anderseni Andersen's Lanternfish
Diaphus bertelseni Bertelsen's Lanternfish
Diaphus brachycephalus Shorthead Lanternfish
Diaphus chrysorhynchus Goldnose Lanternfish
Diaphus coeruleus Blue Lanternfish
Diaphus danae Dana Lanternfish
Diaphus diadematus Crown Lanternfish
Diaphus drachmanni Drachmann's Lanternfish
Diaphus effulgens Headlight Lanternfish
Diaphus fragilis Fragile Lanternfish
Diaphus garmani Garman's Lanternfish
Diaphus hudsoni Hudson's Lanternfish
Diaphus imposter Imposter Lanternfish
Diaphus jenseni Jensen's Lanternfish
Diaphus kapalae Kapala Lanternfish
Diaphus lucidus Spotlight Lanternfish
Diaphus luetkeni Luetken's Lanternfish
Diaphus malayanus Malayan Lanternfish
Diaphus meadi Mead's Lanternfish
Diaphus metopoclampus Bluntnose Lanternfish
Diaphus mollis Soft Lanternfish
Diaphus ostenfeldi Ostenfeld's Lanternfish
Diaphus parri Parr's Lanternfish
Diaphus perspicillatus Flatface Lanternfish
Diaphus phillipsi Phillips' Lanternfish
Diaphus problematicus Problematic Lanternfish
Diaphus regani Regan's Lanternfish
Diaphus splendidus Horned Lanternfish
Diaphus termophilus Warmwater Lanternfish
Diaphus thiollierei Thiolliere's Lanternfish
Diaphus watasei Watase's Lanternfish
Diogenichthys atlanticus Atlantic Lanternfish
Electrona carlsbergi Carlsberg's Lanternfish
Electrona paucirastra Belted Lanternfish
Electrona risso Risso's Lanternfish
Electrona subaspera Rough Lanternfish
Gonichthys barnesi Barnes' Lanternfish
Gymnoscopelus hintonoides False-midas Lanternfish
Gymnoscopelus microlampas Minispotted Lanternfish
Gymnoscopelus piabilis Southern Blacktip Lanternfish
Hygophum hanseni Hansen's Lanternfish
Hygophum hygomii Hygom's Lanternfish
Hygophum macrochir Largefin Lanternfish
Hygophum proximum Firefly Lanternfish
Hygophum reinhardtii Reinhardt's Lanternfish
Lampadena chavesi Chaves' Lanternfish
Lampadena luminosa Luminous Lanternfish
Lampadena notialis Notal Lanternfish
Lampadena speculigera Mirror Lanternfish
Lampadena urophaos Tail-light Lanternfish
Lampanyctodes hectoris Hector's Lanternfish
Lampanyctus alatus Winged Lanternfish
Lampanyctus australis Austral Lanternfish
Lampanyctus festivus Festive Lanternfish
Lampanyctus intricarius Intricate Lanternfish
Lampanyctus lepidolychnus Mermaid Lanternfish
Lampanyctus macdonaldi MacDonald's Lanternfish
Lampanyctus nobilis Noble Lanternfish
Lampanyctus pusillus Pygmy Lanternfish
Lampichthys procerus Blackhead Lanternfish
Lobianchia dofleini Doflein's Lanternfish
Lobianchia gemellarii Gemellar's Lanternfish
Loweina rara Rare Lanternfish
Metelectrona herwigi Herwig Lanternfish
Metelectrona ventralis Flaccid Lanternfish
Myctophum asperum Prickly Lanternfish
Myctophum brachygnathum Shortjaw Lanternfish
Myctophum nitidulum Pearly-spotted Lanternfish
Myctophum obtusirostre Bluntsnout Lanternfish
Myctophum orientale Oriental Lanternfish
Myctophum phengodes Bright Lanternfish
Myctophum selenops Lunar Lanternfish
Myctophum spinosum Spiny Lanternfish
Nannobrachium achirus Cripplefin Lanternfish
Nannobrachium atrum Dusky Lanternfish
Notolychnus valdiviae Topside Lanternfish
Notoscopelus caudispinosus Spinetail Lanternfish
Notoscopelus resplendens Patchwork Lanternfish
Protomyctophum normani Norman's Lanternfish
Protomyctophum parallelum Parallel Lanternfish
Protomyctophum subparallelum Subparallel Lanternfish
Scopelopsis multipunctatus Multispot Lanternfish
Symbolophorus barnardi Barnard's Lanternfish
Symbolophorus boops Spotfin Lanternfish
Symbolophorus evermanni Evermann's Lanternfish
Taaningichthys bathyphilus Deepwater Lanternfish
Triphoturus nigrescens Vagabond Lanternfish

LAMPRIDIFORMES

Lampridiforms can be divided into two main physical types: the Veliferidae and Lampridae are deep-bodied, with a symmetrical caudal fin and strong skeleton; the remaining families are long and ribbon-like, with an asymmetrical caudal fin and weak skeleton. All Lampridiforms have a unique highly protrusible upper jaw for feeding on planktonic prey. They are mainly pelagic fishes, occurring in the upper layers of the open ocean. There are seven families in the order, five of which have been recorded in Australian waters. The remaining two families are distributed worldwide and may also occur here.

VEILFINS

Veliferidae

HIGHFIN VEILFIN
Velifer hypselopterus

The two species of Veilfin are uncommon but widespread, occurring in tropical and temperate waters of the Indian and West Pacific oceans. Both are found in Australian waters, from depths of 30 to 250 m on the continental shelf. Veilfins have a deep, compressed body with tall, sail-like dorsal and anal fins, smooth, fragile scales and a small, highly protrusible mouth without any teeth. They reach about 50 cm in length and are usually found close to the substrate. Virtually nothing is known of their biology.

COMMON VEILFIN
Metavelifer multiradiatus

AUSTRALIAN VELIFERIDAE SPECIES
Metavelifer multiradiatus Common Veilfin
Velifer hypselopterus Highfin Veilfin

OPAHS

Lampridae

OPAH
Lampris guttatus

This family contains two species, both of which are found in Australian waters. Opahs are large, deep-bodied, moderately compressed fishes with minute scales and a small, protrusible mouth without teeth. They are pelagic in the upper layers of the ocean and are usually found well offshore.

The **Opah** is distributed worldwide, except in the polar regions. It is regularly taken by Tuna longline fisheries and is targeted by deepwater anglers off Hawaii. The Opah feeds mainly on squid, but also consumes jellyfishes, pelagic crustaceans and small fishes. It swims by flapping its strong, curved pectoral fins – it has an enormous pectoral muscle for this task, which reaches from the pectoral-fin base, high on the body, down to the ventral profile. The Opah grows to nearly 2 m in length, weighs around 270 kg and is a superlative food fish, with the pectoral muscle being especially prized. The **Southern Moonfish** is found in temperate waters of the southern hemisphere. It is very rarely seen, but is thought to have similar habits to the Opah. It grows to a maximum length of about 1.1 m.

AUSTRALIAN LAMPRIDAE SPECIES

Lampris guttatus Opah
Lampris immaculatus Southern Moonfish

CRESTFISHES

Lophotidae

UNICORN CRESTFISH
Eumecichthys fiski

Crestfishes are found worldwide and the three species in the family are pelagic in mid to deep waters. They have an elongate, compressed body with minute, fragile scales, and a small, protrusible mouth. The forehead is either vertical or projects forward. The dorsal fin is long-based, often with elongate anterior rays, and the anal fin is very small and close to the caudal fin.

Crestfishes have an internal ink sac that can be discharged via the anus, presumably as a defensive function, as in squid and other cephalopods. Crestfishes are rarely encountered, but are occasionally caught on longlines or found washed ashore. The **Unicorn Crestfish** grows to about 1.5 m in length and has been recorded at depths to 1000 m. The **Crested Bandfish** grows to about 2 m in length and is found from the surface to 1000 m or more. Crestfishes feed mainly on squid and small fishes, and probably follow the vertical migrations of their prey towards the surface at night.

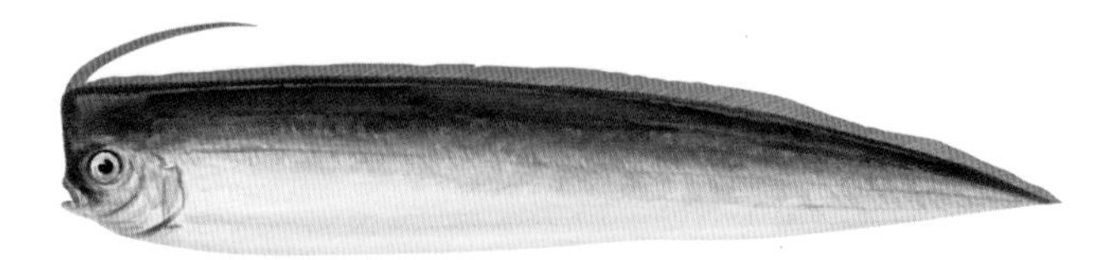

CRESTED BANDFISH
Lophotus lacepede

AUSTRALIAN LOPHOTIDAE SPECIES

Eumecichthys fiski Unicorn Crestfish
Lophotus lacepede Crested Bandfish

RIBBONFISHES

Trachipteridae

SCALLOPED RIBBONFISH (adult)
Zu cristatus

This family is distributed worldwide and contains about 10 species of midwater pelagic fishes. Reaching a maximum length of about 1.7 m, they have a very elongate, compressed, silvery body, which tapers from the large head to the very thin caudal peduncle. Ribbonfishes have no scales and numerous small, bony or cartilaginous tubercles interspersed with pores on the sides of the body. They have large eyes and a highly protrusible mouth. The dorsal fin is long, and in juveniles the anterior portion of the fin forms a high crest.

Ribbonfishes have no ribs and no anal fin, and the caudal fin is directed vertically at a steep angle to the body's horizontal axis. Their body changes quite remarkably with growth, particularly the fins. Most species lose the high crest of the dorsal fin and the upturned caudal fin by the time they reach maturity, and in the **Spotted Ribbonfish** the ventral fins are lost during development. Ribbonfishes are found in the upper layers of the ocean to depths of about 1000 m and they feed on squid, fishes, pelagic crustaceans and worms. Although rarely encountered, they have been observed from submersibles, swimming at an angle of about 45 degrees to the horizontal by undulating the dorsal fin.

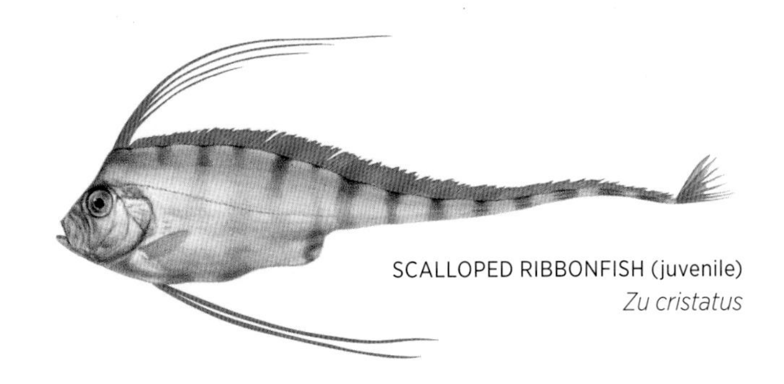

SCALLOPED RIBBONFISH (juvenile)
Zu cristatus

AUSTRALIAN TRACHIPTERIDAE SPECIES

Desmodema polystictum Spotted Ribbonfish
Trachipterus arawatae Southern Ribbonfish
Zu cristatus Scalloped Ribbonfish

OARFISHES

Regalecidae

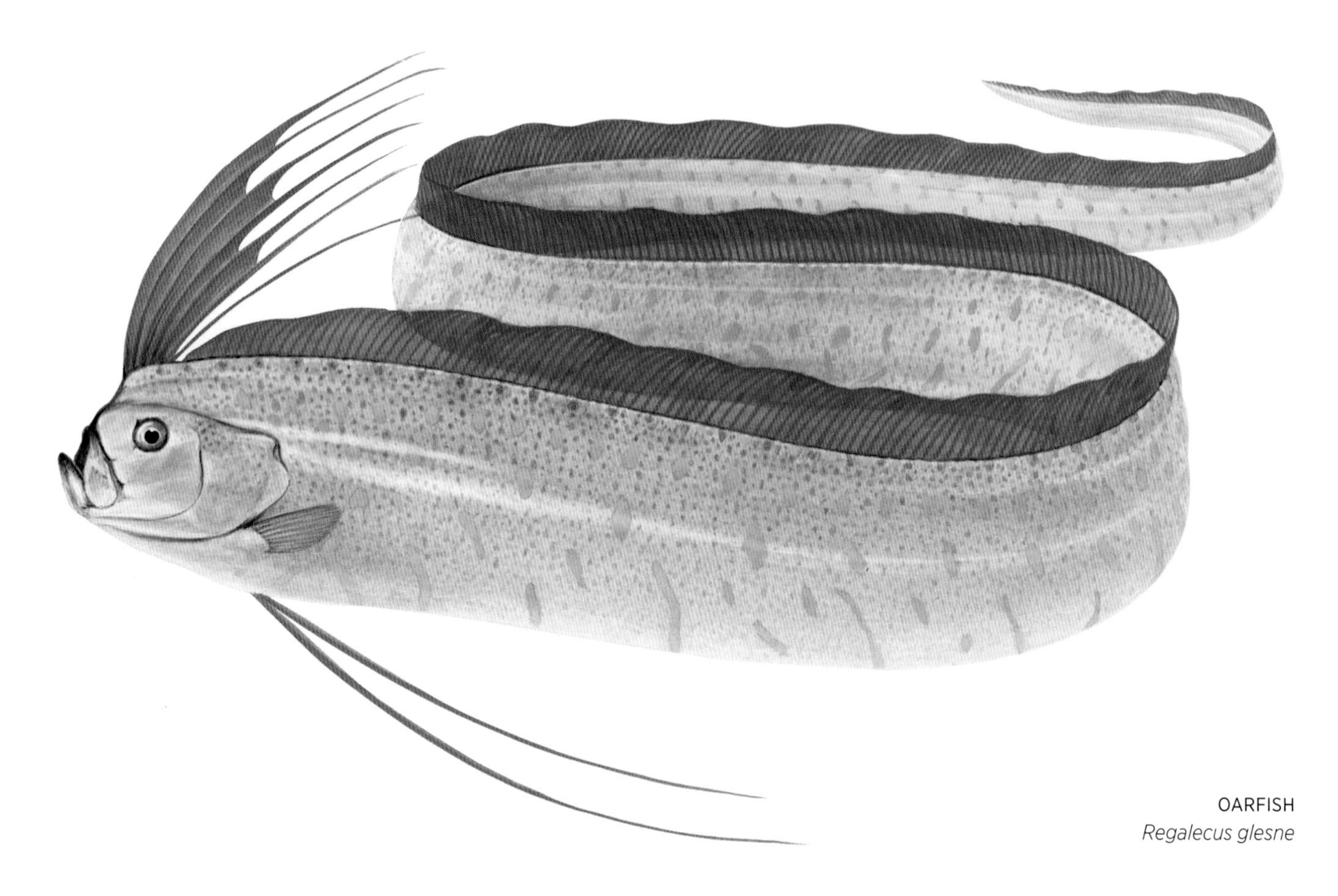

OARFISH
Regalecus glesne

The Oarfish family contains four species, which are pelagic in the upper levels of all the world's oceans. They have an extremely long, compressed, ribbon-like body, with no scales, and are covered with a fragile, silvery layer of pigment. The long dorsal fin has the anterior rays formed into a high crest. There is no anal fin and the caudal fin is reduced to a few trailing filaments or is lost altogether in adults. Each ventral fin consists of a single very long ray with small skin flaps along its length and at the tip. The highly protrusible mouth is vertically oriented and has no teeth.

These rare ocean wanderers are occasionally encountered at the surface and have been taken from as deep as 1000 m, but most records come from strandings after storms at sea. Oarfishes have been observed at depths of 50–200 m, swimming vertically using undulations of the dorsal fin. They feed on planktonic crustaceans, squid and small fishes. The **Oarfish** is the longest of all the bony fishes and can attain 11 m in length. **Benham's Streamerfish** reaches at least 3 m in length and is found in cooler surface waters in the southern hemisphere.

AUSTRALIAN REGALECIDAE SPECIES
Agrostichthys parkeri Benham's Streamerfish
Regalecus glesne Oarfish

POLYMIXIIFORMES

The single family in this order has uncertain relationships to other fishes and was formerly included in the order Beryciformes (p. 351). Polymixiiforms possess two sets of intermuscular bones, similar to those found in much more primitive fishes, but also multiple spines in the fins, as found in more advanced fishes. For other characteristics of the order, see the family Polymixiidae (p. 264).

BEARDFISHES

Polymixiidae

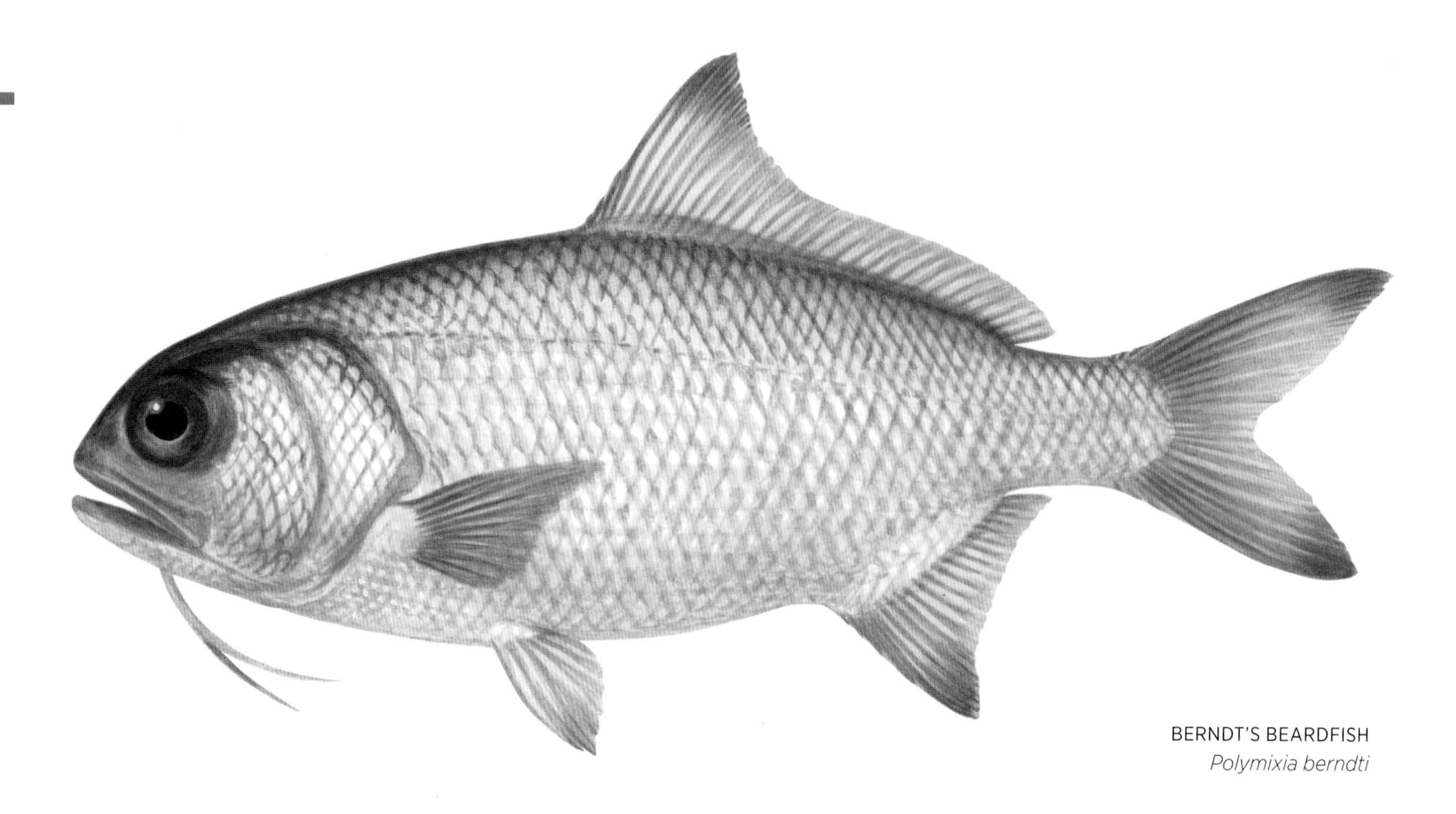

BERNDT'S BEARDFISH
Polymixia berndti

Beardfishes are found in tropical and subtropical waters of the Atlantic, Indian and West Pacific oceans. There are about 10 species in the family. Growing to about 40 cm in length, they have a moderately elongate, compressed body with spiny, silvery scales that extend onto the cheeks and lower jaw. They have a single long dorsal fin with 4–6 spines and an anal fin with four spines. There is a pair of long barbels beneath the chin, set well back from the tip of the lower jaw. The head is large and rounded, the eyes are large, and the slightly inferior mouth bears villiform teeth.

Beardfishes are predominantly bottom dwelling, occurring from depths of 20 to 600 m, usually over sandy or muddy substrates, and are quite common in some areas of the world. The two Australian species are found only in northern waters and feed on benthic crustaceans, invertebrates, squid and small fishes.

AUSTRALIAN POLYMIXIIDAE SPECIES
Polymixia berndti Berndt's Beardfish
Polymixia busakhini Busakhin's Beardfish

GADIFORMES

Gadiforms are bottom-dwelling fishes with elongate bodies that either taper to a point or to a small caudal fin. They have 1–3 dorsal fins and 1–2 anal fins, and the ventral fins are placed well forward on the body, directly beneath or anterior to the pectoral fins. Gadiforms have no true spines in the fins, though some have hardened rays. Many have sensory chin barbels, all have smooth or spiny cycloid scales, and most have a swim bladder. There are nine families in the order, seven of which are found in Australian waters.

CODLETS

Bregmacerotidae

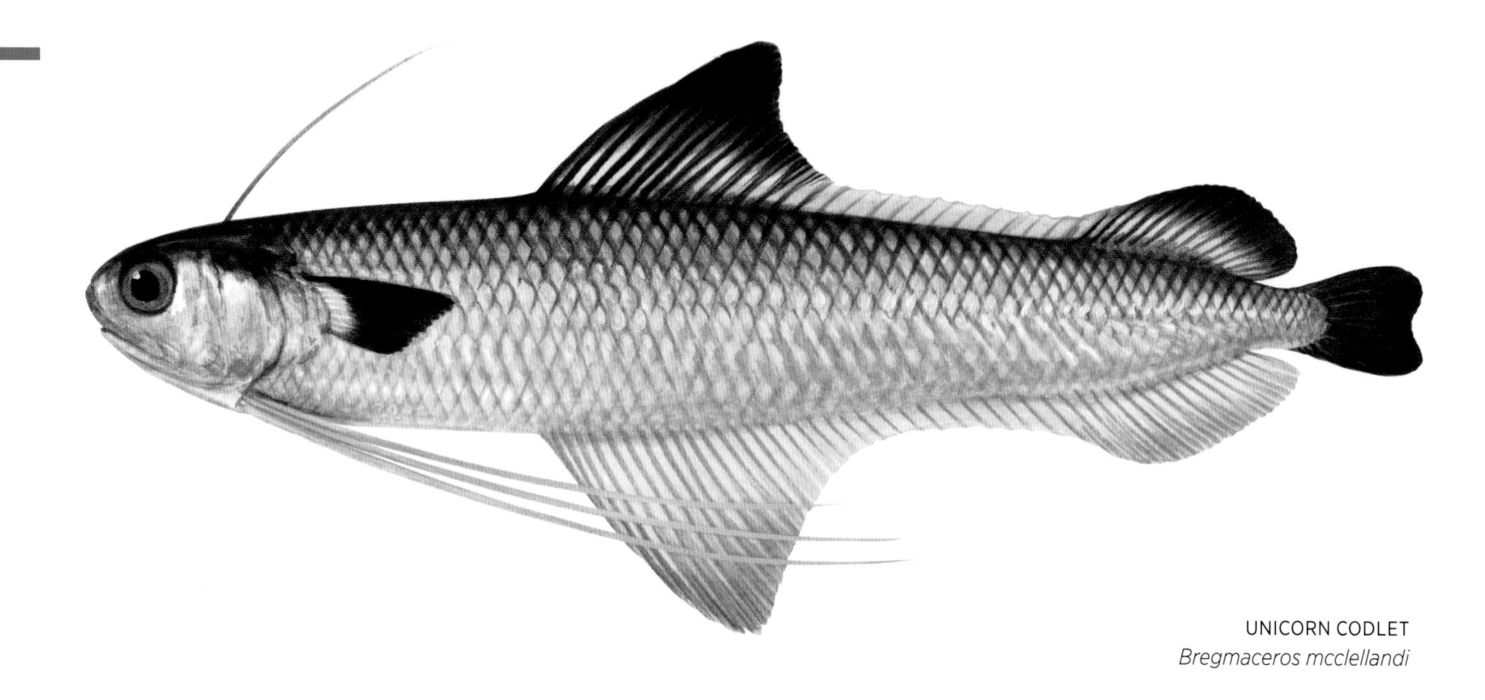

UNICORN CODLET
Bregmaceros mcclellandi

Codlets are found circumglobally in temperate and tropical waters. There are about 15 species in the family. These elongate fishes reach about 10 cm in length and have smooth, fragile scales and a small head. The jaws reach to behind the eyes and bear rows of villiform teeth, and such teeth are also present on the roof of the mouth. The first dorsal fin consists of a single long ray and is well separated from the second dorsal fin, which, like the anal fin, has a much lower central section. The ventral fins are attached under the throat, with the outer three rays formed into long trailing filaments.

Codlets are usually pelagic in open waters and occur in schools. They are found from the surface down to about 1000 m, but most occur in the upper 300 m of the water column and some are found in shallow coastal waters, even entering estuaries. They feed mainly on zooplankton and pelagic crustaceans. The **Unicorn Codlet** is found throughout South-East Asia and northern Australia, in shallow and brackish coastal waters as well as offshore to 2000 m. It is commercially fished in some areas.

AUSTRALIAN BREGMACEROTIDAE SPECIES

Bregmaceros atlanticus Antenna Codlet
Bregmaceros japonicus Japanese Codlet
Bregmaceros lanceolatus Lance Codlet
Bregmaceros mcclellandi Unicorn Codlet
Bregmaceros nectabanus Australian Codlet
Bregmaceros rarisquamosus Bigeye Codlet

EUCLA COD

Euclichthyidae

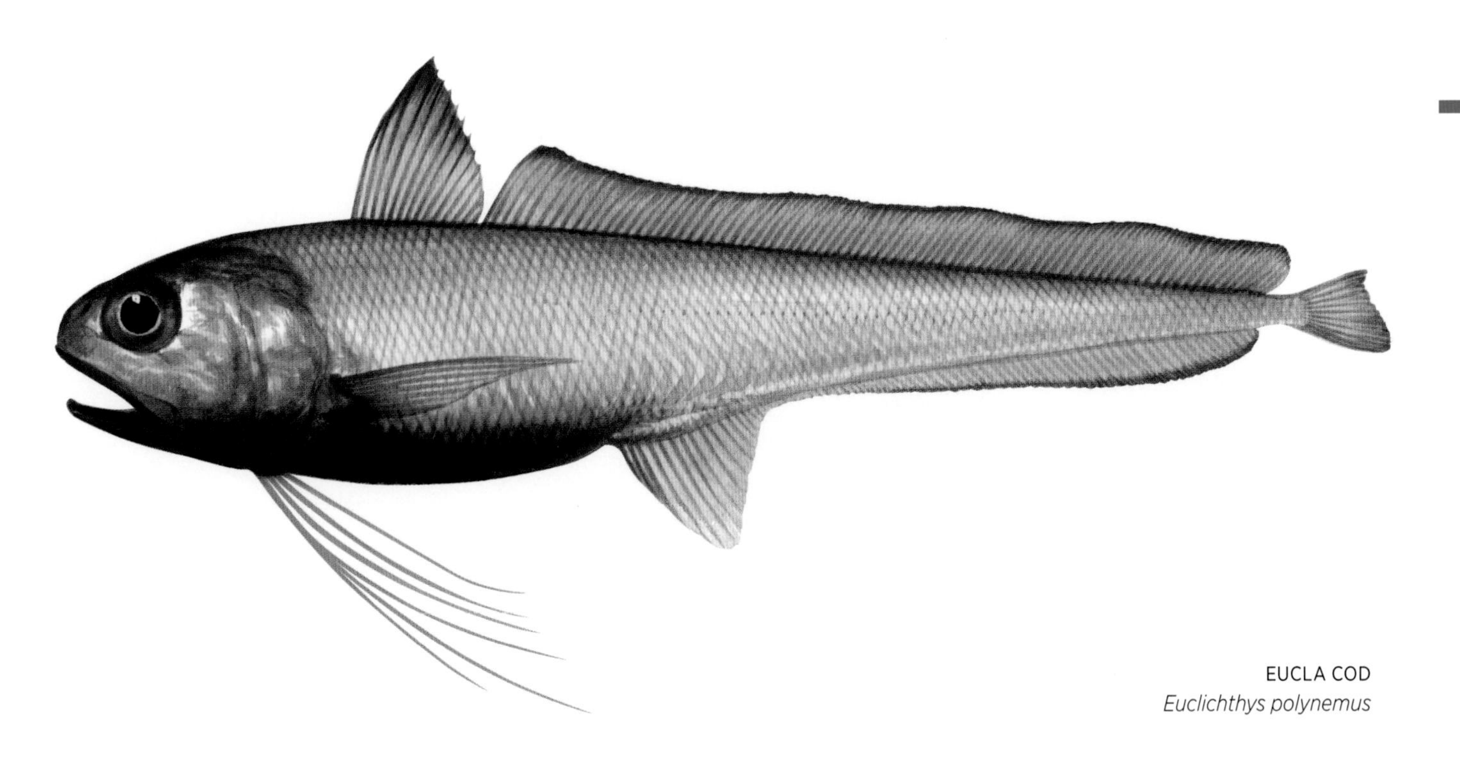

EUCLA COD
Euclichthys polynemus

The single known species in this family is found from the south-west Pacific, around New Zealand and southern Australia, to the continental shelf off the north-west Australian coast. It occurs at depths from 250 to 800 m. There are thought to be other, as yet undescribed species in the same area. The **Eucla Cod** has two nearly continous dorsal fins, the first short-based and tall, the second long-based and lower. The anal fin has a high anterior lobe and a long, low rear section. The ventral fins have four separate long, filamentous rays, the outermost branched into three filaments, and there is a small caudal fin. The Eucla Cod is pelagic near the bottom and reaches a maximum length of about 35 cm. It is caught by deepwater commercial trawls, but is not particularly abundant.

AUSTRALIAN EUCLICHTHYIDAE SPECIES

Euclichthys polynemus Eucla Cod

WHIPTAILS

Macrouridae

KAIYOMARU WHIPTAIL
Coelorinchus kaiyomaru

The family of Whiptails contains over 350 species and members are found in all the world's seas, from the Arctic to the Antarctic. They are bottom dwellers in deep water, usually found from 200 to 2000 m, though some species occur at depths as great as 6000 m. They have a large head and short body, with a long, flattened tail that tapers to a point. Whiptails are one of the dominant families of deepwater bottom-dwelling fishes; in the Pacific they constitute the greatest proportion of vertebrate biomass in deep water. They are sometimes divided into four subfamilies:

BATHYGADINAE Members of this group have two dorsal fins: the first is tall with a long, spiny second ray; the second is long and much taller than the anal fin. They have a large, terminal mouth lined with bands of small teeth. Growing to a maximum length of about 60 cm, they feed mainly on pelagic crustaceans.

MACROUROIDINAE These fishes have a large, soft, rounded head with an inferior mouth, no chin barbels, and a single long, low dorsal fin.

TRACHYRINCINAE These Whiptails have a long, pointed snout with an inferior mouth, and spiny scales and scutes along the dorsal and anal fin bases. Most have a small chin barbel. This group includes the **Unicorn Whiptail**.

MACROURINAE This is by far the largest group in the family. All members have two dorsal fins with a gap between them: the first fin is short-based with tall, spiny anterior rays; the second is long and much lower than the anal fin. The morphology of the mouth varies between species, from large and terminal to small and inferior, with dentition ranging from large fangs to small, villiform teeth. Many species have light-producing organs on the ventral surface. They are bottom dwellers, found from several hundred metres to abyssal depths of more than 6000 m. They grow to a maximum length of between 12 cm and 1.5 m. As the differing mouth structures suggest, the diets of these fishes are extremely varied – ranging from benthic invertebrates such as polychaete worms, holothurians, molluscs and crustaceans, to zooplankton, cephalopods and fishes. Most species tend to concentrate on a narrow range of prey types and these can usually be deduced from the mouth structure: those with a large, terminal mouth tend to be carnivorous, preying on fishes and large decapod crustaceans; those with a long snout and small, protrusible, ventral mouth patrol the substrate with head down, sensing prey with the snout and picking up small invertebrates. In the northern hemisphere, several species in this group are important commercially. In Australian waters they are usually only taken as bycatch, though sometimes in significant numbers.

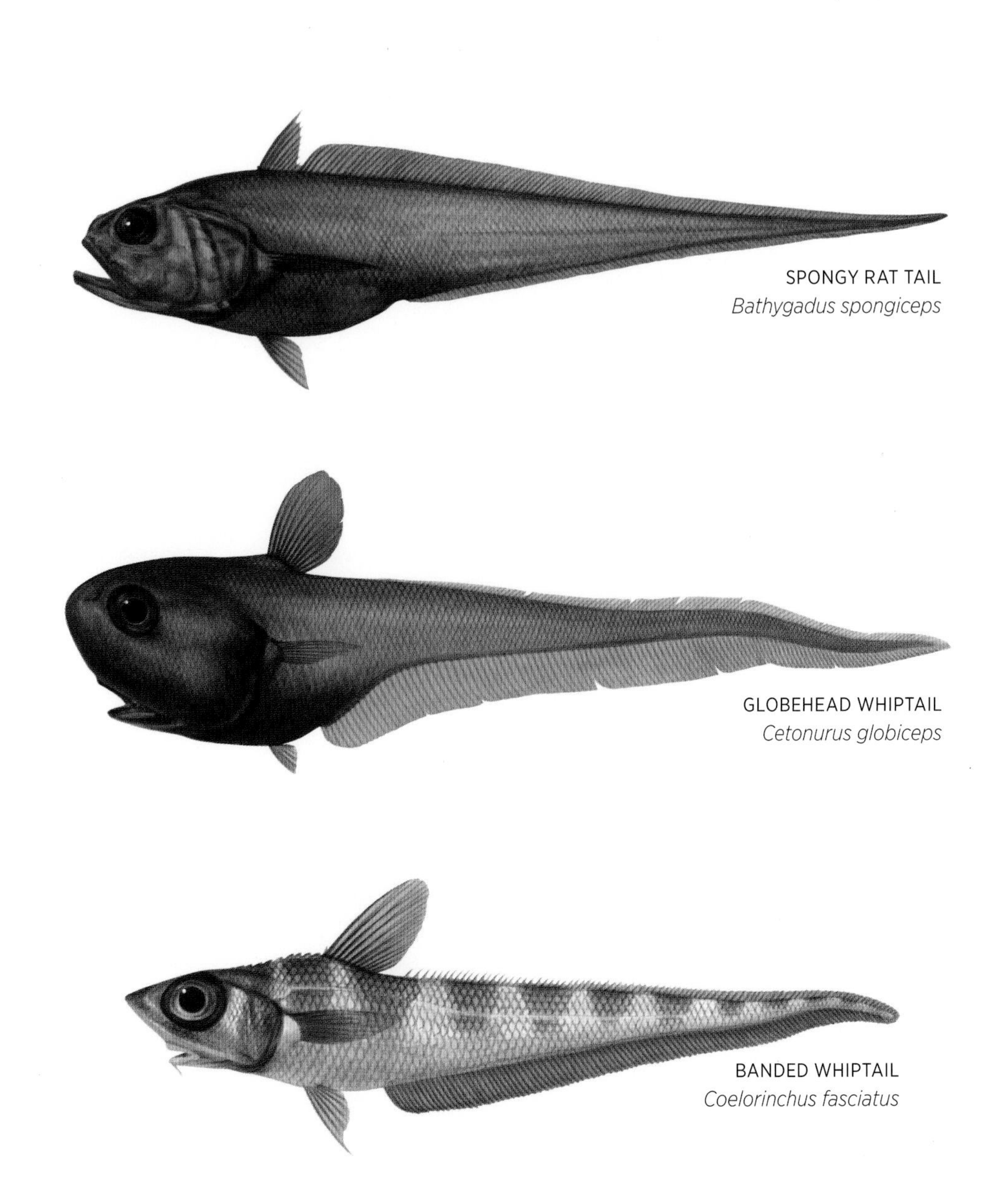

SPONGY RAT TAIL
Bathygadus spongiceps

GLOBEHEAD WHIPTAIL
Cetonurus globiceps

BANDED WHIPTAIL
Coelorinchus fasciatus

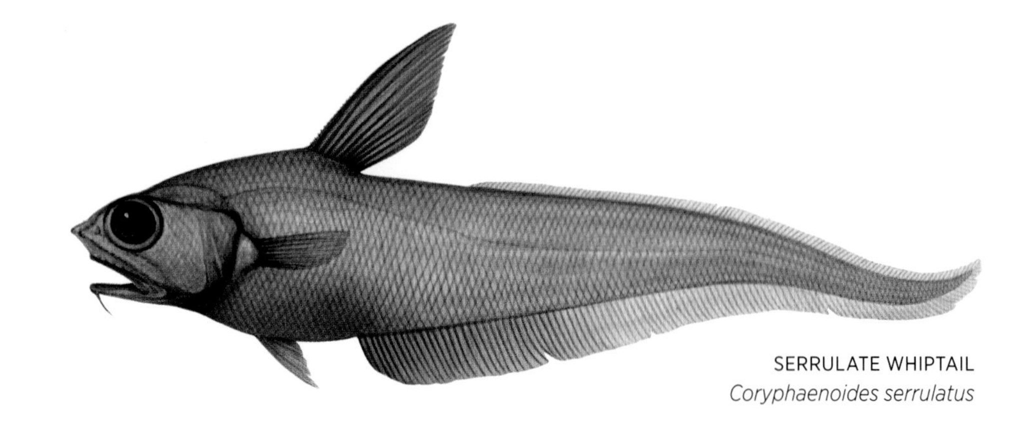

SERRULATE WHIPTAIL
Coryphaenoides serrulatus

TOOTHED WHIPTAIL
Lepidorhynchus denticulatus

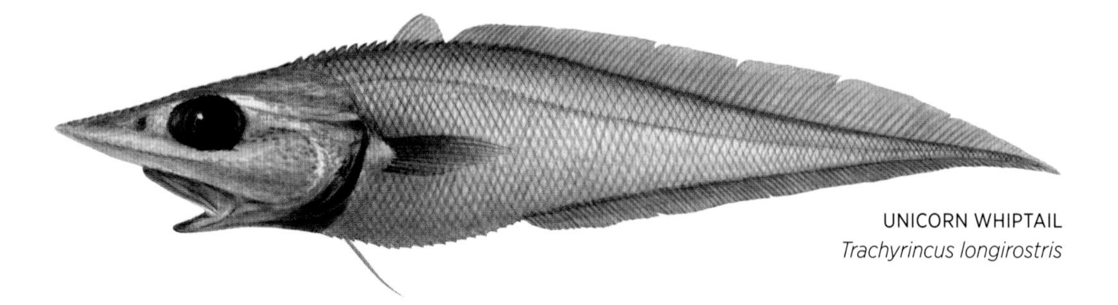

UNICORN WHIPTAIL
Trachyrincus longirostris

AUSTRALIAN MACROURIDAE SPECIES

Asthenomacrurus victoris Victory Whiptail
Bathygadus cottoides Codhead Rat Tail
Bathygadus furvescens Blackfin Rat Tail
Bathygadus spongiceps Spongy Rat Tail
Cetonurichthys subinflatus Smallpore Whiptail
Cetonurus globiceps Globehead Whiptail
Coelorinchus acanthiger Spottyface Whiptail
Coelorinchus acutirostris Spearnose Whiptail

Coelorinchus amydrozosterus Faintbanded Whiptail
Coelorinchus anatirostris Duckbill Whiptail
Coelorinchus argentatus Silver Whiptail
Coelorinchus australis Southern Whiptail
Coelorinchus charius Graceful Whiptail
Coelorinchus fasciatus Banded Whiptail
Coelorinchus gaesorhynchus Javelin Whiptail
Coelorinchus goobala Goobala Whiptail
Coelorinchus innotabilis Notable Whiptail
Coelorinchus kaiyomaru Kaiyomaru Whiptail
Coelorinchus kermadecus Kermadec Whiptail
Coelorinchus lasti Roughsnout Whiptail
Coelorinchus macrorhynchus Bigsnout Whiptail
Coelorinchus maculatus Blotch Whiptail
Coelorinchus matamua Blueband Whiptail
Coelorinchus maurofasciatus Falseband Whiptail
Coelorinchus mayiae False Silver Whiptail
Coelorinchus mirus Gargoyle Fish
Coelorinchus mycterismus Pinocchio Whiptail
Coelorinchus parallelus Spiny Whiptail
Coelorinchus pardus Leopard Whiptail
Coelorinchus parvifasciatus Little Whiptail
Coelorinchus semaphoreus Semaphore Whiptail
Coelorinchus sereti Short-tooth Whiptail
Coelorinchus shcherbachevi False Duckbill Whiptail
Coelorinchus smithi False Graceful Whiptail
Coelorinchus spathulata Spatulate Whiptail
Coelorinchus supernasutus Supanose Whiptail
Coelorinchus thurla Thurla Whiptail
Coelorinchus trachycarus Rough-head Whiptail
Coryphaenoides dossenus Humpback Whiptail
Coryphaenoides fernandezianus Fernandez Whiptail
Coryphaenoides filicauda Humphead Whiptail
Coryphaenoides grahami Graham's Whiptail
Coryphaenoides mcmillani McMillan's Whiptail
Coryphaenoides murrayi Abyssal Whiptail
Coryphaenoides rudis Bighead Whiptail
Coryphaenoides serrulatus Serrulate Whiptail
Coryphaenoides striaturus Striate Whiptail
Coryphaenoides subserrulatus Longray Whiptail
Cynomacrurus piriei Dogtooth Whiptail
Gadomus aoteanus Filamentous Rat Tail
Gadomus introniger Blackmouth Rat Tail
Gadomus pepperi Blacktongue Rat Tail
Haplomacrourus nudirostris Nakedsnout Whiptail
Hymenocephalus adelscotti Celebration Whiptail
Hymenocephalus aeger Plaintail Whiptail
Hymenocephalus aterrimus Blackest Whiptail
Hymenocephalus gracilis Delicate Whiptail
Hymenocephalus kuronumai Kuronuma's Whiptail
Hymenocephalus longibarbis Longbarb Whiptail
Hymenocephalus megalops Bigeye Whiptail
Hymenocephalus nascens Origin Whiptail
Idiolophorhynchus andriashevi Pineapple Whiptail
Kumba gymnorhynchus Smoothsnout Whiptail
Kuronezumia bubonis Bulbous Whiptail
Kuronezumia leonis Snubnose Whiptail
Kuronezumia pallida Pallid Whiptail
Lepidorhynchus denticulatus Toothed Whiptail
Lucigadus microlepis Smallfin Whiptail
Lucigadus nigromaculatus Blackspot Whiptail
Lucigadus ori Bronze Whiptail
Macrouroides inflaticeps Inflated Whiptail
Macrourus carinatus Ridgescale Whiptail
Malacocephalus laevis Smooth Whiptail
Mataeocephalus acipenserinus Sturgeon Whiptail
Mataeocephalus kotlyari Kotlyar's Whiptail
Mesobius antipodum Black Whiptail
Mesobius berryi Berry's Whiptail
Nezumia coheni Cohen's Whiptail
Nezumia kapala Kapala Whiptail
Nezumia leucoura Whitetail Whiptail
Nezumia merretti Merret's Whiptail
Nezumia namatahi Namatahi Whiptail
Nezumia propinqua Aloha Whiptail
Nezumia soela Soela Whiptail
Nezumia spinosa Sawspine Whiptail
Nezumia wularnia Wularni Whiptail
Odontomacrurus murrayi Largefang Whiptail
Pseudonezumia pusilla Tiny Whiptail
Sphagemacrurus pumiliceps Dwarf Whiptail
Sphagemacrurus richardi Richard's Whiptail
Squalogadus modificatus Tadpole Whiptail
Trachonurus gagates Velvet Whiptail
Trachonurus sentipellis Shaggy Whiptail
Trachonurus villosus Furry Whiptail
Trachonurus yiwardaus Yiwarda Whiptail
Trachyrincus longirostris Unicorn Whiptail
Ventrifossa gomoni Pale Smiling Whiptail
Ventrifossa johnboborum Snoutscale Whiptail
Ventrifossa macropogon Longbeard Whiptail
Ventrifossa nigrodorsalis Spinnaker Whiptail
Ventrifossa paxtoni Thinbarbel Whiptail
Ventrifossa sazonovi Dark Smiling Whiptail

MORID CODS AND DEEPSEA CODS

Moridae

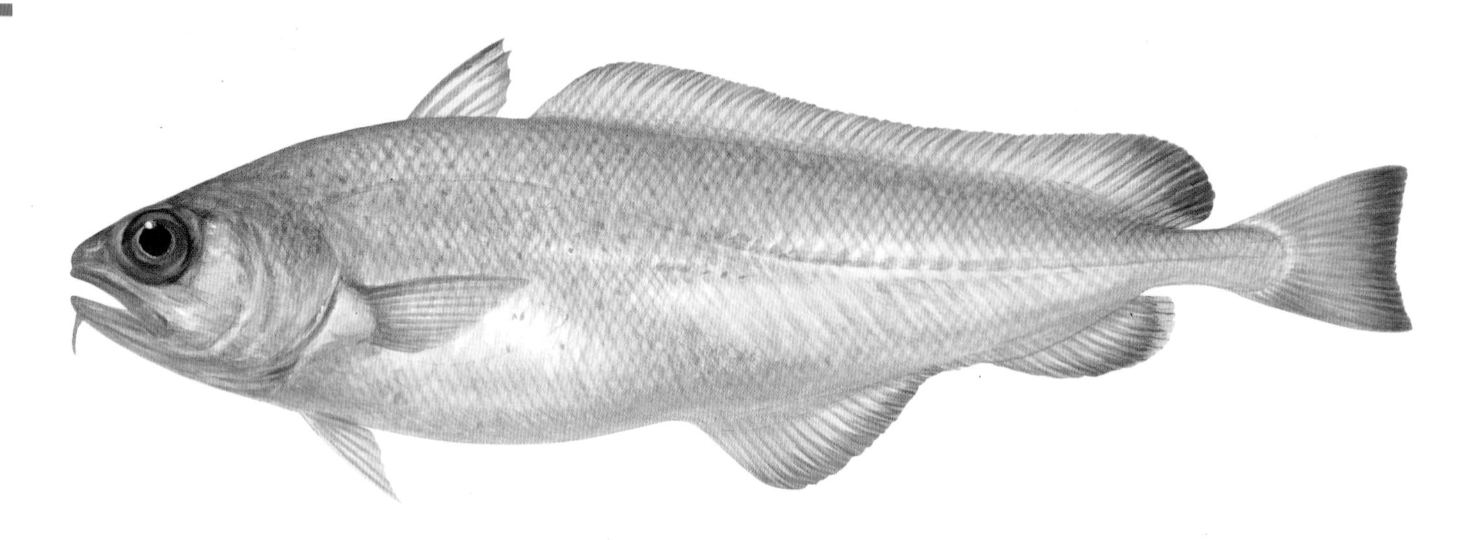

RIBALDO
Mora moro

This family is large and diverse, with more than 100 species found worldwide. Morid Cods and Deepsea Cods have an elongate body that tapers to a narrow caudal peduncle with a distinct caudal fin. They have 1–3 dorsal fins and 1–2 anal fins, and most have small ventral fins placed well forward on the body, anterior to the pectoral fins. They have a moderate to large mouth with small, fine teeth, and almost all have a chin barbel. Some deepwater species have a light-producing organ on the belly, between the anal and ventral fins.

Several species, including the Beardies, **Bastard Red Cod** and **Bearded Rock Cod**, are reasonably common on shallow reefs along the southern coast of Australia, but they are not often seen due to their nocturnal habits and preference for sheltering in caves and crevices during the day. The **Red Cod** is an important commercial species. It attains 80 cm in length and is found on shallow, protected reefs and sandy bottoms in south-eastern Australia, to depths of 170 m. Another species taken commercially is the **Ribaldo**, which grows to about 70 cm and is found worldwide. It occurs in Australia's southern waters from 400 to 900 m. Many other species are taken as bycatch by deepwater trawl fisheries. The **Giant Cod**, found at depths from 900 to 1100 m, is the largest member of the family, reaching a maximum length of around 2 m. Like most Morids and Deepsea Cods, it is a bottom-dwelling predator of small fishes, cephalopods and crustaceans. The **Violet Cod** is found in cold waters in both the southern and northern hemispheres, and is one of the most abundant of the deepwater fishes found on the continental slope. Feeding mainly on benthic invertebrates, it is pelagic near the bottom, in waters as deep as 3000 m.

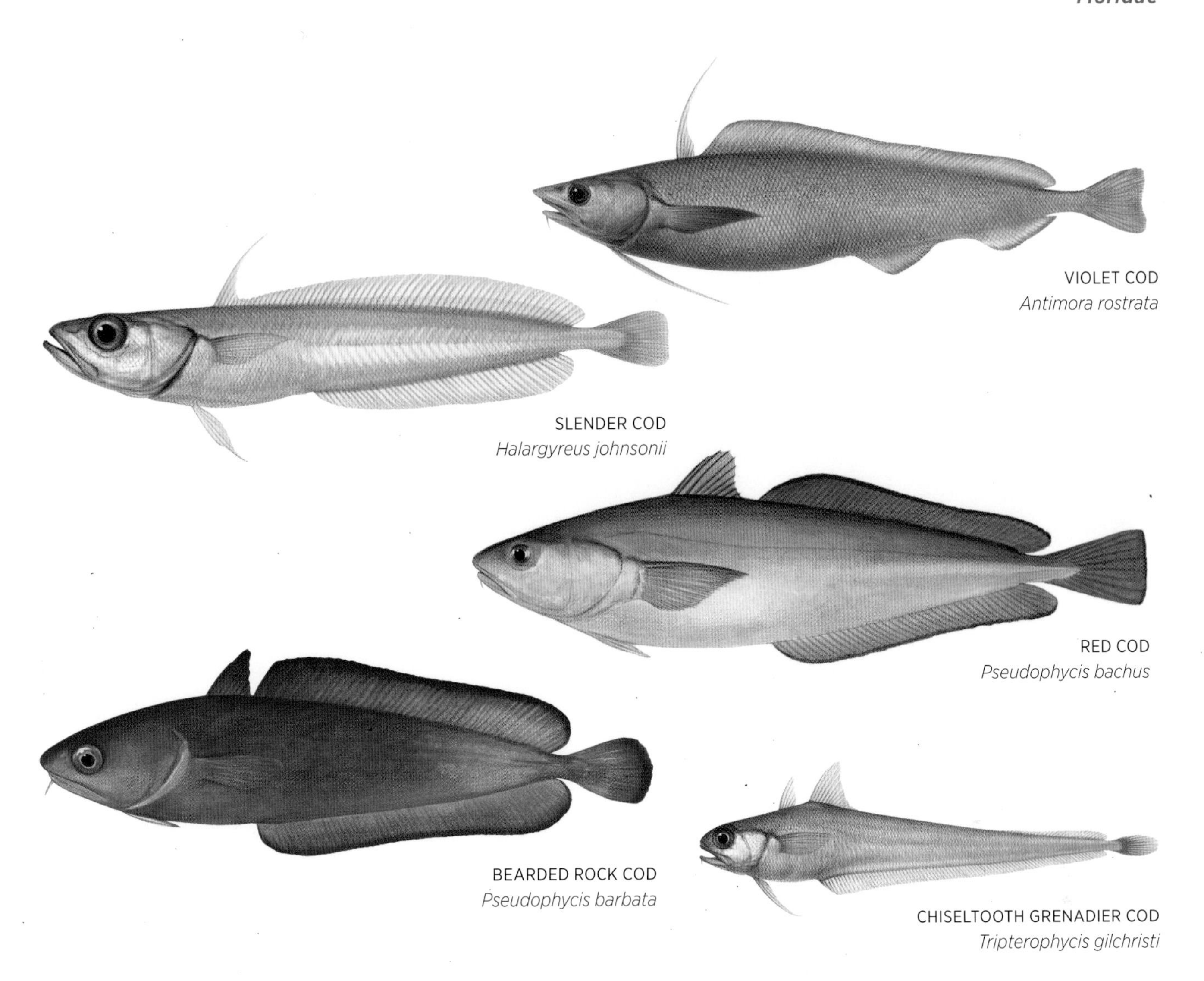

VIOLET COD
Antimora rostrata

SLENDER COD
Halargyreus johnsonii

RED COD
Pseudophycis bachus

BEARDED ROCK COD
Pseudophycis barbata

CHISELTOOTH GRENADIER COD
Tripterophycis gilchristi

AUSTRALIAN MORIDAE SPECIES

Antimora rostrata Violet Cod
Eeyorius hutchinsi Finetooth Beardie
Gadella jordani Jordan's Cod
Gadella macrura Longtail Cod
Guttigadus globiceps Fathead Cod
Guttigadus globosus Tadpole Cod
Guttigadus kongi Austral Cod
Halargyreus johnsonii Slender Cod
Laemonema robustum Robust Cod
Lepidion inosimae Giant Cod
Lepidion microcephalus Smallhead Cod
Lepidion schmidti Schmidt's Cod
Lotella phycis Slender Beardie
Lotella rhacina Largetooth Beardie
Mora moro Ribaldo
Notophycis marginata Forkbeard Cod
Physiculus longifilis Filament Cod
Physiculus luminosa Luminous Cod
Physiculus nigrescens Darktip Cod
Physiculus roseus Rosy Cod
Physiculus therosideros Scalyfin Cod
Pseudophycis bachus Red Cod
Pseudophycis barbata Bearded Rock Cod
Pseudophycis breviuscula Bastard Red Cod
Tripterophycis gilchristi Chiseltooth Grenadier Cod
Tripterophycis svetovidovi Brown Grenadier Cod

PELAGIC CODS
Melanonidae

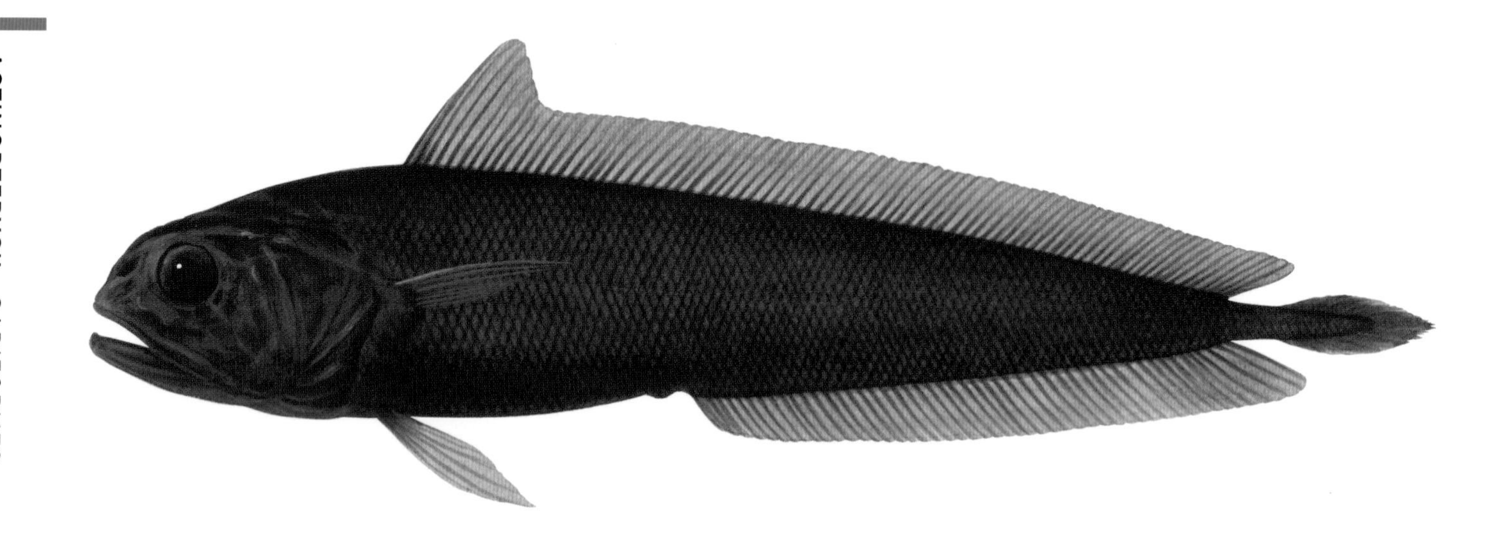

ARROWTAIL COD
Melanonus zugmayeri

The two species in this family are found circumglobally in cool temperate and sub-Antarctic southern hemisphere waters. They are pelagic in mid or bottom waters at depths from 600 to 3500 m. Growing to about 15 cm in length, they have a single long, low dorsal fin and similar anal fin, both of which are separate from the caudal fin. They have a blunt, rounded snout with no chin barbel, fleshy ridges on the head, and a large mouth. The **Pelagic Cod** has numerous small teeth in the jaws, while the **Arrowtail Cod** has larger canines. Neither species is abundant and they are rarely encountered, even in deepwater trawls. Very little is known of their biology.

AUSTRALIAN MELANONIDAE SPECIES
Melanonus gracilis Pelagic Cod
Melanonus zugmayeri Arrowtail Cod

HAKES

Merlucciidae

BLUE GRENADIER
Macruronus novaezelandiae

Hakes are distributed in the Atlantic, Southern and East Pacific oceans, with most species found in the North Atlantic. There are approximately 20 species in the family. Hakes have a large head with a V-shaped ridge on the dorsal surface, a long compressed body with small scales, and a large mouth with numerous sharp teeth on the jaws and anterior roof. They have no chin barbel, two well-developed dorsal fins, and either a long tapering caudal fin (as in the **Blue Grenadier**) or a truncate caudal fin (as in the **Southern Hake**).

The Blue Grenadier is an important commercial species in southern Australian waters, and usually occurs from depths of 200 to 700 m. During the day the Blue Grenadier forms dense schools near the bottom, and at night it migrates upward to within about 50 m of the surface, where it feeds on Lanternfishes, squid, krill and other pelagic crustaceans. In Australia most Blue Grenadiers are caught off the Tasmanian coast, particularly over their spawning grounds off the west coast. Adult females produce about one million eggs each. The eggs drift in the water column and hatch into pelagic larvae before settling on the bottom – usually well offshore but sometimes in estuaries and bays in southern Tasmania, where juveniles are often found. The Blue Grenadier reaches a maximum length of about 110 cm and may live for up to 25 years. The Southern Hake is occasionally found off southern Australia but occurs mainly around New Zealand and South America, where it is commercially fished.

AUSTRALIAN MERLUCCIIDAE SPECIES
Macruronus novaezelandiae Blue Grenadier
Merluccius australis Southern Hake

PHYCID HAKES

Phycidae

NEW ZEALAND ROCKLING
Gaidropsarus novaezealandiae

The family of Phycid Hakes contains about 25 species and is predominantly found in the northern hemisphere. Phycid Hakes are most diverse in the temperate North Atlantic Ocean, but several species occur in temperate waters in the southern hemisphere. The **New Zealand Rockling** can be identified by its two nasal barbels and one chin barbel. The elongate first dorsal fin ray is followed by a series of small, fleshy filaments, and the caudal fin is rounded. In New Zealand it is found from shallow coastal waters (even rock pools) to depths of more than 500 m and feeds on small fishes and invertebrates. It is known in Australia from a relatively small number of larvae taken off southern Tasmania, and these may in fact be a different species. The relationships within the family are not well understood and even the number of valid species is unclear.

AUSTRALIAN PHYCIDAE SPECIES

Gaidropsarus novaezealandiae New Zealand Rockling

OPHIDIIFORMES

Ophidiiforms have elongate bodies that taper to a point or to a small caudal fin. They usually have small cycloid scales, though scales are absent in some. They have long, low, spineless dorsal and anal fins. The small ventral fins have only 1–2 rays and are placed well forward, usually beneath the throat, or they may be absent altogether. Several families reproduce using internal fertilisation and give birth to live young. There are five families in the order, all of which occur in Australian waters.

PEARLFISHES

Carapidae

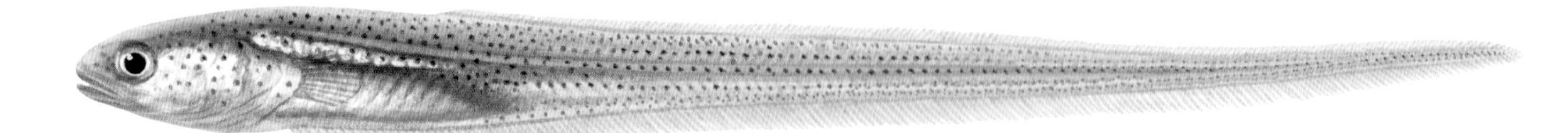

STAR PEARLFISH
Carapus mourlani

About 31 species of Pearlfish are distributed around the world in tropical and subtropical waters. They are small, elongate fishes, with a tapering tail and a semi-translucent, scaleless body. They have long, low dorsal and anal fins that are continuous with the tail.

Some members of this family are free-living in deep water down to 2000 m. Others are commensal, living within the body of another animal, in shallow waters of tropical reefs. The host animals are usually sea cucumbers, or molluscs such as large clams and pearl shells, but starfishes, sponges and tunicates may also be utilised. For some species of Pearlfish the host animal serves only as shelter, while others actively parasitise the host, feeding on its internal tissue.

Pearlfish larvae go through two developmental stages. In the first stage the larvae (called vexillifers) are planktonic, and their bodies are ribbon-like and transparent, with a long, trailing filament before the dorsal fin. The vexillifers develop into cylindrical-bodied, bottom-dwelling 'tenuis' larvae, which have an extremely long tail that gradually diminishes as they develop into adults. Pearlfishes find a host at this second larval stage and enter tail first (through the animal's anus in the case of sea cucumbers). They are usually found in male/female pairs inside the host. The **Star Pearlfish** is common in tropical waters of Australia, and is often found inhabiting the Pin Cushion Starfish (*Culcita* spp.). The **Cucumber Pearlfish** is usually found inside sea cucumbers.

CUCUMBER PEARLFISH
Onuxodon margaritiferae

AUSTRALIAN CARAPIDAE SPECIES

Carapus mourlani Star Pearlfish
Carapus sluiteri Sluiter's Pearlfish
Echiodon anchipterus Closefin Pearlfish
Echiodon rendahli Messmate Fish
Encheliophis boraborensis Pinhead Pearlfish
Encheliophis gracilis Graceful Pearlfish
Encheliophis homei Silver Pearlfish
Encheliophis vermicularis Worm Pearlfish
Encheliophis vermiops Pygmy Pearlfish
Eurypleuron cinereus Wide-rib Pearlfish
Onuxodon fowleri Fowler's Pearlfish
Onuxodon margaritiferae Cucumber Pearlfish
Onuxodon parvibrachium Oyster Pearlfish
Pyramodon lindas Blackedge Pearlfish
Pyramodon punctatus Dogtooth Pearlfish
Pyramodon ventralis Pallid Pearlfish

CUSKS AND LINGS

Ophidiidae

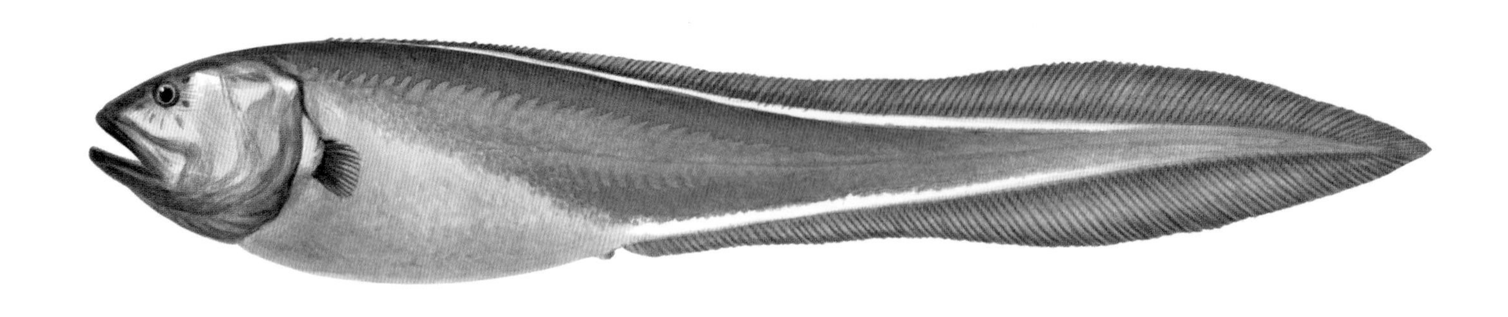

VIOLET CUSK
Brotulotaenia crassa

Cusks and Lings are found throughout the Atlantic, Indian and Pacific oceans. The family contains 220 or more species. They have an elongate body with small scales, a long tapering tail, and dorsal and anal fins continuous with the caudal fin. All (except the **Bearded Cusk**) lack barbels on the chin. The ventral fins are reduced to a few filamentous rays placed well forward under the chin or are absent altogether.

Cusks and Lings are widely distributed, ranging from shallow, rocky reefs to abyssal depths. One Cusk (*Abyssobrotula galatheae*) was captured off Puerto Rico at a depth of 8370 m – the greatest depth at which a fish has ever been caught. Many other species in this family are found at great depths, predominantly in tropical zones, where they feed on small benthic invertebrates such as crustaceans and polychaete worms. Those species inhabiting the shallows live in caves and crevices and are usually nocturnal, emerging at night to feed on crustaceans and small fishes.

The family contains several species that are important commercially. In Australia the **Pink Ling** is taken by trawl at depths of 100–600 m along the southern coast. It can grow to 2 m and weigh up to 20 kg, and is a slow-growing species – a 135 cm specimen was determined to be 21 years old. The Pink Ling feeds mainly on crustaceans such as prawns, as well as on fishes. The **Rock Ling**, which is endemic to Australia, is similar, and occurs on shallow rocky reefs and in seagrass beds and estuaries along the south coast. It is often caught by shore anglers and is also taken by fisheries. Another endemic species, the **Tusk**, occurs in deep waters off southern Australia and is also taken commercially. The Bearded Cusk has barbels on the snout and chin, and is found from shallow reefs to deeper offshore waters in northern Australia.

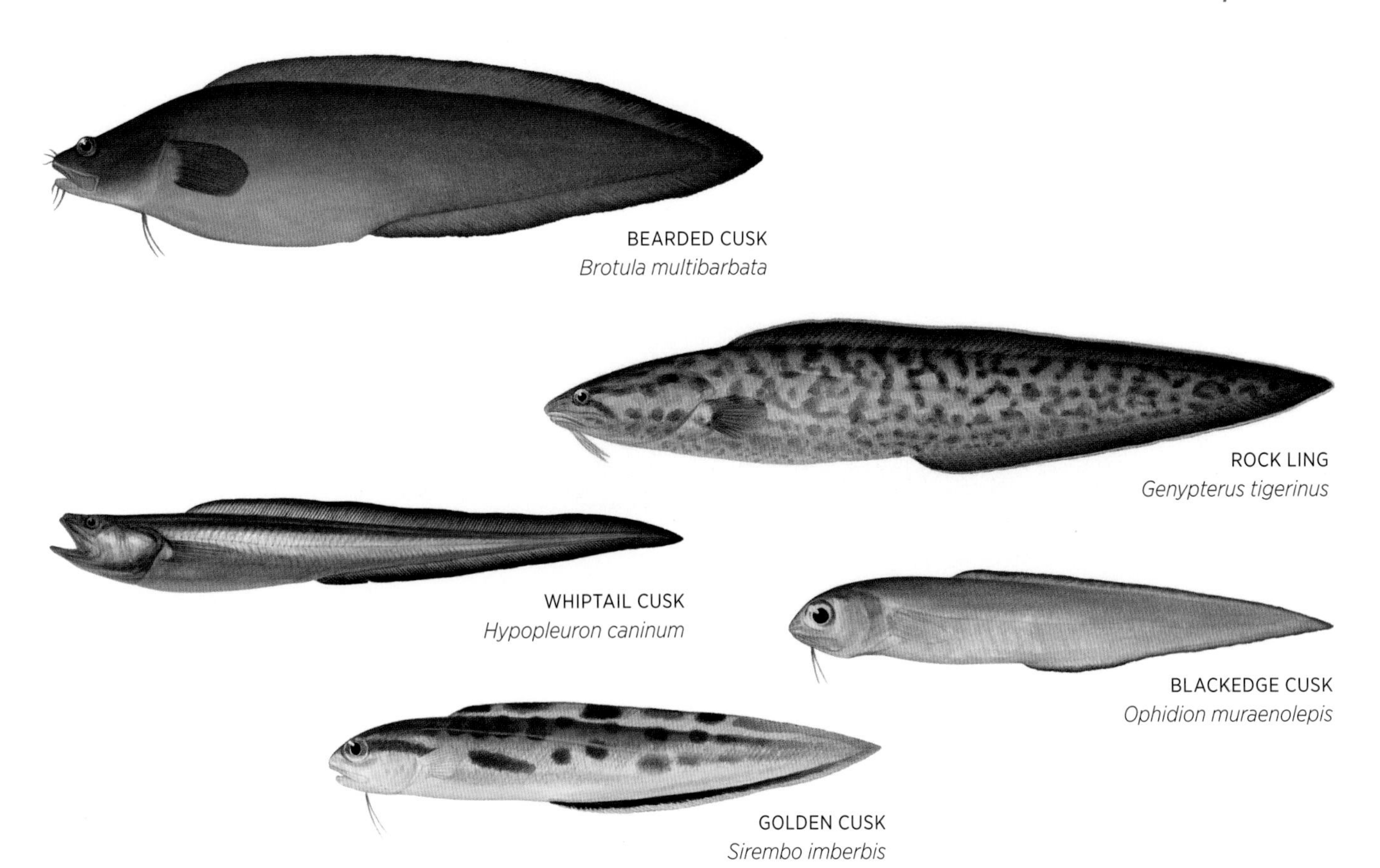

BEARDED CUSK
Brotula multibarbata

ROCK LING
Genypterus tigerinus

WHIPTAIL CUSK
Hypopleuron caninum

BLACKEDGE CUSK
Ophidion muraenolepis

GOLDEN CUSK
Sirembo imberbis

AUSTRALIAN OPHIDIIDAE SPECIES

Bassozetus compressus Abyssal Assfish
Bassozetus galatheae Galathea Assfish
Bassozetus glutinosus Glutin Assfish
Bassozetus robustus Robust Assfish
Brotula multibarbata Bearded Cusk
Brotulotaenia crassa Violet Cusk
Brotulotaenia nigra Dark Cusk
Dannevigia tusca Tusk
Epetriodus freddyi Needletooth Cusk
Genypterus blacodes Pink Ling
Genypterus tigerinus Rock Ling
Glyptophidium japonicum Japanese Cusk
Glyptophidium lucidum Sculptured Cusk
Homostolus acer Filament Cusk
Hoplobrotula armata Armoured Cusk
Hypopleuron caninum Whiptail Cusk
Lamprogrammus brunswigi Brunswig's Cusk
Lamprogrammus niger Black Cusk
Lamprogrammus shcherbachevi Scaleline Cusk
Microbrotula polyactis Many-ray Cusk
Microbrotula queenslandica Queensland Cusk
Monomitopus garmani Garman's Cusk
Monomitopus vitiazi Spearcheek Cusk
Neobythites australiensis Australian Cusk
Neobythites bimaculatus Twospot Cusk
Neobythites franzi Franz's Cusk
Neobythites longipes Longray Cusk
Neobythites macrops Spotfin Cusk
Neobythites nigriventris Blackbelly Cusk
Neobythites pallidus Pale Cusk
Neobythites purus Pure Cusk
Neobythites soelae Soela Cusk
Neobythites unimaculatus Onespot Cusk
Ophidion genyopus Ravenous Cusk-eel
Ophidion muraenolepis Blackedge Cusk
Ophidion smithi Smith's Cusk
Pycnocraspedum squamipinne Pelagic Cusk
Sirembo imberbis Golden Cusk
Sirembo jerdoni Brownbanded Cusk
Sirembo metachroma Chameleon Cusk
Spottobrotula amaculata Lined Cusk
Xyelacyba myersi Gargoyle Cusk

BROTULAS AND CUSKS

Bythitidae

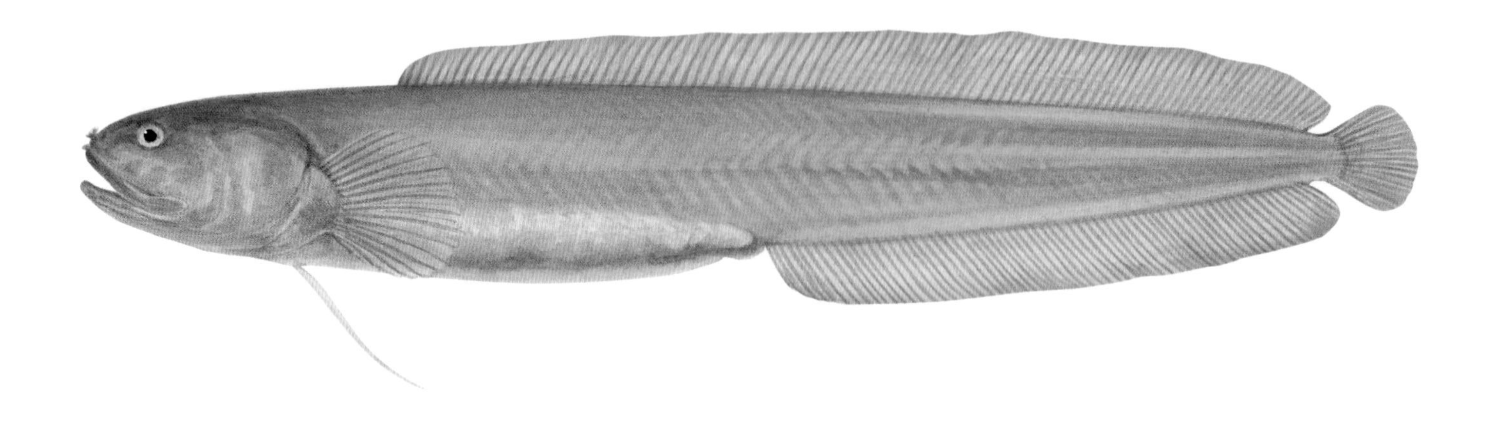

EASTERN YELLOW BLINDFISH
Dermatopsis macrodon

Brotulas and Cusks are found throughout the Atlantic, Indian and Pacific oceans, from the shallows down to 2600 m or more, with several species occurring in fresh water. There are about 107 species in the family. Brotulas have an elongate body, usually with small cycloid scales, though some species are scaleless. They have long dorsal and anal fins that are continuous with or end just before the caudal fin, and ventral fins reduced to a single filamentous ray. They usually have a strong opercular spine and very small eyes – the eyes are vestigial in some freshwater cave-dwelling species. Some species, such as the **Slender Blindfish**, have complex sensory papillae and pores on the snout and head to compensate for this reduction in eye size.

Fertilisation in Brotulas and Cusks is internal (males have a copulatory organ located just anterior to the anal fin) and females give birth to live young. Members of the family range in size from tiny 6 cm long shallow-water species such as the **Southern Pygmy Blindfish**, to 70 cm long deepwater species such as the **Brown Brotula**. Most species are small and secretive, living deep within crevices and caves, and are very rarely seen.

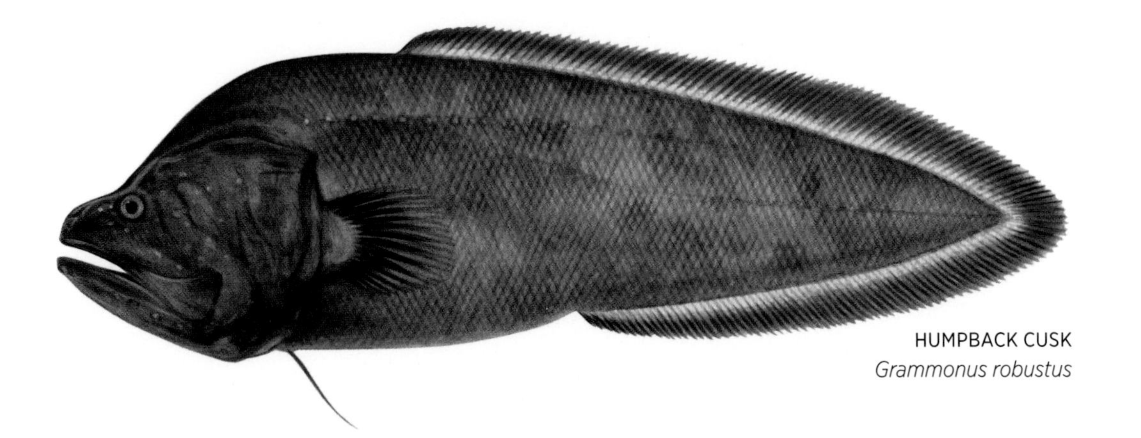

HUMPBACK CUSK
Grammonus robustus

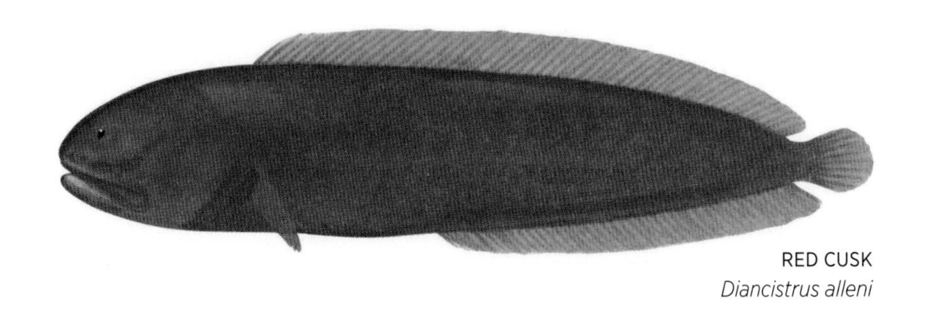

RED CUSK
Diancistrus alleni

AUSTRALIAN BYTHITIDAE SPECIES

Acarobythites larsonae Larson's Cusk
Beaglichthys macrophthalmus Beagle Cusk
Brosmolus longicaudus Longtail Cusk
Brosmophyciops pautzkei Slimy Cusk
Cataetyx niki Brown Brotula
Dactylosurculus gomoni Southern Pygmy Blindfish
Dermatopsis macrodon Eastern Yellow Blindfish
Dermatopsis multiradiatus Slender Blindfish
Diancistrus alleni Red Cusk
Diancistrus longifilis Twinhook Cusk
Dinematichthys dasyrhynchus Shaggy Cusk
Dinematichthys megasoma Robust Cusk
Dinematichthys riukiuensis Bigeye Cusk
Diplacanthopoma kreffti Deepbody Cusk
Dipulus caecus Orange Eelpout
Grammonus robustus Humpback Cusk
Hastatobythites arafurensis Spinyhead Cusk
Melodichthys paxtoni Baldhead Cusk
Monothrix mizolepis Smalleye Cusk
Monothrix polylepis Littoral Cusk
Saccogaster tuberculata Bagbelly Cusk

APHYONIDS

Aphyonidae

GELATINOUS BLINDFISH
Aphyonus gelatinosus

Aphyonids are deepwater inhabitants of the Atlantic, Indian and Pacific oceans, living on or near the bottom, from depths of about 1000 to 6000 m. The family consists of about 22 species. Like the Brotulas (Bythitidae, p. 282), males have a copulatory organ for internal fertilisation and females give birth to live young. They have an elongate, soft body with no scales and loose gelatinous skin that is largely transparent. The long dorsal and anal fins are continuous with the caudal fin, and the ventral fins are reduced to filamentous rays. Aphyonids have a large mouth and very small or vestigial eyes, but unlike the Brotulas they do not possess sensory pores on the head.

Aphyonids grow to around 20 cm in length and feed predominantly on very small benthic invertebrates, mainly crustaceans such as copepods. Some species have fangs in the anterior of the mouth, which indicates larger prey may also be taken. Aphyonids are poor swimmers and probably drift passively when feeding. They may prey on the crustaceans that congregate to feed on animal carcasses that have fallen to the sea floor.

AUSTRALIAN APHYONIDAE SPECIES
Aphyonus gelatinosus Gelatinous Blindfish
Barathronus maculatus Spotted Gelatinous Cusk

FALSE BROTULAS

Parabrotulidae

FALSE CUSK
Parabrotula plagiophthalmus

The three species in this family of bottom-dwelling, deepwater fishes are found scattered throughout the Atlantic, Indian and Pacific oceans. They have a scaleless, eel-like body with a long, low dorsal fin that originates at about mid-body and is continuous with the caudal fin and anal fin. False Brotulas have small pectoral fins and no ventral fins, and the lower jaw projects beyond the upper. They have small, fine teeth in the jaws and are probably predators of small benthic crustaceans and invertebrates. There are no sensory pores on the head, but fine sensory hairs (neuromasts) are present on the head and body. They grow to about 6 cm in length and, like the Brotulas (Bythitidae, p. 282), give birth to live young. Their biology and relationships with other families are not well known.

AUSTRALIAN PARABROTULIDAE SPECIES

Parabrotula plagiophthalmus False Cusk

BATRACHOIDIFORMES

The single family in this order is closely related to the Lophiiformes (p. 291). Batrachoidiforms have a rounded body (usually without scales), two dorsal fins, and ventral fins placed well forward beneath the throat. They have a swim bladder but lack rib bones. For more characteristics of the order see the family Batrachoididae (p. 288).

FROGFISHES

Batrachoididae

BANDED FROGFISH
Halophryne diemensis

Frogfishes are found in temperate and tropical coastal waters of the Atlantic, Indian and Pacific oceans. There are about 78 species in this family of small, stout-bodied fishes. They take their name from the fact that some species make a croaking noise when taken from the water. Frogfishes have two dorsal fins: the first has three spines, which in some species can give an extremely painful sting; the second is long and tall, without spines. The ventral fins, placed well forward beneath the throat, have a short spine, one thick, fleshy ray and one branched ray. The pectoral and caudal fins are broad and rounded. The body and fins are covered with loose skin, which in some species is embedded with small cycloid scales. Frogfishes have a large mouth surrounded by fleshy, branched tentacles (these are also present on the head and body of some species), and strong spines on the operculum, usually hidden beneath skin.

Most species live on sandy or muddy bottoms where they can bury themselves in the substrate, or on reefs in crevices or beneath rocks, from the intertidal zone down to about 750 m. Frogfishes are well camouflaged and, with their dorsally placed eyes and broad caudal fin, are efficient ambush predators. They conceal themselves, then lunge out to snap up crustaceans and small fishes. They also actively forage for invertebrates and molluscs.

The **Plainfin Frogfish** is distinctive in that it possesses rows of small photophores beneath the head; it is one of very few fishes found in shallow coastal waters to have such bioluminescence. Most of the Australian Frogfish species inhabit northern waters, though the **Pinkhead Frogfish** is found in southern Western Australia. The five *Batrachomoeus* species are all endemic to Australia.

DAHL'S FROGFISH
Batrachomoeus dahli

WESTERN FROGFISH
Batrachomoeus occidentalis

OCELLATE FROGFISH
Halophryne ocellatus

AUSTRALIAN BATRACHOIDIDAE SPECIES

Batrachomoeus dahli Dahl's Frogfish
Batrachomoeus dubius Eastern Frogfish
Batrachomoeus occidentalis Western Frogfish
Batrachomoeus rubricephalus Pinkhead Frogfish
Batrachomoeus trispinosus Threespine Frogfish
Halophryne diemensis Banded Frogfish
Halophryne ocellatus Ocellate Frogfish
Halophryne queenslandiae Sculptured Frogfish
Porichthys notatus Plainfin Frogfish

LOPHIIFORMES

Fishes in this order are bottom dwelling or pelagic and found mainly in deep water, though several families occur in coastal waters. Almost all have a modified first dorsal spine (the illicium) that forms a fishing apparatus, with a lure at its tip (the esca) that is used to entice prey close enough to be captured. The form of the esca is an important distinguishing characteristic and in many species it contains symbiotic bioluminescent bacteria in a special chamber.

Lophiiforms have no scales, and the skin is either covered with spines and spinules or is smooth and naked. The gill opening is reduced to a small hole near the pectoral-fin base. Benthic species have leg-like pectoral fins that they use to 'walk' along the bottom. The order contains 11 families of ceratioid Anglerfishes: deepwater pelagic species that have a greatly reduced skeleton, and dwarf males that attach to, and in some cases fuse permanently with and parasitise, the females. There are 18 families in the order, 16 of which occur in Australian waters. The remaining two families are widely distributed in deep waters and may also occur here.

GOOSEFISHES

Lophiidae

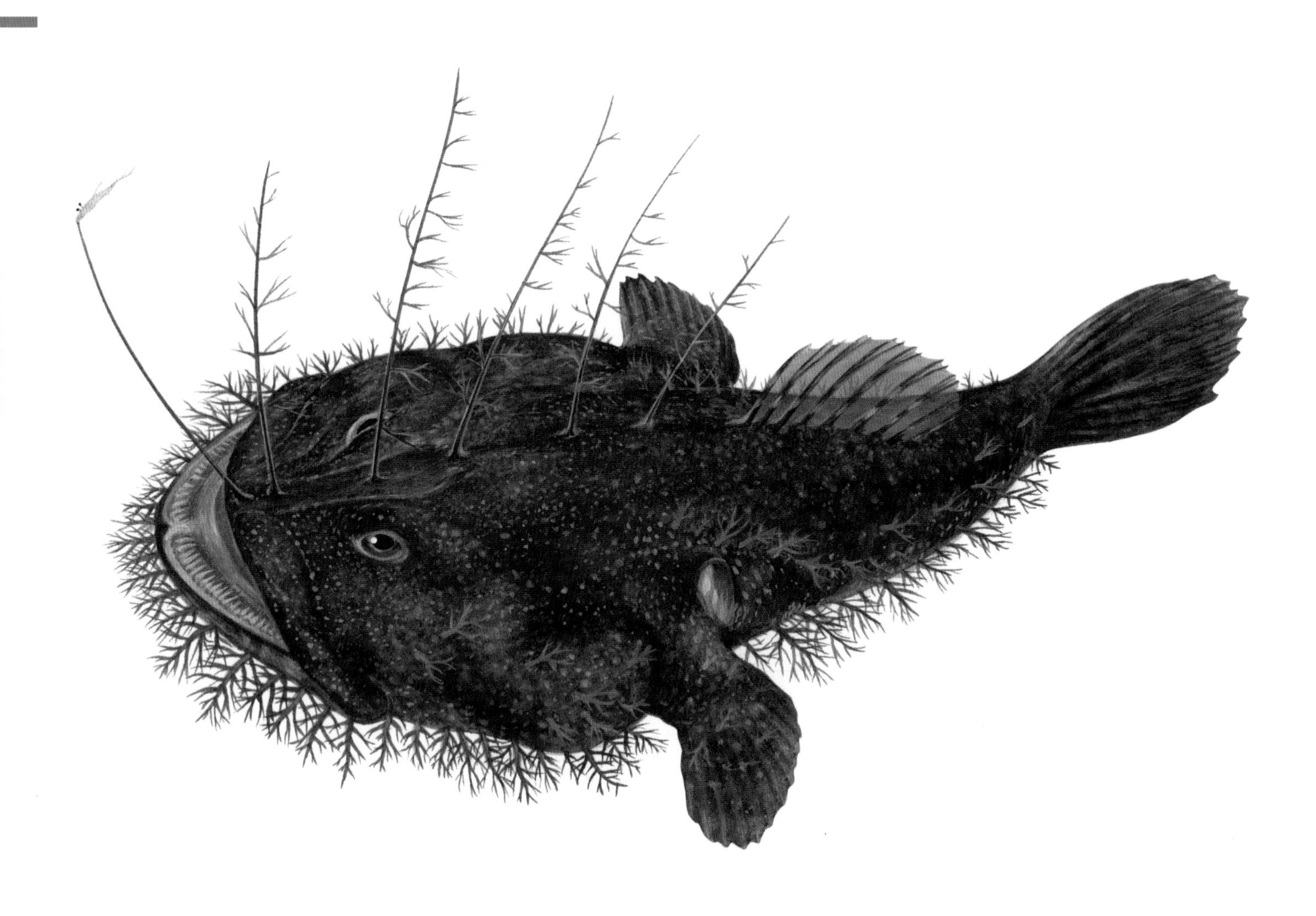

GOOSEFISH
Lophiodus naresi

Goosefishes are found in all oceans of the world, usually in deep water from about 100 m down to at least 1000 m. There are about 25 species in the family. The larger species may reach nearly 2 m in length, but most are less than 1 m. These bizarre-looking fishes have an enormous, flattened head with bony ridges and spines, and an extremely wide mouth lined with sharp, inward-hinging teeth. The mouth and head are fringed with fleshy tendrils that extend onto the body and tail. They have a flabby, scaleless body with loose, thin skin, and leg-like pectoral fins with fleshy rays that are used for 'walking' on the bottom. The gill openings are behind the pectoral fins, and the small ventral fins are beneath the head.

Goosefishes have a series of separate dorsal-fin spines on the head and anterior part of the body. The first spine forms the illicium and esca, which are used to attract the fishes on which it preys. The Goosefish lies camouflaged on the bottom, with the esca dangling above its head, then when a fish is attracted close enough the Goosefish surges upwards to engulf the prey in its cavernous mouth. In some species of Goosefish the females can produce over a million eggs; these are laid in a long, floating, veil-like band that can measure up to 12 m long and 1.5 m wide. Goosefishes are caught commercially in some areas, being prized for their dense, fine-grained flesh, but are not commonly taken in Australian waters.

AUSTRALIAN LOPHIIDAE SPECIES

Lophiodes infrabrunneus Shortspine Goosefish
Lophiodes mutilus Smooth Goosefish
Lophiodes naresi Goosefish
Lophiomus setigerus Broadhead Goosefish

ANGLERFISHES
Antennaridae

SARGASSUM FISH
Histrio histrio

Anglerfishes are widely distributed and occur in all the world's oceans, mainly in tropical waters. The family contains about 42 species, with a great diversity found in temperate waters around southern Australia. Anglerfishes have a rounded, almost globe-shaped body, with no scales and variously developed fleshy papillae and tentacles on the skin. The gill openings are reduced to small holes behind the pectoral fin. They have a large, almost vertical mouth that is lined with villiform teeth. The first dorsal-fin spine is separate from the rest and is developed into an illicium with a fleshy esca at its tip. The second and third dorsal spines are covered with skin and are separate from the soft-rayed second dorsal fin. The pectoral fins are leg-like with a distinct 'elbow' and fleshy rays, and the small ventral fins are placed well forward beneath the head.

One species, the **Sargassum Fish**, is pelagic, inhabiting floating clumps of sargassum weed, and is widely distributed in all oceans of the world. Most Anglerfishes, though, inhabit reefs and weedy areas from the shallows down to 150 m or more. They have highly variable colour patterns and are usually extremely well camouflaged and difficult to see under water. The **Tasselled Anglerfish** and **Glover's Anglerfish**, both found in southern Australian waters, are excellent examples, with long, branched, weed-like filaments all over the body. Brightly coloured species are usually found in areas with similarly coloured sponges.

Anglerfishes are ambush predators – they rest immobile on the substrate and use their esca (which can resemble a small fish, worm or amphipod) as a lure to attract other fishes, which they engulf with a lightning-fast lunge. They have an extremely distensible stomach, which allows them to swallow prey that are longer than themselves. Most species grow to 10–20 cm, with the largest attaining 40 cm in length. Female Anglerfishes lay gelatinous egg masses that are usually guarded by the male. The males of some species cover the eggs with their bodies, while others hold them in a pocket formed by curving the tail forward to the pectoral fin.

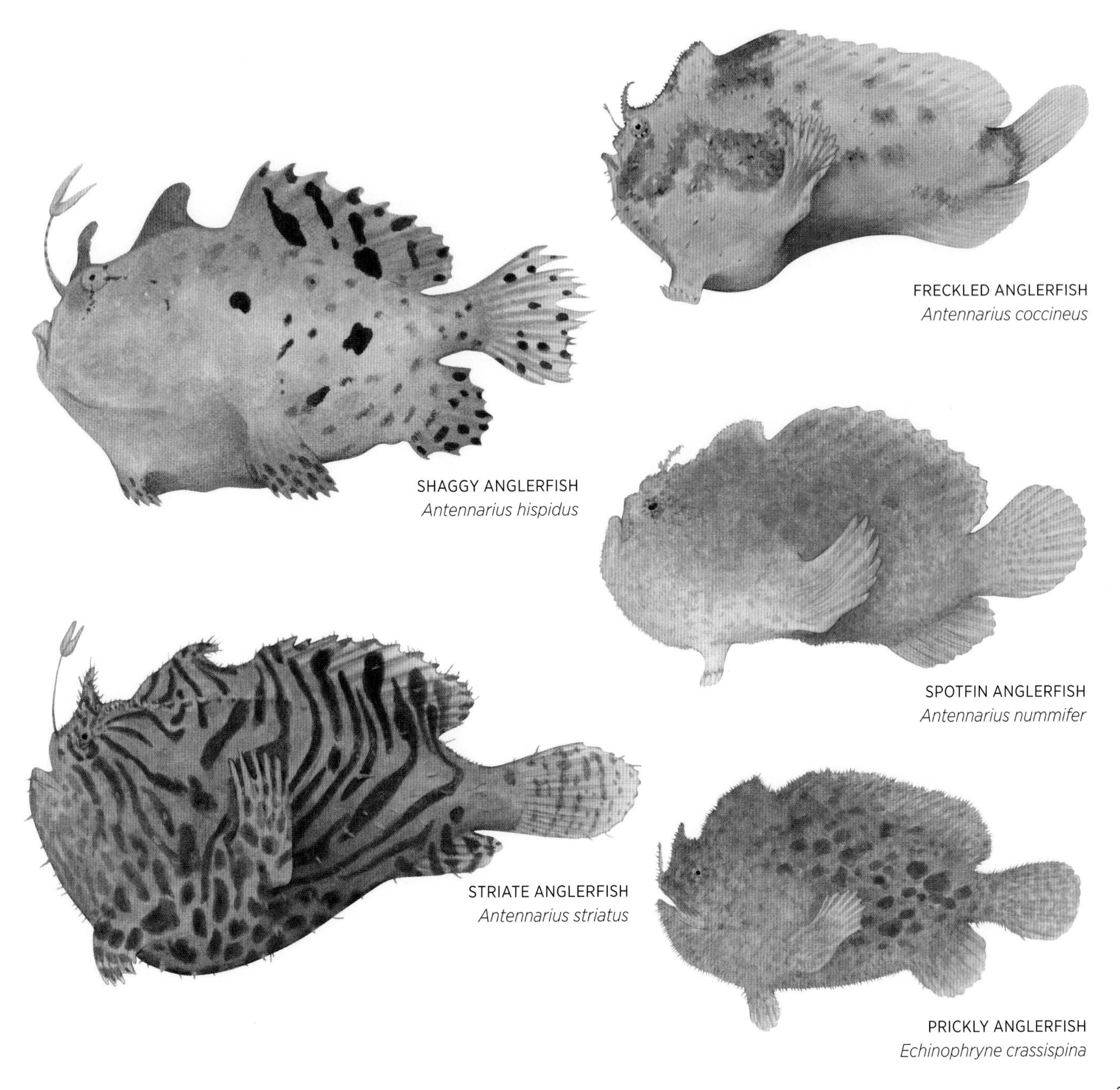

FRECKLED ANGLERFISH
Antennarius coccineus

SHAGGY ANGLERFISH
Antennarius hispidus

SPOTFIN ANGLERFISH
Antennarius nummifer

STRIATE ANGLERFISH
Antennarius striatus

PRICKLY ANGLERFISH
Echinophryne crassispina

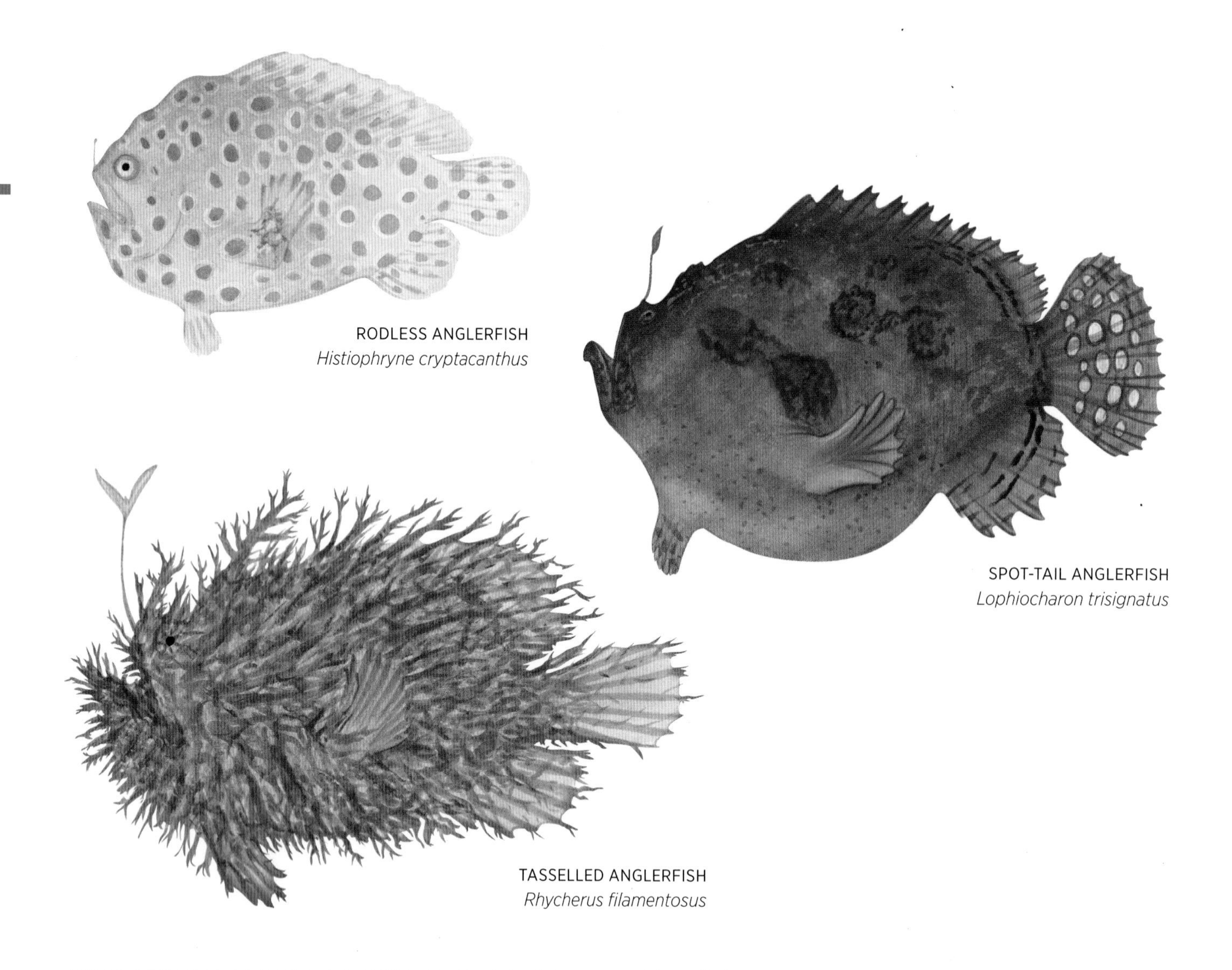

RODLESS ANGLERFISH
Histiophryne cryptacanthus

SPOT-TAIL ANGLERFISH
Lophiocharon trisignatus

TASSELLED ANGLERFISH
Rhycherus filamentosus

AUSTRALIAN ANTENNARIDAE SPECIES

Allenichthys glauerti Glauert's Anglerfish
Antennarius analis Tailjet Anglerfish
Antennarius coccineus Freckled Anglerfish
Antennarius commerson Giant Anglerfish
Antennarius dorehensis New Guinean Anglerfish
Antennarius hispidus Shaggy Anglerfish
Antennarius maculatus Warty Anglerfish
Antennarius nummifer Spotfin Anglerfish
Antennarius pictus Painted Anglerfish
Antennarius rosaceus Rosy Anglerfish
Antennarius striatus Striate Anglerfish
Antennatus tuberosus Tuberculate Anglerfish
Echinophryne crassispina Prickly Anglerfish
Echinophryne mitchellii Spinycoat Anglerfish
Echinophryne reynoldsi Sponge Anglerfish
Histiophryne bougainvilli Smooth Anglerfish
Histiophryne cryptacanthus Rodless Anglerfish
Histrio histrio Sargassum Fish
Kuiterichthys furcipilis Rough Anglerfish
Lophiocharon hutchinsi Hutchins' Anglerfish
Lophiocharon trisignatus Spot-tail Anglerfish
Phyllophryne scortea Whitespotted Anglerfish
Rhycherus filamentosus Tasselled Anglerfish
Rhycherus gloveri Glover's Anglerfish
Tathicarpus butleri Blackspot Anglerfish

HUMPBACK ANGLERFISH

Tetrabrachiidae

HUMPBACK ANGLERFISH
Tetrabrachium ocellatum

The single species in this family is restricted to the west and north coasts of Australia, southern New Guinea and the Molucca Islands of Indonesia. The **Humpback Anglerfish** has a globe-shaped head with eyes set close together on the dorsal surface and a small, vertical mouth with rows of small, curved teeth. The moderately elongate, tapering body has a distinctly humped back. The first dorsal spine is small, without an esca, and the second spine is larger and fringed with tendrils. The pectoral fins are divided into two parts, with the upper ray of the lower part attached to the body by a membrane. The gill opening is in the form of a small pore behind the pectoral fin. The Humpback Anglerfish has smooth skin, without spinules or fleshy papillae. Growing to about 8 cm in length, it inhabits deeper offshore waters, usually over soft substrate, and is occasionally taken by trawlers. Because it lacks an illicium and esca, it probably does not rely on luring prey and perhaps instead uses stealth, burying itself in the soft substrate and simply waiting for suitable prey to pass within striking distance.

AUSTRALIAN TETRABRACHIIDAE SPECIES

Tetrabrachium ocellatum Humpback Anglerfish

BOSCHMA'S ANGLERFISH

Lophichthyidae

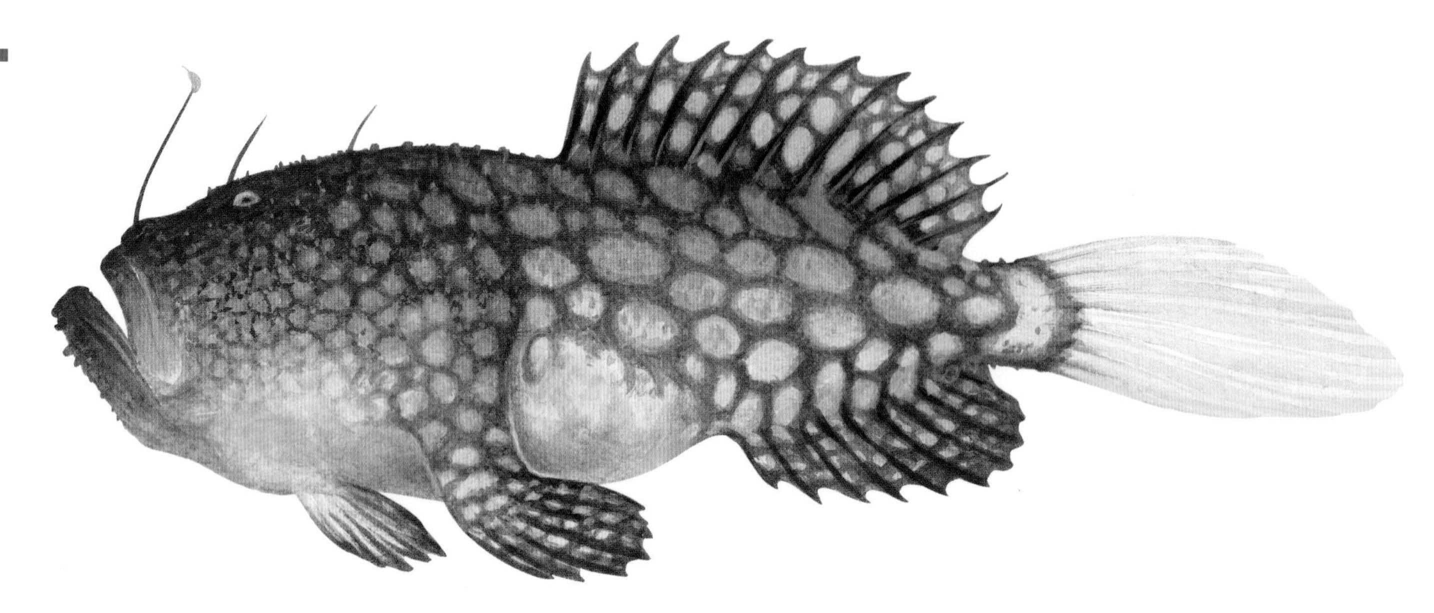

BOSCHMA'S ANGLERFISH
Lophichthys boschmai

The only species in this family is the small and quite rare **Boschma's Anglerfish**, which is found only in the Arafura Sea between Australia and New Guinea. It has an elongate, compressed body and reaches a maximum length of about 7.5 cm. It has spiny skin on the dorsal and lateral surfaces, and numerous small skin flaps over the body. The mouth is large and vertical, with small, curved teeth. The first dorsal spine is modified into an illicium with an esca at its tip, the second and third dorsal spines are separate, and the soft dorsal and anal fins are long-based. The gill opening is in the form of a small tube, which is above and posterior to the pectoral fin. Virtually nothing is known of the biology of Boschma's Anglerfish, though it is likely that it utilises the same feeding strategy as other Anglerfishes – resting immobile on the bottom and using its lure to attract prey within striking distance.

AUSTRALIAN LOPHICHTHYIDAE SPECIES
Lophichthys boschmai Boschma's Anglerfish

HANDFISHES

Brachionichthyidae

RED HANDFISH
Sympterichthys politus

Handfishes are found only in southern Australia, most commonly around Tasmania. There are four known species in the family and a number of others as yet unnamed. They are related to the Anglerfishes (Antennariidae, p. 294) and, like them, the first dorsal spine (which is situated on the tip of the snout) forms an illicium with an esca at its tip. The second and third dorsal spines, however, are joined by a membrane to form a distinct fin. Handfishes have a large head and a tapering body, which are scaleless and either smooth or covered with small spines or papillae. The gill openings are reduced to small pores behind the base of the pectoral fins. The pectoral fins are particularly well developed, with a distinct 'elbow', and are used to 'walk' along the substrate. The small ventral fins are placed beneath the head.

Handfishes are bottom dwellers, found in estuaries and coastal waters down to 100 m or more. They feed on small fishes, and benthic invertebrates such as molluscs and crustaceans. Their distribution is restricted, as they move around very little as adults, and juveniles disperse from their place of hatching only by walking across the bottom to another suitable habitat. All Handfishes are classified as threatened species and are completely protected in Tasmanian waters.

AUSTRALIAN HANDFISH
Brachionichthys australis

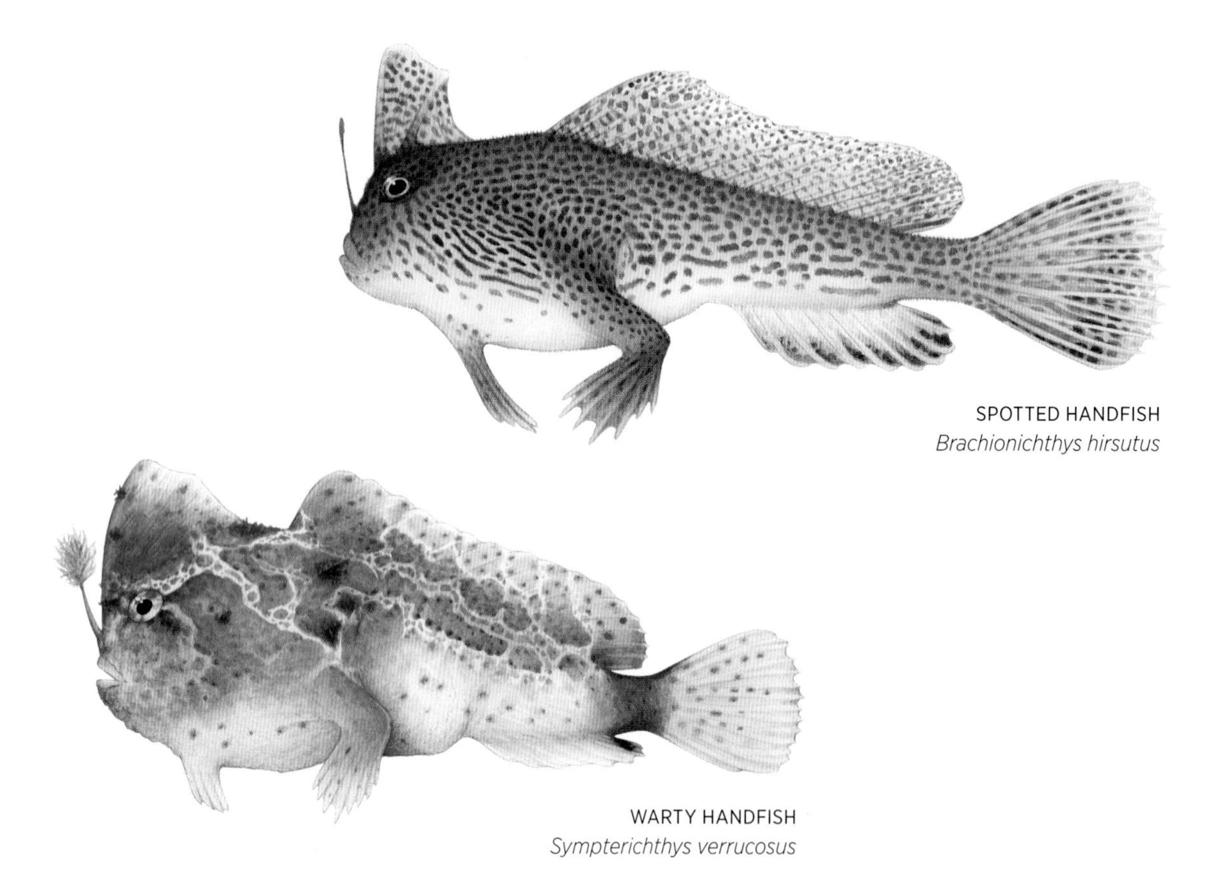

SPOTTED HANDFISH
Brachionichthys hirsutus

WARTY HANDFISH
Sympterichthys verrucosus

AUSTRALIAN BRACHIONICHTHYIDAE SPECIES

Brachionichthys australis Australian Handfish
Brachionichthys hirsutus Spotted Handfish
Sympterichthys politus Red Handfish
Sympterichthys unipennis Smooth Handfish
Sympterichthys verrucosus Warty Handfish

COFFINFISHES

Chaunacidae

TASSELLED COFFINFISH
Chaunax fimbriatus

Coffinfishes are found in deep waters of the Atlantic, Indian and Pacific oceans, from depths of 100 m to more than 2000 m. There are at least 14 species in the family and several more are as yet undescribed. They have a flabby, balloon-like inflatable body and loose skin that is densely covered with small spines. The flattened head has large eyes and a wide, almost vertical mouth with numerous villiform teeth. The single unattached dorsal spine is modified into an illicium (the other dorsal spines are buried beneath the skin). The esca is in the form of a dense bunch of short tentacles, which can be depressed into a hollow between the eyes. The pectoral fins are formed into short, horizontal paddles, with small gill openings above and behind them. There are small ventral fins beneath the head and the anal fin has a distinctly short base. A prominent network of sensory canals is present on the head and body, and these contain sensory filaments (neuromasts). Large, spiny scales are arranged at regular intervals along each side of the canals to protect the sensitive neuromasts. Most Coffinfishes have a pale-pink or reddish body, some with spots and streaks, and reach a maximum length of about 40 cm. They live on soft-sediment bottoms, mainly below depths of 200 m. Very little is known of their biology.

AUSTRALIAN CHAUNACIDAE SPECIES

Bathychaunax melanostomus Tadpole Coffinfish
Chaunax endeavouri Furry Coffinfish
Chaunax fimbriatus Tasselled Coffinfish
Chaunax penicillatus Pencil Coffinfish

BATFISHES

Ogcocephalidae

LONGNOSE SEABAT
Malthopsis lutea

There are about 68 species of Batfish and they are distributed worldwide, in tropical and warm temperate waters. They are found right around Australia. Most species occur from 1500 to 3000 m, though a few are found in coastal waters as shallow as 20 m and one species has been found as deep as 4000 m. Batfishes have an extremely flattened, disc-shaped or triangular body that is covered with bony plates and spines. The spines are often branched and project outwards around the edges of the body. They have a wide, horizontal mouth with small, fine, conical teeth. There is a very short globular illicium on the snout, which can be depressed into a forward-facing hollow. Many species have a distinct bony, horn-like rostrum. The second dorsal fin is small and placed well back on the tail, and the anal fin is greatly reduced. The pectoral fins are long and leg-like, and, along with the ventral fins situated beneath the head, are used for walking along the bottom. The small gill openings are located on the dorsal surface of the membrane that attaches the pectoral fin.

Batfishes attain a maximum length of about 40 cm, with most growing to around 20 cm. They are very poor swimmers and remain on the substrate almost all the time. The glandular esca may produce a chemical that attracts prey, or it may be used as a visual lure, as in other Anglerfishes. Batfishes feed on a variety of benthic invertebrates – such as small crustaceans, polychaete worms and molluscs, as well as small fishes – which indicates that they do not depend on the lure alone to obtain food.

SMOKY SEABAT
Halieutaea fumosa

AUSTRALIAN OGCOCEPHALIDAE SPECIES

Coelophrys brevipes Balloon Seabat
Dibranchus japonicus Japanese Seabat
Halicmetus reticulatus Marbled Seabat
Halieutaea brevicauda Shortfin Seabat
Halieutaea coccinea Scarlet Seabat
Halieutaea fumosa Smoky Seabat
Halieutaea stellata Starry Seabat
Malthopsis lutea Longnose Seabat

FANFIN ANGLERS

Caulophrynidae

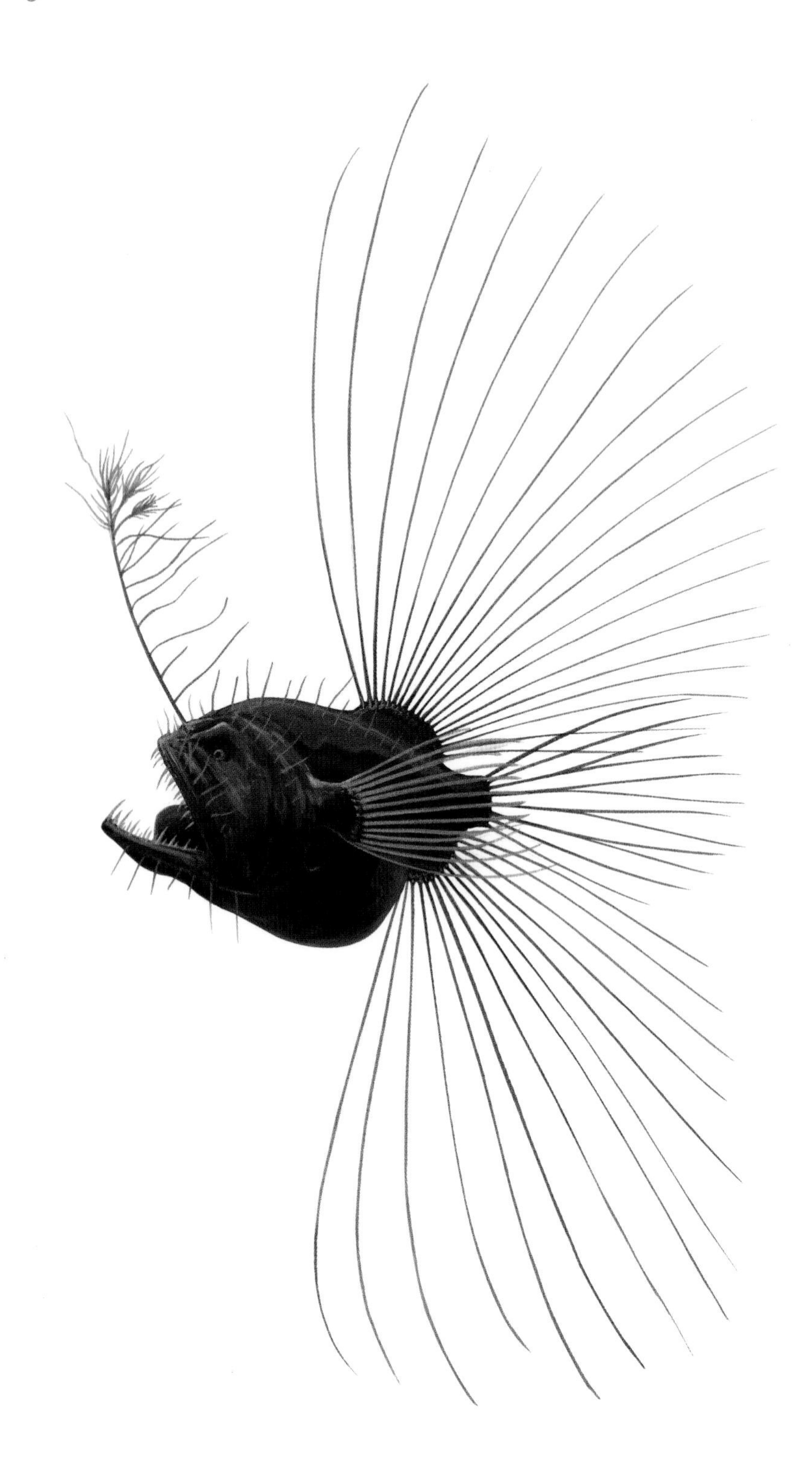

FANFIN ANGLER
Caulophryne jordani

Fanfin Anglers are pelagic in deep waters from depths of 500 m to at least 3000 m, and are found in the Atlantic, Indian and Pacific oceans. There are about five species in the family. Female Fanfin Anglers have no ventral fins, and the rays of the caudal, dorsal and anal fins are extremely elongate, often longer than the body length. There is also a network of long, hair-like sensory filaments on the body and head. Females have an illicium that bears many filaments along its length, and a branched filamentous esca. They have minute eyes and the elongate fin rays and various filaments on the body probably serve as a sensitive vibration detection system for locating prey. The illicium, which lacks a light organ at its tip, would presumably be ineffective as a lure in the darkness of deep waters, so may instead serve as a sensory organ. The diet of female Fanfin Anglers is unknown as the few specimens that have been captured have all had empty stomachs.

Females reach about 20 cm in length, while males are much smaller, growing to only about 2 cm. Males do not possess the illicium and esca, or the elongate fin rays, and have large eyes and very large olfactory organs, both of which are used to find prospective mates. Once a male has located a female, it attaches itself to her permanently, usually on the belly, using specialised denticular teeth. The tissues of the male and female actually fuse together and a connection is established between their circulatory systems. This adaptation ensures reproductive success in an environment where mates can be difficult to locate. Free-living males do not feed once mature, relying on energy stores in their large liver. Once attached to a female, they obtain nutrients from her through blood exchange.

AUSTRALIAN CAULOPHRYNIDAE SPECIES

Caulophryne jordani Fanfin Angler

SPINY SEADEVIL

Neoceratiidae

SPINY SEADEVIL
(female with male attached)
Neoceratias spinifer

The **Spiny Seadevil** is the only species in this family. It is pelagic in mid to bottom waters from depths of 1700 m to at least 2500 m and is found in the Atlantic, Indian and Pacific oceans. Females reach a maximum length of about 10 cm and have a slender, elongate body with smooth skin and a large, horizontal mouth. They are notable for the absence of an illicium and esca, and for their unusual teeth. Along the outside of the jaws they have several rows of very long, movable, fang-like teeth, which are hooked at the tips. As they lack a lure, it is thought that they may feed on gelatinous animals such as salps by swimming rapidly through congregations of prey with their teeth extended laterally.

The males are much smaller than the females, reaching about 1.8 cm in length, and permanently attach themselves to females using their specialised denticular teeth. It is probable that neither males nor females develop functional reproductive tissue until the male attaches to the female. Once the male attaches, the tissues of the two animals fuse and the male subsists on the blood supply of the female; the testes and ovaries then develop. A female may have several males permanently attached to her body. Few specimens have been encountered and the biology of the Spiny Seadevil is largely unknown.

AUSTRALIAN NEOCERATIIDAE SPECIES

Neoceratias spinifer Spiny Seadevil

BLACK SEADEVILS

Melanocetidae

HUMPBACK BLACKDEVIL (female and male)
Melanocetus johnsonii

Melanocetidae

Black Seadevils are pelagic in mid to bottom waters of the Atlantic, Indian and Pacific oceans, from depths of 600 m to more than 2500 m. There are about six species in the family. Females have a smooth-skinned, flabby, globe-shaped body, with no spines on the head or any dermal filaments, and an enormous vertical mouth with numerous large, curved fangs. The illicium is short, with a simple globe-shaped bioluminescent esca at its tip. The anal fin is very short-based and has 3–5 rays, the dorsal fin base is much longer and has 13–16 rays, and the ventral fins are absent. Females reach about 13 cm in length and have very small eyes beneath a layer of transparent skin. A network of low papillae form a sensory lateral line system.

Females are ambush predators, and are thought to spend most of their time drifting passively at depth, using the luminescent lure to attract prey and then seizing it with a sudden lunge; the powerful suction produced as they open their enormous mouth helps to draw the prey towards them. Females also have a greatly distensible stomach and are known to prey on fishes that are larger than themselves. Several specimens have been found to have died after having a large fish become stuck in their jaws – the prey too big to swallow but the Black Seadevil's inward-hinging teeth preventing its release.

The males are much smaller and more elongate than the females, usually reaching only 4 cm in length. They lack the fang-like teeth of females and instead have fused groups of denticular teeth. They lack an illicium and esca but have large eyes and nasal olfactory organs. The olfactory organs are thought to be used to locate females by tracing their pheromones, and the denticular teeth to attach to the female's body. Males are free-living – they attach to a female only until spawning has taken place, and do not become parasitic. The males may not feed at all as adults, relying on stored energy until finding a mate, then spawning and subsequently perishing. The **Humpback Blackdevil** is one of the more widely distributed and common deepsea ceratioid Anglerfishes, and numerous specimens have been collected from around the world.

AUSTRALIAN MELANOCETIDAE SPECIES

Melanocetus johnsonii Humpback Blackdevil
Melanocetus murrayi Deepsea Blackdevil

FOOTBALLFISHES

Himantolophidae

PRICKLY FOOTBALLFISH
Himantolophus appelii

Footballfishes are found in the Atlantic, Indian and Pacific oceans, and are pelagic in midwater from depths of 250 m to about 4000 m. There are about 18 species in the family. Females have a globe-shaped body that bears numerous, widely spaced spines on the skin and warty dermal papillae on the snout and chin. The mouth is smaller than that of most other deepsea Anglerfishes, with a much thicker, protruding lower jaw. Female Footballfishes have numerous fine teeth in the jaws, and minute eyes. The second dorsal and anal fins are very short-based, and are close to the caudal fin. Females also have a short, robust illicium with an ornate, fleshy, bioluminescent esca at the tip, consisting of a bulb and several branched tentacles. In the **Prickly Footballfish** this lure resembles a small squid. Like most deepsea Anglerfishes, Footballfishes can control the amount of light emitted from the esca by expanding or contracting the black pigment around it.

Female Footballfishes grow to a maximum length of about 45 cm, while the males reach only about 4 cm in length. Males lack the illicium and esca, and are more elongate in shape. They have large eyes and well-developed olfactory organs that are presumably used to locate females by tracing their pheromone signals. Their teeth are reduced to a few denticles, which they use to attach themselves to the body of a female prior to mating. Males are much rarer than females and, since they do not attach to the females permanently, it is not known to which species the male specimens that have been found belong.

AUSTRALIAN HIMANTOLOPHIDAE SPECIES

Himantolophus appelii Prickly Footballfish

DOUBLE ANGLERS

Diceratiidae

HORNED TWO-ROD ANGLER
Diceratias bispinosus

The six known species of Double Angler are distributed throughout the Atlantic, Indian and Pacific oceans, and are pelagic in mid to bottom waters from 300 to 2000 m. Females have a globe-shaped body covered with small, close-set dermal spinules, and two prominent bony spines on the dorsal surface of the head. The illicium is attached either near the tip of the snout or much further towards the posterior of the head, depending on the species. The esca varies from a simple bulb shape to branched and filamentous. Immediately behind the illicium, female Double Anglers have a second, smaller dorsal spine, which is also modified and bears a bulbous light organ at the tip. In larger specimens this second spine is often hidden beneath the skin. Female Double Anglers reach a maximum length of about 27 cm and have a large mouth, angled upwards, with numerous curved, fang-like teeth. They consume a wide variety of prey, including other fishes, crustaceans, worms, cnidarians, ctenophores and echinoderms; many of these are benthic animals, indicating that Double Anglers live closer to the bottom than many other deepsea Anglers.

Male Double Anglers are much smaller than the females and lack the illicium and esca. They have well-developed eyes, separate denticular teeth in the jaws, and small olfactory organs. Males are not parasitic on females and use their denticular teeth only to attach to the female until mating. Males are very rare and little is known about them – more than 200 female Double Anglers have been collected, but only one male (free-living and immature) has ever been captured.

AUSTRALIAN DICERATIIDAE SPECIES

Diceratias bispinosus Horned Two-rod Angler

DREAMERS
Oneirodidae

LONGHEAD DREAMER
Chaenophryne longiceps

Oneirodidae

Dreamers are midwater pelagic fishes, and are found in all the oceans of the world, from depths of 300 to 2000 m. Containing about 62 species, this is by far the largest and most diverse family of deepsea Anglerfishes. Female Dreamers have smooth skin and globe-shaped to elongate fusiform bodies, with some species almost spherical. Their large mouths vary from almost horizontal to almost vertical and contain many short, curved teeth. Most have a pair of horn-like spines on the dorsal surface of the head and many also have strong spines on the lower jaw. The illicium is on the end of a pterygiophore that projects from the head between the eyes. Depending on the species, the illicium can be very short, with the tip barely extending beyond the dorsal surface, or it can be longer than the body. The esca is a bulbous, fleshy, bioluminescent lure with simple or many-branched filaments. Different-shaped escas may attract different prey, but most Dreamers seem to feed quite opportunistically on a wide variety of prey, including pelagic crustaceans, squid and fishes.

Female Dreamers reach a maximum length of about 30 cm, with most measuring 10–15 cm. A female Dreamer has been filmed at a depth of more than 1400 m, drifting and swimming slowly, constantly changing its orientation: upside down, vertical with head up or down, and horizontal. It was perhaps exposing its lure to all points of the compass, but we can only speculate. The males reach only about 2 cm in length and are more elongate than females. They lack an illicium and esca, and have large eyes and well-developed olfactory organs. Males are free-living and are not parasitic on the females, attaching to them only until mating has been accomplished. It is difficult to match male specimens with females of the same species, and males of some species have never been identified.

AUSTRALIAN ONEIRODIDAE SPECIES

Chaenophryne draco Smooth Dreamer
Chaenophryne longiceps Longhead Dreamer
Dolopichthys pullatus Lobed Dreamer
Oneirodes kreffti Krefft's Dreamer
Oneirodes sabex Rough Dreamer
Oneirodes whitleyi Whitley's Dreamer

SEADEVILS
Ceratiidae

TRIPLEWART SEADEVIL
Cryptopsaras couesii

Seadevils are distributed throughout the Atlantic, Indian and Pacific oceans. There are four species in the family, all of which have been recorded in Australian waters. They are some of the most common and well known of the deepsea Anglerfishes, occurring mainly at depths of 400–2000 m. Females have an elongate, compressed head and body, with no spines on the head, and skin covered in small dermal spinules. They have a large, almost vertical mouth with short, sharp teeth that are longer in the lower jaw than the top jaw. The illicium is mounted on the tip of a movable, elongate pterygiophore that can slide back and forth in a groove on top of the skull. The posterior end of the sliding pterygiophore projects from the back of the head. The esca is in the form of a fleshy globe with simple filaments. Just anterior to the soft dorsal fin there are several bulbous, fleshy dorsal fin rays bearing light organs.

Female Seadevils reach a maximum length of about 85 cm in the **Longray Seadevil**. Male Seadevils are much smaller than females and lack the illicium, esca and light organs. They have fused denticles instead of teeth, very large, bowl-shaped eyes, and poorly developed nasal olfactory organs. Once a free-living male has found a female, it uses its denticular teeth to attach itself to her permanently, usually on the belly. The tissues of the two animals then fuse and a connection is established between their circulatory systems. Neither the females nor the free-living males develop functioning reproductive ovaries and testes until the male has attached to the female. Free-living males grow to only about 2 cm in length. They do not feed after the larval stage and die if they are unable to find and attach to a female. Once attached, however, the parasitic male continues to grow and can reach up to 14 cm in length. Females often have more than one male attached – eight males have been found on a single large female **Triplewart Seadevil**.

AUSTRALIAN CERATIIDAE SPECIES

Ceratias holboelli Longray Seadevil
Ceratias tentaculatus Southern Seadevil
Ceratias uranoscopus Stargazing Seadevil
Cryptopsaras couesii Triplewart Seadevil

WHIPNOSE ANGLERS

Gigantactinidae

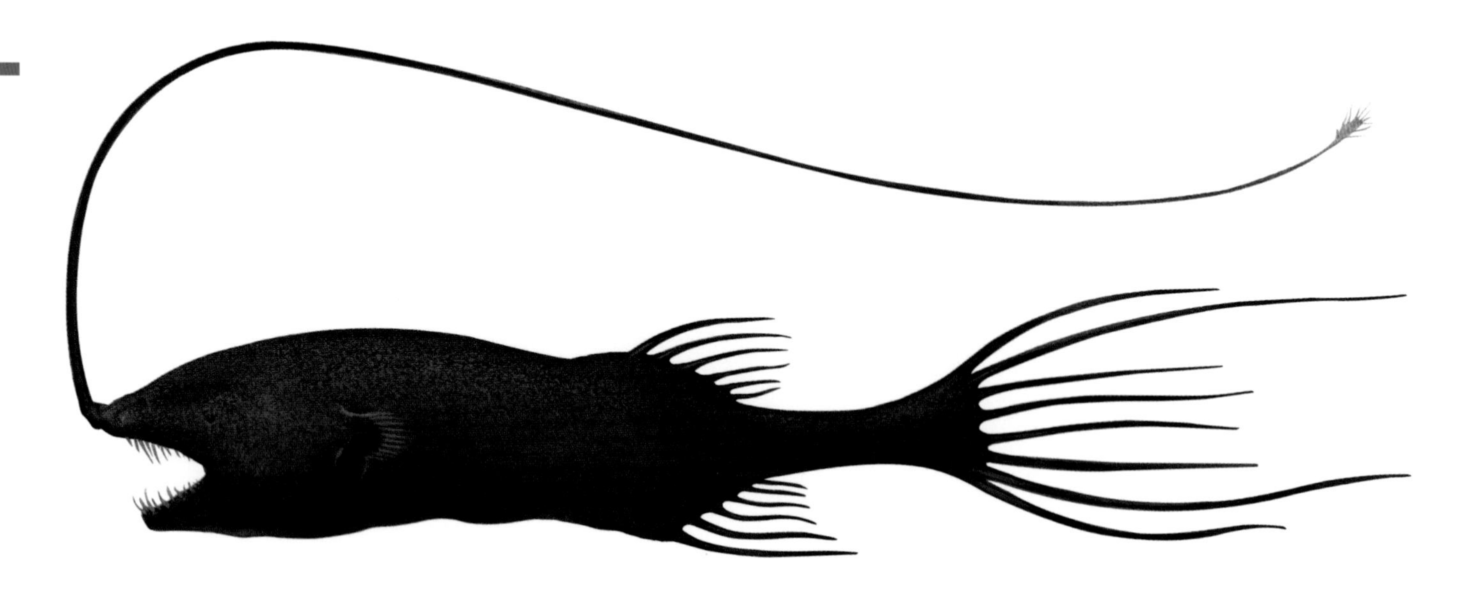

GIGANTIC WHIPNOSE
Gigantactis gargantua

The 22 species of Whipnose Angler are found in the Atlantic, Indian and Pacific oceans. They are pelagic in midwater from depths of about 1000 to 2500 m. Females, which grow to about 40 cm in length, have a much more elongate body than most other deepsea Anglerfishes. They have a very long caudal peduncle and their skin is covered with small dermal spinules. They have a small head, and very small eyes that degenerate in large adults of some species. The illicium is extremely elongate, up to five times the length of the body, with a small bioluminescent esca bearing various appendages and filaments. The illicium is mounted on a large pterygiophore that projects out in front of the jaws and forms the tip of the pointed snout. Most females have a large, almost horizontal mouth with many thin, curved teeth. The longer teeth on the outside of the jaws extend laterally and the lower jaw can be rapidly folded inward like a trap, to hold prey against the curved teeth of the upper jaw, which then move it into the stomach.

Female Whipnose Anglers probably feed by attracting prey items with their bioluminescent lure, which they vibrate through movement of the pterygiophore, then engulfing them with a sudden powerful lunge forward. They have been observed at great depth, drifting upside down just above the bottom with the lure held in front of them: presumably this strategy is designed to attract benthic invertebrates from the sediment. However, the diet of Whipnose Anglers is not well known. Several species have only a few very small teeth, and numerous glands in the mouth that may be bioluminescent.

Male Whipnose Anglers are much smaller than the females, lack the illicium and esca, and have minute eyes and large, well-developed nasal olfactory organs. Males have small, separate denticular teeth that they use to attach to the female for reproduction; they do not attach permanently and are not parasitic. Little is known about the males and although many different species of free-living males have been recorded it is not clear to which female each corresponds.

AUSTRALIAN GIGANTACTINIDAE SPECIES

Gigantactis elsmani Elsman's Whipnose
Gigantactis gargantua Gigantic Whipnose
Gigantactis paxtoni Paxton's Whipnose

LEFTVENTS

Linophrynidae

SOFT LEFTVENT ANGLER
Haplophryne mollis

Leftvents are pelagic in midwater from 1000 to 4000 m and are found in the Atlantic, Indian and Pacific oceans. There are 27 species in the family, with many species known only from single specimens. Females have a globe-shaped body with smooth skin, a pair of strong, bony spines on the dorsal surface of the head, and often well-developed spines on the preoperculum and posterior of the jaws. They are unique amongst all vertebrates in having the anus always situated on the left side of the body, as opposed to on the midline. They have an extremely large mouth – in some species the mouth is proportionately the largest of all vertebrates – with dentition ranging from a few long, curved fangs, to numerous short, curved teeth.

The illicium varies greatly between species, from short and thick to long and thin, and the esca is also highly variable. Many species in the genus *Linophryne* have a large, complex, many-branched barbel beneath the chin that bears numerous small light organs. These light organs do not contain symbiotic bacteria, as do the bioluminescent escae of other Anglerfishes; rather, light is produced from photogenic granules of a type not found in any other bioluminescent fish.

Female Leftvents reach a maximum length of about 27 cm. Males are much smaller and more elongate than females, and lack the illicium, esca and chin barbel. They have large, slightly tubular eyes, and large olfactory organs. Males parasitise the females, using their denticular teeth to attach themselves until a perment fusion of tissues takes place, then subsisting on the blood supply of the female. Males and females do not develop functioning reproductive tissue until the male has attached to the female. Males of many species are still unknown.

Among the other deepsea anglerfishes (there are about 160 species), which are all black or dark brown in colour, the **Soft Leftvent Angler** is unique in having no pigment whatsoever in the skin.

THICKBRANCH ANGLER
Linophryne densiramus

AUSTRALIAN LINOPHRYNIDAE SPECIES

Haplophryne mollis Soft Leftvent Angler
Linophryne densiramus Thickbranch Angler
Linophryne indica Headlight Angler

MUGILIFORMES

This order consists of a single family, which is sometimes placed in the order Perciformes (p. 439). Mugiliforms are distinguished from Perciforms, however, in having no connection, either by bone or ligament, between the ventral fins and the cleithrum (the major bone of the pectoral girdle). They have an elongate, fusiform body with ctenoid scales, two dorsal fins (the first with strong spines), ventral fins placed well behind the pectoral fins, and a small terminal mouth. The single family is well represented in Australian waters.

MULLETS

Mugilidae

DIAMONDSCALE MULLET
Liza vaigiensis

Mullets are found in tropical and temperate coastal waters and estuaries throughout the world. There are about 72 species in the family. They have two widely separated dorsal fins, the first with four strong spines, and the pectoral fins are placed high on the body. Mullets have a blunt head and a small mouth with very small teeth (or none at all), and large eyes. Most species possess an adipose eyelid. Their elongate, cylindrical body has no lateral line. They have ctenoid scales, except for the **Sand Mullet**, which has smooth cycloid scales.

Mullets feed predominantly on algae, phytoplankton and organic detritus, and have a muscular stomach and extremely long intestine to break down these hard-to-digest foods. Sand is often ingested along with detritus to help grind material in the stomach. Some small invertebrates are also taken, particularly by the Sand Mullet, which preys mainly on small marine worms. Mullets are schooling fishes, mostly encountered in surface waters, and are a major food source for larger predatory fishes, sharks and dolphins. It is common to see large numbers leaping out of the water in unison as they are pursued by an unseen predator.

In many parts of the world, Mullets are important food fishes for humans. The **Sea Mullet**, which occurs right around Australia, forms large schools in coastal waters and is a valuable commercial species. Found worldwide, it is one of the largest species of Mullet, reaching a maximum length of about 80 cm. In Australia the adults migrate northward each year along the east and west coasts, to spawn off ocean beaches. Larvae are then carried southward by ocean currents and enter estuaries to mature. The catch in Australian waters is mostly taken using beach seines during the Sea Mullet's annual northerly migration.

Another important commercial species is the **Yelloweye Mullet**, which reaches about 40 cm in length. It is common in shallow inshore waters, estuaries, and the lower reaches of rivers in southern Australia, where it is mostly taken by gillnets. Many smaller tropical species are also caught for food, and are widely used as bait for catching larger fishes. The **Pinkeye Mullet** spends most of its life in freshwater systems along the mid-east coast of Australia, moving into estuaries and coastal waters to spawn.

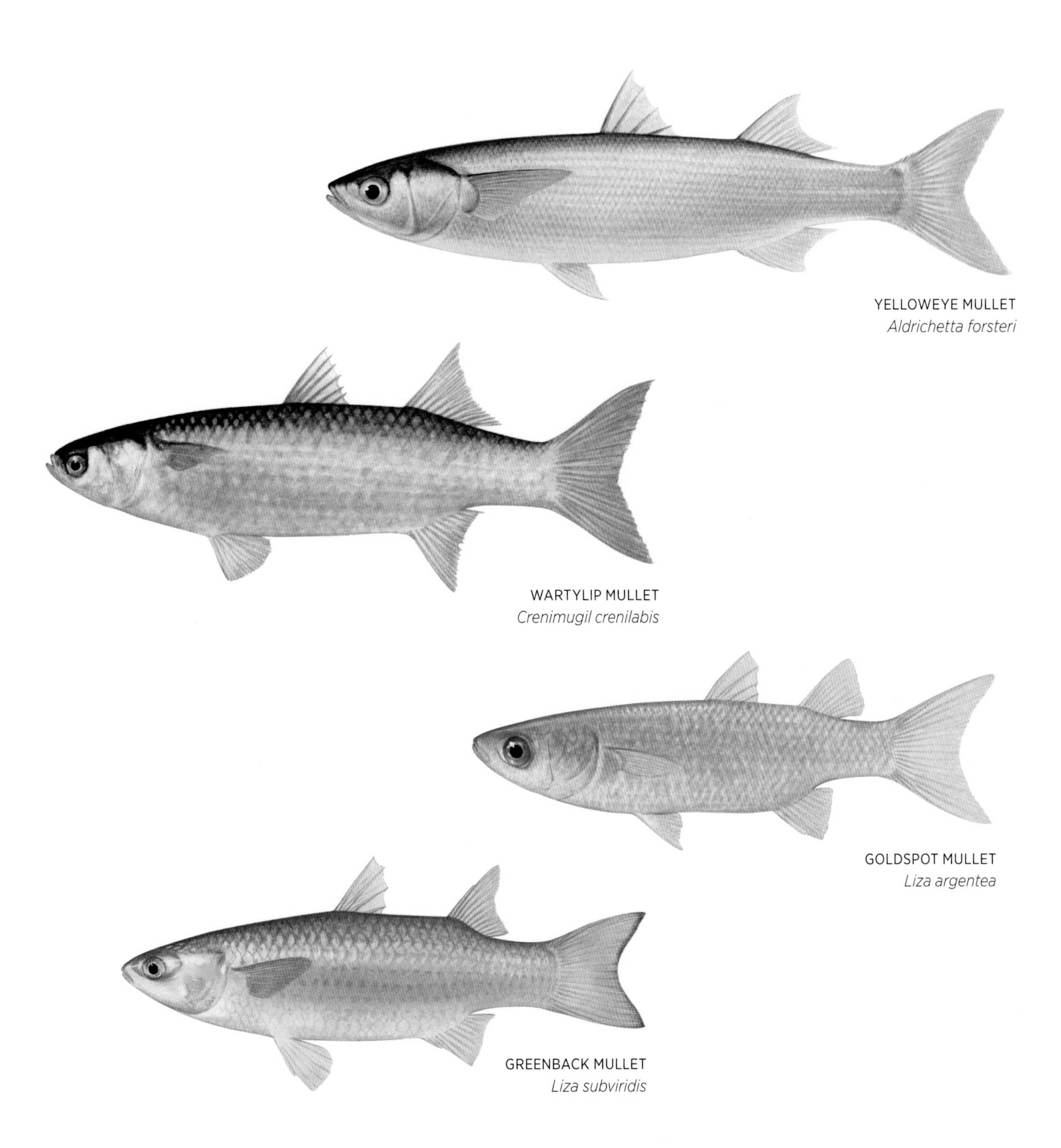

YELLOWEYE MULLET
Aldrichetta forsteri

WARTYLIP MULLET
Crenimugil crenilabis

GOLDSPOT MULLET
Liza argentea

GREENBACK MULLET
Liza subviridis

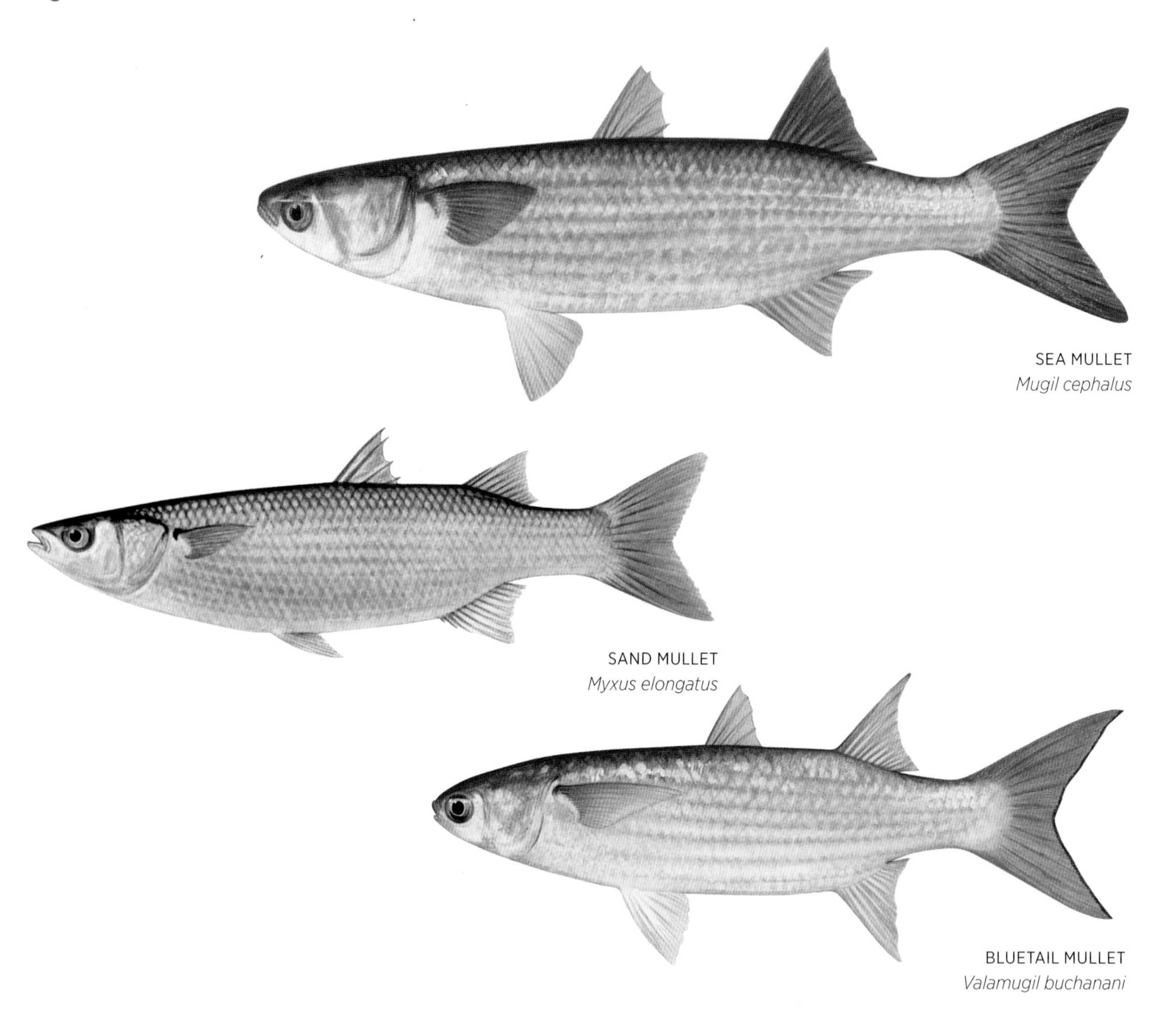

SEA MULLET
Mugil cephalus

SAND MULLET
Myxus elongatus

BLUETAIL MULLET
Valamugil buchanani

AUSTRALIAN MUGILIDAE SPECIES

Aldrichetta forsteri Yelloweye Mullet
Crenimugil crenilabis Wartylip Mullet
Crenimugil heterocheilos Fringelip Mullet
Liza alata Diamond Mullet
Liza argentea Goldspot Mullet
Liza melinoptera Otomebora Mullet
Liza parmata Broadmouth Mullet
Liza subviridis Greenback Mullet
Liza tade Rock Mullet
Liza vaigiensis Diamondscale Mullet
Mugil broussonnetii Broussonnet's Mullet
Mugil cephalus Sea Mullet
Myxus elongatus Sand Mullet
Myxus petardi Pinkeye Mullet
Oedalechilus labiosus Hornlip Mullet
Rhinomugil nasutus Popeye Mullet
Valamugil buchanani Bluetail Mullet
Valamugil cunnesius Roundhead Mullet
Valamugil engeli Kanda Mullet
Valamugil georgii Fantail Mullet
Valamugil seheli Bluespot Mullet
Valamugil speigleri Speigler's Mullet

ATHERINIFORMES

Atheriniforms are small, silvery fishes with an elongate, compressed body. They have two dorsal fins, the first with flexible spines and the second usually with one spine and soft rays. The pectoral fins are high on the body, and the ventral fins are usually posterior to the pectoral fins. There are six families in the order, five of which occur in Australian waters.

SURF SARDINES

Notocheiridae

SURF SARDINE
Iso rhothophilus

Surf Sardines are found in the Indo-West Pacific and around southern South America, most commonly occurring in the surf zone, off rocky headlands and ocean beaches. There are six species in the family. Reaching a maximum length of about 7.5 cm, they have a compressed, translucent, silvery body and are deep-chested. There is a fleshy keel on the ventral profile, a silvery, highly reflective area covering the internal organs, and a silvery stripe running along each side of the body. They have no lateral line, and small scales only on the lateral and dorsal surfaces of the body. The pectoral fins are placed very high on the body, almost on the dorsal surface. The small mouth has a single row of minute teeth in the anterior part of the jaws.

The **Surf Sardine** is found off the south-west and south-east coasts of Australia, usually off rocky headlands in the surf zone but occasionally entering the mouths of rivers. It is a common prey of Australian Salmons (Arripidae, p. 558). Surf Sardines are extremely delicate and although relatively common are rarely encountered due to the turbulent environments they normally frequent. They can survive such conditions because their small size allows them to become part of the turbulence instead of being torn apart by it.

AUSTRALIAN NOTOCHEIRIDAE SPECIES

Iso rhothophilus Surf Sardine

RAINBOWFISHES

Melanotaeniidae

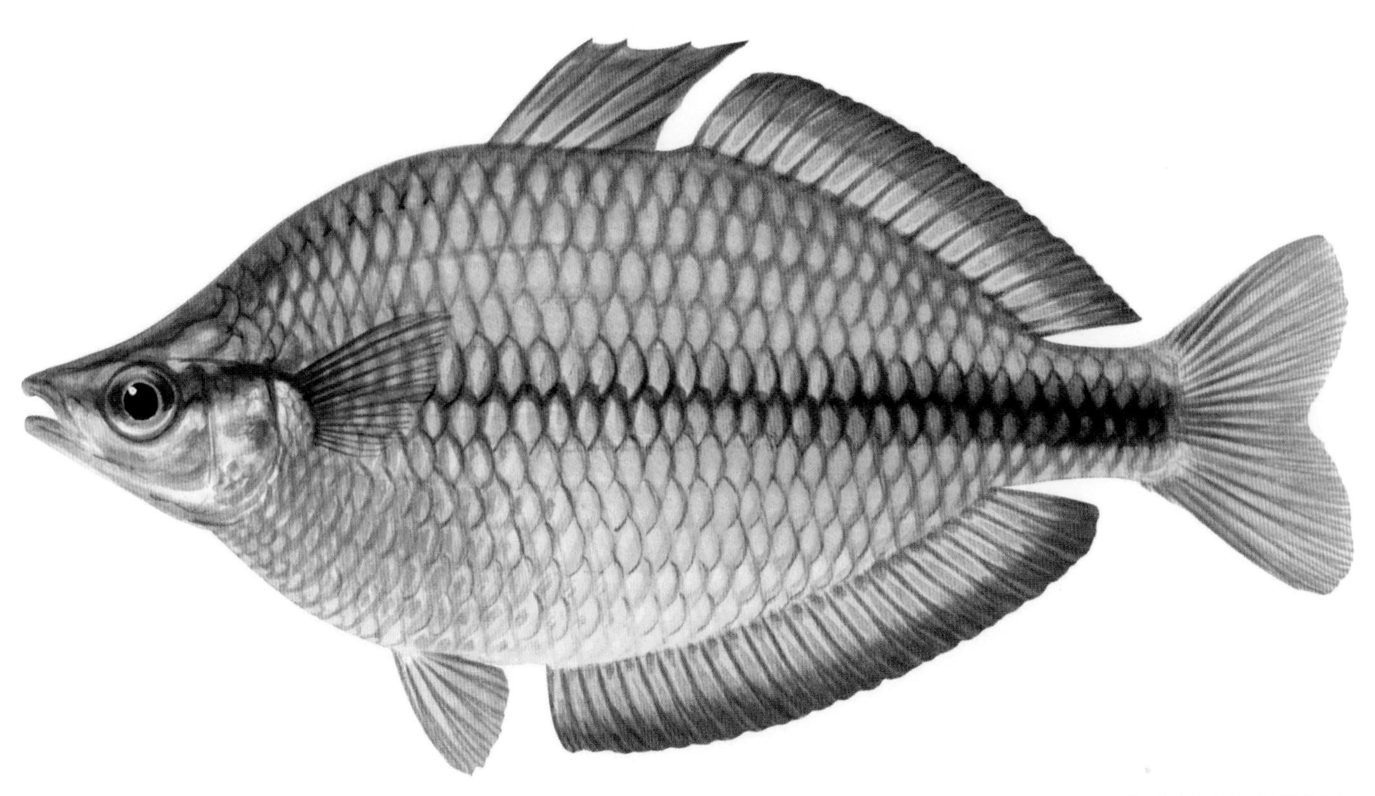

BANDED RAINBOWFISH
Melanotaenia trifasciata

Rainbowfishes are small, brightly coloured freshwater fishes. They are found only in New Guinea and Australia, and many species are common to both areas. There are 66 species in the family, with the greatest diversity found in New Guinea. Rainbowfishes reach a maximum length of about 12 cm and have a compressed body with large scales and a weakly developed or absent lateral line. There are two dorsal fins, the first with 3–7 spines, and a long-based anal fin. They have a small mouth with fine, conical teeth in the jaws. Males are generally larger and more colourful than females, and have a deeper body and more ornate fins.

Rainbowfishes are found in a wide variety of environments, from fast-flowing streams to muddy waterholes. In Australia they are found mainly in waters in the northern half of the continent. The **Murray River Rainbowfish**, however, occurs throughout the Murray-Darling drainage system, right to the mouth of the river on the south coast. Some species of Rainbowfish have very restricted distribution, while others are widespread and exhibit a variety of regional differences in colour pattern. They occur in small groups or large schools and feed on algae, aquatic insects and their larvae, and small crustaceans. Eggs are laid in aquatic vegetation, a few at a time, all year round. Rainbowfishes are relatively easy to breed and have become popular aquarium fishes.

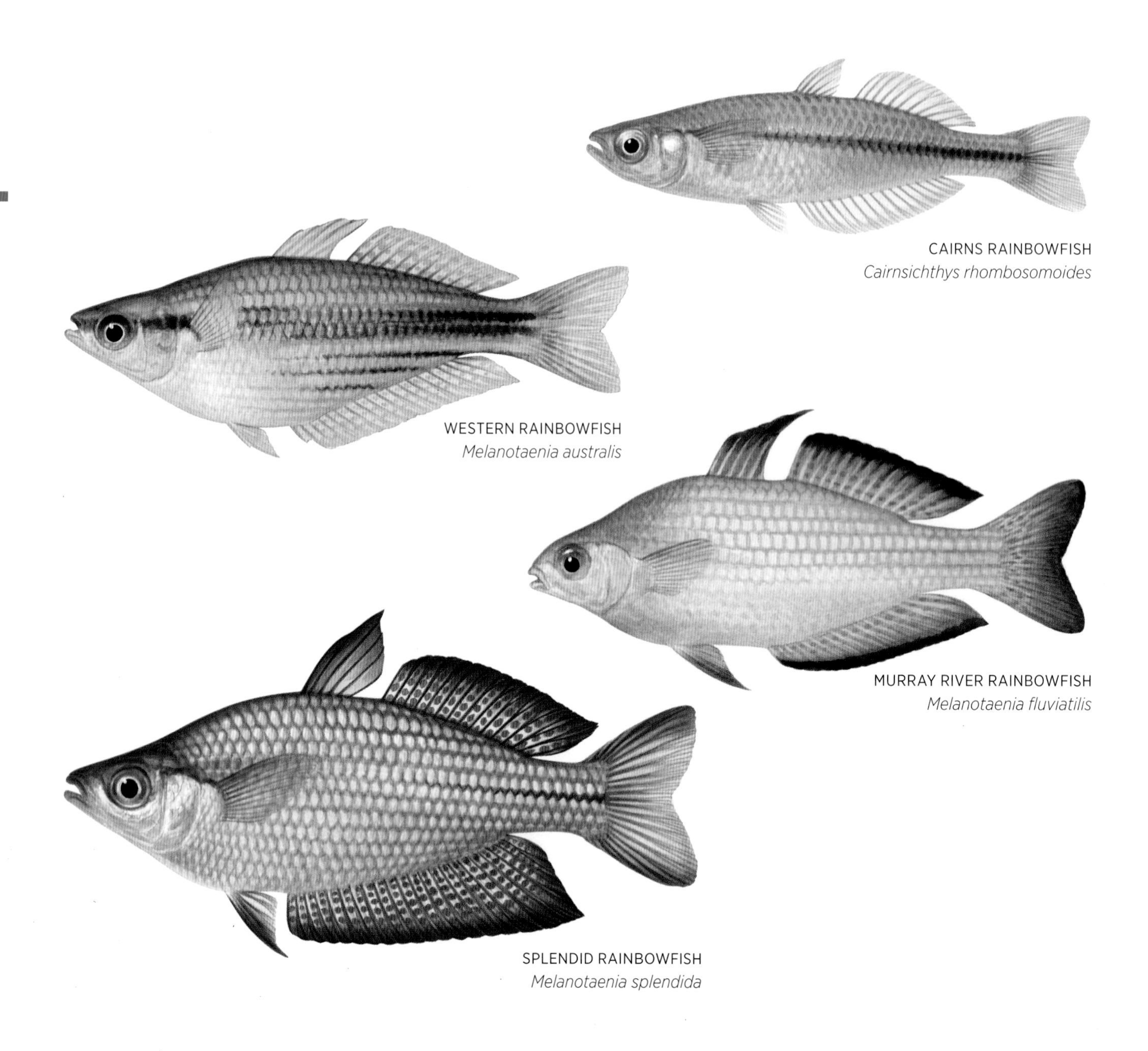

CAIRNS RAINBOWFISH
Cairnsichthys rhombosomoides

WESTERN RAINBOWFISH
Melanotaenia australis

MURRAY RIVER RAINBOWFISH
Melanotaenia fluviatilis

SPLENDID RAINBOWFISH
Melanotaenia splendida

AUSTRALIAN MELANOTAENIIDAE SPECIES

Cairnsichthys rhombosomoides Cairns Rainbowfish
Iriatherina werneri Threadfin Rainbowfish
Melanotaenia australis Western Rainbowfish
Melanotaenia duboulayi Crimsonspotted Rainbowfish
Melanotaenia eachamensis Lake Eacham Rainbowfish
Melanotaenia exquisita Exquisite Rainbowfish
Melanotaenia fluviatilis Murray River Rainbowfish
Melanotaenia gracilis Slender Rainbowfish
Melanotaenia maccullochi McCulloch's Rainbowfish
Melanotaenia nigrans Blackbanded Rainbowfish
Melanotaenia pygmaea Pygmy Rainbowfish
Melanotaenia solata Northern Rainbowfish
Melanotaenia splendida Splendid Rainbowfish
Melanotaenia trifasciata Banded Rainbowfish
Melanotaenia utcheensis Utchee Rainbowfish
Rhadinocentrus ornatus Ornate Rainbowfish

BLUE EYES

Pseudomugilidae

SPOTTED BLUE EYE
Pseudomugil gertrudae

Blue Eyes are closely related to the Rainbowfishes (Melanotaeniidae, p. 323) and share roughly the same distribution, being found only in New Guinea and northern Australia. There are about 17 species in the family. They have a compressed body with large scales and no lateral line. There are two dorsal fins, the first with soft rays instead of spines, and a moderately long anal fin, also with soft rays instead of spines. Many species have large, ornate fins, often brightly coloured, and the iris of the eye is usually a striking blue colour. Their small size and bright colours make them popular aquarium fishes.

Blue Eyes form schools in the lower reaches of rivers and in brackish estuaries around the north and east coasts of Australia. Most reach a maximum length of about 3 cm, the largest species growing to about 7 cm. They breed all year round when conditions are favourable, and feed on microscopic crustaceans, insects and other tiny invertebrates. Several species have very restricted distribution and are endangered due to habitat degradation.

PACIFIC BLUE EYE
Pseudomugil signifer

REDFIN BLUE EYE
Scaturiginichthys vermeilipinnis

AUSTRALIAN PSEUDOMUGILIDAE SPECIES

Pseudomugil cyanodorsalis Blueback Blue Eye
Pseudomugil gertrudae Spotted Blue Eye
Pseudomugil inconspicuus Inconspicuous Blue Eye
Pseudomugil mellis Honey Blue Eye
Pseudomugil signifer Pacific Blue Eye
Pseudomugil tenellus Delicate Blue Eye
Scaturiginichthys vermeilipinnis Redfin Blue Eye

TUSKED SILVERSIDE

Dentatherinidae

TUSKED SILVERSIDE
Dentatherina merceri

The **Tusked Silverside** is the only member of this family. It is found in waters from the Philippines to northern Australia and into the West Pacific. Similar in appearance to the Hardyheads (Atherinidae, p. 328), it differs in having large, lateral projections of bone extending out from the roof of the mouth to beneath the eyes, as well as a large, flattened, tusk-like structure on the anterior of the upper jaw. It also has a distinctive caudal peduncle that is very long and slender. The Tusked Silverside grows to about 5 cm long and frequents inshore waters around islands and coral reefs. Very little is known of its biology.

AUSTRALIAN DENTATHERINIDAE SPECIES
Dentatherina merceri Tusked Silverside

HARDYHEADS

Atherinidae

COMMON HARDYHEAD
Atherinomorus vaigiensis

There are about 165 species of Hardyhead and they are distributed worldwide, in freshwater and coastal marine habitats. Some 50 species occur in fresh water, including many of the Australian species. Hardyheads have elongate, cylindrical to compressed bodies with large, cycloid scales and no lateral line. There are two dorsal fins, the first with flexible spines and the second with one spine and soft rays. The anal fin has one spine and soft rays. They have a small mouth that is usually highly protrusible, and most have a broad silvery stripe running along each side of the body. Hardyheads are omnivorous, feeding on micro-invertebrates, algae and phytoplankton.

Australia has a large number of freshwater Hardyheads of the genus *Craterocephalus*; many of these have very restricted distribution in arid areas and some species are found in only a single small lake or river. The **Dalhousie Hardyhead** and **Glover's Hardyhead**, for example, are found only in the Dalhousie Springs in South Australia. The **Flyspecked Hardyhead**, in contrast, is common and widespread in rivers and streams in northern and eastern Australia.

Marine Hardyhead species often form dense schools in coastal waters and are an important food source for predators. Schools often move into very shallow water in their attempts to avoid large predators, with several different species commonly aggregating in the same school. Hardyheads reach a maximum length of about 10 cm. They are not generally caught for food and are not considered commercially important, but are widely used as bait for catching other fishes.

ENDRACHT HARDYHEAD
Atherinomorus endrachtensis

DESERT HARDYHEAD
Craterocephalus eyresii

FEW-RAY HARDYHEAD
Craterocephalus pauciradiatus

FLYSPECKED HARDYHEAD
Craterocephalus stercusmuscarum

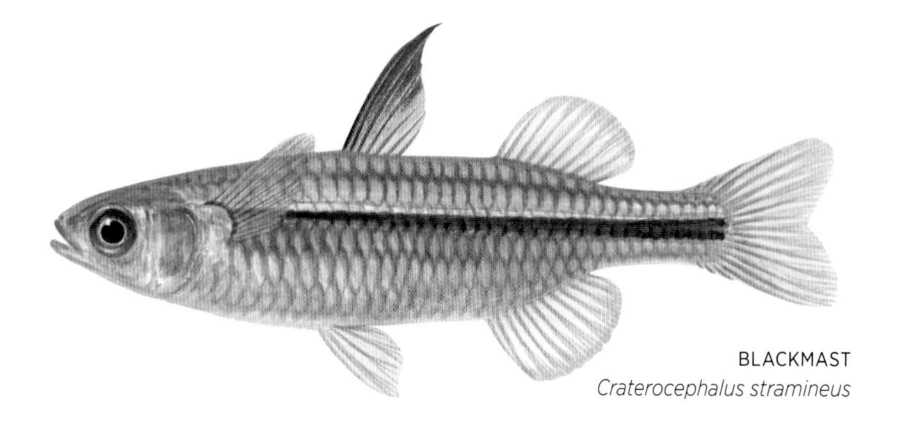

BLACKMAST
Craterocephalus stramineus

WESTERN HARDYHEAD
Leptatherina wallacei

AUSTRALIAN ATHERINIDAE SPECIES

Atherinason hepsetoides Smallscale Hardyhead
Atherinomorus capricornensis Capricorn Hardyhead
Atherinomorus endrachtensis Endracht Hardyhead
Atherinomorus lacunosus Slender Hardyhead
Atherinomorus vaigiensis Common Hardyhead
Atherinosoma elongata Elongate Hardyhead
Atherinosoma microstoma Smallmouth Hardyhead
Atherion elymus Prickleface Hardyhead
Atherion maccullochi McCulloch's Hardyhead
Craterocephalus amniculus Darling Hardyhead
Craterocephalus capreoli North-west Hardyhead
Craterocephalus centralis Finke Hardyhead
Craterocephalus cuneiceps Deep Hardyhead
Craterocephalus dalhousiensis Dalhousie Hardyhead
Craterocephalus eyresii Desert Hardyhead
Craterocephalus fluviatilis Murray Hardyhead
Craterocephalus fulvus Tawny Hardyhead
Craterocephalus gloveri Glover's Hardyhead
Craterocephalus helenae Drysdale Hardyhead
Craterocephalus honoriae Estuarine Hardyhead
Craterocephalus lentiginosus Freckled Hardyhead
Craterocephalus marianae Mariana's Hardyhead
Craterocephalus marjoriae Silverstreak Hardyhead
Craterocephalus mugiloides Spotted Hardyhead
Craterocephalus munroi Munro's Hardyhead
Craterocephalus pauciradiatus Few-ray Hardyhead
Craterocephalus stercusmuscarum Flyspecked Hardyhead
Craterocephalus stramineus Blackmast
Hypoatherina barnesi Barnes' Hardyhead
Hypoatherina ovalaua Fijian Hardyhead
Hypoatherina temminckii Samoan Hardyhead
Hypoatherina tropicalis Tropical Hardyhead
Kestratherina brevirostris Shortsnout Hardyhead
Kestratherina esox Pikehead Hardyhead
Leptatherina presbyteroides Silver Fish
Leptatherina wallacei Western Hardyhead
Stenatherina panatela Panatela Hardyhead

BELONIFORMES

Beloniforms all have a fixed, non-protrusible upper jaw. They have an elongate, silvery body, and the dorsal and anal fins are situated well towards the posterior of the body. There are no spines in the fins. The lateral line runs along the ventral margin of the body. In most the lower jaw, or both jaws, is elongated to form a beak. Beloniforms are pelagic in surface waters, with some occurring in fresh water. There are five families in the order, four of which occur in Australian waters.

FLYINGFISHES

Exocoetidae

FLYINGFISH
Cheilopogon sp.

Flyingfishes are found worldwide in tropical and temperate surface waters. The 52 species in the family are primarily pelagic in open waters, but are also frequently encountered near the coast. Flyingfishes occur in both northern and southern Australian waters. They have an elongate, cylindrical body with large, fragile scales and a small mouth. They reach a maximum length of about 45 cm. The pectoral fins, and often also the ventral fins, are greatly enlarged. It is these stiff, wing-like fins that allow Flyingfishes to 'fly' above the surface of the water. They are not able to flap the fins but can glide for considerable distances – several hundred metres or more. The lower lobe of the caudal fin is longer than the upper lobe and continues to provide them with propulsion as they leave the water, causing a telltale string of small splashes on the surface. The 'wings' are folded flat against the body whilst the fish swims under water. Their flights may take them several metres above the surface and this ability to 'fly' allows them to escape predators such as Tunas and Mackerels (Scombridae, p.717). Flyingfishes are commonly seen in tropical Australian waters and feed mainly on small planktonic organisms. Although common and widely distributed, they are difficult to capture and are not well known. Most records come from strandings on the decks of boats.

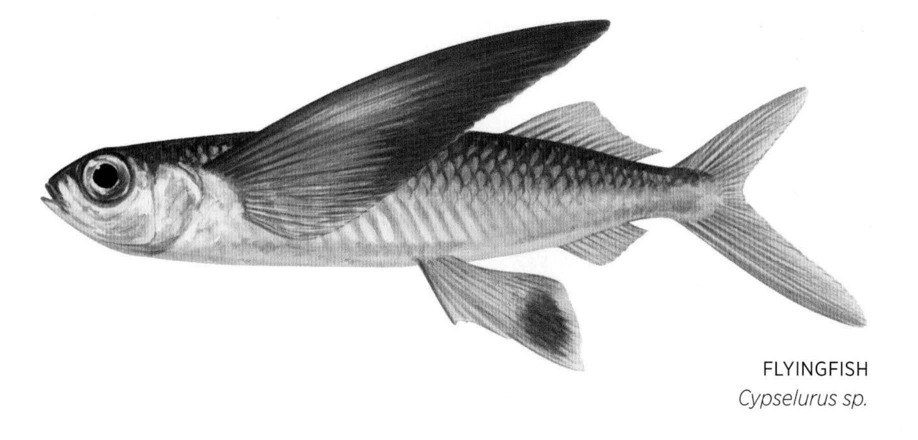

FLYINGFISH
Cypselurus sp.

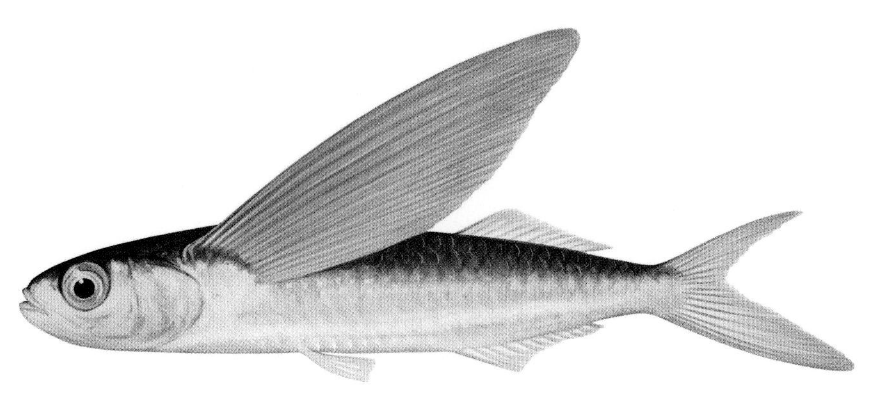

COSMOPOLITAN FLYINGFISH
Exocoetus volitans

AUSTRALIAN EXOCOETIDAE SPECIES

Cheilopogon abei Abe's Flyingfish
Cheilopogon arcticeps Bearhead Flyingfish
Cheilopogon atrisignis Glider Flyingfish
Cheilopogon cyanopterus Margined Flyingfish
Cheilopogon furcatus Spotfin Flyingfish
Cheilopogon heterurus Piebald Flyingfish
Cheilopogon intermedius Intermediate Flyingfish
Cheilopogon katoptron Indonesian Flyingfish
Cheilopogon nigricans Leaping Flyingfish
Cheilopogon pinnatibarbatus Tallfin Flyingfish
Cheilopogon spilonotopterus Stained Flyingfish
Cheilopogon spilopterus Manyspot Flyingfish
Cheilopogon suttoni Sutton's Flyingfish
Cheilopogon unicolor Limpid-wing Flyingfish
Cypselurus angusticeps Narrowhead Flyingfish
Cypselurus hexazona Darkbar Flyingfish
Cypselurus naresii Pharao Flyingfish
Cypselurus oligolepis Largescale Flyingfish
Cypselurus poecilopterus Yellow-wing Flyingfish
Exocoetus gibbosus Oceanic Flyingfish
Exocoetus monocirrhus Barbel Flyingfish
Exocoetus volitans Cosmopolitan Flyingfish
Hirundichthys oxycephalus Bony Flyingfish
Hirundichthys rondeletii Rondelet's Flyingfish
Hirundichthys speculiger Mirrorwing Flyingfish
Oxyporhamphus micropterus Beaked Flyingfish
Parexocoetus brachypterus Sailfin Flyingfish
Parexocoetus mento African Flyingfish

GARFISHES

Hemiramphidae

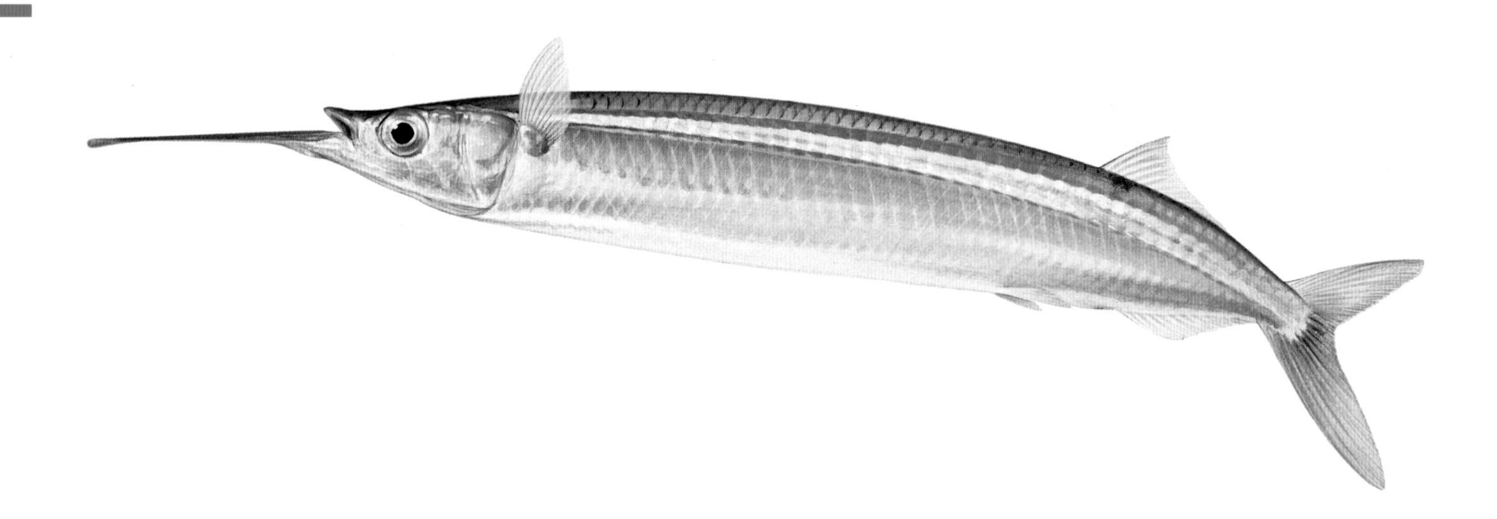

SOUTHERN GARFISH
Hyporhamphus melanochir

About 109 species of Garfish are found worldwide in tropical and temperate surface waters, both coastal and oceanic, with many species occurring in estuarine and freshwater environments. Garfishes have elongate, translucent, silvery to greenish-blue bodies with large, very fragile and easily shed scales. In most species the lower jaw is formed into a long spike with a reddish-orange tip. They have small dorsal and anal fins towards the posterior of the body, and pectoral fins set high on the body. There is a large, forked caudal fin, with the lower lobe often larger than the upper.

Garfishes feed predominantly on algae, though small invertebrates and zooplankton are also consumed. They lack a stomach, and instead have pharyngeal teeth with which they begin to break down their food. They have a long, voluminous intestine that completes the process of digestion and are well known to anglers for the copious amounts of half-digested material they can excrete when captured.

Garfishes are usually encountered just under the surface of the water, where in the shifting and flickering light the reflectiveness, colour and translucence of their bodies help to hide them from predators. They are powerful swimmers and are capable of tail-walking along the surface when startled. The **Longfin Garfish** has large, wing-like pectoral fins, similar to those of the closely related Flyingfishes (Exocoetidae, p. 332), though not as well developed, and is capable of short flights. *Zenarchopterus* species are found in estuarine and freshwater environments in northern Australian waters and males have a modified anal fin used for internal fertilisation. Garfishes occur in small to large schools and are excellent food fishes. They are taken commercially for human consumption and are also widely used as bait for catching larger fishes. They are popular targets for recreational anglers.

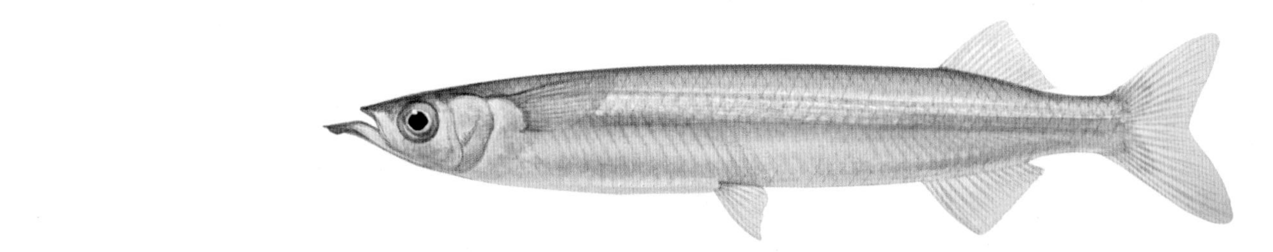

SNUBNOSE GARFISH
Arrhamphus sclerolepis

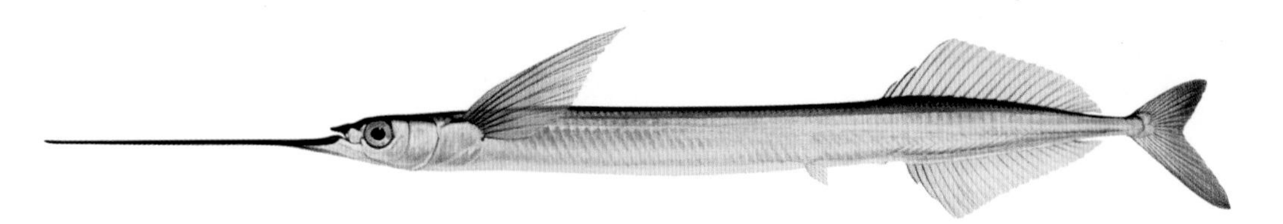

LONGFIN GARFISH
Euleptorhamphus viridis

BLACKBARRED GARFISH
Hemiramphus far

THREE-BY-TWO GARFISH
Hemiramphus robustus

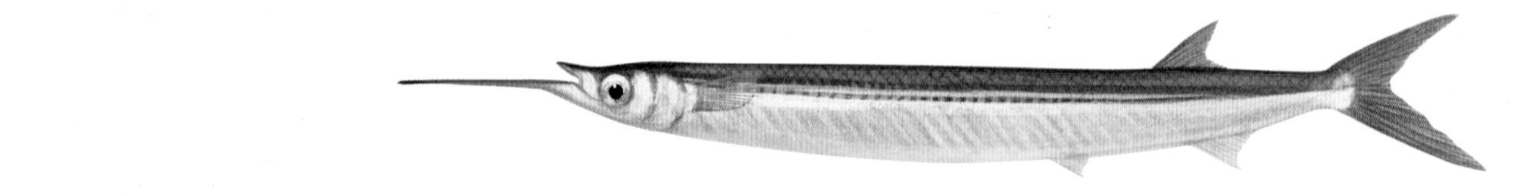

SLENDER GARFISH
Hyporhamphus dussumieri

LONGTAIL GARFISH
Hyporhamphus quoyi

NORTHERN RIVER GARFISH
Zenarchopterus buffonis

AUSTRALIAN HEMIRAMPHIDAE SPECIES

Arrhamphus sclerolepis Snubnose Garfish
Euleptorhamphus viridis Longfin Garfish
Hemiramphus far Blackbarred Garfish
Hemiramphus robustus Three-by-two Garfish
Hyporhamphus affinis Tropical Garfish
Hyporhamphus australis Eastern Sea Garfish
Hyporhamphus dussumieri Slender Garfish
Hyporhamphus melanochir Southern Garfish
Hyporhamphus neglectissimus Neglected Garfish
Hyporhamphus quoyi Longtail Garfish
Hyporhamphus regularis River Garfish
Rhynchorhamphus georgii Duckbill Garfish
Zenarchopterus buffonis Northern River Garfish
Zenarchopterus caudovittatus Longjaw River Garfish
Zenarchopterus dispar Spoonfin River Garfish
Zenarchopterus gilli Shortnose River Garfish
Zenarchopterus novaeguineae Fly River Garfish
Zenarchopterus rasori Short River Garfish

LONGTOMS

Belonidae

BARRED LONGTOM
Ablennes hians

Longtoms are found worldwide in tropical and temperate surface waters. There are about 34 species in the family, some of which are restricted to fresh water. They have an extremely elongate, silvery body that is greenish-blue on the dorsal surface. Both the upper and lower jaws are elongated to form a beak, and are lined with long, needle-sharp teeth. Small juveniles have short equal-sized jaws and as they mature the lower jaw elongates first, followed by the upper – reflecting their evolutionary relationship to the Flyingfishes (Exocoetidae, p. 332) and Garfishes (Hemiramphidae, p. 334). Longtoms have pectoral fins set high on the body, ventral fins at about mid-body, a single long-based dorsal fin and similar anal fin, both set well back on the body, and a large forked caudal fin. They can reach over 1.5 m in length, though most grow to 50–80 cm. They are voracious predators of other fishes. Their silvery flanks and greenish-blue back provide effective camouflage from predators and prey in the flickering reflections of the water just beneath the surface. They are powerful swimmers and are capable of performing large leaps out of the water and long tail-walks on the surface if startled. There have been instances of people being severely injured by large Longtoms leaping over boats at high speed.

In Australia Longtoms are mostly encountered in northern waters. Young adults often form multi-species aggregations, while large specimens are usually more solitary. The **Slender Longtom** is occasionally found around the south coast and one species, the **Freshwater Longtom**, occurs in northern rivers. A traditional indigenous fishing method, still practised in the Torres Strait, uses a lure of spiderweb, which entangles the needle-like teeth of the Longtom. Longtoms are good food fishes, though many people are disinclined to eat them due to their conspicuous green bones.

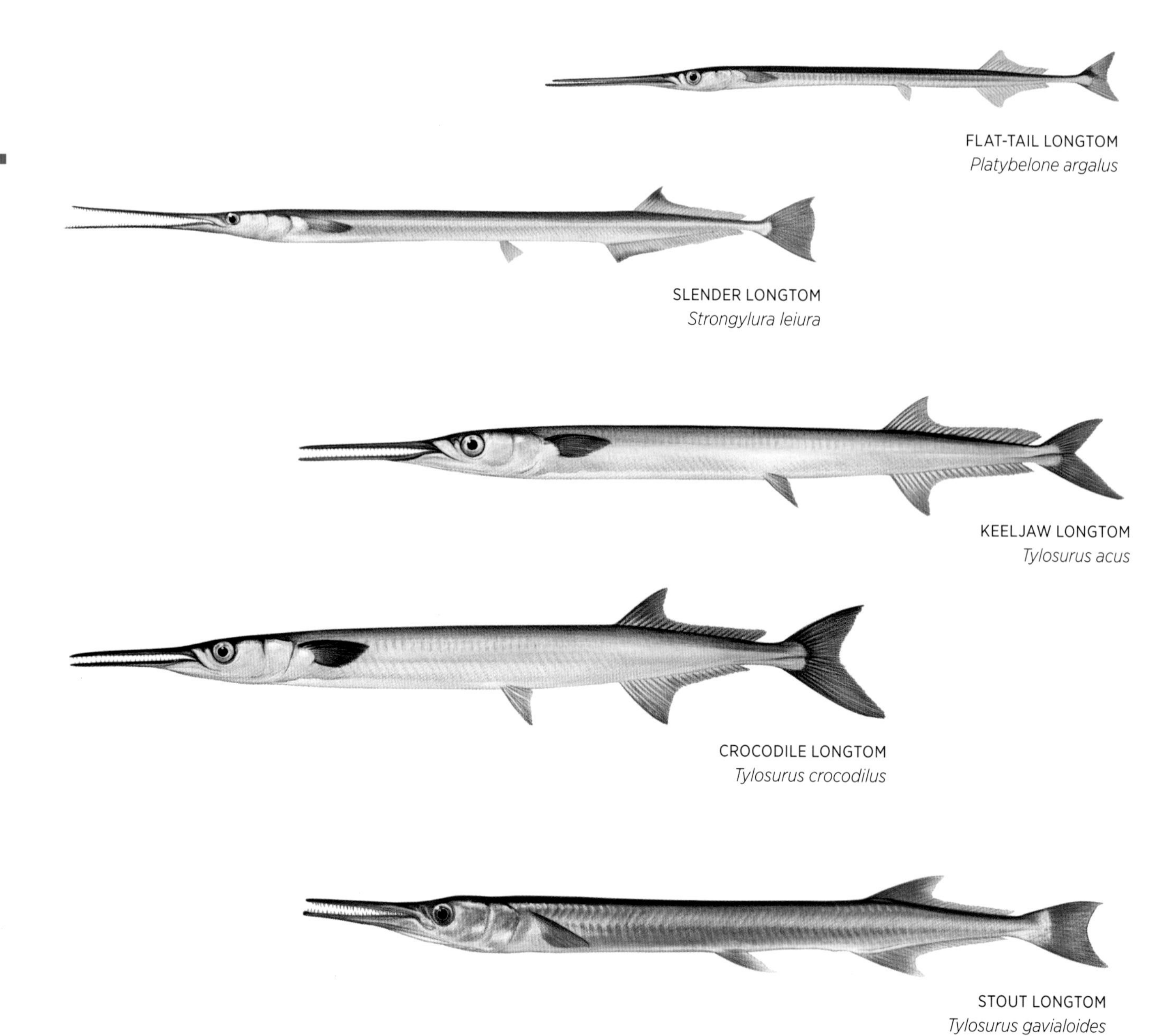

FLAT-TAIL LONGTOM
Platybelone argalus

SLENDER LONGTOM
Strongylura leiura

KEELJAW LONGTOM
Tylosurus acus

CROCODILE LONGTOM
Tylosurus crocodilus

STOUT LONGTOM
Tylosurus gavialoides

AUSTRALIAN BELONIDAE SPECIES

Ablennes hians Barred Longtom
Platybelone argalus Flat-tail Longtom
Strongylura incisa Reef Longtom
Strongylura krefftii Freshwater Longtom
Strongylura leiura Slender Longtom
Strongylura strongylura Blackspot Longtom
Strongylura urvillii Urville's Longtom
Tylosurus acus Keeljaw Longtom
Tylosurus crocodilus Crocodile Longtom
Tylosurus gavialoides Stout Longtom
Tylosurus punctulatus Spongyjaw Longtom

SAURIES

Scomberesocidae

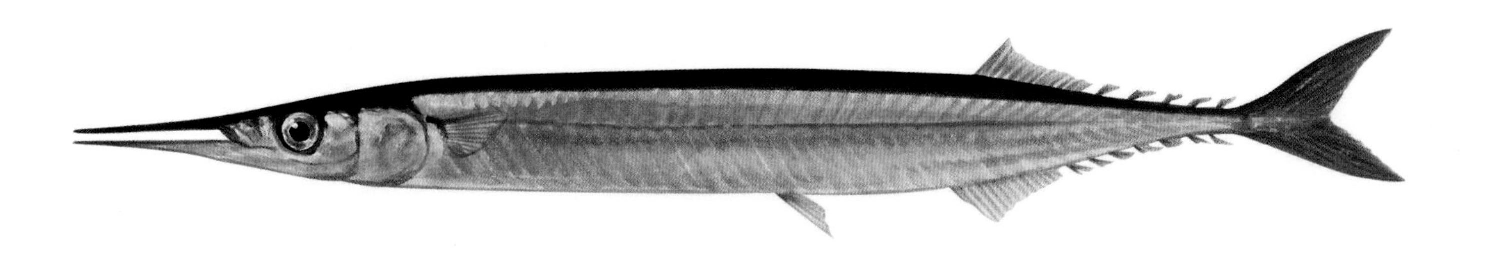

KING GAR
Scomberesox saurus

The four species of Saury are found worldwide in tropical and temperate surface waters. They have an elongate, cylindrical body with fragile scales, and most have both jaws formed into a long, thin beak with minute teeth and a small mouth opening. The single dorsal and anal fins are at the rear of the body, and each is followed by a series of 5–7 small finlets. Sauries have a forked caudal fin and small ventral fins at about mid-body. They usually inhabit oceanic offshore waters, feeding on small pelagic crustaceans and other zooplankton, but are occasionally found inshore, sometimes even entering estuaries.

The **King Gar** occurs off the southern half of Australia and reaches about 45 cm in length. It forms large schools and is preyed on by pelagic fishes such as Tunas (Scombridae, p. 717). It is reputed to be an excellent food fish but does not occur in sufficient concentrations to support a commercial fishery. Another very similar species, the Dwarf Saury (*Scomberesox simulans*), which grows to only 13 cm, may also occur in southern Australian waters as its distribution is circumglobal in the southern hemisphere; specimens may have been mistaken for juvenile King Gars.

AUSTRALIAN SCOMBERESOCIDAE SPECIES
Scomberesox saurus King Gar

CYPRINODONTIFORMES

This is a large group of predominantly freshwater fishes, many of which have adapted to extreme conditions such as salinity and seasonal drying of their environment. Cyprinodontiforms have a compressed body with a single dorsal fin and a rounded or truncate caudal fin. They have sensory pores on the dorsal surface of the head, and a small mouth with a protrusible upper jaw. Usually they are found close to the surface, feeding on insects and material that has fallen into the water. Many species use internal fertilisation, with females giving birth to live young. There are 10 families in the order, two of which have been introduced into Australian waters.

PUPFISHES

Cyprinodontidae

AMERICAN FLAGFISH
Jordanella floridae

Pupfishes occur mainly in fresh water but are also found in coastal marine environments. They are distributed in the waters of North America and South America, North Africa and the Mediterranean. There are more than 100 species of these small fishes, with the largest reaching a maximum length of about 20 cm. Many species have a high tolerance to variations in salinity and temperature, and are found in very diverse environments. Members of this family inhabit lakes in the Andes mountains of South America and hold the record for the highest altitude at which fishes occur. Pupfishes have a deep, compressed body with a single dorsal fin, and an upturned mouth with numerous small teeth. The **American Flagfish**, native to the swamps and streams of Florida, is a common aquarium fish that has been released into the wild in Queensland.

AUSTRALIAN CYPRINODONTIDAE SPECIES

Jordanella floridae American Flagfish

MOSQUITOFISHES

Poeciliidae

SWORDTAIL
Xiphophorus hellerii

Native to fresh and brackish waters in North America, northern South America and Africa, this large family contains more than 300 species. Most species measure from 5 to 10 cm in length, the largest attaining about 20 cm. They have a single dorsal fin and a rounded caudal fin. The males are smaller and more colourful than the females, and in many species the anal fin is modified to form a gonopodium (a structure for internal fertilisation of the female). The family displays a wide range of reproductive strategies. Some species lay fertilised eggs, while others are livebearers. In the latter, the method of nourishing the developing embryos varies widely from species to species – from simple internal incubation of yolk-containing eggs, to the use of specialised internal structures similar to the placenta found in mammals.

The **Mosquitofish** was introduced into Australia in the 1920s as a means of controlling the aquatic larvae of mosquitos. The strategy has had the opposite effect, as the Mosquitofish actually prefers to feed on the invertebrate predators of mosquito larvae. The Mosquitofish is now abundant in most Australian river systems, particularly in the south-east, and has even colonised small, permanent trickles around isolated bores in the arid interior. The other members of this family occurring in Australia are popular aquarium fishes that have been accidentally or intentionally released into the wild; small breeding populations of these species are found mainly in Queensland.

MOSQUITOFISH (male)
Gambusia holbrooki

MOSQUITOFISH (female)
Gambusia holbrooki

AUSTRALIAN POECILIIDAE SPECIES

Gambusia dominicensis Dominican Mosquitofish
Gambusia holbrooki Mosquitofish
Phalloceros caudimaculatus Speckled Mosquitofish
Poecilia latipinna Sailfin Molly
Poecilia reticulata Guppy
Xiphophorus hellerii Swordtail
Xiphophorus maculatus Platy

STEPHANOBERYCIFORMES

Stephanoberyciforms are highly adapted, rare deepwater fishes. They have a large head and rounded body, very thin skull bones, and a single dorsal fin (often with weak spines). The body is usually flabby, either scaleless or with large fragile scales, or sometimes covered with small spines or prickles. Luminous tissue is sometimes present on the body. Most species are rarely encountered and little is known of them, as they are found at great depths in the open ocean. There are eight families in the order, four of which are known to occur in Australian waters, although other families of this order have wide distribution and may also occur here.

BIGSCALES

Melamphaidae

CRESTED BIGSCALE
Poromitra crassiceps

Bigscales are small deepsea fishes that are pelagic in bottom waters from depths of 800 to 5000 m. They have a wide distribution and are found in all oceans except the Arctic. There are 36 species in the family, with the largest growing to about 18 cm in length. Bigscales have a large head with prominent, very thin bones and large mucus-filled cavities covered by thin skin. They take their name from their large scales, which are fragile, easily shed and almost always lost in trawl-caught specimens. The single dorsal fin has 1–3 thin spines, and there are 3–4 spines on the dorsal and ventral profile of the tail at the origin of the caudal fin. They have a large mouth with small, villiform teeth and, from what little is known of their habits, appear to feed on small pelagic crustaceans and gelatinous organisms. Some are known to make vertical migrations towards the surface at night. Their bodies are fragile, with lightly calcified bones, indicating a slow-moving, low-energy mode of life. They occur in deep waters right around the coast of Australia.

AUSTRALIAN MELAMPHAIDAE SPECIES

Melamphaes longivelis Eyebrow Bigscale
Melamphaes suborbitalis Shoulderspine Bigscale
Poromitra crassiceps Crested Bigscale
Scopeloberyx microlepis Smallerscale Bigscale
Scopelogadus beanii Bean's Bigscale
Scopelogadus mizolepis Ragged Bigscale
Sio nordenskjoldii Nordenskjold's Bigscale

REDMOUTH WHALEFISHES

Rondeletiidae

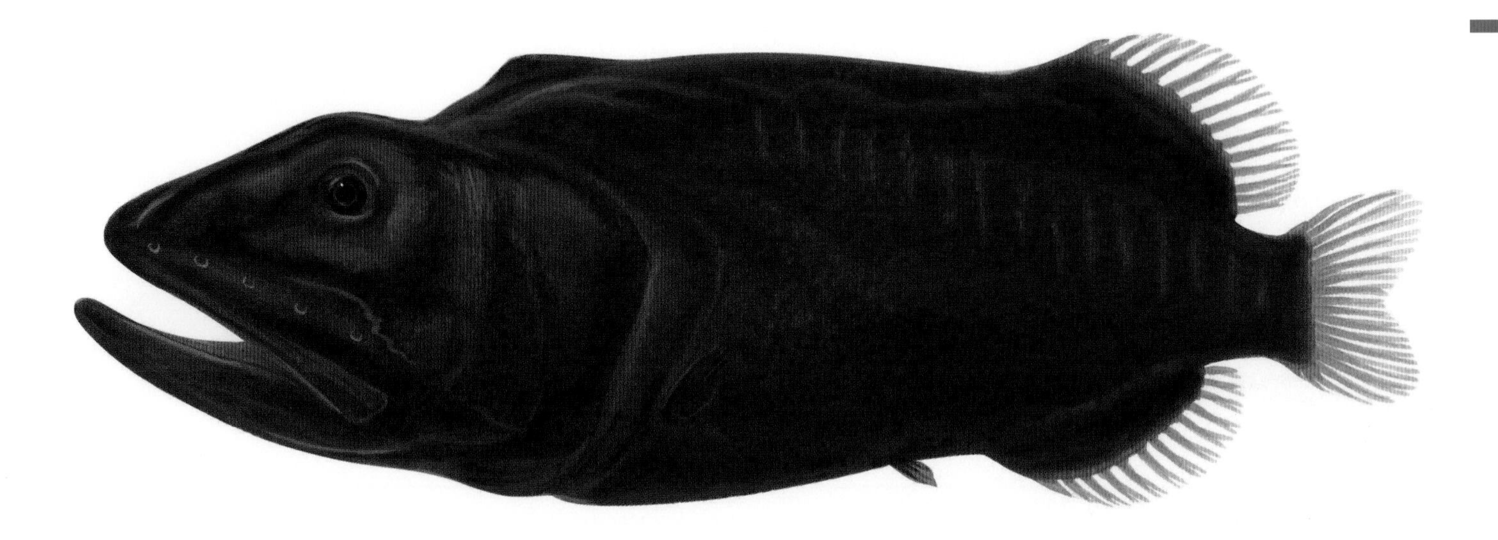

COMMON REDMOUTH WHALEFISH
Rondeletia loricata

The two species in this family are found wordwide in tropical and temperate waters, usually at depths of about 500–1500 m. They are small, strangely proportioned fishes, with a smooth-skinned body, large angular head, long snout and small eyes. They somewhat resemble whales in shape (hence their name), but reach a maximum length of only about 12 cm. Their large mouth reaches back to below the eyes and bears small, close-set teeth. The dorsal and anal fins are placed well back on the body, near the caudal fin. They have no scales and the lateral line is in the form of a series of vertical rows of sensory pores. The **Common Redmouth Whalefish** makes vertical migrations towards the surface at night to feed, and preys on small pelagic crustaceans such as copepods.

AUSTRALIAN RONDELETIIDAE SPECIES
Rondeletia loricata Common Redmouth Whalefish

REDVELVET WHALEFISH

Barbourisiidae

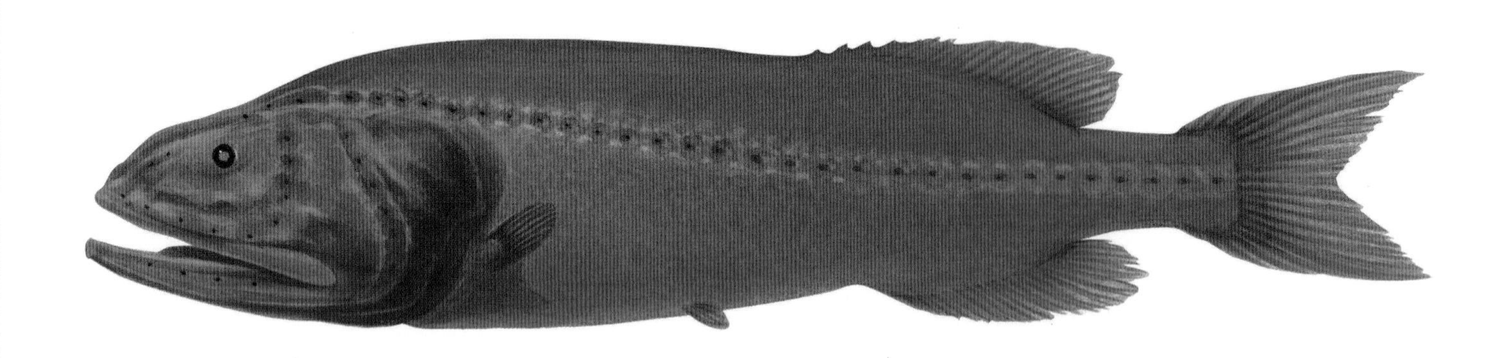

REDVELVET WHALEFISH
Barbourisia rufa

The **Redvelvet Whalefish** is the only species in this family and is found worldwide in tropical and temperate waters, from depths of about 500 to 1500 m. It has a long snout, and a mouth resembling that of a whale, with large jaws reaching to behind the eyes and bands of fine villiform teeth. The bright red–orange skin is densely covered with small spines, giving a velvety feel. The Redvelvet Whalefish has a lateral line that consists of a large tube with numerous pores. Juveniles are pelagic in midwater while adults, which can reach up to about 40 cm in length, are usually found much closer to the bottom. Adults are thought to feed on pelagic crustaceans, but as the species is rarely encountered very little else is known of its biology. It has been recorded in north-eastern Australian waters.

AUSTRALIAN BARBOURISIIDAE SPECIES

Barbourisia rufa Redvelvet Whalefish

FLABBY WHALEFISHES

Cetomimidae

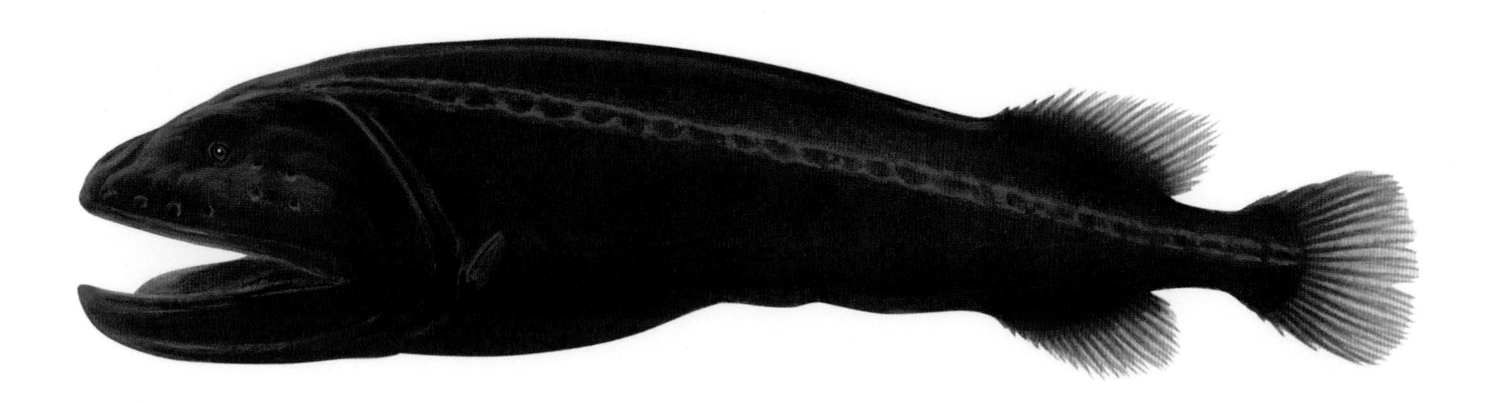

PARR'S COMBTOOTH WHALEFISH
Gyrinomimus parri

There are about 20 species of Flabby Whalefish found worldwide, occurring in deep water from 1000 to 4000 m. They have an elongate, flabby body with no abdominal ribs, no scales and an enormous mouth that reaches to well behind the small eyes. The dorsal and anal fins are set far back on the body, and there are no ventral fins. Flabby Whalefishes have a very well-developed lateral line that consists of a broad tube embedded with large scales and perforated at regular intervals by large pores.

Most species in the family are known from only a very few specimens that have been taken by deepwater trawls; the specimens are usually in poor condition due to the fishes' fragile bodies and the great depths at which they live. Females reach a maximum length of about 30 cm, while males are much smaller, growing to 3–5 cm. Flabby Whalefishes are thought to feed mainly on pelagic crustaceans, though their extremely distensible stomach indicates that they are capable of taking much larger prey.

AUSTRALIAN CETOMIMIDAE SPECIES

Gyrinomimus parri Parr's Combtooth Whalefish
Rhamphocetichthys savagei Birdsnout Whalefish

BERYCIFORMES

The order Beryciformes is closely related to the Stephanoberyciformes (p. 345), but Beryciforms are generally found in shallower waters and have a stronger skeleton and more strongly developed spines in the fins. They have a large head and mouth, large eyes and spiny scales. The single dorsal fin is sometimes deeply notched and there are spines in the dorsal, anal and ventral fins. Most species are carnivorous and bottom dwelling. There are seven families in the order, all of which occur in Australian waters.

FANGTOOTHS

Anoplogastridae

FANGTOOTH
Anoplogaster cornuta

The two species of Fangtooth are found in the Atlantic, Indian and Pacific oceans. They are reasonably common in deep water from about 100 to 3000 m. They have a short, compressed body with minute prickly scales. Their very large, bony head has mucus-filled cavities covered by thin skin. The skull has many serrated, bony ridges but no prominent spines, except in the juvenile stage. The jaws are very large, with enormous fang-like teeth; the largest teeth at the front of the lower jaw fit into sockets in the upper jaw. The single dorsal fin has no spines. The **Fangtooth** occurs individually, in small groups, or in larger schools, and preys on fishes and crustaceans. It has been recorded in deep waters right around the coast of Australia.

AUSTRALIAN ANOPLOGASTRIDAE SPECIES
Anoplogaster cornuta Fangtooth

SPINYFINS
Diretmidae

DISCFISH
Diretmus argenteus

Spinyfins are found throughout the world's tropical and temperate waters, in midwater to depths of about 2000 m. They have an oval, compressed body with ctenoid scales that extend onto the cheeks. The large head has many bony ridges and mucus-filled cavities covered by thin skin. There is no lateral line and the ventral profile has a keel of scutes. Both dorsal and anal fins have no spines, but small spines are present at the base of each ray and along its length. The ventral fin has one strong, serrated spine.

The **Black Spinyfin** grows to about 40 cm in length and the **Discfish** to about 30 cm. Both species feed mainly on small pelagic crustaceans and other zooplankton. Juveniles are found in the upper layers of the ocean, while adults are found at greater depths with increasing size. Spinyfins occur right around the coast of Australia.

AUSTRALIAN DIRETMIDAE SPECIES

Diretmichthys parini Black Spinyfin
Diretmus argenteus Discfish

FLASHLIGHTFISHES

Anomalopidae

TWOFIN FLASHLIGHTFISH
Anomalops katoptron

The eight species of Flashlightfish are found in tropical waters of the Indo-West Pacific, East Pacific and Caribbean. They have an oval, compressed body with small ctenoid scales. The single dorsal fin is either continuous or has a deep notch between the spines and the soft rays. There are 4–6 spines in the dorsal fin and 2–3 spines in the anal fin.

Flashlightfishes are named for the large light organ beneath each eye, which contains symbiotic luminescent bacteria. Members of the family either have a black membrane that can operate as a shutter to 'turn off' the light, or are able to rotate the light organ to achieve the same effect. By day Flashlightfishes shelter in deep crevices or caves, emerging at night to hunt for small crustaceans and other planktonic organisms. The **Twofin Flashlightfish** is one of the largest species in the family, growing to around 27 cm in length. It is usually found near steep dropoffs, descending into deeper waters by day and rising at night to hunt over the reef. Both Australian species are found only in northern waters.

AUSTRALIAN ANOMALOPIDAE SPECIES

Anomalops katoptron Twofin Flashlightfish
Photoblepharon palpebratum Onefin Flashlightfish

PINEAPPLEFISHES

Monocentridae

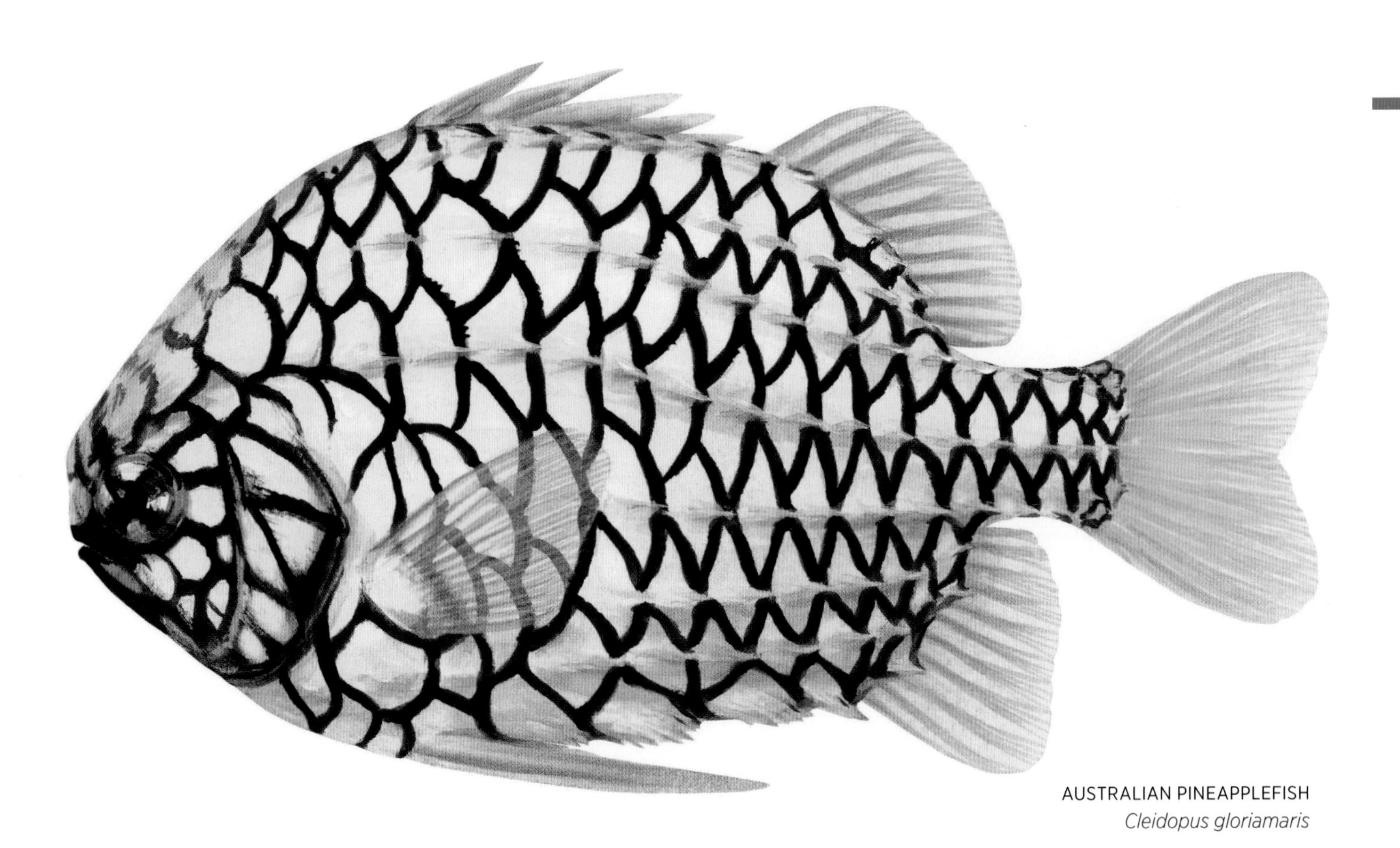

AUSTRALIAN PINEAPPLEFISH
Cleidopus gloriamaris

Pineapplefishes are found in tropical and temperate coastal waters of the Indo-Pacific, from depths of 30 to 300 m. There are about four species in the family. They have a short, rounded body with large, rigid, plate-like scales that are black-edged and form a distinctive pattern. There is a spine at the centre of each scale and these form spiky longitudinal rows. Pineapplefishes have 4–7 large, strong spines in the first dorsal fin and a similar strong spine in the ventral fin. There is a luminous light organ on the lower jaw.

In the **Australian Pineapplefish**, which is endemic to Australia's south-west and south-east coasts, this light organ is a bright red–orange colour, the reddish light emitted perhaps serving to attract the small crustaceans on which it feeds. This species grows to a maximum length of about 30 cm and is often seen by divers, either sheltering in caves on deeper coastal reefs or emerging at night to feed. The widely distributed **Japanese Pineapplefish** is found in more northerly Australian waters and is smaller in size, reaching about 20 cm. It has smaller, paired light organs and a less-defined scale pattern. It forms small schools on coastal and offshore reefs.

AUSTRALIAN MONOCENTRIDAE SPECIES

Cleidopus gloriamaris Australian Pineapplefish
Monocentris japonica Japanese Pineapplefish

ROUGHIES

Trachichthyidae

ORANGE ROUGHY
Hoplostethus atlanticus

The family of Roughies contains about 35 species, distributed mainly in temperate waters of the Atlantic, Indian and Pacific oceans. They have elongate to oval, compressed bodies and a large bony head. The head bears serrated ridges between mucus-filled cavities that are covered by skin, and large jaws with bands of minute teeth. Roughies have a row of scutes along the belly, and the scales on the body may be very small and ctenoid, as in the **Sandpaper Fish**, or thick and spiny, as in the **Southern Roughy**. The single dorsal fin has 3–8 spines and the anal fin has 2–3 spines. There are small spines before the leading edges of the caudal fin and most species have large rearward-pointing spines on the head. Some have bioluminescent tissue along the ventral surface. The family contains many pelagic species that form large schools close to the bottom at depths from 100 to 1500 m. Some of these species are commercially targeted.

In Australia the **Orange Roughy** has been intensively fished in southern offshore waters. This species, which reaches about 50 cm in length, is slow-growing and can live to be over 100 years old, making it extremely vulnerable to overfishing. The southern Australian fishery that targeted Orange Roughy rapidly declined after a few seasons of massive catches and the species is now listed as threatened.

Several species in the family, such as the **Violet Roughy**, **Southern Roughy** and **Western Roughy,** occur on shallow to deep coastal reefs along the southern coast of Australia and are often seen by divers. They shelter beneath ledges during the day and emerge at night to feed on crustaceans and small fishes.

SLENDER ROUGHY
Optivus elongatus

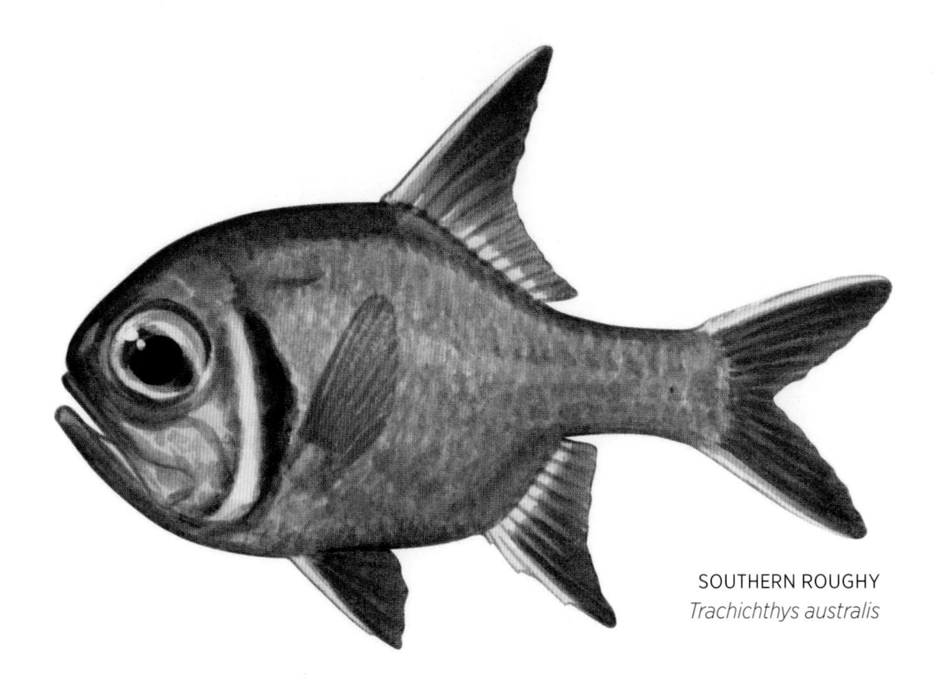

SOUTHERN ROUGHY
Trachichthys australis

AUSTRALIAN TRACHICHTHYIDAE SPECIES

Aulotrachichthys novaezelandicus New Zealand Roughy
Aulotrachichthys pulsator Golden Roughy
Gephyroberyx darwinii Darwin's Roughy
Hoplostethus atlanticus Orange Roughy
Hoplostethus gigas Giant Sawbelly
Hoplostethus intermedius Blacktip Sawbelly
Hoplostethus latus Palefin Sawbelly
Hoplostethus melanopus Smallscale Sawbelly
Hoplostethus shubnikovi Metavay Sawbelly
Optivus agastos Violet Roughy
Optivus agrammus Western Roughy
Optivus elongatus Slender Roughy
Paratrachichthys macleayi Sandpaper Fish
Sorosichthys ananassa Little Pineapplefish
Trachichthys australis Southern Roughy

ALFONSINOS AND REDFISHES

Berycidae

BIGHT REDFISH
Centroberyx gerrardi

Alfonsinos and Redfishes are found in the Atlantic, Indian and West Pacific oceans, on coastal reefs and offshore in deeper waters. There are about nine species in the family. They have an oval, compressed body with spiny scales, 4–7 spines in the single dorsal fin and usually four in the anal fin. There is a single spine in each ventral fin and a series of small spines before the leading edges of the forked caudal fin. Their large, rounded head bears bony ridges and mucus-filled cavities covered with skin. They have large eyes, and large jaws with bands of minute teeth. Most species are silvery-red in colour. The **Imperador** and **Alfonsino** are found worldwide and are pelagic in bottom waters from 180 to 1000 m. They are sometimes taken in commercial trawls off the southern half of Australia.

Centroberyx species occur in small schools over deep reefs and muddy bottoms in Australia's southern half. They have an iridescent, silver–red body and large, intensely coloured red or yellow eyes. The **Bight Redfish** reaches a maximum length of about 65 cm, and has sculptured ridges and exposed bones on the head and cheeks, and enormous brilliant-red eyes. Like others in the family, it is prized as a food fish and is targeted by commercial trawlers and recreational anglers alike. The **Redfish** (also known as the Nannygai) is also commercially trawled off south-eastern Australia. It occurs to depths of about 450 m, schooling near the bottom by day and rising in the water column at night to feed near the surface. It is a slow-growing species and stocks have rapidly declined in areas where it has been targeted. Alfonsinos and Redfishes are carnivorous, preying on crustaceans, small fishes and squid.

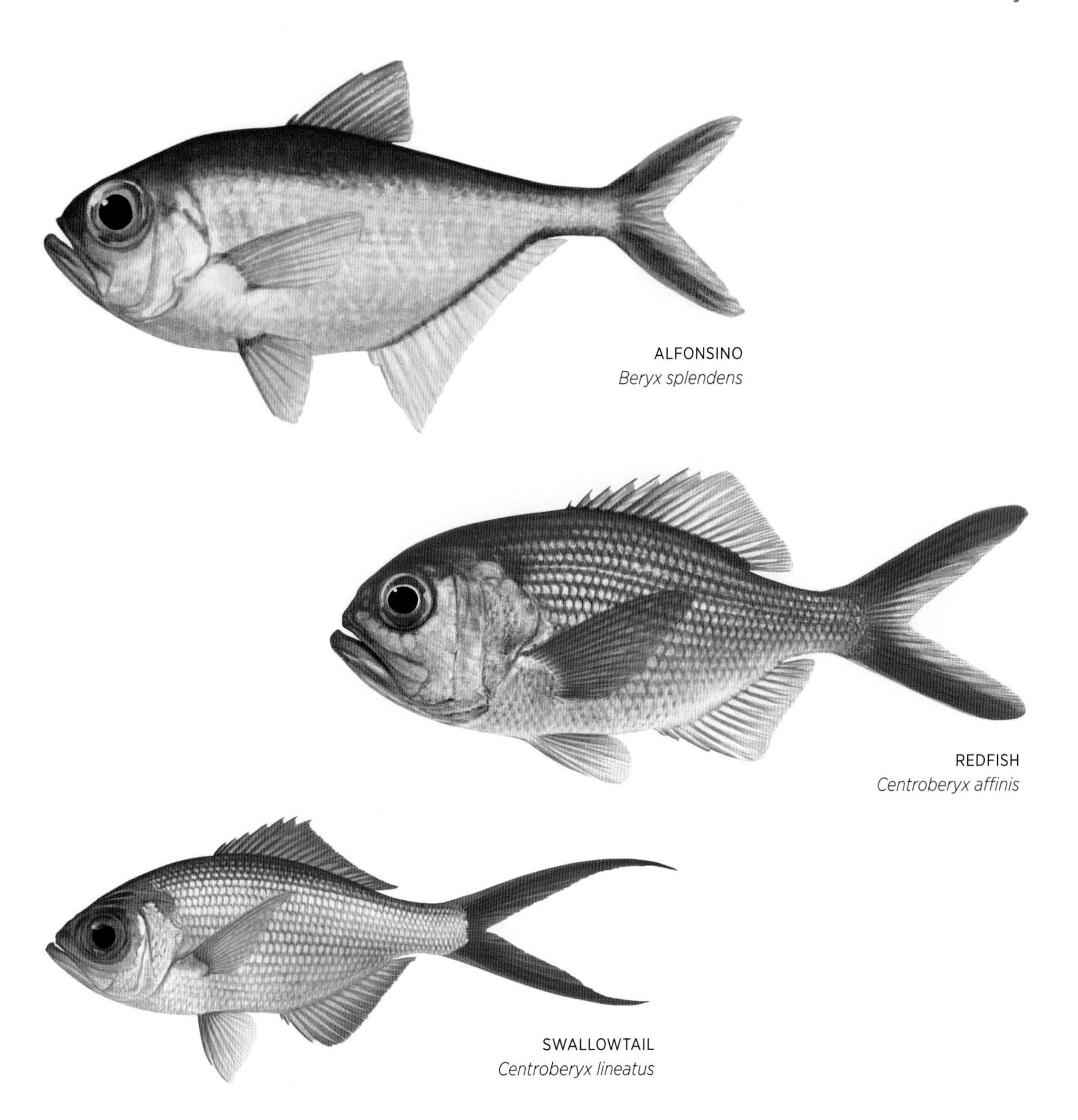

ALFONSINO
Beryx splendens

REDFISH
Centroberyx affinis

SWALLOWTAIL
Centroberyx lineatus

AUSTRALIAN BERYCIDAE SPECIES

Beryx decadactylus Imperador
Beryx splendens Alfonsino
Centroberyx affinis Redfish
Centroberyx australis Yelloweye Redfish
Centroberyx gerrardi Bight Redfish
Centroberyx lineatus Swallowtail

SOLDIERFISHES AND SQUIRRELFISHES

Holocentridae

SABRE SQUIRRELFISH
Sargocentron spiniferum

Soldierfishes and Squirrelfishes are found in coastal tropical waters worldwide, usually on reefs, from the shallows down to about 200 m. There are about 78 species in the family. They have an oval, compressed body with large, spiny, ctenoid scales. The dorsal fin is either continuous or deeply notched, and has 10–13 spines. The anal fin has four spines and the ventral fins each have one spine. The head bears bony ridges, and the preoperculum and operculum often have large spines. Mostly reddish in colour, Soldierfishes and Squirrelfishes have large eyes and are nocturnal feeders, sheltering under ledges and in crevices and caves during the day.

Soldierfishes have a bluntly rounded head and Australian species have no spines on the preoperculum. Squirrelfishes have a more pointed snout and bear a strong, venomous spine on the preoperculum, which can inflict a painful wound. Usually one of the anal-fin spines is also long and robust.

Soldierfishes and Squirrelfishes are common inhabitants of northern Australian reefs, and are frequently encountered in the open at night when they emerge to hunt. Soldierfishes feed mainly on zooplankton in the water column. Squirrelfishes have a similar diet but also take benthic crustaceans and small fishes. The **Sabre Squirrelfish** is one of the largest members of the family, reaching a maximum length of about 50 cm, but most species are no larger than about 30 cm.

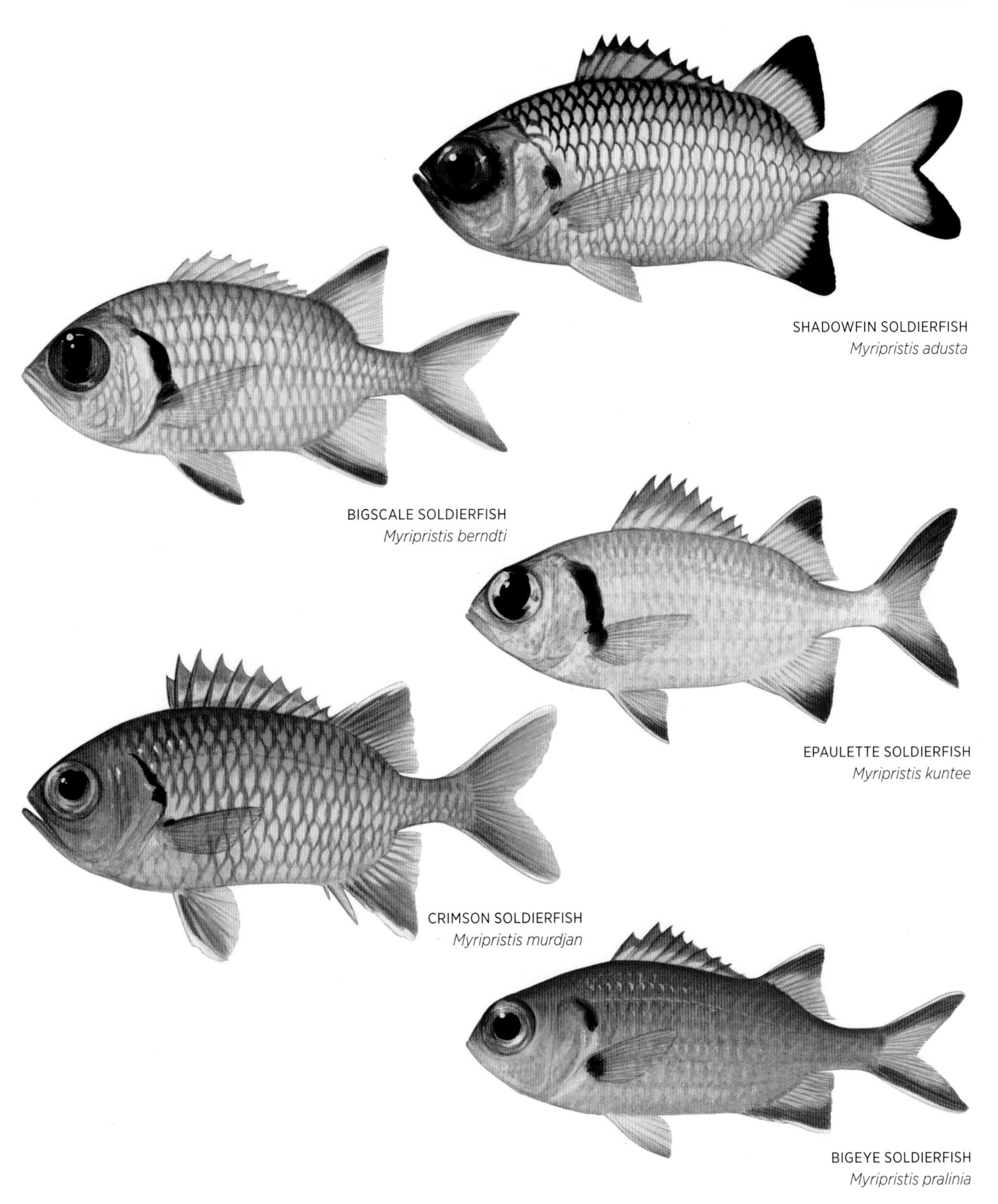

SHADOWFIN SOLDIERFISH
Myripristis adusta

BIGSCALE SOLDIERFISH
Myripristis berndti

EPAULETTE SOLDIERFISH
Myripristis kuntee

CRIMSON SOLDIERFISH
Myripristis murdjan

BIGEYE SOLDIERFISH
Myripristis pralinia

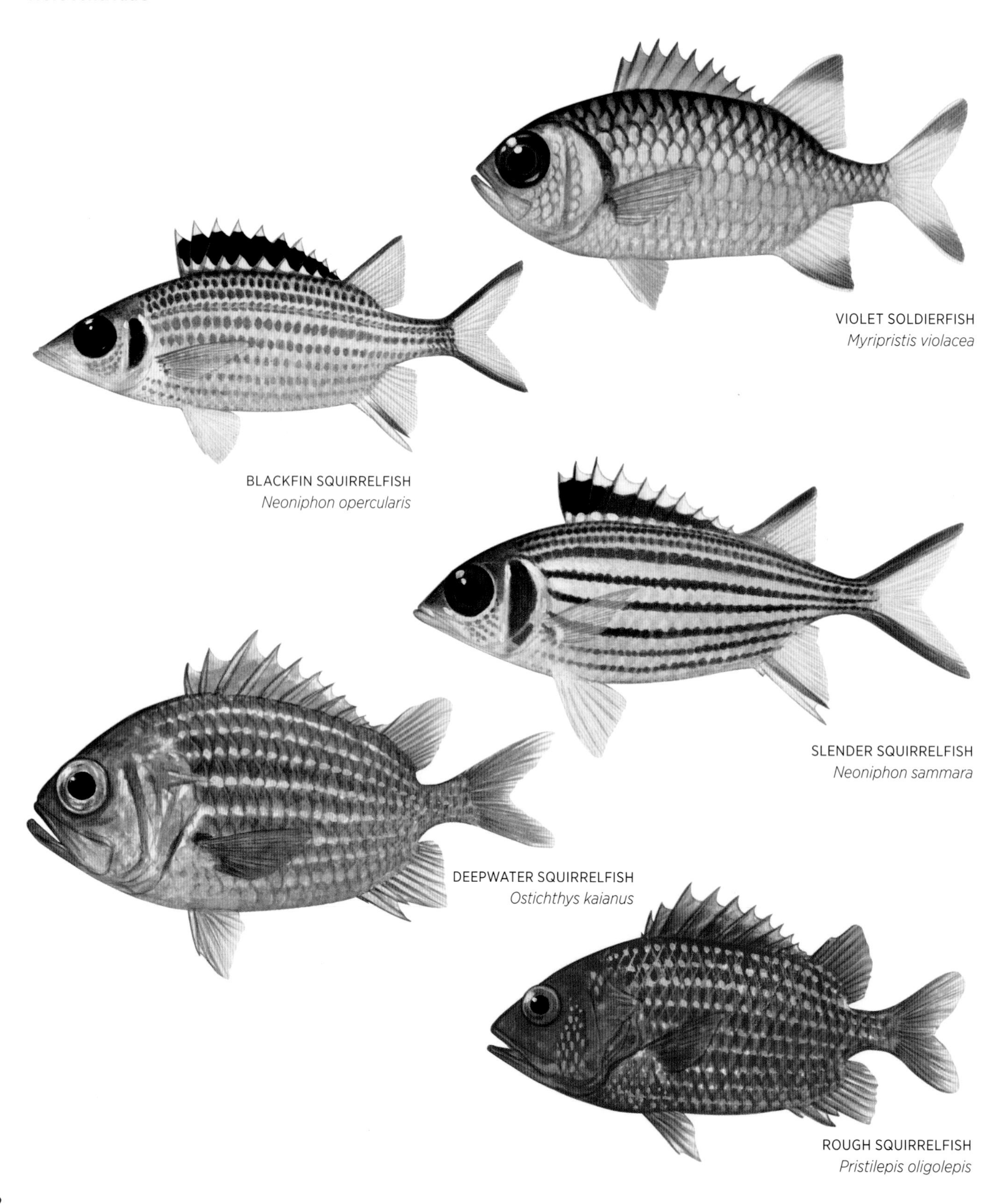

VIOLET SOLDIERFISH
Myripristis violacea

BLACKFIN SQUIRRELFISH
Neoniphon opercularis

SLENDER SQUIRRELFISH
Neoniphon sammara

DEEPWATER SQUIRRELFISH
Ostichthys kaianus

ROUGH SQUIRRELFISH
Pristilepis oligolepis

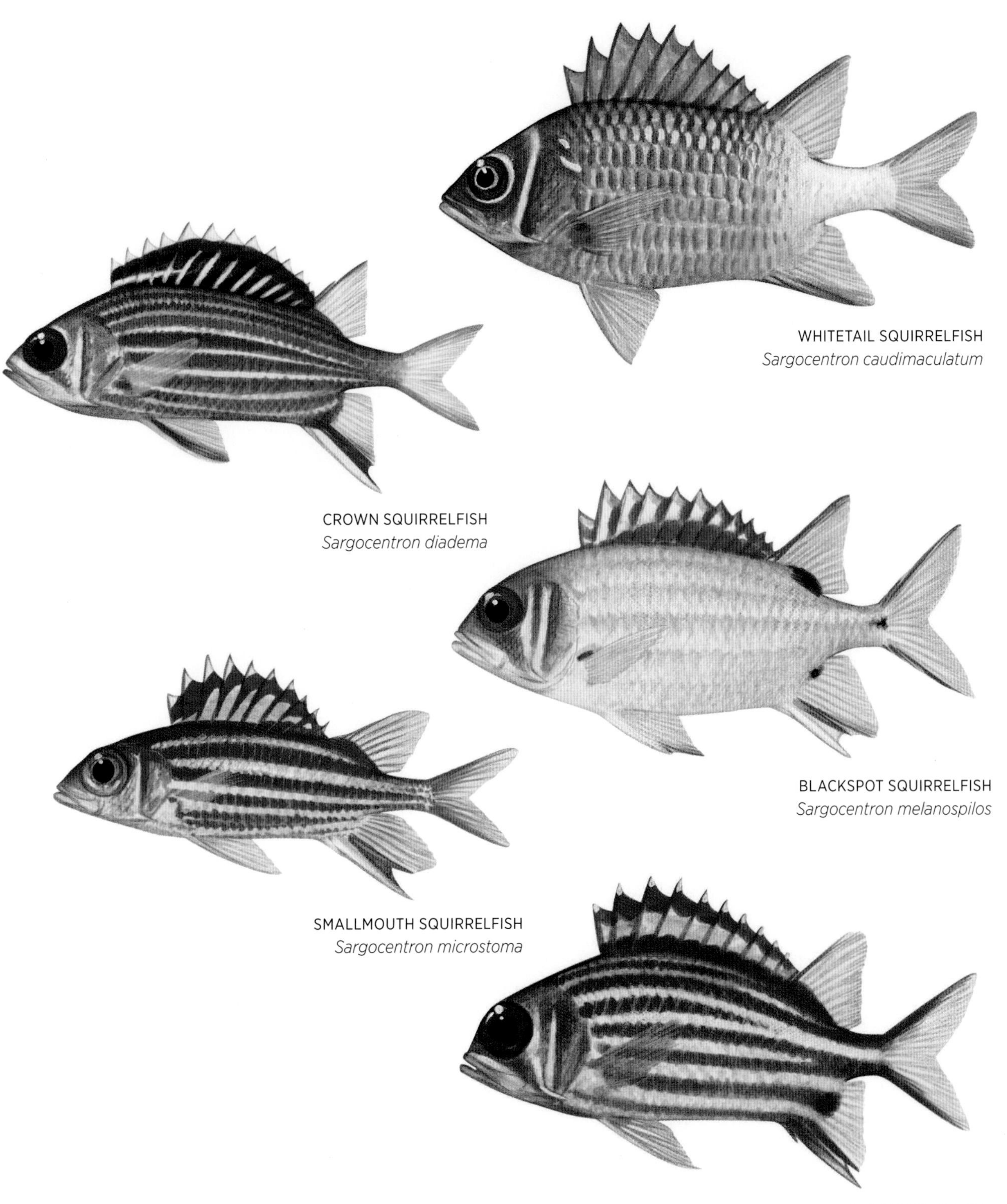

WHITETAIL SQUIRRELFISH
Sargocentron caudimaculatum

CROWN SQUIRRELFISH
Sargocentron diadema

BLACKSPOT SQUIRRELFISH
Sargocentron melanospilos

SMALLMOUTH SQUIRRELFISH
Sargocentron microstoma

RED SQUIRRELFISH
Sargocentron rubrum

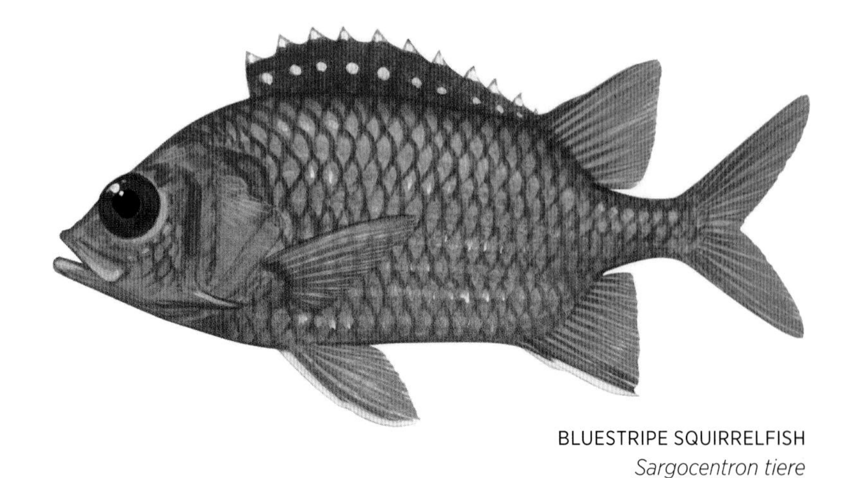

BLUESTRIPE SQUIRRELFISH
Sargocentron tiere

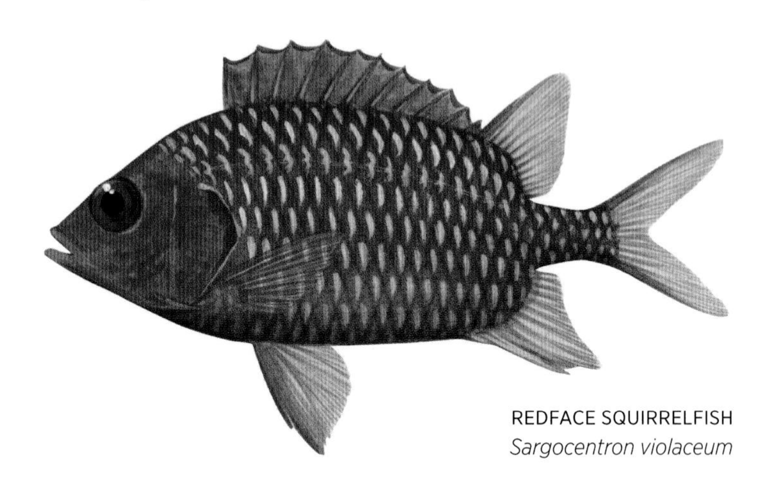

REDFACE SQUIRRELFISH
Sargocentron violaceum

AUSTRALIAN HOLOCENTRIDAE SPECIES

Myripristis adusta Shadowfin Soldierfish
Myripristis berndti Bigscale Soldierfish
Myripristis botche Blacktip Soldierfish
Myripristis chryseres Yellowfin Soldierfish
Myripristis hexagona Doubletooth Soldierfish
Myripristis kuntee Epaulette Soldierfish
Myripristis murdjan Crimson Soldierfish
Myripristis pralinia Bigeye Soldierfish
Myripristis violacea Violet Soldierfish
Myripristis vittata Whitetip Soldierfish
Neoniphon argenteus Silver Squirrelfish
Neoniphon aurolineatus Yellowstriped Squirrelfish
Neoniphon opercularis Blackfin Squirrelfish
Neoniphon sammara Slender Squirrelfish
Ostichthys acanthorhinus Spinesnout Squirrelfish
Ostichthys japonicus Giant Squirrelfish
Ostichthys kaianus Deepwater Squirrelfish
Plectrypops lima Roughscale Soldierfish
Pristilepis oligolepis Rough Squirrelfish
Sargocentron caudimaculatum Whitetail Squirrelfish
Sargocentron cornutum Horned Squirrelfish
Sargocentron diadema Crown Squirrelfish
Sargocentron ittodai Samurai Squirrelfish
Sargocentron lepros Spiny Squirrelfish
Sargocentron melanospilos Blackspot Squirrelfish
Sargocentron microstoma Smallmouth Squirrelfish
Sargocentron praslin Brownspot Squirrelfish
Sargocentron punctatissimum Speckled Squirrelfish
Sargocentron rubrum Red Squirrelfish
Sargocentron spiniferum Sabre Squirrelfish
Sargocentron tiere Bluestripe Squirrelfish
Sargocentron tiereoides Pink Squirrelfish
Sargocentron violaceum Redface Squirrelfish

ZEIFORMES

Zeiforms have highly compressed, oval to circular bodies and a large head, and the mouth is usually highly protrusible. Most are found in deep water, though some occur in coastal shallows. They have well-developed spines in the fins, and the dorsal, anal and pectoral fin rays are unbranched. Many have bony plates called bucklers around the dorsal and ventral profiles. There are six families in the order, all of which occur in Australian waters.

LOOKDOWN DORIES

Cyttidae

KING DORY
Cyttus traversi

The three species of Lookdown Dory are found in temperate waters of the southern hemisphere, off South Africa, Australia and New Zealand. Named for their steep dorsal profile, they have silvery, oval to circular bodies with small, rough, ctenoid scales. There are no enlarged bony bucklers on the dorsal and ventral profiles, although in the **King Dory** the ventral profile does bear a single row of larger overlapping scales. The single dorsal fin has 8–10 spines, which are elongated in the **Silver Dory**, and the anal fin has two spines. Lookdown Dories have large eyes and a large, highly protrusible mouth with narrow bands of villiform teeth.

Lookdown Dories occur near the bottom from depths of about 50 to 1000 m. They are frequently taken by deepwater trawl fisheries off southern Australia, usually from depths of about 250–500 m. The King Dory is the largest member of the family, reaching about 65 cm in length, while the Silver Dory grows to about 50 cm. Both these species are taken in significant quantities, usually as bycatch. The **New Zealand Dory** is also taken, but as it only grows to a maximum size of about 40 cm it is too small to be commercially significant. Lookdown Dories feed on benthic and pelagic crustaceans and small fishes.

SILVER DORY
Cyttus australis

AUSTRALIAN CYTTIDAE SPECIES
Cyttus australis Silver Dory
Cyttus novaezealandiae New Zealand Dory
Cyttus traversi King Dory

OREODORIES

Oreosomatidae

SPIKEY OREODORY
Neocyttus rhomboidalis

Oreodories are found throughout deeper waters of the southern hemisphere and are most common off South Africa and southern Australia. The nine species in the family are bottom dwellers, occurring from depths of about 300 to 1600 m. They have deep, compressed, oval to rhomboid bodies with small, rough scales, a large head and large eyes, and a moderate-sized mouth with minute teeth. There are 2–4 spines in the anal fin, one spine in the ventral fin, and 5–8 strong spines in the single dorsal fin, often with one spine enlarged into a very strong spike. The **Smooth Oreodory** is the largest species in the family, reaching a maximum length of about 70 cm.

Oreodories are high-quality food fishes and several species, particularly the Smooth Oreodory and **Black Oreodory**, are frequently taken in commercial quantities by deepwater trawlers off south-eastern Australia. They are mostly caught as bycatch by trawlers targeting Orange Roughy (Trachichthyidae, p. 356). Oreodories are usually found near seamounts or rocky, deepwater reefs. They feed on benthic and pelagic crustaceans, squid, small fishes, and gelatinous planktonic animals such as salps. Oreodories are very slow growing and are reported to live to over 100 years of age, making them extremely vulnerable to overfishing.

SMOOTH OREODORY
Pseudocyttus maculatus

AUSTRALIAN OREOSOMATIDAE SPECIES

Allocyttus niger Black Oreodory
Allocyttus verrucosus Warty Oreodory
Neocyttus psilorhynchus Rough Oreodory
Neocyttus rhomboidalis Spikey Oreodory
Oreosoma atlanticum Oxeye Oreodory
Pseudocyttus maculatus Smooth Oreodory

SMOOTH DORIES

Parazenidae

ROSY DORY
Cyttopsis rosea

The three species of Smooth Dory are found in deep waters of the Atlantic and West Pacific oceans. They have an elongate, oval body with small scales, a large head and large eyes. There are two dorsal fins, the first with 6–8 spines; the anal fin has one spine and the ventral fins have no spines. Smooth Dories have a large, highly protrusible mouth with small conical teeth in the jaws.

The **Rosy Dory** grows to about 30 cm in length and occurs in schools near the bottom, from depths of about 150 to 730 m, where it feeds on pelagic crustaceans and small fishes. It is found right around the coast of Australia and northward to Japan, though it is not common anywhere in its range. The **Little Dory** has similar habits, reaches about 18 cm in length and occurs off northern Australia.

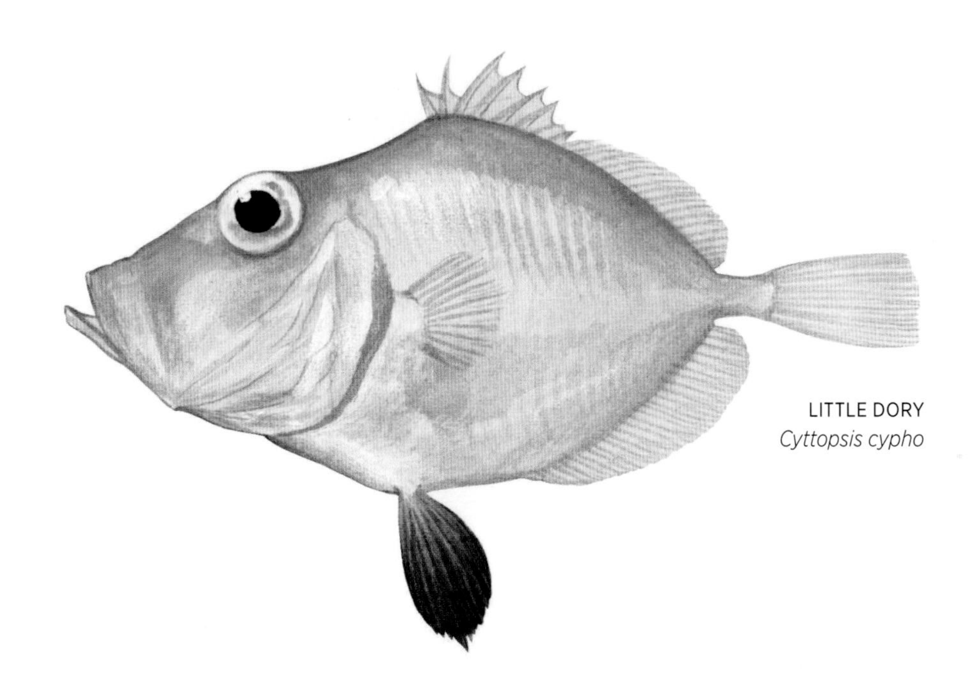

LITTLE DORY
Cyttopsis cypho

AUSTRALIAN PARAZENIDAE SPECIES

Cyttopsis cypho Little Dory
Cyttopsis rosea Rosy Dory

DWARF DORIES

Zeniontidae

FALSE DORY
Cyttomimus affinis

Dwarf Dories are found in the Atlantic, Indian and Pacific oceans, near the bottom in deep water from 150 to 1000 m. There are four species in the family. They have elongate, oval to rhomboid, compressed bodies with rounded to square scales. The single dorsal fin has 6–7 spines and the anal fin has 1–2 small, weak spines. They have bony ridges at the base of the dorsal and anal fins, and a large, serrated spine in each ventral fin. The eyes are very large and the highly protrusible jaws have bands of minute teeth. Dwarf Dories grow to a maximum length of about 12 cm.

The **False Dory** is found off northern Australia and through the West Pacific to Japan. Several other species are also thought to occur in Australian waters. The relationships between Dwarf Dories and other Zeiforms, and between species within this family, are uncertain and very little is known of their biology.

AUSTRALIAN ZENIONTIDAE SPECIES
Cyttomimus affinis False Dory

TINSELFISHES

Grammicolepididae

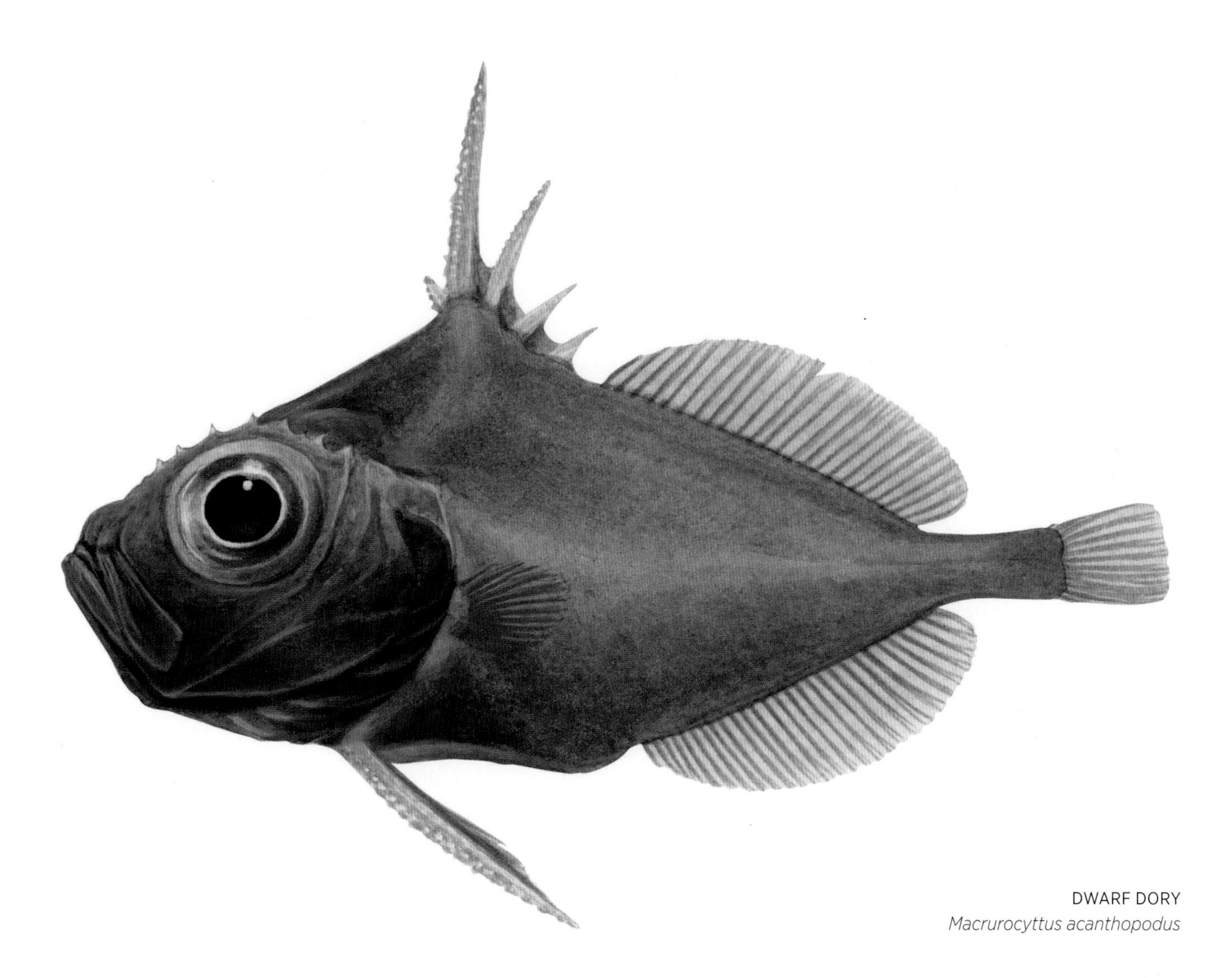

DWARF DORY
Macrurocyttus acanthopodus

The three species of Tinselfish are found in scattered localities worldwide, except in Arctic and Antarctic waters. They have deep, oval to diamond-shaped compressed bodies, with narrow, vertically elongated scales extending onto the head, and a moderate to small mouth with minute teeth. The dorsal fin has 5–7 spines, the anal fin usually has two spines and the ventral fin has one spine. At the base of the dorsal and anal fins there is a row of small spines. In the **Dwarf Dory** the second dorsal spine and the ventral-fin spines are very long and strongly serrated, and the anal-fin spines are absent. The Dwarf Dory has been found at depths of around 900 m, from northern Australia to the Philippines. It is the only species of Tinselfish recorded in Australian waters and is quite different from all other members of the family. In fact, the Dwarf Dory was formerly placed in its own family. It is rare and virtually nothing is known of its biology.

AUSTRALIAN GRAMMICOLEPIDIDAE SPECIES

Macrurocyttus acanthopodus Dwarf Dory

DORIES

Zeidae

JOHN DORY
Zeus faber

The five species of Dory are widely distributed in the Atlantic, Indian and Pacific oceans, from shallow coastal waters down to 700 m or more. They have a very deep, almost circular, highly compressed body, either with very small scales, as in the **John Dory**, or with none at all, as in the **Mirror Dory**. The dorsal fin has 8–10 spines with elongate filaments extending from the tips, the anal fin has 1–4 spines, and there are no spines in the ventral fins. Along the base of the dorsal and anal fins there are rows of large bucklers, each with a central spine.

Both Australian species are highly prized food fishes. The Mirror Dory forms schools in deeper offshore waters, from depths of about 50 to 600 m, around the southern half of Australia and over the north-west shelf. It grows to around 70 cm in length and is caught commercially off New South Wales and Victoria. The slightly smaller John Dory has a similar distribution in Australia but is found in shallower waters, often in coastal bays of only a few metres depth and rarely deeper than about 200 m. It is also taken commercially, though usually only as bycatch in south-eastern trawl fisheries. It is often caught by recreational anglers using whole fish as bait. Both species are bottom dwelling and prey on fishes and crustaceans. They actively stalk their prey, shooting out their large, highly protrusible mouth to engulf the target when it is within reach.

MIRROR DORY
Zenopsis nebulosus

AUSTRALIAN ZEIDAE SPECIES
Zenopsis nebulosus Mirror Dory
Zeus faber John Dory

GASTEROSTEIFORMES

Members of this diverse order exhibit a wide range of body forms, from extremely flattened and encased in bony armour, to elongate and smooth skinned. Most have an elongate snout with a small mouth at the tip and bony dermal plates developed to varying degrees. Gasterosteiforms are shallow-water fishes, with many species occurring in brackish and fresh waters. There are 11 families in the order, seven of which occur in Australian waters. The order is often separated into two groups, with the families found in Australia comprising one group (Syngnathiformes) and the remaining four families, which occur only in the northern hemisphere, another group (Gasterosteiformes).

SEAMOTHS

Pegasidae

SLENDER SEAMOTH
Pegasus volitans

The five species of Seamoth are found only in tropical and temperate waters of the Indo-West Pacific. They are small, bottom-dwelling fishes with a broad, flattened body and head. The body is covered with bony plates, which take the form of jointed rings on the long tail so that it can remain flexible. There is a bony rostrum with a small, protrusible mouth on the underside. The large, wing-like pectoral fins are held out horizontally from the body.

Seamoths are found right around the coast of Australia, on sandy or muddy bottoms or in weed beds, from the shallows down to about 50 m. The **Little Dragonfish** is a tropical species that periodically sheds its skin to rid itself of encrusting organisms. The **Sculptured Seamoth** is found in southern Australian waters and is occasionally encountered in shallow bays and estuaries. It is difficult to spot, as it is capable of rapidly changing colour to match its surroundings and will also often bury itself in the sand. The **Slender Seamoth**, which grows to about 15 cm in length, is another tropical species. It is usually found in muddy estuaries and shallow bays down to about 30 m. Seamoths feed on minute benthic invertebrates, using their ventral fins to propel themselves slowly over the substrate and picking up prey with their protrusible jaws.

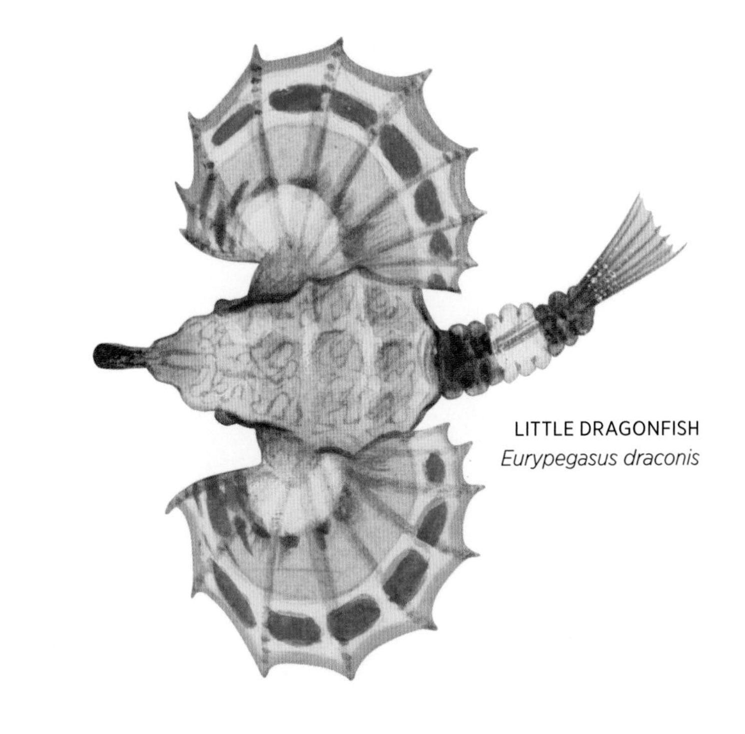

LITTLE DRAGONFISH
Eurypegasus draconis

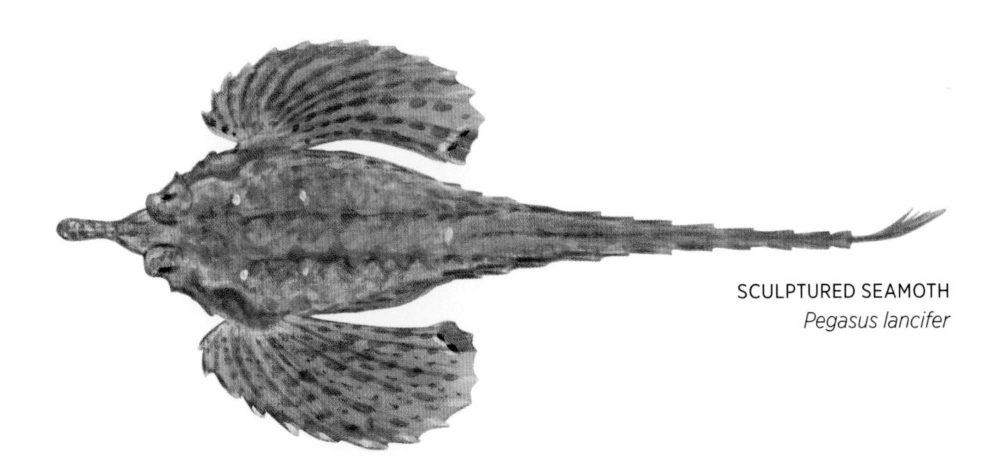

SCULPTURED SEAMOTH
Pegasus lancifer

AUSTRALIAN PEGASIDAE SPECIES

Eurypegasus draconis Little Dragonfish
Pegasus lancifer Sculptured Seamoth
Pegasus volitans Slender Seamoth

GHOSTPIPEFISHES

Solenostomidae

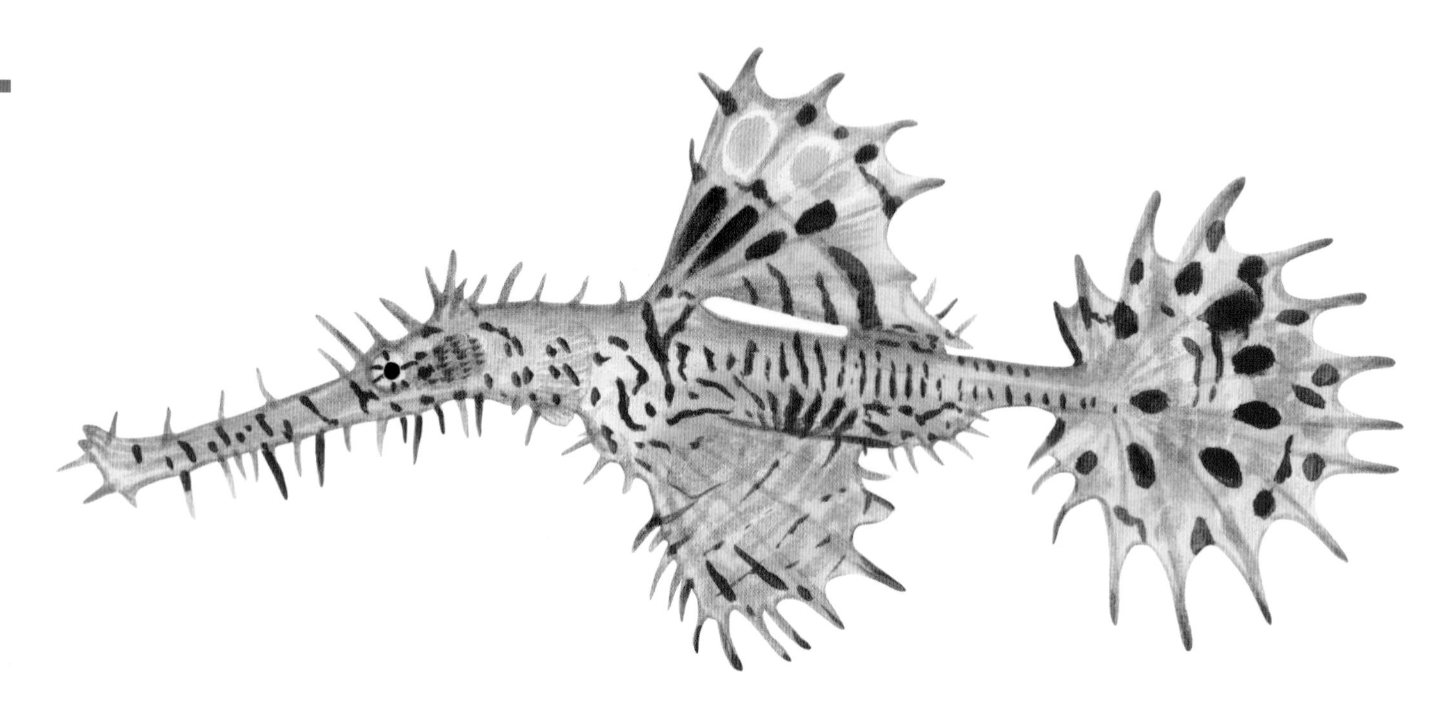

ORNATE GHOSTPIPEFISH
Solenostomus paradoxus

Ghostpipefishes are found in tropical waters of the Indo-West Pacific. There are about six species in the family. They have a rigid, compressed body that is encased in bony plates, and a long, tube-like snout with a tiny mouth at the tip. There are two dorsal fins, the first with five long spines and the second positioned on a projecting bony base opposite the anal fin. They have small pectoral fins, and large ventral fins that each have a single spine.

Female Ghostpipefishes hold their ventral fins together like cupped hands to form a brood pouch in which their eggs are incubated. Ghostpipefishes inhabit a variety of environments in protected coastal waters such as estuaries and shallow bays. They are found on coral reefs and over sand near rubble areas, usually near structures such as weed beds or sessile invertebrates. These fishes are beautifully patterned with camouflaging dermal appendages, and their colours and forms can vary greatly within a single species.

The **Halimeda Ghostpipefish** provides one of the most remarkable examples of camouflage in this family. It lives amongst Halimeda algae and its size, colour and shape exactly match those of the algae – it will even mimic the encrusting growths found on the algae and position itself precisely to look like just another branch. The **Ornate Ghostpipefish** imitates the crinoids it lives amongst, with vivid red and orange colouring and numerous spiky dermal appendages. Other Ghostpipefishes are green–brown all over and shelter amongst various types of algae and seagrass. The **Roughsnout Ghostpipefish** has long, hairy filaments to camouflage it in algae-covered rubble. Ghostpipefishes reach a maximum size of about 15 cm. They move along slowly, close to the substrate and usually with head down, feeding on minute crustaceans such as mysid shrimps and copepods.

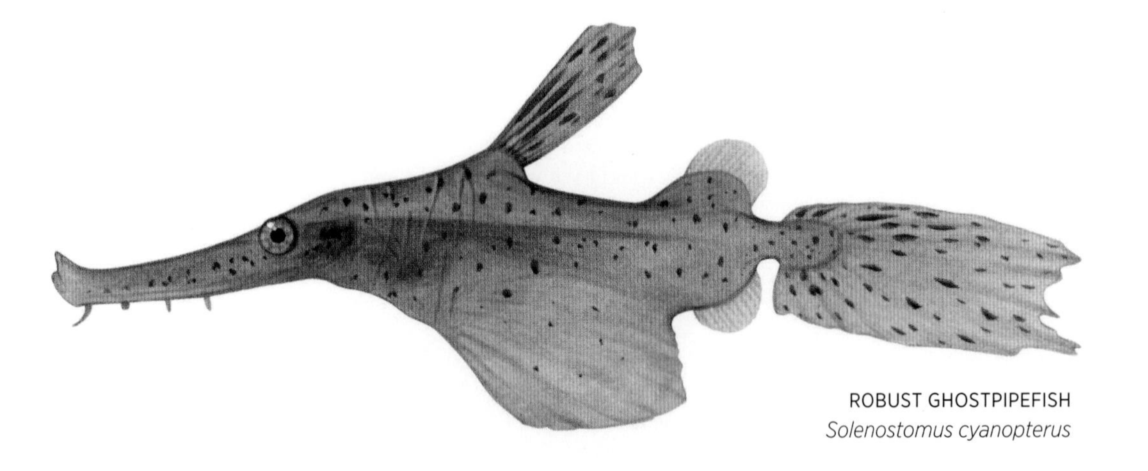

ROBUST GHOSTPIPEFISH
Solenostomus cyanopterus

HALIMEDA GHOSTPIPEFISH
Solenostomus halimeda

AUSTRALIAN SOLENOSTOMIDAE SPECIES

Solenostomus armatus Longtail Ghostpipefish
Solenostomus cyanopterus Robust Ghostpipefish
Solenostomus halimeda Halimeda Ghostpipefish
Solenostomus leptosoma Delicate Ghostpipefish
Solenostomus paegnius Roughsnout Ghostpipefish
Solenostomus paradoxus Ornate Ghostpipefish

PIPEFISHES AND SEAHORSES

Syngnathidae

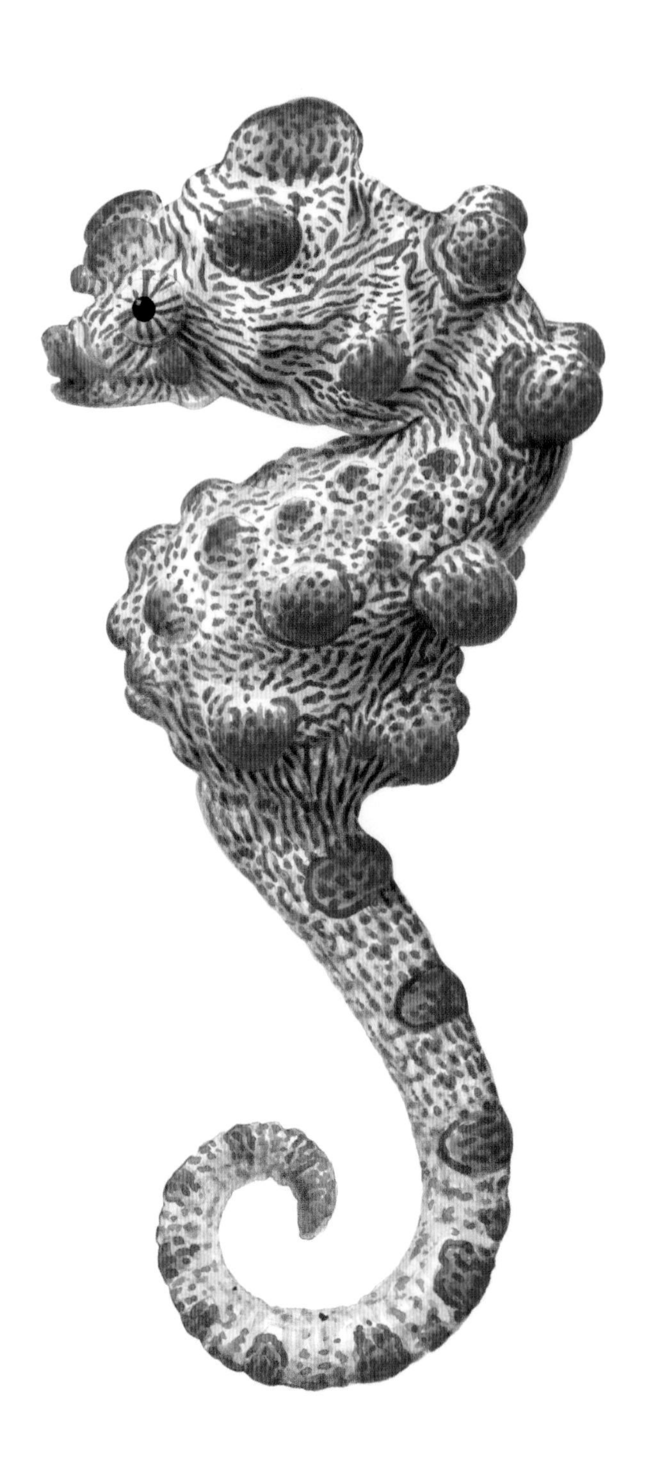

PYGMY SEAHORSE
Hippocampus bargibanti

Pipefishes and Seahorses are widely distributed in warm temperate and tropical waters around the world. There are about 230 species in the family. Several species of Pipefish, including a few in Australia, are found in fresh water, inhabiting brackish estuaries and occasionally entering the lower reaches of rivers. Members of the family have an elongate body encased in bony rings and the snout is formed into a bony tube with a small mouth at the tip. The single dorsal fin has no spines, the ventral fins are absent and the anal fin is very small or absent. They feed on microscopic invertebrates, mostly crustaceans such as mysid shrimps and copepods, either picked off the substrate or taken from the water column. The family can be divided into two subfamilies:

HIPPOCAMPINAE All Seahorses are now recognised as belonging to the genus *Hippocampus*. They have a long prehensile tail that is used to hold onto the substrate, and no caudal fin. The head is held at an angle to the body. There are about 60 species of Seahorse around the world and they are quite common in some areas. They are usually found in shallow waters amongst seagrass or algal growths, on reefs or in sponge gardens. Each species is usually restricted to a particular habitat, although they may be wide ranging geographically. Several species are found in deep water and are only known from a few trawl-caught specimens. Seahorses are usually well camouflaged. **Pygmy Seahorses** that live amongst gorgonian corals, for example, have the same colouring and lumpy protuberances as the coral. All male Seahorses have a pouch beneath the belly into which the female deposits the eggs. The male incubates the eggs until they hatch, the young emerging as fully developed miniature Seahorses. In the case of the **Potbelly Seahorse**, males compete for the attention of a female by inflating their brood pouches. This species, reaching up to 35 cm in length, is one of the largest of the Seahorses. Most Seahorses are 5–15 cm in length. The Pygmy Seahorse is one of the smallest, at about 2 cm.

SYNGNATHINAE Pipefishes generally have a long, slender body, either straight or slightly curved, with the head in line with the body. Some species have a well-developed caudal fin. The tail is sometimes flexible but is not prehensile as in Seahorses. Like Seahorses, male Pipefishes incubate the eggs, but they do not have a well-formed pouch. Instead, the eggs are attached under the belly and upper tail, and the skin may grow over them to some extent.

There are species in the Syngnathinae that are somewhere between Seahorses and Pipefishes in form. The Pipehorses resemble a straightened-out Seahorse, and the Seadragons are similar to Pipehorses but with leafy appendages. There are two spectacular Seadragons found in southern Australian waters: the **Leafy Seadragon**, which has large, branching leafy appendages on stalks along its body and tail, and the **Common Seadragon**, which has more simple leafy appendages. Both are found in seagrass and weed beds around the southern coast and their popularity as aquarium fishes has seen breeding programs established to supply them for the trade. Both species are superbly camouflaged and can be very difficult to spot underwater, particularly as juveniles.

Some Pipefishes, such as the **Bluestripe Pipefish**, have a well-developed caudal fin, are more mobile, and are brightly coloured rather than camouflaged. They are usually found on reefs in tropical waters and swim freely near the bottom, though never venturing too far from sheltering caves or crevices. The majority of Pipefishes, however, remain on the bottom, lying on the substrate or holding onto something with their flexible tail. Most Pipefishes display drab brown and green colouring, to match their preferred habitat of weed beds and algae-covered reef and rubble. Most grow to be 10–20 cm in length, but some can reach quite a large size. The **Pallid Pipehorse**, which inhabits northern Australian offshore waters, has a maximum length of about 50 cm, while the **Brushtail Pipefish**, found in southern seagrass beds, can grow to 65 cm.

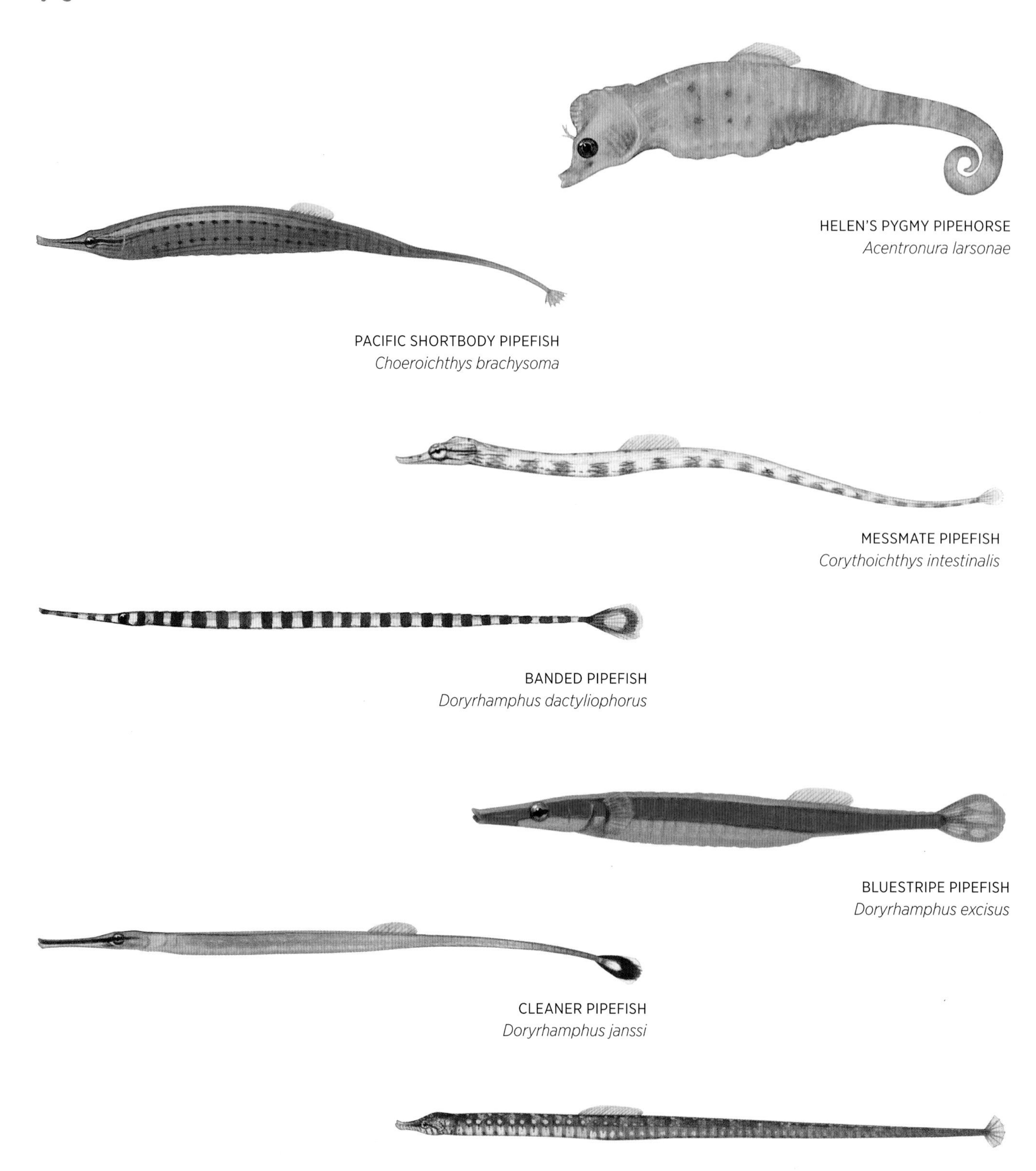

HELEN'S PYGMY PIPEHORSE
Acentronura larsonae

PACIFIC SHORTBODY PIPEFISH
Choeroichthys brachysoma

MESSMATE PIPEFISH
Corythoichthys intestinalis

BANDED PIPEFISH
Doryrhamphus dactyliophorus

BLUESTRIPE PIPEFISH
Doryrhamphus excisus

CLEANER PIPEFISH
Doryrhamphus janssi

LADDER PIPEFISH
Festucalex scalaris

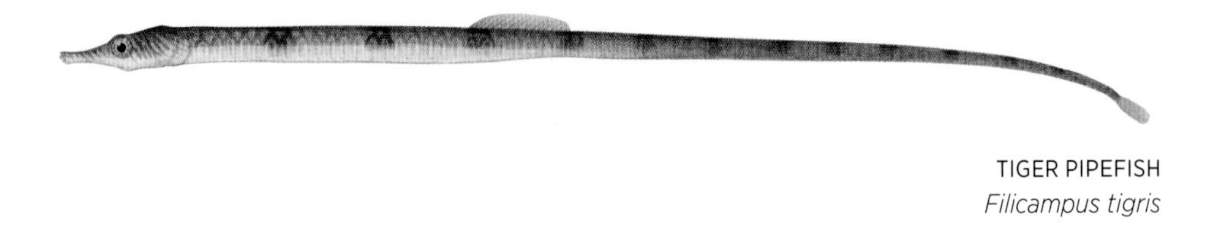

TIGER PIPEFISH
Filicampus tigris

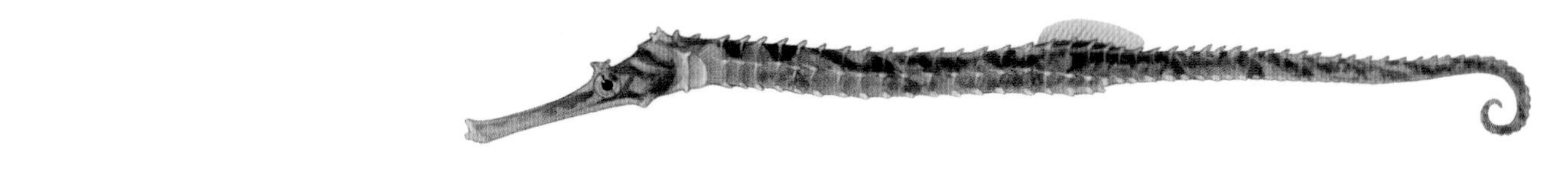

RIBBONED PIPEHORSE
Haliichthys taeniophorus

BEADY PIPEFISH
Hippichthys penicillus

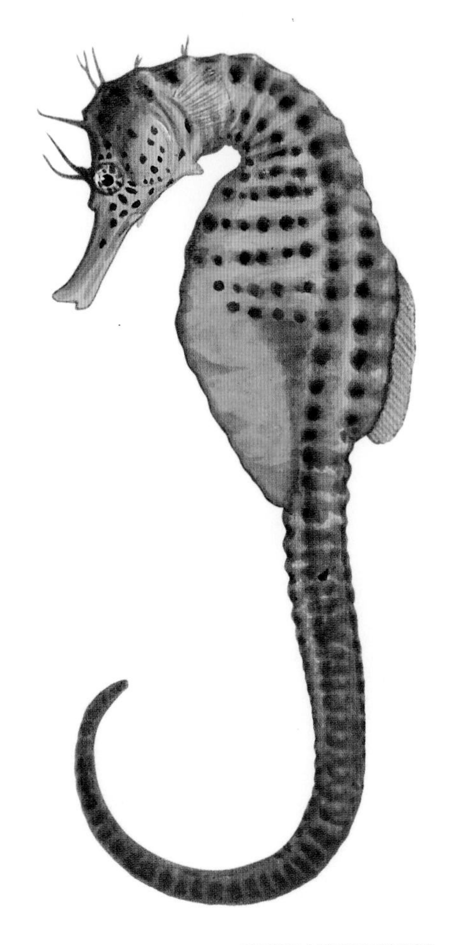

BIGBELLY SEAHORSE
Hippocampus abdominalis

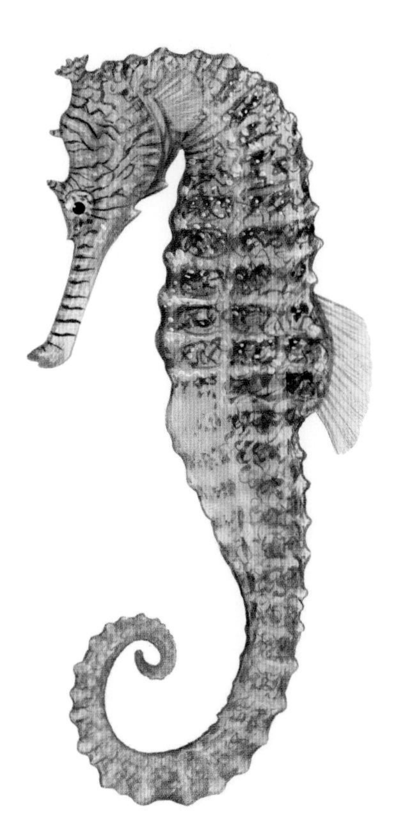

WESTERN SPINY SEAHORSE
Hippocampus angustus

Syngnathidae

SHORTHEAD SEAHORSE
Hippocampus breviceps

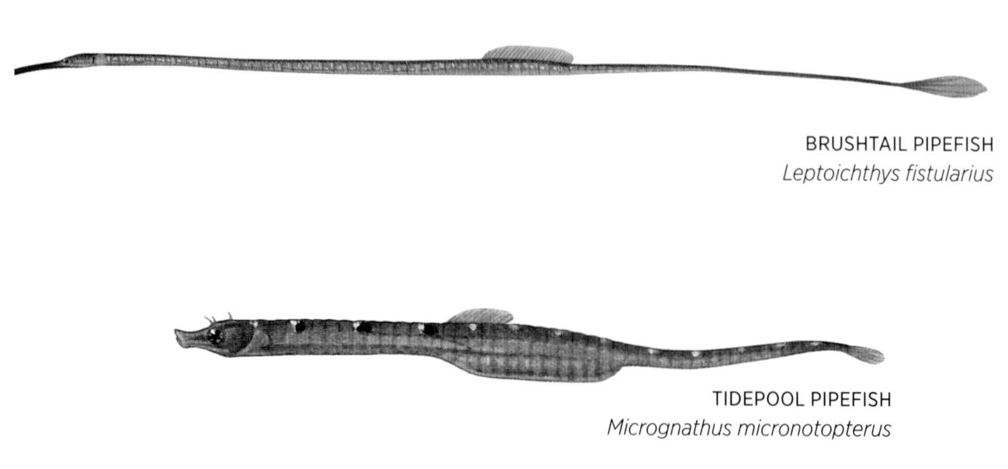

BRUSHTAIL PIPEFISH
Leptoichthys fistularius

TIDEPOOL PIPEFISH
Micrognathus micronotopterus

LEAFY SEADRAGON
Phycodurus eques

COMMON SEADRAGON
Phyllopteryx taeniolatus

GUNTHER'S PIPEHORSE
Solegnathus lettiensis

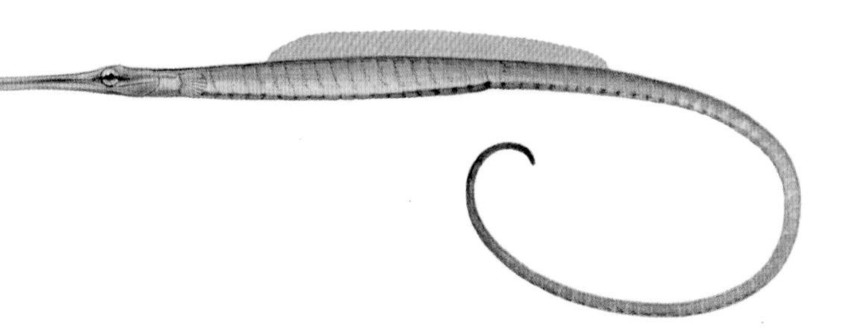

WIDEBODY PIPEFISH
Stigmatopora nigra

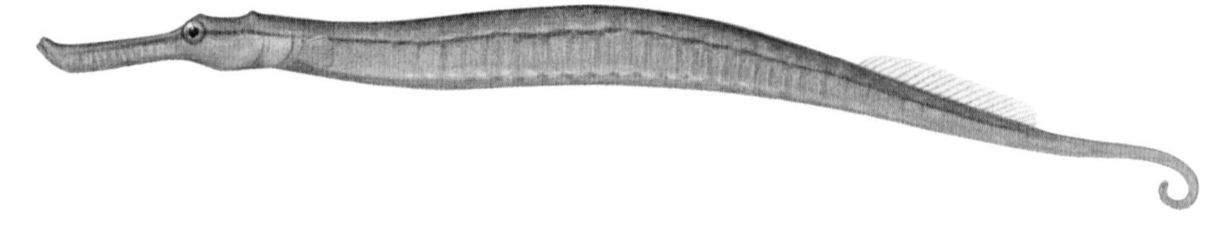

DOUBLE-END PIPEHORSE
Syngnathoides biaculeatus

BENTSTICK PIPEFISH
Trachyrhamphus bicoarctatus

PORT PHILLIP PIPEFISH
Vanacampus phillipi

AUSTRALIAN SYNGNATHIDAE SPECIES

Acentronura australe Southern Pygmy Pipehorse
Acentronura larsonae Helen's Pygmy Pipehorse
Acentronura tentaculata Shortpouch Pygmy Pipehorse
Bhanotia fasciolatus Barbed Pipefish
Bulbonaricus brauni Braun's Pughead Pipefish
Bulbonaricus davaoensis Davao Pughead Pipefish
Campichthys galei Gale's Pipefish
Campichthys tricarinatus Three-keel Pipefish
Campichthys tryoni Tryon's Pipefish
Choeroichthys brachysoma Pacific Shortbody Pipefish
Choeroichthys cinctus Barred Shortbody Pipefish
Choeroichthys latispinosus Muiron Pipefish
Choeroichthys sculptus Sculptured Pipefish
Choeroichthys suillus Pigsnout Pipefish
Corythoichthys amplexus Redbanded Pipefish
Corythoichthys flavofasciatus Reticulate Pipefish
Corythoichthys haematopterus Reeftop Pipefish
Corythoichthys intestinalis Messmate Pipefish
Corythoichthys ocellatus Ocellate Pipefish
Corythoichthys paxtoni Paxton's Pipefish
Corythoichthys schultzi Schultz's Pipefish
Cosmocampus banneri Rough-ridge Pipefish

Cosmocampus darrosanus Whiteface Pipefish
Cosmocampus howensis Lord Howe Pipefish
Cosmocampus maxweberi Maxweber's Pipefish
Doryrhamphus dactyliophorus Banded Pipefish
Doryrhamphus excisus Bluestripe Pipefish
Doryrhamphus janssi Cleaner Pipefish
Doryrhamphus negrosensis Flagtail Pipefish
Dunckerocampus pessuliferus Yellowbanded Pipefish
Festucalex cinctus Girdled Pipefish
Festucalex gibbsi Gibbs' Pipefish
Festucalex scalaris Ladder Pipefish
Filicampus tigris Tiger Pipefish
Halicampus boothae Booth's Pipefish
Halicampus brocki Tasselled Pipefish
Halicampus dunckeri Ridgenose Pipefish
Halicampus grayi Mud Pipefish
Halicampus macrorhynchus Whiskered Pipefish
Halicampus mataafae Samoan Pipefish
Halicampus nitidus Glittering Pipefish
Halicampus spinirostris Spinysnout Pipefish
Haliichthys taeniophorus Ribboned Pipehorse
Heraldia nocturna Upside-down Pipefish
Hippichthys cyanospilos Bluespeckled Pipefish
Hippichthys heptagonus Madura Pipefish
Hippichthys parvicarinatus Short-keel Pipefish
Hippichthys penicillus Beady Pipefish
Hippichthys spicifer Bellybar Pipefish
Hippocampus abdominalis Bigbelly Seahorse
Hippocampus alatus Winged Seahorse
Hippocampus angustus Western Spiny Seahorse
Hippocampus bargibanti Pygmy Seahorse
Hippocampus biocellatus False-eye Seahorse
Hippocampus bleekeri Potbelly Seahorse
Hippocampus breviceps Shorthead Seahorse
Hippocampus colemani Coleman's Seahorse
Hippocampus dahli Lowcrown Seahorse
Hippocampus grandiceps Bighead Seahorse
Hippocampus hendriki Eastern Spiny Seahorse
Hippocampus jugumus Collar Seahorse
Hippocampus kampylotrachelos Smooth Seahorse
Hippocampus minotaur Bullneck Seahorse
Hippocampus montebelloensis Monte Bello Seahorse
Hippocampus multispinus Northern Spiny Seahorse
Hippocampus planifrons Flatface Seahorse
Hippocampus procerus Highcrown Seahorse
Hippocampus queenslandicus Queensland Seahorse
Hippocampus semispinosus Halfspine Seahorse
Hippocampus subelongatus West Australian Seahorse
Hippocampus taeniopterus Common Seahorse
Hippocampus tristis Sad Seahorse
Hippocampus tuberculatus Knobby Seahorse
Hippocampus whitei White's Seahorse
Hippocampus zebra Zebra Seahorse
Histiogamphelus briggsii Crested Pipefish
Histiogamphelus cristatus Rhino Pipefish
Hypselognathus horridus Shaggy Pipefish
Hypselognathus rostratus Knifesnout Pipefish
Idiotropiscis lumnitzeri Sydney's Pygmy Pipehorse
Kaupus costatus Deepbody Pipefish
Kimblaeus bassensis Trawl Pipefish
Leptoichthys fistularius Brushtail Pipefish
Lissocampus caudalis Smooth Pipefish
Lissocampus fatiloquus Prophet's Pipefish
Lissocampus runa Javelin Pipefish
Maroubra perserrata Sawtooth Pipefish
Micrognathus andersonii Anderson's Pipefish
Micrognathus brevirostris Thorntail Pipefish
Micrognathus micronotopterus Tidepool Pipefish
Micrognathus natans Offshore Pipefish
Microphis brachyurus Short-tail Pipefish
Microphis manadensis Manado Pipefish
Mitotichthys meraculus Western Crested Pipefish
Mitotichthys mollisoni Mollison's Pipefish
Mitotichthys semistriatus Halfbanded Pipefish
Mitotichthys tuckeri Tucker's Pipefish
Nannocampus pictus Painted Pipefish
Nannocampus subosseus Bonyhead Pipefish
Notiocampus ruber Red Pipefish
Phoxocampus belcheri Black Rock Pipefish
Phoxocampus diacanthus Paleblotched Pipefish
Phycodurus eques Leafy Seadragon
Phyllopteryx taeniolatus Common Seadragon
Pugnaso curtirostris Pugnose Pipefish
Siokunichthys breviceps Softcoral Pipefish
Solegnathus dunckeri Duncker's Pipehorse
Solegnathus hardwickii Pallid Pipehorse
Solegnathus lettiensis Gunther's Pipehorse
Solegnathus robustus Robust Pipehorse
Solegnathus spinosissimus Spiny Pipehorse
Stigmatopora argus Spotted Pipefish
Stigmatopora nigra Widebody Pipefish
Stigmatopora olivacea Gulf Pipefish
Stipecampus cristatus Ringback Pipefish
Syngnathoides biaculeatus Double-end Pipehorse
Trachyrhamphus bicoarctatus Bentstick Pipefish
Trachyrhamphus longirostris Straightstick Pipefish
Urocampus carinirostris Hairy Pipefish
Vanacampus margaritifer Mother-of-pearl Pipefish
Vanacampus phillipi Port Phillip Pipefish
Vanacampus poecilolaemus Longsnout Pipefish
Vanacampus vercoi Verco's Pipefish

TRUMPETFISHES
Aulostomidae

TRUMPETFISH
Aulostomus chinensis

The three species of Trumpetfish are found in tropical waters of the Atlantic and Indo-Pacific. They have long, cylindrical to slightly compressed bodies and their long snout is formed into a laterally compressed tube with a small, highly protrusible mouth at its tip. There is a small, fleshy barbel on the tip of the lower jaw. The first dorsal fin is formed by a series of isolated spines, the second dorsal fin and the anal fin are positioned well to the rear of the body, and the ventral fins are at about mid-body.

The **Trumpetfish** grows to about 80 cm in length and is a common inhabitant of coral reefs in northern Australia. It occurs in two colour forms: overall yellow, and pale grey–brown with black and white markings. There is a small black spot on the upper caudal fin, which is a false eye used to confuse predators.

Trumpetfishes are predators of small fishes and crustaceans, and employ a number of strategies to capture their prey. They are often seen lying along the backs of larger, non-predatory fishes such as Puffers (Tetraodontidae, p. 773) and large Parrotfishes (Scaridae, p. 634), moving in unison with them to camouflage their approach towards prey. On occasion they may drift in a vertical position over coral heads (presenting a very small silhouette to any fishes or crustaceans sheltering there), then dart downwards to inhale unwary prey. At other times they move around the reef in a horizontal position near the substrate.

AUSTRALIAN AULOSTOMIDAE SPECIES

Aulostomus chinensis Trumpetfish

FLUTEMOUTHS

Fistulariidae

SMOOTH FLUTEMOUTH
Fistularia commersonii

The four species of Flutemouth are found in tropical waters of the Atlantic and Indo-Pacific, usually close to reefs but sometimes in open water. They have a very long, slightly flattened body and a long, tube-like snout with a small, highly protrusible mouth at the tip. The single dorsal fin and the anal fin are both short-based, tall and placed well towards the rear of the body. There is a long, trailing filament extending from the centre of the forked caudal fin. The **Rough Flutemouth** has a row of bony plates along the dorsal midline.

The **Smooth Flutemouth** is commonly encountered over reefs and sandy bottoms in northern Australian waters, and is often seen in small groups hovering close to the bottom. Like the Rough Flutemouth, it preys on small fishes, invertebrates, squid and crustaceans. Large specimens of these two species can reach nearly 2 m in length, and despite their small mouth they are sometimes taken on surprisingly large lures by anglers trolling near reefs. Although their movements are normally languid, they are capable of unexpected bursts of speed.

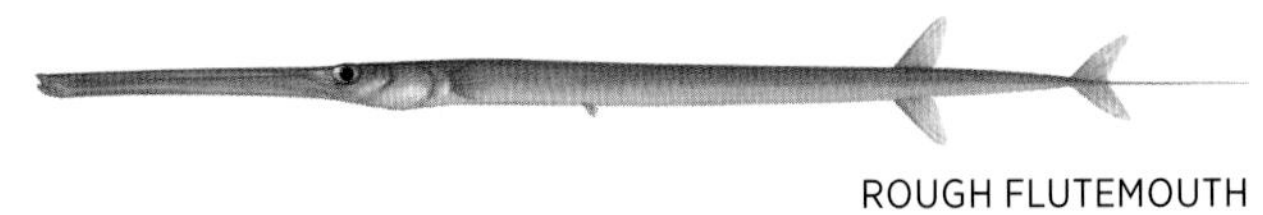

ROUGH FLUTEMOUTH
Fistularia petimba

AUSTRALIAN FISTULARIIDAE SPECIES

Fistularia commersonii Smooth Flutemouth
Fistularia petimba Rough Flutemouth

BELLOWSFISHES

Macroramphosidae

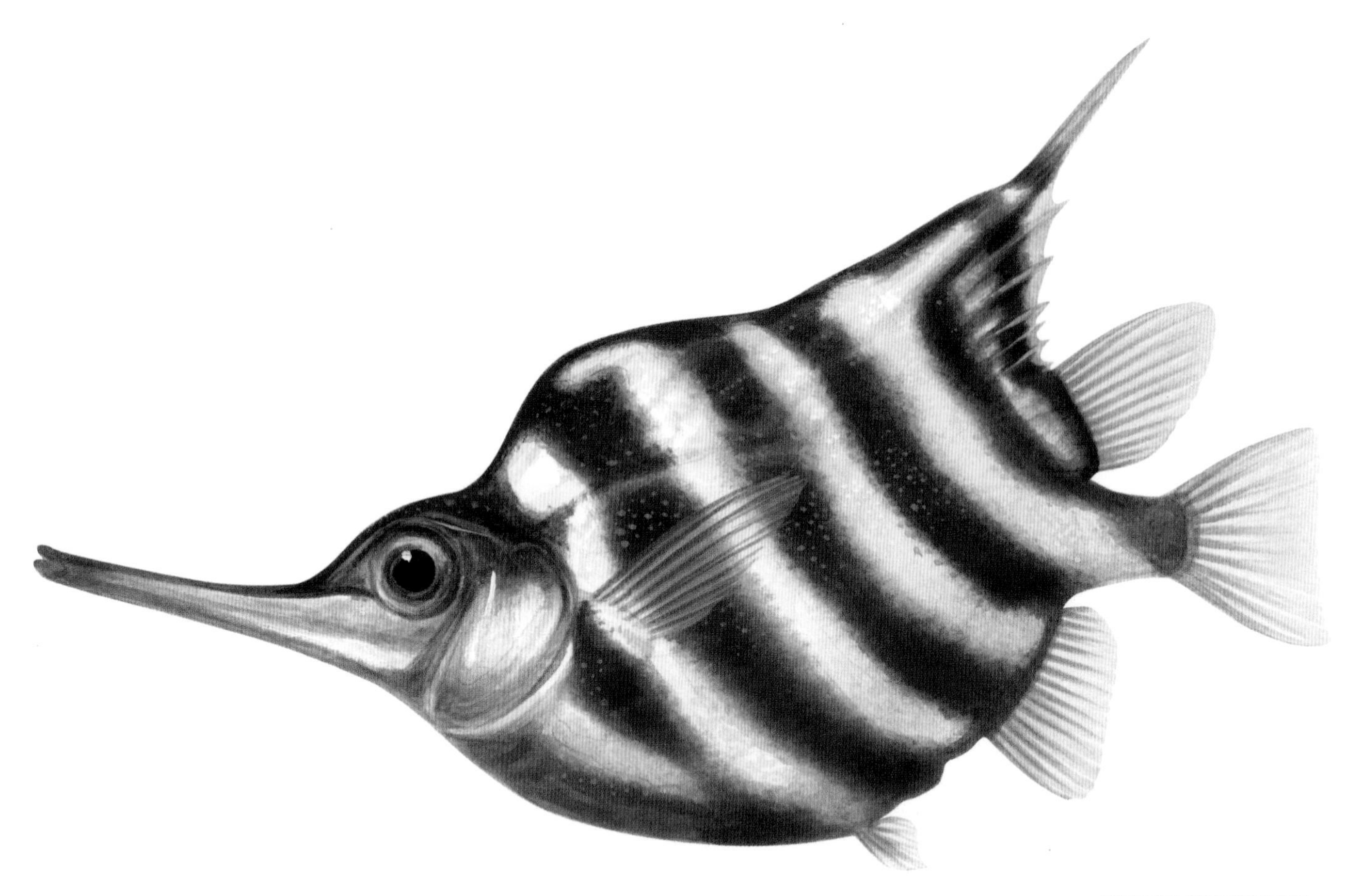

BANDED BELLOWSFISH
Centriscops humerosus

Bellowsfishes are found in temperate and tropical waters of the Atlantic, Indian and Pacific oceans. The family contains about 11 species. They have very compressed, oval to elongate bodies and skin covered by small denticles. A series of bony plates forms ridges along each side of the body, above the pectoral fins. The head is formed into a long, bony snout with a small toothless mouth at its tip. The second spine of the dorsal fin is greatly enlarged and angled backwards, and there is often a distinct hump on the back.

Bellowsfishes are usually found near the bottom in deep water, from about 50 m down to at least 1000 m for the **Banded Bellowsfish**. Some species, such as the **Little Bellowsfish**, form huge schools in shallower tropical waters. Bellowsfishes are often taken in large numbers in trawls and can be a nuisance as their large dorsal spine easily becomes entangled in the net. Bellowsfishes reach a maximum length of about 30 cm and feed on small benthic invertebrates. They occur right around the Australian coast, though they are more common in southern waters.

COMMON BELLOWSFISH
Macroramphosus scolopax

AUSTRALIAN MACRORAMPHOSIDAE SPECIES

Centriscops humerosus Banded Bellowsfish
Macroramphosus gracilis Little Bellowsfish
Macroramphosus scolopax Common Bellowsfish
Notopogon lilliei Crested Bellowsfish
Notopogon xenosoma Orange Bellowsfish

RAZORFISHES

Centriscidae

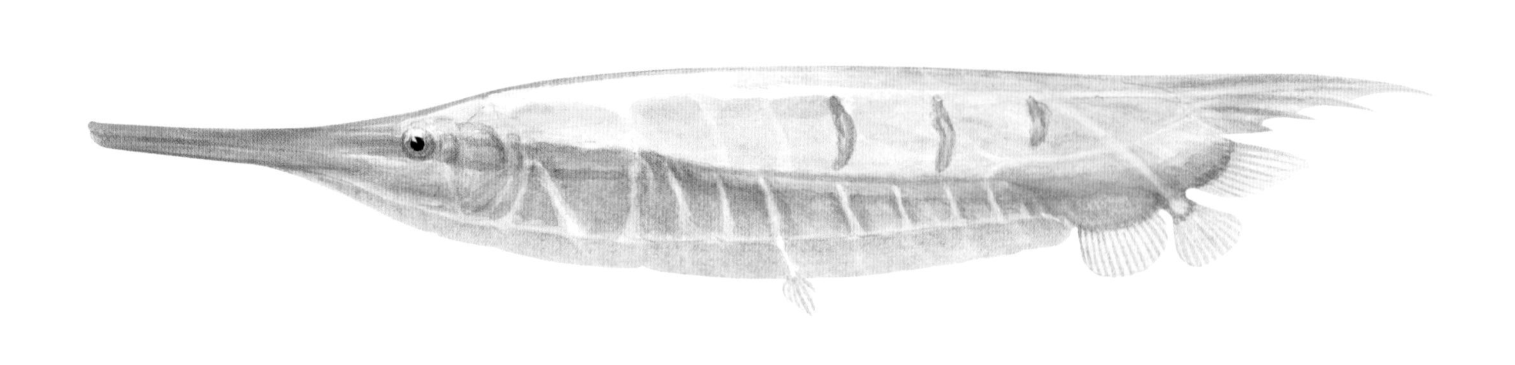

SMOOTH RAZORFISH
Centriscus cristatus

Razorfishes are found in tropical waters of the Indo-Pacific and there are about 15 species in the family. They have a thin, blade-like body encased by fine, rigid, bony plates. The head is elongated into a flattened, tube-like snout with a tiny mouth at its tip. The dorsal fin has three spines, the first of which is large, angled backwards in line with the dorsal profile, and hinged in some species. The soft dorsal rays and the caudal fin are displaced onto the ventral profile, close to the anal fin. These modifications allow Razorfishes to swim in a vertical position with the head downwards.

Razorfishes occur in schools over seagrass and reefs in sheltered waters. They often take refuge between branching corals or, when they are smaller, between the long spines of *Diadema* sea urchins. They are remarkable to watch under water, the schools shifting and veering as one, and moving surprisingly rapidly, with light reflecting off their silvery blade-like bodies. Razorfishes reach a maximum length of about 15 cm and feed on very small invertebrates, mainly crustaceans, which they pick off the substrate with their long snout.

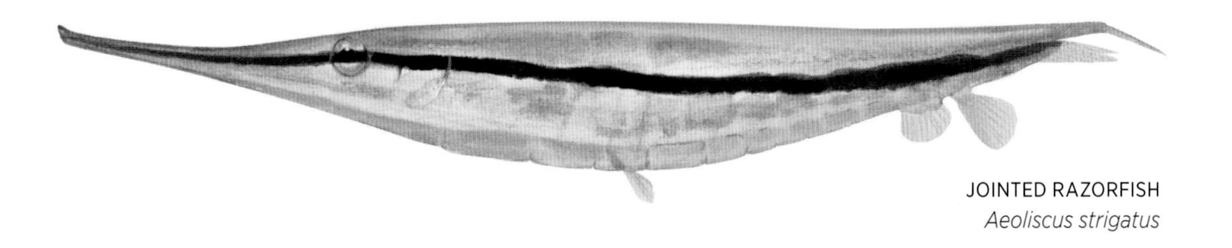

JOINTED RAZORFISH
Aeoliscus strigatus

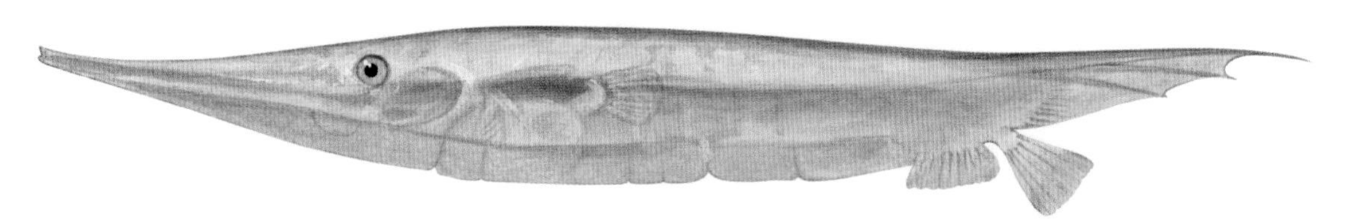

GROOVED RAZORFISH
Centriscus scutatus

AUSTRALIAN CENTRISCIDAE SPECIES

Aeoliscus strigatus Jointed Razorfish
Centriscus cristatus Smooth Razorfish
Centriscus scutatus Grooved Razorfish

SYNBRANCHIFORMES

Synbranchiforms have an elongate, eel-like body and display a number of remarkable adaptations. They have lost many features common to most bony fishes, including the dorsal, anal, pectoral and ventral fins. Most are capable of breathing air and several are known to move about extensively on land. The single family in the order occurs in northern Australia.

SWAMP EELS

Synbranchidae

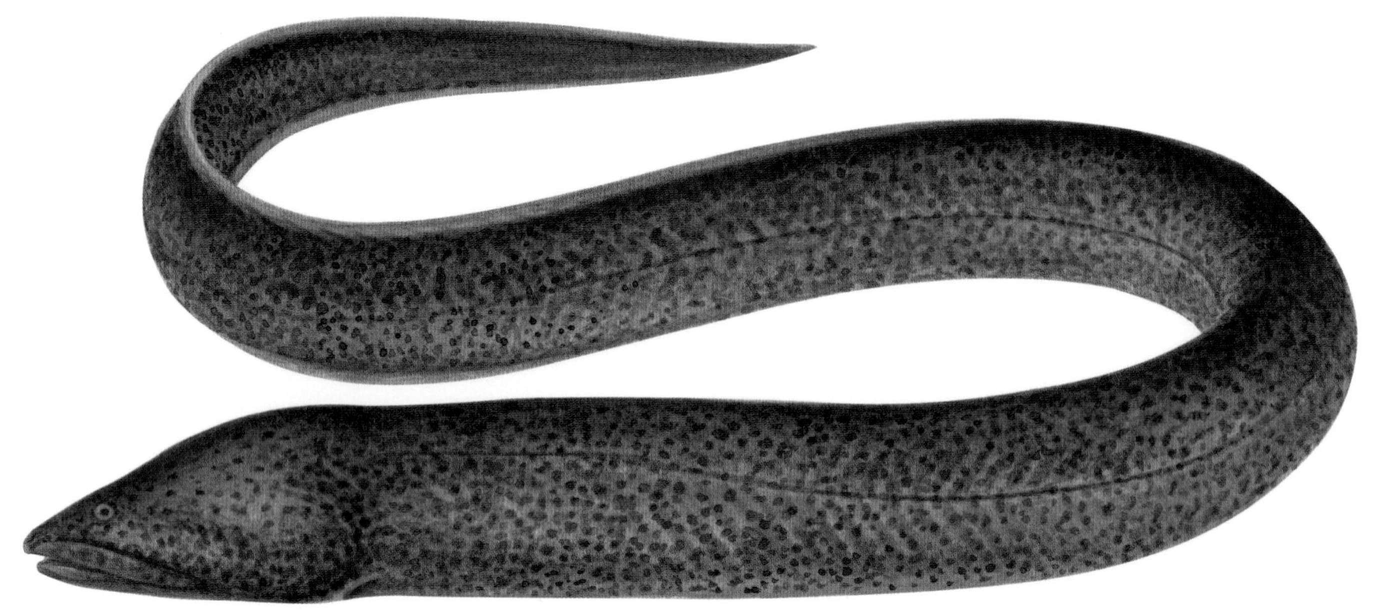

ONEGILL EEL
Ophisternon bengalense

Despite their name and appearance, Swamp Eels are not related to the true eels (Anguilliformes, p. 153). About 17 species of Swamp Eel are distributed around the world in tropical and subtropical freshwater environments. They have no pectoral or ventral fins, and the dorsal and anal fins are each reduced to a fold of skin. In Australian species the caudal fin is also absent. The gill openings are joined into a single pore or slit beneath the throat. Most species have no scales and very small eyes; in some species the eyes are vestigial.

In the **Blind Cave Eel** the eyes have disappeared altogether. This remarkable fish lives in an underground aquifer beneath North West Cape in Western Australia and is occasionally sighted in wells in the area. Other species inhabit swamps and slow-moving waters in northern and north-eastern Australia, often burrowing into soft sediment. Most have air-breathing capabilities, though these are developed to varying extents, and some can survive short periods out of the water. The **Belut** has a pair of lung-like sacs and relies on these for respiration, as the gills have become vestigial. It grows to about 90 cm in length and is quite common in rice paddies and swamps throughout South-East Asia, where it is used for food. It is capable of moving around on land during the wet season and has even been reported to feed while out of the water. In Australia the Belut is known only from a few sightings in Queensland, as is the **Onegill Eel**. The **Swamp Eel** is more widespread, occurring in the lower reaches of rivers, estuaries and swamps across northern Australia.

AUSTRALIAN SYNBRANCHIDAE SPECIES

Monopterus albus Belut
Ophisternon bengalense Onegill Eel
Ophisternon candidum Blind Cave Eel
Ophisternon gutturale Swamp Eel

SCORPAENIFORMES

This large and diverse order contains about 1500 species. Scorpaeniforms have a large, bony head with prominent spines, and all possess a bony ridge on the cheek beneath each eye that attaches to the preoperculum. They have rounded caudal and pectoral fins, the latter often enlarged and deeply incised between the rays. Many species have extremely venomous fin spines. This order is in a constant state of flux, with much disagreement on the relationships of many families and where they should be placed. There are 33 families of Scorpaeniforms recognised in this book, 21 of which occur in Australian waters.

FLYING GURNARDS

Dactylopteridae

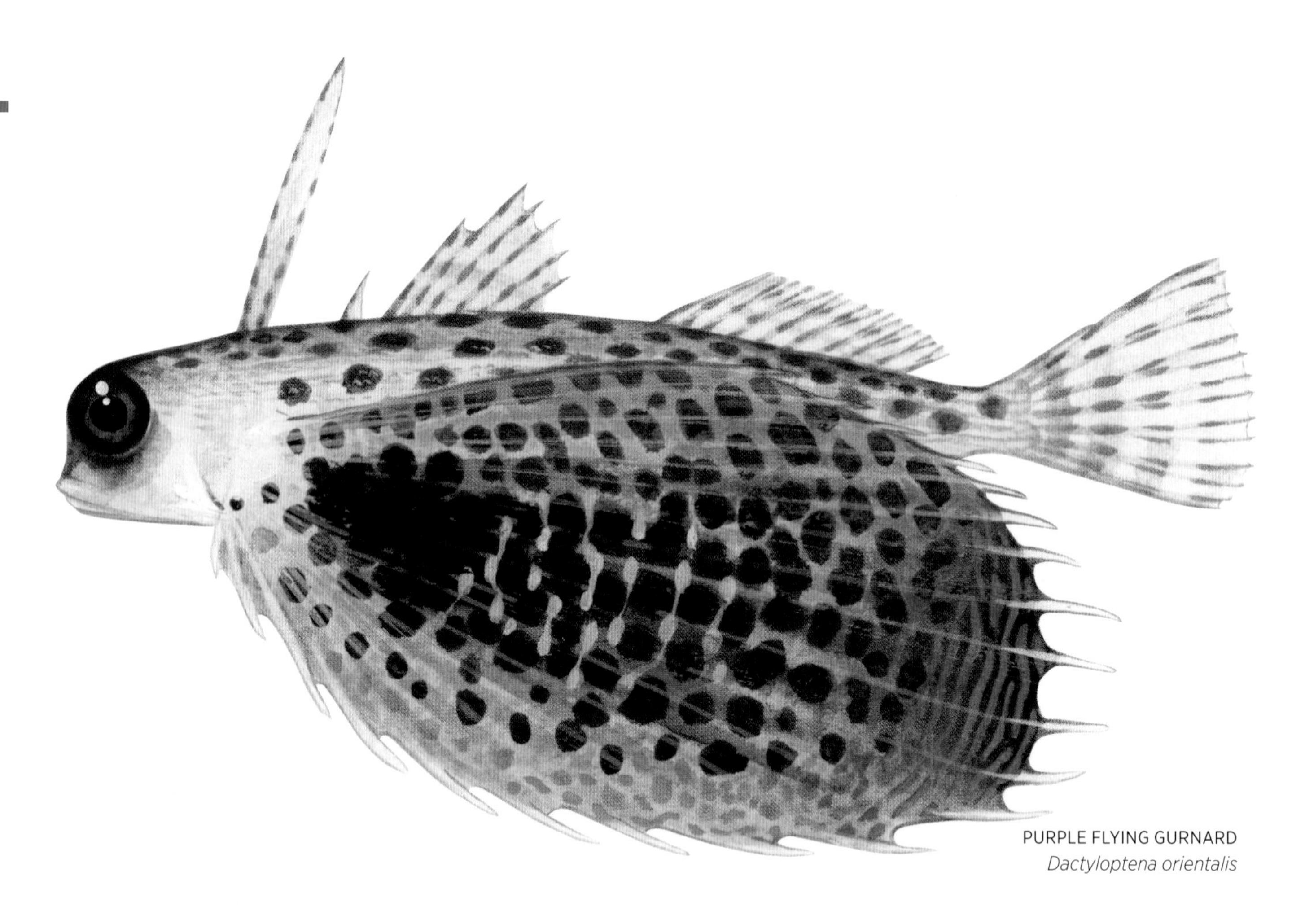

PURPLE FLYING GURNARD
Dactyloptena orientalis

Flying Gurnards are found in tropical and temperate coastal waters worldwide, and there are about seven species in the family. They are named for their large, wing-like pectoral fins, but they are unable to actually 'fly' and always remain close to the bottom. They have a blunt, rounded head that is encased in bony armour with spines and ridges, and a very long spine on the angle of the preoperculum. The body bears spiny scales, each with a central keel, and the pectoral fins are enormously enlarged. The dorsal fin is composed of 1–2 separate long spines followed by two fins, the first of which has five spines. The short anal fin has no spines. They have one spine in each ventral fin, which they use to 'walk' along the bottom.

Flying Gurnards are found over sandy bottoms in northern Australian coastal and offshore waters, down to about 200 m. They live close to the substrate and feed on small benthic crustaceans and invertebrates. The **Purple Flying Gurnard** is occasionally encountered by divers at night over sandy bottoms near reefs, gliding slowly with the large pectoral fins extended, or walking on the ventral fins. The pectoral fins are folded flat against the body during fast swimming.

AUSTRALIAN DACTYLOPTERIDAE SPECIES

Dactyloptena orientalis Purple Flying Gurnard
Dactyloptena papilio Largespot Flying Gurnard
Dactyloptena peterseni Onespine Flying Gurnard
Dactyloptena tiltoni Plain Helmet Gurnard

OCEAN PERCHES

Sebastidae

BIGEYE OCEAN PERCH
Helicolenus barathri

Ocean Perches are found worldwide in temperate waters and are particularly diverse in the North Pacific. There are about 128 species in the family. They have oval to elongate bodies and a large, bony head. There are several spines on top of the head and sometimes one beneath the eye – they generally have less-developed spines than other Scorpaeniforms. The preoperculum has a spiny edge and the body is covered with ctenoid scales. The dorsal fin has 12–13 spines, the anal fin three spines, and the ventral fins one spine; all the spines have a venom gland at the base and are capable of giving a painful sting.

Ocean Perches occur in southern Australian waters, from shallow coastal reefs to depths of more than 1000 m. The **Bigeye Ocean Perch** is taken by commercial trawlers off the southern coast, usually as bycatch, and is an excellent food fish. Until recently it was thought that the **Reef Ocean Perch**, which occurs in shallower waters than the Bigeye Ocean Perch, was a different colour form of the same species. The taxonomy of the family is still unclear and the east- and west-coast populations of Reef Ocean Perch may be separate species. The **Deepsea Ocean Perch** also appears in commercial catches from deepwater trawlers and is found from 400 to 1000 m. Ocean Perches are bottom dwelling and mainly prey on crustaceans, fishes and squid. Instead of laying eggs like most other fishes, they give birth to larvae about 1 mm in length.

REEF OCEAN PERCH
Helicolenus percoides

AUSTRALIAN SEBASTIDAE SPECIES

Helicolenus barathri Bigeye Ocean Perch
Helicolenus percoides Reef Ocean Perch
Plectrogenium nanum Bigeye Scorpionfish
Sebastiscus marmoratus False Kelpfish
Trachyscorpia eschmeyeri Deepsea Ocean Perch

DEEPWATER SCORPIONFISHES

Setarchidae

DEEPWATER SCORPIONFISH
Setarches guentheri

The five species of Deepwater Scorpionfish are found worldwide in tropical and temperate waters, from depths of about 150–2000 m. They have an oval, compressed body with thin, fragile scales and a large bony head with delicate spines. The dorsal fin has 11–13 spines and the anal fin 2–3 spines; the spines each have a venom gland at the base and can give an extremely painful sting. Deepwater Scorpionfishes have a well-developed lateral line in the form of a continuous groove or trough covered with thin membranous scales.

The **Black Scorpionfish** is pelagic in bottom waters and feeds on small crustaceans and amphipods. The other Australian species are bottom dwelling but may rise up through the water column to feed. Deepwater Scorpionfishes occasionally appear in commercial trawl nets but are not abundant enough to be commercially important. They reach a maximum length of about 25 cm and occur right around the coast of Australia.

AUSTRALIAN SETARCHIDAE SPECIES

Ectreposebastes imus Black Scorpionfish
Lioscorpius longiceps Slender Scorpionfish
Lioscorpius trifasciatus Tripleband Scorpionfish
Setarches guentheri Deepwater Scorpionfish
Setarches longimanus Red Deepwater Scorpionfish

GURNARD PERCHES

Neosebastidae

COMMON GURNARD PERCH
Neosebastes scorpaenoides

Gurnard Perches are found in the western Pacific and southeast Indian oceans. There are about 18 species in the family. They have a large, bony head with strong spines and ridges, large eyes and jaws, and rough ctenoid scales on the body that extend onto the head. The dorsal fin has 13 spines, some of which may be very long and curved in the anterior part of the fin. The anal fin has three spines and the ventral fin one spine. All the fin spines have venom glands at the base and can give excruciatingly painful stings. Many Gurnard Perches have large, beautifully patterned pectoral fins with fleshy anterior rays, which are used to prop up the fish as it rests on the bottom.

Gurnard Perches are common in coastal waters around Australia, occurring over reef, weed and sandy bottoms, from the shallows down to 500 m or more. They are ambush predators of small fishes and crustaceans, their mottled colour patterns and large fins providing effective camouflage. Gurnard Perches are excellent food fishes and several species are occasionally caught by anglers. However, great care must be taken when handling them to avoid the venomous spines. The nocturnal **Little Gurnard Perch** inhabits shallow waters off the south and west coasts of Australia. During the day it buries itself in sand with just the dorsal surface exposed, often resulting in painful stings to waders.

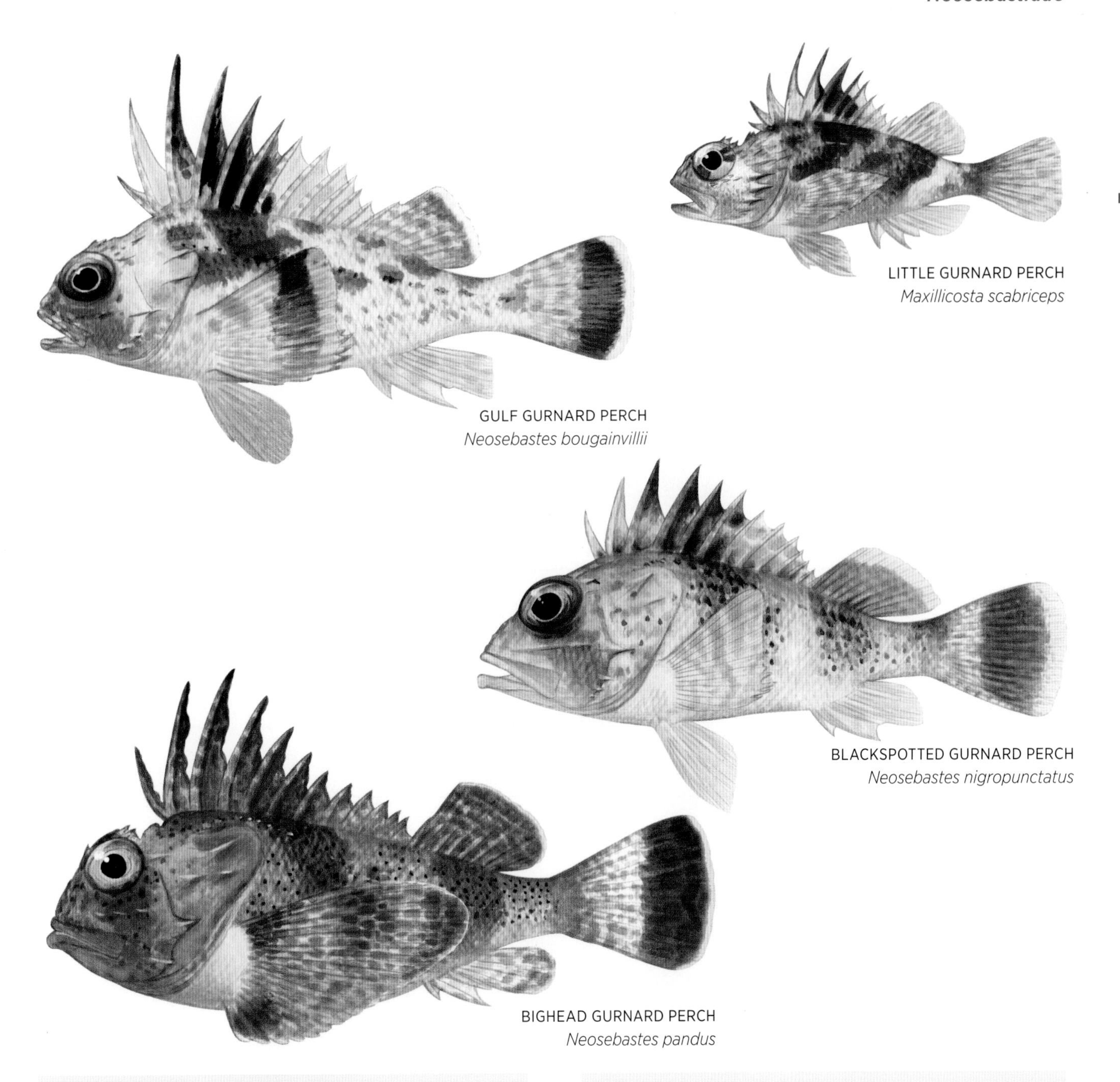

LITTLE GURNARD PERCH
Maxillicosta scabriceps

GULF GURNARD PERCH
Neosebastes bougainvillii

BLACKSPOTTED GURNARD PERCH
Neosebastes nigropunctatus

BIGHEAD GURNARD PERCH
Neosebastes pandus

AUSTRALIAN NEOSEBASTIDAE SPECIES

Maxillicosta lopholepis Bigeye Gurnard Perch
Maxillicosta raoulensis Red Little Gurnard Perch
Maxillicosta scabriceps Little Gurnard Perch
Maxillicosta whitleyi Whitley's Gurnard Perch
Neosebastes bougainvillii Gulf Gurnard Perch
Neosebastes entaxis Orangebanded Gurnard Perch
Neosebastes incisipinnis Incised Gurnard Perch
Neosebastes johnsoni Johnson's Gurnard Perch
Neosebastes longirostris Longsnout Gurnard Perch
Neosebastes nigropunctatus Blackspotted Gurnard Perch
Neosebastes pandus Bighead Gurnard Perch
Neosebastes scorpaenoides Common Gurnard Perch
Neosebastes thetidis Thetis Fish

LIONFISHES AND SCORPIONFISHES

Scorpaenidae

COMMON LIONFISH
Pterois volitans

Lionfishes and Scorpionfishes are found throughout the world's tropical and temperate oceans, and the family contains about 185 species. They have a large, bony head with strong spines and ridges. There are ctenoid scales on the body, which usually do not extend onto the head. The dorsal fin has 12–13 spines, often greatly elongated; the anal fin usually has three spines; and the ventral fins each have one spine. All spines have associated venom glands at the base and can inflict extremely painful wounds. Many species formerly contained in the Scorpaenidae have now been classified in their own families. The remaining species are divided into two subfamilies.

PTEROINAE The Lionfishes have remarkably elongated fin spines and rays, often with trailing filaments and flaps, and are popular aquarium fishes. Many species have elongate fringed antennae over the eyes and various dermal flaps around the mouth. They inhabit reefs in tropical waters and prey on small fishes and crustaceans. Small groups of **Common Lionfish** will often hunt cooperatively with their enormous fins outspread, herding small fishes then making lightning-fast lunges and snapping them up in their large, protrusible jaws. Most Lionfishes hunt at night near the bottom, and they are often seen by divers sheltering in caves and under ledges during the day. Stings from Lionfish spines are extremely painful.

SCORPAENINAE Scorpionfishes are common inhabitants of reef areas right around the Australian coast. They are bottom dwelling and often brightly coloured, with many spines, filaments and flaps of skin on the head, body and fins to aid in camouflage. The **Weedy Scorpionfish**, found in tropical waters, is an extreme example, resembling a crinoid or feather star, with ornate filaments all over the head, body and fins, making it very difficult to spot under water. Most Scorpionfishes are ambush predators, resting camouflaged and immobile on the bottom until unwary small fishes approach and then snapping them up with a sudden rapid lunge. They are rarely spotted by divers, so well do they blend with their surroundings. They also possess venom glands associated with the spines and can inflict agonising stings. The largest species reach about 30 cm in length, but most grow to only 10–20 cm.

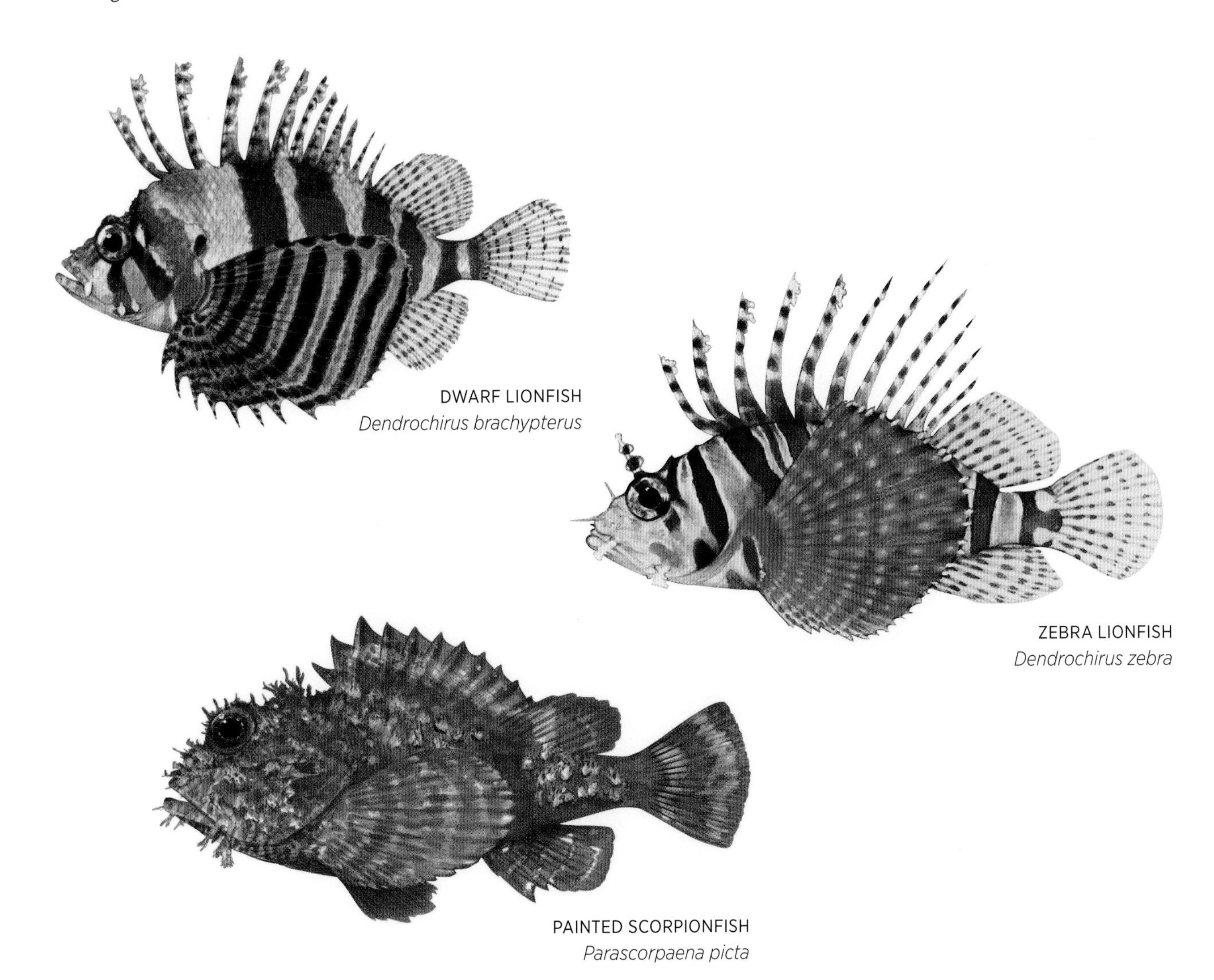

DWARF LIONFISH
Dendrochirus brachypterus

ZEBRA LIONFISH
Dendrochirus zebra

PAINTED SCORPIONFISH
Parascorpaena picta

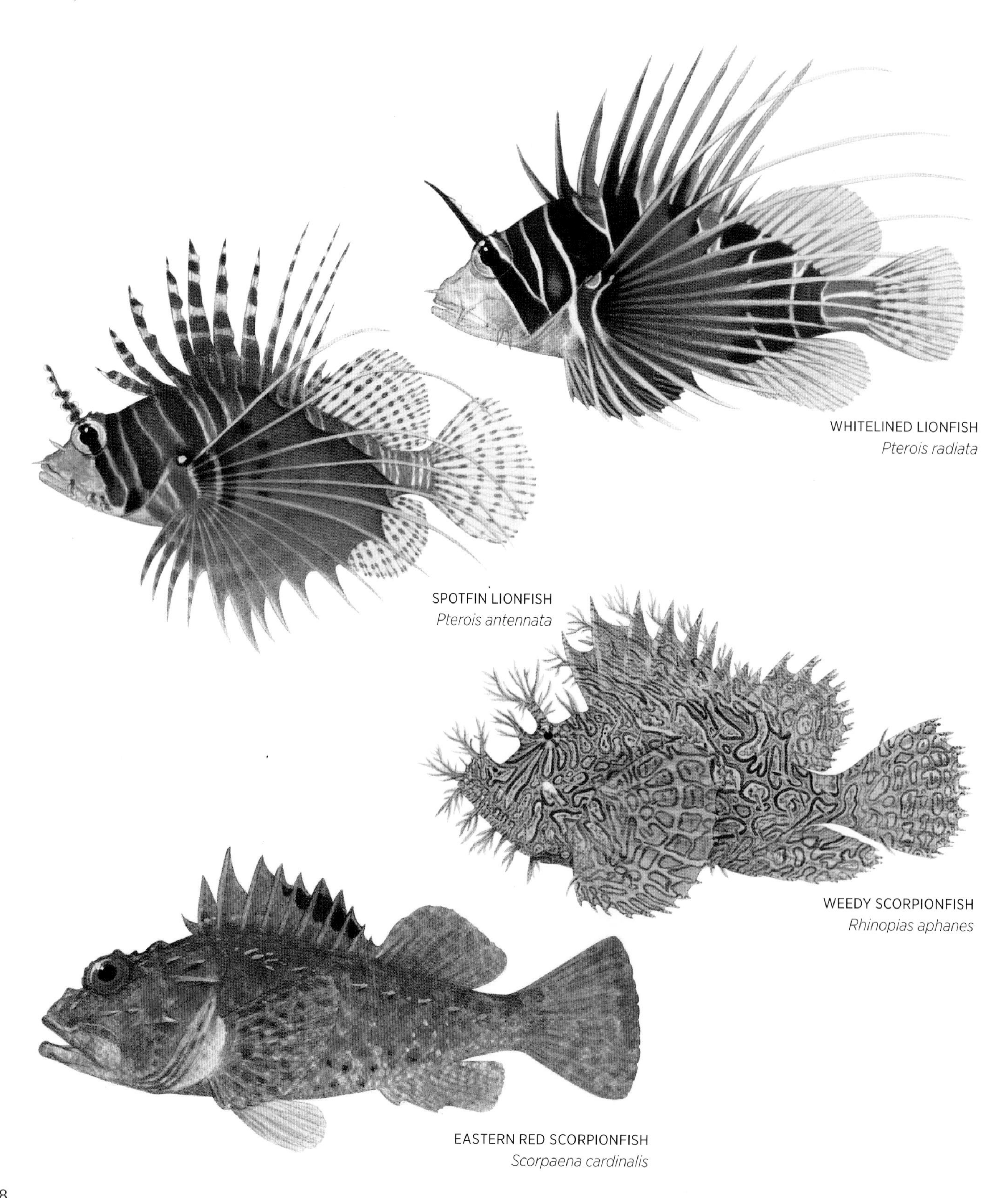

WHITELINED LIONFISH
Pterois radiata

SPOTFIN LIONFISH
Pterois antennata

WEEDY SCORPIONFISH
Rhinopias aphanes

EASTERN RED SCORPIONFISH
Scorpaena cardinalis

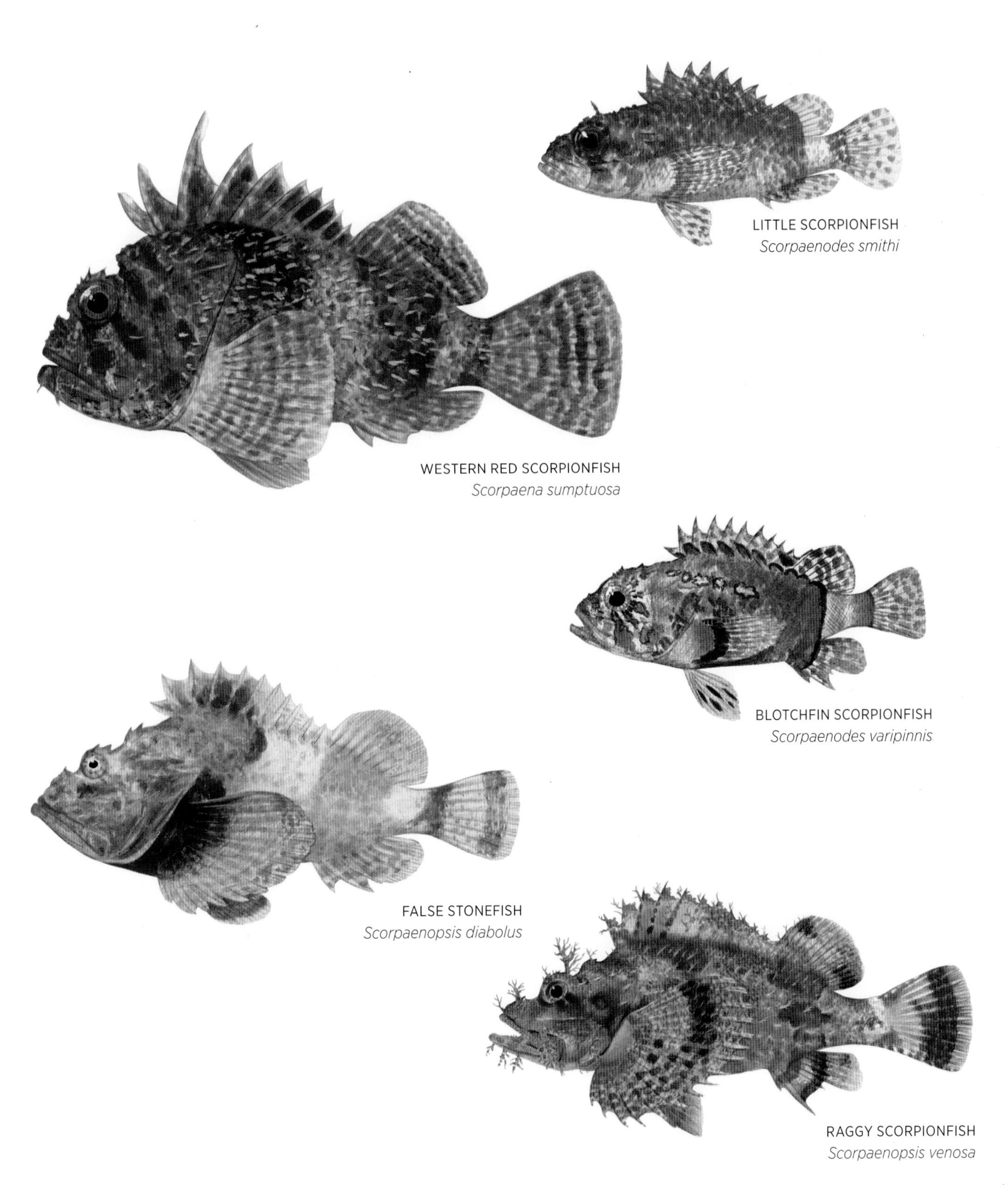

LITTLE SCORPIONFISH
Scorpaenodes smithi

WESTERN RED SCORPIONFISH
Scorpaena sumptuosa

BLOTCHFIN SCORPIONFISH
Scorpaenodes varipinnis

FALSE STONEFISH
Scorpaenopsis diabolus

RAGGY SCORPIONFISH
Scorpaenopsis venosa

YELLOWSPOTTED SCORPIONFISH
Sebastapistes cyanostigma

LEAF SCORPIONFISH
Taenianotus triacanthus

AUSTRALIAN SCORPAENIDAE SPECIES

Dendrochirus brachypterus Dwarf Lionfish
Dendrochirus zebra Zebra Lionfish
Ebosia bleekeri Cockscomb Lionfish
Neomerinthe amplisquamiceps Orange Scorpionfish
Neomerinthe megalepis Largescale Scorpionfish
Neomerinthe procurva Curvedspine Scorpionfish
Parascorpaena aurita Golden Scorpionfish
Parascorpaena maculipinnis Spotfin Scorpionfish
Parascorpaena mcadamsi Ocellate Scorpionfish
Parascorpaena mossambica Mozambique Scorpionfish
Parascorpaena picta Painted Scorpionfish
Pteroidichthys godfreyi Godfrey's Scorpionfish
Pterois antennata Spotfin Lionfish
Pterois lunulata Dragon's Beard Fish
Pterois mombasae African Lionfish
Pterois radiata Whitelined Lionfish
Pterois russelii Plaintail Lionfish
Pterois volitans Common Lionfish
Rhinopias aphanes Weedy Scorpionfish
Scorpaena bulacephala Bullhead Scorpionfish
Scorpaena cardinalis Eastern Red Scorpionfish
Scorpaena cookii Cook's Scorpionfish
Scorpaena grandisquamis Bigscale Scorpionfish
Scorpaena izensis Izu Scorpionfish
Scorpaena neglecta Neglected Scorpionfish
Scorpaena papillosa Southern Red Scorpionfish
Scorpaena sumptuosa Western Red Scorpionfish
Scorpaenodes albaiensis Longfinger Scorpionfish
Scorpaenodes guamensis Guam Scorpionfish
Scorpaenodes hirsutus Hairy Scorpionfish
Scorpaenodes kelloggi Dwarf Scorpionfish
Scorpaenodes littoralis Cheekspot Scorpionfish
Scorpaenodes minor Minor Scorpionfish
Scorpaenodes parvipinnis Coral Scorpionfish
Scorpaenodes scaber Pygmy Scorpionfish
Scorpaenodes smithi Little Scorpionfish
Scorpaenodes steenei Steene's Scorpionfish
Scorpaenodes varipinnis Blotchfin Scorpionfish
Scorpaenopsis cirrosa Bearded Scorpionfish
Scorpaenopsis cotticeps Sculpin Scorpionfish
Scorpaenopsis diabolus False Stonefish
Scorpaenopsis furneauxi Furneaux Scorpionfish
Scorpaenopsis insperatus Sydney Scorpionfish
Scorpaenopsis macrochir Humpback Scorpionfish
Scorpaenopsis neglecta Yellowfin Scorpionfish
Scorpaenopsis obtusa Shortsnout Scorpionfish
Scorpaenopsis oxycephala Smallscale Scorpionfish
Scorpaenopsis papuensis Papua Scorpionfish
Scorpaenopsis possi Poss' Scorpionfish
Scorpaenopsis venosa Raggy Scorpionfish
Sebastapistes cyanostigma Yellowspotted Scorpionfish
Sebastapistes mauritiana Spineblotch Scorpionfish
Sebastapistes strongia Barchin Scorpionfish
Sebastapistes tinkhami Darkspotted Scorpionfish
Taenianotus triacanthus Leaf Scorpionfish

WASPFISHES

Apistidae

LONGFIN WASPFISH
Apistus carinatus

Apistidae

The three species in this family are found throughout the shallow coastal waters of the Indo-Pacific and all occur in northern Australian waters. They have an elongate body with small scales, 14–16 spines in the dorsal fin, 3–4 spines in the anal fin and 3–5 small sensory tentacles beneath the head. There is a large black spot in the centre of the dorsal fin.

Waspfishes are named for the painful stings that can be inflicted by their venomous dorsal spines. They inhabit sandy and muddy bottoms from depths of about 15 to 60 m, and are frequently found in the nets of prawn trawlers in northern Australian waters. Waspfishes remain buried in the substrate during the day and hunt at night for small benthic invertebrates, fishes and crustaceans, using their sensory tentacles to detect prey in the sediment. The large pectoral fins have brightly coloured inner surfaces and when a Waspfish is alarmed or threatened these fins may be opened suddenly to confuse predators. The **Longfin Waspfish** is the largest member of the family, reaching a maximum length of about 20 cm.

SHORTFIN WASPFISH
Apistops caloundra

AUSTRALIAN APISTIDAE SPECIES
Apistops caloundra Shortfin Waspfish
Apistus carinatus Longfin Waspfish
Cheroscorpaena tridactyla Humpback Waspfish

FORTESCUES AND WASPFISHES

Tetrarogidae

EASTERN FORTESCUE
Centropogon australis

Fortescues and Waspfishes are found in the Indo-West Pacific, mostly at shallow to moderate depths in coastal waters, with one species occurring in fresh water. There are about 35 species in this family. They have a compressed body with small scales deeply embedded in the skin, and a large, blunt head with bony ridges and spines. They lack the sensory tentacles beneath the head found in Waspfishes of the family Apistidae (p. 411). Their single long dorsal fin originates on the head and has 12–18 spines. There are usually three spines in the anal fin and one spine in the ventral fins. All the spines are venomous and capable of inflicting painful stings.

Members of this family are small, bottom-dwelling fishes that are very common over sandy bottoms, weed beds and reefs in many areas around the Australian coast. Some species, such as the **Soldier**, are abundant in very shallow waters and are often responsible for stings to waders and swimmers. Roguefishes and species such as the **Yellow Waspfish** inhabit deeper offshore waters over reefs or sandy bottoms, and are a common bycatch of prawn trawlers. The strangely shaped **Goblinfish** is a nocturnal inhabitant of southern Australian reefs. The widespread and abundant Fortescues also inhabit shallow waters, and are often found in seagrass beds and sheltered bays. The **Bullrout** is the only member of the family found in fresh water and it is common in slow-flowing areas of east-coast rivers. Waspfishes and Fortescues feed on small fishes, crustaceans and benthic invertebrates, and reach a maximum length of about 80 cm, though most species grow to only 10–20 cm.

COCKATOO WASPFISH
Ablabys taenianotus

WESTERN FORTESCUE
Centropogon latifrons

YELLOW WASPFISH
Cottapistus cottoides

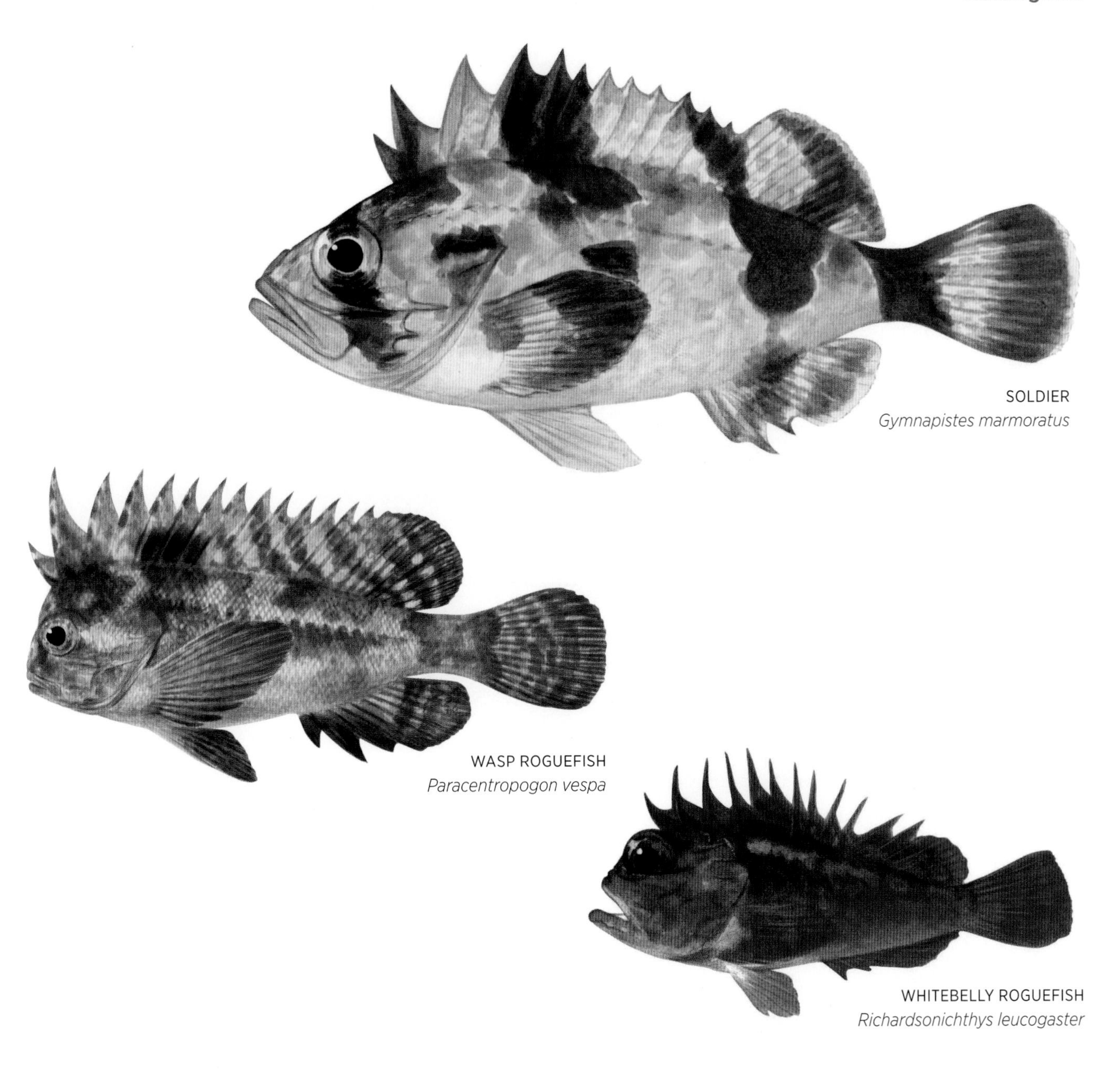

SOLDIER
Gymnapistes marmoratus

WASP ROGUEFISH
Paracentropogon vespa

WHITEBELLY ROGUEFISH
Richardsonichthys leucogaster

AUSTRALIAN TETRAROGIDAE SPECIES

Ablabys taenianotus Cockatoo Waspfish
Centropogon australis Eastern Fortescue
Centropogon latifrons Western Fortescue
Centropogon marmoratus Marbled Fortescue
Cottapistus cottoides Yellow Waspfish
Glyptauchen panduratus Goblinfish
Gymnapistes marmoratus Soldier
Liocranium praepositum Blackspot Waspfish
Neocentropogon aeglefinis Onespot Waspfish
Neocentropogon trimaculatus Threespot Waspfish
Notesthes robusta Bullrout
Paracentropogon longispinus Whiteface Roguefish
Paracentropogon vespa Wasp Roguefish
Richardsonichthys leucogaster Whitebelly Roguefish

STONEFISHES, STINGERFISHES AND STINGFISHES

Synanceiidae

REEF STONEFISH
Synanceia verrucosa

There are about 30 species in this family, and they are distributed in tropical and subtropical waters of the Indo-Pacific region, ranging from the intertidal zone to deeper offshore reefs. They have a moderately compressed body with no scales, rough, warty skin, and a large, bony head with strong ridges and spines. The dorsal fin is long and has 12–18 spines, and the anal fin has 2–3 spines. All spines have venom glands associated with them and the neurotoxin they can inject is one of the most dangerous known to humankind. Stings are capable of causing agonising pain in humans and fatalities are not uncommon.

The Stonefishes, which can reach up to 60 cm in length, are particularly dangerous due to their large size and habit of resting, perfectly camouflaged as algae-covered rocks, in shallow waters, where they can be inadvertently stepped on. They have large, fleshy pectoral fins, which help break up the outline of the fish, and small eyes on the dorsal surface. The **Reef Stonefish** is usually found amongst rocks covered with sessile invertebrates and coralline algae, on or near coral reefs, and is often mottled with bright colours to match its surroundings perfectly. Under water it is extremely difficult to see, even at close range.

Stingerfishes have a more elongate body than Stonefishes and are found over rubble and broken ground near reefs, as well as on sandy and muddy bottoms, in northern Australian waters. They have long dorsal spines, also capable of inflicting excruciating stings. The pectoral fins are large, with the lowermost 2–3 rays separate and used for 'walking' along the substrate. The pectoral fins are often brightly coloured on the interior surface and are used to startle potential predators.

Stingfishes reach about 10–12 cm in length and are found in northern Australian waters over sandy and muddy bottoms down to about 50 m. They have large pectoral fins with only the lowermost ray separate, and also use this pectoral-fin adaptation to 'walk' along the bottom. Most members of this family are ambush predators of crustaceans and small fishes. All should be treated with extreme caution.

DARUMA STINGER
Erosa daruma

SPOTTED STINGERFISH
Inimicus sinensis

PLUMBSTRIPED STINGFISH
Minous versicolor

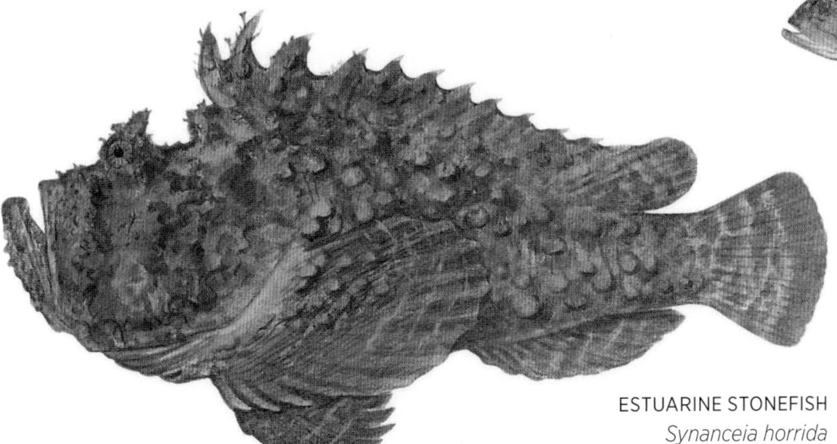

ESTUARINE STONEFISH
Synanceia horrida

AUSTRALIAN SYNANCEIIDAE SPECIES
Erosa daruma Daruma Stinger
Erosa erosa Pacific Monkeyfish
Inimicus caledonicus Demon Stingerfish
Inimicus didactylus Longsnout Stingerfish
Inimicus sinensis Spotted Stingerfish
Minous coccineus Yellowfin Stingfish
Minous trachycephalus Striped Stingfish
Minous versicolor Plumbstriped Stingfish
Synanceia horrida Estuarine Stonefish
Synanceia verrucosa Reef Stonefish

CROUCHERS

Caracanthidae

SPOTTED CROUCHER
Caracanthus maculatus

The four species of Croucher are found only in tropical waters of the Indo-Pacific region. They have an oval, compressed body, a large head with low bony ridges, spines on the edge of the preoperculum and operculum, and a small mouth. The body is densely covered with small papillae, giving it a velvety appearance. The dorsal fin has 6–8 spines, the anal fin two spines and the ventral fins one spine. The spines are venomous, but as Crouchers only reach a maximum of about 5 cm in length they are not dangerous.

All species inhabit branching corals such as *Acropora* and *Pocillopora,* in quite shallow waters and areas of moderate surge. Although reasonably common in northern Australian waters, they are rarely seen and are usually only found when entire colonies of coral are removed from the water. Crouchers feed on small invertebrates and it is probable that they never leave the coral heads in which they live.

AUSTRALIAN CARACANTHIDAE SPECIES

Caracanthus maculatus Spotted Croucher
Caracanthus unipinnis Coral Croucher

VELVETFISHES

Aploactinidae

SOUTHERN VELVETFISH
Aploactisoma milesii

Velvetfishes are distributed in the Indo-West Pacific and are particularly diverse in Indonesian and Australian waters. There are about 37 species in the family. They have an elongate, compressed body, usually densely covered with small spinules that give a velvety appearance, though a few species have smooth skin. The head has rows of rounded bony knobs instead of spines. The long dorsal fin originates on the head and has 12–16 spines; the first 3–5 spines form an almost separate fin, with a deep notch before the remainder of the fin. The anal fin has either 1–2 very small spines or none at all, and the ventral fins usually have one spine. All the fin rays are unbranched. Some species have venom glands associated with the spines.

Velvetfishes inhabit rough and broken bottoms, weed beds, sponge gardens and reefs, from the shallows down to about 100 m. Many are rare and known from only a few specimens. The **Southern Velvetfish**, however, is quite common in shallow waters around the southern coast of Australia, particularly in deep estuaries and sponge beds. It grows to about 20 cm, but is not often seen due to its excellent camouflage. Most species are no larger than 5–10 cm and inhabit more northern waters, and occasionally appear in the nets of prawn trawlers.

Aploactinidae

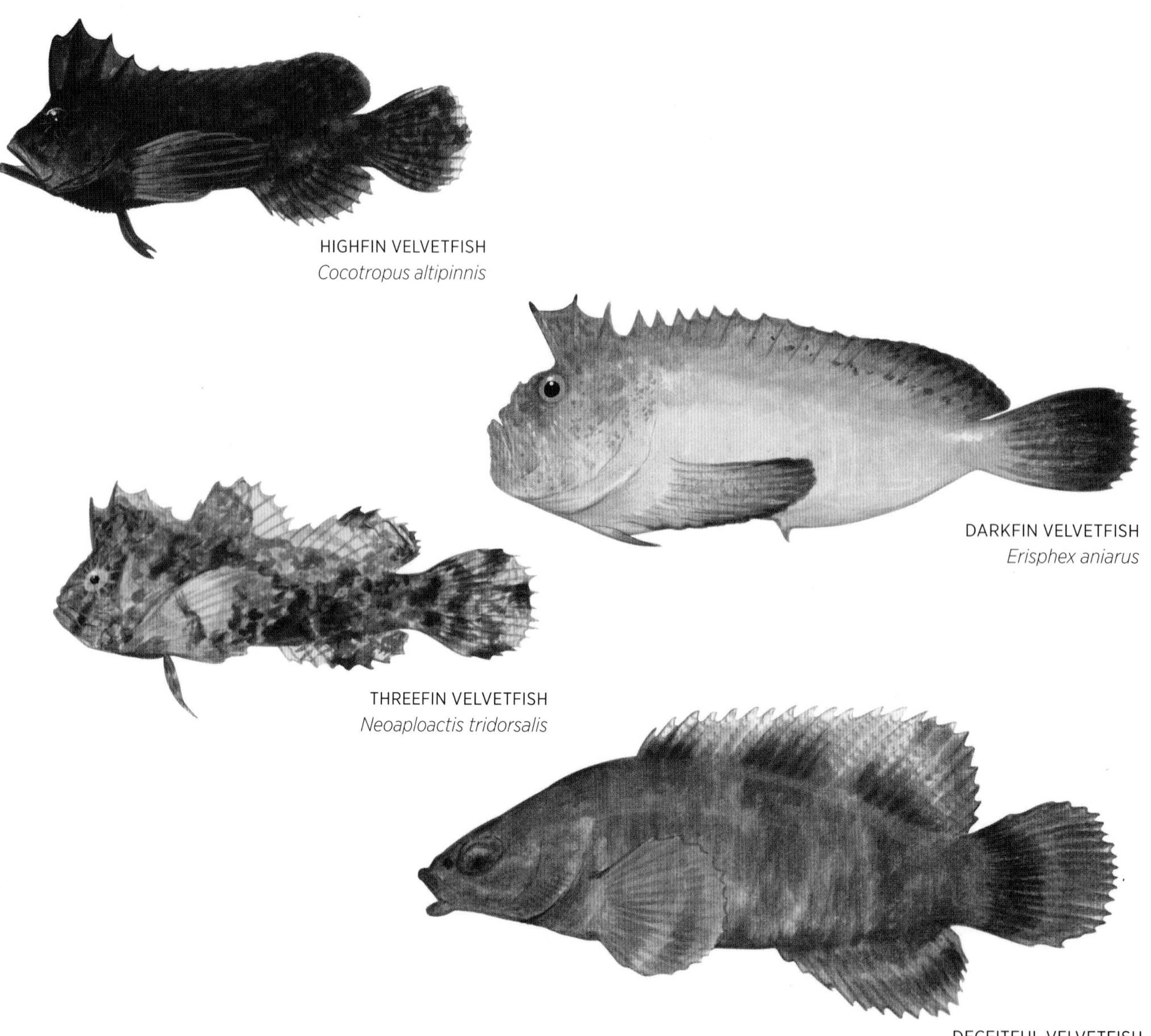

HIGHFIN VELVETFISH
Cocotropus altipinnis

DARKFIN VELVETFISH
Erisphex aniarus

THREEFIN VELVETFISH
Neoaploactis tridorsalis

DECEITFUL VELVETFISH
Peristrominous dolosus

AUSTRALIAN APLOACTINIDAE SPECIES

Acanthosphex leurynnis Wasp-spine Velvetfish
Adventor elongatus Sandpaper Velvetfish
Aploactis aspera Dusky Velvetfish
Aploactisoma milesii Southern Velvetfish
Bathyaploactis curtisensis Port Curtis Mossback
Bathyaploactis ornatissimus Ornate Velvetfish
Cocotropus altipinnis Highfin Velvetfish
Cocotropus microps Patchwork Velvetfish
Erisphex aniarus Darkfin Velvetfish
Kanekonia queenslandica Deep Velvetfish
Matsubarichthys inusitatus Rare Velvetfish
Neoaploactis tridorsalis Threefin Velvetfish
Paraploactis intonsa Bearded Velvetfish
Paraploactis pulvinus Pillow Velvetfish
Paraploactis trachyderma Mossback Velvetfish
Peristrominous dolosus Deceitful Velvetfish
Pseudopataecus taenianotus Longfin Velvetfish
Xenaploactis cautes Rough Velvetfish

PROWFISHES
Pataecidae

RED INDIAN FISH
Pataecus fronto

Prowfishes are found only in coastal waters of the southern half of Australia. They have an elongate, highly compressed body with no scales, and smooth or warty skin. The single dorsal fin, which has 18–26 spines, originates on the front of the steep head and runs the entire length of the body. The anal fin has 4–11 spines, there are no ventral fins and all fin rays are unbranched.

Prowfishes inhabit sheltered coastal waters and are found on reefs, amongst vegetation and sponge gardens, and in deeper offshore areas over broken bottoms. The **Red Indian Fish** grows to about 30 cm in length on deeper reefs and sponge gardens, and is often found in craypots off the west coast. Its thin, leaf-like body has a tall, curved dorsal fin, that bears a resemblance to a traditional Native American feather headdress. The **Whiskered Prowfish** also appears occasionally in craypots and has a fringe of small papillae beneath the head. The **Warty Prowfish** inhabits sponge gardens and sheltered reef areas in shallow waters and has the remarkable ability of being able to shed its skin. It will regularly shrug off the outer layer of skin in a single piece, leaving a perfect empty replica of itself. As a sedentary species, this is perhaps a means by which it can prevent algae and microinvertebrates accumulating on the skin. Prowfishes are not often encountered under water and their biology is poorly known. They feed mainly on small crustaceans.

WARTY PROWFISH
Aetapcus maculatus

AUSTRALIAN PATAECIDAE SPECIES

Aetapcus maculatus Warty Prowfish
Neopataecus waterhousii Whiskered Prowfish
Pataecus fronto Red Indian Fish

RED VELVETFISH

Gnathanacanthidae

RED VELVETFISH
Gnathanacanthus goetzeei

The **Red Velvetfish**, the sole member of this family, is found only around Tasmania and along the southern coast of Australia. It has an elongate, compressed body, and a large head and mouth. The body is densely covered with small spinules that give it a velvety texture. The dorsal fin is deeply notched, giving it the appearance of being two fins, and the first part has 7–8 spines. The anal fin has three spines and the large ventral fins have one spine. There are two spines on the upper edge of the operculum. All fin spines are venomous and can give extremely painful stings. The pectoral fins are large and fan-like and the rays are unbranched in all the fins. Red Velvetfishes are found on shallow reefs down to about 50 m, usually amongst weed, and feed on small crustaceans and cephalopods. Although quite common in some areas, they are rarely seen due to their nocturnal habits.

AUSTRALIAN GNATHANACANTHIDAE SPECIES

Gnathanacanthus goetzeei Red Velvetfish

PIGFISHES

Congiopodidae

WHITENOSE PIGFISH
Perryena leucometopon

Pigfishes are found only in temperate waters of the southern hemisphere and the family contains about nine species. They have an elongate body with no scales and often granular skin, and a large, bulbous head. The tip of the snout is enlarged and faintly resembles the snout of a pig. The gill openings are reduced to single small slits just above the base of each large pectoral fin. In Australian species, the single long dorsal fin has 15–21 spines, the anal fin has 0–3 spines and the ventral fins have one spine.

The **Whitenose Pigfish** has venomous spines that can give a painful sting and its skin is formed into vertical folds along its sides. The **Southern Pigfish**, which grows to about 35 cm in length, does not have venom glands associated with the spines. Pigfishes inhabit rocky offshore reefs to depths of at least 200 m. They are slow moving and sedentary by habit, and feed on small benthic invertebrates. One South African species is reported to shed its skin, like the Warty Prowfish (Pataecidae, p. 421), and others may do the same.

AUSTRALIAN CONGIOPODIDAE SPECIES

Congiopodus leucopaecilus Southern Pigfish
Perryena leucometopon Whitenose Pigfish

GURNARDS

Triglidae

RED GURNARD
Chelidonichthys kumu

Gurnards are found worldwide in tropical and temperate waters, and there are about 100 species in the family. They have an elongate, cylindrical body with small to minute scales that are often embedded in the skin. The large head is encased in bony armour, with prominent spines on the preoperculum and operculum, and often forward-projecting spines on the snout above the mouth. They have two dorsal fins, the first with 7–11 spines, and often bony plates and spines along the base of both fins. The anal fin has a single spine or none at all. The pectoral fins are greatly enlarged into broad wings that are often brightly coloured on the inner surface. The lower three rays of each pectoral fin are developed into separate fleshy fingers; these are used to 'walk' along the bottom and to feel in the substrate for the benthic invertebrates on which Gurnards feed.

Gurnards are bottom dwellers, usually found over sandy or muddy substrates, and some species occur in large numbers. They reach a maximum length of about 1 m, though most species are 10–50 cm in length. They are capable of producing grunting or chirping sounds by expelling gas from the swim bladder. Gurnards are generally regarded as excellent food fishes and in some areas are important commercially. Both the **Red Gurnard** and the **Latchet** are taken by trawlers in southern Australian waters, usually as bycatch but occasionally in considerable numbers. Both are found over sandy substrates from depths of about 20 to 200 m, with the Latchet preferring deeper waters in this range. Many other species of Gurnard are found in quite deep water, below 200 m, and are rarely encountered, only occasionally appearing in the nets of deep-water trawlers.

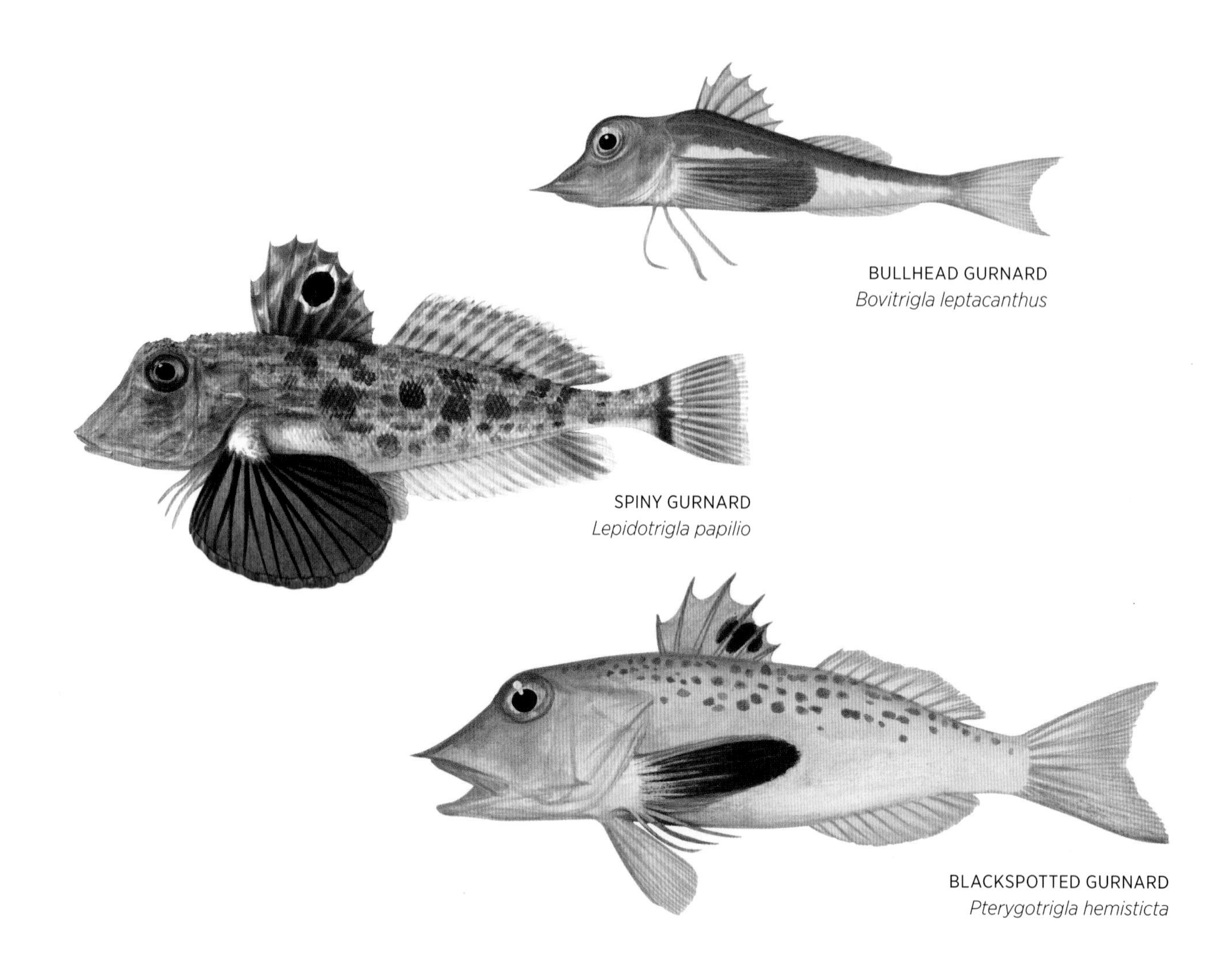

BULLHEAD GURNARD
Bovitrigla leptacanthus

SPINY GURNARD
Lepidotrigla papilio

BLACKSPOTTED GURNARD
Pterygotrigla hemisticta

AUSTRALIAN TRIGLIDAE SPECIES

Bovitrigla leptacanthus Bullhead Gurnard
Chelidonichthys kumu Red Gurnard
Lepidotrigla argus Eye Gurnard
Lepidotrigla calodactyla Drab Longfin Gurnard
Lepidotrigla grandis Little Red Gurnard
Lepidotrigla larsoni Swordtip Gurnard
Lepidotrigla modesta Cocky Gurnard
Lepidotrigla mulhalli Roundsnout Gurnard
Lepidotrigla papilio Spiny Gurnard
Lepidotrigla punctipectoralis Finspot Gurnard
Lepidotrigla russelli Smooth Gurnard
Lepidotrigla spiloptera Spot-wing Gurnard
Lepidotrigla spinosa Shortfin Gurnard
Lepidotrigla umbrosa Blackspot Gurnard
Lepidotrigla vanessa Butterfly Gurnard
Pterygotrigla andertoni Painted Latchet
Pterygotrigla draiggoch Dragon Gurnard
Pterygotrigla elicryste Dwarf Gurnard
Pterygotrigla hemisticta Blackspotted Gurnard
Pterygotrigla hoplites Swordspine Gurnard
Pterygotrigla macrorhynchus Longnose Gurnard
Pterygotrigla pauli Yellowspotted Gurnard
Pterygotrigla polyommata Latchet
Pterygotrigla robertsi Shortspine Gurnard
Pterygotrigla ryukyuensis Ryukyu Gurnard
Pterygotrigla soela Soela Gurnard

ARMOURED GURNARDS

Peristediidae

JAGGEDHEAD GURNARD
Gargariscus prionocephalus

Armoured Gurnards are found in deep waters of the Atlantic, Indian and Pacific oceans, and there are about 36 species in the family. They are similar to the Gurnards (Triglidae, p. 425), differing mainly in having rows of interlocking, bony plates along each side of the body. Like the Gurnards, they have the head encased in bony armour and have forward-projecting, flattened spines on the snout above the mouth. The lateral edges of the head armour are flattened and spread out into broad flanges that are sometimes quite ornate and usually have spines on the rear corners. The pectoral fins are not as enlarged as those of the Gurnards.

Armoured Gurnards have similar habits to the Gurnards, being bottom dwelling over sand or mud and feeding on benthic invertebrates that they detect with their finger-like pectoral rays. They differ, however, in having ornate sensory barbels around the jaws to help locate their prey, and are generally found in deeper waters, from about 50 m to at least 600 m. The larger species reach a maximum length of about 70 cm.

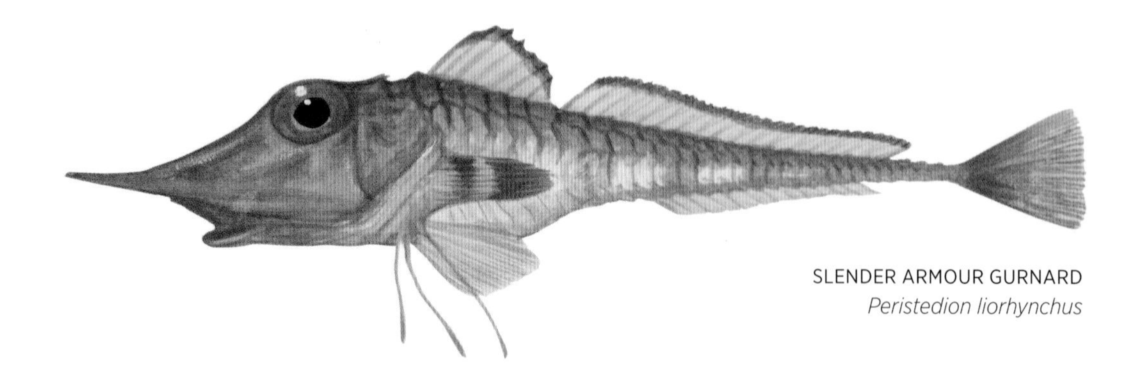

SLENDER ARMOUR GURNARD
Peristedion liorhynchus

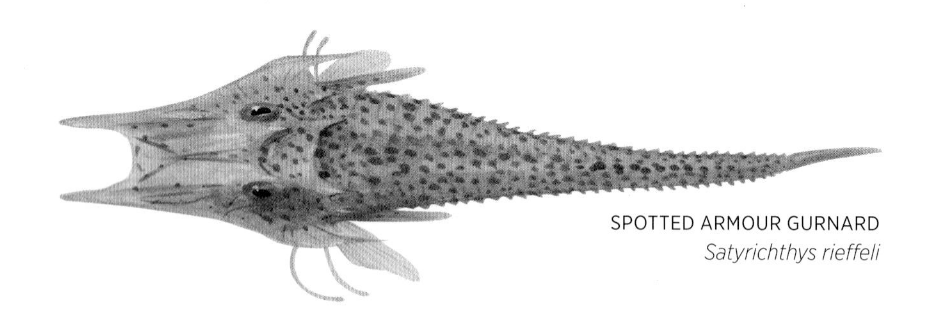

SPOTTED ARMOUR GURNARD
Satyrichthys rieffeli

AUSTRALIAN PERISTEDIIDAE SPECIES

Gargariscus prionocephalus Jaggedhead Gurnard
Peristedion liorhynchus Slender Armour Gurnard
Peristedion picturatum Robust Armour Gurnard
Satyrichthys lingi Ling's Armour Gurnard
Satyrichthys moluccense Blackfin Armour Gurnard
Satyrichthys rieffeli Spotted Armour Gurnard
Satyrichthys welchi Robust Armour Gurnard

DEEPWATER FLATHEADS

Bembridae

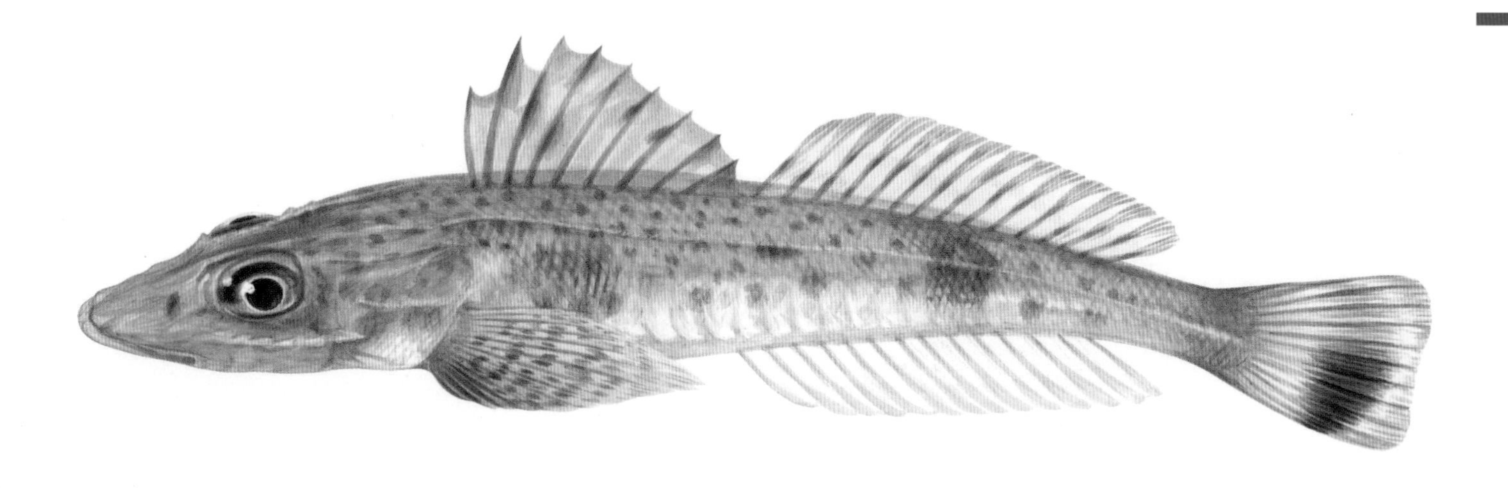

LONGFIN FLATHEAD
Bembras longipinnis

There are about 10 species of Deepwater Flathead and they are found throughout the deeper continental waters of the Indo-Pacific, from depths of about 150 to 650 m. They have an elongate, tapering, rounded body with small ctenoid scales and a flattened, spiny head. There is a strong, bony ridge beneath each large, dorsally directed eye, and the large mouth bears fine teeth on the jaws. There are two dorsal fins, the first with 6–12 spines; the anal fin has 0–3 spines and the ventral fins one spine.

Deepwater Flatheads occur in northern Australian waters. They reach about 30 cm in length and are very similar in appearance to the Flatheads (Platycephalidae, p. 430). They have the same bottom-dwelling habits as the Flatheads, albeit in deeper waters, and also prey on small fishes and benthic crustaceans.

AUSTRALIAN BEMBRIDAE SPECIES
Bembras longipinnis Longfin Flathead
Bembras macrolepis Bigscale Flathead
Bembras megacephala Greenspotted Flathead

FLATHEADS

Platycephalidae

YELLOWTAIL FLATHEAD
Platycephalus endrachtensis

There are 65 species in the family of Flatheads, mostly found in the Indo-Pacific region. They are particularly diverse in Australian waters, where there are a number of endemic species. They have an elongate, cylindrical body with small ctenoid scales, a large flattened head with spiny ridges, and a pair of sharp spines on the angle of the preoperculum. The mouth is large and the lower jaw projects beyond the upper. They have two dorsal fins, the first with 8–10 spines, and the first spine is very small and almost separate from the rest of the fin. There is a long-based anal fin with no spines, and the ventral fins have one spine.

Flatheads are widespread in coastal waters right around Australia, from shallow beaches and estuaries to reefs and deeper offshore waters. They are popular targets for anglers and commercial fishers alike, and are excellent food fishes. The **Dusky Flathead**, which grows to 1.2 m in length and can weigh 15 kg, is one of the largest species and is a major recreational fishing target in south-eastern Australia. Like most Flatheads it is a voracious predator of small fishes and crustaceans, and will readily attack lures. Strict regulations and conservation measures have been put in place to safeguard populations. The **Tiger Flathead**, **Southern Sand Flathead** and **Deepwater Flathead** are targeted commercially off southern Australia, while various other species are caught incidentally, particularly in estuaries along the southern coast. Flatheads are ambush predators. They often lie partially buried in sand or mud, well camouflaged with mottled and spotted colouring, then lunge upward to engulf passing fishes. They will also actively forage for small crustaceans and invertebrates. Flatheads should be handled with care: if the fish is held by the body it will thrash vigorously from side to side and the razor-sharp spines on the preoperculum can inflict nasty gashes.

MUD FLATHEAD
Ambiserrula jugosa

FRINGE-EYE FLATHEAD
Cymbacephalus nematophthalmus

NORTHERN SAND FLATHEAD
Platycephalus arenarius

DUSKY FLATHEAD
Platycephalus fuscus

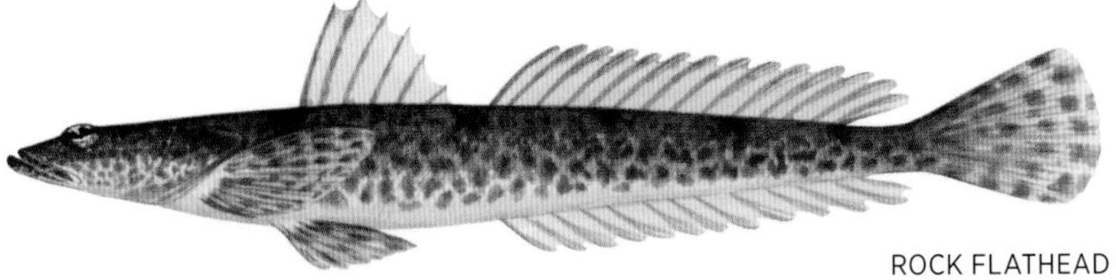
ROCK FLATHEAD
Platycephalus laevigatus

LONGSPINE FLATHEAD
Platycephalus longispinis

TIGER FLATHEAD
Platycephalus richardsoni

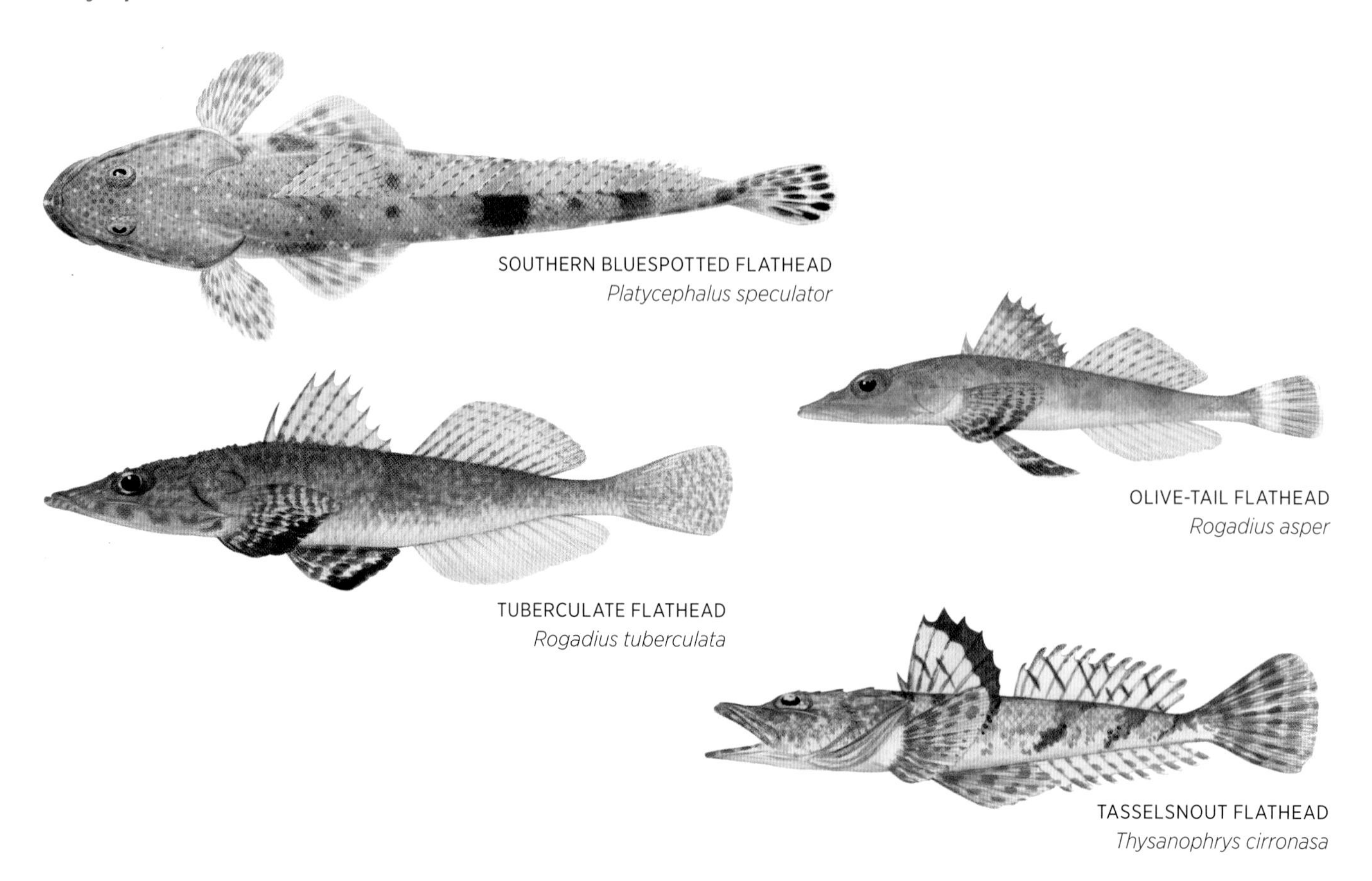

SOUTHERN BLUESPOTTED FLATHEAD
Platycephalus speculator

OLIVE-TAIL FLATHEAD
Rogadius asper

TUBERCULATE FLATHEAD
Rogadius tuberculata

TASSELSNOUT FLATHEAD
Thysanophrys cirronasa

AUSTRALIAN PLATYCEPHALIDAE SPECIES

Ambiserrula jugosa Mud Flathead
Cociella hutchinsi Brownmargin Flathead
Cymbacephalus bosschei Smalleye Flathead
Cymbacephalus nematophthalmus Fringe-eye Flathead
Cymbacephalus staigeri Northern Rock Flathead
Elates ransonnetii Dwarf Flathead
Inegocia harrisii Harris' Flathead
Inegocia japonica Rusty Flathead
Kumococius rodericensis Whitefin Flathead
Leviprora inops Longhead Flathead
Onigocia bimaculata Twospot Flathead
Onigocia macrolepis Notched Flathead
Onigocia oligolepis Shortsnout Flathead
Onigocia pedimacula Broadband Flathead
Onigocia spinosa Midget Flathead
Platycephalus arenarius Northern Sand Flathead
Platycephalus aurimaculatus Toothy Flathead
Platycephalus bassensis Southern Sand Flathead
Platycephalus caeruleopunctatus Bluespotted Flathead
Platycephalus chauliodous Bigtooth Flathead
Platycephalus conatus Deepwater Flathead
Platycephalus endrachtensis Yellowtail Flathead
Platycephalus fuscus Dusky Flathead
Platycephalus indicus Bartail Flathead
Platycephalus laevigatus Rock Flathead
Platycephalus longispinis Longspine Flathead
Platycephalus marmoratus Marbled Flathead
Platycephalus richardsoni Tiger Flathead
Platycephalus speculator Southern Bluespotted Flathead
Ratabulus diversidens Freespine Flathead
Rogadius asper Olive-tail Flathead
Rogadius patriciae Blackbanded Flathead
Rogadius pristiger Thorny Flathead
Rogadius serratus Serrate Flathead
Rogadius tuberculata Tuberculate Flathead
Suggrundus macracanthus Bigspine Flathead
Sunagocia arenicola Broadhead Flathead
Sunagocia otaitensis Fringelip Flathead
Sunagocia sainsburyi Sainsbury's Flathead
Thysanophrys celebica Celebes Flathead
Thysanophrys chiltonae Longsnout Flathead
Thysanophrys cirronasa Tasselsnout Flathead
Thysanophrys papillaris Smallknob Flathead
Thysanophrys papillolabium Fringe Flathead

GHOST FLATHEADS

Hoplichthyidae

DEEPSEA FLATHEAD
Hoplichthys haswelli

Ghost Flatheads are distributed in deep waters of the Indo-Pacific, to depths of about 1500 m. There are about 10 species in the family. They have an elongate, flattened body with no scales and a series of large bony plates, each bearing a spine, along the lateral line. The large head is extremely flattened and broad, with bony ridges and spines. They have two dorsal fins, the first with 5–6 spines; no spines in the anal fin; and one spine in each of the ventral fins. The ventral fins are spaced far apart, on the edges of the body. The pectoral fins have the lower 3–4 rays separated from the rest of the fin and these are probably used, as in the Gurnards (Triglidae, p. 425), to detect benthic invertebrates in the sand and mud bottoms over which they live.

Ghost Flatheads have a large mouth with fine, needle-like teeth and feed on benthic invertebrates, crustaceans and small fishes. The **Deepsea Flathead** is one of the largest species, reaching about 45 cm in length, and often appears in the nets of deepwater trawlers off the south coast of Australia. The other Australian species are found mainly in deeper northern waters. Ghost Flatheads are not considered worthwhile targeting commercially, even though they are quite abundant in some areas, as they are too small to process.

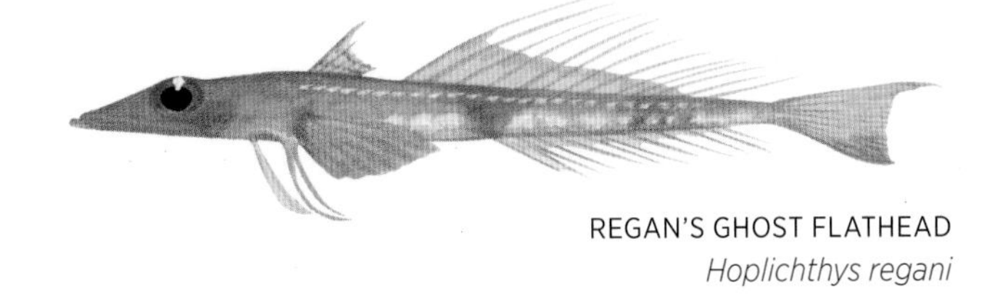

REGAN'S GHOST FLATHEAD
Hoplichthys regani

AUSTRALIAN HOPLICHTHYIDAE SPECIES

Hoplichthys citrinus Lemon Ghost Flathead
Hoplichthys filamentosus Longray Ghost Flathead
Hoplichthys haswelli Deepsea Flathead
Hoplichthys ogilbyi Ogilby's Ghost Flathead
Hoplichthys regani Regan's Ghost Flathead

SCULPINS

Cottidae

DWARF SCULPIN
Antipodocottus elegans

This is a very large family containing about 275 species. Almost all of them occur in the northern hemisphere, with the majority in temperate coastal waters of the North Pacific. Sculpins have a stout, tapered body, a large head with knobs and spines, large eyes on the dorsal surface, and fan-like pectoral fins. The body may be scaleless, partially or fully covered with bony scales, or covered with small spinules. Sculpins have a long dorsal fin that is deeply notched between the spiny and soft-rayed parts, a long anal fin with no spines, and ventral fins with one spine.

Several deepwater species are found around Australia and New Zealand. The **Dwarf Sculpin** is known only from a few specimens taken off the coast of New South Wales at depths of about 400 m, and the **Galathea Sculpin** from specimens taken at about 600 m in the south-west Pacific. Both are small bottom-dwelling fishes, growing to about 5–6 cm in length. Virtually nothing is known of their biology.

AUSTRALIAN COTTIDAE SPECIES

Antipodocottus elegans Dwarf Sculpin
Antipodocottus galatheae Galathea Sculpin

BLOBFISHES

Psychrolutidae

SMOOTH-HEAD BLOBFISH
Psychrolutes marcidus

There are about 29 species of Blobfish and they are found in the Atlantic, Indian and Pacific oceans, from shallow water down to more than 2800 m. They have a large, bulbous, slightly flattened head with bony arches beneath the skin, and a short, tapering body. They have no scales and usually loose, flabby skin, though some species have bony plates with prickles. The dorsal fin has a low spiny portion, often buried in the skin; the anal fin has no spines; and the ventral fins have one very small spine.

The Australian Blobfishes are aptly named as they all have loose flabby skin, a very soft body and a 'blobby' appearance. The **Macquarie Blobfish** and **Smooth-head Blobfish** are bottom dwelling in southern waters and found over soft sand or mud from depths of 500 to 1200 m. The **Western Blobfish** is known only from the North-West Shelf near Rowley Shoals, at depths of 350–700 m. The Macquarie Blobfish is the largest of the Australian species, reaching a maximum length of about 27 cm. Very little is known of the biology of Blobfishes.

AUSTRALIAN PSYCHROLUTIDAE SPECIES

Ebinania macquariensis Macquarie Blobfish
Psychrolutes marcidus Smooth-head Blobfish
Psychrolutes occidentalis Western Blobfish

SNAILFISHES

Liparidae

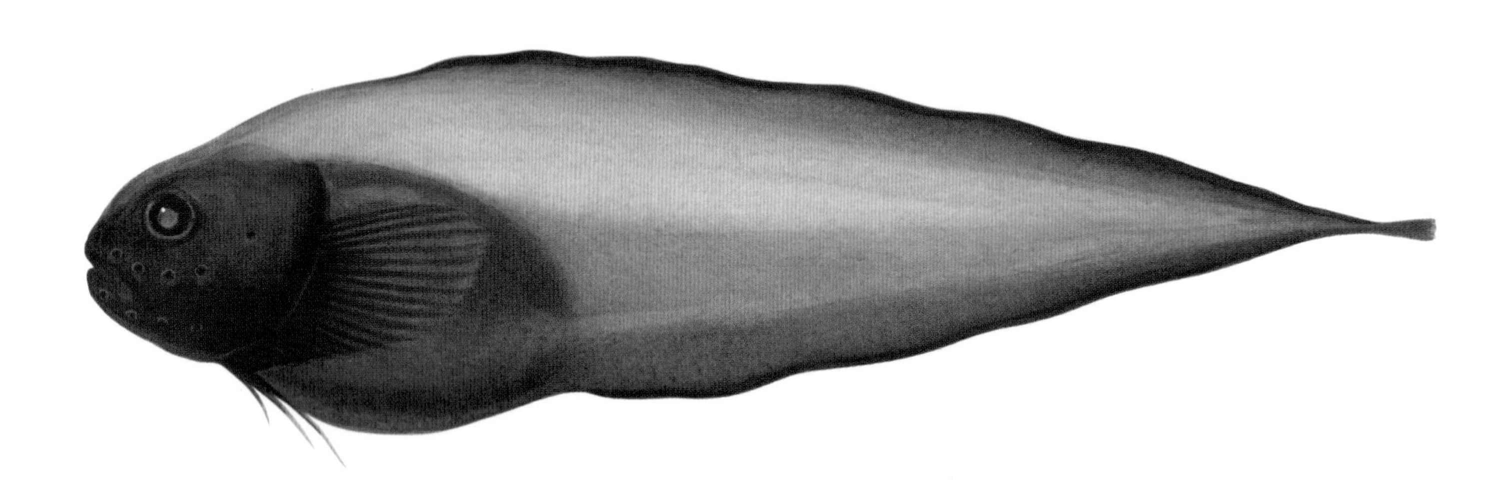

SNAILFISH
Paraliparis sp.

This large family contains more than 195 species. They inhabit an extremely broad range of environments, from the Antarctic to the Arctic and from shallow tide pools to depths of more than 7000 m. Snailfishes are particularly diverse in the Antarctic, North Pacific and North Atlantic oceans. They have a blunt, rounded head and tapering, tadpole-like body with no scales and soft, jelly-like skin. The long dorsal and anal fins have no spines. In many species the ventral fins are joined to form a sucking disc beneath the chest; this adaptation occurs in species that inhabit shallow surge areas, but is absent in all the deepwater *Paraliparis* species. Snailfishes were unknown in Australian waters until deepwater trawls discovered many species living on seamounts off south-eastern Australia at depths of 1000 m or more. Very little is understood of their biology, with many species known from only a few specimens and some species seemingly restricted to a single seamount. Snailfishes reach a maximum length of about 50 cm, but most grow no larger than 10–20 cm.

AUSTRALIAN LIPARIDAE SPECIES

Careproctus paxtoni Blunt-tooth Snailfish
Paraliparis anthracinus Coalskin Snailfish
Paraliparis ater Sooty Snailfish
Paraliparis atrolabiatus Darklip Snailfish
Paraliparis auriculatus Smallcheek Snailfish
Paraliparis australiensis Australian Snailfish
Paraliparis avellaneus Nutty Snailfish
Paraliparis badius Dark Brown Snailfish
Paraliparis brunneocaudatus Browntail Snailfish
Paraliparis brunneus Brown Snailfish
Paraliparis coracinus Black Snailfish
Paraliparis costatus Black Ribbed Snailfish
Paraliparis csiroi Loweye Snailfish
Paraliparis delphis Dolphin Snailfish
Paraliparis dewitti Brown Ribbed Snailfish
Paraliparis eastmani Thickskin Snailfish
Paraliparis gomoni Squarechin Snailfish
Paraliparis hobarti Palepore Snailfish
Paraliparis impariporus Unipore Snailfish
Paraliparis infeliciter Badluck Snailfish
Paraliparis labiatus Biglip Snailfish
Paraliparis lasti Rusty Snailfish
Paraliparis obtusirostris Bluntsnout Snailfish
Paraliparis piceus Tarred Snailfish
Paraliparis plagiostomus Sharkmouth Snailfish
Paraliparis retrodorsalis Shortfin Snailfish
Paraliparis tasmaniensis Tasmanian Snailfish
Psednos balushkini Palemouth Snailfish
Psednos nataliae Darkgill Snailfish
Psednos whitleyi Bigcheek Snailfish

PERCIFORMES

This is by far the largest order of fishes, containing more than 10 000 species divided into about 20 sub-orders. Perciforms display a vast range of adaptations and morphologies, and it is difficult to find anatomical features common to all. Their classification is the subject of much ongoing debate. Almost all have ventral fins with one spine and five rays, and the caudal fin with 17 rays. Most have spines in the dorsal and anal fins, and possess ctenoid scales and four pairs of gills. Usually the ventral fins are close to the pectoral fins, with the supporting structures of both fins attached to each other either by ligaments or bones. There are about 160 families in the order, 120 of which occur in Australian waters.

GLASSFISHES

Ambassidae

GIANT GLASSFISH
Parambassis gulliveri

Glassfishes are distributed from East Africa through India and South-East Asia to Australia, and there are about 41 species in the family. They are predominantly freshwater fishes, though a few species occur in shallow coastal marine waters and brackish estuarine environments. They have elongate to oval compressed bodies and thin, fragile cycloid scales. The body is transparent in some species, with the vertebral column and internal organs clearly visible. They have a deeply notched dorsal fin with 7–9 spines and 8–11 rays, while the anal fin has three spines and 8–11 rays.

Glassfishes are small schooling fishes, with a number of species locally abundant in certain northern Australian rivers. They feed nocturnally on small fishes and invertebrates such as crustaceans, insects and larvae, sheltering by day in vegetation or between sunken branches. Glassfishes lay small adhesive eggs amongst aquatic vegetation. The **Longspine Glassfish**, **Vachell's Glassfish** and **Estuary Glassfish** are marine species found in north- and east-coast mangroves, estuaries and tidal creeks. The remaining species inhabit fresh water and are found in northern and eastern Australian river systems. The deep-bodied **Giant Glassfish** is the largest member of the family, reaching about 24 cm in length, and is found in the lower reaches of a number of northern Australian rivers.

LONGSPINE GLASSFISH
Ambassis interrupta

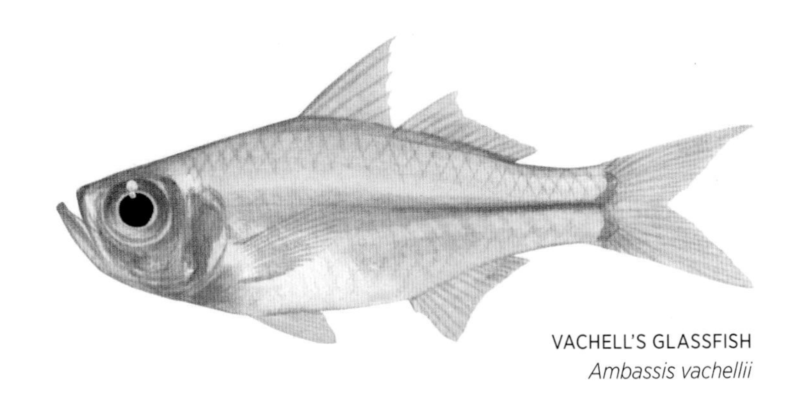

VACHELL'S GLASSFISH
Ambassis vachellii

AUSTRALIAN AMBASSIDAE SPECIES

Ambassis agassizii Agassiz's Glassfish
Ambassis agrammus Sailfin Glassfish
Ambassis elongatus Elongate Glassfish
Ambassis gymnocephalus Barehead Glassfish
Ambassis interrupta Longspine Glassfish
Ambassis jacksoniensis Port Jackson Glassfish
Ambassis macleayi Macleay's Glassfish
Ambassis marianus Estuary Glassfish
Ambassis miops Flagtail Glassfish
Ambassis nalua Scalloped Glassfish
Ambassis vachellii Vachell's Glassfish
Denariusa australis Pennyfish
Parambassis gulliveri Giant Glassfish

SAND BASSES AND BARRAMUNDI

Latidae

BARRAMUNDI
Lates calcarifer

Sand Basses and Barramundi are found in fresh waters in Africa and Australia, and coastal marine and estuarine environments of the Indo-Pacific. This small family contains about 10 species, including one of Australia's iconic fishes, the **Barramundi**. In all members of this family the dorsal fin is in two parts or deeply notched, the first part with 8–9 spines. The anal fin has three spines. They have an elongate, compressed body with large cycloid scales, and a large mouth.

The Barramundi occurs throughout northern Australia in coastal waters and tidal streams, venturing far up rivers. Adults spawn in the mouths of tidal rivers and creeks, and the larvae, which need brackish water to develop, move into estuaries and coastal lagoons. Juveniles travel upstream into swamps, creeks and freshwater lagoons to mature before moving downstream again to spawn. The Barramundi undergoes sex reversal at about six years of age: after functioning as males in spawning seasons for several years, they become female. Large Barramundi are almost all females, while smaller ones are almost all males. Females can reach up to 2 m in length, but specimens this size are no longer encountered. The Barramundi is an aggressive predator of crustaceans and fishes such as Mullets (Mugilidae, p. 318) and Hardyheads (Atherinidae, p. 328).

The Barramundi is hugely popular as a sportfish and food fish, and is targeted by anglers in freshwater creeks, tidal rivers and coastal waters across the northern half of Australia. Commercial gillnetting of the species is restricted to tidal reaches of rivers and coastal waters, as stocks are under constant pressure. The Barramundi is now extensively farmed and stocked as juveniles in many man-made lakes.

The **Sand Bass** and small **Spiky Bass** are marine species. The Sand Bass is widely distributed in the Indo-West Pacific and reaches about 50 cm in length. It prefers rocky and weedy areas, and coral reefs, and is nocturnal, sheltering in crevices and caves during the day. The Spiky Bass is restricted to north-western Australia and is found in shallow inshore waters.

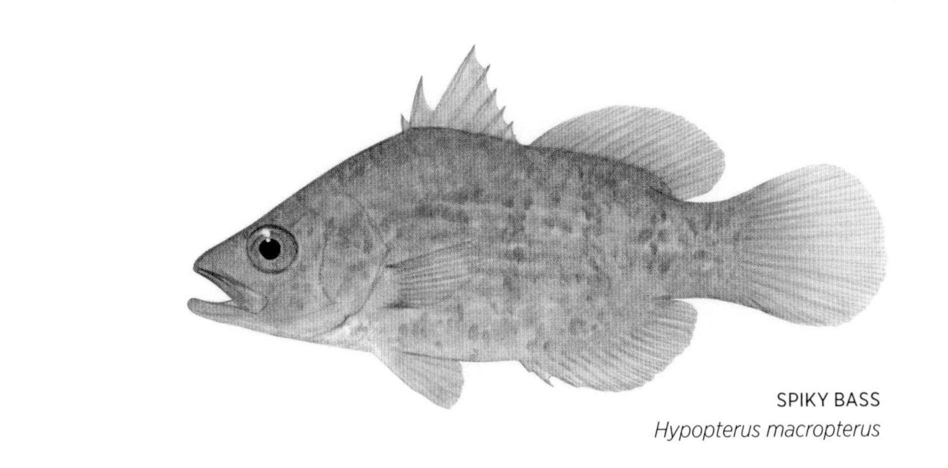

SPIKY BASS
Hypopterus macropterus

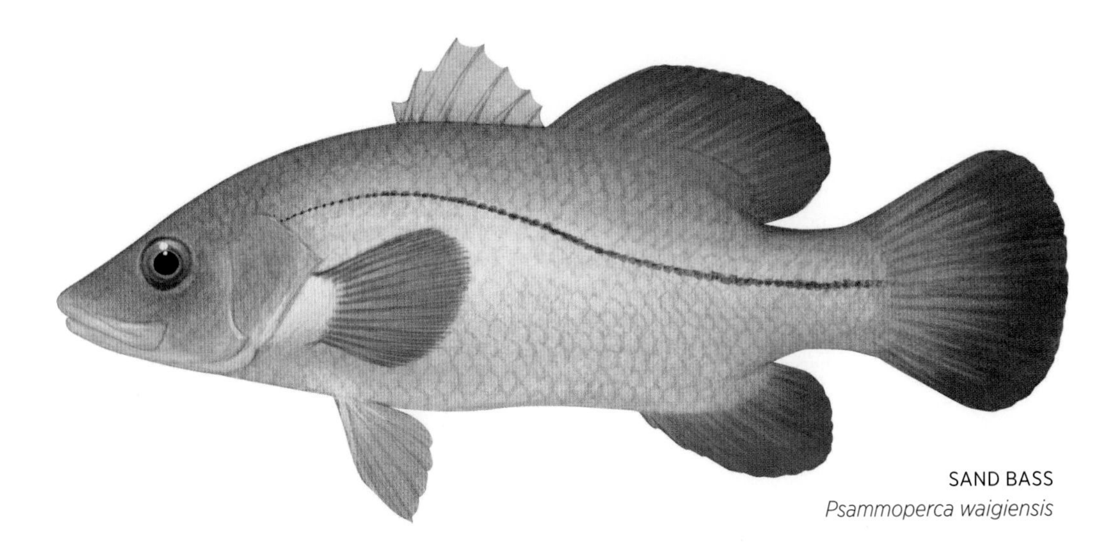

SAND BASS
Psammoperca waigiensis

AUSTRALIAN LATIDAE SPECIES

Hypopterus macropterus Spiky Bass
Lates calcarifer Barramundi
Psammoperca waigiensis Sand Bass

TEMPERATE BASSES

Moronidae

JAPANESE SEAPERCH
Lateolabrax japonicus

The six species of Temperate Bass are found in freshwater and marine environments in North America, Europe, North Africa and Asia. They have an elongate, slightly compressed body with small scales, and a large mouth with slightly projecting lower jaws. In the **Japanese Seaperch** the dorsal fin is deeply notched, with 12–15 spines and 12–14 rays, and the anal fin has three spines.

The Japanese Seaperch is found in the western Pacific from Japan to northern Australia. It inhabits inshore waters over rocky reefs where there is plenty of water movement, and feeds on small fishes and crustaceans. Juveniles have large black spots scattered over the dorsal surface that are usually lost as they mature. Large adults can reach up to 1 m in length and are a plain silvery colour. The Japanese Seaperch is prized as a food fish in Japan but is not common in Australian waters.

AUSTRALIAN MORONIDAE SPECIES
Lateolabrax japonicus Japanese Seaperch

TEMPERATE PERCHES

Percichthyidae

MURRAY COD
Maccullochella peelii

Temperate Perches are found mainly in fresh waters, but occasionally brackish waters, of South America and Australia. There are about 34 species in the family. They are particularly diverse in southern Australia, where they form a major component of the freshwater fauna. The family has an ancient lineage that can be traced back to Gondwana, some 150 million years ago. This is reflected in its distribution, which is restricted to the southern hemisphere, and the diversity of species now found in the long-isolated waters of southern Australia. Temperate Perches display a diverse range of forms, with body shapes ranging from elongate to oval, rounded to compressed. They have small scales and a large mouth. In most the dorsal fin is notched with 7–12 spines, the anal fin has three spines, and the anterior rays of the ventral fins are elongated. Temperate Perches can be divided into three subfamilies, which are sometimes considered to be separate families:

GADOPSINAE Blackfishes are small, elongate freshwater fishes found only in rivers and streams in south-eastern Australia and Tasmania. They have a single long dorsal fin that is not deeply notched. They are territorial and remain in one small stretch of a stream or river throughout their life. Blackfishes are nocturnal predators, feeding on insects, crustaceans and other invertebrates, as well as small fishes. They reach a maximum length of about 30 cm.

NANNOPERCINAE Pygmy Perches are small freshwater fishes found only in coastal drainages in southern Australia. They reach a maximum length of about 10 cm and feed mainly on insects, larvae and microscopic invertebrates. Several species are endangered due to their restricted range and habitat loss.

PERCICHTHYINAE The group of Freshwater Basses and Cods includes an iconic Australian freshwater fish, the **Murray Cod**, as well as a number of other species that are highly sought-after as sport or food fishes. The Murray Cod can reach 1.8 m in length and weigh over 100 kg. It was once common throughout the Murray-Darling river system, but despite numerous restocking programs, numbers have drastically declined due to overfishing and habitat degradation. It feeds on virtually anything that moves, including fishes, turtles, crustaceans and even small water birds. Murray Cods occupy a territorial range in a given stretch of river and rarely move outside that area. There are several subspecies in east-coast drainages and the **Clarence River Cod** and **Trout Cod** are quite similar. These two species require near-pristine, clear, fast-flowing streams – habitats that are now rare in Australia. Both now have very restricted ranges and are listed as endangered. Restocking programs have built up the numbers of these fishes somewhat. The **Golden Perch** is widely distributed in inland waters in eastern Australia and is another species keenly sought by anglers. It can reach about 75 cm in length and weigh over 20 kg, and is far more common than many of the other species in this family due to its tolerance of high temperatures and poor water quality. The **Estuary Perch** and **Australian Bass** are also highly sought-after as sportfishes; they can be found in south-eastern coastal drainages and seasonally in estuaries. The **Nightfish** is a small, nocturnal species that grows to about 10 cm in length and is found in south-western rivers. The **Bloomfield River Cod** occurs only in a small stretch of the Bloomfield River in North Queensland.

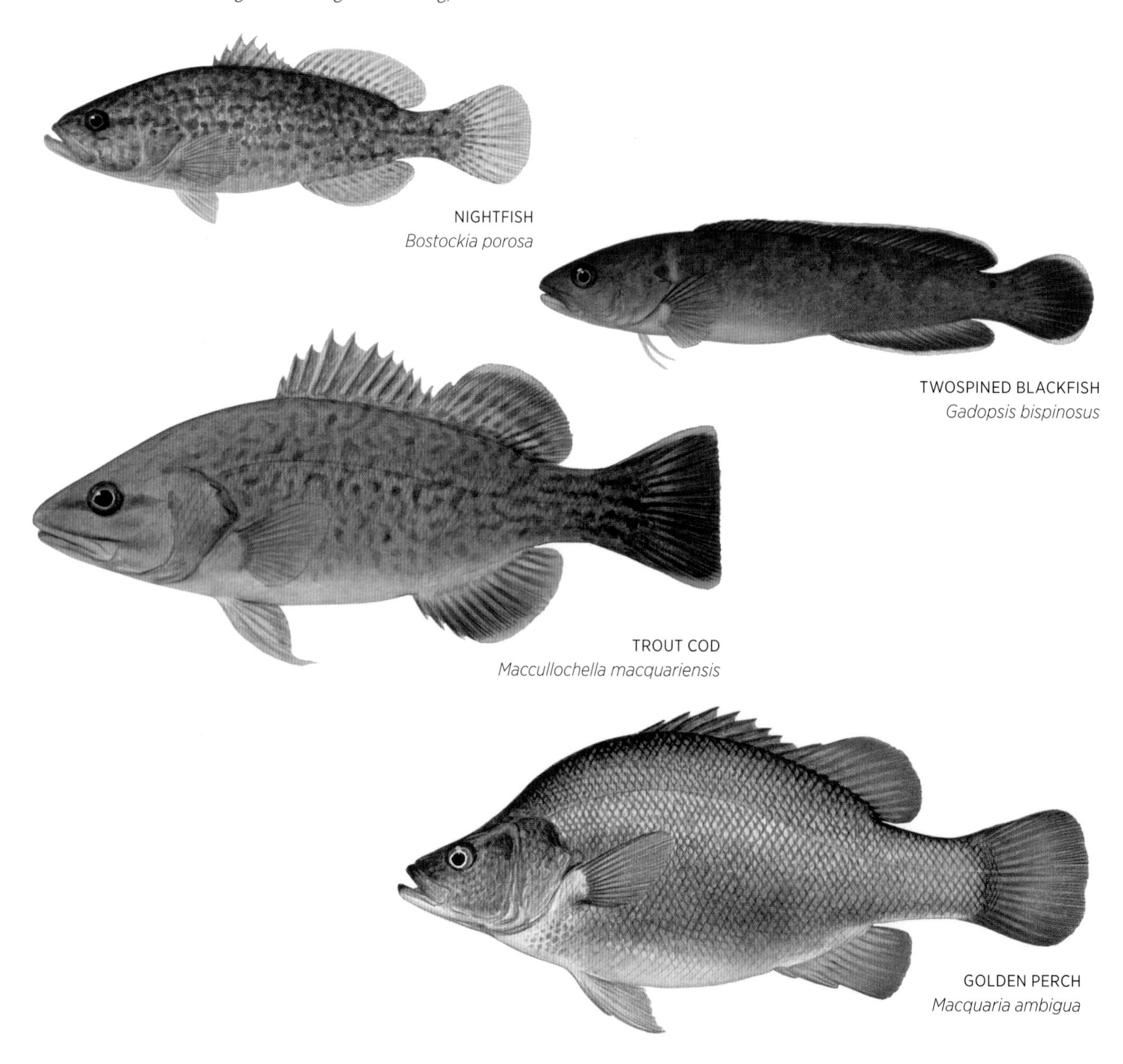

NIGHTFISH
Bostockia porosa

TWOSPINED BLACKFISH
Gadopsis bispinosus

TROUT COD
Maccullochella macquariensis

GOLDEN PERCH
Macquaria ambigua

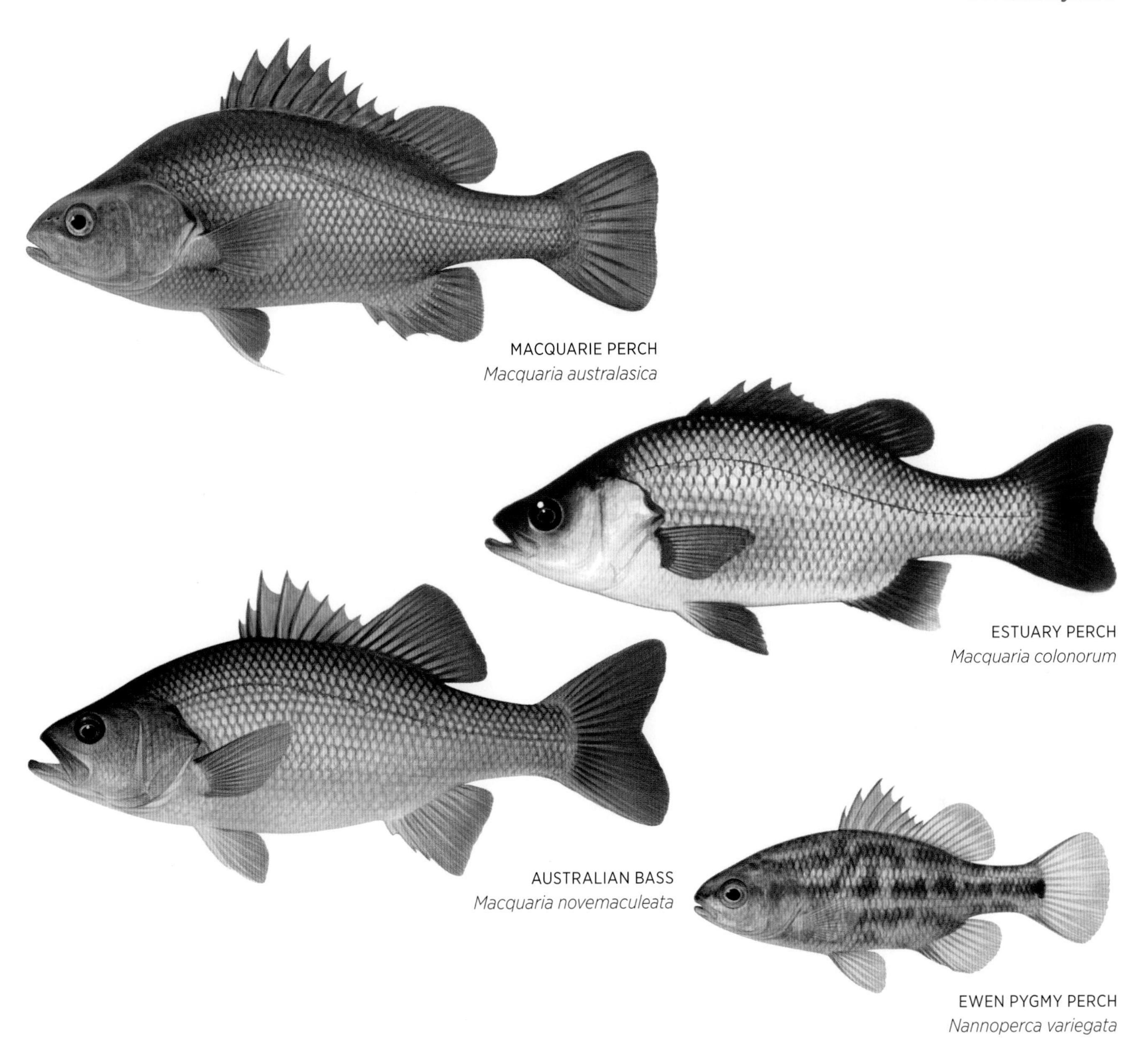

MACQUARIE PERCH
Macquaria australasica

ESTUARY PERCH
Macquaria colonorum

AUSTRALIAN BASS
Macquaria novemaculeata

EWEN PYGMY PERCH
Nannoperca variegata

AUSTRALIAN PERCICHTHYIDAE SPECIES

Bostockia porosa Nightfish
Edelia vittata Western Pygmy Perch
Gadopsis bispinosus Twospined Blackfish
Gadopsis marmoratus River Blackfish
Guyu wujalwujalensis Bloomfield River Cod
Maccullochella ikei Clarence River Cod
Maccullochella macquariensis Trout Cod
Maccullochella peelii Murray Cod
Macquaria ambigua Golden Perch
Macquaria australasica Macquarie Perch
Macquaria colonorum Estuary Perch
Macquaria novemaculeata Australian Bass
Nannatherina balstoni Balston's Pygmy Perch
Nannoperca australis Southern Pygmy Perch
Nannoperca obscura Yarra Pygmy Perch
Nannoperca oxleyana Oxleyan Pygmy Perch
Nannoperca variegata Ewen Pygmy Perch

TEMPERATE SEABASSES

Acropomatidae

GLOWBELLY SEABASS
Acropoma japonica

Temperate Seabasses are widely distributed in the Atlantic, Indian and Pacific oceans, mainly in deeper offshore waters. There are about 30 species in the family. They have elongate to oval, compressed bodies, small to large cycloid scales, and a large mouth with the lower jaw projecting beyond the upper. The dorsal fin is deeply notched, with 7–10 spines in the first part, and the anal fin has 2–3 spines. The eyes are large and there are several flattened spines on the upper edge of the operculum. Temperate Seabasses have the anus placed near the ventral fins, much further forward than other Perciforms. *Acropoma* species, such as the **Glowbelly Seabass**, have light organs along the belly that contain symbiotic bioluminescent bacteria. Temperate Seabasses are mostly small, around 15–20 cm in length. In Australia they occur mainly in the north, inhabiting bottom waters from about 50 to 600 m. The **Rosy Seabass** is the largest species in the family, reaching about 40 cm in length, and is taken commercially as a food fish in South-East Asia.

AUSTRALIAN ACROPOMATIDAE SPECIES

Acropoma japonica Glowbelly Seabass
Apogonops anomalus Threespine Cardinalfish
Doederleinia berycoides Rosy Seabass
Malakichthys elegans Splendid Seabass
Malakichthys griseus Grey Seabass
Malakichthys levis Smooth Seabass
Malakichthys mochizuki Mochizuki's Seabass
Synagrops analis Threespine Seabass
Synagrops japonicus Japanese Seabass
Synagrops philippinensis Sharptooth Seabass
Synagrops serratospinosus Roughspine Seabass

WRECKFISHES

Polyprionidae

HAPUKU
Polyprion oxygeneios

The six species of Wreckfish are found in temperate waters of the Atlantic, Indian and Pacific oceans. They are extremely large, some species reaching up to 2.5 m in length, and are usually found in deeper waters, from 50 to 600 m. They have elongate, rounded to slightly compressed bodies, with small ctenoid scales. Their large mouth bears villiform teeth in the jaws, and the lower jaw projects slightly beyond the upper. The dorsal fin has 11–12 strong spines and the anal fin has three spines.

The **Bass Groper** is found on rocky reefs off the south coast of Australia from depths of 100 to 900 m, and can reach 2 m in length and over 100 kg in weight. It has a distinctive small bony crest on the head and a short, thickset body. The **Hapuku** has a similar distribution, though it also occasionally appears off north-western Australia. The Hapuku is caught commercially by longline in the Tasman Sea, where it is quite common. It has has a more elongate form than the Bass Groper and can reach a maximum length of about 1.5 m and weigh 70 kg. Both species are highly esteemed food fishes and are commonly taken by dropline fisheries off south-eastern Australia.

BASS GROPER
Polyprion americanus

AUSTRALIAN POLYPRIONIDAE SPECIES

Polyprion americanus Bass Groper
Polyprion oxygeneios Hapuku

SLOPEFISHES

Symphysanodontidae

TYPICAL SLOPEFISH
Symphysanodon typus

Slopefishes are found in tropical and warm temperate waters of the Atlantic, Indian and Pacific oceans, and there are about 10 species in the family. They have an elongate, compressed body, large eyes and two flat spines on the operculum. The long dorsal fin has nine spines and 10 rays, the anal fin has three spines and 7–8 rays, and the caudal fin is deeply forked.

Slopefishes reach a maximum length of about 20 cm and are pelagic in bottom waters, feeding on zooplankton. Most species are known from only a few specimens caught in deep water. The **Typical Slopefish** is found from depths of 50 to 400 m, from the waters of the Philippines through Indonesia, across northern Australia to Hawaii. Very little is known of the biology of this family.

AUSTRALIAN SYMPHYSANODONTIDAE SPECIES
Symphysanodon typus Typical Slopefish

ROCKCODS, GROUPERS, BASSLETS, SEAPERCHES AND CORAL TROUTS

Serranidae

GOLDSPOTTED ROCKCOD
Epinephelus coioides

The Serranidae is a very large and diverse family with over 475 species. They are found in a wide range of environments, in tropical and temperate waters of all the oceans, with a few species found in fresh water. Most are associated with reefs and rocky bottoms, and occur from the shallows down to 300 m or more. All have three spines on the operculum, a serrated edge to the lower preoperculum, a continuous or notched dorsal fin with 7–13 spines (most with 10 spines or less), and three spines in the anal fin. The entire upper jaw is visible when the mouth is closed; the rear of it does not slide beneath the suborbital bones. The family can be divided into three subfamilies:

ANTHIINAE This group includes small colourful fishes such as the Fairy Basslets, **Butterfly Perch** and **Barber Perch**. They are mainly zooplankton feeders and are commonly found in small to large schools in midwater, over dropoffs and on reef edges where there is plenty of current. At the approach of danger the whole school will sink down to shelter in the reef. Schools contain a few highly colourful males along with harems of less-colourful females. These species are hermaphroditic and, should the need arise, a large female from the harem will undergo sex reversal to replace a missing male, over a period of several weeks completely changing form, colour and reproductive function. Other species within this group, such as the Orange Perches, are more solitary and slightly larger in size, but also feed on zooplankton. Such species are found near deepwater reefs and are seen only occasionally in trawl nets. They do not seem to undergo sex reversal.

Wirrahs and Seaperches are bottom dwellers, found on rocky reefs in southern Australian waters and feeding principally on benthic crustaceans. Wirrahs grow to about 50 cm in length and are very shy, never venturing far from sheltering crevices and caves. They are often caught by anglers, but are not considered good food fishes and are usually released. Seaperches are solitary benthic predators of small fishes and crustaceans. They are frequently seen by divers on rocky reefs in southern waters, resting in the open with head held up, constantly swivelling their remarkably mobile eyes to survey their surroundings; they have thickened rays in the lower part of the pectoral fin, which they use to support themselves in this position. They seem quite curious, often allowing divers to approach closely. The colourful **Harlequin Fish** is a large, solitary species that grows to about 75 cm in length and frequents south-coast rocky reefs. It is an inquisitive fish, easily approached by divers as it rests on the bottom, and is often taken by anglers, who appreciate it as a fine food fish. The **Breaksea Cod**, a south-west-coast species, is found on reefs down to about 100 m and is another excellent food fish.

SERRANINAE This group is not well represented in Australian waters. It contains about 80 species of Dwarf Bass and Hamlet, which are found in coastal waters of the Atlantic and East Pacific. They are colourful bottom-dwelling fishes, most growing to about 10–30 cm in length. The **Pearly Perchlet** is one of a small number of species in this group from the tropical Indo-Pacific. It can be found over sand or silt bottoms, usually at depths of 100–200 m though sometimes in shallow water, resting on its ventral fins on the substrate. It occasionally appears in the nets of deepwater trawls.

EPINEPHELINAE Rockcods and Groupers are mostly bottom-dwelling fishes, usually associated with reefs, rocky bottoms or weed beds to depths of about 200 m. They are found right around the coast of Australia. All members of this group are thought to be sequential hermaphrodites, growing to maturity as females and then changing to males. Generally solitary fishes, many species form spawning aggregations at preferred sites, where males defend small patches of territory and attract females. Rockcods and Groupers have a large mouth and powerful jaws, and are aggressive predators of other fishes and crustaceans. They feed by making a sudden lunge from a position of ambush and engulfing their prey whole. They range in size from the small **Birdwire Rockcod**, which grows to about 30 cm, to the giant **Queensland Groper**, which can reach up to 3 m. Large individuals establish a home range that they defend against others; if left undisturbed they can live for many decades in the same location. Several other species of Rockcod can grow to around 2 m in length, including the **Goldspotted Rockcod**, **Blackspotted Rockcod** and **Potato Rockcod**, all of which are found in northern Australian waters.

Many Rockcods and Groupers are highly sought-after food fishes of excellent quality. Several species are targeted commercially, while many other species are taken as bycatch by trawl, trap and line fisheries. The Coral Trouts, distributed in coastal waters around the northern half of Australia, are among the finest food fishes in Australian waters and are heavily fished by amateurs and professionals alike, particularly in the waters of the Great Barrier Reef. They will vigorously attack lures or baits, which makes them relatively easy to catch and vulnerable to overfishing. Overfishing has occurred in many areas, for many species of Rockcod and Grouper. Most species are slow growing and territorial, and in areas heavily frequented by fishers all the large specimens have been removed, leaving only juveniles and small females, which cannot reproduce and repopulate the area without the larger males.

Within the Epinephelinae there are two groups that are sometimes placed in their own subfamilies or families. The Reef Basslets are shy, reef-dwelling fishes that inhabit caves and crevices, and feed on small benthic invertebrates. The **Rainfordia** is the largest, growing to a maximum length of 15 cm. The other group is known as the Soapfishes, due to their ability to exude a toxic substance from their skin that can cause a soap-like froth in the water. Presumably this bitter-tasting substance serves to deter predators, though I have watched a **Lined Soapfish** repeatedly rubbing its body along the substrate in various spots, perhaps to mark a territory. The toxin is quite powerful and Soapfishes held in an aquarium with other species can kill them if stimulated to release the toxin. The Podges, also part of the Soapfish group, grow to 8–10 cm in length and they too are capable of producing the toxin. Soapfishes reach a maximum size of about 30 cm.

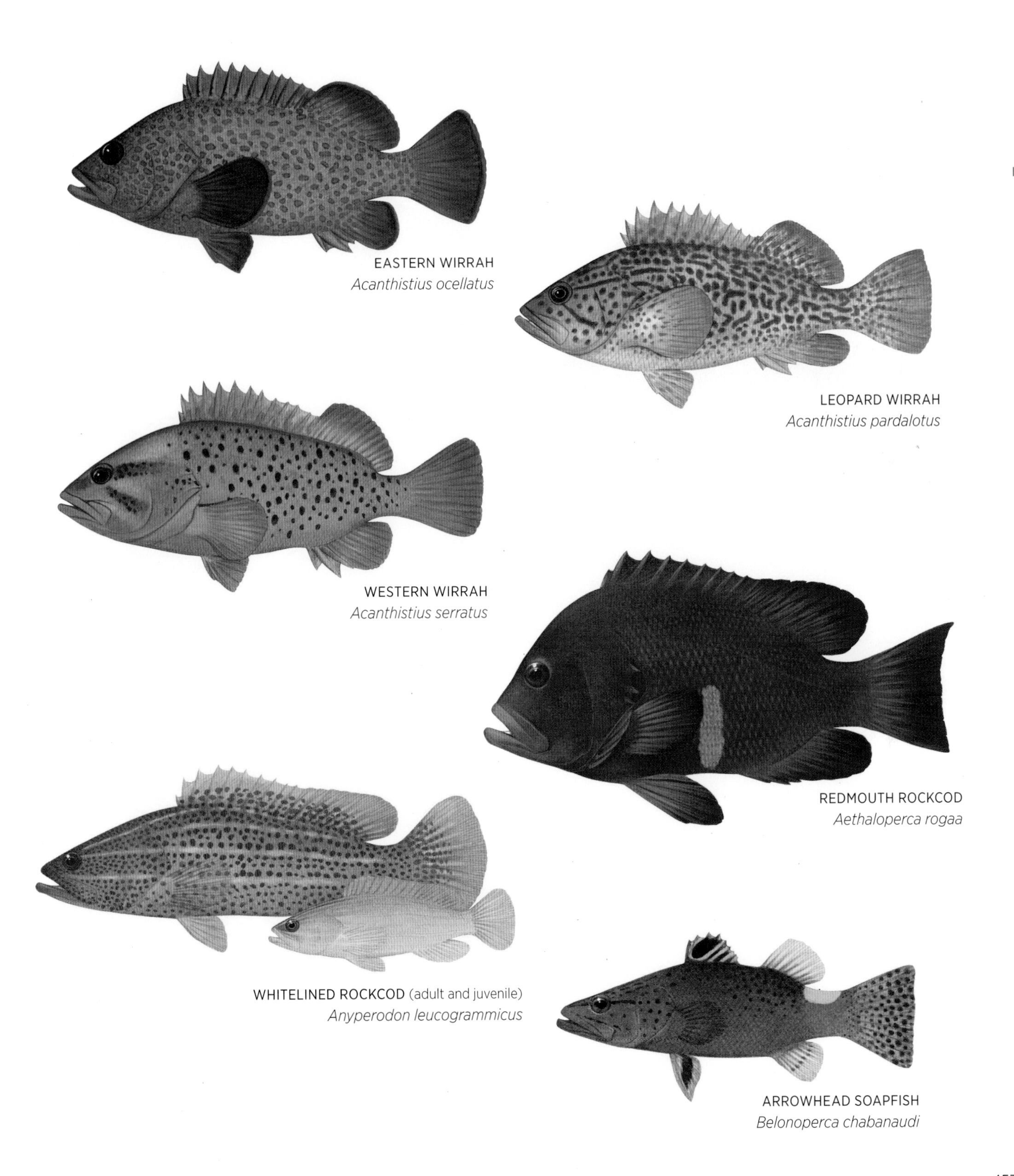

EASTERN WIRRAH
Acanthistius ocellatus

LEOPARD WIRRAH
Acanthistius pardalotus

WESTERN WIRRAH
Acanthistius serratus

REDMOUTH ROCKCOD
Aethaloperca rogaa

WHITELINED ROCKCOD (adult and juvenile)
Anyperodon leucogrammicus

ARROWHEAD SOAPFISH
Belonoperca chabanaudi

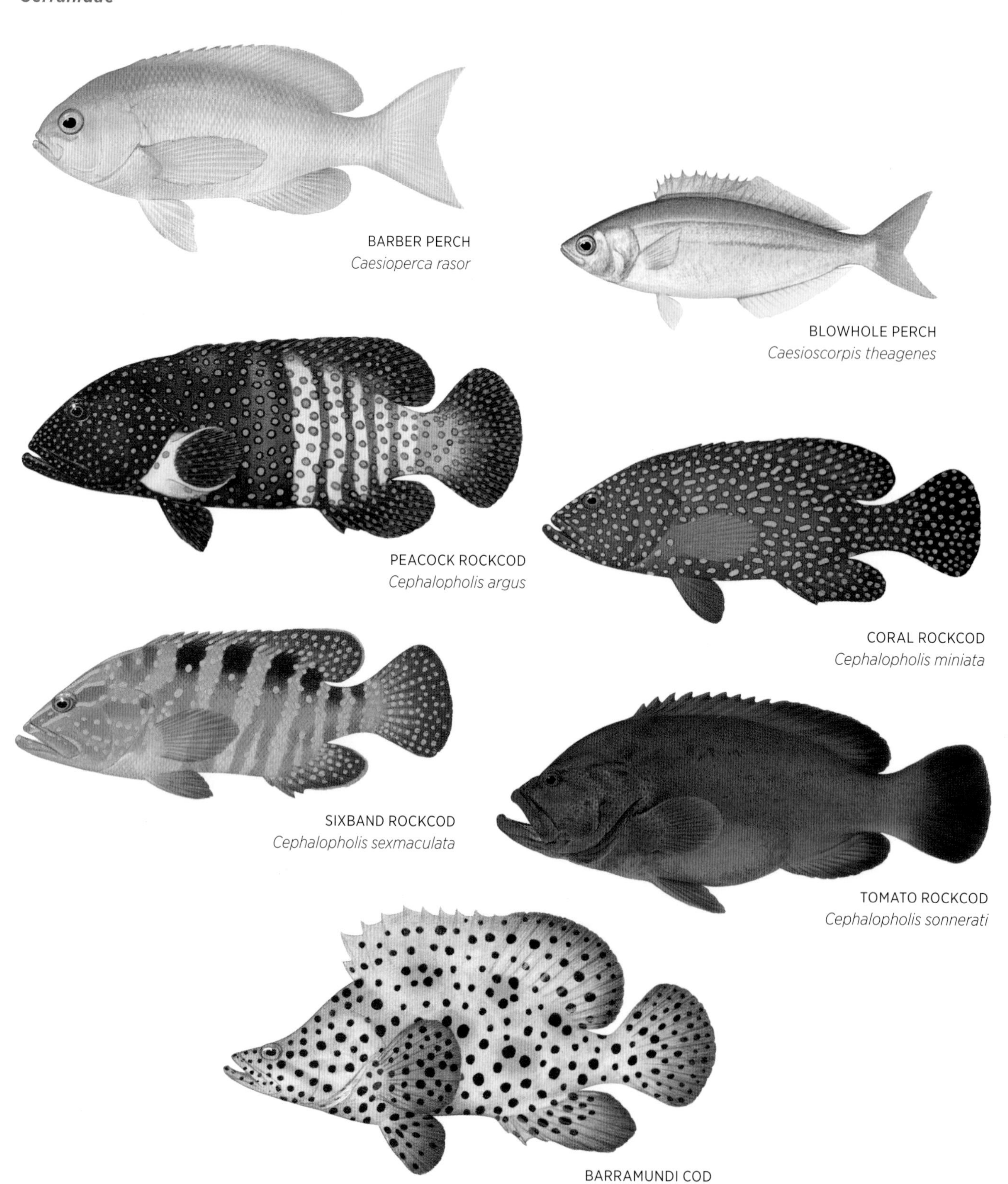

BARBER PERCH
Caesioperca rasor

BLOWHOLE PERCH
Caesioscorpis theagenes

PEACOCK ROCKCOD
Cephalopholis argus

CORAL ROCKCOD
Cephalopholis miniata

SIXBAND ROCKCOD
Cephalopholis sexmaculata

TOMATO ROCKCOD
Cephalopholis sonnerati

BARRAMUNDI COD
Chromileptes altivelis

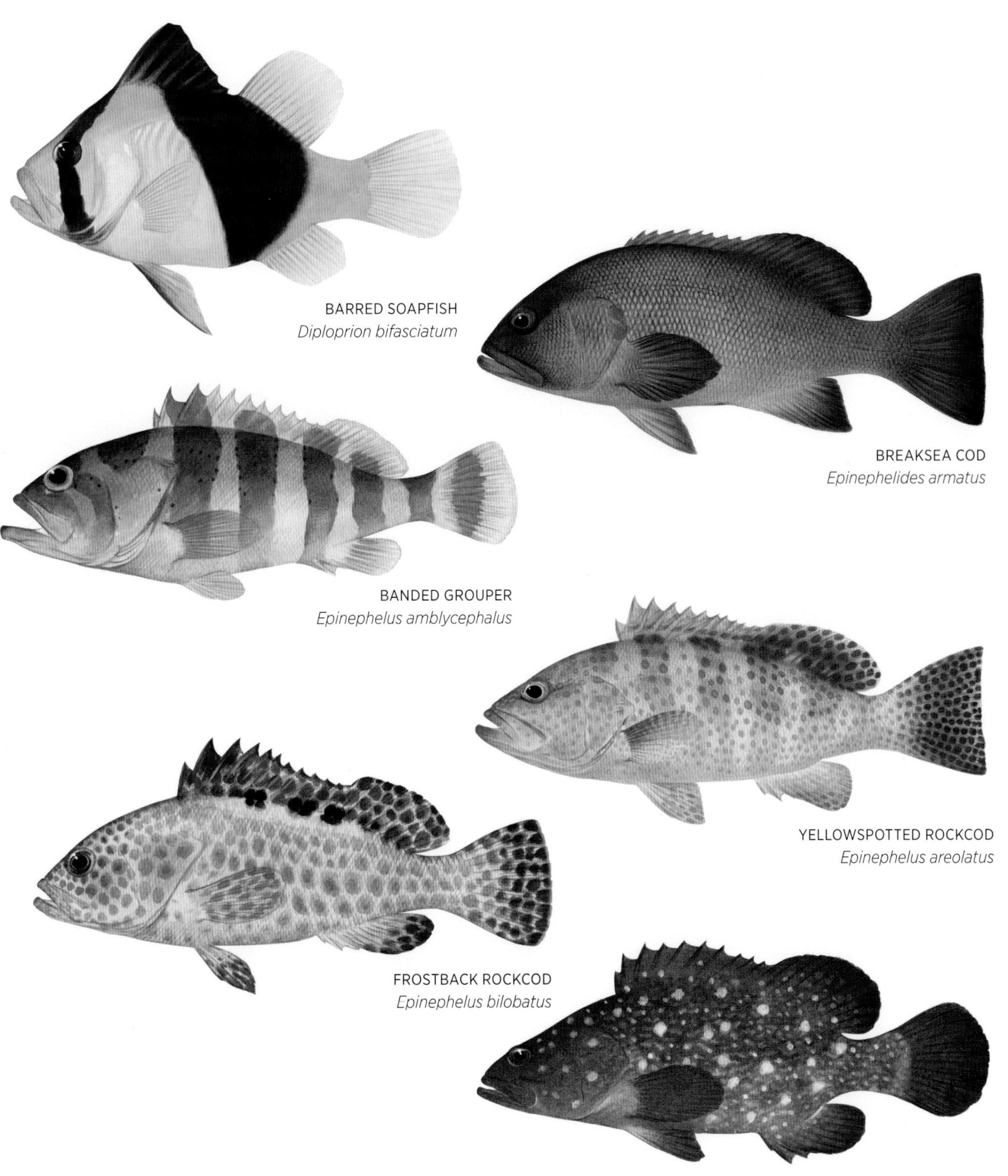

BARRED SOAPFISH
Diploprion bifasciatum

BREAKSEA COD
Epinephelides armatus

BANDED GROUPER
Epinephelus amblycephalus

YELLOWSPOTTED ROCKCOD
Epinephelus areolatus

FROSTBACK ROCKCOD
Epinephelus bilobatus

WHITESPOTTED GROUPER
Epinephelus coeruleopunctatus

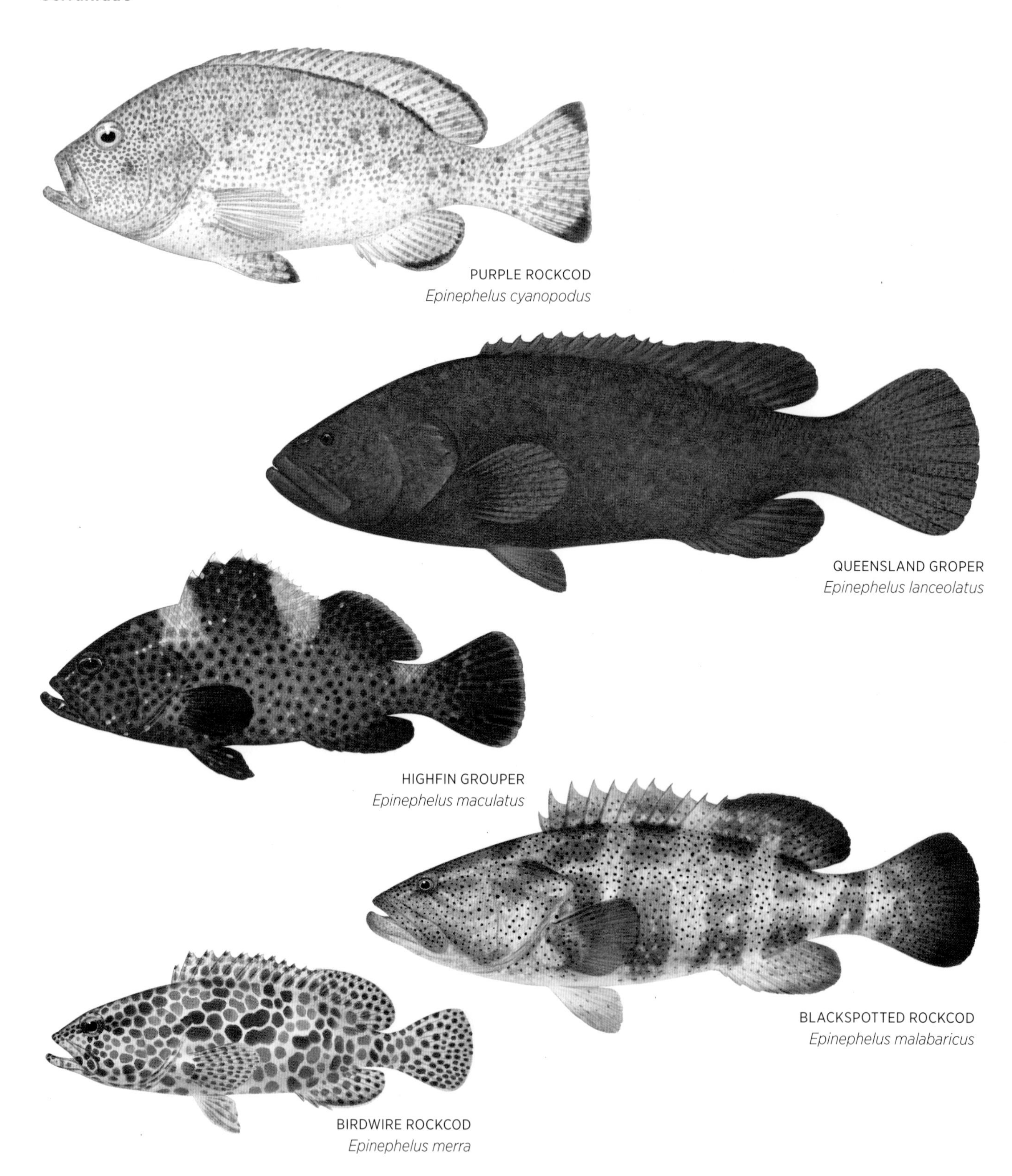

PURPLE ROCKCOD
Epinephelus cyanopodus

QUEENSLAND GROPER
Epinephelus lanceolatus

HIGHFIN GROUPER
Epinephelus maculatus

BLACKSPOTTED ROCKCOD
Epinephelus malabaricus

BIRDWIRE ROCKCOD
Epinephelus merra

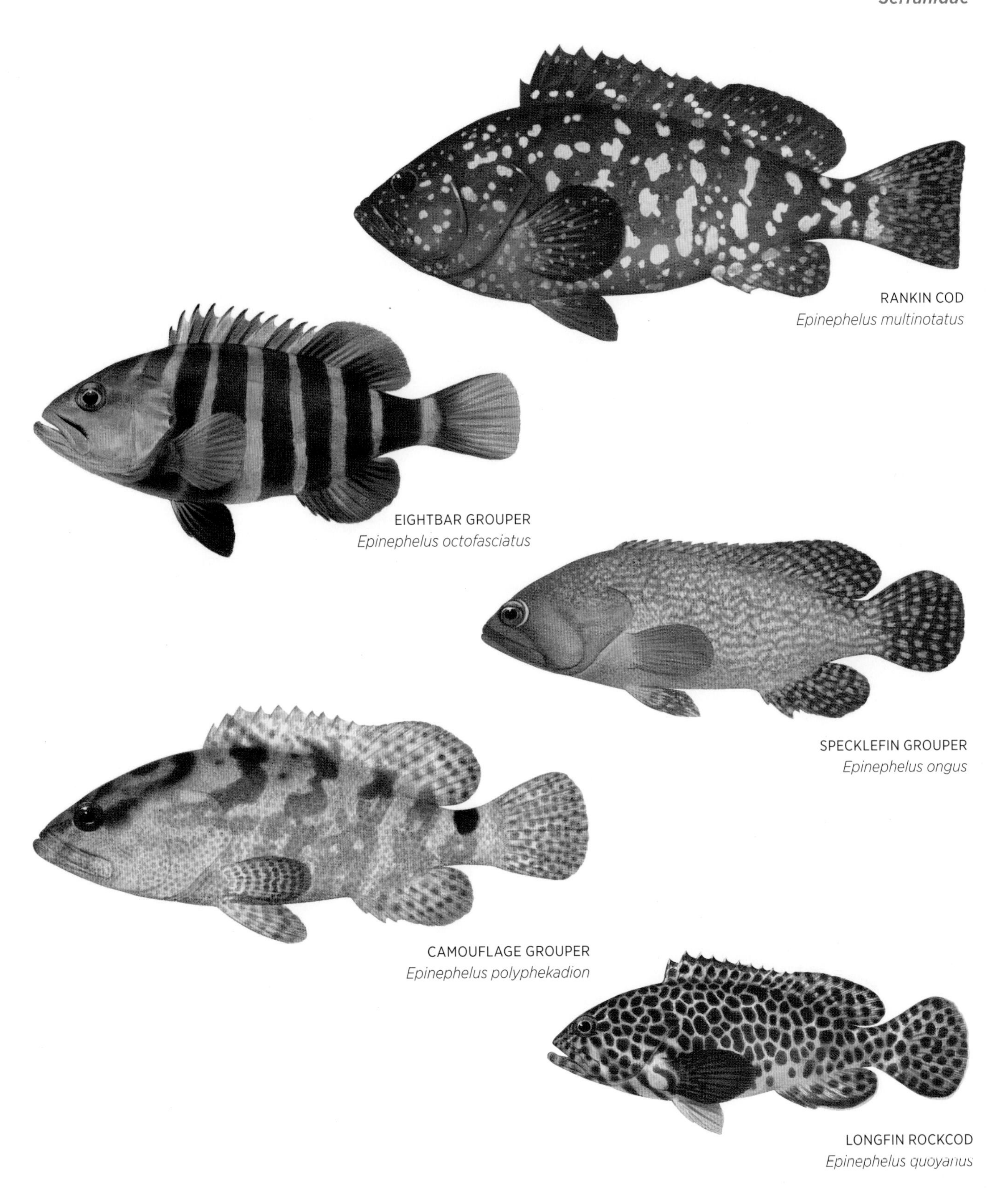

RANKIN COD
Epinephelus multinotatus

EIGHTBAR GROUPER
Epinephelus octofasciatus

SPECKLEFIN GROUPER
Epinephelus ongus

CAMOUFLAGE GROUPER
Epinephelus polyphekadion

LONGFIN ROCKCOD
Epinephelus quoyanus

CHINAMAN ROCKCOD
Epinephelus rivulatus

POTATO ROCKCOD
Epinephelus tukula

THINSPINE GROUPER
Gracila albomarginata

LINED SOAPFISH
Grammistes sexlineatus

BANDED SEAPERCH
Hypoplectrodes nigroruber

HARLEQUIN FISH
Othos dentex

CITRON PERCHLET
Plectranthias megalophthalmus

COMMON CORAL TROUT
Plectropomus leopardus

BARCHEEK CORAL TROUT
Plectropomus maculatus

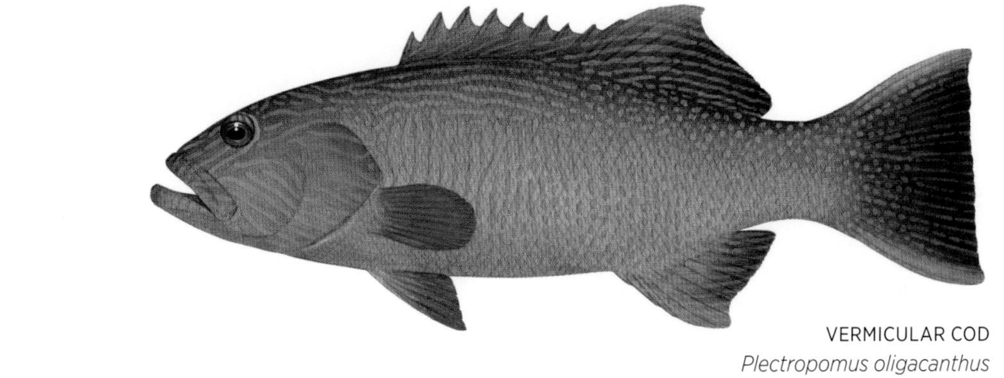

VERMICULAR COD
Plectropomus oligacanthus

PINK BASSLET (male and female)
Pseudanthias hypselosoma

FAIRY BASSLET (male and female)
Pseudanthias dispar

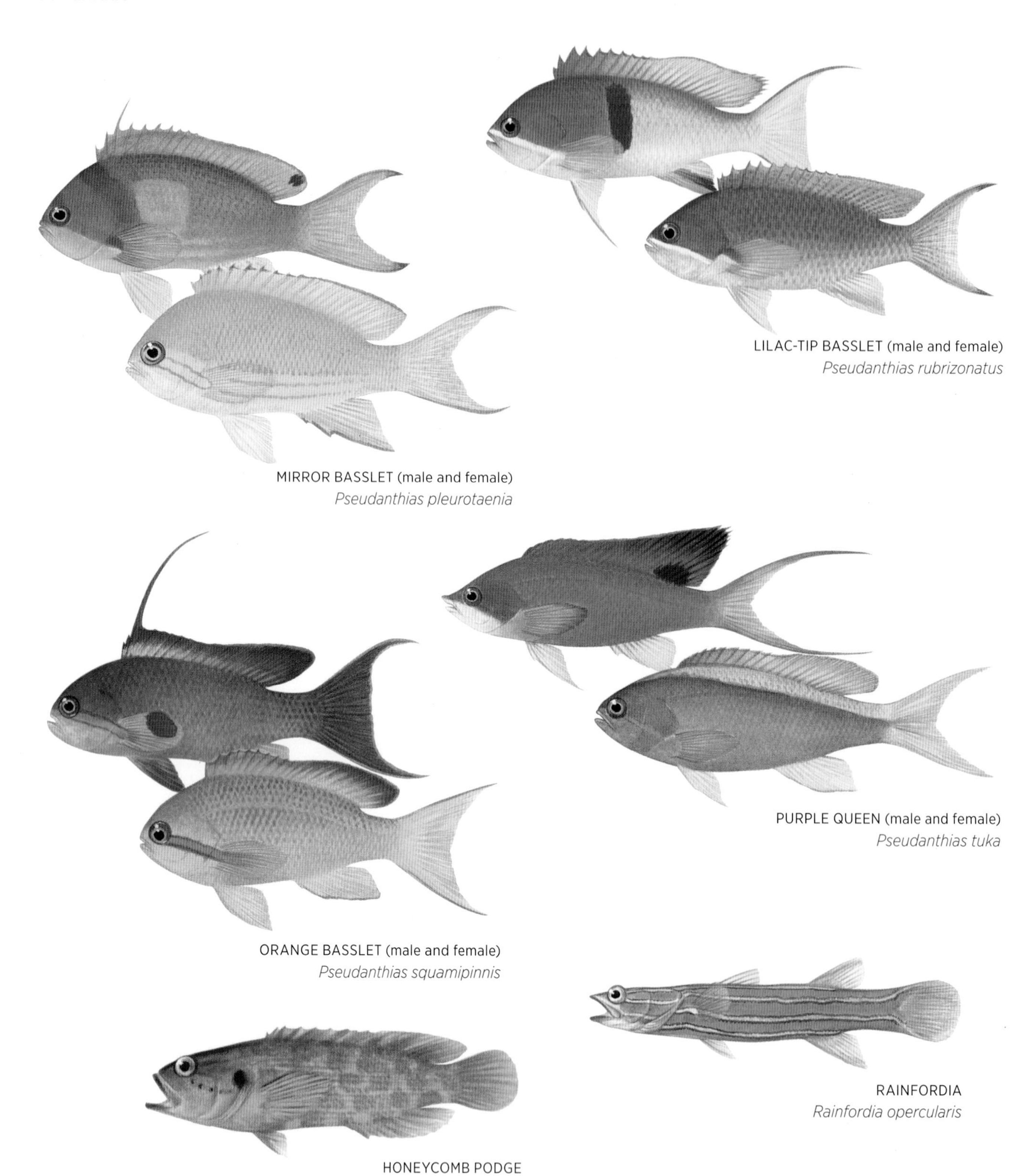

LILAC-TIP BASSLET (male and female)
Pseudanthias rubrizonatus

MIRROR BASSLET (male and female)
Pseudanthias pleurotaenia

PURPLE QUEEN (male and female)
Pseudanthias tuka

ORANGE BASSLET (male and female)
Pseudanthias squamipinnis

RAINFORDIA
Rainfordia opercularis

HONEYCOMB PODGE
Pseudogramma polyacanthum

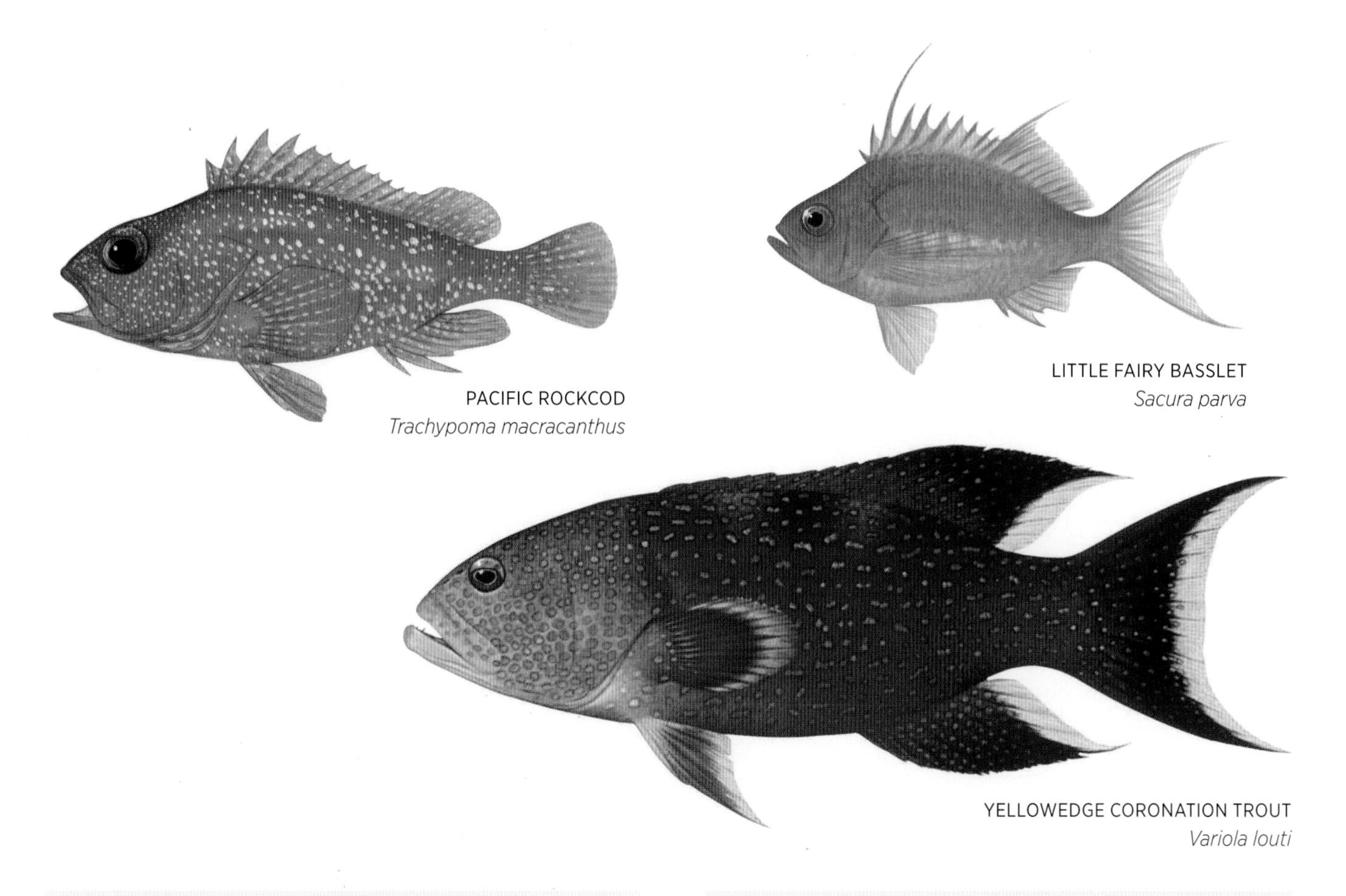

PACIFIC ROCKCOD
Trachypoma macracanthus

LITTLE FAIRY BASSLET
Sacura parva

YELLOWEDGE CORONATION TROUT
Variola louti

AUSTRALIAN SERRANIDAE SPECIES

Acanthistius cinctus Yellowbanded Wirrah
Acanthistius ocellatus Eastern Wirrah
Acanthistius pardalotus Leopard Wirrah
Acanthistius paxtoni Orangelined Wirrah
Acanthistius serratus Western Wirrah
Aethaloperca rogaa Redmouth Rockcod
Anyperodon leucogrammicus Whitelined Rockcod
Aporops bilinearis Blotched Podge
Aulacocephalus temminckii Goldribbon Cod
Belonoperca chabanaudi Arrowhead Soapfish
Caesioperca lepidoptera Butterfly Perch
Caesioperca rasor Barber Perch
Caesioscorpis theagenes Blowhole Perch
Caprodon krasyukovae Krasyukova's Perch
Caprodon longimanus Longfin Perch
Caprodon schlegelii Sunrise Perch
Cephalopholis argus Peacock Rockcod
Cephalopholis boenak Brownbarred Rockcod
Cephalopholis cyanostigma Bluespotted Rockcod
Cephalopholis formosa Bluelined Rockcod
Cephalopholis leopardus Leopard Rockcod
Cephalopholis microprion Dot-head Rockcod
Cephalopholis miniata Coral Rockcod
Cephalopholis sexmaculata Sixband Rockcod
Cephalopholis sonnerati Tomato Rockcod
Cephalopholis spiloparaea Strawberry Rockcod
Cephalopholis urodeta Flagtail Rockcod
Chelidoperca margaritifera Pearly Perchlet
Chromileptes altivelis Barramundi Cod
Diploprion bifasciatum Barred Soapfish
Epinephelides armatus Breaksea Cod
Epinephelus amblycephalus Banded Grouper
Epinephelus areolatus Yellowspotted Rockcod
Epinephelus bilobatus Frostback Rockcod
Epinephelus bleekeri Duskytail Grouper
Epinephelus coeruleopunctatus Whitespotted Grouper
Epinephelus coioides Goldspotted Rockcod
Epinephelus corallicola Coral Grouper
Epinephelus cyanopodus Purple Rockcod
Epinephelus daemelii Black Rockcod
Epinephelus darwinensis Darwin Grouper
Epinephelus epistictus Dotted Grouper

Epinephelus ergastularius Banded Rockcod
Epinephelus fasciatus Blacktip Rockcod
Epinephelus fuscoguttatus Flowery Rockcod
Epinephelus heniochus Threeline Rockcod
Epinephelus hexagonatus Wirenet Rockcod
Epinephelus howlandi Blacksaddle Rockcod
Epinephelus lanceolatus Queensland Groper
Epinephelus latifasciatus Striped Grouper
Epinephelus macrospilos Snubnose Grouper
Epinephelus maculatus Highfin Grouper
Epinephelus magniscuttis Speckled Grouper
Epinephelus malabaricus Blackspotted Rockcod
Epinephelus melanostigma Oneblotch Grouper
Epinephelus merra Birdwire Rockcod
Epinephelus miliaris Netfin Grouper
Epinephelus morrhua Comet Grouper
Epinephelus multinotatus Rankin Cod
Epinephelus octofasciatus Eightbar Grouper
Epinephelus ongus Specklefin Grouper
Epinephelus perplexus Puzzling Grouper
Epinephelus polyphekadion Camouflage Grouper
Epinephelus polystigma Whitedotted Grouper
Epinephelus quoyanus Longfin Rockcod
Epinephelus radiatus Radiant Rockcod
Epinephelus retouti Redtip Grouper
Epinephelus rivulatus Chinaman Rockcod
Epinephelus septemfasciatus Convict Grouper
Epinephelus sexfasciatus Sixbar Grouper
Epinephelus spilotoceps Foursaddle Grouper
Epinephelus stictus Blackdotted Grouper
Epinephelus tauvina Greasy Rockcod
Epinephelus timorensis Yellowspotted Grouper
Epinephelus trophis Plump Grouper
Epinephelus tukula Potato Rockcod
Epinephelus undulatostriatus Maori Rockcod
Gracila albomarginata Thinspine Grouper
Grammistes sexlineatus Lined Soapfish
Grammistops ocellatus Ocellate Soapfish
Hypoplectrodes annulatus Blackbanded Seaperch
Hypoplectrodes cardinalis Red Seaperch
Hypoplectrodes jamesoni Jameson's Seaperch
Hypoplectrodes maccullochi Halfbanded Seaperch
Hypoplectrodes nigroruber Banded Seaperch
Hypoplectrodes wilsoni Spotty Seaperch
Lepidoperca brochata Fangtooth Perch
Lepidoperca filamenta Western Orange Perch
Lepidoperca magna Sharphead Perch
Lepidoperca occidentalis Slender Orange Perch
Lepidoperca pulchella Eastern Orange Perch
Lepidoperca tasmanica Tasmanian Perch
Liopropoma mitratum Headband Perch
Liopropoma multilineatum Yellow Reef Basslet
Liopropoma susumi Pinstripe Reef Basslet
Luzonichthys waitei Pygmy Basslet
Othos dentex Harlequin Fish
Plectranthias alleni Allen's Perchlet
Plectranthias japonicus Japanese Perchlet
Plectranthias lasti Trawl Perchlet
Plectranthias longimanus Spot-tail Perchlet
Plectranthias megalophthalmus Citron Perchlet
Plectranthias nanus Dwarf Perchlet
Plectranthias pallidus Pale Perchlet
Plectranthias robertsi Filamentous Perchlet
Plectranthias wheeleri Spotted Perchlet
Plectranthias winniensis Redblotch Perchlet
Plectropomus areolatus Passionfruit Coral Trout
Plectropomus laevis Bluespotted Coral Trout
Plectropomus leopardus Common Coral Trout
Plectropomus maculatus Barcheek Coral Trout
Plectropomus oligacanthus Vermicular Cod
Pseudanthias bicolor Yellowback Basslet
Pseudanthias caesiopercula Greycheek Basslet
Pseudanthias cooperi Red Basslet
Pseudanthias dispar Fairy Basslet
Pseudanthias engelhardi Barrier Reef Basslet
Pseudanthias fasciata Redstripe Basslet
Pseudanthias georgei George's Basslet
Pseudanthias huchtii Pacific Basslet
Pseudanthias hypselosoma Pink Basslet
Pseudanthias lori Lori's Basslet
Pseudanthias luzonensis Luzon Basslet
Pseudanthias pascalus Sailfin Queen
Pseudanthias pictilis Painted Basslet
Pseudanthias pleurotaenia Mirror Basslet
Pseudanthias rubrizonatus Lilac-tip Basslet
Pseudanthias sheni Shen's Basslet
Pseudanthias smithvanizi Princess Basslet
Pseudanthias squamipinnis Orange Basslet
Pseudanthias tuka Purple Queen
Pseudanthias ventralis Longfin Basslet
Pseudogramma astigmum Spotless Podge
Pseudogramma polyacanthum Honeycomb Podge
Rainfordia opercularis Rainfordia
Sacura parva Little Fairy Basslet
Saloptia powelli Golden Grouper
Selenanthias analis Pearlspot Fairy Basslet
Serranocirrhitus latus Swallowtail Basslet
Suttonia lineata Freckleface Podge
Trachypoma macracanthus Pacific Rockcod
Triso dermopterus Oval Rockcod
Variola albimarginata White-edge Coronation Trout
Variola louti Yellowedge Coronation Trout

SPINYCHEEK SEABASSES

Ostracoberycidae

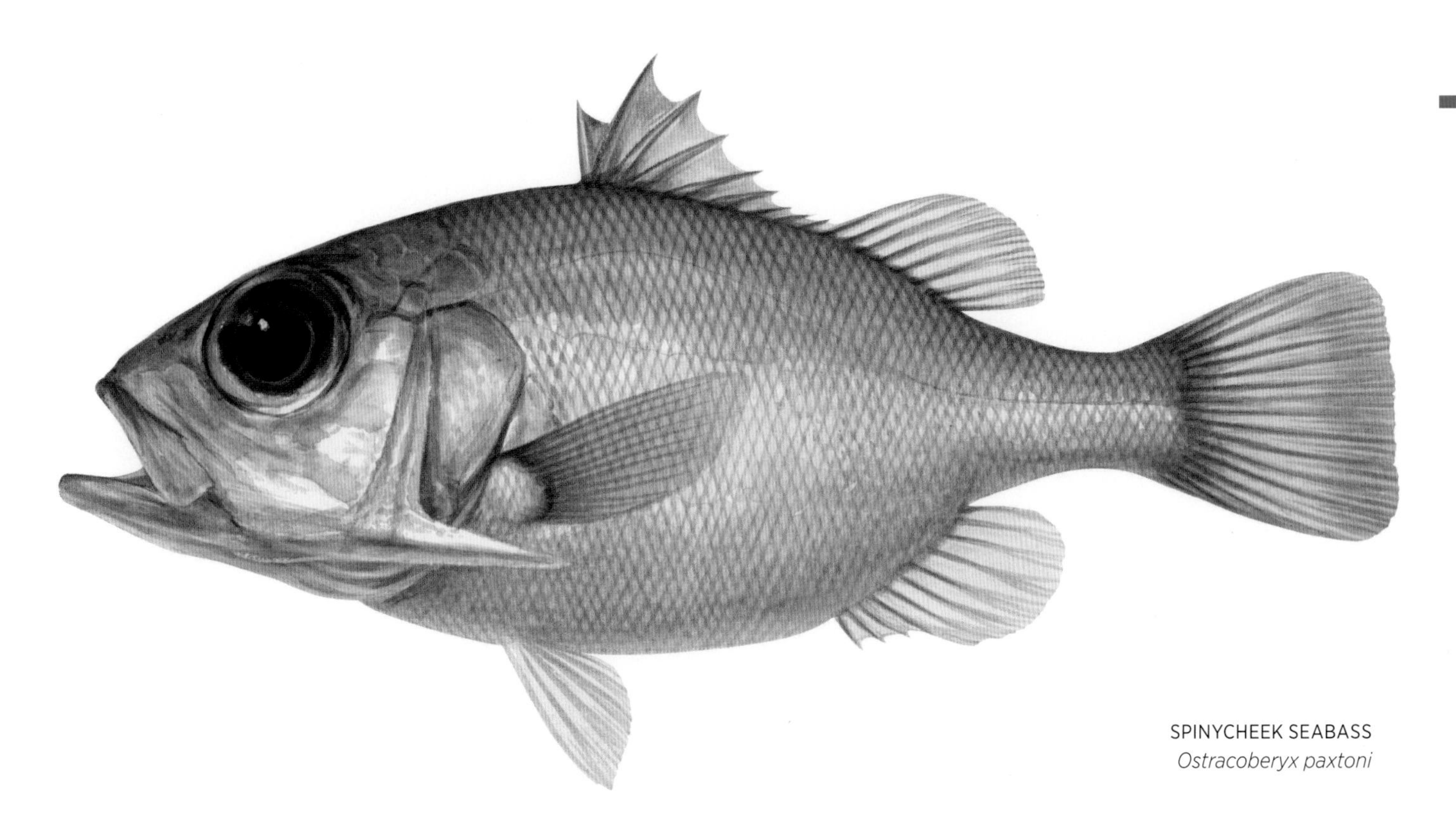

SPINYCHEEK SEABASS
Ostracoberyx paxtoni

The three species of Spinycheek Seabass are found in the Indian and western Pacific oceans. They are characterised by the very large spine on the angle of the preoperculum, which extends past the base of the pectoral fin. They have two dorsal fins, the first with nine spines and 9–10 rays; the anal fin has three spines and seven rays. The **Spinycheek Seabass** is found off the east coast of Australia, near the bottom in deep water from about 300 to 700 m, and grows to about 16 cm in length. It is occasionally encountered in deep trawl nets, but virtually nothing is known of its biology.

AUSTRALIAN OSTRACOBERYCIDAE SPECIES
Ostracoberyx paxtoni Spinycheek Seabass

SPLENDID PERCHES

Callanthiidae

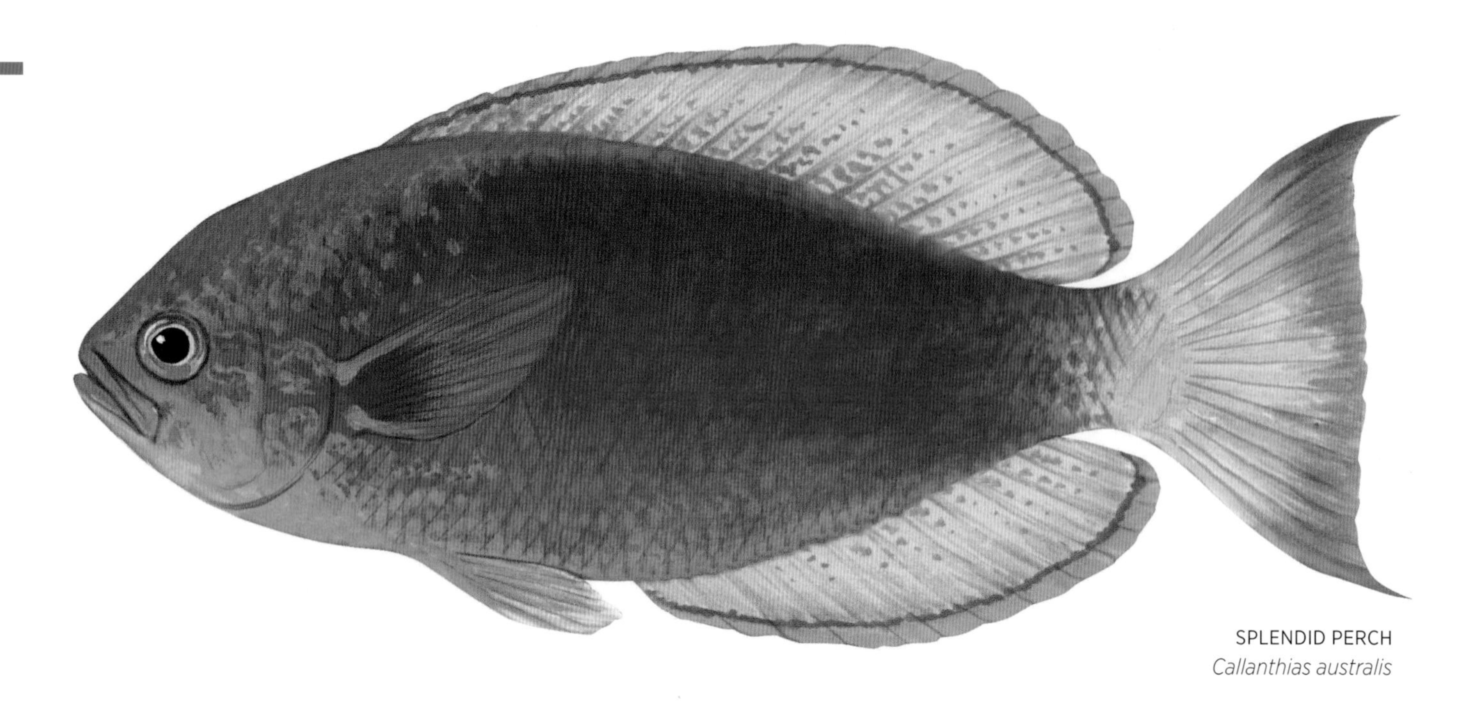

SPLENDID PERCH
Callanthias australis

Splendid Perches are found in the eastern Atlantic, Indian and Pacific oceans, and there are about 12 species in the family. They are small fishes, reaching about 30 cm in length, brightly coloured and usually found in deep water from about 50 to 400 m. They have an oval, compressed body, a single dorsal fin with 11 spines and 9–11 rays, an anal fin with three spines and 9–11 rays, and there are 1–2 flat spines on an operculum. The lateral line runs distinctively along the base of the dorsal fin.

Splendid Perches are hermaphroditic, with the smaller, less colourful females capable of changing sex to become larger, more colourful males. They form small to large aggregations, each group usually with a few males and many females, and feed on zooplankton.

The **Rosy Perch** is found in deep waters in Australia, mainly off the east coast and around Tasmania, and is occasionally seen in deepwater trawls. The **Splendid Perch** is distributed around New South Wales and the south coast of Australia, from about 25 m down to 200 m or more, and may be quite abundant over some deeper reefs. It prefers steep dropoffs and pinnacles, with crevices and caves in which it can shelter at night or when danger threatens. The brilliantly coloured males become even more magnificent during breeding times, with bright-purple patches on the head and body that they can switch on and off.

AUSTRALIAN CALLANTHIIDAE SPECIES

Callanthias allporti Rosy Perch
Callanthias australis Splendid Perch

FALSE SCORPIONFISH

Centrogeniidae

FALSE SCORPIONFISH
Centrogenys vaigiensis

The **False Scorpionfish** is the only member of this family and it is found in the eastern Indian and western Pacific oceans. It resembles a small Scorpionfish (Scorpaenidae, p. 406) but does not have numerous spines and ridges on the head, and the fin spines are not venomous. It has a single dorsal fin with 13–14 spines and 9–11 rays, a short-based anal fin with three spines and five rays, and a single strong spine on the operculum. Growing to about 20 cm in length, the False Scorpionfish is found over shallow, rocky reefs and rubble areas in coastal waters of northern Australia, occasionally entering estuaries. It feeds on small fishes and crustaceans.

AUSTRALIAN CENTROGENIIDAE SPECIES
Centrogenys vaigiensis False Scorpionfish

DOTTYBACKS AND EEL BLENNIES

Pseudochromidae

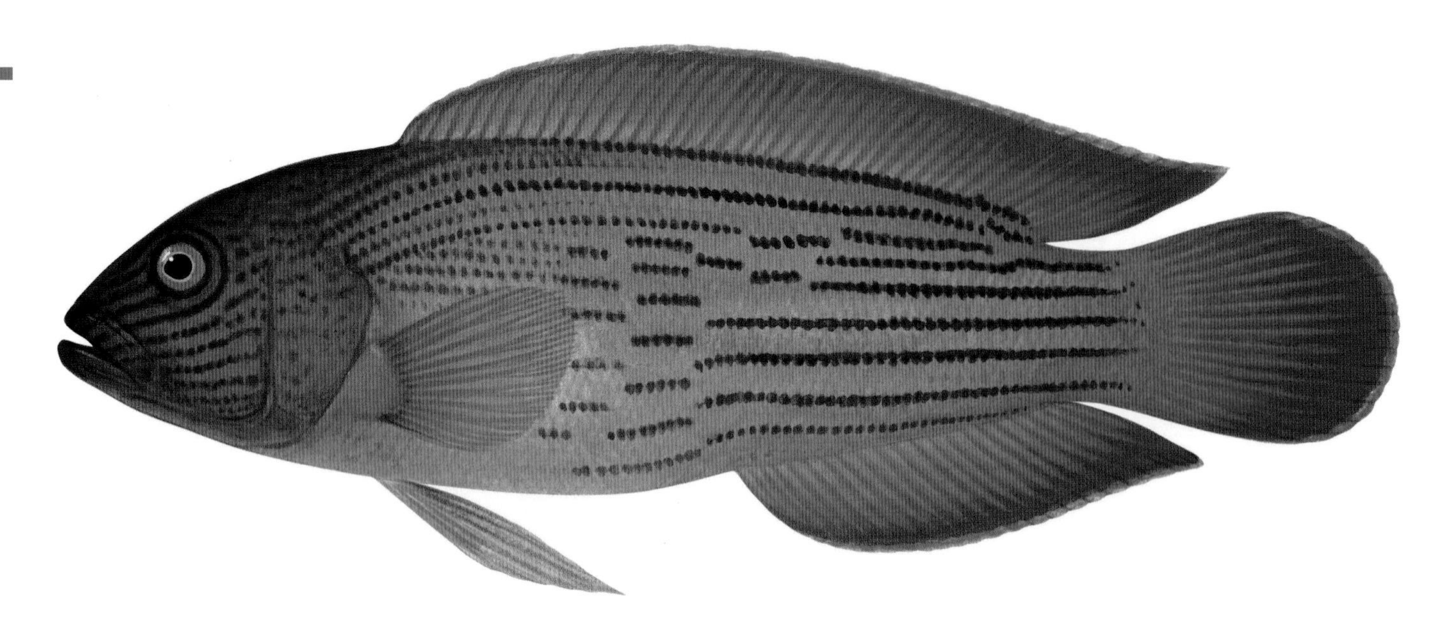

FIRETAIL DOTTYBACK
Labracinus cyclophthalmus

This large family contains about 120 species, and they are found in tropical waters of the Indo-Pacific. The Australian species can be divided into two groups: the Eel Blennies and the Dottybacks.

Eel Blennies have an elongate, eel-like body, a long-based dorsal fin with one spine and 32–79 rays, and a long anal fin with no spines and 26–66 rays. Some species have no ventral fins, but when present they are placed well forward, beneath the throat. They have a single strong spine on the upper edge of the operculum. Eel Blennies are small, cryptic, bottom-dwelling fishes, found over rubble areas and reefs in northern Australian coastal waters, hiding in crevices and caves and amongst weed. They reach a maximum length of about 50 cm, but most grow to 10–20 cm.

Dottybacks are small, often brightly coloured fishes, with an elongate, compressed body. The single long-based dorsal fin has 1–3 thin spines and 21–37 rays, and the anal fin has 1–3 small spines and 11–21 rays. In most Dottybacks the lateral line is in two parts, the first part following the dorsal profile high on the body to the rear of the dorsal fin, the second part on the rear midline of the body, up to the caudal fin. In *Pseudoplesiops* species, the lateral line is reduced to a single scale with a sensory pore. Dottybacks are shy inhabitants of coral reefs and rubble areas, keeping to the shelter of caves and crevices most of the time, but quite active and on the move during the day. Female Dottybacks lay a ball of adhesive eggs, which the male guards aggressively. Male and female Dottybacks often have quite different colour patterns and some species are known to be hermaphroditic, with sex reversal from female to male. The **Lined Dottyback** is one of the largest species, reaching about 25 cm in length, but most are around 10 cm. Dottybacks are carnivorous, preying on zooplankton and small benthic invertebrates such as crustaceans, molluscs and worms.

BLUESPOTTED DOTTYBACK
Assiculus punctatus

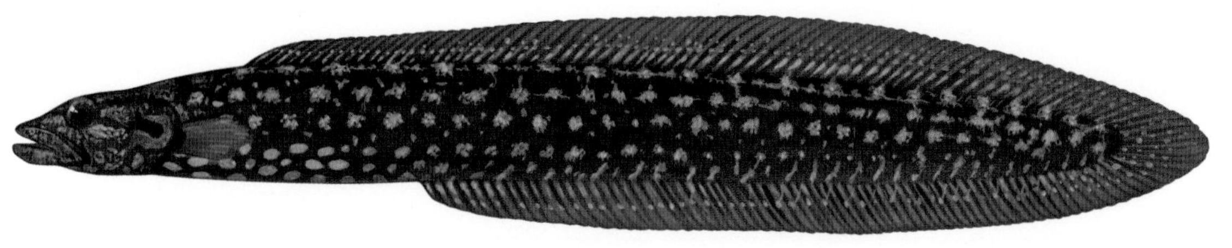

CARPET EEL BLENNY
Congrogadus subducens

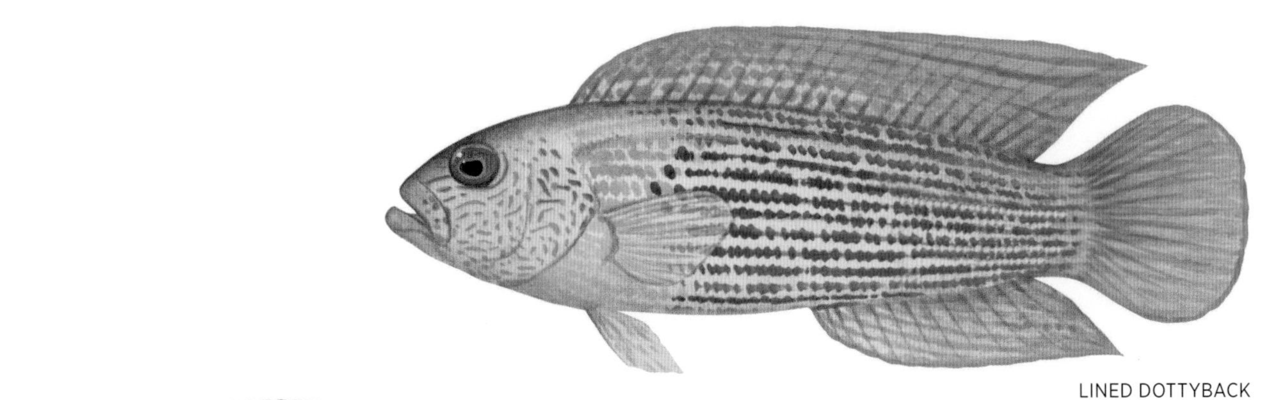

LINED DOTTYBACK
Labracinus lineatus

MULTICOLOUR DOTTYBACK
Ogilbyina novaehollandiae

DUSKY DOTTYBACK
Pseudochromis fuscus

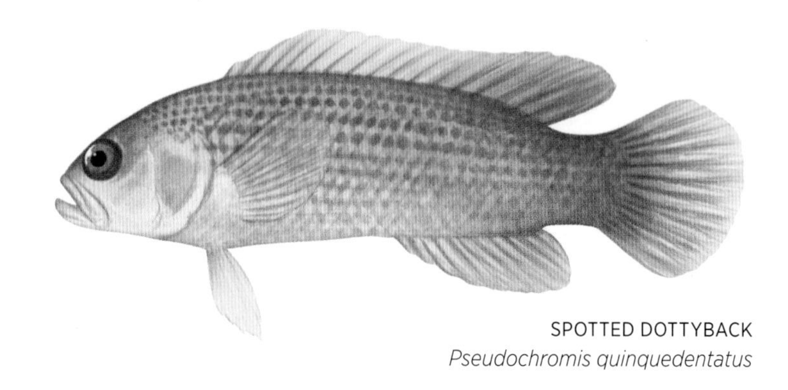

SPOTTED DOTTYBACK
Pseudochromis quinquedentatus

AUSTRALIAN PSEUDOCHROMIDAE SPECIES

Amsichthys knighti Bugeye Dottyback
Assiculoides desmonotus Bondfin Dottyback
Assiculus punctatus Bluespotted Dottyback
Blennodesmus scapularis Ocellate Eel Blenny
Congrogadus amplimaculatus Largespot Eel Blenny
Congrogadus malayanus Malayan Eel Blenny
Congrogadus spinifer Spiny Eel Blenny
Congrogadus subducens Carpet Eel Blenny
Congrogadus winterbottomi Pilbara Eel Blenny
Cypho purpurascens Lavender Dottyback
Labracinus cyclophthalmus Firetail Dottyback
Labracinus lineatus Lined Dottyback
Lubbockichthys multisquamatus Finescale Dottyback
Ogilbyina novaehollandiae Multicolour Dottyback
Ogilbyina queenslandiae Queensland Dottyback
Ogilbyina velifera Longtail Dottyback
Pictichromis coralensis Bicoloured Dottyback
Pictichromis paccagnellae Royal Dottyback
Pseudochromis andamanensis Sunset Dottyback
Pseudochromis bitaeniatus Slender Dottyback
Pseudochromis cyanotaenia Yellowhead Dottyback
Pseudochromis flammicauda Orangetail Dottyback
Pseudochromis fuscus Dusky Dottyback
Pseudochromis howsoni Howson's Dottyback
Pseudochromis jamesi Spot-tail Dottyback
Pseudochromis marshallensis Marshall Dottyback
Pseudochromis paranox Midnight Dottyback
Pseudochromis quinquedentatus Spotted Dottyback
Pseudochromis reticulatus Reticulate Dottyback
Pseudochromis splendens Splendid Dottyback
Pseudochromis tapeinosoma Blackmargin Dottyback
Pseudochromis wilsoni Yellowfin Dottyback
Pseudoplesiops annae Anne's Dottyback
Pseudoplesiops howensis Lord Howe Dottyback
Pseudoplesiops immaculatus Immaculate Dottyback
Pseudoplesiops rosae Rose Dottyback
Pseudoplesiops typus Ringeye Dottyback
Pseudoplesiops wassi Fleckfin Dottyback

LONGFINS

Plesiopidae

EASTERN BLUE DEVIL
Paraplesiops bleekeri

Longfins are found in the Indo-West Pacific and there are about 46 species in the family. They have elongate to oval compressed bodies, a single long-based dorsal fin with 10–13 spines (the membrane between the spines is often deeply incised) and 6–7 rays. The anal fin has three spines and 8–9 rays. Spiny Basslets are an exception, with 17–26 spines and only 2–6 rays in the dorsal fin, and 7–16 spines and 2–6 rays in the anal fin. Longfins have a lateral line in two or three parts: the first part is along the dorsal profile, the second is along the midline of the rear of the body, and there is sometimes a third along the ventral profile above the anal fin. Some have only a single short section of lateral line on the anterior of the body. They have elongate ventral fins with one spine and 2–4 rays, with the first ray usually thickened and quite prominent. Most are small, reef-dwelling fishes, keeping to the shelter of caves and crevices.

Blue Devils are spectacular residents of rocky southern Australian reefs and are often seen by divers under ledges or in caves. They grow to about 30 cm in length and are popular fishes for larger aquariums. Hulafishes are schooling plankton feeders, found in small groups or occasionally large aggregations on southern Australian coastal reefs down to about 30 m. Several different species may school together. Hulafishes grow to about 15 cm in length and are named for their undulating swimming motion.

The small Scissortails have an elongate body with a long, forked caudal fin and are found in caves along the Great Barrier Reef. The **Comet** is another favourite with aquarium enthusiasts. Its broad white-spotted fins and prominent black eye spot mimic a Moray Eel (Muraenidae, p. 158); adaptations to deter predators. To complete the illusion, it will even position its head in a crevice with the tail projecting, the gap between the anal fin and caudal fin resembling the mouth of the Moray.

Various other Longfins are found on coral and rocky reefs in northern Australian waters, though most are rarely seen due to their secretive habits. Some species of Longfin, such as the Scissortails, are known to be mouthbrooders, the male holding the mass of adhesive eggs in its mouth until they hatch. Others, such as the Blue Devils, attach their eggs to the substrate and guard them until hatching. Longfins feed on zooplankton, small fishes, crustaceans and other benthic invertebrates.

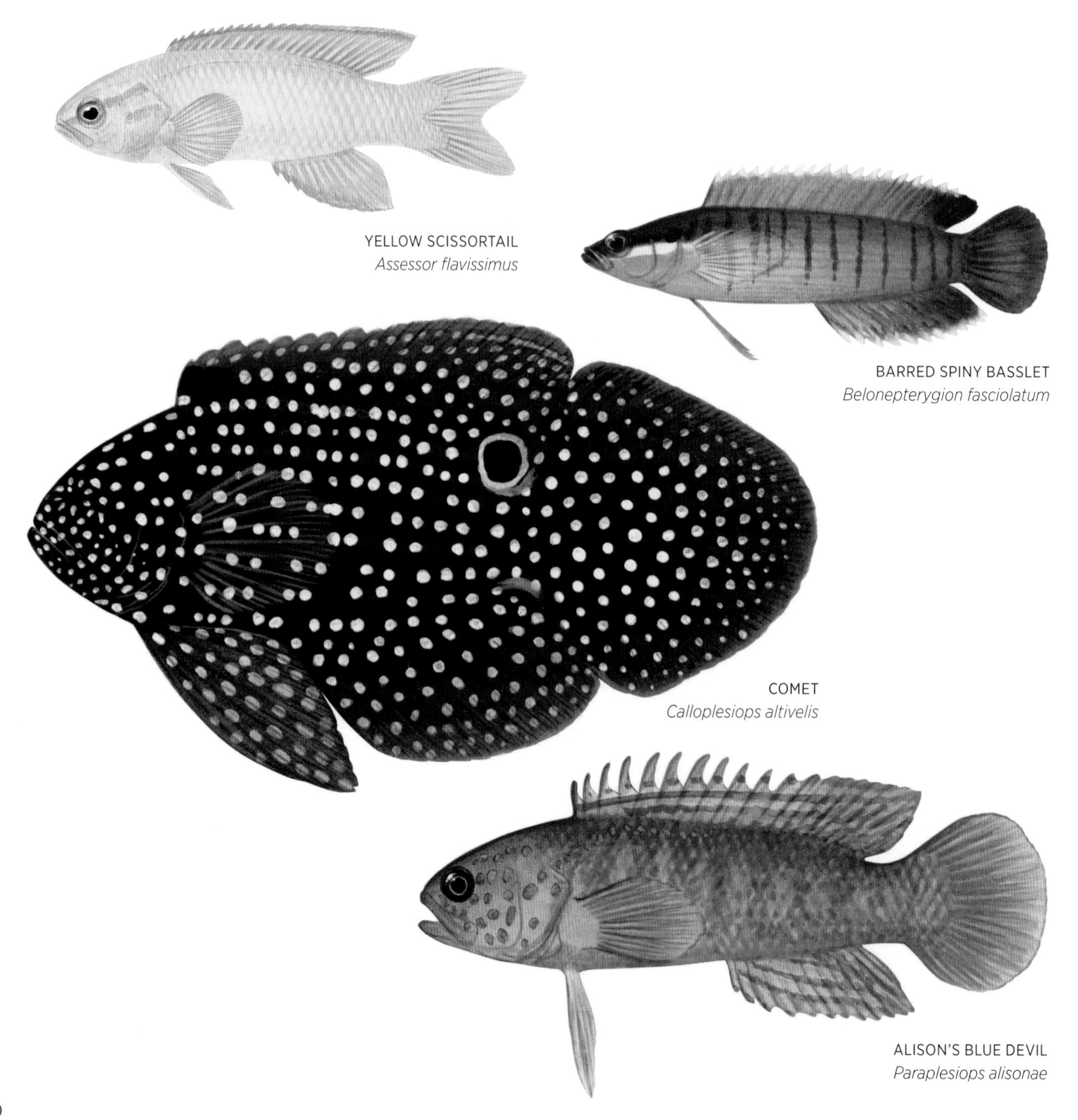

YELLOW SCISSORTAIL
Assessor flavissimus

BARRED SPINY BASSLET
Belonepterygion fasciolatum

COMET
Calloplesiops altivelis

ALISON'S BLUE DEVIL
Paraplesiops alisonae

CORAL DEVIL
Plesiops coeruleolineatus

SOUTHERN HULAFISH
Trachinops caudimaculatus

BLUELINED HULAFISH
Trachinops brauni

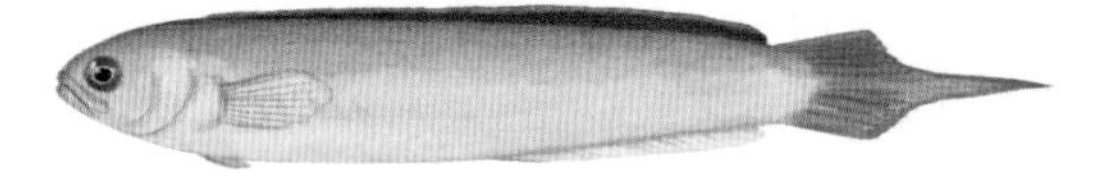

YELLOWHEAD HULAFISH
Trachinops noarlungae

AUSTRALIAN PLESIOPIDAE SPECIES

Assessor flavissimus Yellow Scissortail
Assessor macneilli Blue Scissortail
Beliops xanthokrossos Southern Longfin
Belonepterygion fasciolatum Barred Spiny Basslet
Calloplesiops altivelis Comet
Fraudella carassiops Carp Prettyfin
Paraplesiops alisonae Alison's Blue Devil
Paraplesiops bleekeri Eastern Blue Devil
Paraplesiops meleagris Southern Blue Devil
Paraplesiops poweri Northern Blue Devil
Paraplesiops sinclairi Western Blue Devil
Plesiops coeruleolineatus Coral Devil
Plesiops genaricus Cheekveil Longfin
Plesiops gracilis Threadfin Longfin
Plesiops insularis Island Longfin
Plesiops verecundus Redtip Longfin
Steeneichthys plesiopsus Steene's Prettyfin
Trachinops brauni Bluelined Hulafish
Trachinops caudimaculatus Southern Hulafish
Trachinops noarlungae Yellowhead Hulafish
Trachinops taeniatus Eastern Hulafish

BEARDED EEL BLENNIES

Notograptidae

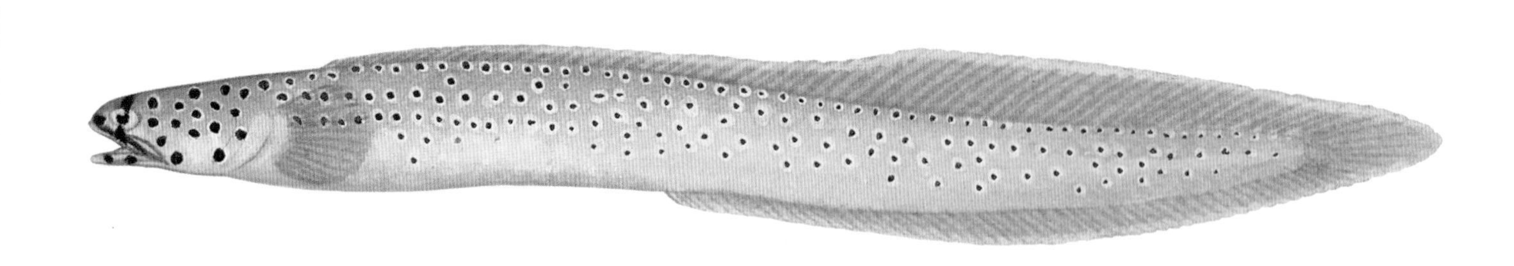

SPOTTED EEL BLENNY
Notograptus guttatus

Bearded Eel Blennies are found only in waters of northern Australia and New Guinea. There are about three species in the family. They have an eel-like body, a small barbel on the chin and a large mouth that reaches behind the posterior margin of the eye. The long-based dorsal and anal fins are continuous with the caudal fin, and the ventral fins have one small spine and two rays.

The **Shark Bay Eel Blenny** is known from only a few specimens found in Shark Bay, Western Australia. The **Spotted Eel Blenny** is more widely distributed in northern Australian waters and is found in crevices and caves in coastal reefs. Bearded Eel Blennies are cryptic, small fishes, reaching a maximum length of about 14 cm. They feed on small crustaceans, with some species appearing to specialise in alpheid shrimps. They are rarely encountered and little is known of their biology.

SHARK BAY EEL BLENNY
Notograptus gregoryi

AUSTRALIAN NOTOGRAPTIDAE SPECIES

Notograptus gregoryi Shark Bay Eel Blenny
Notograptus guttatus Spotted Eel Blenny

JAWFISHES

Opistognathidae

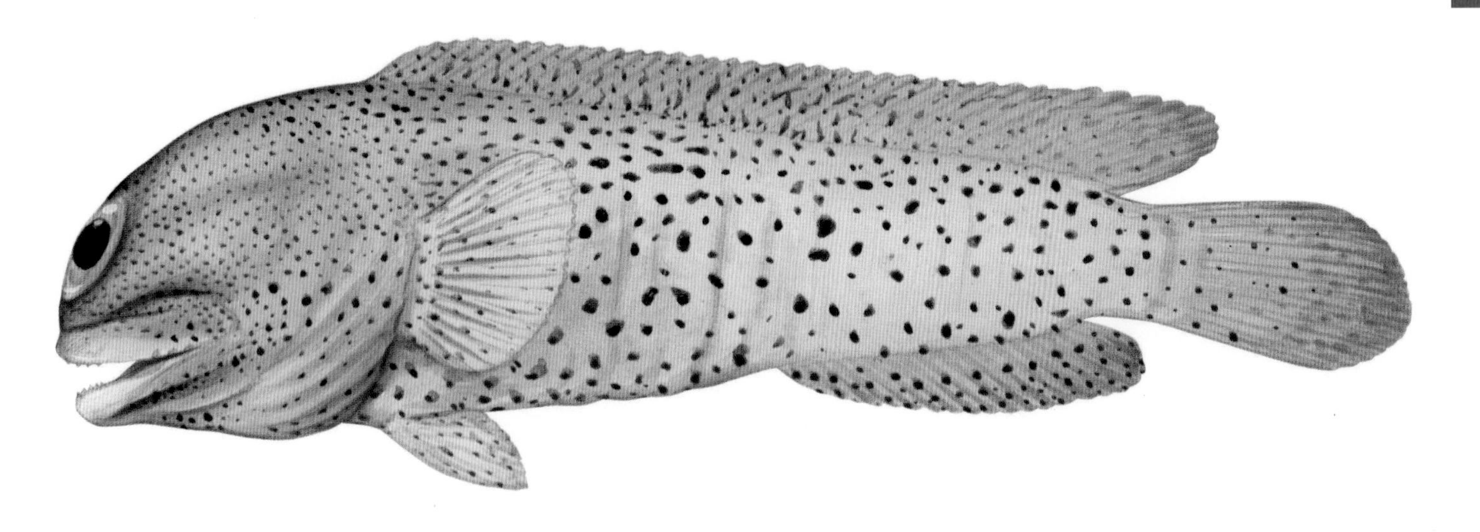

FINESPOTTED JAWFISH
Opistognathus punctatus

There are about 60 species of Jawfish distributed around the world in warm temperate and tropical coastal waters. They have a tapering, compressed body with a large head and large jaws, sometimes with a long rearward extension to the upper jaw. They have a single long dorsal fin with 9–12 spines and 12–22 rays, the anal fin has 2–3 spines and 10–21 rays, and the ventral fins are anterior to the pectoral fins. In the **Harlequin Jawfish** and **Yellow Jawfish**, the anterior dorsal-fin spines are unusually forked at their tips. The lateral line is high on the body and ends at about the middle of the dorsal fin.

Growing to a maximum length of about 50 cm, Jawfishes use their large, powerful jaws to dig burrows, which they often line with stones and shells. They make short forays into the open to feed on benthic and planktonic invertebrates, retreating tail first into the burrow at any sign of danger. Their eyes are large and placed well forward and dorsally, giving these fishes an excellent view of their surroundings with only a very small part of their head projecting from the burrow. Jawfishes are mouthbrooders, the male holding a ball of eggs in its mouth until hatching. Males will leave the eggs in the burrow when making short feeding excursions. Each egg bears filaments that hold the mass together. Australian Jawfishes usually live in small colonies over sandy and rubble-strewn areas on northern reefs. They are found mainly in shallow water down to about 30 m, although some occur as deep as 300 m. They are uncommon and not well known, and there may well be more species yet to be described. Jawfishes are occasionally seen by divers, usually with only the head protruding from the burrow, and are infrequently taken by anglers. They have powerful jaws with small, sharp teeth and can give a strong bite, so they should be handled carefully.

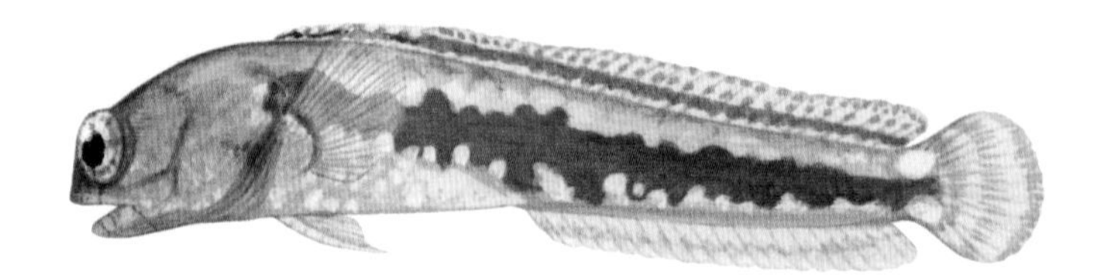

ABROLHOS JAWFISH
Opistognathus alleni

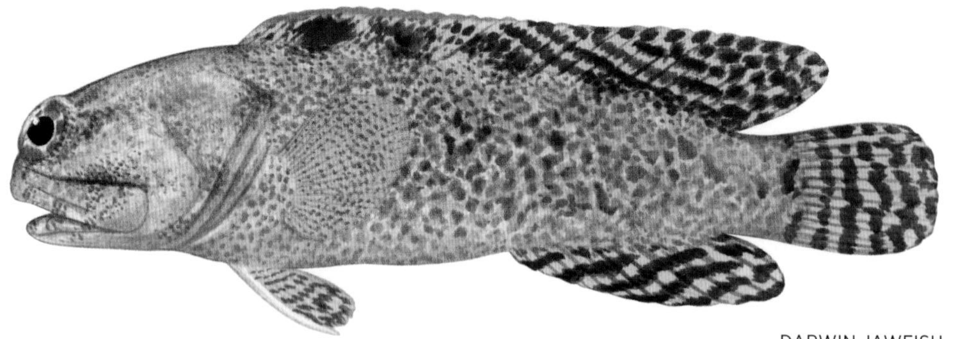

DARWIN JAWFISH
Opistognathus darwiniensis

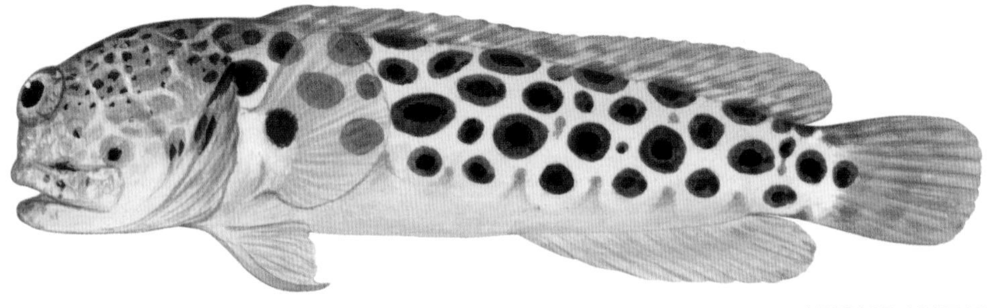

LEOPARD JAWFISH
Opistognathus reticulatus

AUSTRALIAN OPISTOGNATHIDAE SPECIES

Opistognathus alleni Abrolhos Jawfish
Opistognathus castelnaui Castelnau's Jawfish
Opistognathus darwiniensis Darwin Jawfish
Opistognathus elizabethensis Elizabeth Reef Jawfish
Opistognathus eximius Harlequin Smiler
Opistognathus inornatus Black Jawfish
Opistognathus jacksoniensis Southern Smiler
Opistognathus latitabundus Blotched Jawfish
Opistognathus macrolepis Bigscale Jawfish
Opistognathus papuensis Papuan Jawfish
Opistognathus punctatus Finespotted Jawfish
Opistognathus reticeps Reticulated Jawfish
Opistognathus reticulatus Leopard Jawfish
Opistognathus seminudus Halfnaked Jawfish
Opistognathus stigmosus Coral Sea Jawfish
Opistognathus verecundus Bashful Jawfish
Stalix flavida Yellow Jawfish
Stalix histrio Harlequin Jawfish

BANJOFISH

Banjosidae

BANJOFISH
Banjos banjos

The **Banjofish**, the only member of this family, is found in the western Pacific and around the northern half of Australia. It is deep-bodied and compressed, the dorsal fin has 10 large strong spines and 12 rays, and the anal fin has three spines (the second very long) and seven rays. The Banjofish grows to about 30 cm in length and is found in deeper coastal waters, from about 50 to 400 m, usually over sand. It is occasionally caught by anglers and often appears in trawl nets.

AUSTRALIAN BANJOSIDAE SPECIES
Banjos banjos Banjofish

FRESHWATER PERCHES

Percidae

REDFIN PERCH
Perca fluviatilis

Freshwater Perches are northern-hemisphere fishes, with more than 200 species found in the freshwater systems of North America, Europe and Asia. The **Redfin Perch** is a typical member of the family, with a slightly elongate, compressed body, strong ctenoid scales and a large head and jaws. It has two dorsal fins, the first with 13–17 spines and the second with 13–16 rays, and a single flattened spine on the operculum.

The Redfin Perch was the first freshwater fish to be introduced to Australia from Europe. Brought over in the 1860s as an angling species, it is now widespread in the southern half of Australia, in rivers, streams and dams. The Redfin Perch has become a popular target for freshwater anglers, as it can reach up to 50 cm in length, fights hard and is an excellent food fish. However, it is also a voracious predator of small fishes and crustaceans, and in some areas is thought to be responsible for drastic reductions in numbers of small native fishes. In addition, the species produces a large number of eggs that are distasteful to other fishes, allowing the Redfin Perch to breed very successfully. In some areas this has resulted in extremely large populations of Redfin Perch and the elimination of most other species.

AUSTRALIAN PERCIDAE SPECIES

Perca fluviatilis Redfin Perch

BIGEYES

Priacanthidae

LUNARTAIL BIGEYE
Priacanthus hamrur

Bigeyes are found throughout the tropical waters of the Atlantic, Indian and Pacific oceans, and there are about 18 species in the family. They have an oval, compressed body with spiny scales, a large oblique mouth and very large, reflective eyes. The single long dorsal fin has 10 spines and 11–15 rays, and the anal fin has three spines and 10–16 rays. The large ventral fins are attached to the body by a membrane.

Most species are a magnificent silvery-red colour, which they can alter at will, changing from red to a dull silver in an instant. Their brightly coloured large eyes possess an internal reflective layer, which makes the eyes shine brilliantly under a light at night. This feature may serve to amplify light within the eye, improving night vision.

Bigeyes feed nocturnally, principally on zooplankton, small fishes and cephalopods. They are usually seen by divers over reef areas in small groups, though some species occur in aggregations over open ground in deeper waters and these are frequently taken in large numbers by offshore trawlers. Bigeyes are occasionally taken by anglers at night and are considered good food fishes. The **Spotted Bigeye**, **Purplespotted Bigeye** and **Lunartail Bigeye** appear quite often at fish markets. In Australia most Bigeyes are found only in northern waters, though the **Blotched Bigeye**, which has a long pelagic larval stage, is seen at times along the southern coast. Bigeyes reach a maximum length of about 50 cm, though most attain no more than 30 cm.

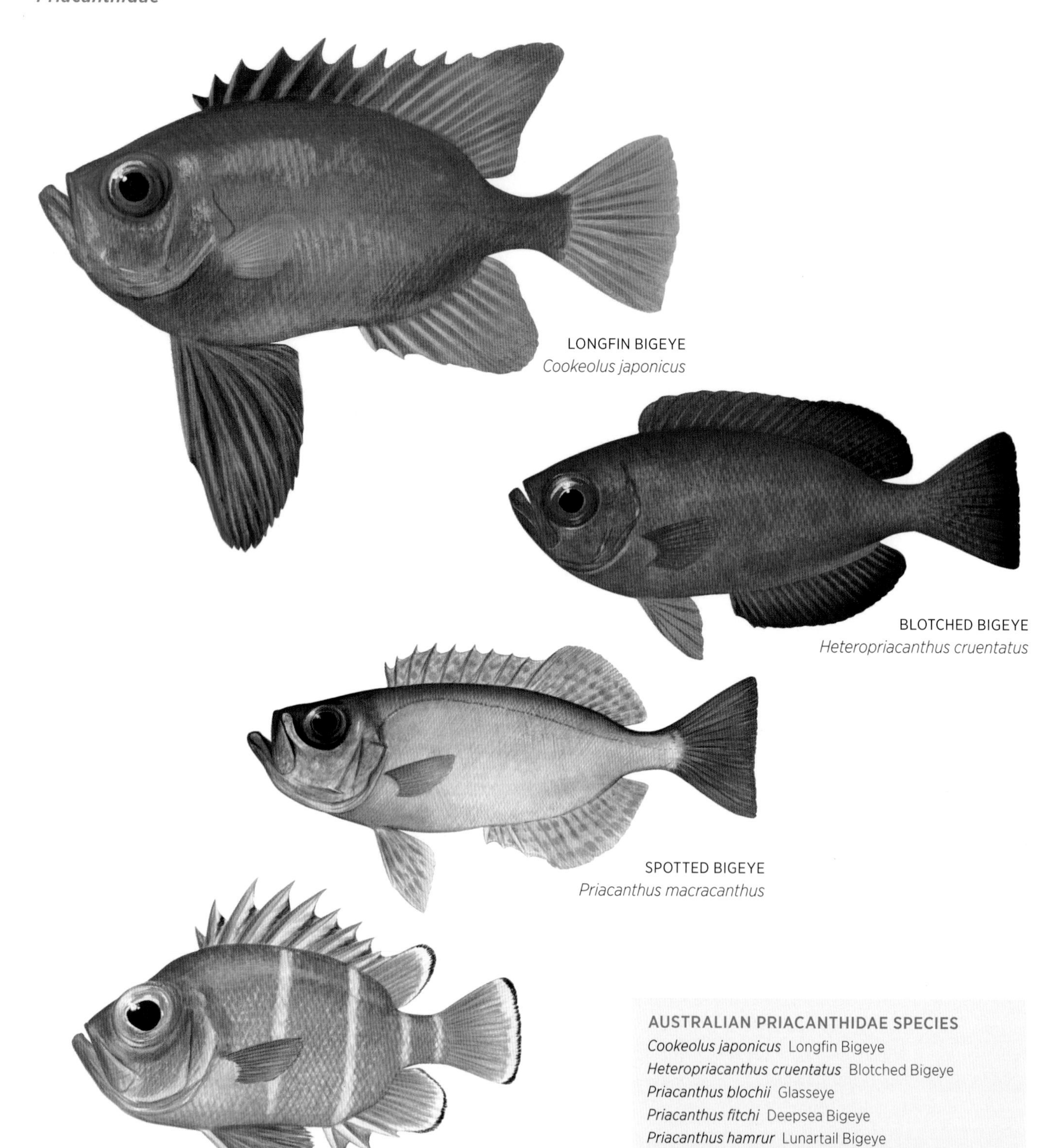

AUSTRALIAN PRIACANTHIDAE SPECIES

Cookeolus japonicus Longfin Bigeye
Heteropriacanthus cruentatus Blotched Bigeye
Priacanthus blochii Glasseye
Priacanthus fitchi Deepsea Bigeye
Priacanthus hamrur Lunartail Bigeye
Priacanthus macracanthus Spotted Bigeye
Priacanthus sagittarius Arrowfin Bigeye
Priacanthus tayenus Purplespotted Bigeye
Pristigenys niphonia Whiteband Bigeye

CARDINALFISHES

Apogonidae

TIGER CARDINALFISH
Cheilodipterus macrodon

Cardinalfishes are found in tropical and warm temperate waters around the world, with a few species found in brackish and fresh waters. This large family contains about 250 species, and they are particularly diverse in the Indo-Pacific and around coral reefs in northern Australia. They have oval to elongate bodies with large scales, a large mouth and large eyes, and two tall dorsal fins, the first with 6–8 spines, the second with 8–14 rays. The anal fin has two spines and 8–18 rays.

During the day Cardinalfishes are usually found in caves, under ledges or amongst branching corals. They emerge from shelter at night to hunt zooplankton and benthic invertebrates. Though the family is widespread, most species are restricted to a specific habitat type, such as between the branches of a certain species of coral, and are found nowhere else. Some species, such as the **Southern Cardinalfish** and the **Western Gobbleguts**, frequent estuaries and weedbeds in sheltered waters around the south coast of Australia. Others are found in deep waters offshore and are known only from specimens taken by trawlers.

The **Mouth Almighty** is a freshwater species found in streams, lakes and swamps across northern Australia. The **Mangrove Cardinalfish** is found in tidal creeks, and occasionally in fresh water, in northern Queensland. The vast majority of species, however, are tropical reef dwellers. They range from large, solitary species such as the **Tiger Cardinalfish**, which is often seen floating almost motionless under ledges and in caves, to schooling species such as the **Slender Cardinalfish**, which forms large aggregations over coral bommies. Many species have horizontal stripes along the body and are difficult to differentiate from each other. Some are more distinctive, such as the superbly patterned **Pyjama Cardinalfish**, a favourite with aquarium enthusiasts. The **Girdled Cardinalfish** is another spectacular species, commonly found amongst branching Acropora corals along the Great Barrier Reef. The **Ringtail Cardinalfish**, found around the northern coast, is also beautifully coloured.

Cardinalfishes are all mouthbrooders, with the male holding the gelatinous mass of eggs in its mouth for several days until hatching. It is quite common to see males sheltering under ledges or in caves with throat bulging and mouth open, crammed with eggs. Most Cardinalfishes are 5–10 cm in length, with the largest species reaching a maximum of about 22 cm.

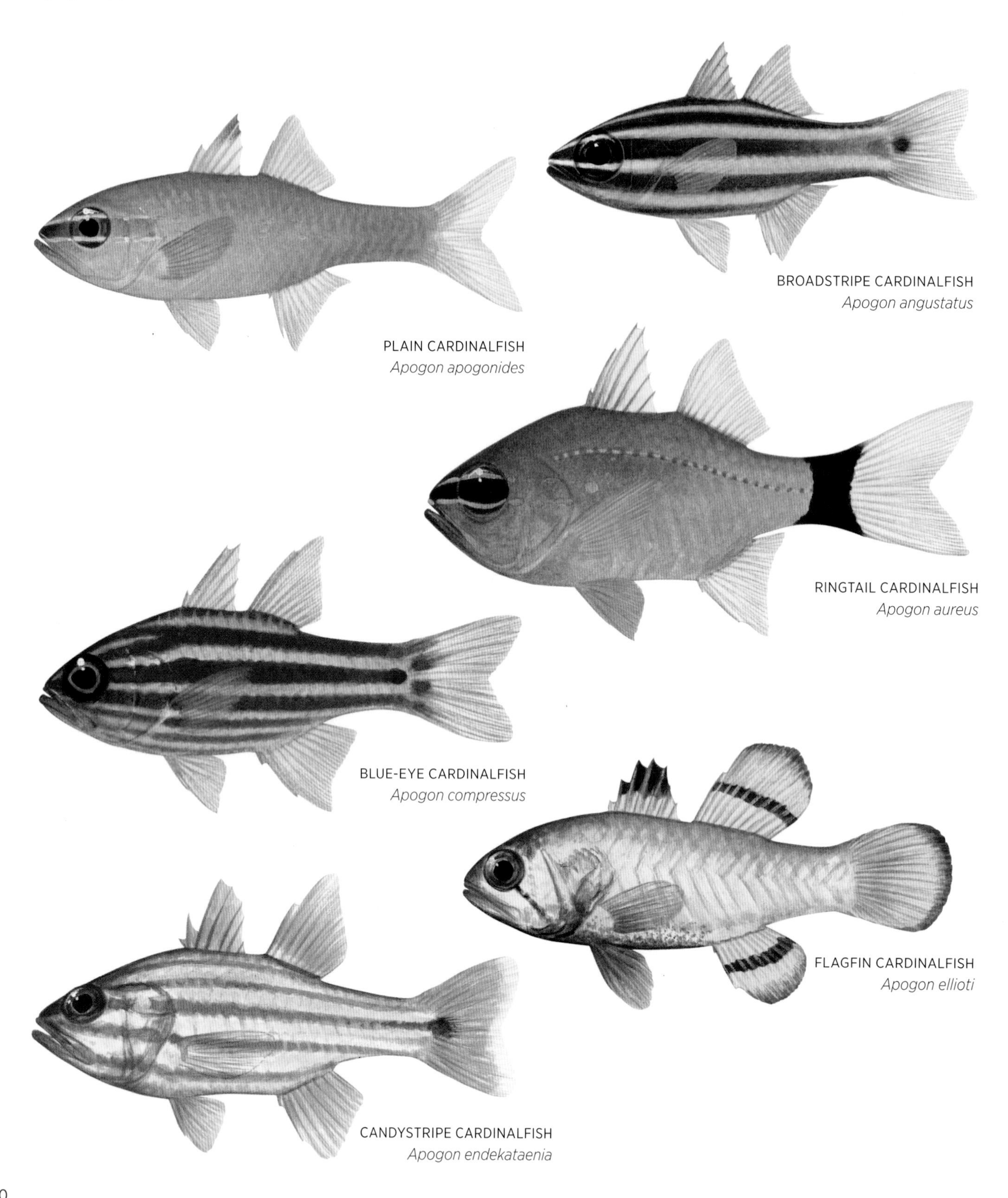

BROADSTRIPE CARDINALFISH
Apogon angustatus

PLAIN CARDINALFISH
Apogon apogonides

RINGTAIL CARDINALFISH
Apogon aureus

BLUE-EYE CARDINALFISH
Apogon compressus

FLAGFIN CARDINALFISH
Apogon ellioti

CANDYSTRIPE CARDINALFISH
Apogon endekataenia

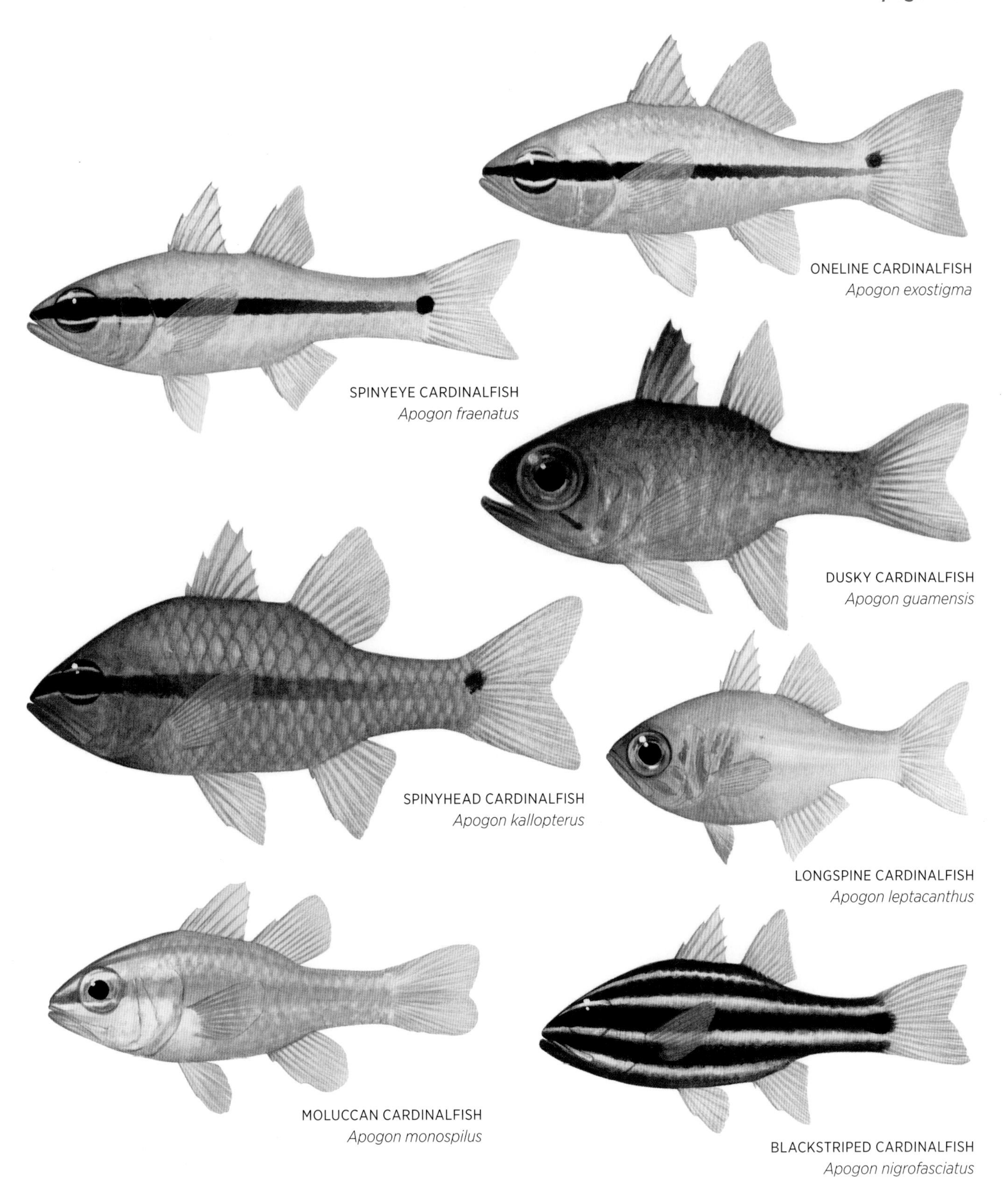

ONELINE CARDINALFISH
Apogon exostigma

SPINYEYE CARDINALFISH
Apogon fraenatus

DUSKY CARDINALFISH
Apogon guamensis

SPINYHEAD CARDINALFISH
Apogon kallopterus

LONGSPINE CARDINALFISH
Apogon leptacanthus

MOLUCCAN CARDINALFISH
Apogon monospilus

BLACKSTRIPED CARDINALFISH
Apogon nigrofasciatus

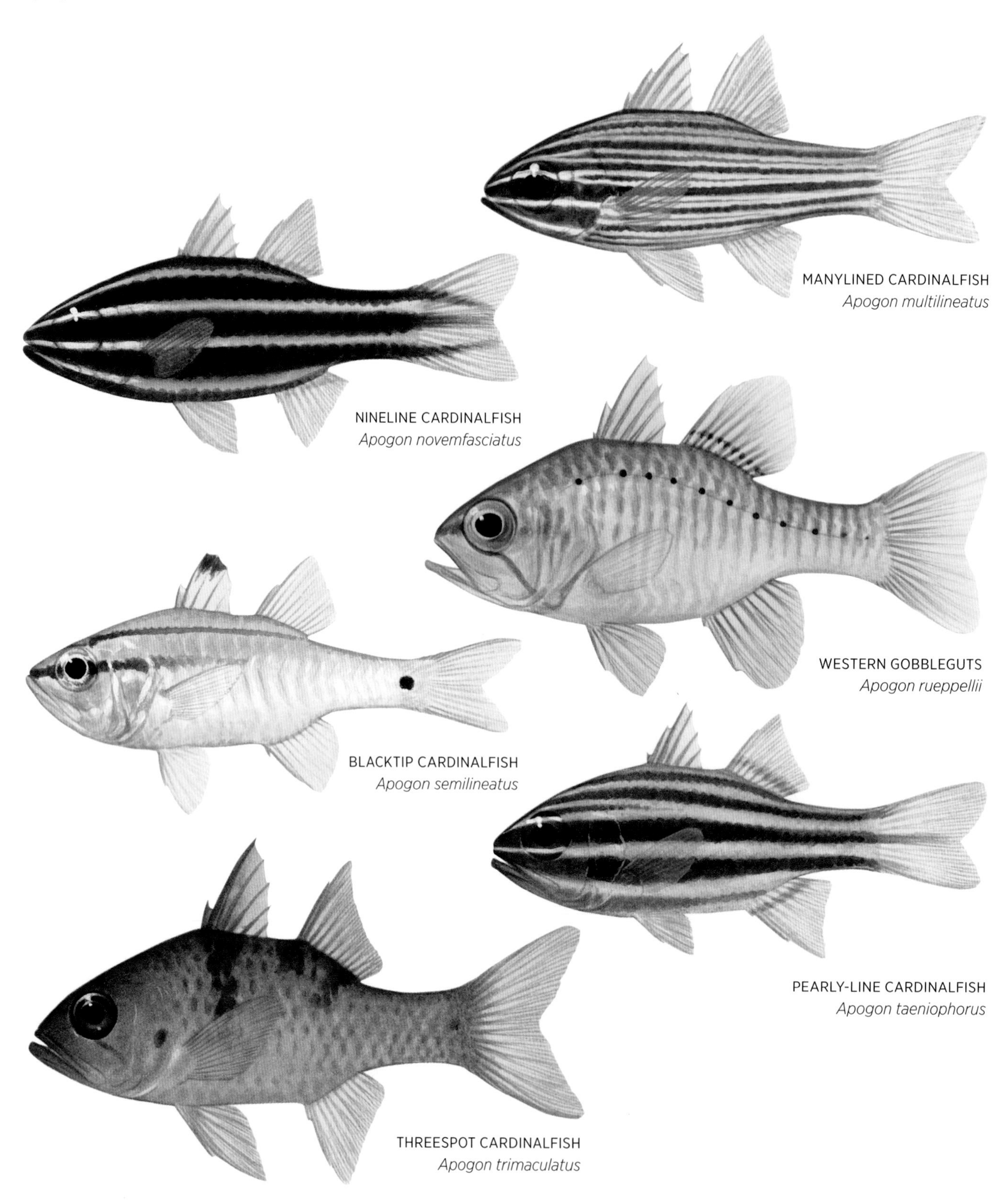

MANYLINED CARDINALFISH
Apogon multilineatus

NINELINE CARDINALFISH
Apogon novemfasciatus

WESTERN GOBBLEGUTS
Apogon rueppellii

BLACKTIP CARDINALFISH
Apogon semilineatus

PEARLY-LINE CARDINALFISH
Apogon taeniophorus

THREESPOT CARDINALFISH
Apogon trimaculatus

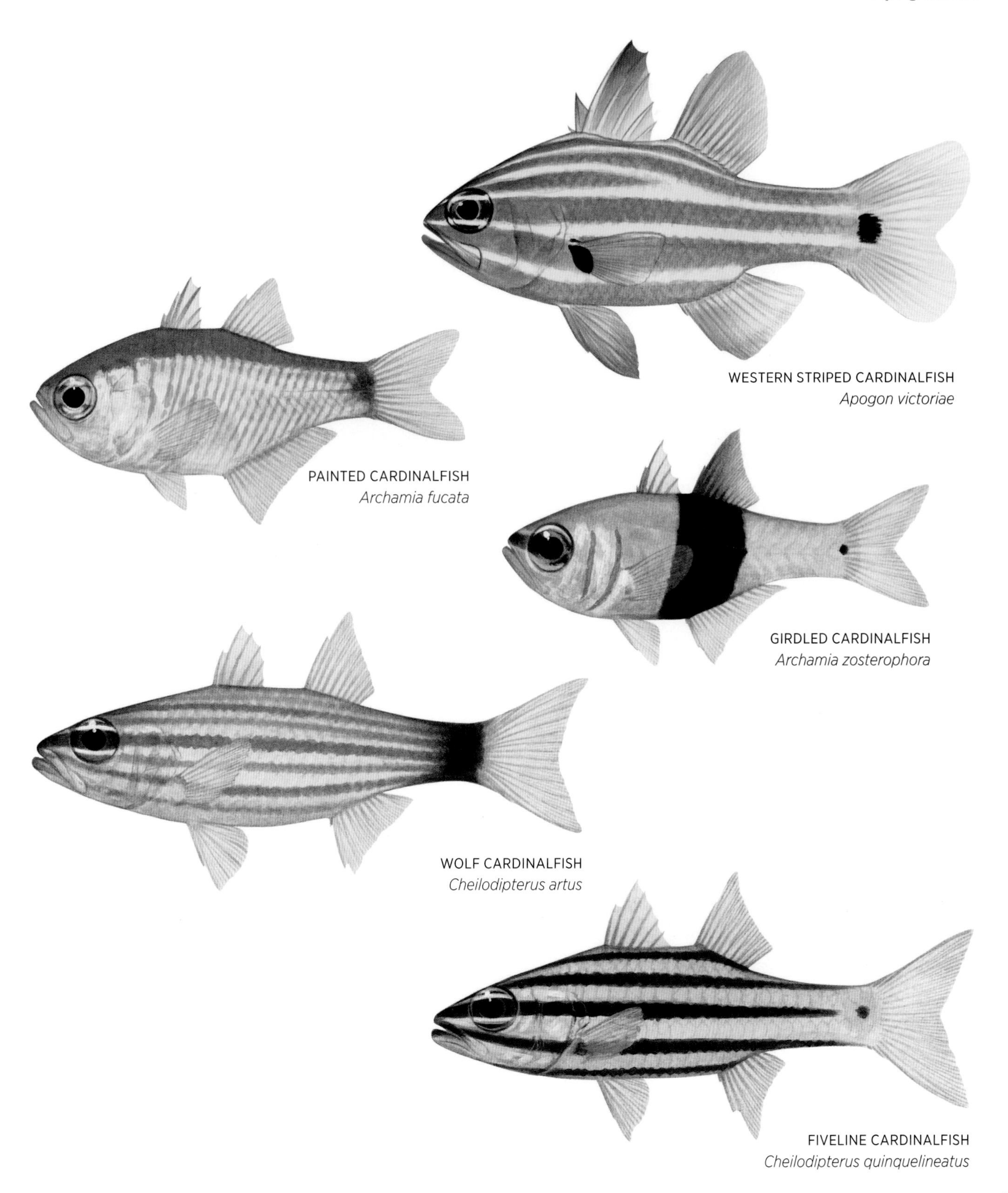

WESTERN STRIPED CARDINALFISH
Apogon victoriae

PAINTED CARDINALFISH
Archamia fucata

GIRDLED CARDINALFISH
Archamia zosterophora

WOLF CARDINALFISH
Cheilodipterus artus

FIVELINE CARDINALFISH
Cheilodipterus quinquelineatus

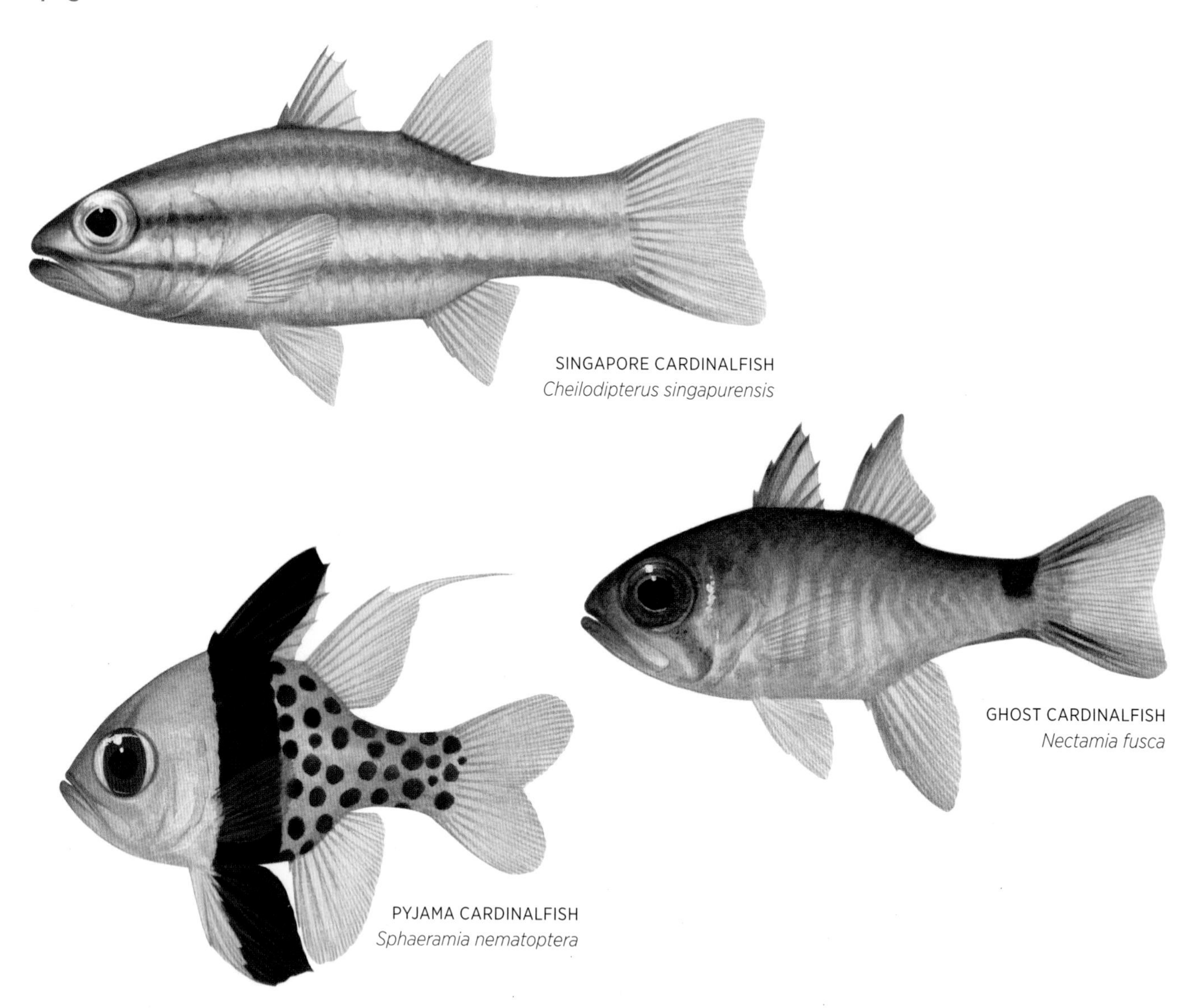

SINGAPORE CARDINALFISH
Cheilodipterus singapurensis

GHOST CARDINALFISH
Nectamia fusca

PYJAMA CARDINALFISH
Sphaeramia nematoptera

AUSTRALIAN APOGONIDAE SPECIES

Apogon albimaculosus Creamspotted Cardinalfish
Apogon angustatus Broadstripe Cardinalfish
Apogon apogonides Plain Cardinalfish
Apogon atrogaster Blackbelly Cardinalfish
Apogon aureus Ringtail Cardinalfish
Apogon brevicaudatus Manyband Cardinalfish
Apogon bryx Offshore Cardinalfish
Apogon capricornis Capricorn Cardinalfish
Apogon carinatus Ocellate Cardinalfish
Apogon cavitiensis Whiteline Cardinalfish
Apogon chrysotaenia Yellowlined Cardinalfish
Apogon cladophilos Shelter Cardinalfish
Apogon compressus Blue-eye Cardinalfish
Apogon cookii Cook's Cardinalfish
Apogon crassiceps Ruby Cardinalfish
Apogon cyanosoma Orangelined Cardinalfish
Apogon doederleini Fourline Cardinalfish
Apogon doryssa Night Cardinalfish
Apogon ellioti Flagfin Cardinalfish
Apogon endekataenia Candystripe Cardinalfish
Apogon evermanni Cave Cardinalfish
Apogon exostigma Oneline Cardinalfish
Apogon fasciatus Striped Cardinalfish
Apogon flavus Yellow Cardinalfish

Apogon fleurieu Bullseye Cardinalfish
Apogon fraenatus Spinyeye Cardinalfish
Apogon fragilis Fragile Cardinalfish
Apogon fuscomaculatus Brownspotted Cardinalfish
Apogon gilberti Gilbert's Cardinalfish
Apogon guamensis Dusky Cardinalfish
Apogon hartzfeldii Silverlined Cardinalfish
Apogon hoevenii Frostfin Cardinalfish
Apogon hyalosoma Mangrove Cardinalfish
Apogon jenkinsi Spotnape Cardinalfish
Apogon kallopterus Spinyhead Cardinalfish
Apogon kiensis Rifle Cardinalfish
Apogon lateralis Pinstripe Cardinalfish
Apogon leptacanthus Longspine Cardinalfish
Apogon limenus Sydney Cardinalfish
Apogon melanopus Monster Cardinalfish
Apogon melas Black Cardinalfish
Apogon monospilus Moluccan Cardinalfish
Apogon multilineatus Manylined Cardinalfish
Apogon neotes Mini Cardinalfish
Apogon nigripinnis Two-eye Cardinalfish
Apogon nigrofasciatus Blackstriped Cardinalfish
Apogon norfolcensis Norfolk Cardinalfish
Apogon novemfasciatus Nineline Cardinalfish
Apogon opercularis Pearlycheek Cardinalfish
Apogon pallidofasciatus Palestriped Cardinalfish
Apogon perlitus Pearly Cardinalfish
Apogon poecilopterus Pearlyfin Cardinalfish
Apogon properuptus Coral Cardinalfish
Apogon quadrifasciatus Bar-striped Cardinalfish
Apogon rhodopterus Twobar Cardinalfish
Apogon rubrimacula Redspot Cardinalfish
Apogon rueppellii Western Gobbleguts
Apogon sangiensis Sangi Cardinalfish
Apogon sealei Cheekbar Cardinalfish
Apogon selas Meteor Cardinalfish
Apogon semilineatus Blacktip Cardinalfish
Apogon semiornatus Halfband Cardinalfish
Apogon septemstriatus Sevenband Cardinalfish
Apogon taeniophorus Pearly-line Cardinalfish
Apogon talboti Flame Cardinalfish
Apogon timorensis Timor Cardinalfish
Apogon trimaculatus Threespot Cardinalfish
Apogon unicolor Big Red Cardinalfish
Apogon unitaeniatus Singlestripe Cardinalfish
Apogon victoriae Western Striped Cardinalfish
Apogon wassinki Kupang Cardinalfish
Apogonichthys ocellatus Ocellate Cardinalfish
Apogonichthys perdix Perdix Cardinalfish
Archamia biguttata Blackspot Cardinalfish
Archamia bleekeri Gon's Cardinalfish
Archamia fucata Painted Cardinalfish
Archamia leai Lea's Cardinalfish
Archamia zosterophora Girdled Cardinalfish
Cercamia eremia Glassy Cardinalfish
Cheilodipterus artus Wolf Cardinalfish
Cheilodipterus intermedius Intermediate Cardinalfish
Cheilodipterus isostigmus Toothy Cardinalfish
Cheilodipterus macrodon Tiger Cardinalfish
Cheilodipterus parazonatus Mimic Cardinalfish
Cheilodipterus quinquelineatus Fiveline Cardinalfish
Cheilodipterus singapurensis Singapore Cardinalfish
Foa brachygramma Harbour Cardinalfish
Fowleria aurita Crosseye Cardinalfish
Fowleria marmorata Ear Cardinalfish
Fowleria punctulata Peppered Cardinalfish
Fowleria vaiulae Dwarf Cardinalfish
Fowleria variegata Variegated Cardinalfish
Glossamia aprion Mouth Almighty
Gymnapogon annona Naked Cardinalfish
Gymnapogon philippinus Philippine Cardinalfish
Gymnapogon urospilotus Tailspot Cardinalfish
Gymnapogon vanderbilti Vanderbilt's Cardinalfish
Neamia octospina Eightspine Cardinalfish
Nectamia fusca Ghost Cardinalfish
Paxton concilians Paxton's Cardinalfish
Pseudamia amblyuroptera Whitejaw Cardinalfish
Pseudamia gelatinosa Gelatinous Cardinalfish
Pseudamia hayashii Hayashi's Cardinalfish
Pseudamia nigra Estuary Cardinalfish
Pseudamiops gracilicauda Slendertail Cardinalfish
Pterapogon mirifica Sailfin Cardinalfish
Rhabdamia cypselurus Schooling Cardinalfish
Rhabdamia gracilis Slender Cardinalfish
Siphamia argyrogaster Silvermouth Siphonfish
Siphamia cephalotes Wood's Siphonfish
Siphamia cuneiceps Wedgehead Siphonfish
Siphamia fuscolineata Crown-of-thorns Cardinalfish
Siphamia guttulatus Speckled Siphonfish
Siphamia majimae Midnight Siphonfish
Siphamia roseigaster Pinkbreast Siphonfish
Siphamia tubulata Pipe Siphonfish
Siphamia versicolor Urchin Cardinalfish
Siphamia zaribae Striped Siphonfish
Sphaeramia nematoptera Pyjama Cardinalfish
Vincentia badia Scarlet Cardinalfish
Vincentia chrysura Golden Cardinalfish
Vincentia conspersa Southern Cardinalfish
Vincentia macrocauda Smooth Cardinalfish
Vincentia novaehollandiae Eastern Gobbleguts
Vincentia punctata Orange Cardinalfish

DEEPSEA CARDINALFISHES

Epigonidae

BIGEYE DEEPSEA CARDINALFISH
Epigonus lenimen

Deepsea Cardinalfishes are found in the Atlantic, Indian and Pacific oceans, near the bottom from depths of about 300 to 2000 m. There are about 12 species in the family and they are very similar to the Cardinalfishes (Apogonidae, p. 479). They differ in having smaller scales, overall drab brown and black colouring, the lateral line extending onto the caudal fin, and the second dorsal and anal fins usually covered with a scaly sheath at the base. Deepsea Cardinalfishes have two dorsal fins, the first with 6–8 spines and the second with one spine and 8–10 rays. The anal fin has two spines and 8–9 rays.

The **Black Deepsea Cardinalfish** reaches a maximum length of about 55 cm, but most species grow to about half that size. Deepsea Cardinalfishes are mostly pelagic, forming schools and feeding on small fishes and zooplankton. They are widely distributed in Australian waters and form an important part of the diet of larger deepwater predatory fishes. Significant numbers are also taken by deepwater trawl fisheries.

AUSTRALIAN EPIGONIDAE SPECIES

Epigonus denticulatus White Deepsea Cardinalfish
Epigonus lenimen Bigeye Deepsea Cardinalfish
Epigonus macrops Luminous Deepsea Cardinalfish
Epigonus occidentalis Western Deepsea Cardinalfish
Epigonus robustus Robust Deepsea Cardinalfish
Epigonus telescopus Black Deepsea Cardinalfish
Rosenblattia robusta Stout Cardinalfish
Sphyraenops bairdianus Triplespine Deepwater Cardinalfish

WHITINGS

Sillaginidae

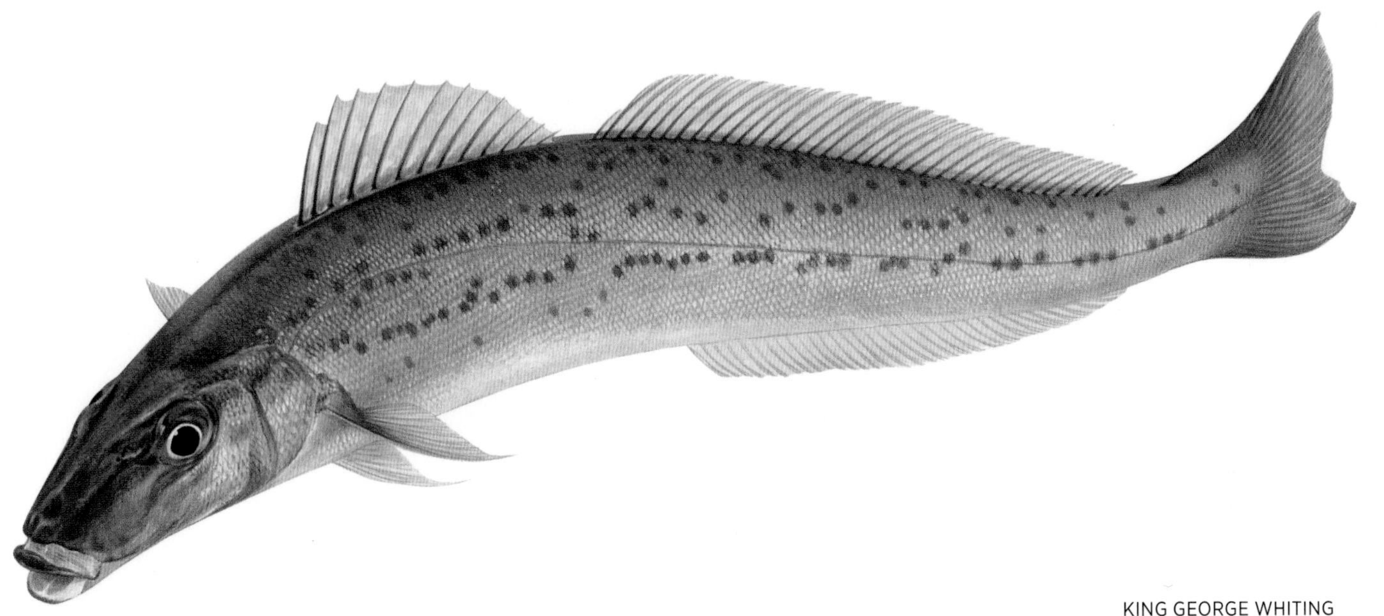

KING GEORGE WHITING
Sillaginodes punctata

Whitings are found in coastal waters of the Indo-West Pacific region, from the intertidal zone down to about 200 m. There are about 31 species in the family. They have an elongate, rounded, tapering body with small scales, and a long, pointed snout with a small, slightly inferior mouth. There are two dorsal fins, the first with 10–13 thin spines and the second with one spine and 16–27 rays. The long-based anal fin has two spines and 14–26 rays.

Whitings are generally found over areas of sand and weed, where they probe the sediment with their sensitive snout in search of small benthic invertebrates such as polychaete worms, molluscs and small crustaceans. They have a complicated swim bladder with distinctive tubes and projections that are thought to aid in sensing vibrations caused by prey items in the substrate. These adaptations also make them very sensitive to sound and they are easily frightened by noise in shallow water.

Whitings are schooling fishes and are common along beaches, in estuaries and over reefs right around the Australian coast. They will often move inshore on rising tides to feed over sand and mudflats in very shallow water, and at times may burrow beneath the sand to escape detection. The **King George Whiting**, found along the southern coast, is one of the largest species, reaching a maximum length of about 70 cm. Like all the Whitings it is a highly sought-after food fish, and it is targeted by commercial and recreational fishers alike, particularly in South Australia. Juvenile King George Whitings inhabit seagrass beds and shallow inshore waters, while larger adults move into deeper waters down to 200 m. Several other species are targeted commercially, including the School Whitings, **Sand Whiting**, **Yellowfin Whiting** and **Trumpeter Whiting**. Whitings are a favourite target of anglers, as they are easily accessible from shore and are superb food fishes. In the summer months they can often be seen along crowded metropolitan beaches, feeding in knee-deep water on invertebrates flushed from the sand by waders.

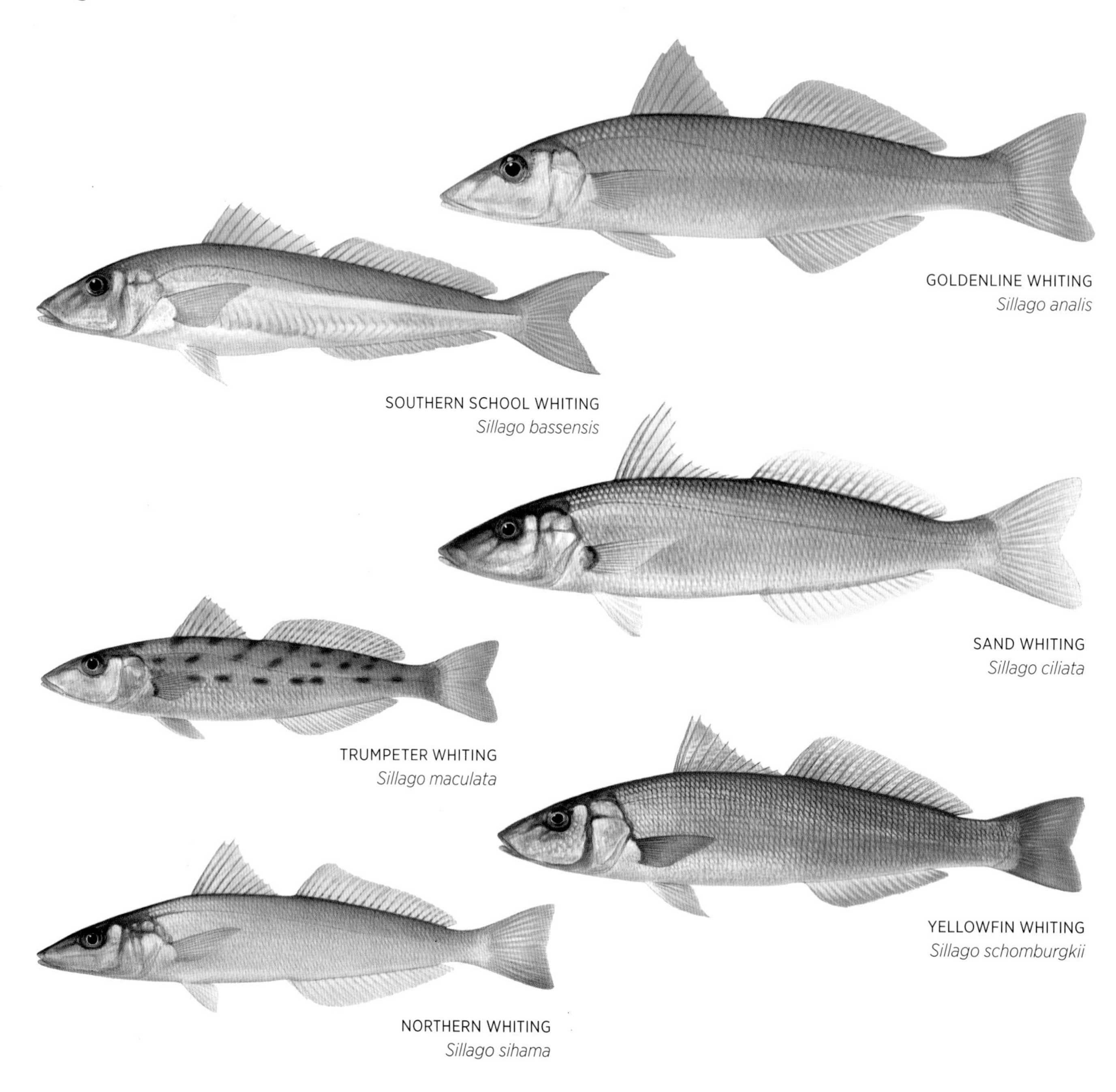

GOLDENLINE WHITING
Sillago analis

SOUTHERN SCHOOL WHITING
Sillago bassensis

SAND WHITING
Sillago ciliata

TRUMPETER WHITING
Sillago maculata

YELLOWFIN WHITING
Sillago schomburgkii

NORTHERN WHITING
Sillago sihama

AUSTRALIAN SILLAGINIDAE SPECIES

Sillaginodes punctata King George Whiting
Sillago analis Goldenline Whiting
Sillago bassensis Southern School Whiting
Sillago burrus Western Trumpeter Whiting
Sillago ciliata Sand Whiting
Sillago flindersi Eastern School Whiting
Sillago ingenuua Bay Whiting
Sillago lutea Mud Whiting
Sillago maculata Trumpeter Whiting
Sillago robusta Stout Whiting
Sillago schomburgkii Yellowfin Whiting
Sillago sihama Northern Whiting
Sillago vittata Western School Whiting

TILEFISHES

Malacanthidae

RED TILEFISH
Branchiostegus japonicus

There are about 40 species of Tilefish and they are found in warmer waters of the Atlantic, Indian and Pacific oceans. They have elongate, rounded to compressed bodies, a single long dorsal fin and a similar long anal fin. Tilefishes can be divided into two subfamilies:

LATILINAE This group includes the *Branchiostegus* species, which have a rounded, blunt head and an elongate, compressed body. The dorsal fin has 6–10 spines, there is a ridge on the dorsal midline anterior to the dorsal fin, and there is no spine on the preoperculum. They are found in deeper coastal waters from 20 to 600 m and are rarely seen. Some are known from only a few specimens, while others appear occasionally in deepwater trawls. They prey on small crustaceans, worms and molluscs.

MALACANTHINAE This group includes the *Hoplolatilus* and *Malacanthus* species, which have an elongate, rounded body, with a rounded or pointed snout. The dorsal fin has 1–10 spines, there is no dorsal ridge before the dorsal fin, and most species have a single enlarged spine on the preoperculum. They are found on steep, rubble-strewn outer reef slopes in northern Australian waters and feed on benthic invertebrates and zooplankton. They construct a burrow, where they shelter, usually in pairs or small groups. As they enlarge their burrow, they deposit the excavated pieces of shell and rubble around the entrance, and large colonies of some species eventually accumulate great mounds of rubble around the burrows they inhabit. Mounds 5 m long, 3 m wide and 1 m high have been reported – a Herculean effort for fishes that are less than 20 cm in length. The mounds themselves become important habitats for other animals, particularly in areas where they are the only hard substrate in a large expanse of sand. The **Green Tilefish** and **Bluehead Tilefish** mostly inhabit barren areas on deep outer-reef slopes, retreating into their burrow if approached, and are not often seen by divers. The **Flagtail Blanquillo** and **Blue Blanquillo** are sometimes encountered in northern Australian waters, swimming over rubble areas near coastal reefs. The Blue Blanquillo is one of the larger and more common species in the family, reaching a maximum length of about 35 cm.

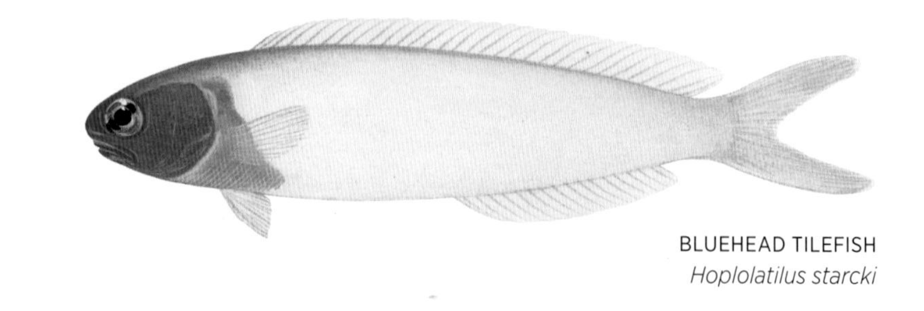

BLUEHEAD TILEFISH
Hoplolatilus starcki

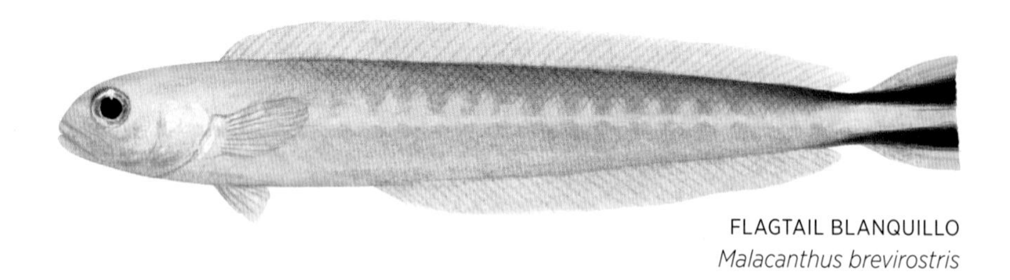

FLAGTAIL BLANQUILLO
Malacanthus brevirostris

BLUE BLANQUILLO
Malacanthus latovittatus

AUSTRALIAN MALACANTHIDAE SPECIES

Branchiostegus australiensis Australian Tilefish
Branchiostegus hedlandensis Port Hedland Tilefish
Branchiostegus japonicus Red Tilefish
Branchiostegus paxtoni Paxton's Tilefish
Branchiostegus sawakinensis Freckled Tilefish
Branchiostegus serratus Australian Barred Tilefish
Branchiostegus wardi Pink Tilefish
Hoplolatilus cuniculus Green Tilefish
Hoplolatilus starcki Bluehead Tilefish
Malacanthus brevirostris Flagtail Blanquillo
Malacanthus latovittatus Blue Blanquillo

FALSE TREVALLY

Lactariidae

FALSE TREVALLY
Lactarius lactarius

The **False Trevally**, the only species in this family, is found in the Indo-West Pacific. It has an oval, compressed body with moderate-sized, fragile scales. It is very similar to the true Trevallies (Carangidae, p. 497), but differs from them in having a broad caudal peduncle that does not have scutes along the lateral line, and in having the second dorsal and anal fins covered with scales. The False Trevally has two dorsal fins, the first with 7–8 spines, and the second with one spine and 19–23 rays. The anal fin has three spines and 25–28 rays.

The False Trevally is found in schools in shallow coastal waters, usually over sand or mud bottoms, down to about 100 m. It grows to a maximum length of about 35 cm and feeds on benthic invertebrates and small fishes. It is an important commercial species in some areas of South-East Asia, but is only occasionally encountered in northern Australian waters.

AUSTRALIAN LACTARIIDAE SPECIES

Lactarius lactarius False Trevally

LONGFIN PIKE

Dinolestidae

LONGFIN PIKE
Dinolestes lewini

The single species in this family is the **Longfin Pike**, which is found only in coastal waters off southern Australia, to depths of about 60 m. It has an elongate, rounded body with small cycloid scales. The large mouth has a series of small teeth along the jaws and large canines anteriorly. It has a deep caudal peduncle and two widely spaced dorsal fins, the first with 4–5 spines, and the second with one spine and 17–19 rays. The anal fin has one spine and 25–28 rays.

The Longfin Pike grows to a maximum length of about 60 cm and is quite common in a variety of habitats in southern waters, from seagrass beds to reefs. In shallow waters it tends to be solitary, but divers often see large aggregations over deeper reefs. The Longfin Pike resembles the Barracudas (Sphyraenidae, p. 711) and like them it preys on other fishes. It differs, however, in having longer-based soft dorsal and anal fins, and is actually more closely related to the Cardinalfishes (Apogonidae, p. 479).

AUSTRALIAN DINOLESTIDAE SPECIES

Dinolestes lewini Longfin Pike

TAILOR
Pomatomidae

TAILOR
Pomatomus saltatrix

The **Tailor** is the only member of this family and is widely distributed around the world in warm temperate coastal waters. It has an elongate, compressed body with small scales, and robust jaws with rows of very sharp teeth. It has two dorsal fins, the first with 8–9 spines and the second with 23–28 rays. The anal fin has three spines and 23–27 rays. The soft dorsal and anal fins have scaly sheaths on their bases.

The Tailor is a schooling fish and is found in coastal waters, usually in areas of high water movement such as coastal reefs and surf beaches, though juveniles are found in estuaries and protected waters. It is a voracious predator of small schooling fishes such as Mullets (Mugilidae, p. 318). In Australia it is widely targeted by recreational anglers, as it is an excellent food fish and will readily attack lures or bait, fighting tenaciously and often leaping clear of the water. When actively feeding it can become extremely aggressive and enter a feeding frenzy, attacking with great ferocity anything that moves, including any bait presented. At other times it will bite quite tentatively.

In North America, where it is known as the Bluefish, the Tailor is a very important commercial species. It is also taken commercially to a limited extent in Queensland and mid-west coast fisheries. In Australian waters Tailor are migratory, moving southward along the west coast in summer, and are eagerly awaited each year by anglers. Occasionally they move all the way around to the south coast, where they will often join schools of Australian Salmons (Arripidae, p. 558). On the east coast, schools move northward to spawning grounds near Fraser Island, Queensland, in late winter and then head southward again in early summer. The species is rare along the south coast and in Tasmanian waters. The Tailor is reported to reach a maximum length of 1.3 m and weight of 14 kg, though any fish over 5 kg is considered a large catch. Their powerful jaws and razor-sharp teeth can inflict a nasty gash, so care should be taken when handling them.

AUSTRALIAN POMATOMIDAE SPECIES
Pomatomus saltatrix Tailor

MAHI MAHI

Coryphaenidae

MAHI MAHI
Coryphaena hippurus

The two species of Mahi Mahi are wide-ranging pelagic fishes, found in tropical and warm temperate oceanic waters around the world. They have an elongate, tapering, compressed body with minute scales, and a blunt, rounded head. The head profile becomes very steep and develops a dorsal hump in large males, which can reach up to 1.5 m in length. They have very long dorsal fins, which originate on the dorsal surface of the head, and long, low anal fins without spines.

Mahi Mahi are pelagic in surface waters, forming large schools when young and becoming more solitary as they grow larger. They occur right around the coast of Australia, usually well offshore, and feed mainly on Flyingfishes (Exocoetidae, p. 332) and Garfishes (Hemiramphidae, p. 334). Mahi Mahi are popular targets for offshore gamefishers, as they take lures, baits and flies, fight hard and leap repeatedly. They are often attracted to floating debris, and in some locations fish-aggregating devices (FADs) – consisting of netting suspended beneath the surface – have been installed to attract them. Upon capture they display brilliant golden colours, dark fins and iridescent blue spots, but they rapidly fade to overall silvery white after death. Mahi Mahi are extremely fast-growing and are excellent food fishes. Aquaculture projects are being developed in Australia and overseas to farm them commercially.

AUSTRALIAN CORYPHAENIDAE SPECIES

Coryphaena equiselis Pompano Mahi Mahi
Coryphaena hippurus Mahi Mahi

COBIA

Rachycentridae

COBIA
Rachycentron canadum

The **Cobia** is the only member of this family and it is found in warm temperate and tropical waters of the Atlantic and Indo-Pacific. It has an elongate, cylindrical body with minute scales, a wide, flattened head and a large mouth with a projecting lower jaw. The dorsal fins consist of 6–9 short, strong, free spines followed by a long-based fin with 1–3 spines and 26–33 rays. The anal fin has 2–3 spines and 23–28 rays. The Cobia is a large, muscular fish, growing to a maximum of about 2 m in length and weighing up to 50 kg.

In Australia it is found in northern coastal waters, usually close to the surface in small groups near reefs but also offshore and sometimes close to the bottom, down to at least 100 m. The Cobia is well known for its habit of closely following Manta Rays (Myliobatidae, p. 136) and large sharks, and in this respect it resembles the closely related Remoras (Echeneidae, p. 496). This association probably serves to camouflage the Cobia's approach to its prey, which is usually fishes, cephalopods and crustaceans. The Cobia is a popular sportfish as it fights tenaciously and is an excellent food fish.

AUSTRALIAN RACHYCENTRIDAE SPECIES
Rachycentron canadum Cobia

REMORAS

Echeneidae

SHARKSUCKER
Echeneis naucrates

There are about eight species of Remora and they are widespread in tropical and temperate waters around the world. Remoras are closely related to the Cobia (Rachycentridae, p. 495) and have a similar elongate, cylindrical body, flattened head and projecting lower jaw. However, the dorsal fin of Remoras is remarkably different. The dorsal-fin spines are modified to form broad lamellae in an oval shape on the head. Muscular contraction of these lamellae creates a powerful suction, by which the Remora attaches itself to large marine animals such as whales, sharks, Marlins (Istiophoridae, p. 723), Manta Rays (Myliobatidae, p. 136) and a variety of other large fishes. Some species, such as the **Whalesucker** and **Marlinsucker**, are, as their names suggest, always found on the same hosts. Others are not host-specific and will attach to any large moving object, including boats and occasionally even divers. The largest species in the family, the **Sharksucker**, can grow to over 1 m in length and is often seen free-swimming over coral reefs. Remoras feed on scraps of food that have been missed or stirred up by their hosts, as well as on any parasitic copepods and other small crustaceans that may infest them.

REMORA
Remora remora

AUSTRALIAN ECHENEIDAE SPECIES

Echeneis naucrates Sharksucker
Phtheirichthys lineatus Louse Fish
Remora australis Whalesucker
Remora brachyptera Spearfish Remora
Remora osteochir Marlinsucker
Remora remora Remora
Remorina albescens White Suckerfish

TREVALLIES

Carangidae

GIANT TREVALLY
Caranx ignobilis

The large family of Trevallies contains about 140 species, which are widely distributed in temperate and tropical waters around the world. There is a great variety of body shapes within the family, from elongate and cylindrical to deep-bodied and very compressed. Trevallies have minute scales, often embedded in the skin, and many species have a series of enlarged scutes along the rear of the lateral line and on the narrow caudal peduncle. They usually have two dorsal fins, the first short-based with 4–8 spines and the second long-based with 17–44 rays. The anal fin usually has two free spines, then one attached spine and 15–39 rays. The caudal fin is deeply forked. Trevallies can be divided into four subfamilies:

TRACHINOTINAE Darts have silvery oval to deep bodies. In Australia they are found mainly in northern coastal waters, usually along sandy beaches in the surf zone and over shallow rocky reefs. They have no scutes along the lateral line. The soft dorsal and anal fins are formed into long lobes, and in the **Common Dart** and **Swallowtail Dart** these may extend as trailing filaments. The **Snubnose Dart** is a shy, deep-bodied, tropical species, sought-after by sportfishers due to its tremendous power and stamina. However, in contrast to many other species of Trevally, Darts are not considered to be good food fishes. They feed mainly on crustaceans such as crabs and shrimps, molluscs, and other benthic invertebrates.

SCOMBEROIDINAE Queenfishes have a long, tapering, compressed body with elongate scales embedded in tough leathery skin, and no scutes along the lateral line. The dorsal fin has 6–7 short free spines before a long-based soft dorsal fin. Posterior to both the soft dorsal fin and the anal fins are a number of semi-detached finlets. Queenfishes have a large mouth and are fast-swimming predators, mainly of other fishes. They are found in shallow coastal waters in northern Australia, down to about 100 m. They occur in small groups and are a favourite of anglers, as they attack lures ferociously and leap repeatedly. The **Giant Queenfish** is the largest of the group and can reach up to 1.2 m in length.

NAUCRATINAE This group includes the **Yellowtail Kingfish**, **Samson Fish**, **Rainbow Runner**, **Pilotfish** and Amberjacks. These all have an elongate, rounded body, with a small first dorsal fin and a long, low second dorsal fin and anal fin. There are no scutes on the lateral line. The Yellowtail Kingfish, Samson Fish and Amberjacks are large, powerful fishes, the largest being the Samson Fish, which can reach up to 1.8 m in length and 55 kg in weight. Both the Samson Fish and the Yellowtail Kingfish are found in southern Australian waters. Juveniles occur inshore and adults in deeper offshore waters, often in large schools. Both species are targeted by anglers due to their immensely powerful fight, though large individuals are not good food fishes. The very similar **Amberjack** is found in more northern waters. The Rainbow Runner is pelagic in offshore waters, but is occasionally encountered over coastal reefs, usually in small schools. The Pilotfish is so named because of its habit of riding the pressure wave in front of large fishes and sharks, where it can feed on scraps from the larger fishes' meals or prey on small invertebrates disturbed by their passage. It occurs worldwide in temperate and tropical waters, and is sometimes found under boats or floating debris.

CARANGINAE Including the Trevallies, Scads, Jack Mackerels and the **Pennantfish**, this is by far the largest and most diverse group in the family. All members possess scutes, developed to varying degrees, along the lateral line and on the caudal peduncle. The **Finny Scad** has very large scutes along the length of the lateral line and is a pelagic schooling species. The other Scads also occur in schools and most have an elongate, rounded body and a number of separate finlets at the rear of the soft dorsal and anal fins. The **Common Jack Mackerel** is an important commercial species in southern Australian waters. It is pelagic in large schools, offshore down to about 400 m. It is caught mainly by purse seine off Tasmania and processed as fish meal for livestock and aquaculture feed. The Trevallies of the genera *Caranx* and *Carangoides* have a deep, oval to elongate body, very small, embedded scales and a deeply forked caudal fin. They are strong swimmers, occurring mostly in midwater or over reefs, and are found right around the coast of Australia. There are numerous species and most are caught by commercial fisheries, either being specifically targeted or taken as incidental catch. Some reach a large size, such as the **Giant Trevally**, which is abundant in northern coastal and offshore waters, and which can reach up to 1.7 m in length and over 30 kg in weight. Like many of the Trevallies, large individuals tend to be solitary, while juveniles occur in schools. Other species are much smaller, such as the **Skipjack Trevally**, which grows to about 25 cm in length. It is found in southern coastal waters, over reefs, along beaches and in estuaries. Almost all Trevallies are excellent food fishes and are keenly sought by anglers. Most occur in schools, a notable example being the **Bigeye Trevally**, which forms spectacular aggregations over deep reefs and bommies, moving slowly during the day and dispersing at night to feed on fishes, crustaceans and cephalopods. Some species, such as the **Fringefin Trevally**, inhabit shallow, turbid waters around the northern coast. Others, such as the **Bluefin Trevally** and **Black Trevally**, are only found in the clear waters of offshore reefs. The spectacular **Diamond Trevally**, which can reach 1.3 m in length, is widely distributed in tropical waters. In juveniles the dorsal and anal fins extend into long, trailing filaments, which are lost in adults.

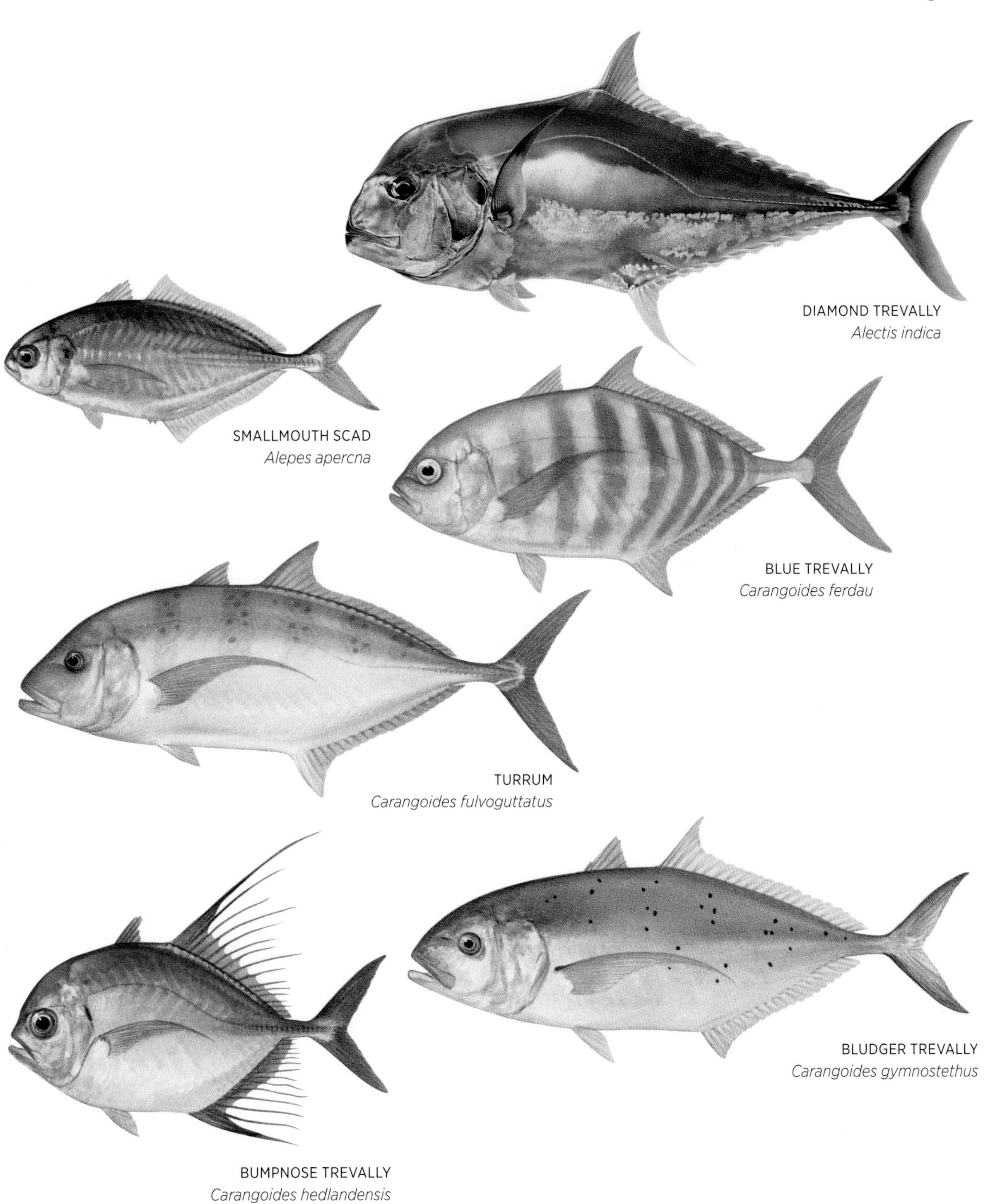

DIAMOND TREVALLY
Alectis indica

SMALLMOUTH SCAD
Alepes apercna

BLUE TREVALLY
Carangoides ferdau

TURRUM
Carangoides fulvoguttatus

BLUDGER TREVALLY
Carangoides gymnostethus

BUMPNOSE TREVALLY
Carangoides hedlandensis

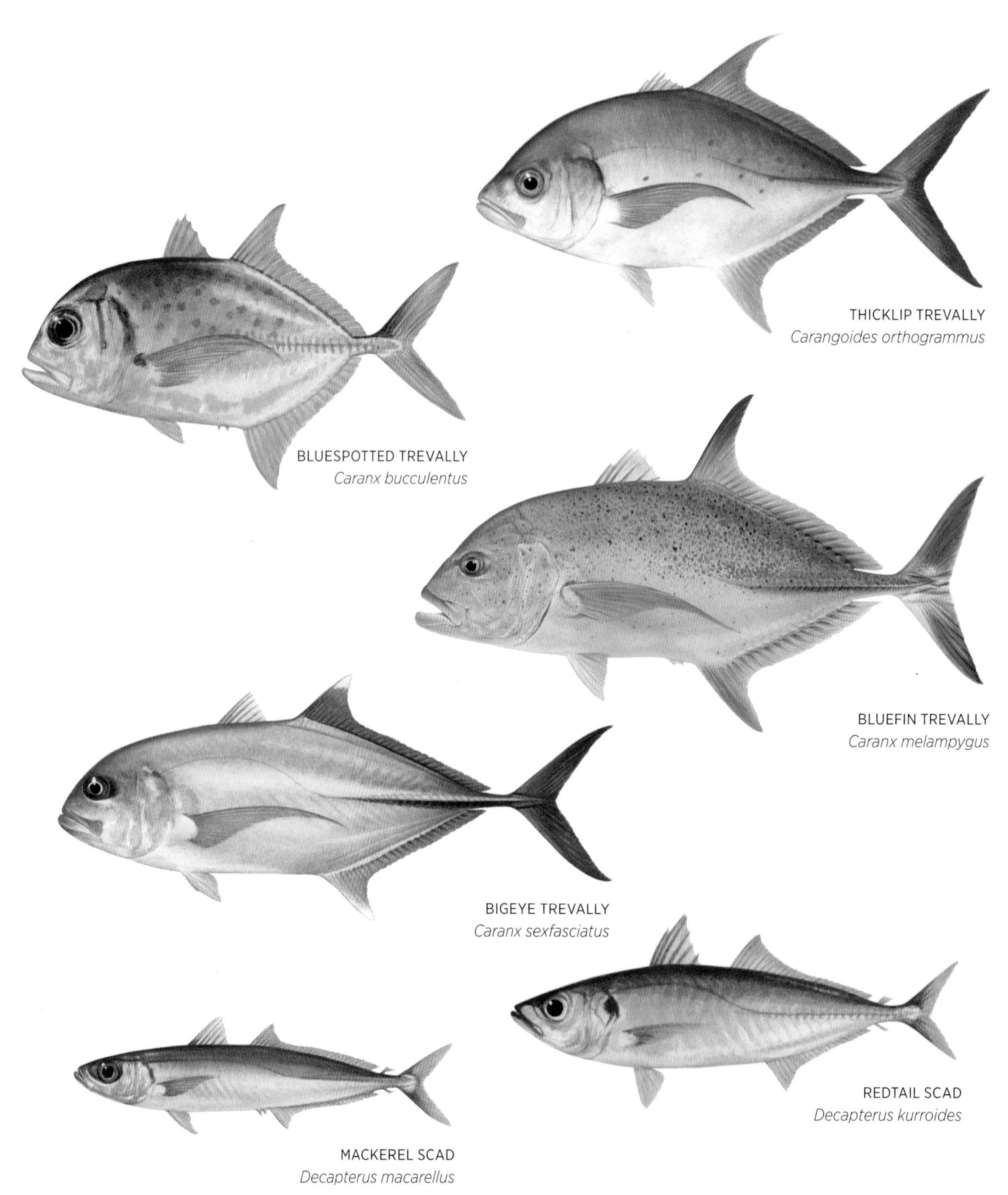

THICKLIP TREVALLY
Carangoides orthogrammus

BLUESPOTTED TREVALLY
Caranx bucculentus

BLUEFIN TREVALLY
Caranx melampygus

BIGEYE TREVALLY
Caranx sexfasciatus

REDTAIL SCAD
Decapterus kurroides

MACKEREL SCAD
Decapterus macarellus

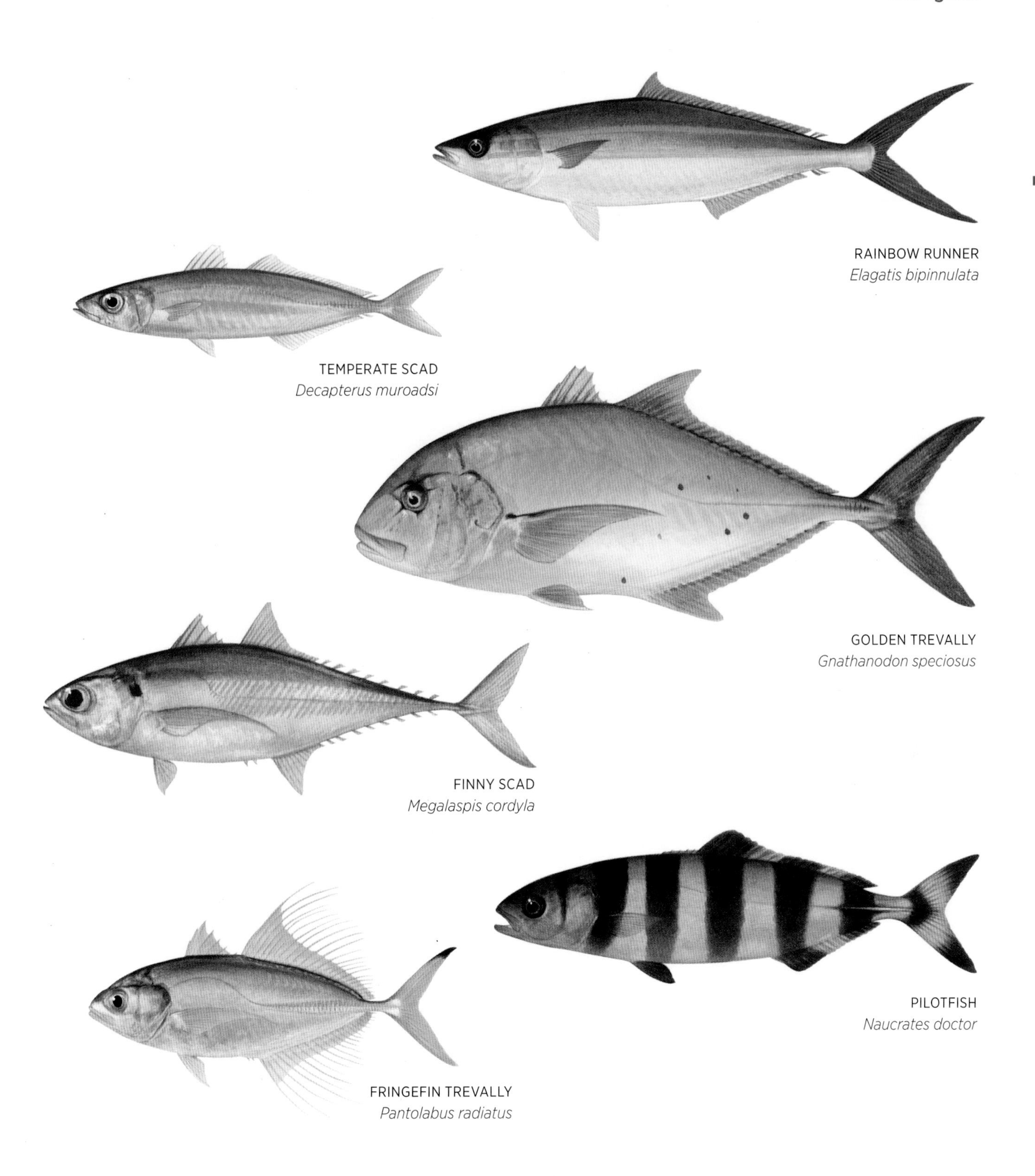

RAINBOW RUNNER
Elagatis bipinnulata

TEMPERATE SCAD
Decapterus muroadsi

GOLDEN TREVALLY
Gnathanodon speciosus

FINNY SCAD
Megalaspis cordyla

PILOTFISH
Naucrates doctor

FRINGEFIN TREVALLY
Pantolabus radiatus

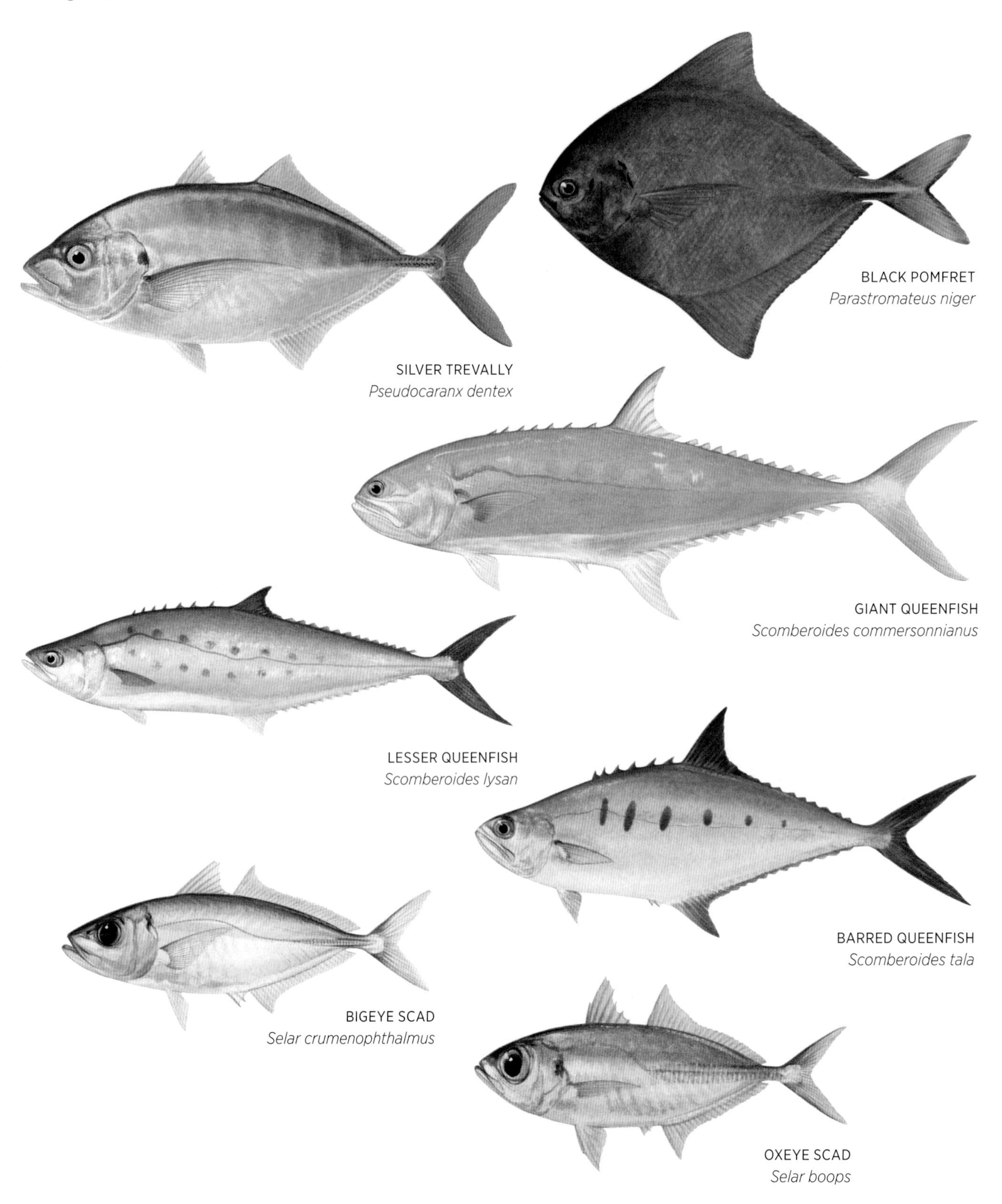

BLACK POMFRET
Parastromateus niger

SILVER TREVALLY
Pseudocaranx dentex

GIANT QUEENFISH
Scomberoides commersonnianus

LESSER QUEENFISH
Scomberoides lysan

BARRED QUEENFISH
Scomberoides tala

BIGEYE SCAD
Selar crumenophthalmus

OXEYE SCAD
Selar boops

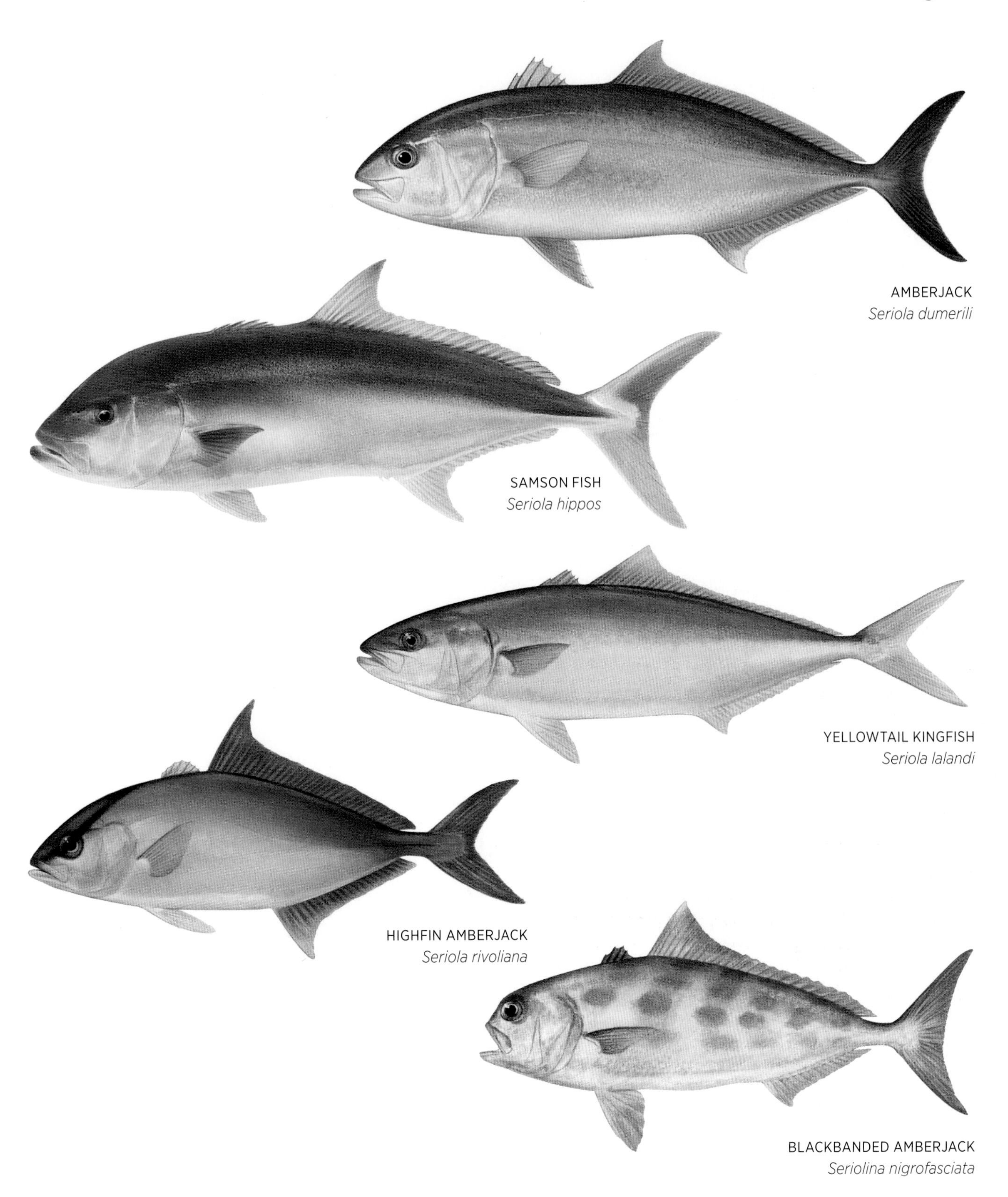

AMBERJACK
Seriola dumerili

SAMSON FISH
Seriola hippos

YELLOWTAIL KINGFISH
Seriola lalandi

HIGHFIN AMBERJACK
Seriola rivoliana

BLACKBANDED AMBERJACK
Seriolina nigrofasciata

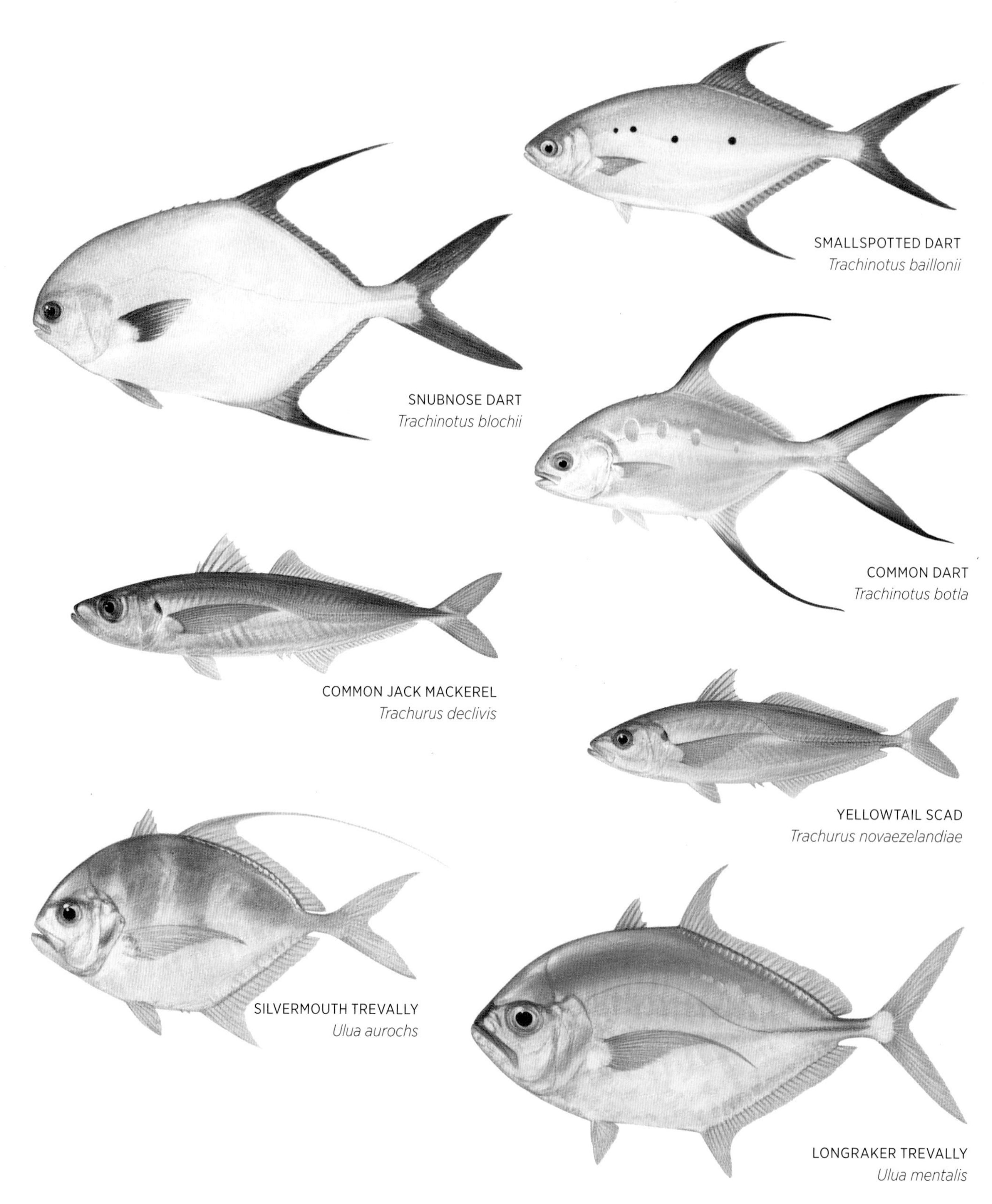

SMALLSPOTTED DART
Trachinotus baillonii

SNUBNOSE DART
Trachinotus blochii

COMMON DART
Trachinotus botla

COMMON JACK MACKEREL
Trachurus declivis

YELLOWTAIL SCAD
Trachurus novaezelandiae

SILVERMOUTH TREVALLY
Ulua aurochs

LONGRAKER TREVALLY
Ulua mentalis

WHITEMOUTH TREVALLY
Uraspis uraspis

AUSTRALIAN CARANGIDAE SPECIES

Alectis ciliaris Pennantfish
Alectis indica Diamond Trevally
Alepes apercna Smallmouth Scad
Alepes vari Herring Scad
Atule mate Barred Yellowtail Scad
Carangoides caeruleopinnatus Onion Trevally
Carangoides chrysophrys Longnose Trevally
Carangoides dinema Shadow Trevally
Carangoides equula Whitefin Trevally
Carangoides ferdau Blue Trevally
Carangoides fulvoguttatus Turrum
Carangoides gymnostethus Bludger Trevally
Carangoides hedlandensis Bumpnose Trevally
Carangoides humerosus Epaulette Trevally
Carangoides malabaricus Malabar Trevally
Carangoides oblongus Coachwhip Trevally
Carangoides orthogrammus Thicklip Trevally
Carangoides plagiotaenia Barcheek Trevally
Carangoides talamparoides Whitetongue Trevally
Caranx bucculentus Bluespotted Trevally
Caranx ignobilis Giant Trevally
Caranx kleinii Razorbelly Trevally
Caranx lugubris Black Trevally
Caranx melampygus Bluefin Trevally
Caranx papuensis Brassy Trevally
Caranx sexfasciatus Bigeye Trevally
Caranx tille Tille Trevally
Decapterus kurroides Redtail Scad
Decapterus macarellus Mackerel Scad
Decapterus macrosoma Slender Scad
Decapterus muroadsi Temperate Scad
Decapterus russelli Indian Scad
Decapterus tabl Rough-ear Scad
Elagatis bipinnulata Rainbow Runner
Gnathanodon speciosus Golden Trevally
Megalaspis cordyla Finny Scad
Naucrates doctor Pilotfish
Pantolabus radiatus Fringefin Trevally
Parastromateus niger Black Pomfret
Pseudocaranx dentex Silver Trevally
Pseudocaranx wrighti Skipjack Trevally
Scomberoides commersonnianus Giant Queenfish
Scomberoides lysan Lesser Queenfish
Scomberoides tala Barred Queenfish
Scomberoides tol Needleskin Queenfish
Selar boops Oxeye Scad
Selar crumenophthalmus Bigeye Scad
Selaroides leptolepis Yellowstripe Scad
Seriola dumerili Amberjack
Seriola hippos Samson Fish
Seriola lalandi Yellowtail Kingfish
Seriola rivoliana Highfin Amberjack
Seriolina nigrofasciata Blackbanded Amberjack
Trachinotus anak Giant Oystercracker
Trachinotus baillonii Smallspotted Dart
Trachinotus blochii Snubnose Dart
Trachinotus botla Common Dart
Trachinotus coppingeri Swallowtail Dart
Trachurus declivis Common Jack Mackerel
Trachurus murphyi Peruvian Jack Mackerel
Trachurus novaezelandiae Yellowtail Scad
Ulua aurochs Silvermouth Trevally
Ulua mentalis Longraker Trevally
Uraspis secunda Cottonmouth Trevally
Uraspis uraspis Whitemouth Trevally

RAZOR MOONFISH

Menidae

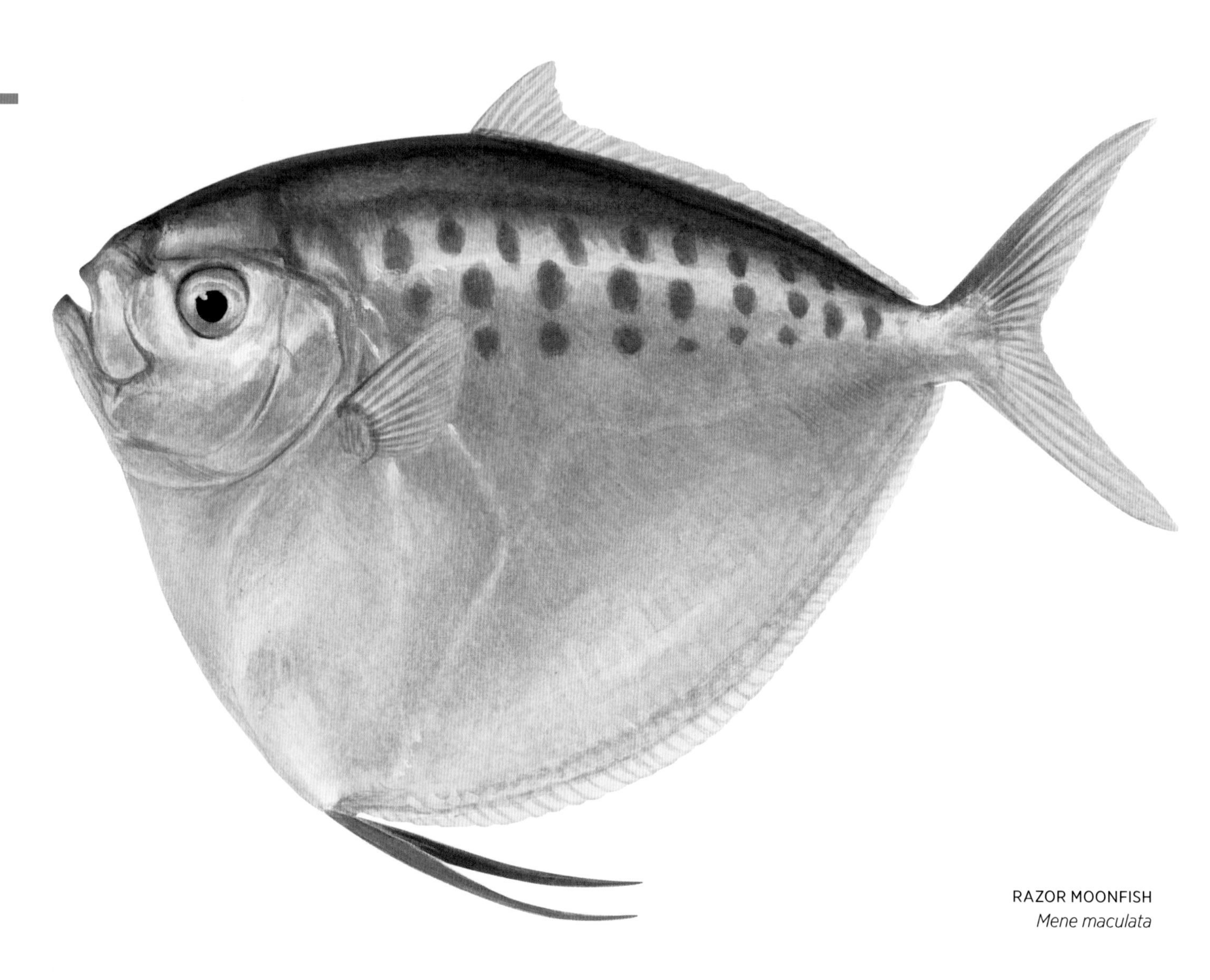

RAZOR MOONFISH
Mene maculata

The **Razor Moonfish**, the only species in this family, is found in the Indo-Pacific, from Africa to Japan. It is a distinctive fish, with a highly compressed, very deep body and an almost straight dorsal profile. Their single long, low dorsal fin has 3–4 spines and 40–45 rays, the anal fin has no spines and 30–33 rays, and the ventral fins have the first ray elongated. It has an extremely protrusible mouth with a broad upper jaw, and grows to about 30 cm in length. The Razor Moonfish forms schools close to the bottom and feeds mainly on small benthic invertebrates. In Australia it is usually found in deeper offshore northern waters.

AUSTRALIAN MENIDAE SPECIES

Mene maculata Razor Moonfish

PONYFISHES

Leiognathidae

COMMON PONYFISH
Leiognathus equulus

Ponyfishes are found in coastal marine and estuarine waters of the Indo-West Pacific, and there are about 30 species in the family. They have an oval, compressed body with slimy skin and minute scales. The single dorsal fin has 8–9 spines, often with some elongated, and 14–16 rays. The anal fin has three spines and 14 rays. The fin spines have a locking mechanism to fix them in a vertical position (making the fish difficult to swallow), as well as a low, scaly sheath around the base into which the fin can be depressed. Ponyfishes have an extremely protrusible mouth and bony ridges on the head. They possess a luminous organ in the throat that is thought to serve as a recognition and schooling signal. Some species have canine teeth in the jaws and feed on small fishes and crustaceans. Others, with an upwardly directed mouth, take zooplankton and phytoplankton from the water column, while those with a downwardly directed mouth take mainly benthic invertebrates.

Ponyfishes occur around northern Australia, in large schools over sand, mud and weed, usually in shallow, turbid coastal waters, tidal creeks and estuaries, though some species occur in deep water down to about 160 m. They are small fishes, the largest in the family being the **Common Ponyfish**, which reaches a maximum length of about 28 cm. However, they are abundant and easily caught with nets in small coastal fisheries, and hence are an important food source in many developing countries.

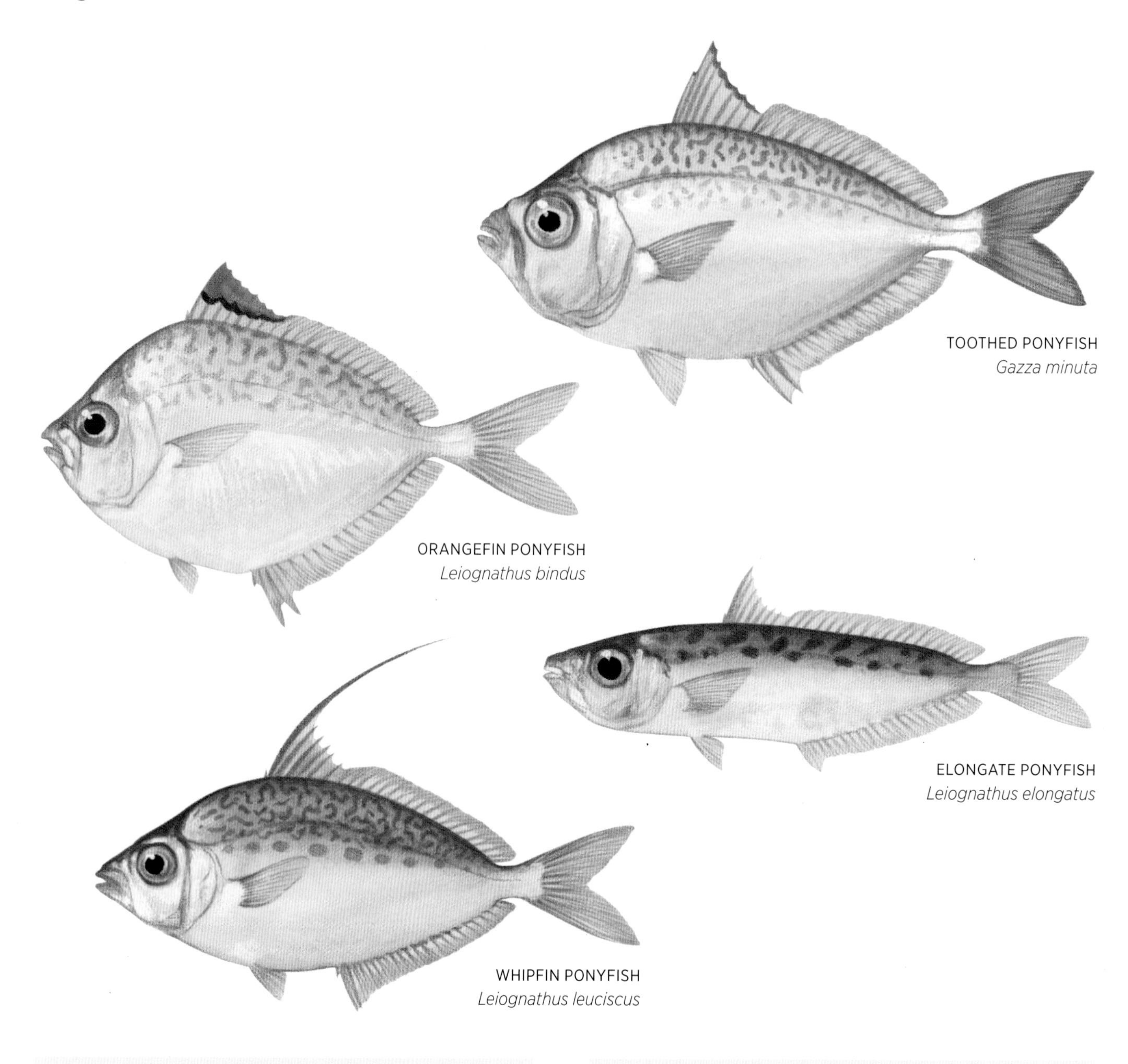

TOOTHED PONYFISH
Gazza minuta

ORANGEFIN PONYFISH
Leiognathus bindus

ELONGATE PONYFISH
Leiognathus elongatus

WHIPFIN PONYFISH
Leiognathus leuciscus

AUSTRALIAN LEIOGNATHIDAE SPECIES

Gazza dentex Ovoid Ponyfish
Gazza minuta Toothed Ponyfish
Gazza rhombea Rhomboid Ponyfish
Leiognathus aureus Golden Ponyfish
Leiognathus bindus Orangefin Ponyfish
Leiognathus blochii Twoblotch Ponyfish
Leiognathus decorus Ornate Ponyfish
Leiognathus elongatus Elongate Ponyfish
Leiognathus equulus Common Ponyfish
Leiognathus fasciatus Threadfin Ponyfish
Leiognathus leuciscus Whipfin Ponyfish
Leiognathus longispinis Longspine Ponyfish
Leiognathus moretoniensis Zigzag Ponyfish
Leiognathus splendens Blacktip Ponyfish
Secutor insidiator Pugnose Ponyfish
Secutor interruptus Deep Pugnose Ponyfish
Secutor megalolepis Bigscale Ponyfish

POMFRETS AND FANFISHES

Bramidae

RAY'S BREAM
Brama brama

There are about 21 species in this family and they are pelagic in offshore waters of the Atlantic, Indian and Pacific oceans. Pomfrets and Fanfishes have oval to elongate, compressed bodies with large scales, and a blunt, rounded head with scales on the upper jaw. Pomfrets of this family are not to be confused with members of the Mondactylidae (p.555), which have a similar body shape and are also known as Pomfrets but are not related.

Fanfishes have large, sail-like dorsal and anal fins that can be depressed into scaly sheaths. They are wide-ranging in oceanic waters and grow to about 50 cm in length. The few specimens that have been recorded have mostly been found stranded or in the stomachs of other fishes.

The Pomfrets, Ray's Breams and **Lesser Bream** are also wide-ranging oceanic species, occurring mainly in the upper 200 m of the water column, but found to at least 1000 m. They have long dorsal and anal fins with the anterior portions elevated or formed into curved lobes, with both fins mostly covered in scales and not depressible. Pomfrets occur in small schools and feed in midwater on pelagic crustaceans, small fishes and cephalopods. Some species are quite large – the **Bigscale Pomfret** can reach 1 m in length – and these are often taken by longline fisheries targeting Tunas (Scombridae, p.717). The Ray's Breams and larger Pomfrets are superb food fishes, though they are not common in Australian markets – allegedly because, of all the bycatch taken by longliners, they are the crew's first choice for eating. There are also several dwarf species of Pomfret found in tropical Pacific waters, which do not grow much larger than 8–10 cm.

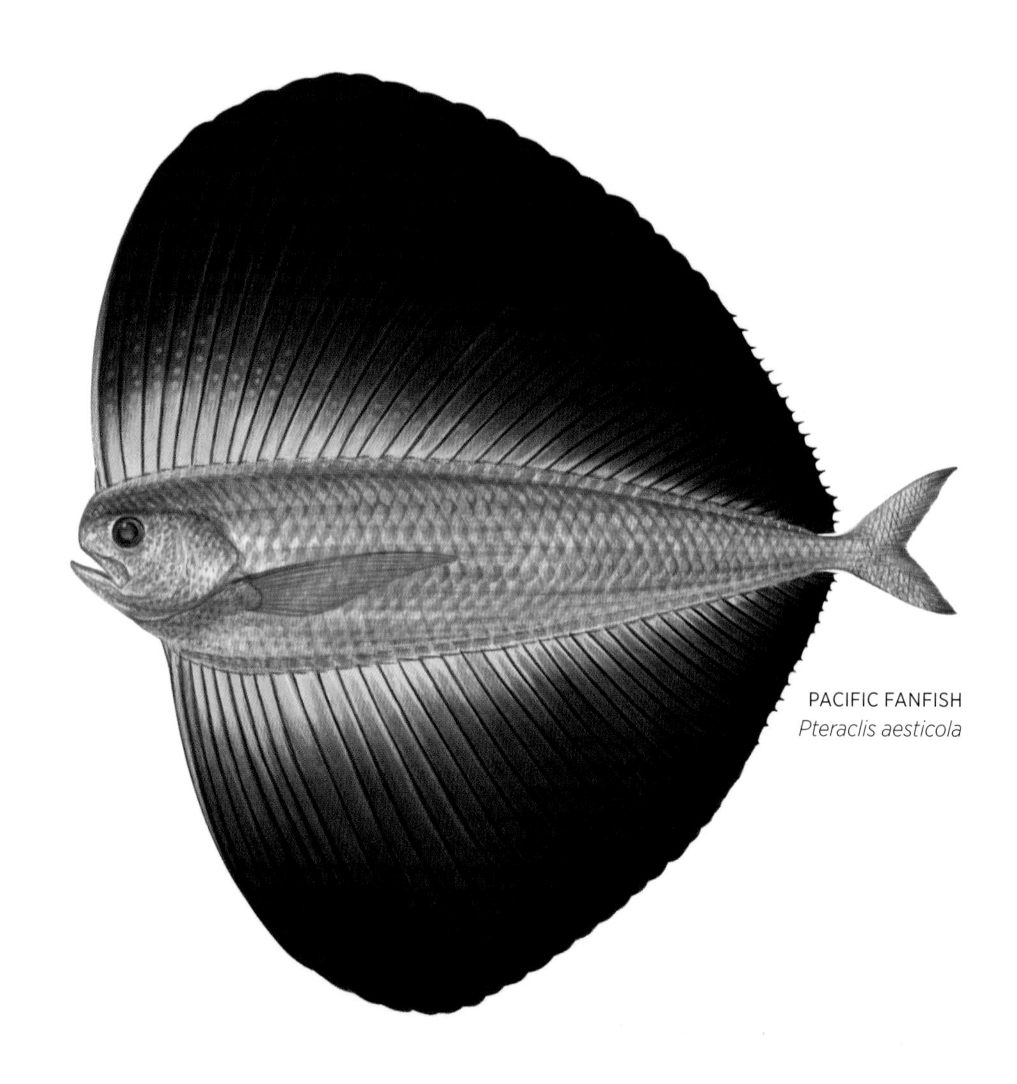

PACIFIC FANFISH
Pteraclis aesticola

AUSTRALIAN BRAMIDAE SPECIES

Brama australis Southern Ray's Bream
Brama brama Ray's Bream
Brama dussumieri Lesser Bream
Brama myersi Myers' Pomfret
Brama orcini Bigbelly Pomfret
Brama pauciradiata Shortfin Pomfret
Eumegistus illustris Brilliant Pomfret
Pteraclis aesticola Pacific Fanfish
Pteraclis velifera Southern Fanfish
Pterycombus petersii Prickly Fanfish
Taractes asper Flathead Pomfret
Taractes rubescens Knifetail Pomfret
Taractichthys longipinnis Bigscale Pomfret
Taractichthys steindachneri Sickle Pomfret
Xenobrama microlepis Golden Pomfret

MANEFISHES

Caristiidae

MANEFISH
Caristius macropus

Caristiidae

Manefishes are distributed throughout the world's oceans, except the North Pacific, and this small family contains about four species. They have a deep, compressed body with fragile, ctenoid scales and a blunt, rounded head with a steep profile and large eyes. The dorsal fin, which originates on the dorsal surface of the head, is large and sail-like, with 28–36 rays. The anal fin is also large, with 17–22 rays, and has a wide sheath of skin over the base. The ventral fins are elongate and fold down into a groove.

The **Manefish** occurs in open waters, from the surface to about 500 m, and is usually found in association with siphonophores (large colonial invertebrates with stinging cells). It reaches at least 32 cm in length and feeds on small fishes and probably zooplankton. It is rarely encountered, but has been recorded from waters off New South Wales, around Tasmania and over the North West Shelf. Very little is known of its biology.

AUSTRALIAN CARISTIIDAE SPECIES

Caristius macropus Manefish

BONNETMOUTHS

Emmelichthyidae

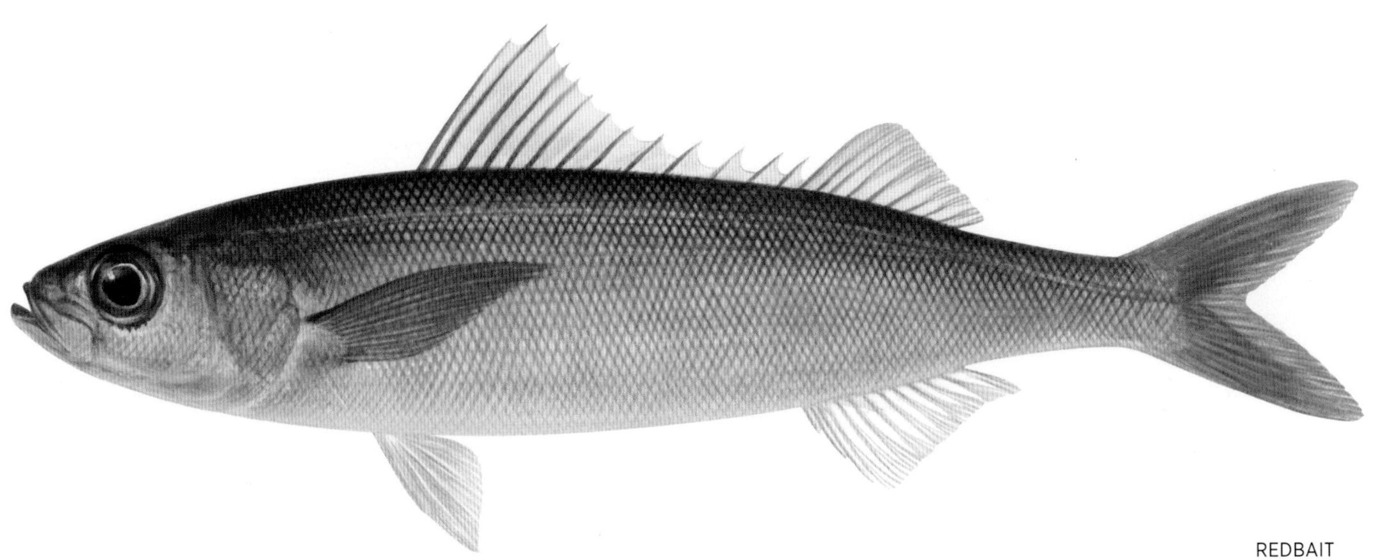

REDBAIT
Emmelichthys nitidus

There are about 15 species of Bonnetmouth and they are found in tropical and warm temperate waters of the Indo-Pacific, South Pacific and eastern Atlantic. They have an elongate, almost cylindrical body with very small ctenoid scales, and two flat spines on the operculum. Their jaws are greatly protrusible and the upper jaw is very broad at the rear and covered with scales. Bonnetmouths have two dorsal fins, the first with 9–13 spines, then 1–3 free spines before the soft dorsal fin, which has 10–12 rays. The anal fin has three spines and 9–10 rays.

Bonnetmouths take their name from the large tube-like extension formed when the jaws protrude. They are usually found close to the bottom, in offshore waters from about 100 to 400 m, though the **Redbait** sometimes occurs in shallower coastal habitats from depths of about 20 to 100 m. Bonnetmouths form schools and feed on small fishes, crustaceans and zooplankton. They reach a maximum length of about 60 cm and are excellent food fishes. The family is widespread but each species seems to have a patchy and quite restricted distribution. It is uncertain whether very similar specimens taken from various locations around the world are of the same or different species.

AUSTRALIAN EMMELICHTHYIDAE SPECIES
Emmelichthys nitidus Redbait
Emmelichthys struhsakeri Golden Redbait
Plagiogeneion macrolepis Bigscale Rubyfish
Plagiogeneion rubiginosus Cosmopolitan Rubyfish

SNAPPERS

Lutjanidae

RED EMPEROR
Lutjanus sebae

Snappers are widely distributed in tropical and warm temperate waters of the Atlantic, Indian and Pacific oceans. The family contains about 105 species, several of which are found in fresh and brackish waters. Many species are important commercially and Snappers are widely utilised as food fishes. They have oval to elongate, compressed bodies with ctenoid scales. Their large mouth usually has well-developed canine teeth, and the upper jaw slips beneath the suborbital bones when the mouth is closed. They have a single dorsal fin, often deeply notched, the first part with 9–12 spines and the second with 9–18 rays. The anal fin has three spines and 7–11 rays, and the ventral fins have a scaly axillary process. Snappers can be divided into four subfamilies:

ETELINAE The Jobfishes and Snappers of the genera *Etelis* and *Pristipomoides* are elongate fishes with deeply forked tails. They are usually associated with deeper reefs in tropical waters and are found down to at least 300 m for the **Ruby Snapper** and **Flame Snapper**. Members of this group prey on fishes, crustaceans and cephalopods, with several species also feeding on zooplankton. They are excellent food fishes and are heavily exploited wherever they occur. The **Goldband Snapper**, found throughout the Indo-Pacific region, is a typical example. Slow-growing and easily caught by line, it is very susceptible to overfishing. The **Green Jobfish** is frequently encountered over reefs in northern Australian waters and is a fast-swimming predator, mainly of fishes.

APSILINAE This group includes Snappers of the genus *Paracaesio*. They have an oval body and are usually found in deep water from 50 to 300 m, over rocky reefs. They feed mainly on zooplankton, as well as on small fishes and crustaceans, and are occasionally caught by deepwater line fishers. **Tang's Snapper** is very similar to the *Lutjanus* species, but has a slightly inferior lower jaw and a fleshy, protruding upper lip. It is found over rocky bottoms from about 90 to 350 m in northern Australian waters, where it feeds on small fishes and invertebrates. It is widespread, though not common, in the Indo-West Pacific.

PARADICHTHYINAE The two species in this group are the **Sailfin Snapper** and the **Chinamanfish**. Both are commonly encountered on northern Australian reefs, where they prey on small fishes, molluscs and crustaceans. The Sailfin Snapper reaches about 60 cm in length and the young are magnificently decorated, with horizontal blue and yellow stripes, and dorsal-fin rays elongated into arched, trailing filaments. Young Chinamanfish also have elongated dorsal fin-rays, but in both species these are lost as the fish matures. The Chinamanfish, which can reach up to 1 m in length, is often a carrier of the toxin that causes ciguatera poisoning, and so should not be eaten.

LUTJANINAE This is the largest group in the family and contains species of the *Lutjanus*, *Pinjalo* and *Macolor* genera. Members of the latter two genera are zooplanktivorous fishes, often found in schools over steep outer-reef slopes and dropoffs. The *Lutjanus* genus contains about 64 species worldwide, with most Australian species occurring over reefs in northern coastal waters. They are strong-swimming predators, feeding nocturnally on fishes, crustaceans, cephalopods and other invertebrates, and often occur in large schools. They have a large mouth and strong jaws with well-developed conical teeth, and some species have large canines. The lower rear edge of the preoperculum always has a shallow to deep notch and the tail is slightly forked to truncate.

Many *Lutjanus* species are important food fishes, though several, including the **Red Bass** and **Paddletail**, are known to often carry ciguatoxin and thus should not be eaten. The **Mangrove Jack** is widespread around the northern half of Australia and juveniles are occasionally carried further south. Young fish inhabit mangroves and estuaries, and occasionally lower reaches of freshwater streams. Adults move into deeper waters over reefs and can reach a maximum length of about 1.2 m. The Mangrove Jack is a popular target for recreational fishers wherever it occurs. The **Golden Snapper** is another species sought-after by anglers across the north of Australia, appreciated for its hard fighting and fine eating qualities. The **Red Emperor** is also heavily targeted by commercial and recreational fishers, on deeper reefs along the north-east and north-west coasts, and is one of the finest of all food fishes in Australian waters. The **Saddletail Snapper** and **Crimson Snapper** are also important commercial species in Australia, and are caught mainly over reefs off the Queensland coast. Many other species are frequently encountered by divers on northern reefs, either in small groups, as with the **Fiveline Snapper**, **Moses' Snapper** and **Stripey Snapper**, or in large schools, as with the **Bluestriped Snapper**. Some species are more solitary, such as the **Maori Snapper**, which is occasionally seen over outer-reef slopes.

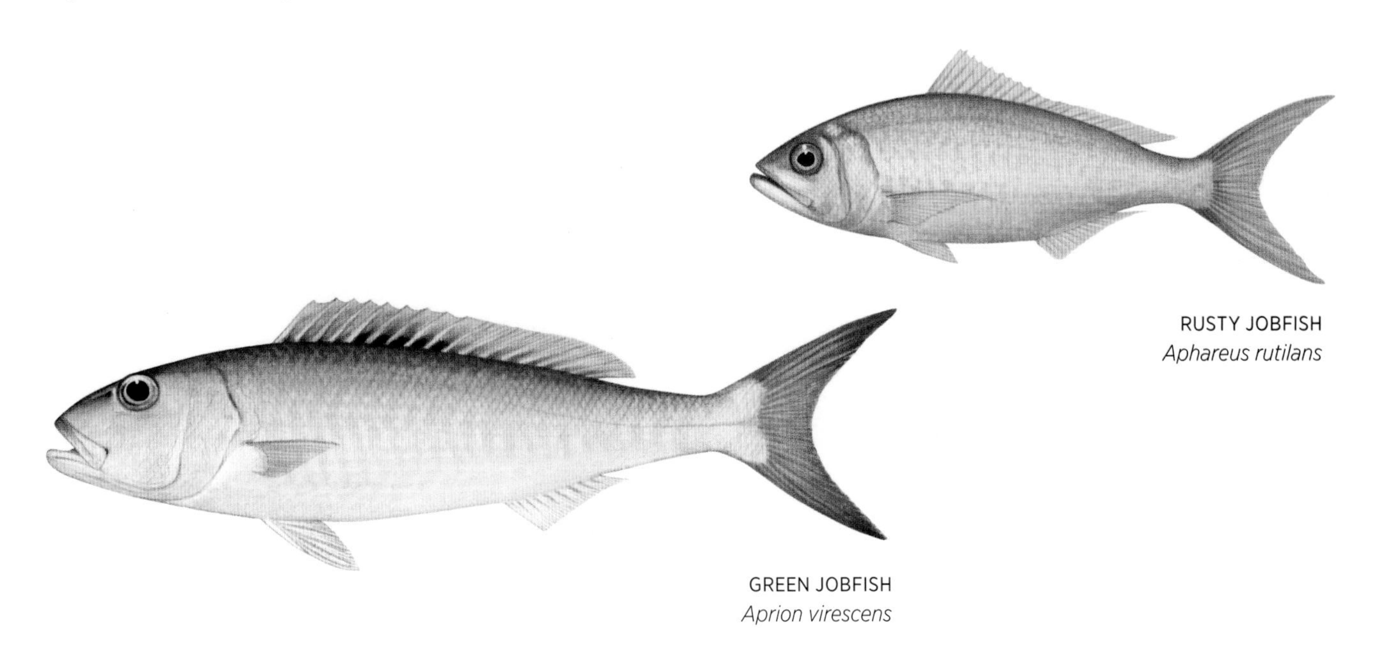

RUSTY JOBFISH
Aphareus rutilans

GREEN JOBFISH
Aprion virescens

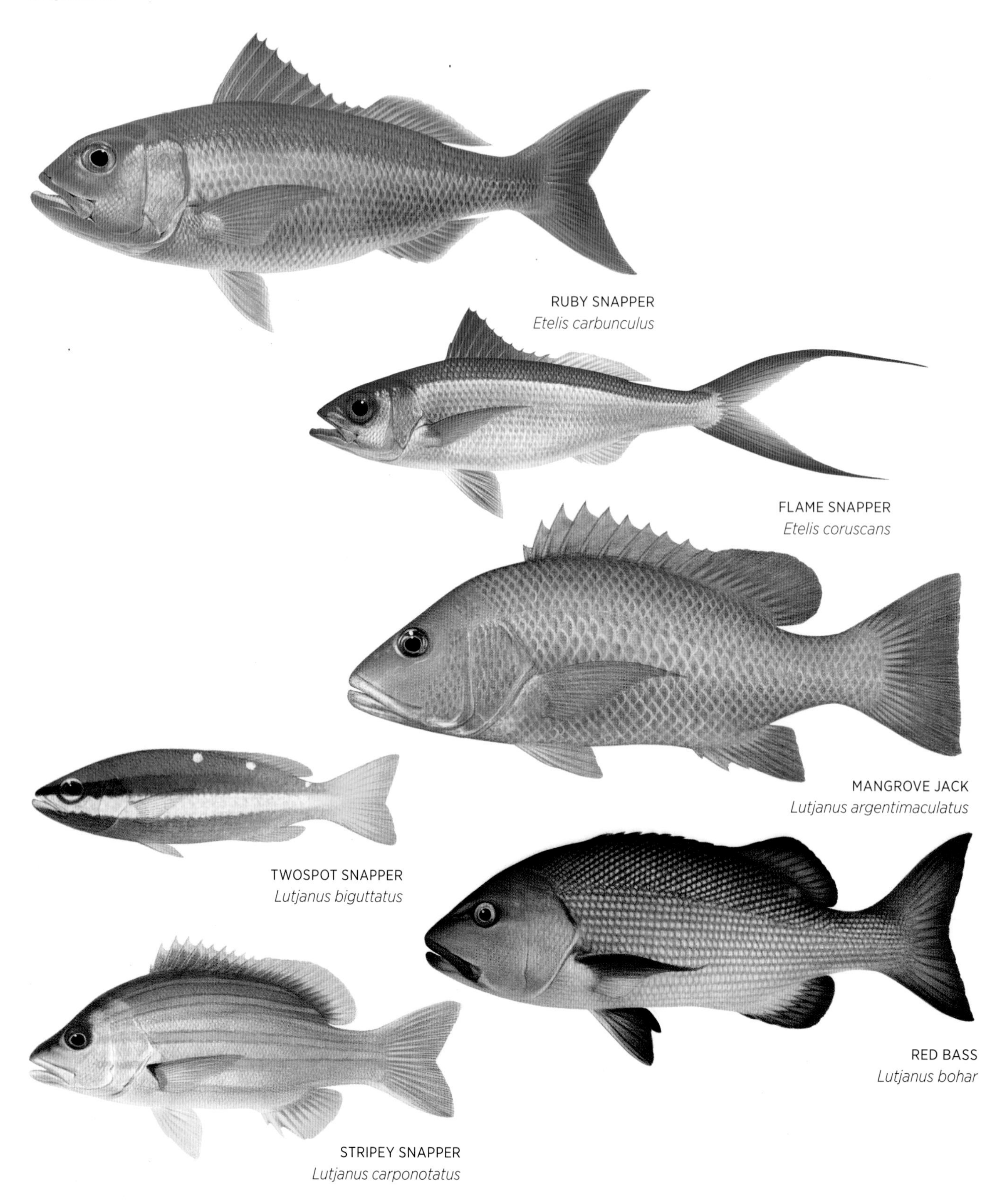

RUBY SNAPPER
Etelis carbunculus

FLAME SNAPPER
Etelis coruscans

MANGROVE JACK
Lutjanus argentimaculatus

TWOSPOT SNAPPER
Lutjanus biguttatus

RED BASS
Lutjanus bohar

STRIPEY SNAPPER
Lutjanus carponotatus

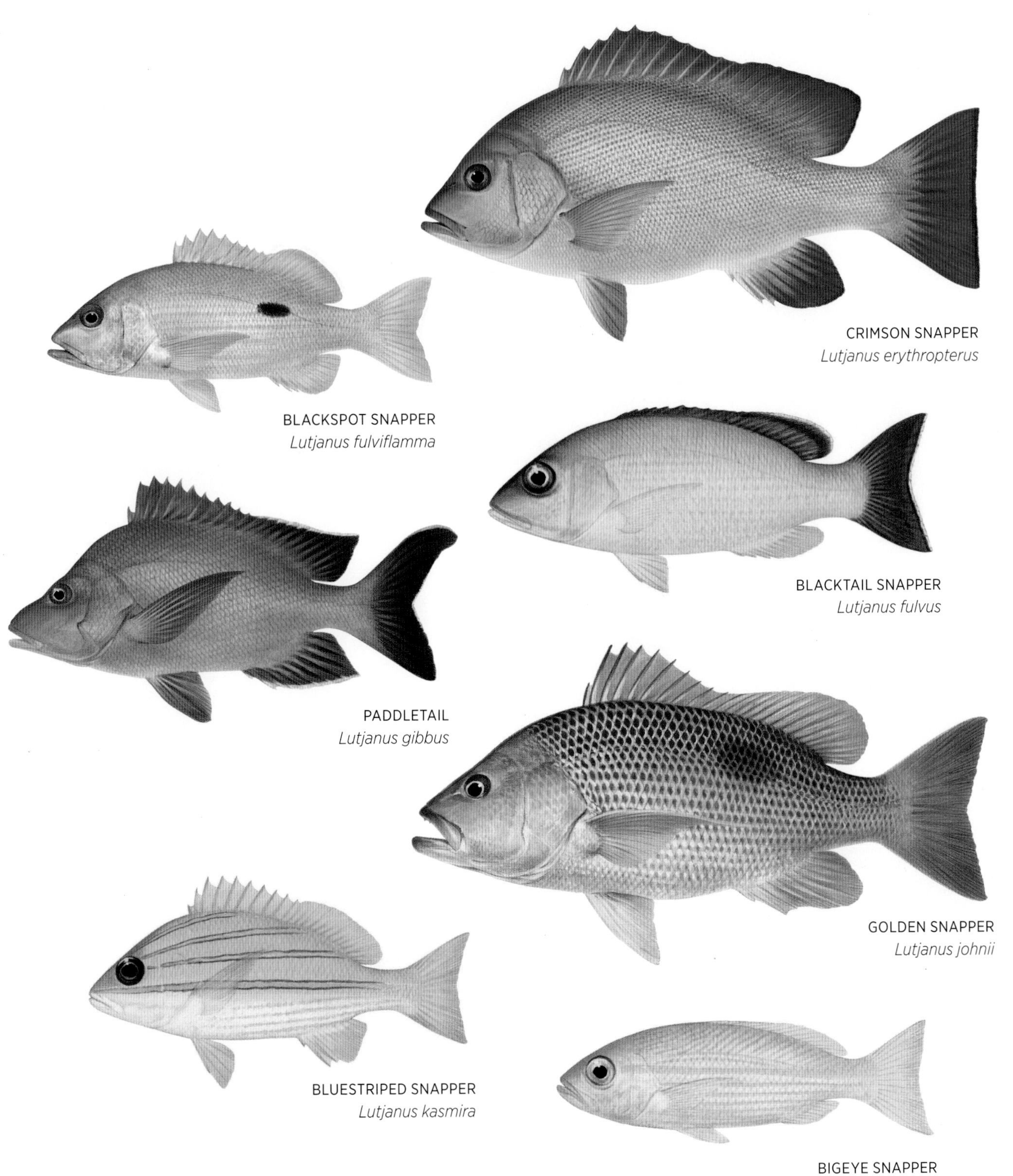

CRIMSON SNAPPER
Lutjanus erythropterus

BLACKSPOT SNAPPER
Lutjanus fulviflamma

BLACKTAIL SNAPPER
Lutjanus fulvus

PADDLETAIL
Lutjanus gibbus

GOLDEN SNAPPER
Lutjanus johnii

BLUESTRIPED SNAPPER
Lutjanus kasmira

BIGEYE SNAPPER
Lutjanus lutjanus

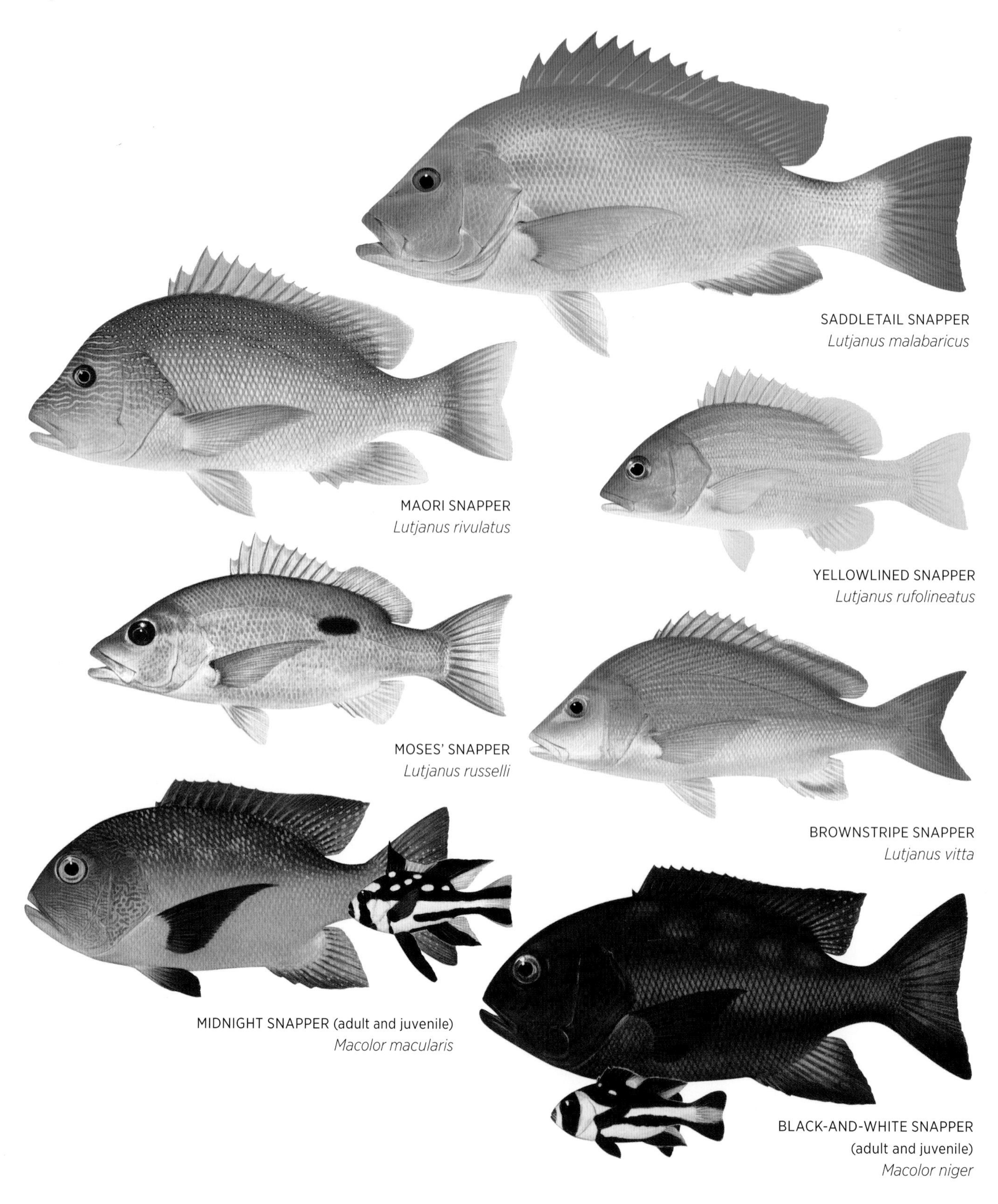

SADDLETAIL SNAPPER
Lutjanus malabaricus

MAORI SNAPPER
Lutjanus rivulatus

YELLOWLINED SNAPPER
Lutjanus rufolineatus

MOSES' SNAPPER
Lutjanus russelli

BROWNSTRIPE SNAPPER
Lutjanus vitta

MIDNIGHT SNAPPER (adult and juvenile)
Macolor macularis

BLACK-AND-WHITE SNAPPER
(adult and juvenile)
Macolor niger

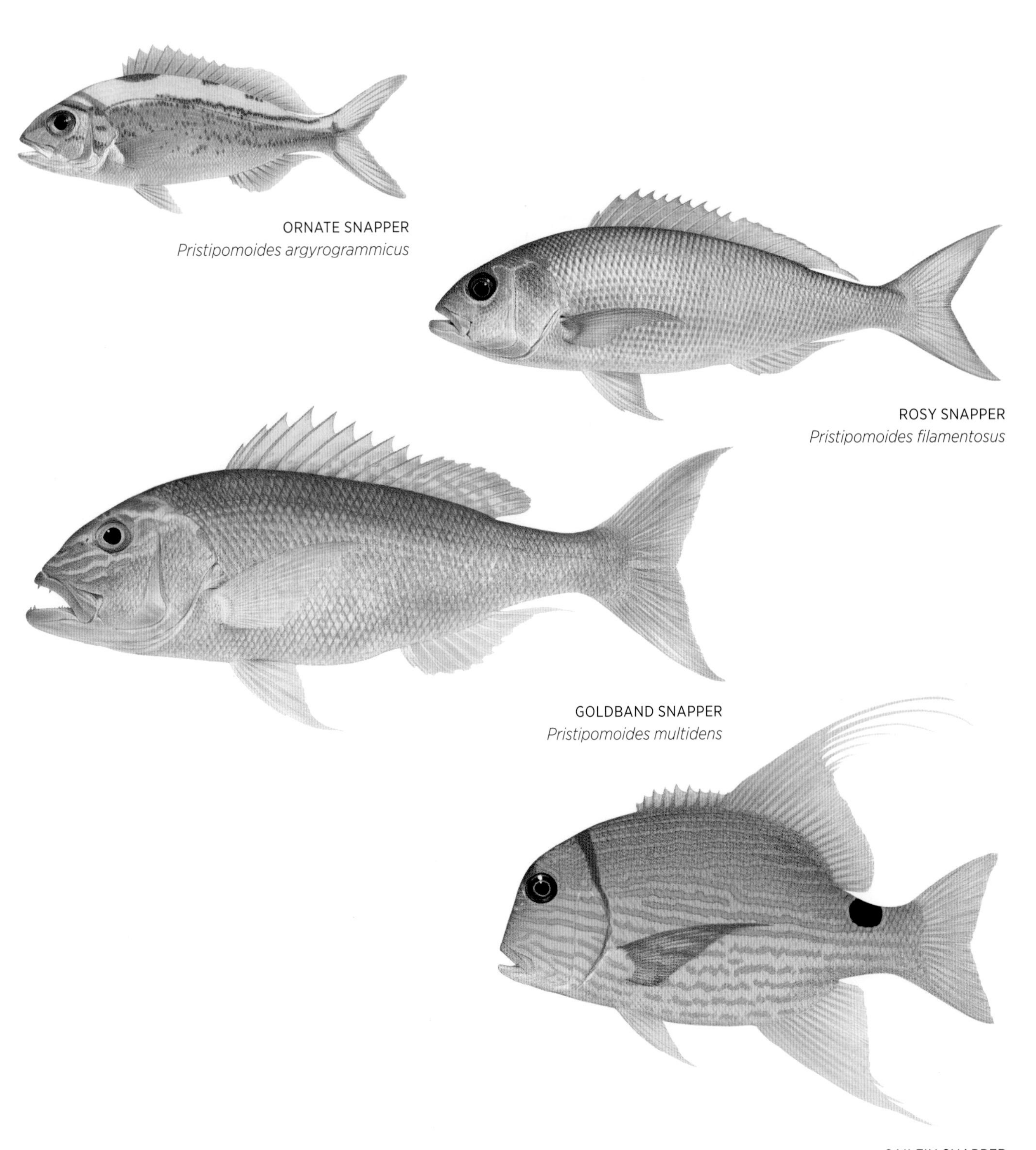

ORNATE SNAPPER
Pristipomoides argyrogrammicus

ROSY SNAPPER
Pristipomoides filamentosus

GOLDBAND SNAPPER
Pristipomoides multidens

SAILFIN SNAPPER
Symphorichthys spilurus

CHINAMANFISH
Symphorus nematophorus

AUSTRALIAN LUTJANIDAE SPECIES

Aphareus furca Smalltooth Jobfish
Aphareus rutilans Rusty Jobfish
Aprion virescens Green Jobfish
Etelis carbunculus Ruby Snapper
Etelis coruscans Flame Snapper
Etelis radiosus Pale Ruby Snapper
Lipocheilus carnolabrum Tang's Snapper
Lutjanus adetii Hussar
Lutjanus argentimaculatus Mangrove Jack
Lutjanus biguttatus Twospot Snapper
Lutjanus bitaeniatus Indonesian Snapper
Lutjanus bohar Red Bass
Lutjanus carponotatus Stripey Snapper
Lutjanus decussatus Checkered Snapper
Lutjanus ehrenbergii Ehrenberg's Snapper
Lutjanus erythropterus Crimson Snapper
Lutjanus fulviflamma Blackspot Snapper
Lutjanus fulvus Blacktail Snapper
Lutjanus gibbus Paddletail
Lutjanus johnii Golden Snapper
Lutjanus kasmira Bluestriped Snapper
Lutjanus lemniscatus Darktail Snapper
Lutjanus lutjanus Bigeye Snapper
Lutjanus malabaricus Saddletail Snapper
Lutjanus monostigma Onespot Snapper
Lutjanus quinquelineatus Fiveline Snapper
Lutjanus rivulatus Maori Snapper
Lutjanus rufolineatus Yellowlined Snapper
Lutjanus russelli Moses' Snapper
Lutjanus sebae Red Emperor
Lutjanus semicinctus Blackbanded Snapper
Lutjanus timoriensis Timor Snapper
Lutjanus vitta Brownstripe Snapper
Macolor macularis Midnight Snapper
Macolor niger Black-and-white Snapper
Paracaesio gonzalesi Vanuatu Snapper
Paracaesio kusakarii Saddleback Snapper
Paracaesio sordida Sordid Snapper
Paracaesio stonei Cocoa Snapper
Paracaesio xanthura False Fusilier
Pinjalo lewisi Red Pinjalo
Pristipomoides argyrogrammicus Ornate Snapper
Pristipomoides auricilla Goldflag Snapper
Pristipomoides filamentosus Rosy Snapper
Pristipomoides flavipinnis Goldeneye Snapper
Pristipomoides multidens Goldband Snapper
Pristipomoides sieboldii Lavender Snapper
Pristipomoides typus Sharptooth Snapper
Pristipomoides zonatus Oblique-banded Snapper
Symphorichthys spilurus Sailfin Snapper
Symphorus nematophorus Chinamanfish

FUSILIERS

Caesionidae

BLUE FUSILIER
Caesio teres

Fusiliers are found in tropical waters of the Indo-West Pacific and there are about 20 species in the family. They have elongate, rounded to slightly compressed bodies with small ctenoid scales. The small mouth has highly protrusible jaws and minute conical teeth. The single dorsal fin has 10–15 spines and 8–22 rays, the anal fin has three spines and 9–13 rays, and the caudal fin is strongly forked.

Fusiliers feed on zooplankton and occur in large schools over reefs and dropoffs in northern Australian waters. During the day they form large feeding aggregations in midwater, often with several species occurring together, and at night they retreat to the shelter of the reef. I have watched an enormous feeding aggregation of up to five different species, forming a dancing cloud of fishes over a large coral bommie. At the high-speed approach of a large Spanish Mackerel (Scombridae, p.717), they separated almost instantly into schools of the same species, then returned to the mixed aggregation as soon as the Mackerel departed. Fusiliers are abundant in many parts of their range and are important food items for larger predatory fishes. They are also important commercially and are taken as food fishes in areas across much of the western Pacific, though not generally in Australia.

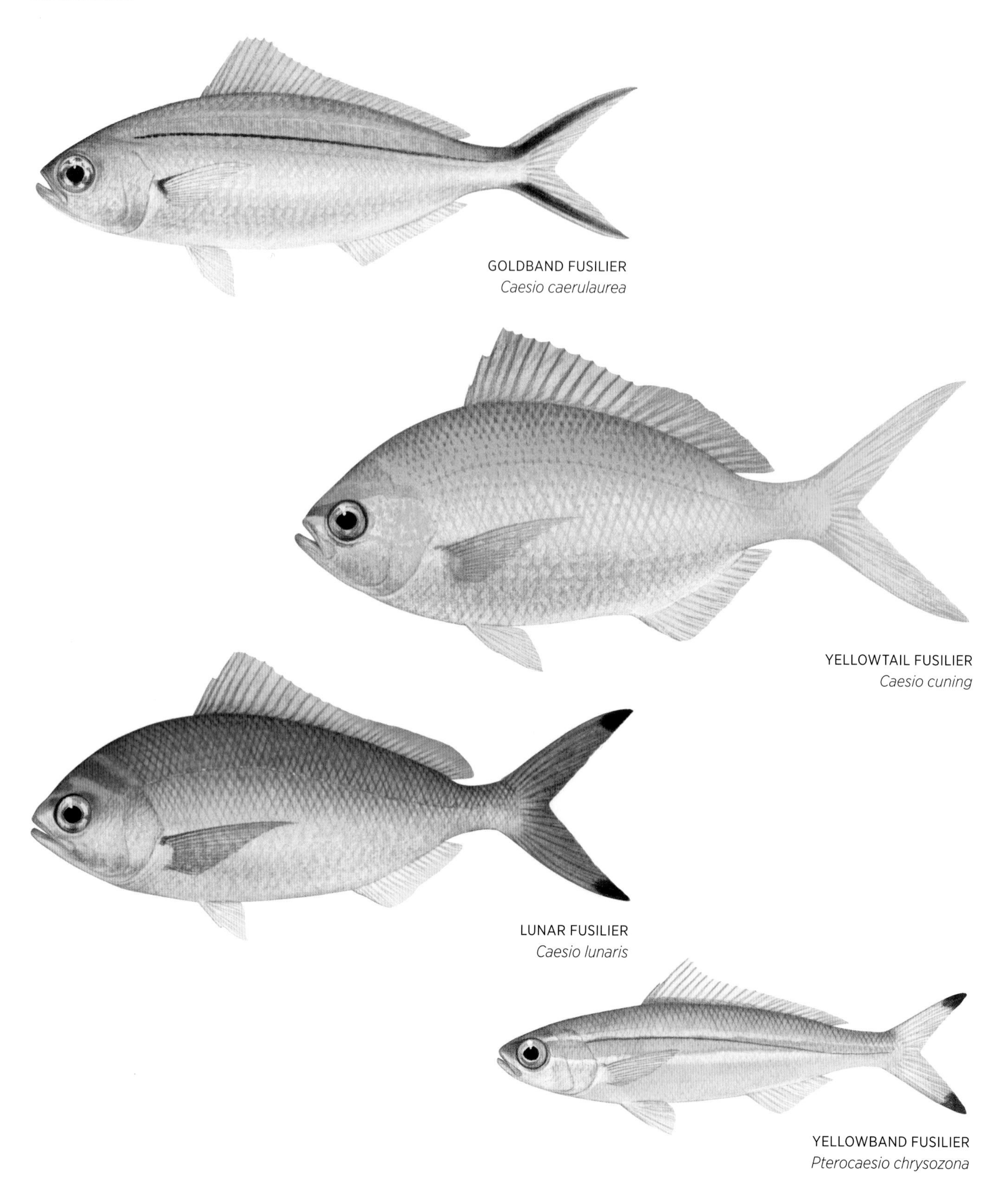

GOLDBAND FUSILIER
Caesio caerulaurea

YELLOWTAIL FUSILIER
Caesio cuning

LUNAR FUSILIER
Caesio lunaris

YELLOWBAND FUSILIER
Pterocaesio chrysozona

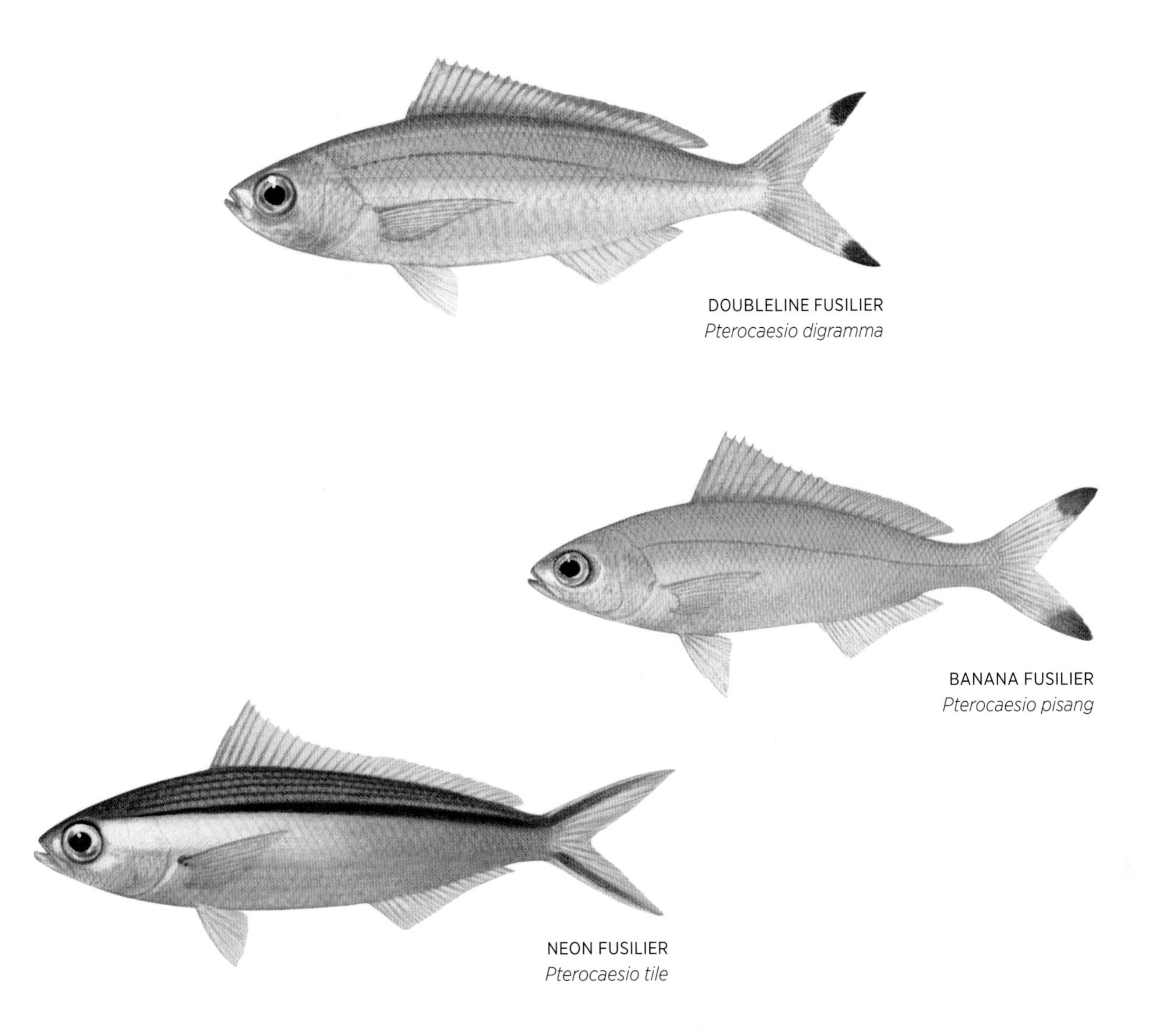

DOUBLELINE FUSILIER
Pterocaesio digramma

BANANA FUSILIER
Pterocaesio pisang

NEON FUSILIER
Pterocaesio tile

AUSTRALIAN CAESIONIDAE SPECIES

Caesio caerulaurea Goldband Fusilier
Caesio cuning Yellowtail Fusilier
Caesio lunaris Lunar Fusilier
Caesio teres Blue Fusilier
Dipterygonotus balteatus Mottled Fusilier
Pterocaesio chrysozona Yellowband Fusilier
Pterocaesio digramma Doubleline Fusilier
Pterocaesio marri Bigtail Fusilier
Pterocaesio pisang Banana Fusilier
Pterocaesio tile Neon Fusilier
Pterocaesio trilineata Threestripe Fusilier

TRIPLEFINS

Lobotidae

TRIPLETAIL
Lobotes surinamensis

Triplefins are found in tropical and subtropical waters around the world, with several species occurring in brackish and freshwater habitats. There are about eight species in the family, but the **Tripletail** is the only one found in Australian waters. It has a deep, heavy-set body with moderate-sized ctenoid scales and a large mouth with rows of small, closely set canines. The edge of the preoperculum is deeply serrated. The single dorsal fin has 11–12 strong spines and 15–16 rays, and the anal fin has three spines and 11–12 rays. The dorsal, anal and caudal fins are all large with broadly rounded rear margins, giving the species its common name.

The Tripletail grows to about 1 m in length, and is found in coastal waters and estuaries in northern Australia. Juveniles, which are mottled brown, can be found in estuaries and shallow bays, where they may float on their sides, imitating dead leaves. The Tripletail feeds mainly on benthic invertebrates and small fishes. It is occasionally found in offshore waters around floating debris and is considered an excellent food fish.

AUSTRALIAN LOBOTIDAE SPECIES
Lobotes surinamensis Tripletail

SILVERBIDDIES

Gerreidae

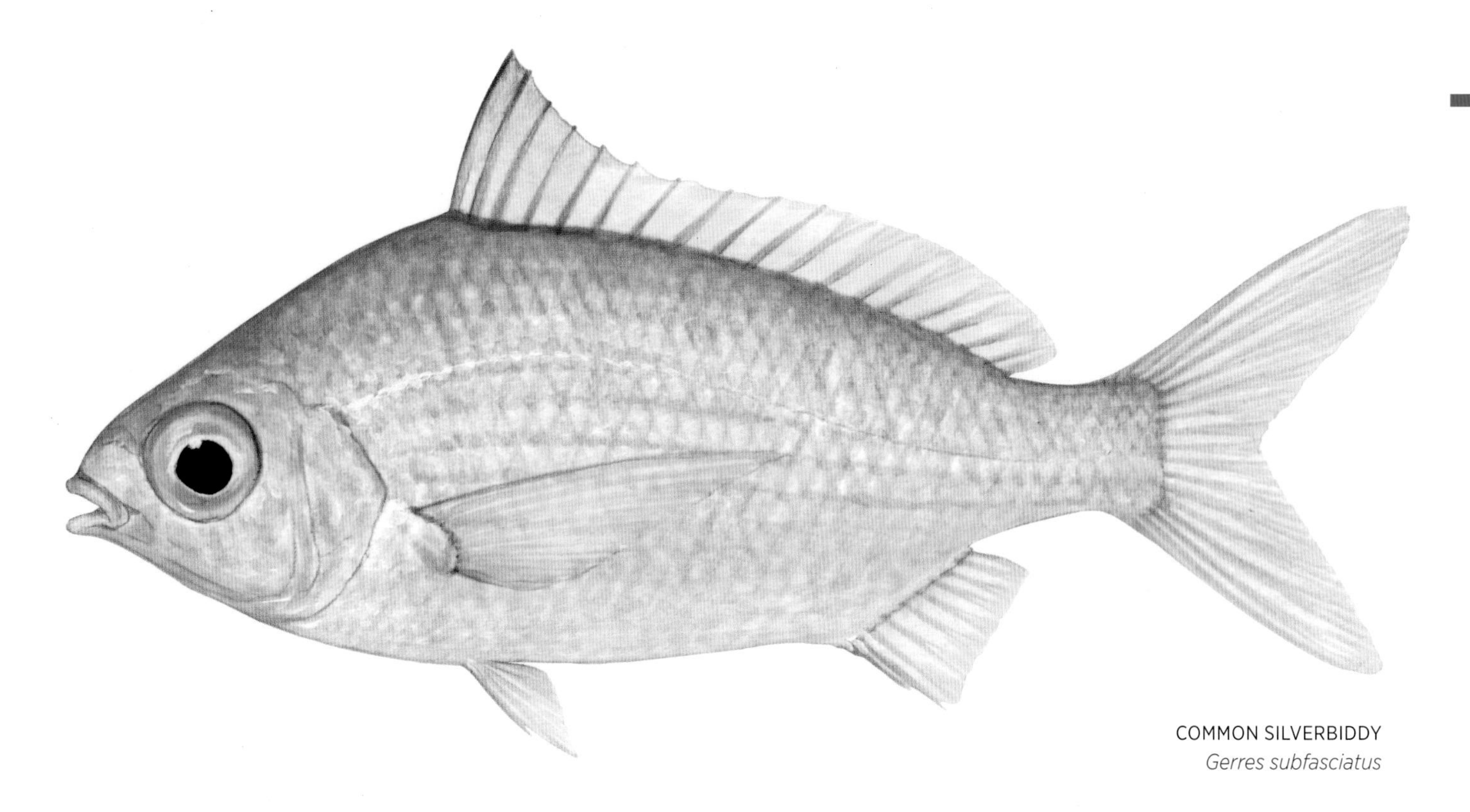

COMMON SILVERBIDDY
Gerres subfasciatus

Silverbiddies are found worldwide in tropical and warm temperate waters. There are about 44 species in the family, several of which are found in fresh and brackish waters. Silverbiddies have an oval, compressed body with large eyes, fragile silvery cycloid or finely ctenoid scales, and a small mouth that is protrusible downwards. They have a single dorsal fin with 9–10 spines, with the anterior spines often elongated, and 9–17 rays. The anal fin has 2–6 spines and 6–18 rays, the pectoral fins are long and pointed, and the caudal fin is forked. The dorsal and anal fins can be lowered into a scaly sheath at their base.

Silverbiddies are small schooling fishes found in shallow coastal bays and estuaries, and occasionally in the lower reaches of freshwater streams. They feed on benthic invertebrates, such as worms and small crustaceans, by taking mouthfuls of sediment with their downwardly directed, tubelike mouth. In Australia Silverbiddies are common in estuaries and protected bays around the north coast, also occurring over sandy bottoms near reefs and in tidal creeks. Most species are found in tropical waters, but the **Silverbelly** is found in southern waters. The Silverbelly reaches about 18 cm in length, and forms small schools over seagrass and sand or mud bottoms, from the shallows down to about 100 m. Silverbiddies are used as bait for catching other fishes and are only occasionally used as food fishes. The largest species reach about 25 cm in length.

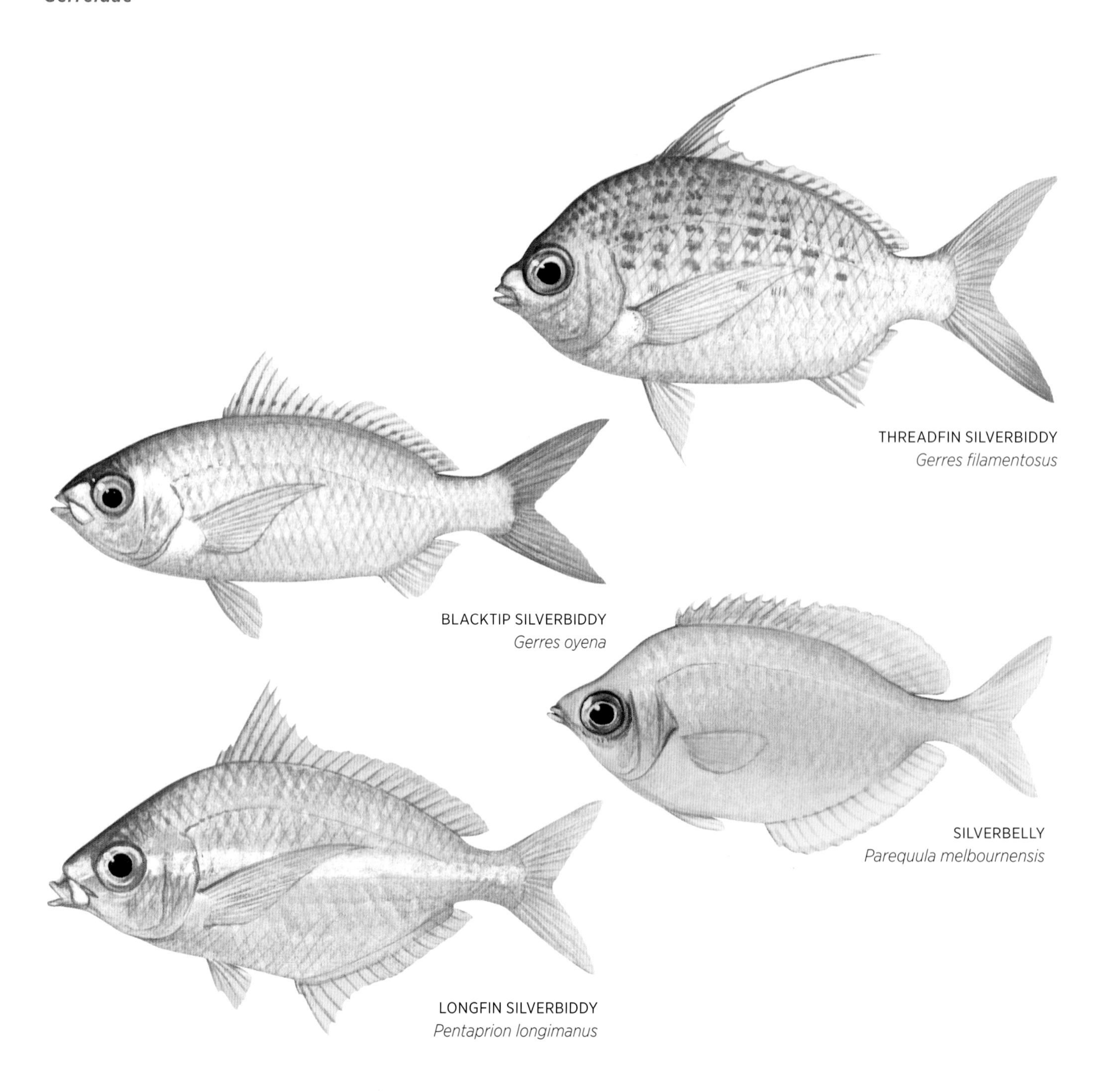

THREADFIN SILVERBIDDY
Gerres filamentosus

BLACKTIP SILVERBIDDY
Gerres oyena

SILVERBELLY
Parequula melbournensis

LONGFIN SILVERBIDDY
Pentaprion longimanus

AUSTRALIAN GERREIDAE SPECIES

Gerres erythrourus Short Silverbiddy
Gerres filamentosus Threadfin Silverbiddy
Gerres kapas Cotton Silverbiddy
Gerres longirostris Strongspine Silverbiddy
Gerres macracanthus Longspine Silverbiddy
Gerres oblongus Slender Silverbiddy
Gerres oyena Blacktip Silverbiddy
Gerres subfasciatus Common Silverbiddy
Parequula melbournensis Silverbelly
Pentaprion longimanus Longfin Silverbiddy

SWEETLIPS AND JAVELINS

Haemulidae

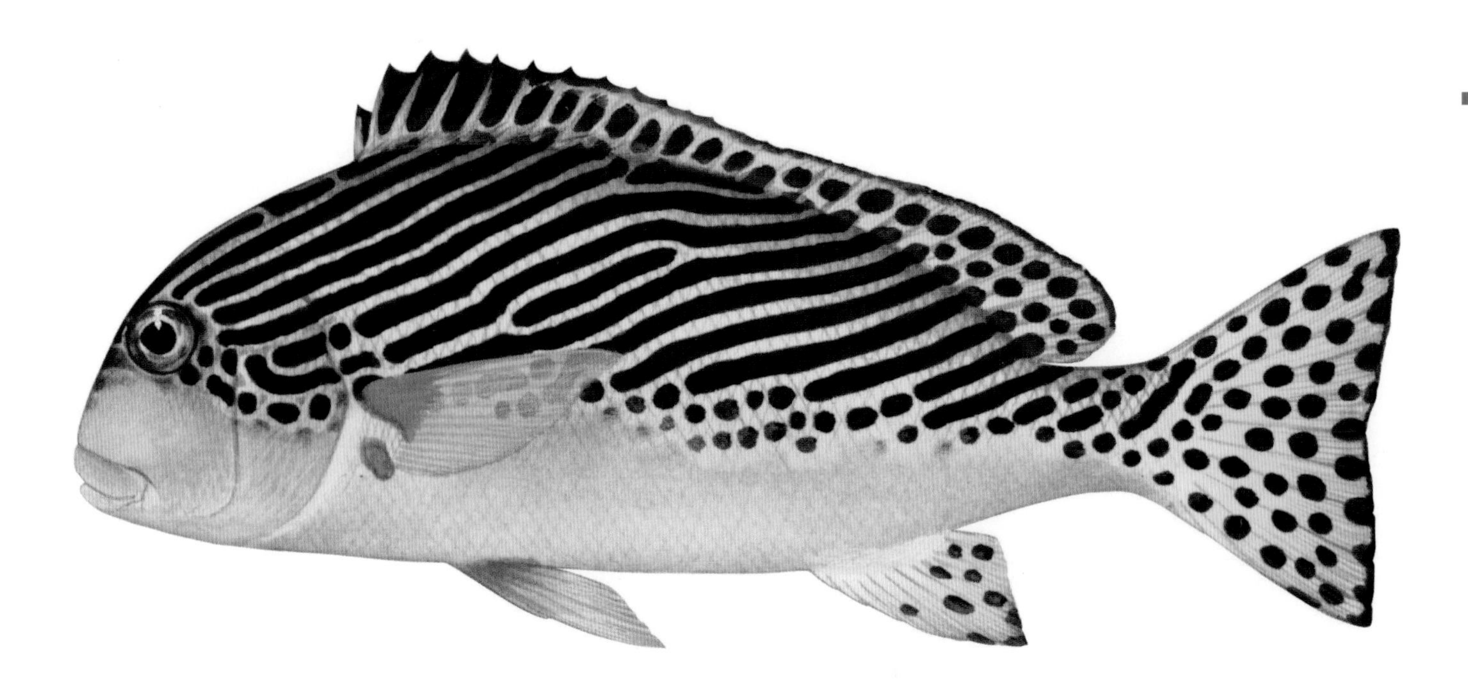

OBLIQUE-BANDED SWEETLIPS
Plectorhinchus lineatus

Sweetlips and Javelins are found in coastal and brackish waters of the Atlantic, Indian and Pacific oceans and there are about 145 species in the family. They have an oval, compressed body with a tall back, small scales, and a small mouth, often with thick fleshy lips. There is a single dorsal fin with 9–14 spines and 11–26 rays. The anal fin has a short base with three spines, the second spine often long and robust, and 6–18 rays. The caudal fin is square to slightly forked in shape.

Sweetlips are similar to Snappers (Lutjanidae, p. 514) in general appearance, differing in having fleshy lips, and scales on the snout and beneath the eyes. They vary greatly in colour between species, from an overall dull silver to strongly patterned with bands or spots, often changing remarkably from juvenile to adult. Juveniles of many species are very brightly coloured and some mimic toxic flatworms or nudibranchs to deter predators. Sweetlips are common over rocky and coral reefs, mainly in northern Australian waters, though the **Gold-spotted Sweetlips** is found in south-western waters and off New South Wales. Some species, such as the **Oblique-banded Sweetlips**, form large schools, though most are solitary or occur in small groups. Sweetlips feed, mainly nocturnally, on benthic invertebrates such as crabs and molluscs, and possess grinding pharyngeal teeth to crush their hard-shelled prey. During the day they are often seen sheltering under overhanging coral or ledges, or in stationary schools. Larger species can reach up to 1 m in length and most are considered excellent food fishes. They are easily approached under water and as a consequence can be overfished by spearing.

Javelins are usually silver, often marked with dark spots or bars. They frequent tropical coastal and offshore waters over sandy and muddy bottoms, and can tolerate the turbid, muddy waters of northern Australian estuaries, shallow bays and tidal creeks. The **Barred Javelin**, which reaches about 80 cm in length, is widespread in the Indo-West Pacific and is a popular target for recreational anglers in northern Australian waters. Like most Javelins, it is an excellent food fish.

PAINTED SWEETLIPS (adult and juvenile)
Diagramma pictum

SPOTTED SWEETLIPS (adult and juvenile)
Plectorhinchus chaetodonoides

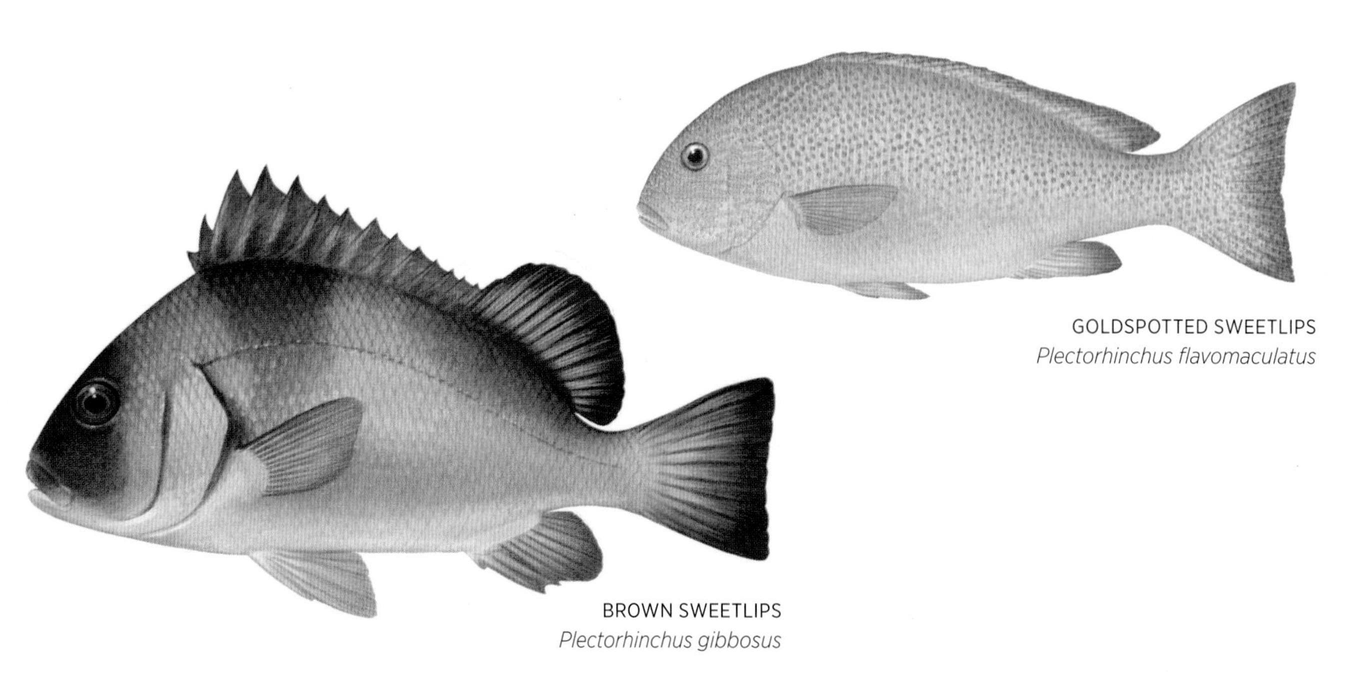

GOLDSPOTTED SWEETLIPS
Plectorhinchus flavomaculatus

BROWN SWEETLIPS
Plectorhinchus gibbosus

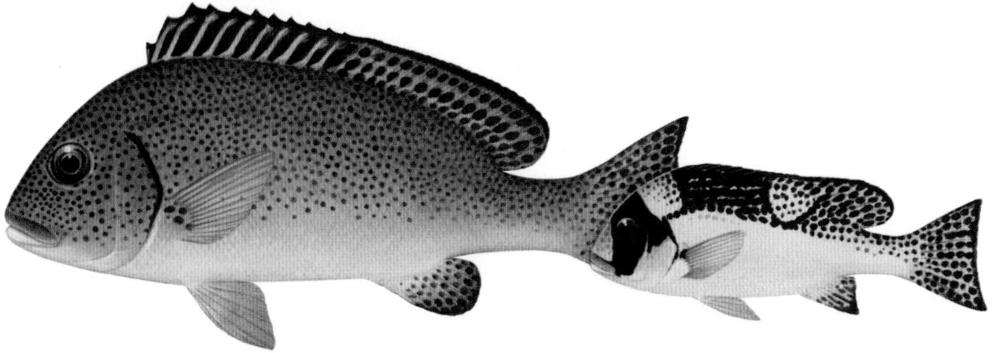

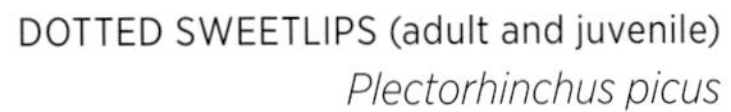

DOTTED SWEETLIPS (adult and juvenile)
Plectorhinchus picus

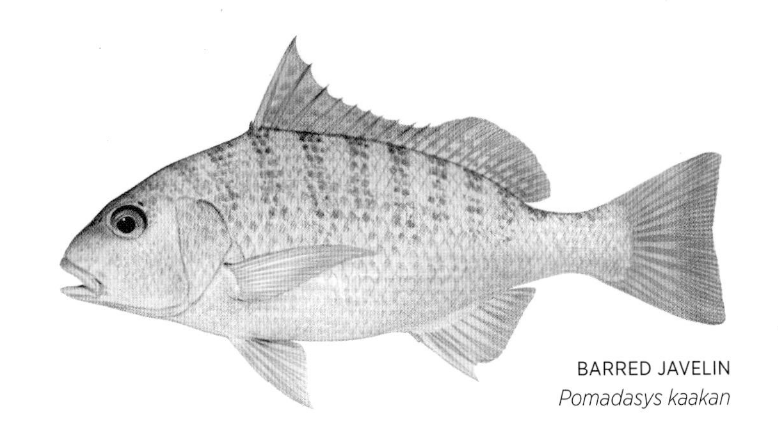

BARRED JAVELIN
Pomadasys kaakan

AUSTRALIAN HAEMULIDAE SPECIES

Diagramma melanacrum Blackfin Sweetlips
Diagramma pictum Painted Sweetlips
Hapalogenys kishinouyei Lined Javelinfish
Plectorhinchus albovittatus Giant Sweetlips
Plectorhinchus chaetodonoides Spotted Sweetlips
Plectorhinchus chrysotaenia Goldlined Sweetlips
Plectorhinchus chubbi Dusky Sweetlips
Plectorhinchus flavomaculatus Goldspotted Sweetlips
Plectorhinchus gibbosus Brown Sweetlips
Plectorhinchus lessonii Striped Sweetlips
Plectorhinchus lineatus Oblique-banded Sweetlips
Plectorhinchus multivittatus Manyline Sweetlips
Plectorhinchus picus Dotted Sweetlips
Plectorhinchus polytaenia Ribbon Sweetlips
Plectorhinchus schotaf Sombre Sweetlips
Plectorhinchus vittatus Oriental Sweetlips
Pomadasys argenteus Silver Javelin
Pomadasys argyreus Bluecheek Javelin
Pomadasys auritus Longhead Grunt
Pomadasys kaakan Barred Javelin
Pomadasys maculatus Blotched Javelin
Pomadasys trifasciatus Black-ear Javelin

THREADFIN BREAMS AND MONOCLE BREAMS

Nemipteridae

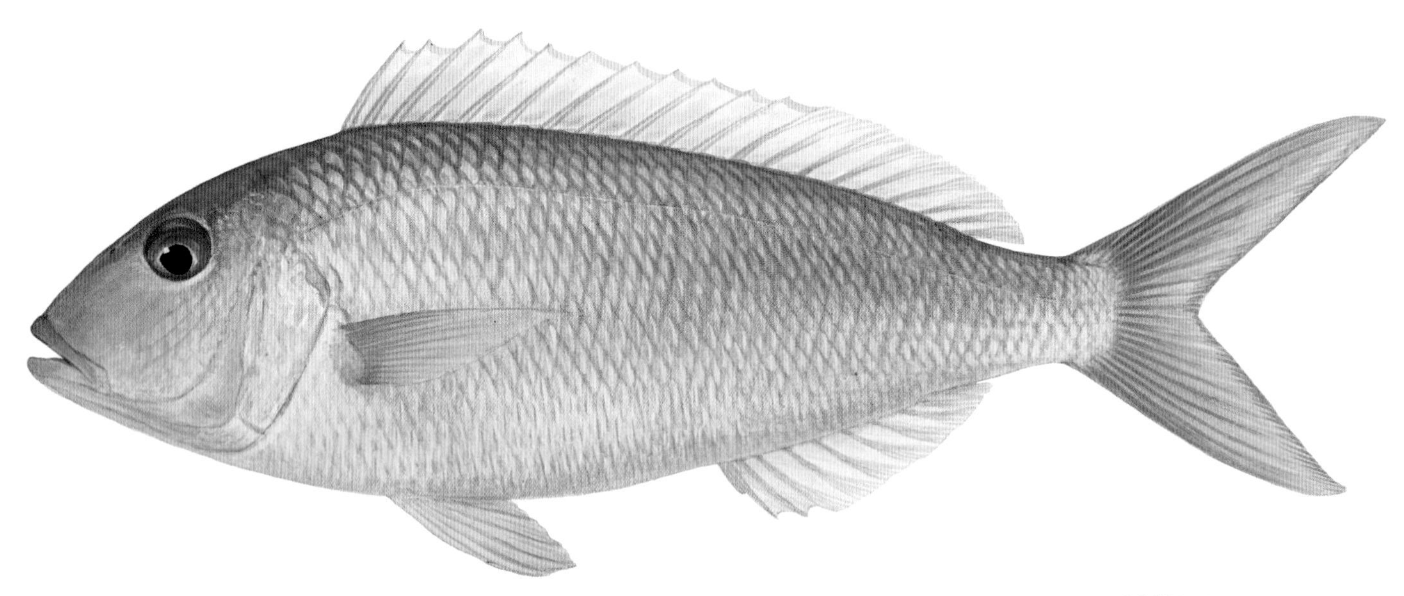

ROSY THREADFIN BREAM
Nemipterus furcosus

Threadfin Breams and Monocle Breams are found in the Indo-West Pacific region and there are about 64 species in the family. They are elongate to oval bodied, with large ctenoid scales and a small mouth with bands of small conical teeth. The single dorsal fin has 10 spines and 8–11 rays, the anal fin has three spines and 7–8 rays, and the caudal fin is slightly to deeply forked, often with a trailing filament on the upper lobe.

Threadfin Breams of the genus *Nemipterus* are found off northern Australia, generally in deeper coastal waters, over sand or mud bottoms down to about 100 m. They occur in large schools, feeding on small benthic invertebrates and some small fishes, and are frequently taken in large numbers by trawlers. Most are silvery-red with yellow markings, and reach a maximum size of 30–40 cm. They are excellent food fishes.

In Australia Monocle Breams are found near reefs over sandy bottoms, mainly in northern waters. They are small, elegant fishes, often beautifully coloured and generally solitary or in small groups. The **Two-line Monocle Bream** is very common on northern Australian coral reefs, often with one or more juveniles accompanying a larger adult. They stay close to the bottom, often hovering in a stationary position over sand patches for long periods, then making quick darting movements to pick up small benthic invertebrates. The abundant **Western Butterfish** occurs down the west coast of Australia to Geographe Bay, and forms small schools over seagrass and sandy bottoms in shallow coastal waters. Several Monocle Breams, such as the *Parascolopsis* species, occur in deeper offshore waters down to about 100 m, but most are restricted to coastal reefs. Monocle Breams reach a maximum length of about 30 cm and are good food fishes.

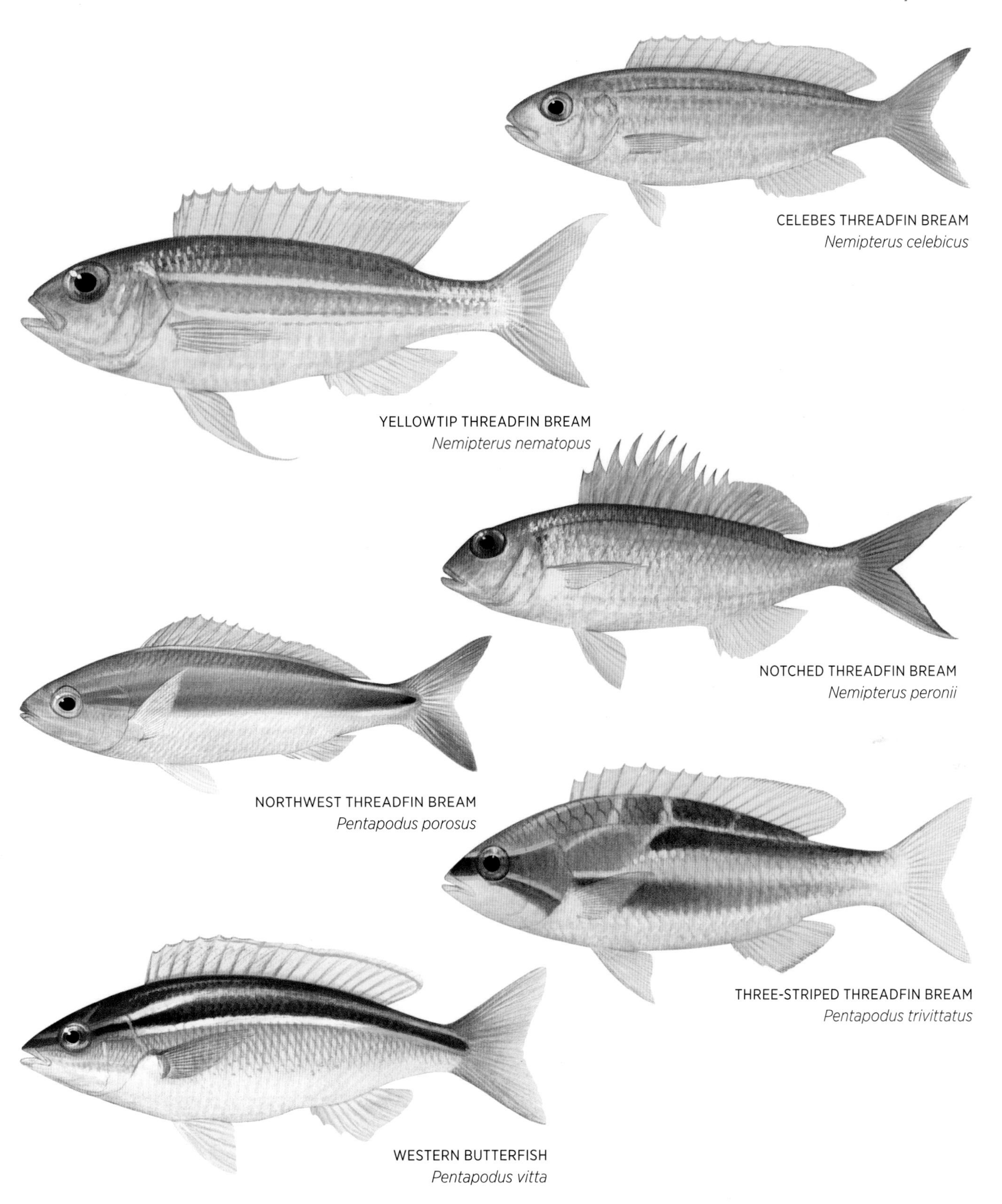

CELEBES THREADFIN BREAM
Nemipterus celebicus

YELLOWTIP THREADFIN BREAM
Nemipterus nematopus

NOTCHED THREADFIN BREAM
Nemipterus peronii

NORTHWEST THREADFIN BREAM
Pentapodus porosus

THREE-STRIPED THREADFIN BREAM
Pentapodus trivittatus

WESTERN BUTTERFISH
Pentapodus vitta

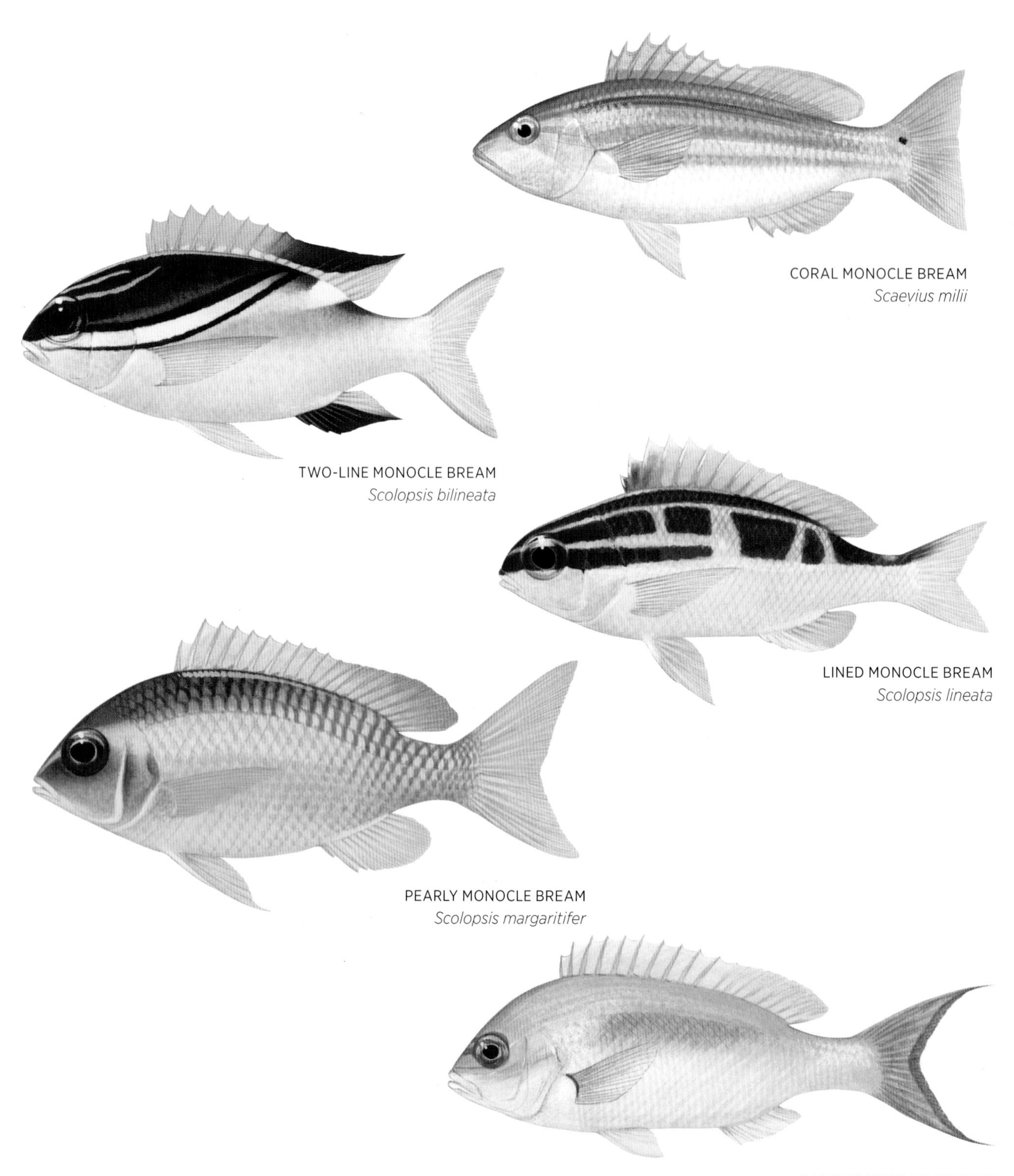

CORAL MONOCLE BREAM
Scaevius milii

TWO-LINE MONOCLE BREAM
Scolopsis bilineata

LINED MONOCLE BREAM
Scolopsis lineata

PEARLY MONOCLE BREAM
Scolopsis margaritifer

RAINBOW MONOCLE BREAM
Scolopsis monogramma

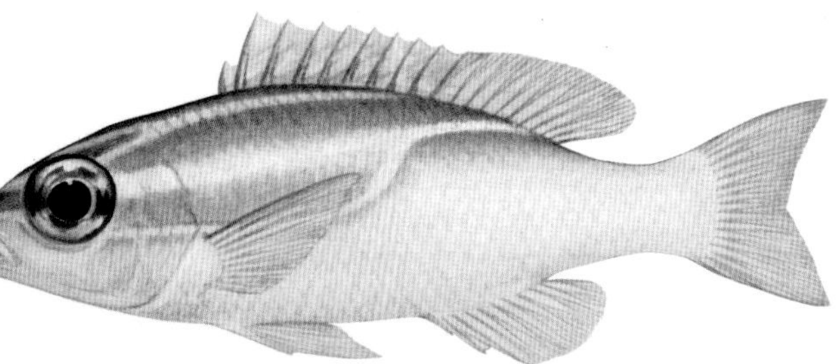

THREELINE MONOCLE BREAM
Scolopsis trilineatus

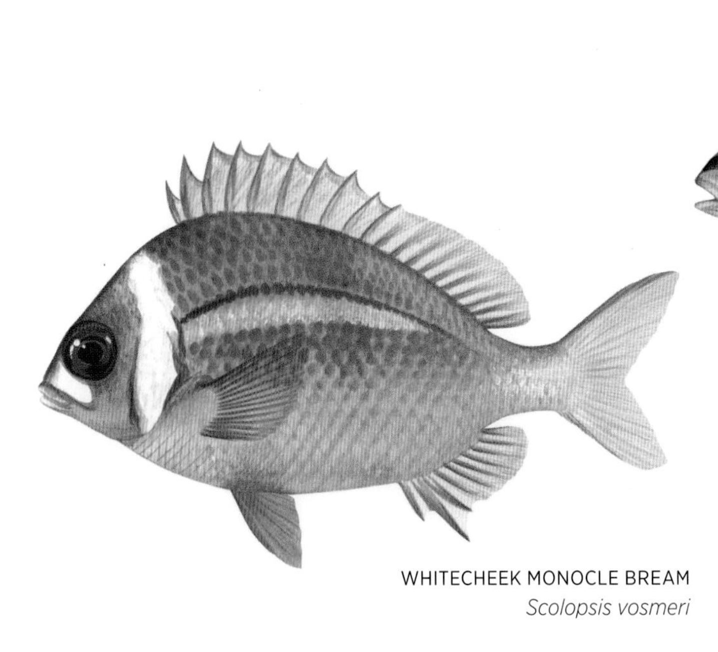

WHITECHEEK MONOCLE BREAM
Scolopsis vosmeri

OBLIQUE-BAR MONOCLE BREAM
Scolopsis xenochrous

AUSTRALIAN NEMIPTERIDAE SPECIES

Nemipterus aurifilum Yellowlip Threadfin Bream
Nemipterus balinensis Bali Threadfin Bream
Nemipterus balinensoides Dwarf Threadfin Bream
Nemipterus bathybius Yellowbelly Threadfin Bream
Nemipterus celebicus Celebes Threadfin Bream
Nemipterus furcosus Rosy Threadfin Bream
Nemipterus hexodon Ornate Threadfin Bream
Nemipterus isacanthus Teardrop Threadfin Bream
Nemipterus marginatus Red-filament Threadfin Bream
Nemipterus nematopus Yellowtip Threadfin Bream
Nemipterus peronii Notched Threadfin Bream
Nemipterus tambuloides Fiveline Threadfin Bream
Nemipterus theodorei Theodore's Threadfin Bream
Nemipterus virgatus Golden Threadfin Bream
Nemipterus zysron Slender Threadfin Bream
Parascolopsis inermis Redbelt Monocle Bream
Parascolopsis tanyactis Longray Monocle Bream
Parascolopsis tosensis Yellowstripe Monocle Bream
Pentapodus aureofasciatus Yellowstripe Threadfin Bream
Pentapodus emeryii Purple Threadfin Bream
Pentapodus nagasakiensis Japanese Threadfin Bream
Pentapodus paradiseus Paradise Threadfin Bream
Pentapodus porosus Northwest Threadfin Bream
Pentapodus trivittatus Three-striped Threadfin Bream
Pentapodus vitta Western Butterfish
Scaevius milii Coral Monocle Bream
Scolopsis affinis Bridled Monocle Bream
Scolopsis bilineata Two-line Monocle Bream
Scolopsis lineata Lined Monocle Bream
Scolopsis margaritifer Pearly Monocle Bream
Scolopsis monogramma Rainbow Monocle Bream
Scolopsis taenioptera Redspot Monocle Bream
Scolopsis trilineatus Threeline Monocle Bream
Scolopsis vosmeri Whitecheek Monocle Bream
Scolopsis xenochrous Oblique-bar Monocle Bream

EMPERORS AND SEABREAMS

Lethrinidae

LONGNOSE EMPEROR
Lethrinus olivaceus

Emperors and Seabreams are found in the coastal waters of tropical West Africa and the Indo-West Pacific. There are about 40 species in the family and many are important commercial food fishes. They have large, oval to elongate, compressed bodies with finely ctenoid scales and large eyes. Their large mouth bears strong canines in the front of the jaws, and the upper jaw slides beneath the suborbital bones when the mouth is closed. The single dorsal fin has 10 spines and 9–10 rays, the anal fin has three spines and 8–10 rays, and the caudal fin is slightly to strongly forked.

Seabreams occur near reefs in coastal waters of northern Australia. The **Mozambique Seabream** is found down to 200 m or more, while some species, such as the **Goldspot Seabream**, form large schools over shallow reefs. Others are solitary or occur in small groups. The **Robinson's Seabream**, which is an excellent food fish, reaches about 80 cm in length and is found on offshore reefs from 50 to 150 m.

Emperors are found in coastal waters of northern Australia, over reefs and nearby sandy areas, and near the bottom offshore, down to 150 m or more. They often occur in schools, and prey on small fishes, squid, crustaceans, echinoderms and other invertebrates. Some have molariform teeth in the jaws for crushing hard-shelled prey. Emperors are superb food fishes and are targeted by commercial fisheries and recreational anglers alike. The **Spangled Emperor** is common throughout its range, on inshore reefs and along beaches, as well as offshore down to about 150 m, and is a favourite of both boating and shore-based anglers. The **Longnose Emperor**, which can reach 1 m in length, is another popular angling target found over deeper coastal reefs, as are the **Redthroat Emperor** and **Grass Emperor**. The **Variegated Emperor** and **Threadfin Emperor** are smaller species, reaching about 25 cm in length, and are common in seagrass and weed beds around northern Australia. The **Orangespotted Emperor** is a large, solitary inhabitant of outer reef slopes, where it feeds mainly on echinoderms and crustaceans.

Emperors and Seabreams are sequential hermaphrodites, functioning as fully reproductive females initially, then changing sex from female to male at a certain stage of growth. They form large spawning aggregations and this reproductive strategy ensures the success of the species as a single large male can fertilise numerous small females.

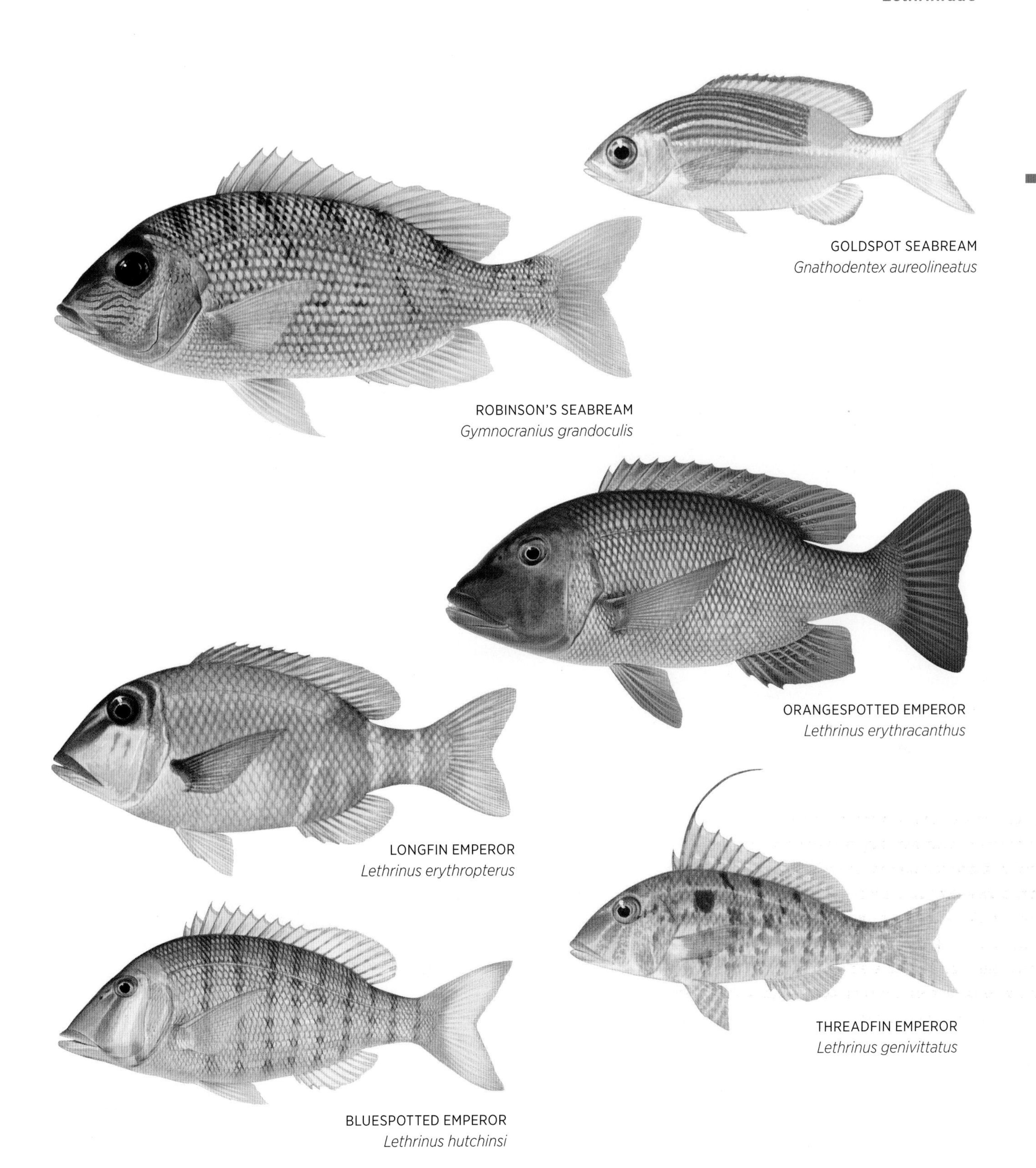

GOLDSPOT SEABREAM
Gnathodentex aureolineatus

ROBINSON'S SEABREAM
Gymnocranius grandoculis

ORANGESPOTTED EMPEROR
Lethrinus erythracanthus

LONGFIN EMPEROR
Lethrinus erythropterus

THREADFIN EMPEROR
Lethrinus genivittatus

BLUESPOTTED EMPEROR
Lethrinus hutchinsi

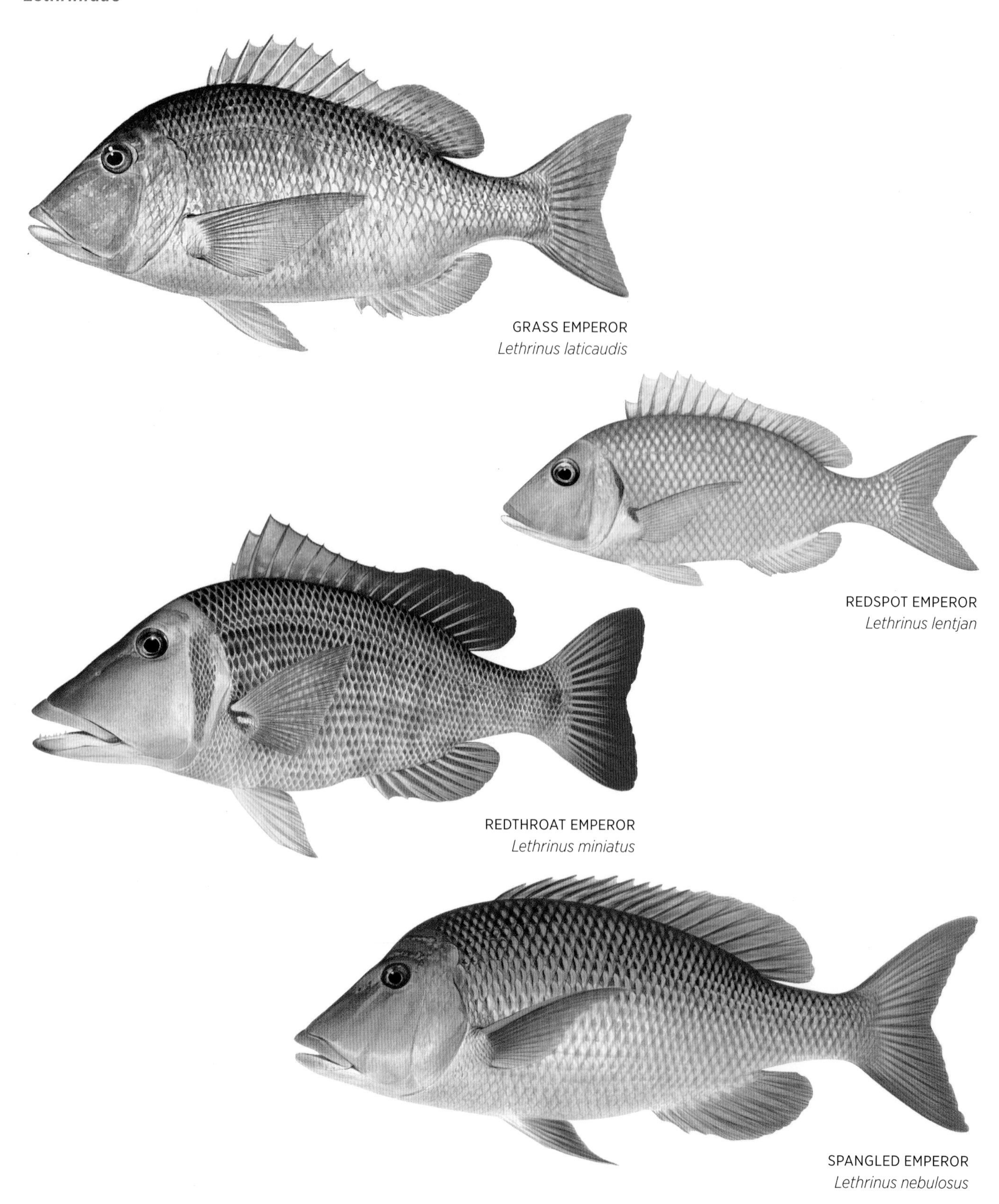

GRASS EMPEROR
Lethrinus laticaudis

REDSPOT EMPEROR
Lethrinus lentjan

REDTHROAT EMPEROR
Lethrinus miniatus

SPANGLED EMPEROR
Lethrinus nebulosus

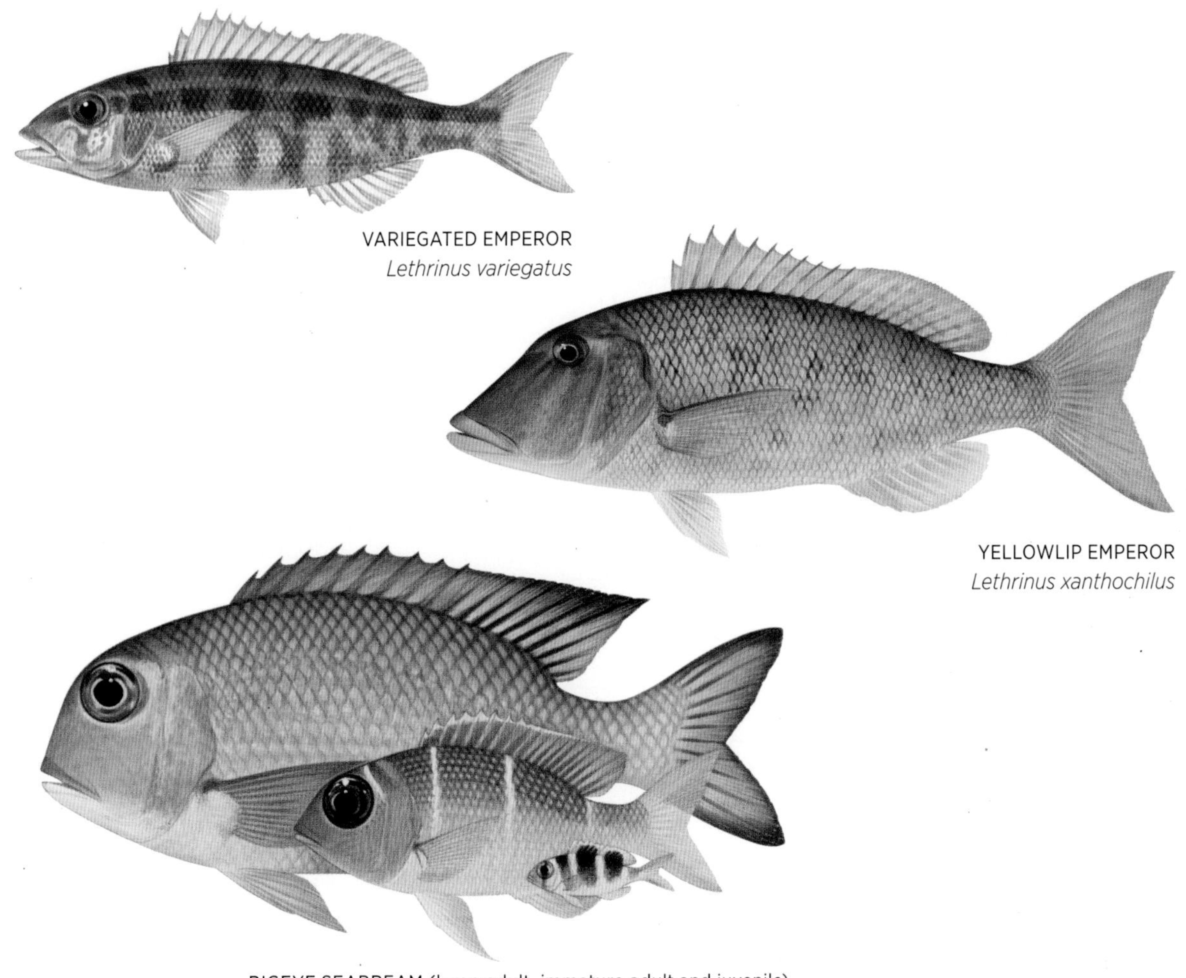

VARIEGATED EMPEROR
Lethrinus variegatus

YELLOWLIP EMPEROR
Lethrinus xanthochilus

BIGEYE SEABREAM (large adult, immature adult and juvenile)
Monotaxis grandoculis

AUSTRALIAN LETHRINIDAE SPECIES

Gnathodentex aureolineatus Goldspot Seabream
Gymnocranius audleyi Collar Seabream
Gymnocranius elongatus Swallowtail Seabream
Gymnocranius euanus Paddletail Seabream
Gymnocranius grandoculis Robinson's Seabream
Gymnocranius griseus Grey Seabream
Gymnocranius microdon Bluespotted Seabream
Lethrinus amboinensis Ambon Emperor
Lethrinus atkinsoni Yellowtail Emperor
Lethrinus erythracanthus Orangespotted Emperor
Lethrinus erythropterus Longfin Emperor
Lethrinus genivittatus Threadfin Emperor
Lethrinus harak Thumbprint Emperor
Lethrinus hutchinsi Bluespotted Emperor
Lethrinus laticaudis Grass Emperor
Lethrinus lentjan Redspot Emperor
Lethrinus microdon Smalltooth Emperor
Lethrinus miniatus Redthroat Emperor
Lethrinus nebulosus Spangled Emperor
Lethrinus obsoletus Orangestriped Emperor
Lethrinus olivaceus Longnose Emperor
Lethrinus ornatus Ornate Emperor
Lethrinus ravus Drab Emperor
Lethrinus rubrioperculatus Spotcheek Emperor
Lethrinus semicinctus Blackblotch Emperor
Lethrinus variegatus Variegated Emperor
Lethrinus xanthochilus Yellowlip Emperor
Monotaxis grandoculis Bigeye Seabream
Wattsia mossambica Mozambique Seabream

BREAMS

Sparidae

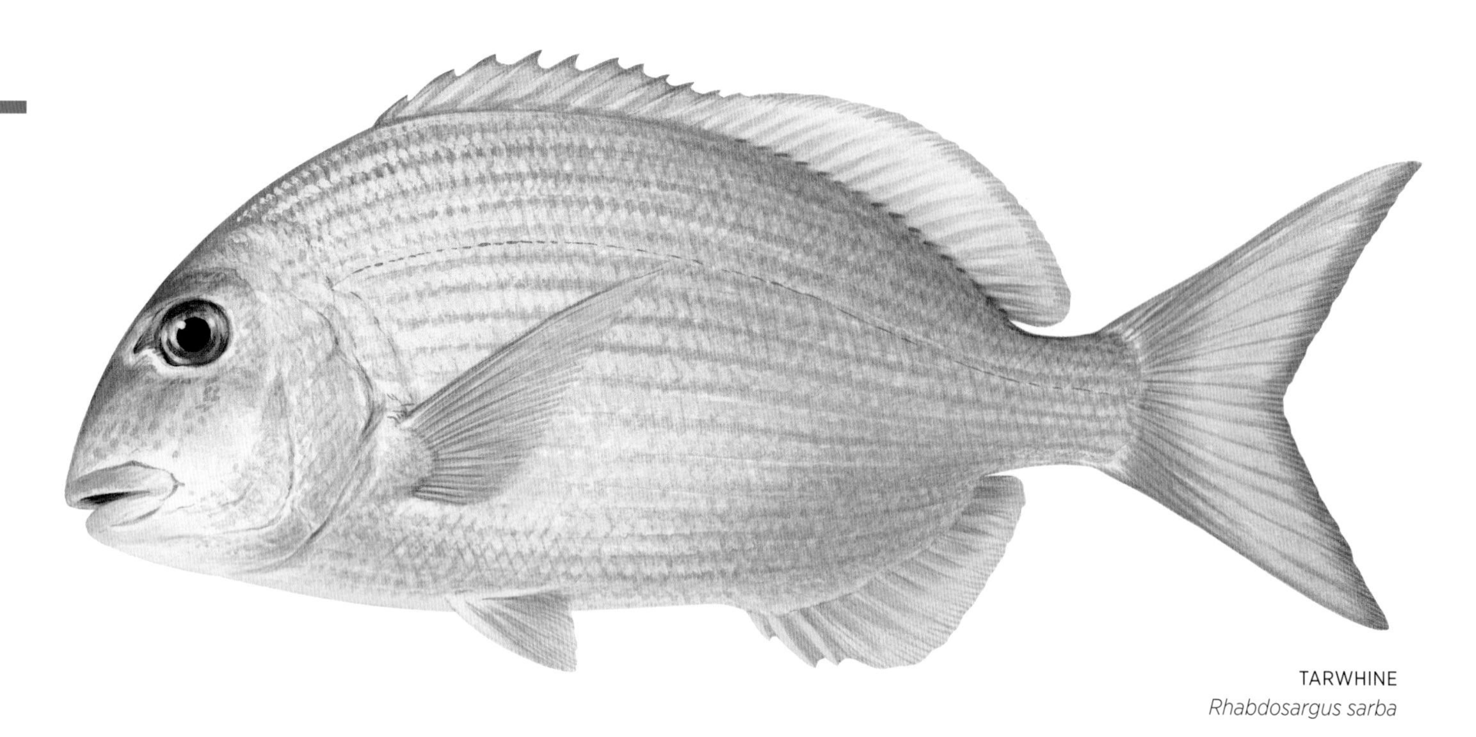

TARWHINE
Rhabdosargus sarba

There are about 115 species of Bream and they are found worldwide in tropical and temperate waters, with some species entering fresh and brackish waters. They have a deep, compressed body with finely ctenoid scales, and a small mouth with anterior canines and rear molariform teeth. The single dorsal fin has 10–13 spines, the anal fin has three spines and 8–14 rays, the ventral fin has a scaly axillary process, and the caudal fin is truncate to slightly forked.

Breams feed mainly on benthic invertebrates such as crustaceans, cephalopods and molluscs, and some also prey on fishes. The family includes the **Snapper**, an excellent food fish that is an extremely important commercial species in Australian waters and which is also keenly sought-after by recreational anglers. The Snapper can reach up to 1.3 m in length and is found, often in large schools, over reefs and rocky bottoms around the southern two-thirds of Australia, down to at least 100 m. Juveniles are common in estuaries around the southern coast and in seagrass beds throughout their range. Large adults develop a hump on the forehead and sometimes a large lump on the snout. Large bony nodules may also develop on the vertebral spines.

Other Breams, including the **Black Bream** and **Yellowfin Bream**, are also important angling and commercial species. Both are common in estuaries and occasionally venture into the lower reaches of freshwater rivers. The Black Bream occurs across the southern coast of Australia, and the Yellowfin Bream in south-eastern coastal waters and estuaries, where it is one of the most sought-after angling species. The **Tarwhine** is found over reefs, in estuaries and along sandy beaches on the south-east and south-west coasts of Australia. It reaches a maximum length of about 80 cm, though most specimens encountered are about 30 cm. The **Frypan Bream**, another excellent food fish, is a resident of deeper offshore reefs in northern Australian waters, and has greatly elongated, filamentous anterior fin spines.

Most members of this family are superb food fishes and heavily targeted wherever they occur. In Australia restrictions such as bag and size limits are now imposed on most Breams, and closed seasons apply in some areas where populations, particularly of Snapper, have been severely depleted.

Different species of Bream show varying degrees of hermaphroditism. The family includes examples of rudimentary hermaphroditism, where juveniles can develop into either male or female; simultaneous hermaphroditism, where individuals have both male and female organs at the same time; and sequential hermaphroditism, where sex reversal occurs.

YELLOWFIN BREAM
Acanthopagrus australis

PIKEY BREAM
Acanthopagrus berda

BLACK BREAM
Acanthopagrus butcheri

FRYPAN BREAM
Argyrops spinifer

SNAPPER (adult)
Pagrus auratus

SNAPPER (juvenile)
Pagrus auratus

AUSTRALIAN SPARIDAE SPECIES

Acanthopagrus australis Yellowfin Bream
Acanthopagrus berda Pikey Bream
Acanthopagrus butcheri Black Bream
Acanthopagrus latus Western Yellowfin Bream
Acanthopagrus palmaris Northwest Black Bream
Argyrops spinifer Frypan Bream
Dentex tumifrons Yellowback Bream
Pagrus auratus Snapper
Rhabdosargus sarba Tarwhine
Sparidentex hasta Sobaity Bream

THREADFINS

Polynemidae

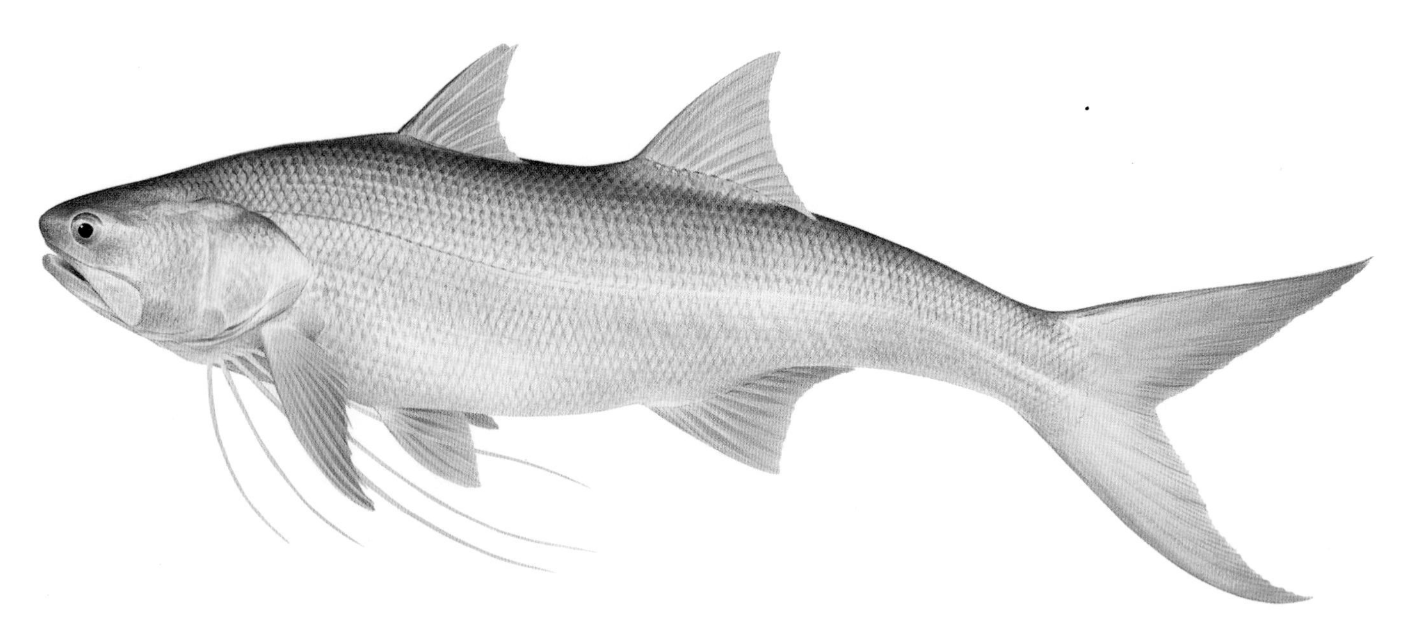

KING THREADFIN
Polydactylus macrochir

About 41 species of Threadfin are found in tropical and subtropical waters worldwide. Many enter the brackish waters of estuaries and some are found in fresh water. They are distinctive fishes, with elongate, rounded to compressed bodies and finely ctenoid scales. The eyes are covered with a clear, fleshy adipose eyelid, and they have a rounded, projecting snout with a large inferior mouth and bands of villiform teeth in the jaws. They have two separate dorsal fins, the first with 7–8 spines and the second with one spine and 11–18 rays. The anal fin has 2–3 spines and 9–30 rays, and the caudal fin is deeply forked. The pectoral fins are divided into two parts: the upper part is a normal fin and the lower part consists of 3–7 separate long, filamentous rays. These rays are used as feelers to locate prey in the often turbid waters Threadfins inhabit. In the small **Streamer Threadfin** the filaments may be longer than the body of the fish.

Threadfins occur in shallow coastal waters of Australia's north, usually over sandy or muddy bottoms, and are common in mangroves, estuaries and tidal creeks. They prey mainly on small crustaceans – the **Australian Threadfin** feeds almost exclusively on prawns – and small fishes are taken by some species. Their diet changes seasonally according to prey abundance. The **Blue Threadfin** is the largest member of the family, reportedly growing to up to 1.8 m in length, but most Threadfins are much smaller. Both the Blue Threadfin and the **King Threadfin** are popular angling targets in northern waters, readily taking baits and fighting hard, and they are excellent food fishes. The **Striped Threadfin** is widespread around the northern half of Australia and reaches about 45 cm in length. It is common around river mouths and along ocean beaches, feeding mainly on crabs and other benthic crustaceans.

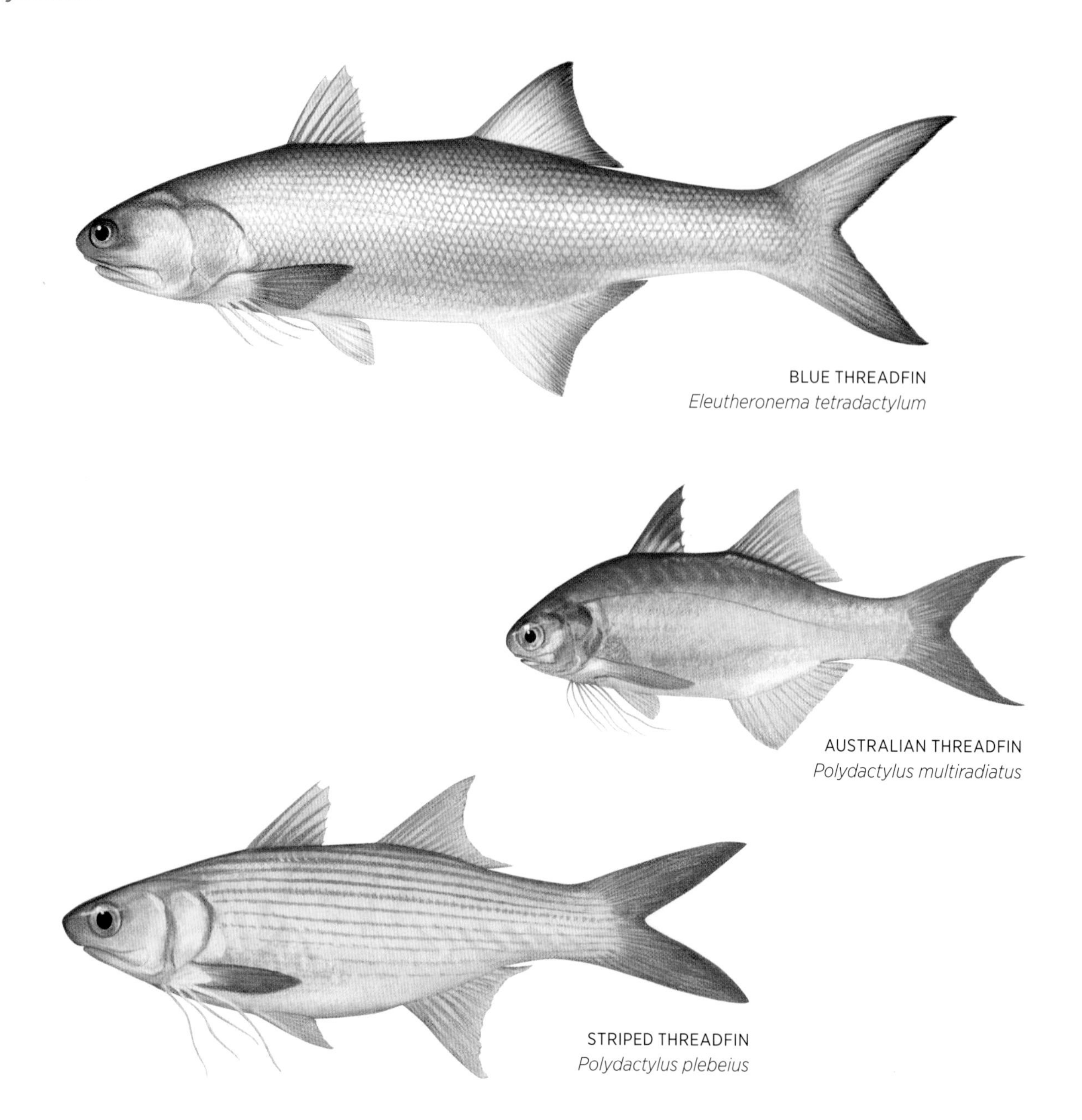

BLUE THREADFIN
Eleutheronema tetradactylum

AUSTRALIAN THREADFIN
Polydactylus multiradiatus

STRIPED THREADFIN
Polydactylus plebeius

AUSTRALIAN POLYNEMIDAE SPECIES

Eleutheronema tetradactylum Blue Threadfin
Parapolynemus verekeri Streamer Threadfin
Polydactylus macrochir King Threadfin
Polydactylus multiradiatus Australian Threadfin
Polydactylus nigripinnis Blackfin Threadfin
Polydactylus plebeius Striped Threadfin

JEWFISHES

Sciaenidae

MULLOWAY
Argyrosomus japonicus

The family of Jewfishes, or Croakers as they are sometimes known, contains about 270 species. They are found worldwide in tropical and warm temperate waters, and are particularly diverse in the eastern Pacific. Several species are found in fresh water in South America. Jewfishes have elongate, rounded to slightly compressed bodies, with cycloid or ctenoid scales that extend onto the fins, and a large mouth with small conical teeth. They have a deeply notched dorsal fin, the anterior part with 9–11 spines and the posterior part with one spine and 23–31 rays. The very short-based anal fin has two spines and 6–9 rays and the caudal fin is square to rounded or diamond-shaped.

Jewfishes take their alternative name of Croaker from the ability to produce croaking or grunting sounds using their muscular swimbladder. This ability is more pronounced in males and is used in schooling and mating behaviour. Jewfishes also have two very well-developed auditory chambers in the skull, each containing a large bony sagitta or otolith. The shape of the swimbladder and otolith are important in identification of many of the very similar small species.

In Australia Jewfishes are found in coastal waters and estuaries, with the majority of species occurring in northern waters. The **Mulloway**, however, is found around the southern half of Australia and is a popular angling target off ocean beaches and in estuaries. It can reach up to 2 m in length and weigh 60 kg. The Mulloway feeds mainly on small fishes and migrates seasonally, following prey abundance. In the north it is replaced by the **Black Jewfish**, which occurs in similar habitats and reaches an equally large size. The **Teraglin** is also found in southern waters, occurring in schools over offshore reefs off New South Wales. It feeds at night, in midwater, preying on small fishes. Numerous smaller species of Jewfish are found around the northern half of Australia, particularly in the far north and Gulf of Carpentaria, most reaching a maximum length of 25–50 cm. These are usually found in shallow inshore waters, over sand or mud bottoms, in mangroves, estuaries and tidal creeks, and feed mainly on benthic crustaceans and other invertebrates.

Sciaenidae

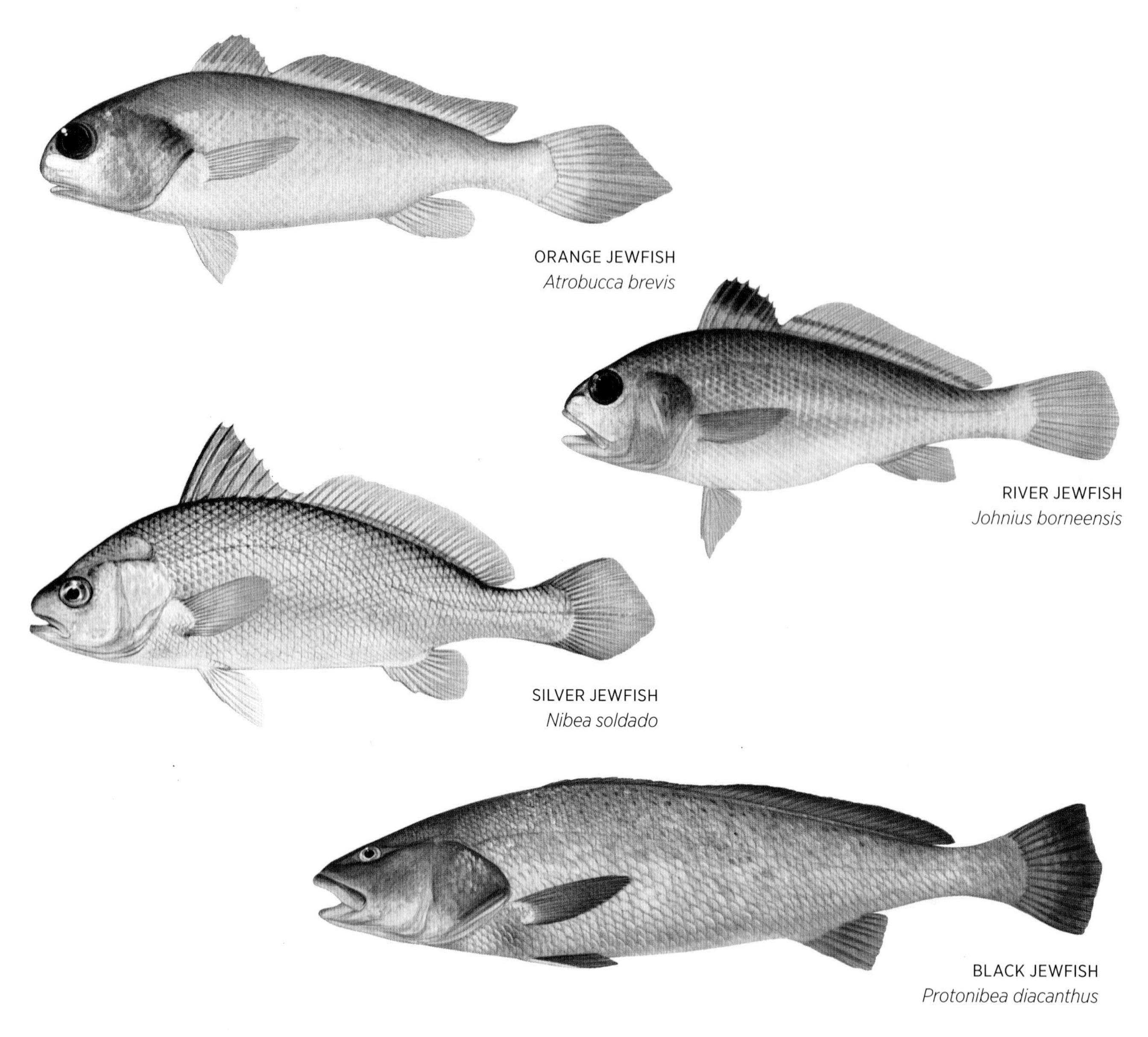

ORANGE JEWFISH
Atrobucca brevis

RIVER JEWFISH
Johnius borneensis

SILVER JEWFISH
Nibea soldado

BLACK JEWFISH
Protonibea diacanthus

AUSTRALIAN SCIAENIDAE SPECIES

Argyrosomus japonicus Mulloway
Atractoscion aequidens Teraglin
Atrobucca brevis Orange Jewfish
Atrobucca nibe Longmouth Jewfish
Austronibea oedogenys Yellowtail Jewfish
Johnius amblycephalus Bearded Jewfish
Johnius australis Little Jewfish
Johnius belangerii Belanger's Jewfish
Johnius borneensis River Jewfish
Johnius laevis Smooth Jewfish
Johnius novaeguineae Paperhead Jewfish
Johnius novaehollandiae Bottlenose Jewfish
Larimichthys pamoides Southern Yellow Jewfish
Nibea leptolepis Smallscale Jewfish
Nibea microgenys Smallmouth Jewfish
Nibea soldado Silver Jewfish
Nibea squamosa Scaly Jewfish
Otolithes ruber Silver Teraglin
Protonibea diacanthus Black Jewfish

GOATFISHES

Mullidae

YELLOWSPOT GOATFISH
Parupeneus indicus

Goatfishes are found worldwide in tropical and temperate waters, and there are about 62 species in the family. They have elongate, rounded to slightly compressed bodies, with large, fragile, finely ctenoid scales. The mouth has thick, fleshy lips and there are two long, highly mobile sensory barbels beneath the chin, which give them their common name. They have two widely separated dorsal fins, the first with 7–8 spines and the second with nine rays. The anal fin has one spine and 6–7 rays, and the caudal fin is strongly forked.

Goatfishes are usually found over sandy bottoms near reef and weed areas. Most feed on small sand-dwelling invertebrates such as worms, molluscs, echinoderms and crustaceans. They search out prey by burrowing into the soft sediment, blowing jets of water into the sand and stirring up clouds of sediment, sometimes until their entire head is buried. They are usually accompanied by a small group of other species, such as Wrasses (Labridae, p. 617), which pounce on any morsels exposed by this feeding activity. The **Goldsaddle Goatfish** is unusual in the family, as it feeds mainly on small fishes. The **Bluespotted Goatfish** is one of the only species found in southern Australian waters. It occurs right around the southern coast, in small schools down to about 40 m, and although growing to only about 30 cm in length, is an excellent food fish.

The **Blackspot Goatfish** and **Yellowstripe Goatfish** are often seen in large, stationary schools over reefs during the day, dispersing at night to feed over nearby sandy areas. Many other species are solitary, or occur in small groups or pairs, over sand and weed areas near tropical reefs. Some are active at night, others during the day. One of the largest species, the **Dot-and-dash Goatfish**, can reach over 50 cm in length, though most species are no larger than about 30 cm. Most Goatfishes, though soft-bodied, are considered good food fishes.

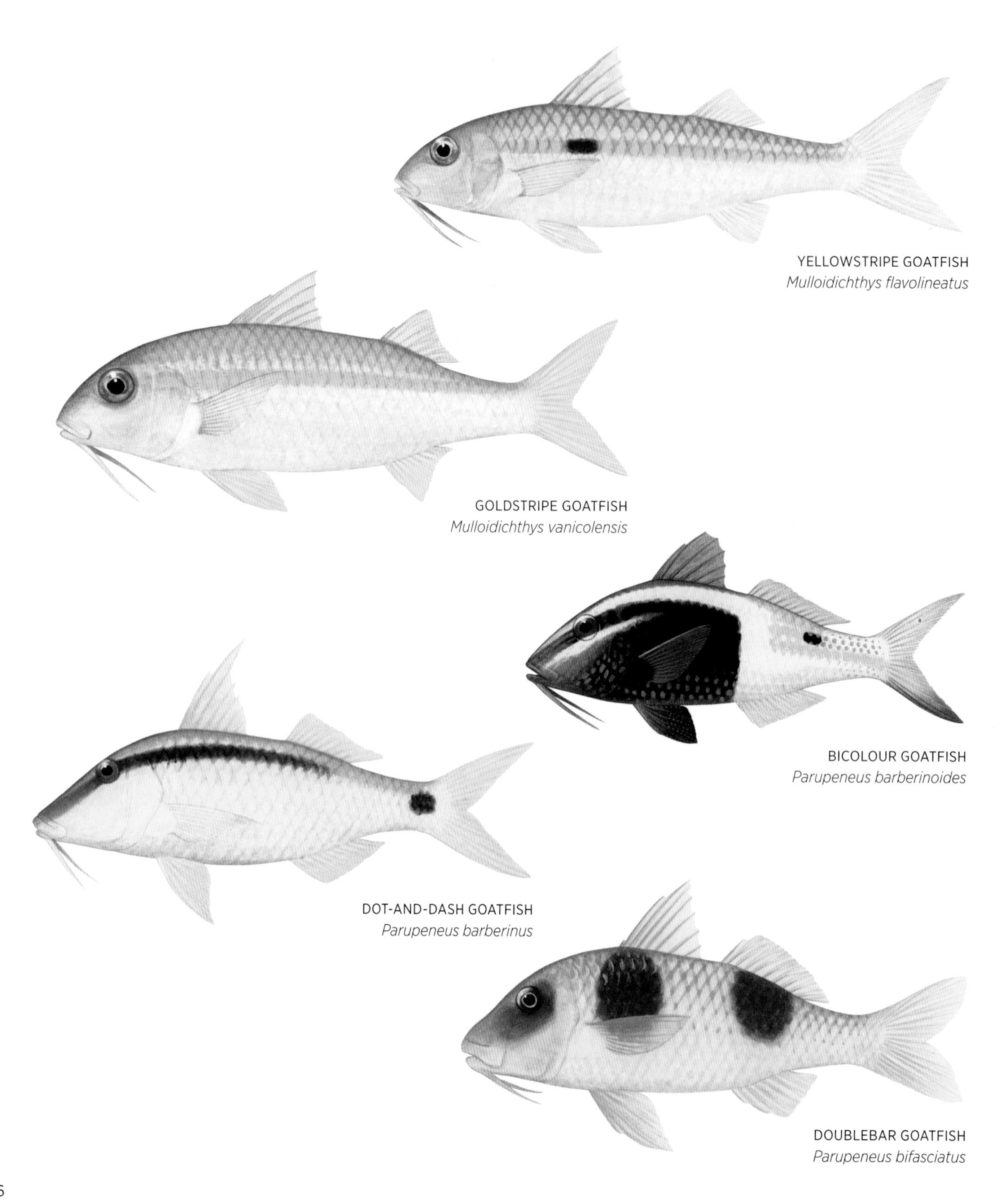

YELLOWSTRIPE GOATFISH
Mulloidichthys flavolineatus

GOLDSTRIPE GOATFISH
Mulloidichthys vanicolensis

BICOLOUR GOATFISH
Parupeneus barberinoides

DOT-AND-DASH GOATFISH
Parupeneus barberinus

DOUBLEBAR GOATFISH
Parupeneus bifasciatus

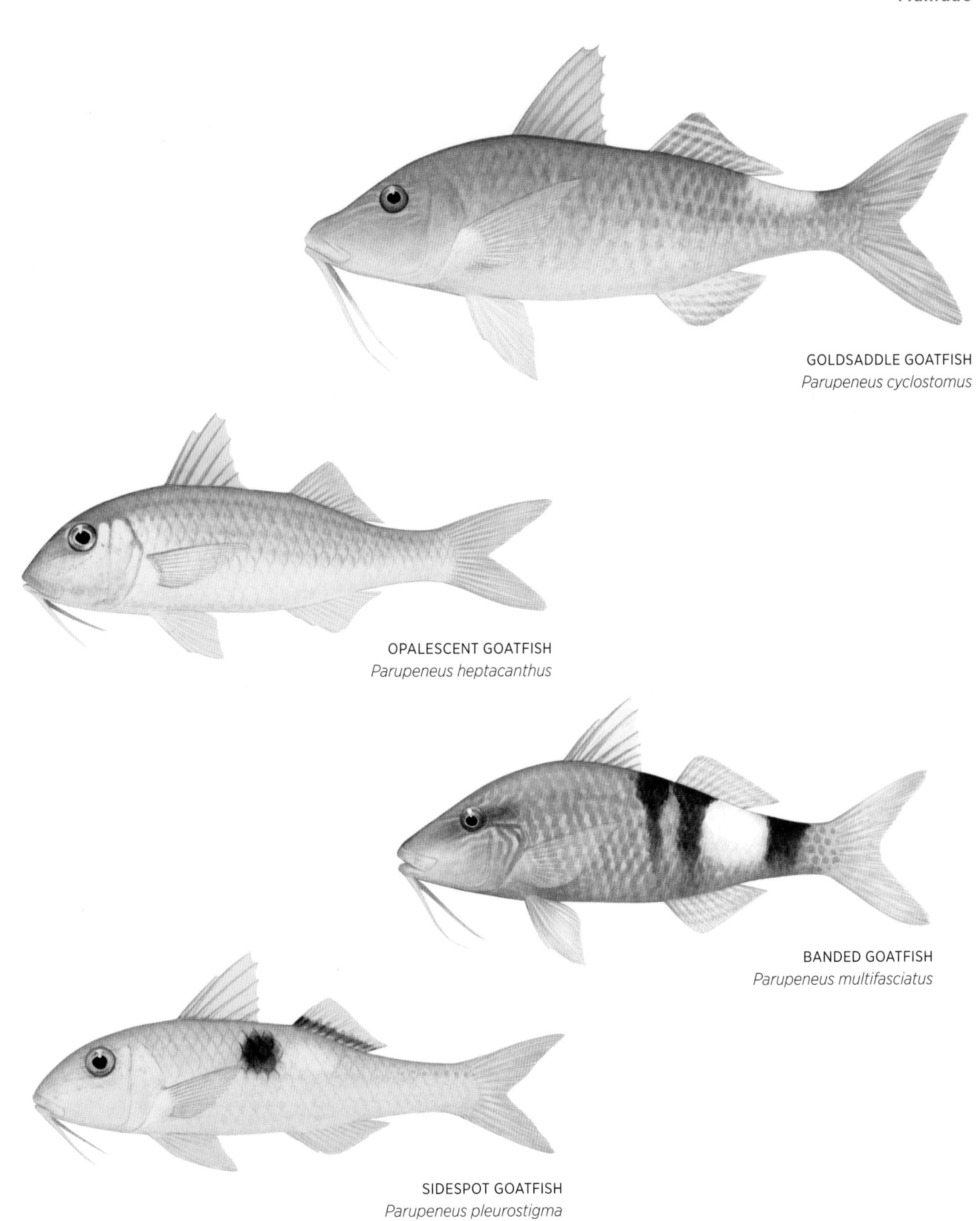

GOLDSADDLE GOATFISH
Parupeneus cyclostomus

OPALESCENT GOATFISH
Parupeneus heptacanthus

BANDED GOATFISH
Parupeneus multifasciatus

SIDESPOT GOATFISH
Parupeneus pleurostigma

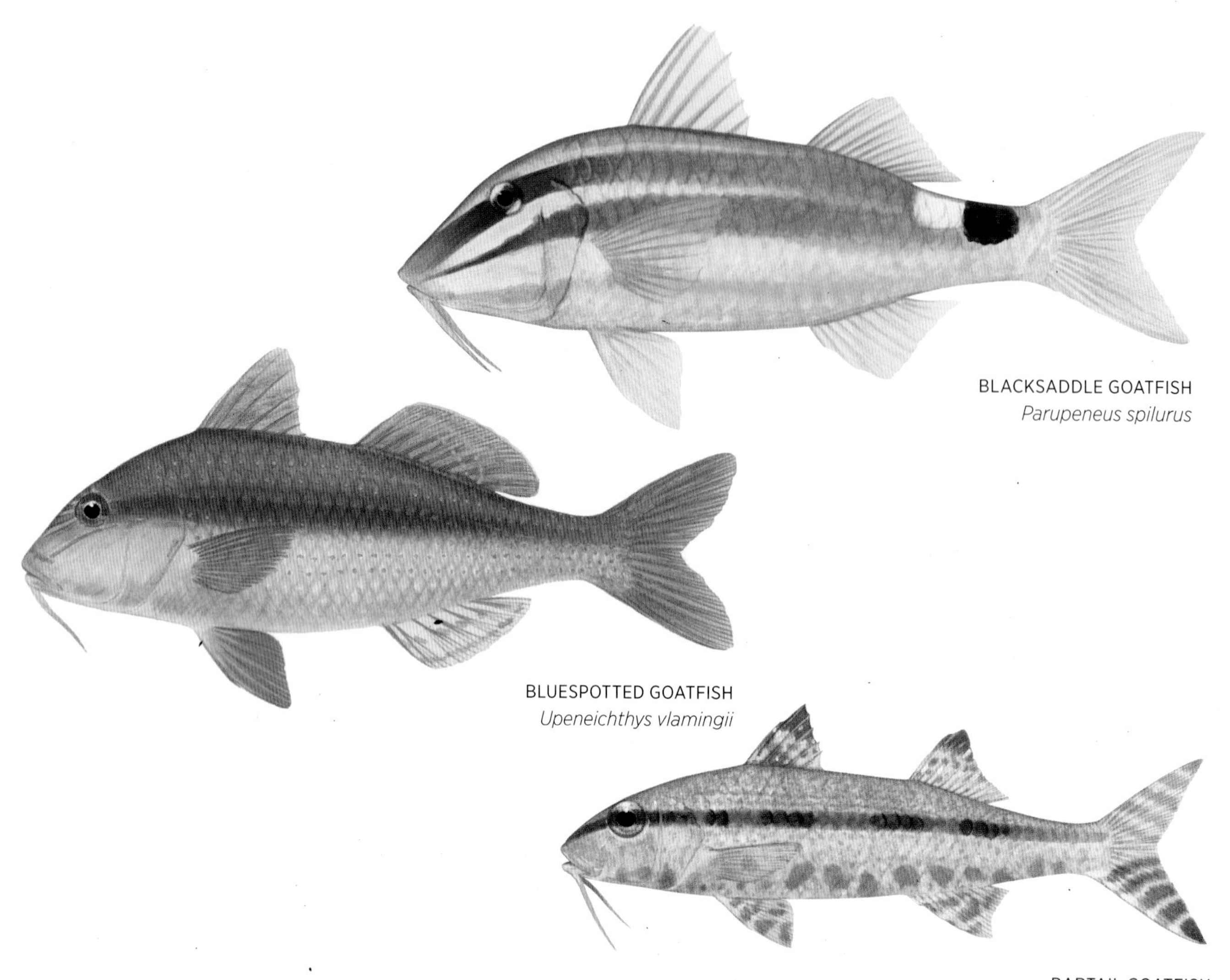

BLACKSADDLE GOATFISH
Parupeneus spilurus

BLUESPOTTED GOATFISH
Upeneichthys vlamingii

BARTAIL GOATFISH
Upeneus tragula

AUSTRALIAN MULLIDAE SPECIES

Mulloidichthys flavolineatus Yellowstripe Goatfish
Mulloidichthys vanicolensis Goldstripe Goatfish
Parupeneus barberinoides Bicolour Goatfish
Parupeneus barberinus Dot-and-dash Goatfish
Parupeneus bifasciatus Doublebar Goatfish
Parupeneus chrysopleuron Rosy Goatfish
Parupeneus ciliatus Diamondscale Goatfish
Parupeneus cyclostomus Goldsaddle Goatfish
Parupeneus heptacanthus Opalescent Goatfish
Parupeneus indicus Yellowspot Goatfish
Parupeneus multifasciatus Banded Goatfish
Parupeneus pleurostigma Sidespot Goatfish
Parupeneus rubescens Blackspot Goatfish
Parupeneus spilurus Blacksaddle Goatfish
Upeneichthys lineatus Bluestriped Goatfish
Upeneichthys stotti Stott's Goatfish
Upeneichthys vlamingii Bluespotted Goatfish
Upeneus asymmetricus Asymmetric Goatfish
Upeneus australiae Australian Goatfish
Upeneus filifer Pennant Goatfish
Upeneus japonicus Japanese Goatfish
Upeneus luzonius Luzon Goatfish
Upeneus moluccensis Goldband Goatfish
Upeneus sulphureus Sunrise Goatfish
Upeneus sundaicus Ochreband Goatfish
Upeneus tragula Bartail Goatfish
Upeneus vittatus Striped Goatfish

BULLSEYES

Pempheridae

BIGSCALE BULLSEYE
Pempheris multiradiata

Bullseyes are found in the Indian and Pacific oceans, with a few species occurring in the Atlantic Ocean. There are about 26 species in the family. Bullseyes have a very deep, compressed body, with fragile cycloid or strong ctenoid scales, large eyes and a rounded head with large oblique mouth. The single short-based dorsal fin has 4–7 spines and 7–12 rays. The anal fin is very long and usually has two spines and 17–45 rays, and the caudal fin is slightly forked. Some species have light organs associated with the digestive tract and along the belly, and the bioluminescence they produce seems to be obtained from their food items.

Bullseyes occur right around the coast of Australia, in small to large schools on reefs down to about 100 m. They shelter in caves and crevices by day, and feed by night on zooplankton, crustaceans and other invertebrates. The **Bigscale Bullseye** reaches about 30 cm in length and is commonly encountered by divers in southern Australian waters, sheltering in deep caves in small groups. The **Rough Bullseye** is very common on west-coast reefs and has similar habits.

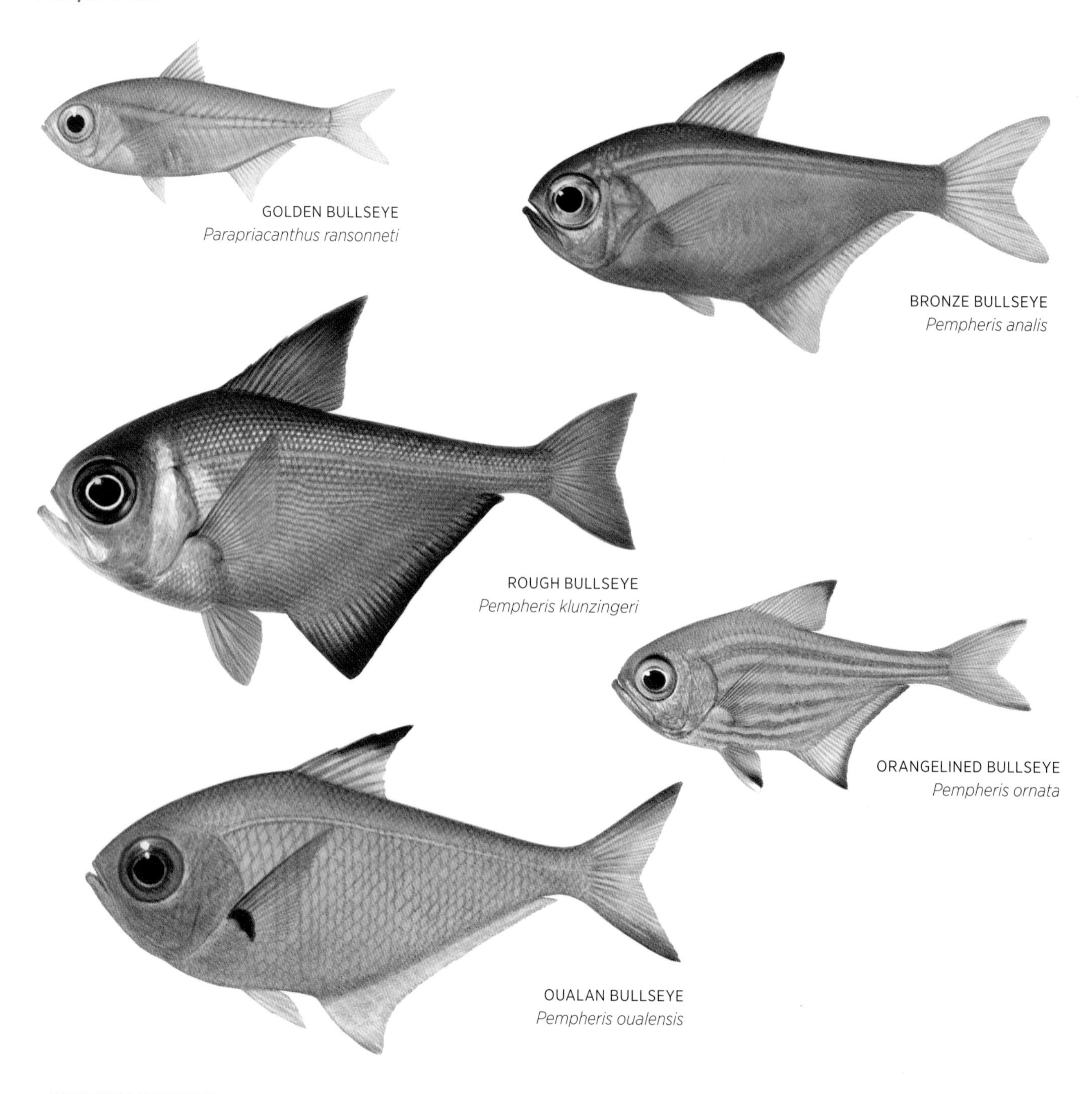

GOLDEN BULLSEYE
Parapriacanthus ransonneti

BRONZE BULLSEYE
Pempheris analis

ROUGH BULLSEYE
Pempheris klunzingeri

ORANGELINED BULLSEYE
Pempheris ornata

OUALAN BULLSEYE
Pempheris oualensis

AUSTRALIAN PEMPHERIDAE SPECIES

Parapriacanthus dispar Deep Bullseye
Parapriacanthus elongatus Elongate Bullseye
Parapriacanthus ransonneti Golden Bullseye
Pempheris affinis Blacktip Bullseye
Pempheris analis Bronze Bullseye
Pempheris compressa Smallscale Bullseye
Pempheris klunzingeri Rough Bullseye
Pempheris multiradiata Bigscale Bullseye
Pempheris ornata Orangelined Bullseye
Pempheris oualensis Oualan Bullseye
Pempheris schwenkii Silver Bullseye
Pempheris ypsilychnus Ypsilon Bullseye

PEARL PERCHES

Glaucosomatidae

WEST AUSTRALIAN DHUFISH
Glaucosoma hebraicum

Pearl Perches are distributed from the eastern Indian Ocean to the western Pacific. There are four species in the family, all of which are found in Australian waters. They have oval to almost circular compressed bodies, with small to moderate-sized ctenoid scales, a large head and large eyes. The large mouth has bands of small pointed teeth in the jaws. They have a single dorsal fin with eight spines and 12–14 rays, the rays being much longer than the spines and sometimes elongated into trailing filaments. The anal fin has three spines and 9–10 rays, and the caudal fin is lunate to slightly forked.

Pearl Perches inhabit rocky reefs and weed areas down to about 100 m, occasionally moving into shallow water, but usually found on deeper reefs. They feed at dawn or dusk, mostly near the bottom, using their large eyes to locate a wide range of invertebrates and small fishes in low light conditions. They will sometimes move up the water column to take midwater prey.

The **West Australian Dhufish** is reported to grow to 1.25 m in length and weigh up to 32 kg, and is one of the most highly sought-after food fishes in Australian waters. It inhabits deep rocky reefs off south-western Australia. Large individuals are territorial and will take up residence in a particular area, moving into shallower waters to spawn in summer. The territorial habits, large size and superb eating qualities of this species have contributed to its being severely overfished in most of its range. The magnificent **Threadfin Pearl Perch** occurs from Cape York across to north-western Australia and has long, trailing filaments on the dorsal and anal fins. It is usually found on rocky reefs at moderate depths, in small to large schools. The **Northern Pearl Perch** is found off the north-west coast and the **Pearl Perch** off the north-east coast, with both species occurring over rocky bottoms in deeper waters down to about 100 m. The Pearl Perch has a prominent, silvery, skin-covered bone just behind the head, which is lacking in the Northern Pearl Perch. Both species reach a maximum length of about 45 cm and are considered superb food fishes.

NORTHERN PEARL PERCH
Glaucosoma buergeri

PEARL PERCH
Glaucosoma scapulare

AUSTRALIAN GLAUCOSOMATIDAE SPECIES

Glaucosoma buergeri Northern Pearl Perch
Glaucosoma hebraicum West Australian Dhufish
Glaucosoma magnificum Threadfin Pearl Perch
Glaucosoma scapulare Pearl Perch

BEACH SALMON

Leptobramidae

BEACH SALMON
Leptobrama muelleri

The **Beach Salmon**, the single species in this family, is found only in the waters of southern New Guinea and northern Australia. It has an elongate, compressed body with small ctenoid scales, large eyes with an adipose eyelid, and a large mouth with broad bands of small, pointed teeth. The Beach Salmon has a single dorsal fin, set well back on the body, with four spines and 16–18 rays. The long-based anal fin has three spines and 26–30 rays, and the caudal fin is truncate. The dorsal and anal fins both have scaly sheaths at the base. The Beach Salmon occurs in shallow inshore waters, estuaries and along sandy beaches, often in schools. It reaches a maximum length of about 45 cm and feeds mainly on small fishes.

AUSTRALIAN LEPTOBRAMIDAE SPECIES

Leptobrama muelleri Beach Salmon

DEEPSEA HERRINGS

Bathyclupeidae

SLENDER DEEPSEA HERRING
Bathyclupea gracilis

There are about seven species of Deepsea Herring and they are found in tropical and subtropical waters of the Atlantic, Indian and western Pacific oceans. They occur in deep water from about 400 to 3000 m and superficially resemble the true Herrings (Clupeidae, p. 184), though they are not related. Deepsea Herrings have an elongate, very compressed body with large, fragile cycloid scales that extend onto the fins. They have large eyes, and a large oblique mouth with fine, villiform teeth. The single, short-based dorsal fin has one weak spine and 8–10 rays, and the long-based anal fin has one weak spine and 24–39 rays. The ventral fins are very small and placed well forward on the body, before the pectoral-fin base, and the pectoral fins are large and pointed. (The true Herrings have no fin spines and the ventral fins are behind the pectoral-fin base.) Deepsea Herrings feed on small pelagic crustaceans. They are rarely encountered and little is known of their biology.

AUSTRALIAN BATHYCLUPEIDAE SPECIES

Bathyclupea gracilis Slender Deepsea Herring

POMFRETS

Monodactylidae

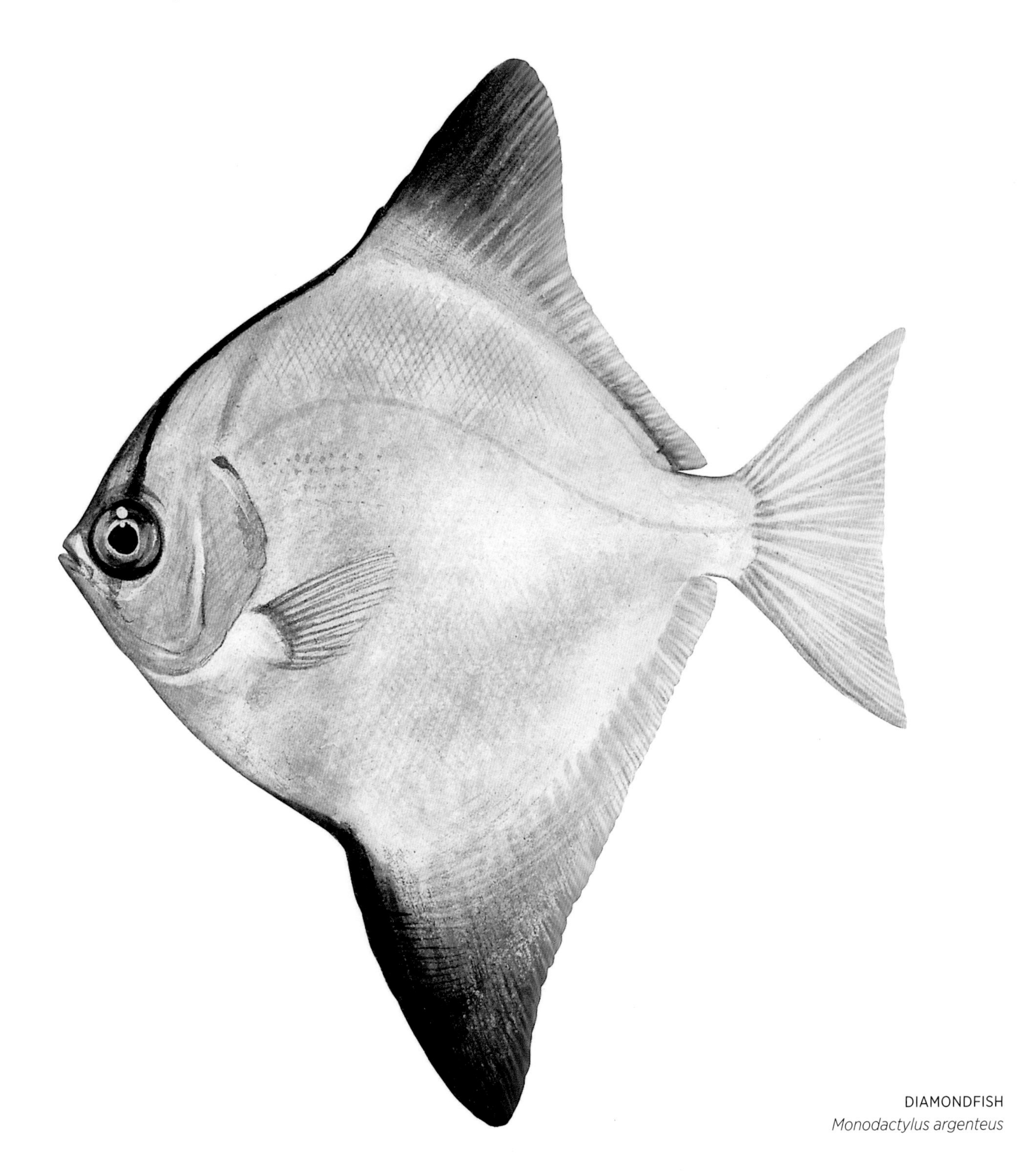

DIAMONDFISH
Monodactylus argenteus

Monodactylidae

Pomfrets occur in tropical and temperate coastal waters of the Indo-Pacific and West Africa, and occasionally in brackish and fresh waters. There are about five species in the familly. Pomfrets have a deep, compressed body, with small, fragile cycloid or ctenoid scales that extend onto the fins, and a small mouth. The single, tall dorsal fin has 5–8 spines and 26–31 rays, the anal fin is similar in shape to the dorsal fin and has three spines and 26–31 rays, and the caudal fin is truncate to slightly forked. Although members of this family have a somewhat similar body shape to Pomfrets of the family Bramidae (p. 509), they are not related.

The **Diamondfish** is a small silvery fish that grows to about 27 cm in length. Adults lack ventral fins, but they are present in small juveniles. The species forms schools in estuaries and protected bays in northern Australian waters. It occasionally enters the lower reaches of freshwater streams and is common around jetties and in estuaries in New South Wales. The **Eastern Pomfret** and **Western Pomfret** are found on the south-east and south-west Australian coasts respectively. They inhabit estuaries and shallow reefs down to about 30 m, where they form small schools close to the shelter of crevices and weed. Juveniles are common in very shallow coastal waters and estuaries. Pomfrets feed on zooplankton and reach a maximum length of about 25 cm.

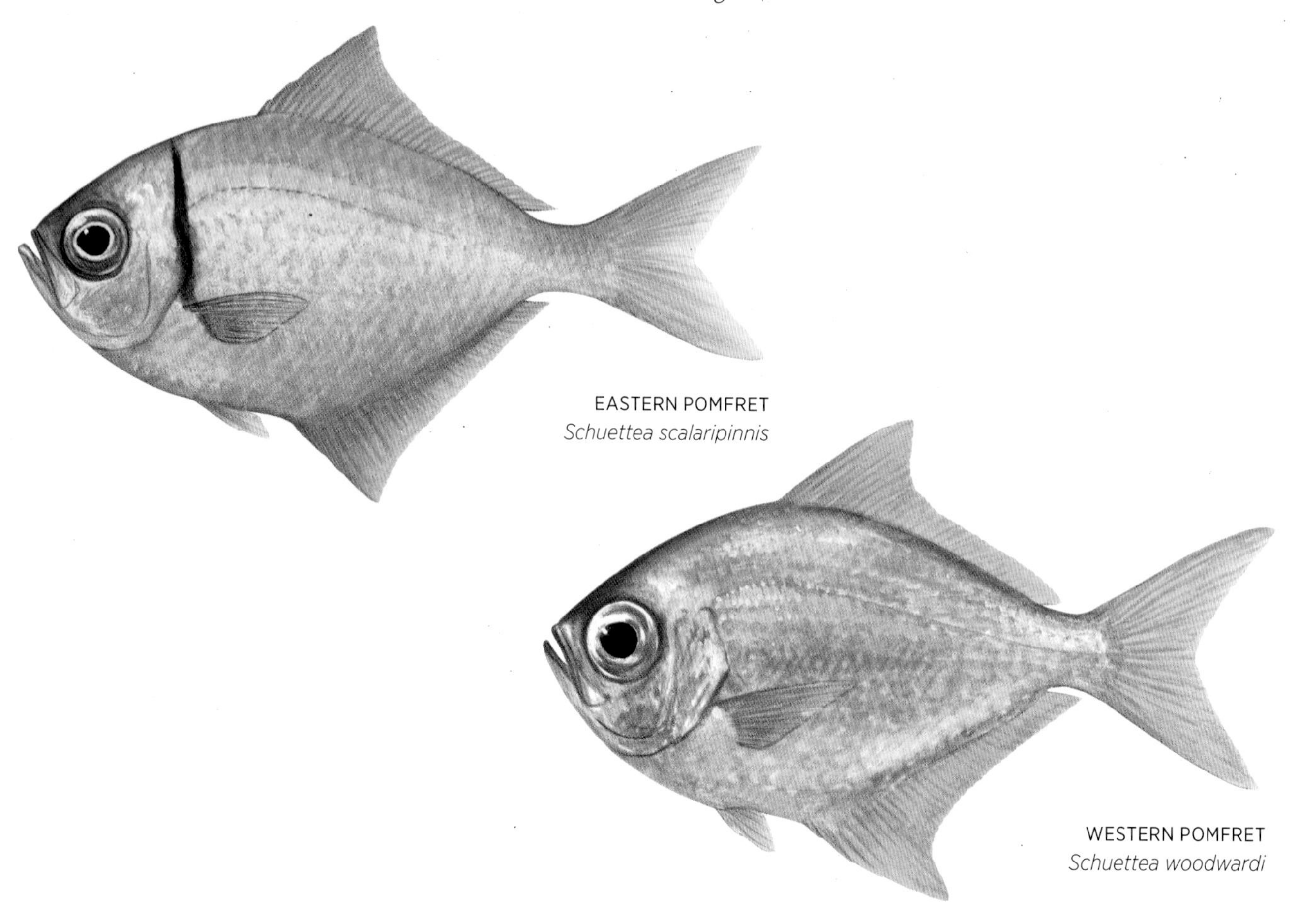

EASTERN POMFRET
Schuettea scalaripinnis

WESTERN POMFRET
Schuettea woodwardi

AUSTRALIAN MONODACTYLIDAE SPECIES
Monodactylus argenteus Diamondfish
Schuettea scalaripinnis Eastern Pomfret
Schuettea woodwardi Western Pomfret

ARCHERFISHES

Toxotidae

BANDED ARCHERFISH
Toxotes jaculatrix

Archerfishes are found in coastal marine and fresh waters from India through South-East Asia to northern Australia, and there are about six species in the family. They have a deep, compressed body with large ctenoid scales that extend onto the fins, and a long straight dorsal profile with a large mouth and large eyes. The single dorsal fin is placed towards the rear of the body and has 4–6 spines and 11–14 rays. The anal fin has three spines and 15–18 rays, and the caudal fin is truncate.

Archerfishes are known for their remarkable ability to shoot jets of water into the air through a groove in the roof of the mouth, to knock down insects flying near the surface of the water or perched on overhanging vegetation. They are astoundingly accurate and are capable of hitting a small insect several metres away.

Archerfishes are widely distributed in northern Australia in mangrove swamps, brackish estuaries and freshwater streams, often well inland. Most inhabit areas with abundant vegetation and consume floating fruit and flower buds, as well as insects, aquatic invertebrates and small fishes. The largest species reach a maximum length of about 40 cm.

AUSTRALIAN TOXOTIDAE SPECIES

Toxotes chatareus Sevenspot Archerfish
Toxotes jaculatrix Banded Archerfish
Toxotes kimberleyensis Kimberley Archerfish
Toxotes lorentzi Primitive Archerfish

AUSTRALIAN SALMONS AND AUSTRALIAN HERRING

Arripidae

WESTERN AUSTRALIAN SALMON (adult)
Arripis truttaceus

The four species in this family are restricted to the coastal waters of southern Australia and New Zealand. They are not related to the true Salmons (Salmonidae, p. 214) or true Herrings (Clupeidae, p. 184). They have elongate, rounded to slightly compressed bodies, with small, finely ctenoid scales, and a moderate-sized mouth with bands of fine teeth. The single dorsal fin has nine spines and 13–19 rays, the anal fin has three spines and 9–10 rays, and the caudal fin is strongly forked.

Australian Salmons form large schools around the southern half of Australia in shallow coastal waters and along ocean beaches, and make annual migrations from the south coast to spawning grounds along the west and east coasts. The **Western Australian Salmon** reaches a maximum length of about 75 cm, while the **Eastern Australian Salmon** is very similar but grows to a maximum size of about 55 cm. Their ranges overlap in south-eastern Australian waters.

Australian Salmons are voracious predators of small fishes and crustaceans. They are important commercial species, and are taken by beach seine net during their seasonal migrations. The Western Australian Salmon migration is eagerly awaited each autumn by anglers on the west coast. During the migration, large schools of Salmon move close to shore along beaches and around rocky headlands, and will readily strike lures and baits. They are tenacious fighters and good food fishes when well prepared.

The **Australian Herring** spawns on the west coast and juveniles are carried southward and eastward by the Leeuwin Current, growing larger in shallow coastal waters and estuaries for several years before migrating westward again in large schools. The Australian Herring is one of the most popular recreational angling fishes in southern waters. It grows to a maximum length of about 40 cm and is abundant close to shore, making it accessible to many anglers, and it is an excellent food fish. The **Giant Kahawai**, a New Zealand species, is very similar to the Australian Salmons and occurs around Lord Howe and Norfolk islands.

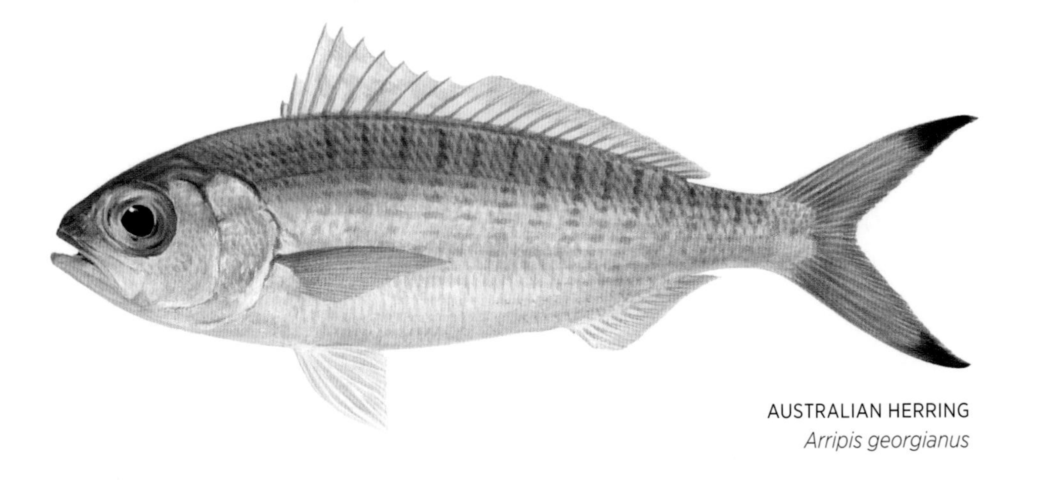

AUSTRALIAN HERRING
Arripis georgianus

WESTERN AUSTRALIAN SALMON (juvenile)
Arripis truttaceus

AUSTRALIAN ARRIPIDAE SPECIES

Arripis georgianus Australian Herring
Arripis trutta Eastern Australian Salmon
Arripis truttaceus Western Australian Salmon
Arripis xylabion Giant Kahawai

DRUMMERS

Kyphosidae

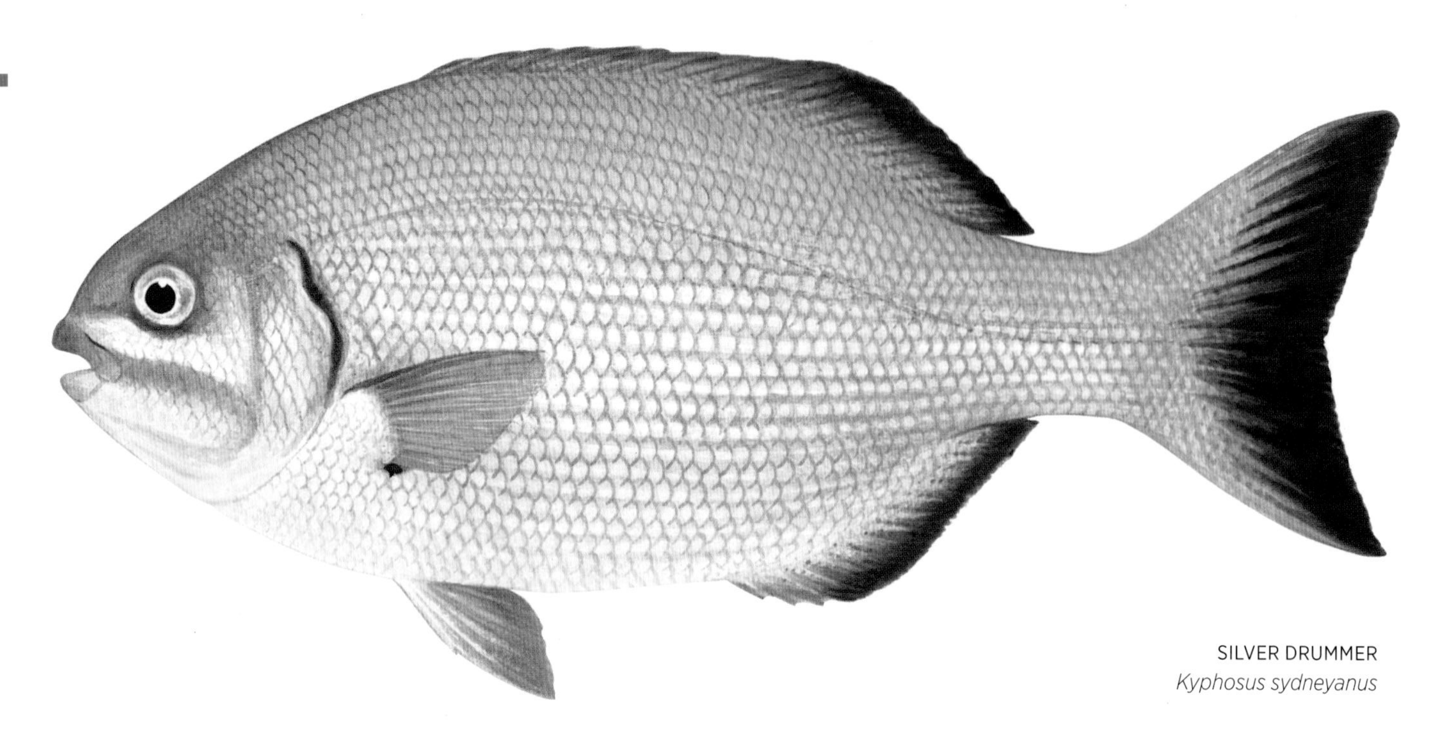

SILVER DRUMMER
Kyphosus sydneyanus

Drummers are found in tropical and temperate waters of the Atlantic, Indian and Pacific oceans. There are about 13 species in the family. They have an oval, compressed body with small ctenoid scales that extend onto the fins. The small mouth has an outer row of incisor teeth in the jaws and bands of minute teeth inside. The rear upper jaw does not slide beneath the suborbital bones and is exposed when the mouth is closed. Drummers have a single long-based dorsal fin with 11 spines and 12–15 rays; the anal fin has three spines and 11–14 rays; and the caudal fin is slightly forked with pointed lobes.

Drummers are large schooling fishes and are mainly herbivorous, feeding predominantly on benthic algae but also preying on small invertebrates such as crustaceans and worms. They occur over rocky and coral reefs, and weed beds right around the Australian coastline, from the shallows down to about 30 m. Juvenile Drummers are often found in oceanic waters beneath floating weed and debris. The **Silver Drummer** is a large species, reaching about 80 cm in length, that forms schools over shallow reefs in Australia's southern half and is a popular angling target in south-eastern Australia. Fast-swimming schools of large Silver Drummers are frequently encountered by divers in shallow areas with high water movement. The other Drummers found in Australia occur in northern waters and often form small schools comprising a number of Drummer species. In some parts of Australia, Drummers are considered good food fishes, while in others they are deemed barely edible – this may be due to the different types of algae being consumed by the fishes.

WESTERN BUFFALO BREAM
Kyphosus cornelii

BRASSY DRUMMER
Kyphosus vaigiensis

AUSTRALIAN KYPHOSIDAE SPECIES

Kyphosus bigibbus Grey Drummer
Kyphosus cinerascens Snubnose Drummer
Kyphosus cornelii Western Buffalo Bream
Kyphosus gibsoni Northern Silver Drummer
Kyphosus pacificus Pacific Drummer
Kyphosus sydneyanus Silver Drummer
Kyphosus vaigiensis Brassy Drummer

NIBBLERS

Girellidae

LUDERICK
Girella tricuspidata

Nibblers are found in the Indo-Pacific and Atlantic oceans, and there are about 17 species in the family. They are similar in appearance to the closely related Drummers (Kyphosidae, p. 560) and are sometimes included in the same family. They differ in having the rear upper jaw covered by the suborbital bones when the mouth is closed, and in having some of the incisor teeth with three lobes. They have an oval, compressed body with small ctenoid scales that extend onto the base of the fins. The single long-based dorsal fin has 13–16 spines and 11–14 rays, the anal fin has three spines and 11–12 rays, and the large caudal fin is truncate to slightly forked.

In Australia Nibblers are found mainly in southern waters, over shallow wave-washed reefs, in estuaries and over seagrass beds. They occur in small groups or individually and are herbivorous, feeding on a variety of benthic algae. The **Luderick** is a popular angling target in south-eastern Australia, growing to a maximum length of about 50 cm, and is found in large schools. The Rock Blackfishes are more solitary and inhabit shallow rocky reefs in areas of high water movement, down to about 30 m. In south-eastern Australia the **Rock Blackfish** is considered a good food fish, but in the south-west the closely related **Western Rock Blackfish** is considered inedible – this is probably due to differences in their respective diets. The **Zebrafish** is common in small groups over weed beds and shallow reefs around the south coast, and also enters estuaries.

ROCK BLACKFISH
Girella elevata

ZEBRAFISH
Girella zebra

AUSTRALIAN GIRELLIDAE SPECIES

Girella cyanea Blue Drummer
Girella elevata Rock Blackfish
Girella tephraeops Western Rock Blackfish
Girella tricuspidata Luderick
Girella zebra Zebrafish

MICROCANTHIDS

Microcanthidae

STRIPEY
Microcanthus strigatus

Microcanthids are mainly confined to temperate waters around Australia and New Zealand, and this small family contains about seven species. Microcanthids have oval to deep, compressed bodies with small ctenoid or cycloid scales that extend onto the fins, and a small mouth with conical, tricuspid or brush-like teeth. The single long-based dorsal fin has 11–12 spines and 16–18 rays, the anal fin has three spines and 16–19 rays, and the caudal fin is slightly forked.

In Australia Microcanthids are found mainly in southern waters, over rocky reefs and in estuaries. The **Mado** is common, in large schools in shallow coastal waters around south-eastern Australia down to about 30 m, often around jetties and in estuaries. The **Footballer Sweep** prefers deeper reefs in the south-west, where it forms small schools. The **Stripey** is found on both east and west coasts, forming small to large groups close to the bottom over rocky substrates. The **Moonlighter** is more solitary and occurs on deeper coastal reefs in the south and south-west. It is the largest species in the family, reaching a maximum length of about 40 cm. Microcanthids are omnivorous, feeding on algae and small invertebrates.

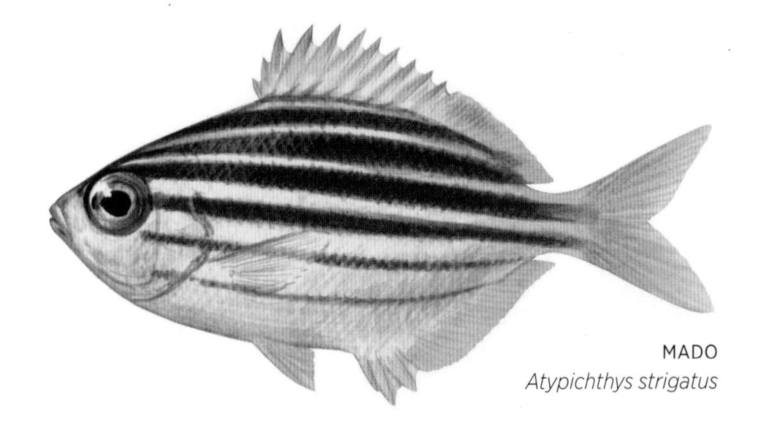

MADO
Atypichthys strigatus

FOOTBALLER SWEEP
Neatypus obliquus

MOONLIGHTER
Tilodon sexfasciatus

AUSTRALIAN MICROCANTHIDAE SPECIES

Atypichthys latus New Zealand Mado
Atypichthys strigatus Mado
Microcanthus strigatus Stripey
Neatypus obliquus Footballer Sweep
Tilodon sexfasciatus Moonlighter

SWEEPS

Scorpididae

BANDED SWEEP
Scorpis georgiana

Sweeps are found only in the Indo-Pacific region, mainly in Australian waters, and there are about seven species in the family. They have a deep, compressed body, with small ctenoid scales that extend onto the fins, and a small mouth with bands of strong teeth, the outer rows of which are enlarged. They have a single tall dorsal fin with 9–10 spines and 26–28 rays, with the rays much longer than the spines. The anal fin has three spines and 27–28 rays, with the rays longer than the spines, and the caudal fin is slightly to strongly forked.

Sweeps are found around southern Australia, over rocky reefs in areas of high water movement. The **Sea Sweep** is one of the largest in the family, reaching a maximum length of about 60 cm. It forms big schools and feeds mainly in midwater on zooplankton. The **Banded Sweep** is found in south-western waters, in loose aggregations or individually, over rocky reefs near crevices and caves, down to 35 m or more. It grows to a maximum length of about 45 cm and is considered a fine food fish. The **Silver Sweep**, also a good food fish, is found in similar habitats in south-eastern waters. Most Sweeps are omnivorous, feeding on algae, small invertebrates and zooplankton.

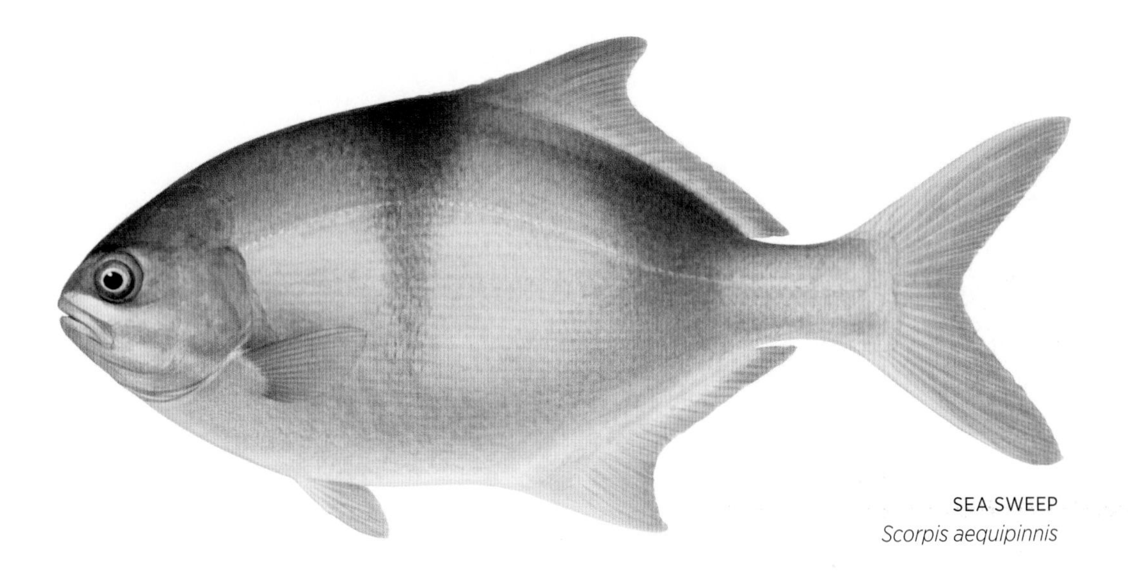

SEA SWEEP
Scorpis aequipinnis

AUSTRALIAN SCORPIDIDAE SPECIES

Bathystethus cultratus Knifefish
Labracoglossa nitida Blue Knifefish
Scorpis aequipinnis Sea Sweep
Scorpis georgiana Banded Sweep
Scorpis lineolata Silver Sweep
Scorpis violacea Violet Sweep

SICKLEFISHES

Drepaneidae

SICKLEFISH
Drepane punctata

There are three species of Sicklefish and they are found in waters around West Africa and in the Indo-Pacific region. They have a deep, compressed body with small ctenoid scales that extend onto the base of the dorsal and anal fins. The forehead is steep, and large adults develop a bony hump on the forehead over the eyes. The small mouth is extremely protrusible, forming a downwardly directed tube, with bands of fine teeth on the jaws. The single dorsal fin has 8–10 spines and 19–22 rays, and the anal fin has three spines and 16–19 rays. The caudal fin is slightly rounded and the pectoral fins are very long and curved.

Sicklefishes are found in northern Australia over sand and mud bottoms and shallow coastal reefs, also entering harbours and estuaries. They reach a maximum length of about 50 cm and feed on small benthic invertebrates such as crustaceans. They are occasionally taken by anglers and prawn trawlers in northern waters, but very little is known of their biology.

AUSTRALIAN DREPANEIDAE SPECIES
Drepane longimana Banded Sicklefish
Drepane punctata Sicklefish

BUTTERFLYFISHES

Chaetodontidae

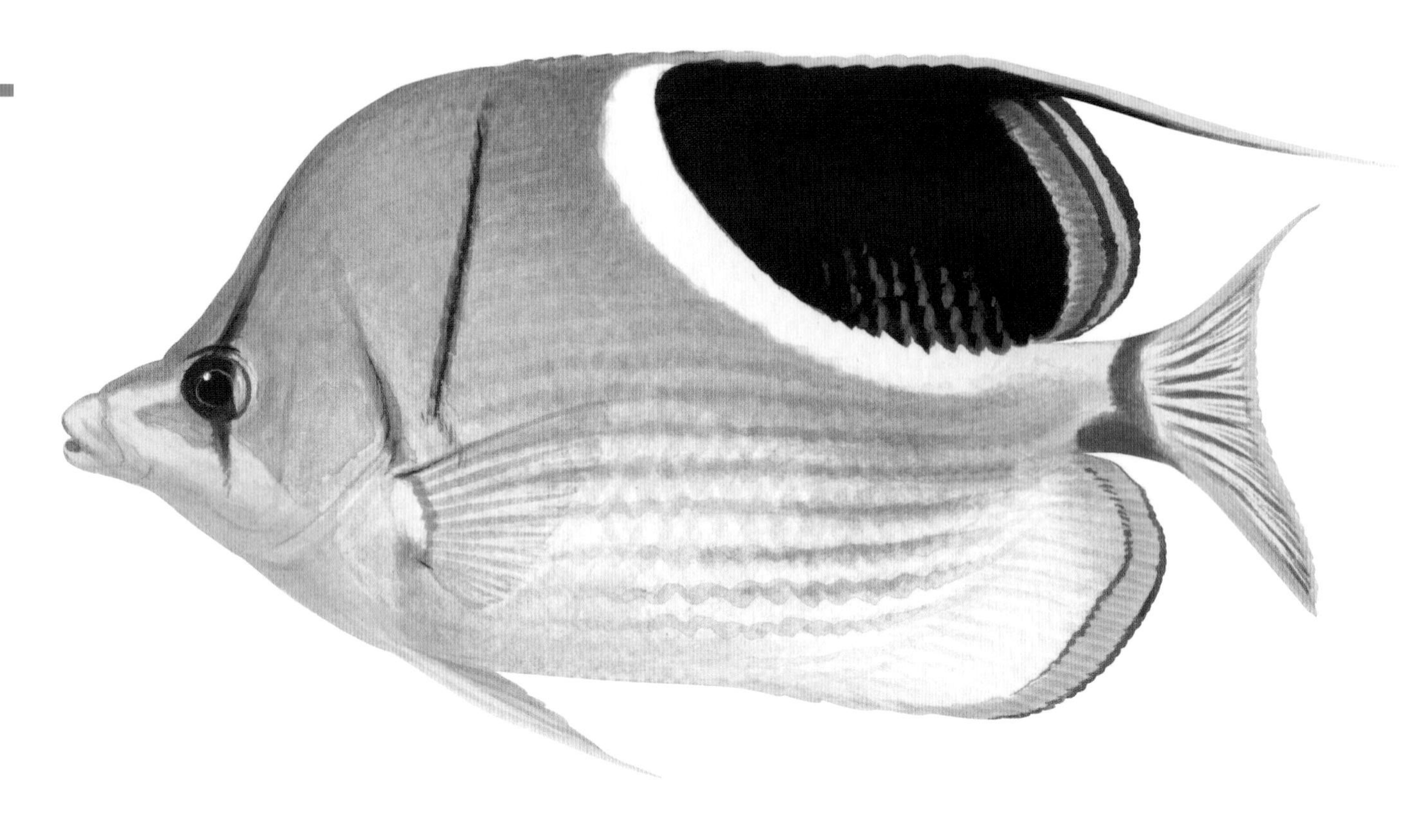

SADDLE BUTTERFLYFISH
Chaetodon ephippium

Butterflyfishes are found in tropical and warm temperate waters around the world, and are particularly diverse in the Indo-Pacific region. This large family contains about 122 species. Butterflyfishes have oval to circular, strongly compressed bodies with small ctenoid scales that extend onto the fins. The small mouth, which contains fine bristle-like teeth, is often formed into an elongate snout. The single long-based dorsal fin has 6–16 spines and 15–30 rays, the anal fin has 3–5 spines and 14–23 rays, and the caudal fin is rounded to slightly forked.

These brightly coloured residents of tropical coral reefs are familiar to divers, being one of the most conspicuous of the reef fish families. Several species, including the Talmas and **Western Butterflyfish**, occur in southern Australian waters over rocky reefs, and many of the tropical species also occasionally appear there, having been carried south as juveniles by ocean currents. Butterflyfishes are found from the shallows down to about 60 m, with a few species, such as the **Tripleband Butterflyfish**, preferring deeper water, down to 200 m or so. They occur individually, in pairs, or in small groups, and feed mainly on coral polyps, algae and small invertebrates, nipping and picking up morsels with their small mouth. However, several species, including the **Schooling Bannerfish** and the **Pyramid Butterflyfish**, form large schools in midwater, where they feed on zooplankton. The **Longnose Butterflyfish** and **Forceps Fish** have an extremely elongate snout, enabling them to pick small invertebrates from amongst branching corals. Many species of Butterflyfish form mating pairs, a relationship that in some is maintained for life. Butterflyfishes are extremely popular aquarium fishes and have been overfished in some areas to supply the aquarium trade.

LORD HOWE BUTTERFLYFISH
Amphichaetodon howensis

TRIANGULAR BUTTERFLYFISH
Chaetodon baronessa

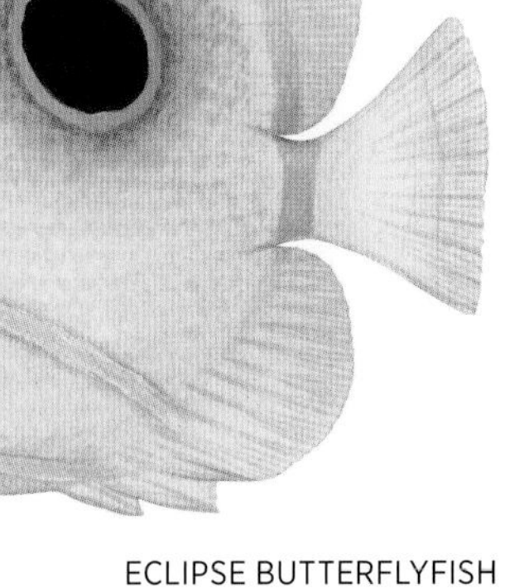

ECLIPSE BUTTERFLYFISH
Chaetodon bennetti

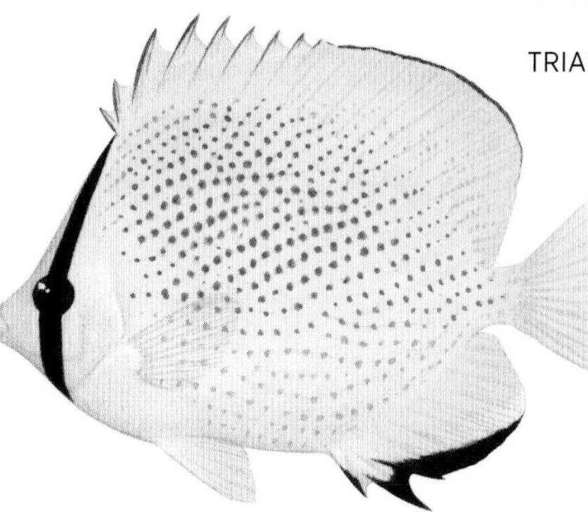

SPECKLED BUTTERFLYFISH
Chaetodon citrinellus

KLEIN'S BUTTERFLYFISH
Chaetodon kleinii

LINED BUTTERFLYFISH
Chaetodon lineolatus

RACCOON BUTTERFLYFISH
Chaetodon lunula

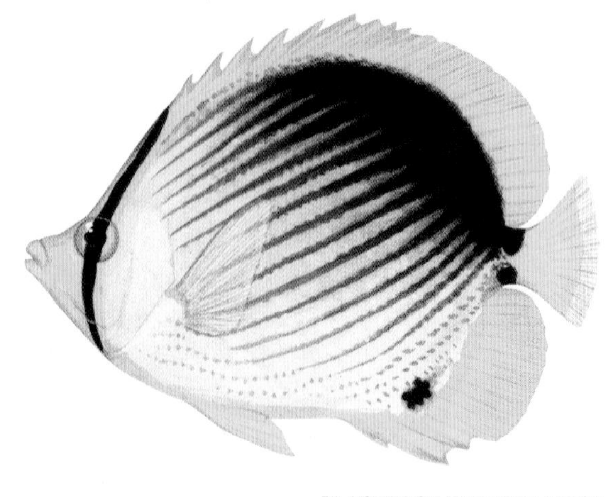

BLACKBACK BUTTERFLYFISH
Chaetodon melannotus

MERTENS' BUTTERFLYFISH
Chaetodon mertensii

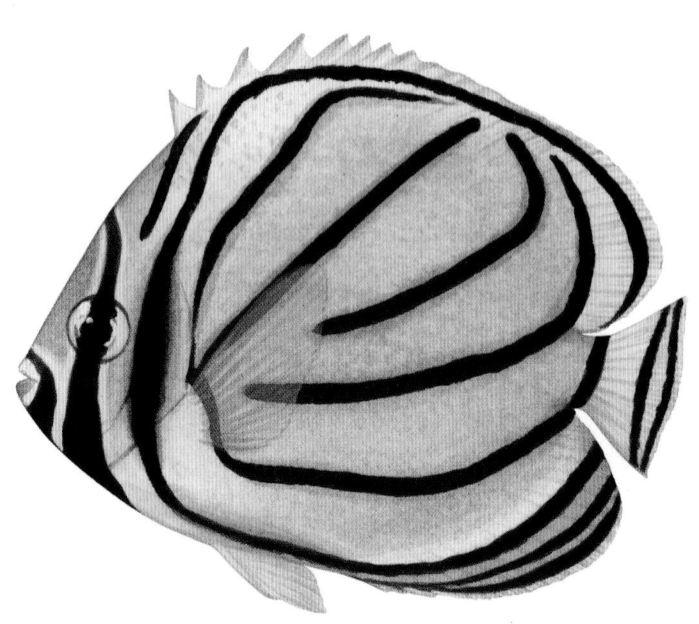

MEYER'S BUTTERFLYFISH
Chaetodon meyeri

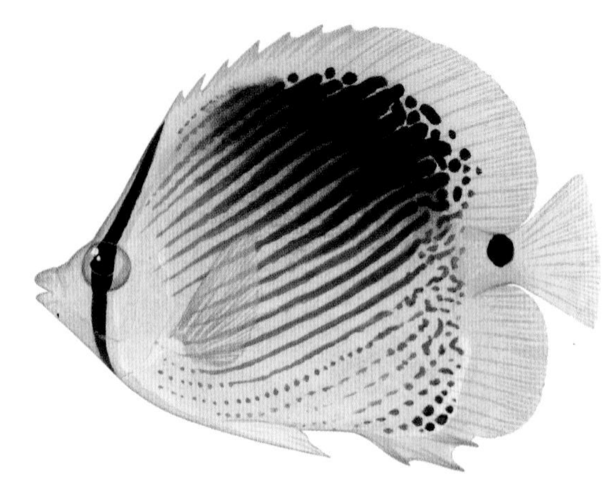

SPOT-TAIL BUTTERFLYFISH
Chaetodon ocellicaudus

ORNATE BUTTERFLYFISH
Chaetodon ornatissimus

SPOTNAPE BUTTERFLYFISH
Chaetodon oxycephalus

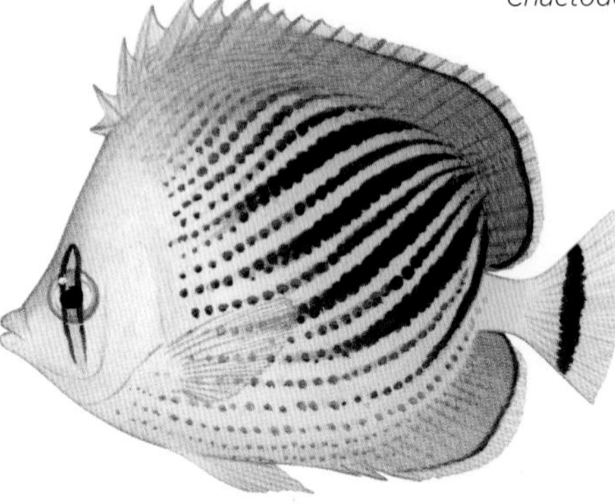

DOT-AND-DASH BUTTERFLYFISH
Chaetodon pelewensis

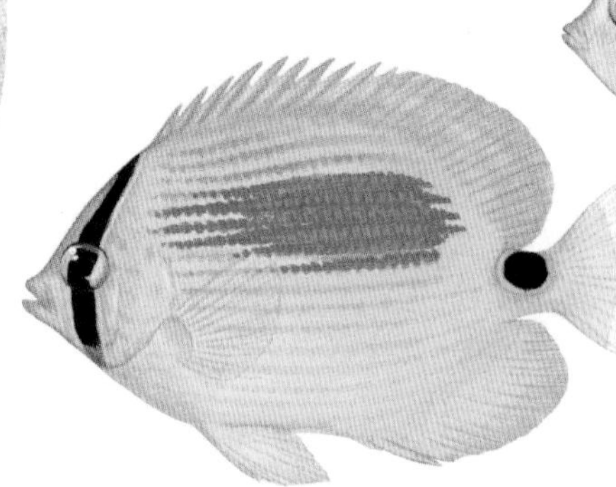

BLUESPOT BUTTERFLYFISH
Chaetodon plebeius

SPOTBANDED BUTTERFLYFISH
Chaetodon punctatofasciatus

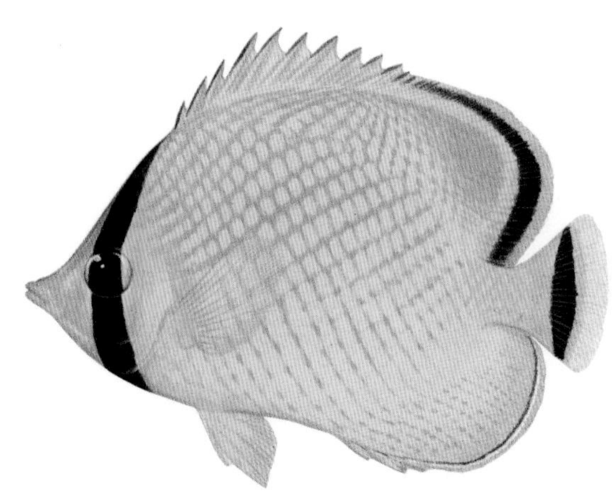

LATTICE BUTTERFLYFISH
Chaetodon rafflesii

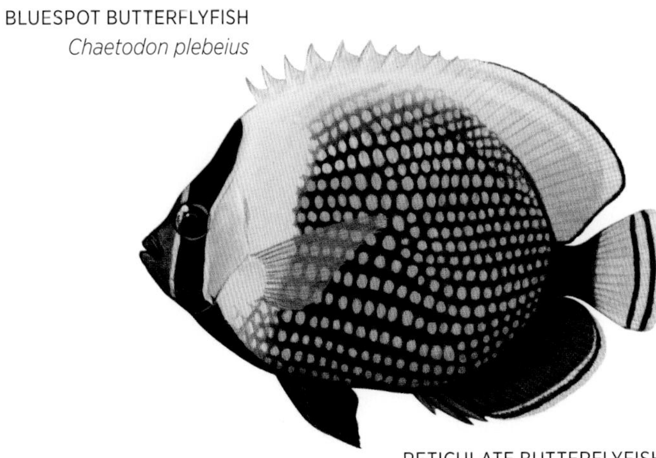

RETICULATE BUTTERFLYFISH
Chaetodon reticulatus

DOTTED BUTTERFLYFISH
Chaetodon semeion

OVALSPOT BUTTERFLYFISH
Chaetodon speculum

CHEVRON BUTTERFLYFISH
Chaetodon trifascialis

TEARDROP BUTTERFLYFISH
Chaetodon unimaculatus

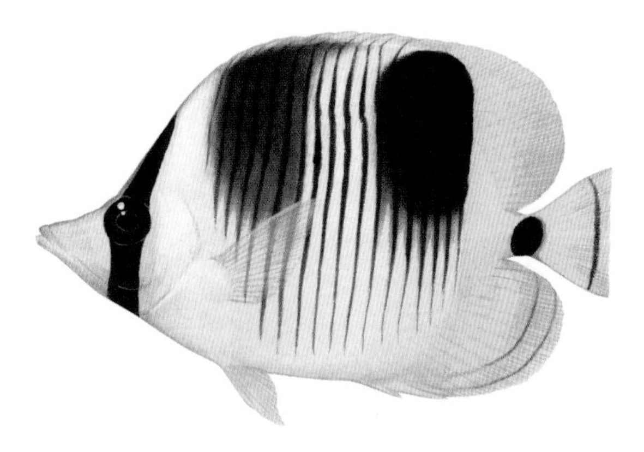

DOUBLESADDLE BUTTERFLYFISH
Chaetodon ulietensis

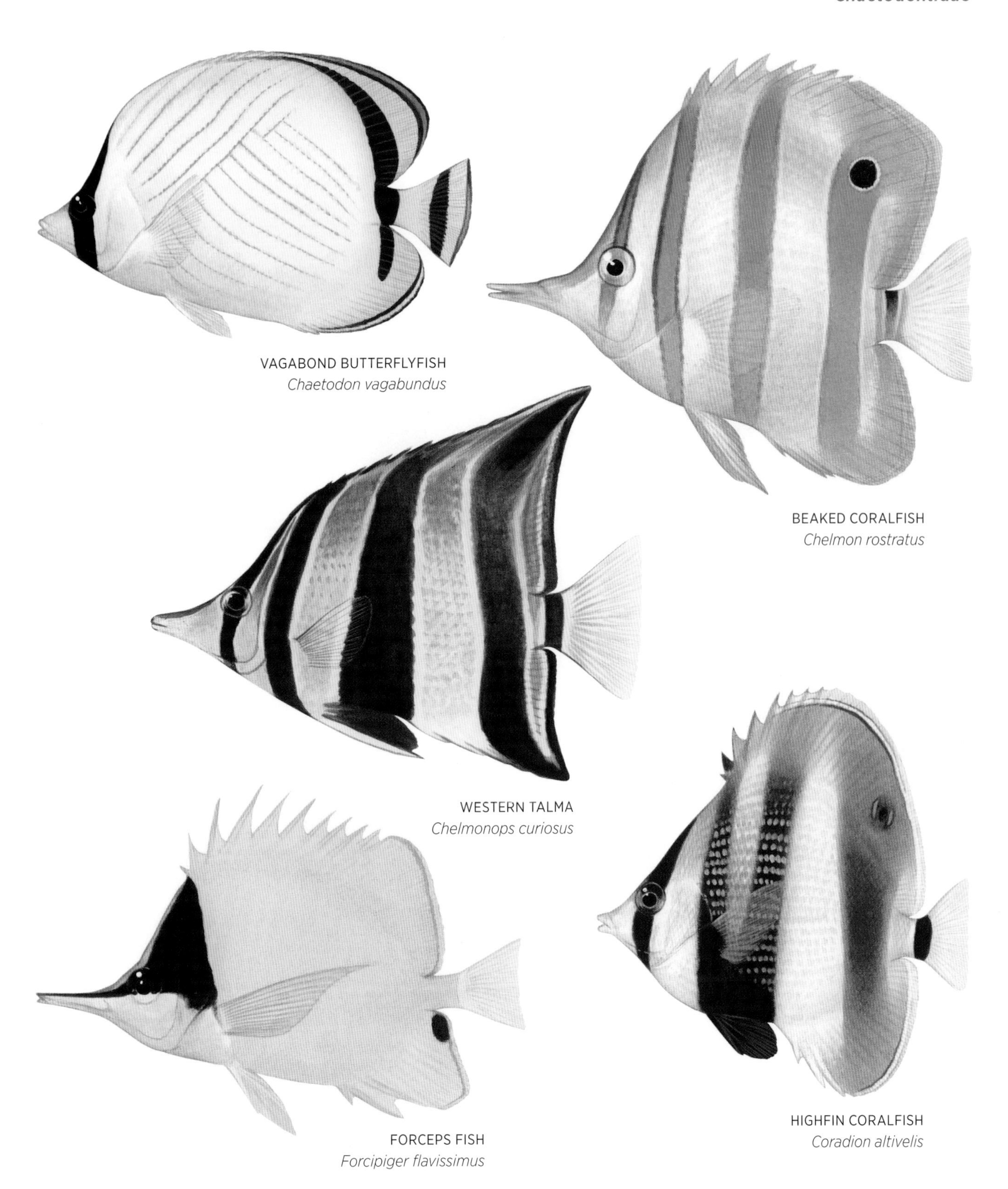

VAGABOND BUTTERFLYFISH
Chaetodon vagabundus

BEAKED CORALFISH
Chelmon rostratus

WESTERN TALMA
Chelmonops curiosus

HIGHFIN CORALFISH
Coradion altivelis

FORCEPS FISH
Forcipiger flavissimus

PYRAMID BUTTERFLYFISH
Hemitaurichthys polylepis

LONGFIN BANNERFISH
Heniochus acuminatus

PENNANT BANNERFISH
Heniochus chrysostomus

MASKED BANNERFISH
Heniochus monoceros

SINGULAR BANNERFISH
Heniochus singularius

HORNED BANNERFISH
Heniochus varius

OCELLATE BUTTERFLYFISH
Parachaetodon ocellatus

AUSTRALIAN CHAETODONTIDAE SPECIES

Amphichaetodon howensis Lord Howe Butterflyfish
Chaetodon adiergastos Philippine Butterflyfish
Chaetodon assarius Western Butterflyfish
Chaetodon aureofasciatus Goldstripe Butterflyfish
Chaetodon auriga Threadfin Butterflyfish
Chaetodon baronessa Triangular Butterflyfish
Chaetodon bennetti Eclipse Butterflyfish
Chaetodon citrinellus Speckled Butterflyfish
Chaetodon decussatus Indian Vagabond Butterflyfish
Chaetodon ephippium Saddle Butterflyfish
Chaetodon flavirostris Dusky Butterflyfish
Chaetodon guentheri Gunther's Butterflyfish
Chaetodon kleinii Klein's Butterflyfish
Chaetodon lineolatus Lined Butterflyfish
Chaetodon lunula Raccoon Butterflyfish
Chaetodon lunulatus Redfin Butterflyfish
Chaetodon melannotus Blackback Butterflyfish
Chaetodon mertensii Mertens' Butterflyfish
Chaetodon meyeri Meyer's Butterflyfish
Chaetodon ocellicaudus Spot-tail Butterflyfish
Chaetodon octofasciatus Eightband Butterflyfish
Chaetodon ornatissimus Ornate Butterflyfish
Chaetodon oxycephalus Spotnape Butterflyfish
Chaetodon pelewensis Dot-and-dash Butterflyfish
Chaetodon plebeius Bluespot Butterflyfish
Chaetodon punctatofasciatus Spotbanded Butterflyfish
Chaetodon rafflesii Lattice Butterflyfish
Chaetodon rainfordi Rainford's Butterflyfish
Chaetodon reticulatus Reticulate Butterflyfish
Chaetodon semeion Dotted Butterflyfish
Chaetodon speculum Ovalspot Butterflyfish
Chaetodon tricinctus Threeband Butterflyfish
Chaetodon trifascialis Chevron Butterflyfish
Chaetodon ulietensis Doublesaddle Butterflyfish
Chaetodon unimaculatus Teardrop Butterflyfish
Chaetodon vagabundus Vagabond Butterflyfish
Chelmon marginalis Margined Coralfish
Chelmon muelleri Muller's Coralfish
Chelmon rostratus Beaked Coralfish
Chelmonops curiosus Western Talma
Chelmonops truncatus Eastern Talma
Coradion altivelis Highfin Coralfish
Coradion chrysozonus Orangebanded Coralfish
Forcipiger flavissimus Forceps Fish
Forcipiger longirostris Longnose Butterflyfish
Hemitaurichthys polylepis Pyramid Butterflyfish
Heniochus acuminatus Longfin Bannerfish
Heniochus chrysostomus Pennant Bannerfish
Heniochus diphreutes Schooling Bannerfish
Heniochus monoceros Masked Bannerfish
Heniochus singularius Singular Bannerfish
Heniochus varius Horned Bannerfish
Parachaetodon ocellatus Ocellate Butterflyfish
Roa australis Tripleband Butterflyfish

ANGELFISHES

Pomacanthidae

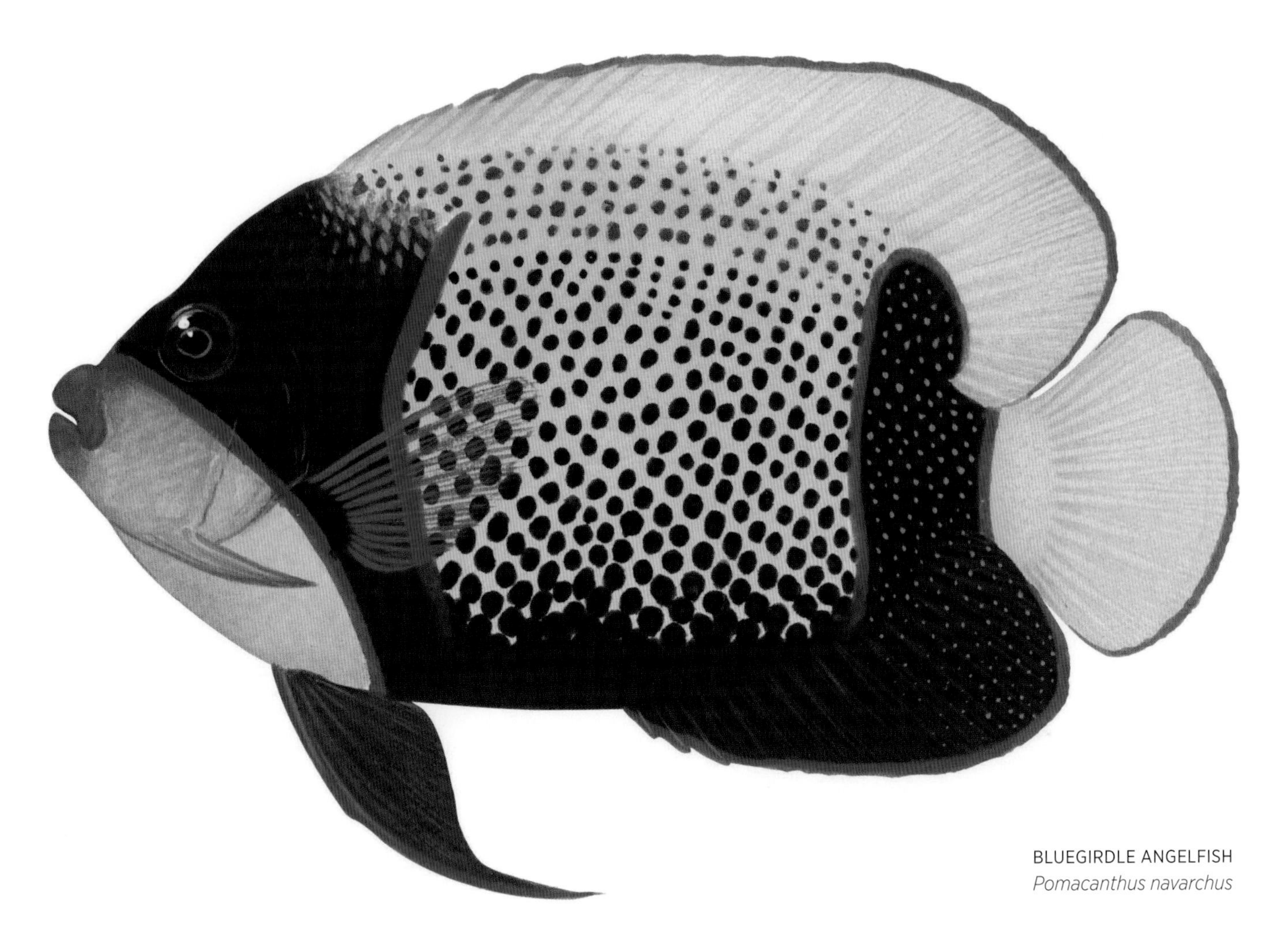

BLUEGIRDLE ANGELFISH
Pomacanthus navarchus

There are about 82 species of Angelfish and they are found in tropical waters of the Atlantic, Indian and Pacific oceans. Angelfishes have oval to circular, compressed bodies with small ctenoid scales that extend onto the fins, a small mouth with bristle-like teeth, and a large, rearward-directed spine on the angle of the preoperculum. The single long-based dorsal fin has 9–15 spines and 15–37 rays, and the anal fin has three spines and 14–25 rays. The caudal fin is rounded to lunate, sometimes with trailing filaments on the upper and lower lobes.

In Australia Angelfishes are generally found on or near coral reefs in northern waters, rarely deeper than about 50 m, though some *Genicanthus* species are found down to at least 75 m and new Angelfishes are still being discovered on even deeper reefs. Diet varies between the different species and may include algae, sponges, zooplankton and small benthic invertebrates. Many species have remarkably beautiful colour patterns, often with radically different juvenile and adult markings.

Juveniles of *Pomacanthus* species have striking patterns, with black, blue and white lines on the body and fins. Adults of the larger *Pomacanthus* species can reach up to about 50 cm in length. Species in this genus are found over reefs and in lagoons where there is abundant coral. Most are quite shy and retreat to the shelter of caves and ledges at the approach of divers. They feed predominantly on sponges and tunicates.

Genicanthus species are usually found in midwater, feeding on zooplankton. The *Centropyge* species, which reach about 10–15 cm in length, are primarily algal feeders and are found close to the substrate, never far from a sheltering crevice or hole. Most Angelfishes are hermaphroditic: a male will defend a territory and a small group of females, and if the male is removed one of the females can change sex to replace it. Angelfishes are popular aquarium fishes but most are difficult to keep successfully, having specific dietary requirements.

THREESPOT ANGELFISH
Apolemichthys trimaculatus

BICOLOR ANGELFISH
Centropyge bicolour

TWOSPINE ANGELFISH
Centropyge bispinosa

WHITETAIL ANGELFISH
Centropyge flavicauda

FLAME ANGELFISH
Centropyge loricula

MIDNIGHT ANGELFISH
Centropyge nox

KEYHOLE ANGELFISH
Centropyge tibicen

PEARLSCALE ANGELFISH
Centropyge vroliki

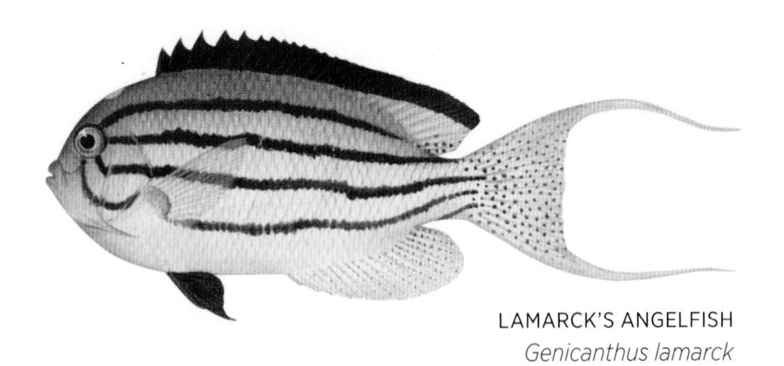

LAMARCK'S ANGELFISH
Genicanthus lamarck

VERMICULATE ANGELFISH
Chaetodontoplus mesoleucus

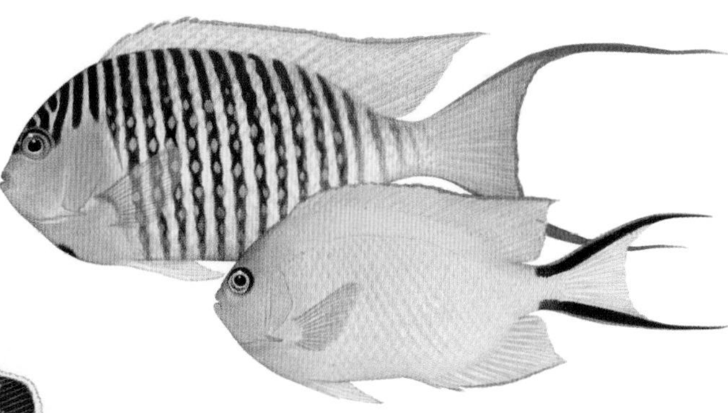

SWALLOWTAIL ANGELFISH (male and female)
Genicanthus melanospilos

MULTIBAR ANGELFISH
Paracentropyge multifasciatus

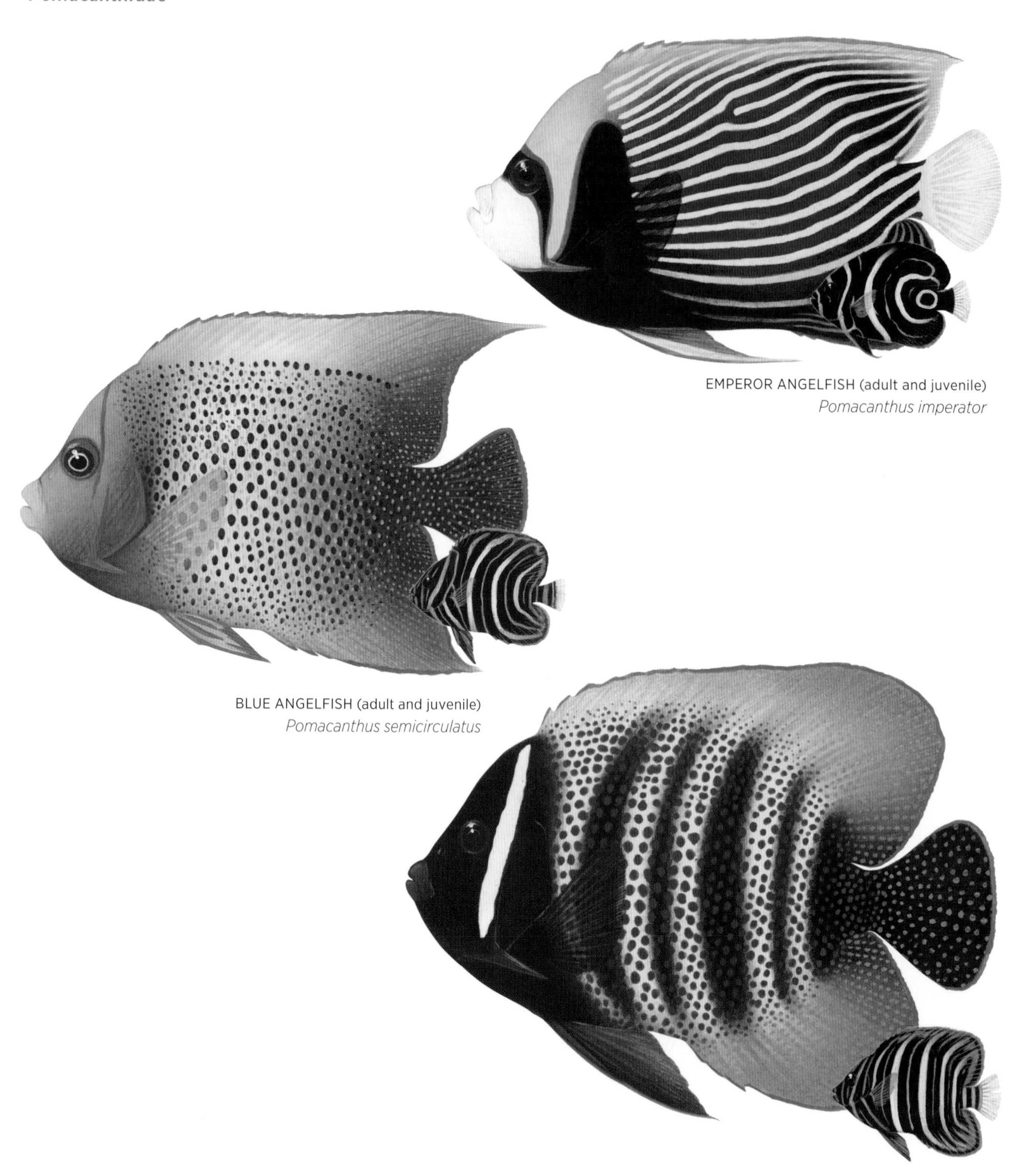

EMPEROR ANGELFISH (adult and juvenile)
Pomacanthus imperator

BLUE ANGELFISH (adult and juvenile)
Pomacanthus semicirculatus

SIXBAND ANGELFISH (adult and juvenile)
Pomacanthus sexstriatus

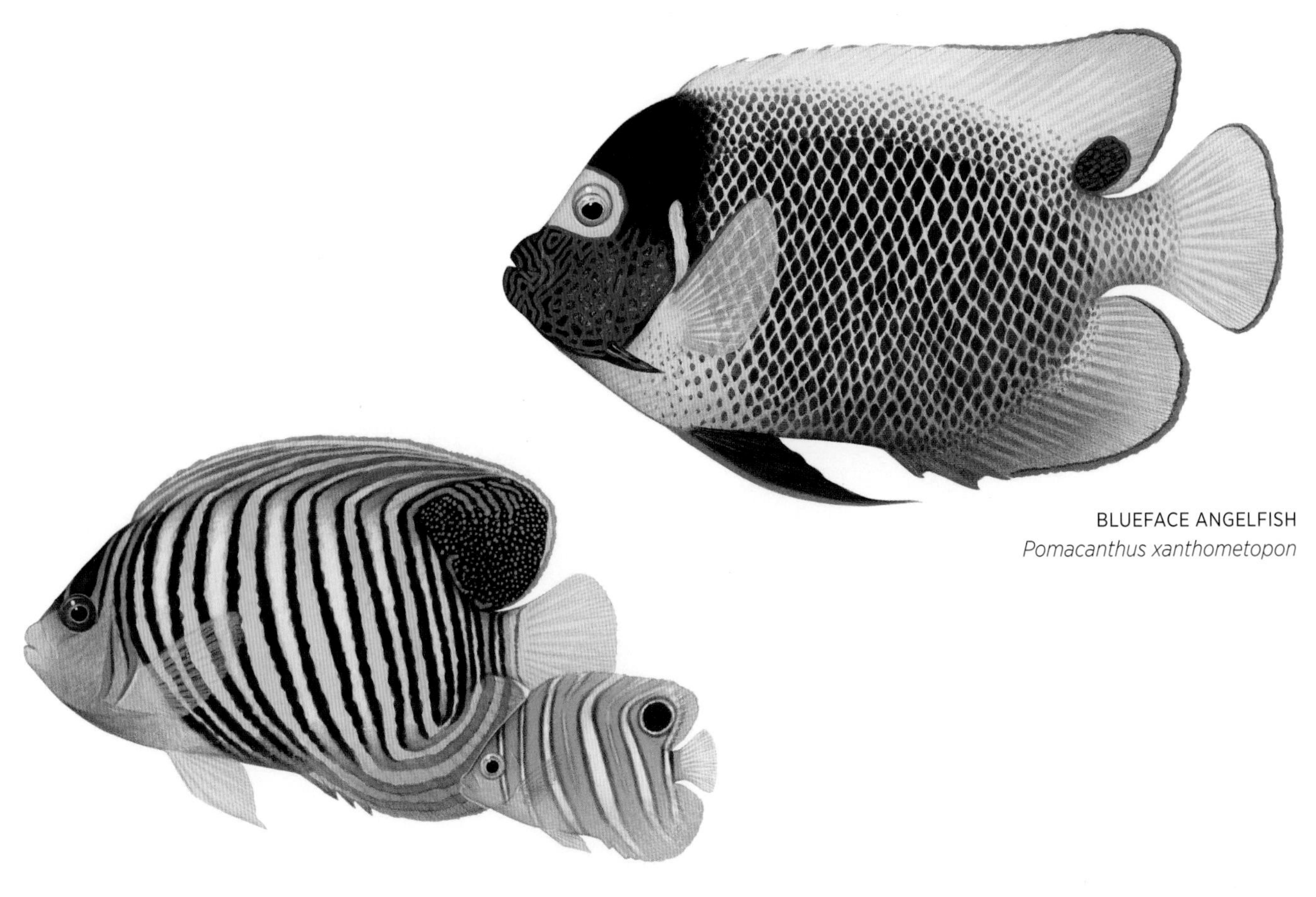

BLUEFACE ANGELFISH
Pomacanthus xanthometopon

REGAL ANGELFISH (adult and juvenile)
Pygoplites diacanthus

AUSTRALIAN POMACANTHIDAE SPECIES

Apolemichthys trimaculatus Threespot Angelfish
Centropyge aurantia Golden Angelfish
Centropyge bicolor Bicolor Angelfish
Centropyge bispinosa Twospine Angelfish
Centropyge eibli Eibl's Angelfish
Centropyge flavicauda Whitetail Angelfish
Centropyge flavissima Lemonpeel Angelfish
Centropyge heraldi Yellow Angelfish
Centropyge loricula Flame Angelfish
Centropyge nox Midnight Angelfish
Centropyge tibicen Keyhole Angelfish
Centropyge vroliki Pearlscale Angelfish
Chaetodontoplus ballinae Ballina Angelfish
Chaetodontoplus conspicillatus Conspicuous Angelfish
Chaetodontoplus duboulayi Scribbled Angelfish
Chaetodontoplus melanosoma Blackvelvet Angelfish
Chaetodontoplus mesoleucus Vermiculate Angelfish
Chaetodontoplus personifer Yellowtail Anglefish
Genicanthus lamarck Lamarck's Angelfish
Genicanthus melanospilos Swallowtail Angelfish
Genicanthus semicinctus Halfbanded Angelfish
Genicanthus watanabei Watanabe's Angelfish
Paracentropyge multifasciatus Multibar Angelfish
Pomacanthus imperator Emperor Angelfish
Pomacanthus navarchus Bluegirdle Angelfish
Pomacanthus semicirculatus Blue Angelfish
Pomacanthus sexstriatus Sixband Angelfish
Pomacanthus xanthometopon Blueface Angelfish
Pygoplites diacanthus Regal Angelfish

OLD WIFE

Enoplosidae

OLD WIFE
Enoplosus armatus

The single species in this family, the **Old Wife**, is found only in southern Australian waters. It is a distinctive fish, with an oval, compressed body covered in minute cycloid scales that extend onto the base of the fins. There are two sharp spines on the angle of the preoperculum. The Old Wife has two well-separated dorsal fins, the first with eight spines and the second with 14–15 rays that extend into a trailing lobe. The anal fin has three spines and 14–15 rays, the caudal fin is slightly lunate, and the ventral fins are very large. The fin spines are venomous and can inflict painful stings.

The Old Wife is often seen in small schools over reefs and seagrass beds, or hanging in small to large groups beneath jetties. It is quite common within its range, occurring from shallow water down to about 100 m. The Old Wife feeds on small benthic crustaceans and other invertebrates.

AUSTRALIAN ENOPLOSIDAE SPECIES

Enoplosus armatus Old Wife

BOARFISHES

Pentacerotidae

YELLOWSPOTTED BOARFISH
Paristiopterus gallipavo

Boarfishes are found in the Indo-Pacific and south-west Atlantic oceans, and there are about 12 species in the family. They have oval to deep, compressed bodies, with small ctenoid scales, a projecting snout, and a small mouth with bands of fine teeth. Their large head bears striated bony plates. They have a single dorsal fin, often deeply notched between the spinous and soft parts, with 4–15 strong spines and 8–29 rays. The anal fin has 2–5 strong spines and 6–17 rays, the ventral fins are large with a strong spine, and the caudal fin is truncate to slightly forked.

Some species, such as the **Short Boarfish**, range onto shallow coastal reefs, but most Boarfishes occur over deeper offshore reefs. In Australia the **Bigspine Boarfish** and **Pelagic Armourhead** are found at depths of about 300–700 m off the southern coast, and the **Longsnout Boarfish**, **Giant Boarfish** and **Yellowspotted Boarfish** are occasionally encountered by anglers or divers in southern waters on deep coastal reefs. The **Threebar Boarfish** occurs from about 40 to 400 m over rocky bottoms in northern waters and is sometimes seen by divers. Usually, however, Boarfishes are only encountered in the nets of deepwater trawlers. The Giant Boarfish can grow to a length of about 90 cm, but most other species are from 30 to 50 cm. Boarfishes feed mainly on benthic invertebrates such as molluscs and crustaceans. They are considered excellent food fishes.

THREEBAR BOARFISH
Histiopterus typus

SHORT BOARFISH
Parazanclistius hutchinsi

LONGSNOUT BOARFISH
Pentaceropsis recurvirostris

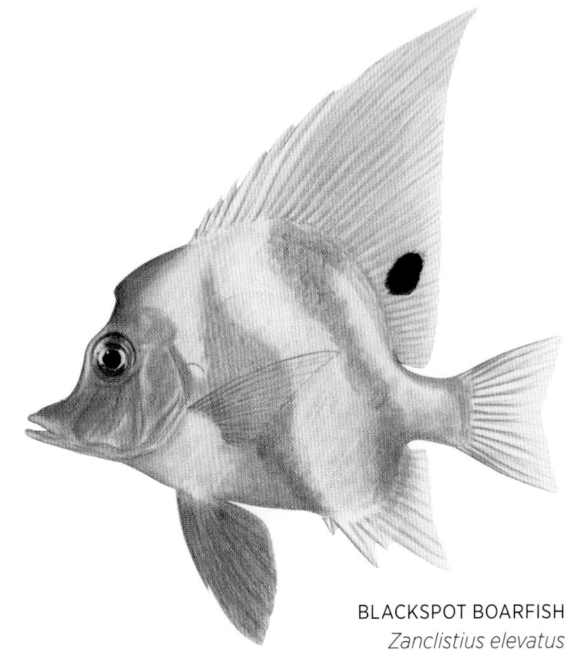

BLACKSPOT BOARFISH
Zanclistius elevatus

AUSTRALIAN PENTACEROTIDAE SPECIES

Evistias acutirostris Striped Boarfish
Histiopterus typus Threebar Boarfish
Parazanclistius hutchinsi Short Boarfish
Paristiopterus gallipavo Yellowspotted Boarfish
Paristiopterus labiosus Giant Boarfish
Pentaceropsis recurvirostris Longsnout Boarfish
Pentaceros capensis Cape Armourhead
Pentaceros decacanthus Bigspine Boarfish
Pseudopentaceros richardsoni Pelagic Armourhead
Zanclistius elevatus Blackspot Boarfish

GRUNTERS

Terapontidae

SOOTY GRUNTER
Hephaestus fuliginosus

Grunters are found in coastal marine, brackish and fresh waters of the Indo-West Pacific. There are about 48 species in the family. They have elongate to oval, compressed bodies, with small ctenoid scales that extend onto the base of the fins and form a scaly sheath into which the fin spines can be depressed. The small mouth has bands of villiform teeth or flattened incisors. Grunters have a single dorsal fin with 11–14 spines and 8–14 rays. The anal fin has three spines and 7–12 rays, and the caudal fin is truncate to slightly forked. They have two sharp spines on the operculum. Muscles attached to the swim bladder enable them to make a drumming or grunting noise.

A few species, including the Striped Grunters and **Sea Trumpeter**, are found in marine habitats such as estuaries, river mouths and shallow coastal waters around southern Australia. The **Yellowtail Grunter** is a marine species that is capable of tolerating extremely high salinity, and it is common in shallow environments such as Shark Bay in Western Australia. The **Crescent Grunter** is abundant in shallow coastal waters around the northern half of Australia, and the **Spinycheek Grunter** and **Largescale Grunter** are found in similar habitats. Most Australian Grunters, however, are found in freshwater habitats such as streams, rivers and lakes in the north of the country. Some have very restricted distribution, such as the **Drysdale Grunter**, which is found only in the Drysdale River in Western Australia. The **Silver Perch**, which can reach 40 cm in length, is widespread in the Murray-Darling drainage, though its numbers have declined dramatically since the introduction of artificial barriers to its upstream spawning migrations. It is a popular angling species and a fine food fish, and is now the subject of extensive aquaculture. The **Spangled Perch**, which reaches about 30 cm in length, is very widespread, and can be found in most freshwater systems, except those along the south coast.

Most freshwater Grunters are carnivorous, feeding on aquatic and terrestrial invertebrates, small fishes, frogs and crustaceans. However, *Pingalla* and *Syncomistes* species are herbivorous, with specially adapted teeth for scraping algae from the substrate. The Sooty Grunters, along with several other very similar species, are sought-after by anglers in northern Australian rivers and are excellent food fishes.

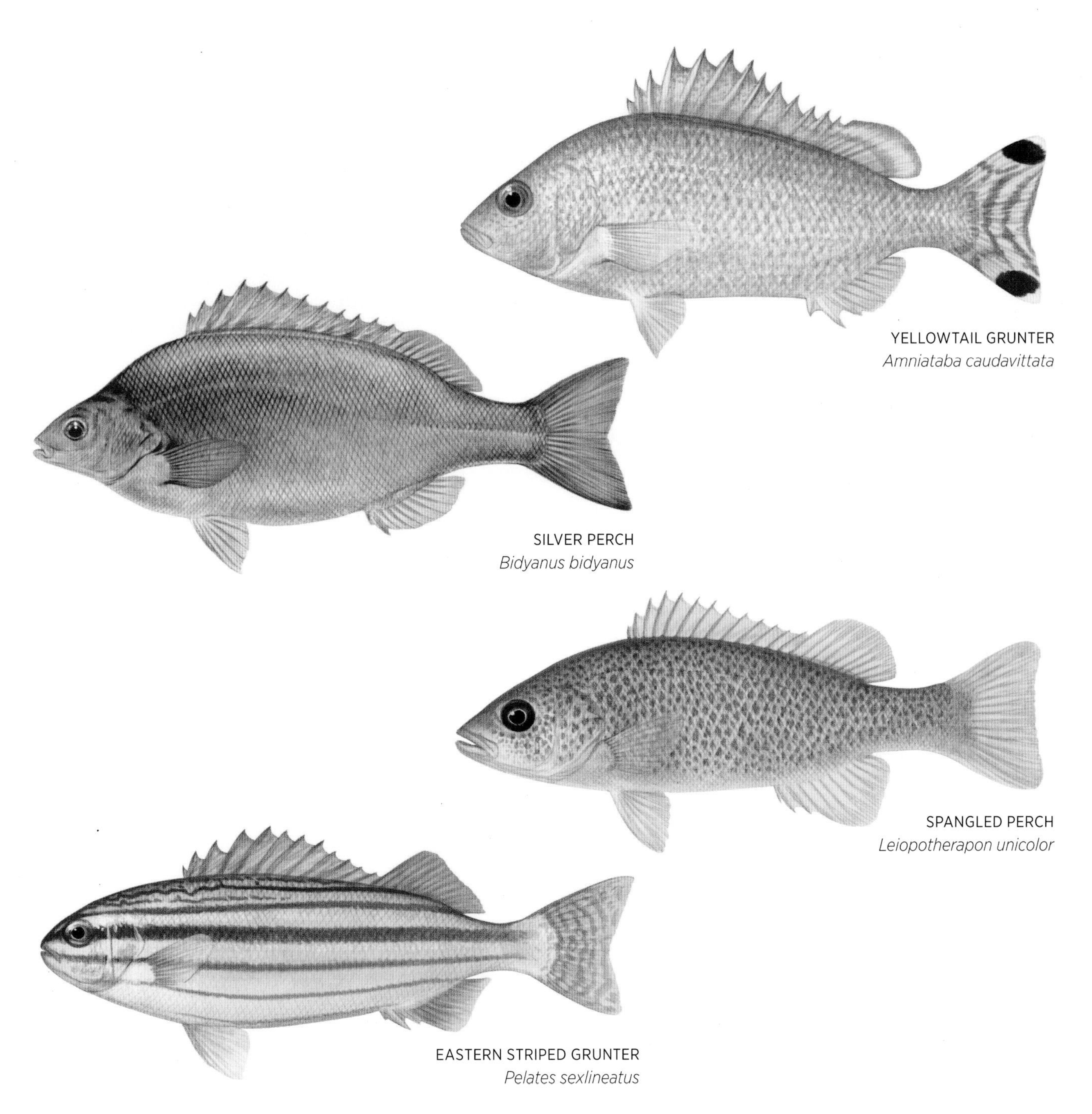

YELLOWTAIL GRUNTER
Amniataba caudavittata

SILVER PERCH
Bidyanus bidyanus

SPANGLED PERCH
Leiopotherapon unicolor

EASTERN STRIPED GRUNTER
Pelates sexlineatus

SEA TRUMPETER
Pelsartia humeralis

DRYSDALE GRUNTER
Syncomistes rastellus

CRESCENT GRUNTER
Terapon jarbua

AUSTRALIAN TERAPONTIDAE SPECIES

Amniataba caudavittata Yellowtail Grunter
Amniataba percoides Barred Grunter
Bidyanus bidyanus Silver Perch
Bidyanus welchi Welch's Grunter
Hannia greenwayi Greenway's Grunter
Hephaestus carbo Coal Grunter
Hephaestus epirrhinos Longnose Sooty Grunter
Hephaestus fuliginosus Sooty Grunter
Hephaestus jenkinsi Western Sooty Grunter
Hephaestus tulliensis Khaki Grunter
Leiopotherapon aheneus Fortescue Grunter
Leiopotherapon macrolepis Kimberley Spangled Perch
Leiopotherapon unicolor Spangled Perch
Mesopristes argenteus Silver Grunter
Pelates octolineatus Western Striped Grunter
Pelates quadrilineatus Fourline Striped Grunter
Pelates sexlineatus Eastern Striped Grunter
Pelsartia humeralis Sea Trumpeter
Pingalla gilberti Gilbert's Grunter
Pingalla lorentzi Lorentz Grunter
Pingalla midgleyi Midgley's Grunter
Scortum barcoo Barcoo Grunter
Scortum hillii Leathery Grunter
Scortum neili Angalarri Grunter
Scortum ogilbyi Gulf Grunter
Scortum parviceps Smallhead Grunter
Syncomistes butleri Sharpnose Grunter
Syncomistes kimberleyensis Kimberley Grunter
Syncomistes rastellus Drysdale Grunter
Syncomistes trigonicus Longnose Grunter
Terapon jarbua Crescent Grunter
Terapon puta Spinycheek Grunter
Terapon theraps Largescale Grunter
Variichthys lacustris Lake Grunter

FLAGTAILS

Kuhliidae

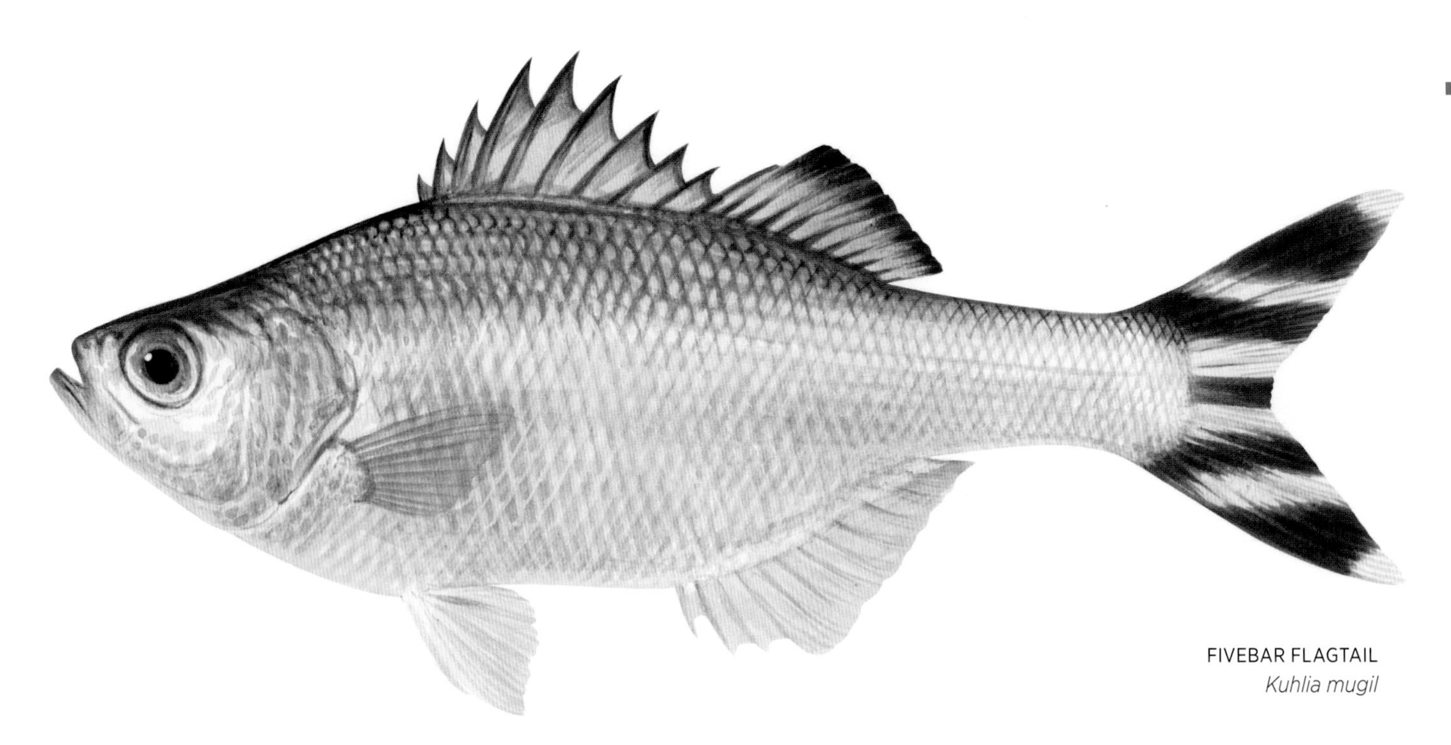

FIVEBAR FLAGTAIL
Kuhlia mugil

Flagtails occur in coastal marine, brackish and fresh waters of the Indo-Pacific, and there are about 10 species in the family. They are small fishes with oval to elongate, compressed bodies and small ctenoid scales that extend onto the base of the fins, forming a scaly sheath. They have a moderate-sized mouth with bands of fine teeth, and two spines on the operculum. The single, notched dorsal fin has 10–12 spines and 9–16 rays, the anal fin has three spines and 9–13 rays, and the caudal fin is slightly forked. Most have conspicuous dark markings on the tail.

The **Jungle Perch** grows to about 45 cm and is found in streams and rivers in north-eastern Australia, preferring clear, fast-flowing water. It feeds on insects, crustaceans, small fishes, and fruit that falls into the water, and is a popular angling fish. Adults occur only in fresh water, though they probably have a marine larval stage, which would contribute to their wide distribution. The other Australian Flagtail species are found in coastal waters in the north, over coral reefs, along beaches and in estuaries. The **Spotted Flagtail** and **Silver Flagtail** also occur in the lower reaches of coastal streams. The **Fivebar Flagtail** is very widespread, occurring from Africa to the eastern Pacific and forming dense schools in surge areas along rocky shore-lines and over reefs. It feeds nocturnally, mainly on pelagic crustaceans, and reaches a maximum length of about 20 cm.

JUNGLE PERCH
Kuhlia rupestris

AUSTRALIAN KUHLIIDAE SPECIES

Kuhlia marginata Spotted Flagtail
Kuhlia mugil Fivebar Flagtail
Kuhlia munda Silver Flagtail
Kuhlia rupestris Jungle Perch

KNIFEJAWS

Oplegnathidae

KNIFEJAW
Oplegnathus woodwardi

Knifejaws are found in temperate waters of the Indo-Pacific, and the family contains about seven species. They have an oval, compressed body with very small, finely ctenoid scales. The strong jaws have teeth fused to form a beak – a structure similar to that found in Parrotfishes (Scaridae, p. 634). The single dorsal fin has 11–12 spines and 11–22 rays, the anal fin has three spines and 11–16 rays, and the caudal fin is slightly forked.

The **Knifejaw** is found in southern Australian waters, over reefs and rocky bottoms from about 50–400 m. It occasionally enters shallow coastal areas but is generally found on deeper reefs. The beak-like teeth in the jaws and their strong pharyngeal teeth are adapted for crushing hard-shelled prey such as molluscs, crustaceans and echinoderms. The Knifejaw is an excellent food fish but is not often encountered.

AUSTRALIAN OPLEGNATHIDAE SPECIES
Oplegnathus woodwardi Knifejaw

HAWKFISHES

Cirrhitidae

FRECKLED HAWKFISH
Paracirrhites forsteri

Hawkfishes are found in tropical waters of the Atlantic, Indian and Pacific oceans, with the majority of the 33 species in the family occurring in the Indo-Pacific region. They have oval to elongate, rounded to slightly compressed bodies with small, cycloid scales. The small mouth bears an outer row of canines and inner bands of villiform teeth, and there is a fringe of small filaments around the rear of the nostrils. The single dorsal fin has 10 spines and 11–17 rays, and the spines have distinctive tufts of short, branched filaments at the tips. The short-based anal fin has three spines and 5–7 rays, and the pectoral fins are large, with the lower rays thickened, unbranched and separate from the other rays. These lower rays are used by Hawkfishes to prop themselves on the substrate, usually on a conspicuous outcrop with a clear view of the surroundings. From such positions they ambush their prey, lunging forward to snap up small fishes and crustaceans.

Many Hawkfishes are brightly coloured, and they are often seen by divers on coral reefs around the north of Australia. The **Longnose Hawkfish** is usually found amongst intricately branched gorgonian corals, its cross-hatched colour pattern providing excellent camouflage. The **Lyretail Hawkfish** prefers deeper waters and, unlike most species in the family, feeds on plankton in midwater. Most Hawkfishes are from 10 to 20 cm in length; the **Ornate Hawkfish** is one of the largest, growing to about 30 cm. Hawkfishes are hermaphroditic: male Hawkfishes defend a territory and a small group of females, and females are able to change sex to replace a missing male.

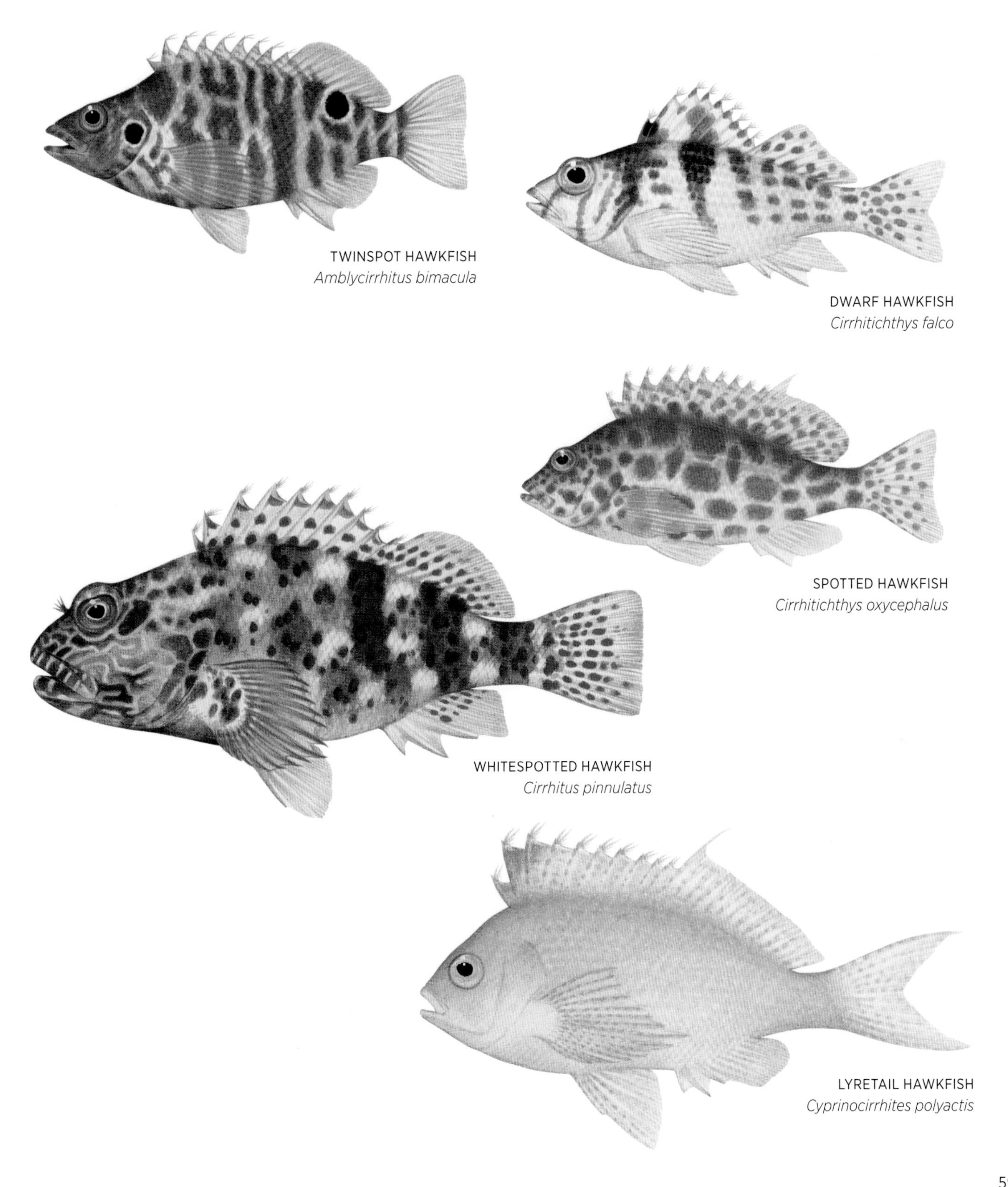

TWINSPOT HAWKFISH
Amblycirrhitus bimacula

DWARF HAWKFISH
Cirrhitichthys falco

SPOTTED HAWKFISH
Cirrhitichthys oxycephalus

WHITESPOTTED HAWKFISH
Cirrhitus pinnulatus

LYRETAIL HAWKFISH
Cyprinocirrhites polyactis

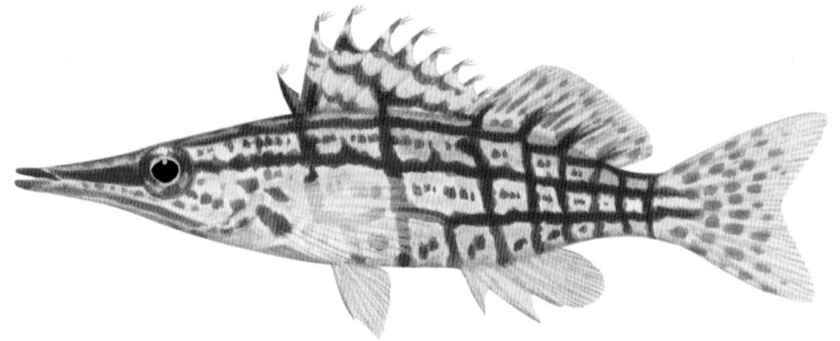

LONGNOSE HAWKFISH
Oxycirrhites typus

RINGEYE HAWKFISH
Paracirrhites arcatus

ORNATE HAWKFISH
Paracirrhites hemistictus

AUSTRALIAN CIRRHITIDAE SPECIES

Amblycirrhitus bimacula Twinspot Hawkfish
Cirrhitichthys aprinus Blotched Hawkfish
Cirrhitichthys falco Dwarf Hawkfish
Cirrhitichthys oxycephalus Spotted Hawkfish
Cirrhitus pinnulatus Whitespotted Hawkfish
Cyprinocirrhites polyactis Lyretail Hawkfish
Neocirrhites armatus Flame Hawkfish
Notocirrhitus splendens Splendid Hawkfish
Oxycirrhites typus Longnose Hawkfish
Paracirrhites arcatus Ringeye Hawkfish
Paracirrhites forsteri Freckled Hawkfish
Paracirrhites hemistictus Ornate Hawkfish

KELPFISHES

Chironemidae

SILVER SPOT
Threpterius maculosus

The five species of Kelpfish are found only in southern waters of Australia and New Zealand. They have an elongate, slightly compressed body with cycloid scales, a moderate-sized mouth with conical or villiform teeth, and tufts of small filaments on the nostrils. The single, notched dorsal fin has 14–16 spines and 15–21 rays, the anal fin has three spines and 6–8 rays, and the pectoral fins are large, with the lower six rays thickened, unbranched and separate from the other rays.

Kelpfishes inhabit weed beds and rocky reefs in shallow coastal waters, often in areas of strong surge, holding themselves in position with their strong pectoral fins. They are reasonably common, but are not often seen due to their secretive nature and excellent camouflage. The **Western Kelpfish** has a tuft of branched filaments at the tip of each dorsal spine and strongly resembles the Hawkfishes (Cirrhitidae, p. 594). Kelpfishes feed on a variety of small benthic invertebrates. The **Eastern Kelpfish** is the largest member of the family, reaching a maximum length of about 40 cm.

AUSTRALIAN CHIRONEMIDAE SPECIES

Chironemus georgianus Western Kelpfish
Chironemus marmoratus Eastern Kelpfish
Chironemus microlepis Smallscale Kelpfish
Threpterius maculosus Silver Spot

MARBLEFISHES

Aplodactylidae

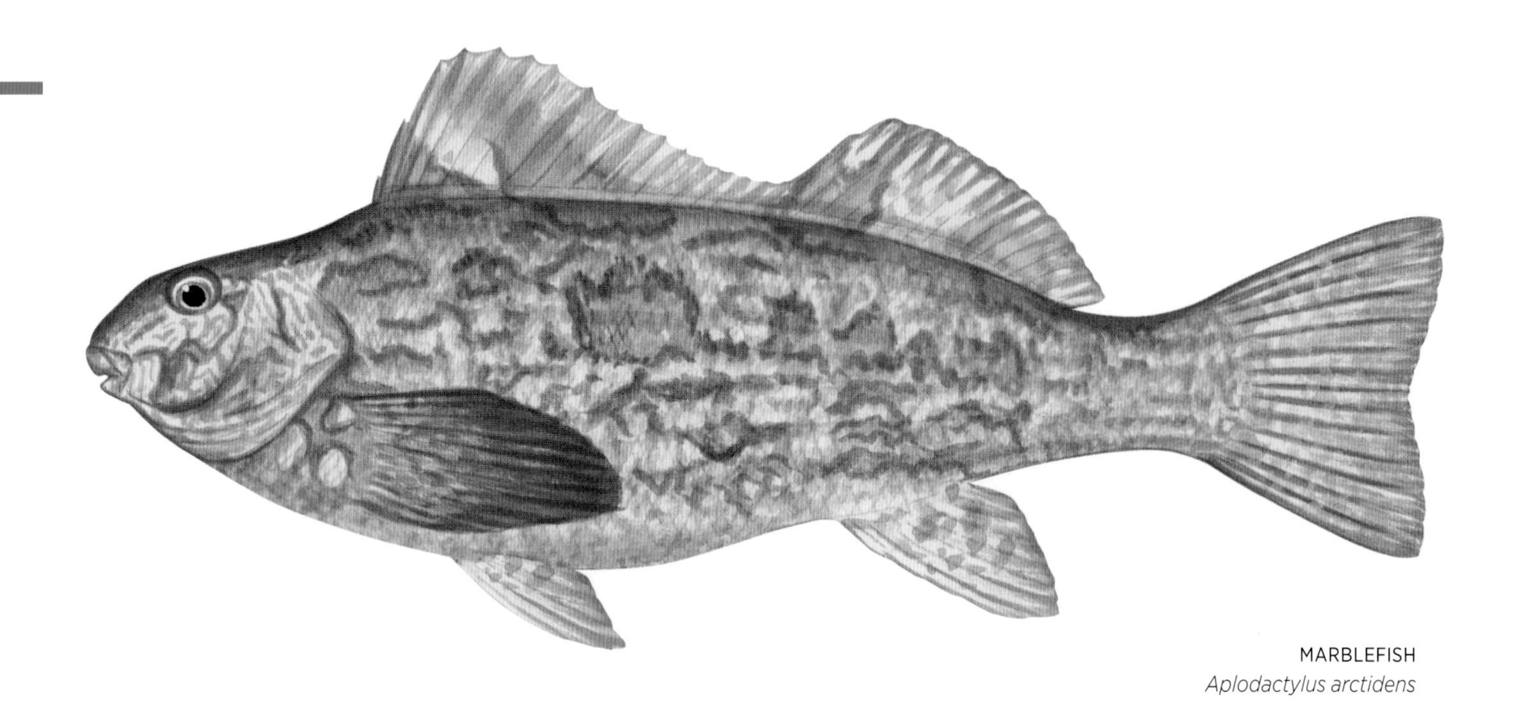

MARBLEFISH
Aplodactylus arctidens

There are about five species of Marblefish and they are found in temperate coastal waters around southern Australia, New Zealand and South America. Marblefishes have an elongate, rounded body with small cycloid scales embedded in the skin, and a single flattened spine on the operculum. They have a bluntly rounded snout above a small, ventrally directed mouth with fleshy lips. The single, deeply notched dorsal fin has 14–23 spines and 16–21 rays, and the short-based, triangular anal fin has three spines and 7–8 rays. The caudal fin is truncate to slightly forked, and the pectoral fins are large, with the lower 5–6 rays thickened, unbranched and separate from the other rays.

Marblefishes are found on shallow rocky reefs and weed beds along the southern coast of Australia. They are well camouflaged and remain close to the bottom, foraging amongst weed and feeding on algae. The **Western Seacarp** is the largest at about 60 cm in length. Marblefishes are occasionally taken by spearfishers but are poor food fishes.

ROCK CALE
Aplodactylus lophodon

AUSTRALIAN APLODACTYLIDAE SPECIES

Aplodactylus arctidens Marblefish
Aplodactylus etheridgii Notch-head Marblefish
Aplodactylus lophodon Rock Cale
Aplodactylus westralis Western Seacarp

MORWONGS

Cheilodactylidae

BLUE MORWONG
Nemadactylus valenciennesi

Morwongs are found in temperate waters of the Atlantic, Indian and Pacific oceans. They occur mainly in the southern hemisphere and there are about 22 species in the family. Morwongs have oval to tapering, compressed bodies, with moderate-sized cycloid scales that form a sheath at the base of the fins. Their small, slightly downwardly directed mouth has thick, fleshy lips and small villiform teeth. The single long dorsal fin has 14–22 spines and 19–39 rays, and the anal fin has three spines and 7–19 rays. The caudal fin is forked and the pectoral fins are large, usually with the lower 4–7 rays thickened, unbranched and elongate.

Most Australian Morwongs are found in coastal waters in the south, over rocky reefs, weed beds or sandy bottoms. The **Jackass Morwong**, an important commercial species, occurs down to about 300 m. Morwongs are usually encountered close to the bottom and prey on benthic invertebrates such as crabs, molluscs and worms, as well as small fishes. The **Dusky Morwong**, which can reach 1.2 m in length, is frequently taken by spearfishers on shallow reefs, though it is a poor food fish. This is also true of the **Redlip Morwong**, which inhabits shallow water and is often encountered resting on the bottom or sheltering in caves. The **Blue Morwong** (also known as the Queen Snapper) can reach up to 1 m in length. It is usually solitary and is found on deeper coastal reefs down to at least 100 m, around the southern coast. It is an excellent food fish and is keenly sought-after by recreational and commercial fishers. The **Red Morwong** forms large schools over reefs in New South Wales and is another fine food fish. The **Crested Morwong** and **Magpie Morwong**, both small species, are found on opposite sides of the continent but are very similar. They are common on shallow, protected coastal reefs and occasionally enter estuaries.

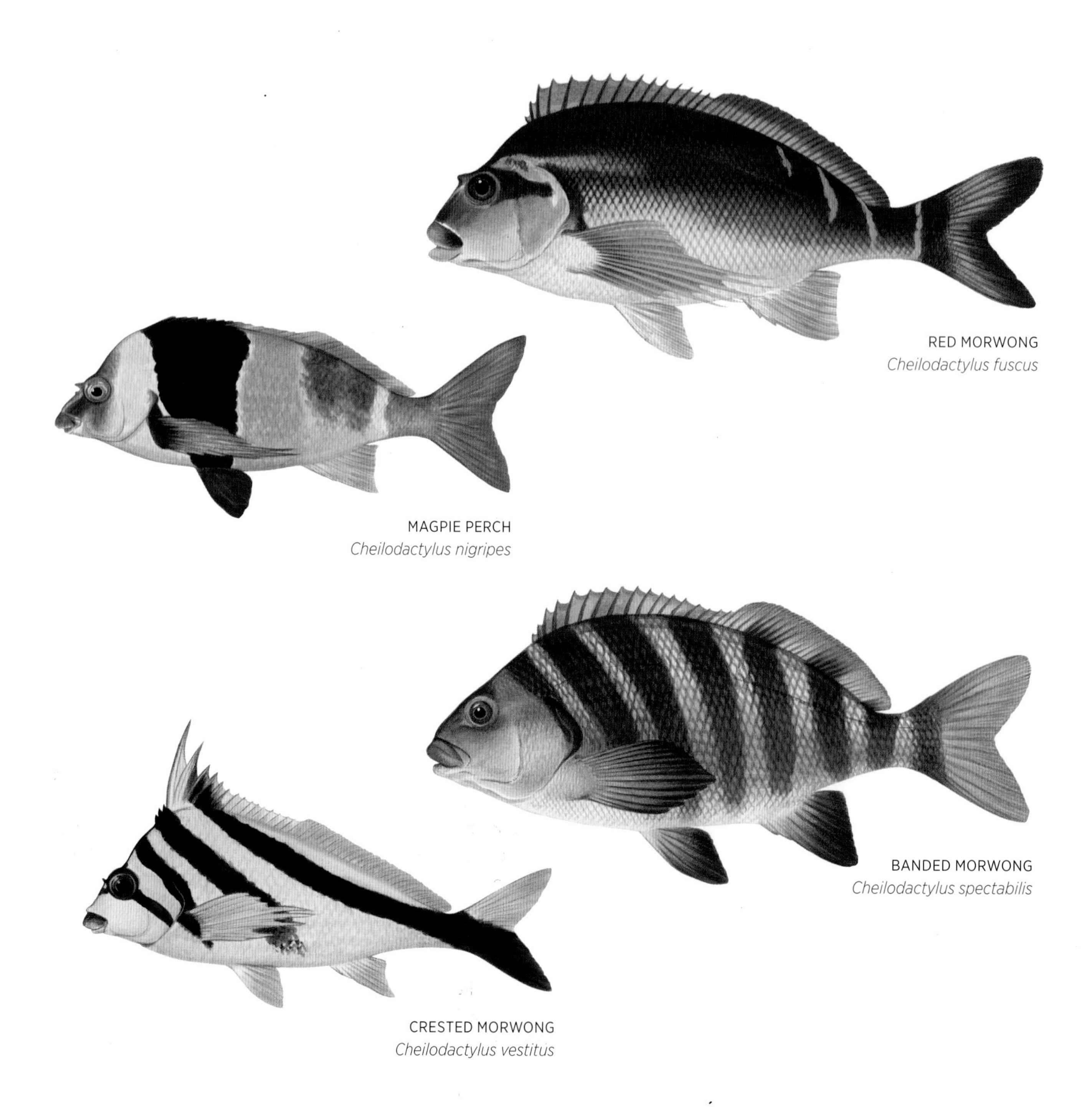

RED MORWONG
Cheilodactylus fuscus

MAGPIE PERCH
Cheilodactylus nigripes

BANDED MORWONG
Cheilodactylus spectabilis

CRESTED MORWONG
Cheilodactylus vestitus

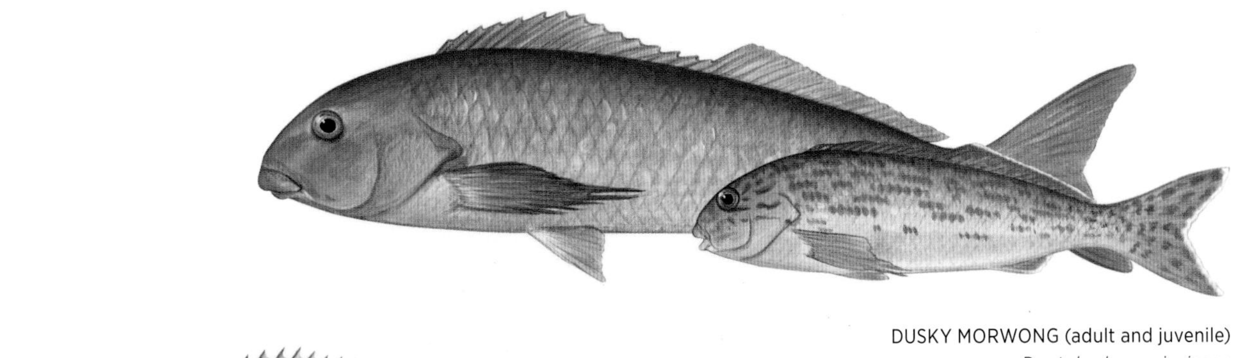

DUSKY MORWONG (adult and juvenile)
Dactylophora nigricans

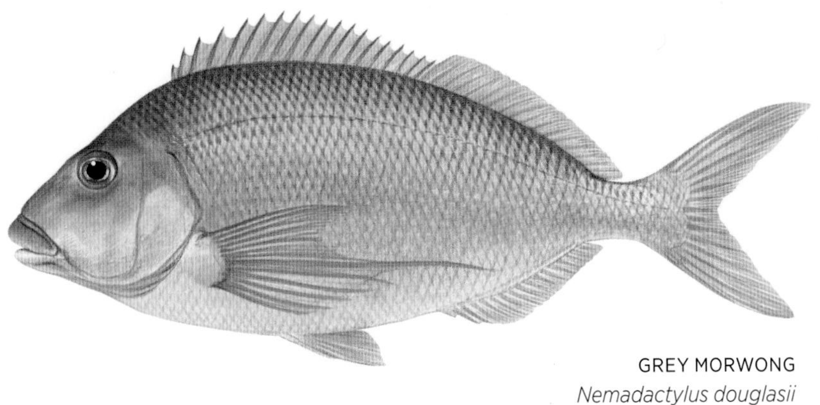

GREY MORWONG
Nemadactylus douglasii

JACKASS MORWONG
Nemadactylus macropterus

AUSTRALIAN CHEILODACTYLIDAE SPECIES

Cheilodactylus ephippium Painted Morwong
Cheilodactylus francisi Blacktip Morwong
Cheilodactylus fuscus Red Morwong
Cheilodactylus gibbosus Magpie Morwong
Cheilodactylus nigripes Magpie Perch
Cheilodactylus rubrolabiatus Redlip Morwong
Cheilodactylus spectabilis Banded Morwong
Cheilodactylus vestitus Crested Morwong
Cheilodactylus vittatus Hawaiian Morwong
Dactylophora nigricans Dusky Morwong
Nemadactylus douglasii Grey Morwong
Nemadactylus macropterus Jackass Morwong
Nemadactylus valenciennesi Blue Morwong

TRUMPETERS

Latridae

STRIPED TRUMPETER
Latris lineata

Trumpeters are found in temperate waters around southern Australia, New Zealand and Chile, and in the South Atlantic Ocean. There are about eight species in the family. They have oval to elongate, compressed bodies with small cycloid scales, and small to moderate mouths, most with fleshy lips. Their single, deeply notched dorsal fin has 14–24 spines and 23–40 rays, the anal fin has three spines and 18–35 rays, and the caudal fin is forked.

The **Real Bastard Trumpeter** is found only off south-eastern Tasmania, where it occurs in schools. It feeds on zooplankton, which it takes in midwater using its highly protrusible mouth. The **Striped Trumpeter**, which grows to a maximum length of about 1.2 m, is found in deeper reefs off southern Australia, down to about 300 m. It is a superb food fish and is highly sought-after by anglers. Populations are severely depleted due to overfishing and it has now disappeared from large parts of its former range. The **Bastard Trumpeter** is another excellent food fish and is found in small to very large schools, mainly in Tasmanian waters. Trumpeters feed on a variety of invertebrates and small fishes.

BASTARD TRUMPETER
Latridopsis forsteri

REAL BASTARD TRUMPETER
Mendosoma lineata

AUSTRALIAN LATRIDAE SPECIES

Latridopsis ciliaris Blue Moki
Latridopsis forsteri Bastard Trumpeter
Latris lineata Striped Trumpeter
Mendosoma lineata Real Bastard Trumpeter

BANDFISHES

Cepolidae

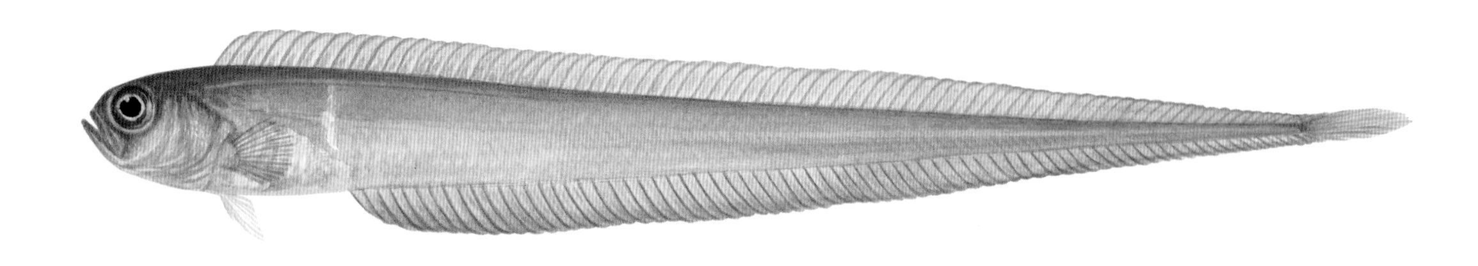

RED BANDFISH
Cepola schlegelii

Bandfishes are found in the eastern Atlantic and Indo-West Pacific, and there are about 19 species in the family. They have elongate to eel-like bodies with minute cycloid scales and a lateral line that closely follows the dorsal profile of the body. The large, oblique mouth has slender, curved teeth and a broad upper jaw. The single long-based dorsal fin has 0–4 spines and 21–89 rays, and the anal fin has 0–1 spine and 13–102 rays. The Bandfishes are divided into two subfamilies:

CEPOLINAE Members of this group have an eel-like body, with the dorsal and anal fins continuous with the caudal fin. They are bottom-dwelling fishes and burrow into soft substrates such as the sediment of estuaries. The **Australian Bandfish**, which grows to about 40 cm in length, lives in small colonies in south-eastern coastal waters and is occasionally seen when it leaves its burrow to feed on zooplankton.

OWSTONIINAE These fishes have an elongate caudal fin that is separate from the dorsal and anal fins, and long ventral fins. They are found in deep waters off northern Australia, down to about 475 m, and also burrow into the substrate.

Bandfishes are rarely encountered and, due to their burrowing, do not often appear in deepwater trawl nets. Little is known about the family and the biology of its species.

AUSTRALIAN CEPOLIDAE SPECIES

Acanthocepola abbreviata Yellowspotted Bandfish
Acanthocepola krusensternii Redspotted Bandfish
Acanthocepola limbata Blackspot Bandfish
Cepola australis Australian Bandfish
Cepola schlegelii Red Bandfish
Owstonia maccullochi McCulloch's Bandfish
Owstonia pectinifer Comb Bandfish
Owstonia totomiensis Short-tail Bandfish

CICHLIDS

Cichlidae

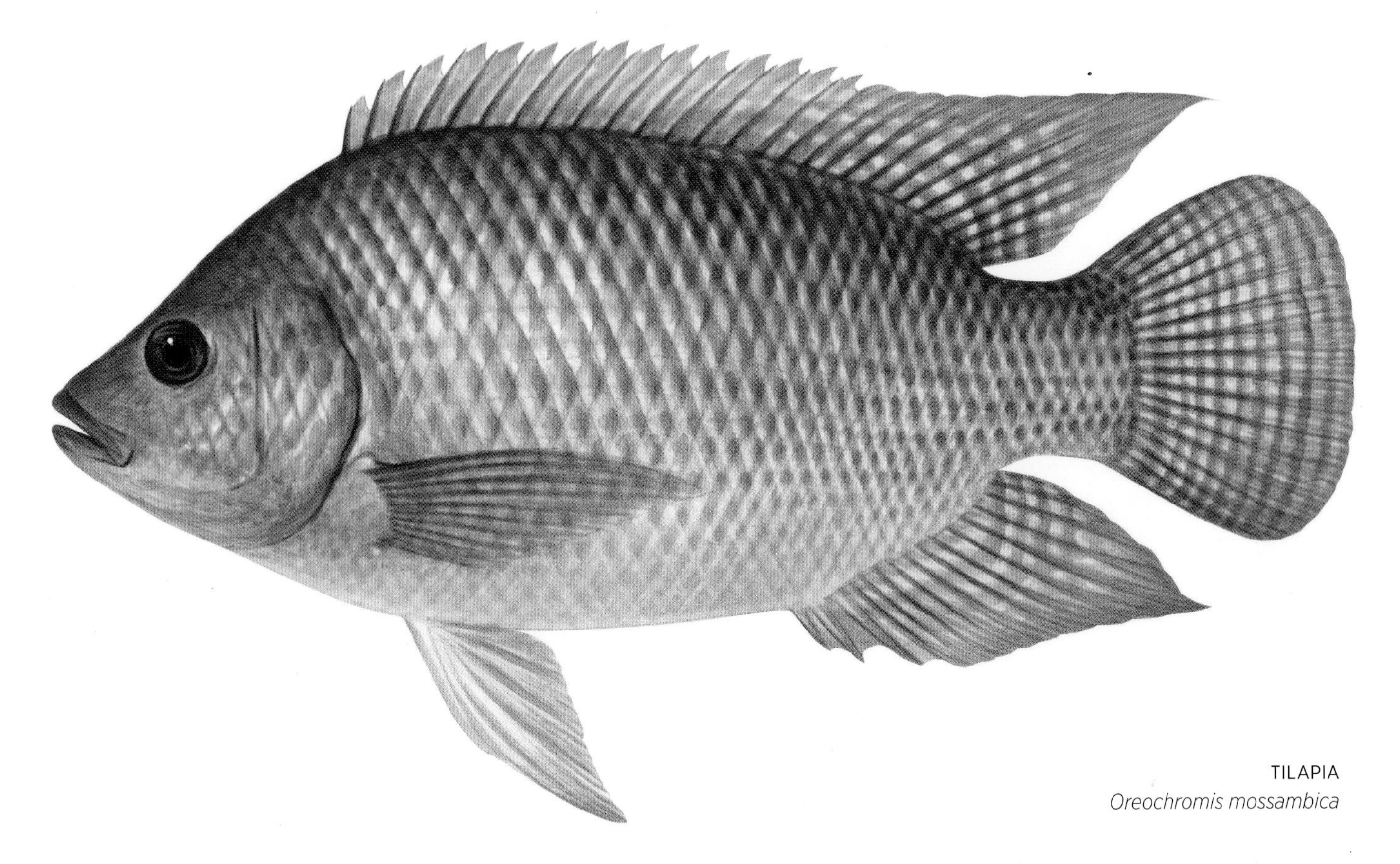

TILAPIA
Oreochromis mossambica

This very large freshwater family contains over 1300 species. They are found around South America, the West Indies, Africa (where the greatest diversity occurs) and scattered localities in the Middle East and India. Their single dorsal fin has 7–25 spines and 5–30 rays, the anal fin usually has three spines and 4–15 rays, and the caudal fin is truncate to rounded. They have deep, oval to almost circular, compressed bodies with large scales. Dentition varies widely between the various species, with adaptations to suit a broad range of feeding specialisations.

The exact number of species of Cichlids is unknown, as new species are constantly being discovered and named while others with limited ranges disappear. An astounding array of Cichlids has developed over a very short geological period. For example, during the 15 000 years or so since Africa's Lake Victoria was last completely dry, 200–400 different species of Cichlid have evolved there, filling every imaginable ecological niche.

Cichlids, many of which are popular aquarium fishes, have been introduced into Australian waters mainly through releases from aquariums, and several species have established breeding populations. The **Tilapia**, which can reach about 35 cm in length, is farmed as a speciality food fish in many parts of the world. It is a hardy fish that is principally herbivorous, and it is now found in a variety of freshwater habitats on the east and west coasts of Australia.

AUSTRALIAN CICHLIDAE SPECIES

Aequidens pulchrus Blue Acara
Cichlasoma nigrofasciatum Convict Cichlid
Cichlasoma octofasciatum Jack Dempsey
Oreochromis mossambica Tilapia
Tilapia mariae Spotted Tilapia
Tilapia zillii Zille's Cichlid

DAMSELFISHES

Pomacentridae

GOLDEN DAMSEL
Amblyglyphidodon aureus

Damselfishes are found throughout the world's tropical and warm temperate waters, but are most diverse in the Indo-Pacific region. There are about 350 species in the family. They have deep, oval to slightly elongate, compressed bodies, with moderately large ctenoid scales that extend onto the base of the fins. Their small mouth has conical or incisor teeth. The single dorsal fin has 8–17 spines and 10–21 rays, the anal fin usually has two spines and 10–16 rays, and the caudal fin is truncate to forked.

In Australia Damselfishes are found mostly in tropical waters on coral reefs, though several species occur around the south coast, including some of the Pullers and Scalyfins. Damselfishes display a great range of behaviours and feeding strategies. Some species are herbivorous and aggressively guard a small patch of filamentous algae, chasing off any other fishes that intrude into their territory and even attacking divers. Others form large schools, feeding on zooplankton in midwater, and these are among the most abundant of all reef fishes.

Anemonefishes are well known for their habit of living amongst sea anemones. They live in small groups within the shelter of the anemone's stinging tentacles, producing a covering of mucus to protect themselves from stings. It was thought that this relationship was commensal, with the Anemonefishes deriving protection from the anemones. However, experiments found that when Anemonefishes were removed, host anemones were quickly consumed by Butterflyfishes (Chaetodontidae, p.570), suggesting that the relationship is mutually beneficial.

Some species of Damselfish, such as the small Humbugs, are closely associated with coral heads, hovering above them to feed on zooplankton and quickly retreating to the shelter of their branches at any hint of danger. Others species inhabit sandy areas and prey on small benthic invertebrates. Damselfishes lay batches of adhesive eggs on the substrate and guard them against predators. Juveniles of many species are brilliantly coloured, often yellow with electric-blue markings, and frequently look quite different from the adults.

BANDED SERGEANT
Abudefduf septemfasciatus

SCISSORTAIL SERGEANT
Abudefduf sexfasciatus

BLACKSPOT SERGEANT
Abudefduf sordidus

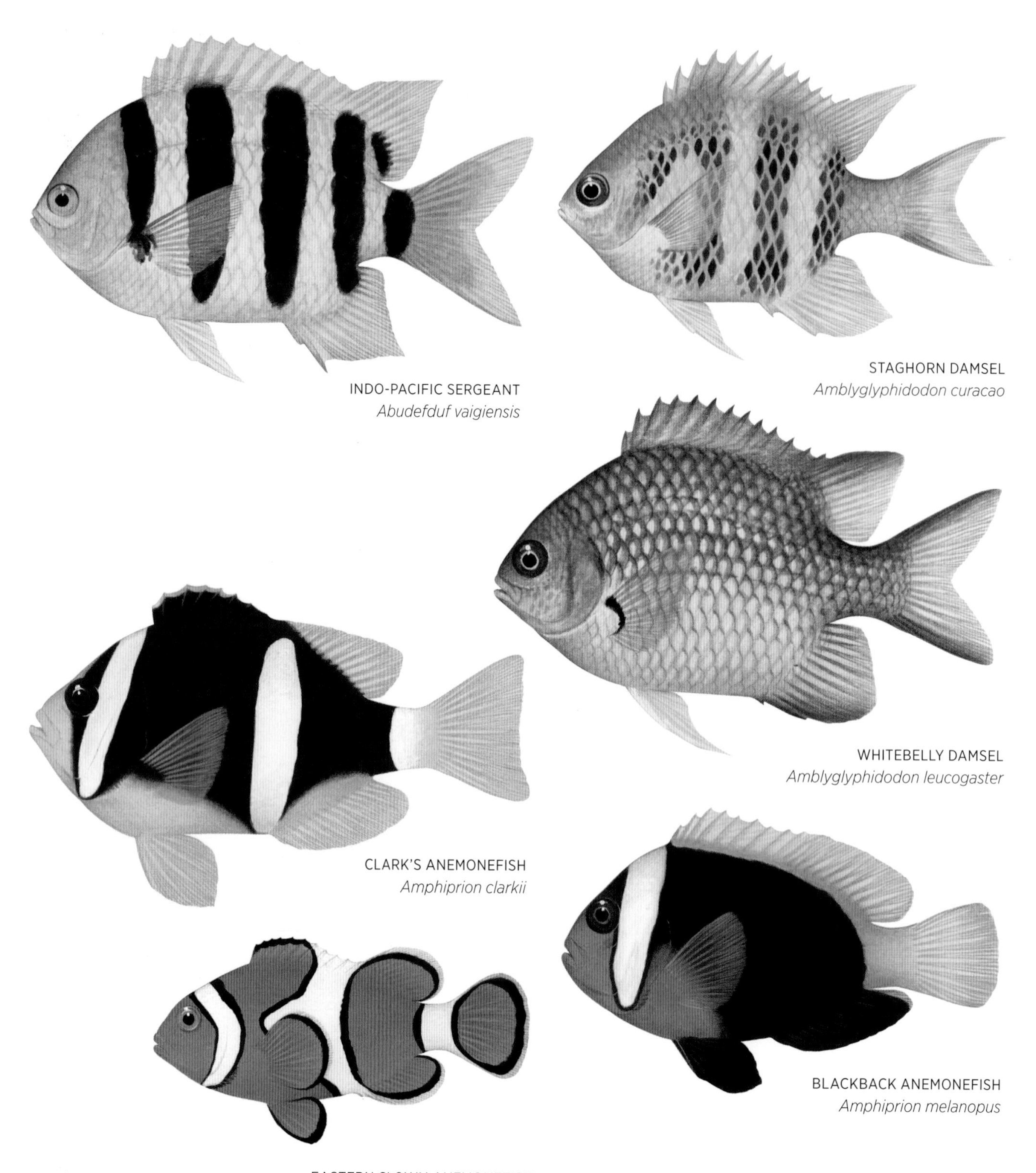

INDO-PACIFIC SERGEANT
Abudefduf vaigiensis

STAGHORN DAMSEL
Amblyglyphidodon curacao

WHITEBELLY DAMSEL
Amblyglyphidodon leucogaster

CLARK'S ANEMONEFISH
Amphiprion clarkii

BLACKBACK ANEMONEFISH
Amphiprion melanopus

EASTERN CLOWN ANEMONEFISH
Amphiprion percula

PINK ANEMONEFISH
Amphiprion perideraion

SADDLEBACK ANEMONEFISH
Amphiprion polymnus

BIGLIP DAMSEL
Cheiloprion labiatus

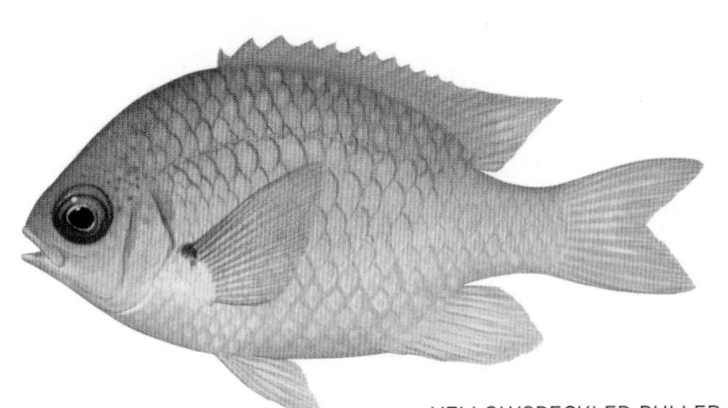

YELLOWSPECKLED PULLER
Chromis alpha

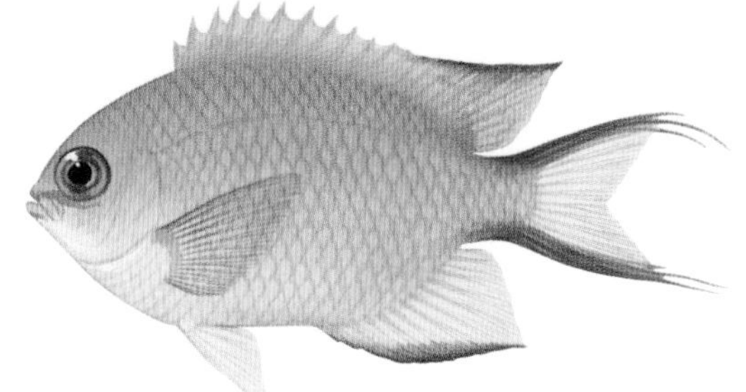

AMBON PULLER
Chromis amboinensis

BLACKAXIL PULLER
Chromis atripectoralis

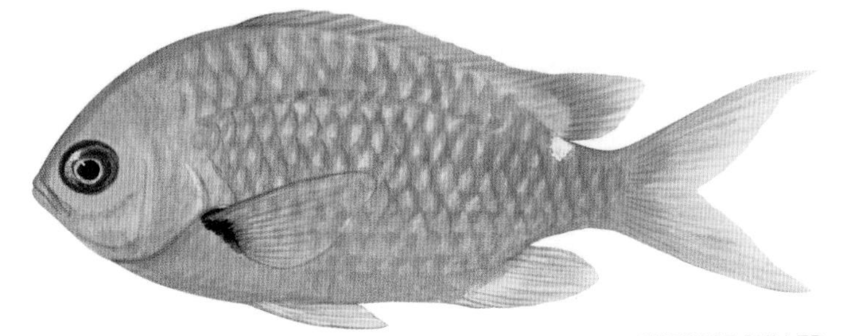

ONESPOT PULLER
Chromis hypsilepis

LINED PULLER
Chromis lineata

WEBER'S PULLER
Chromis weberi

PALE-TAIL PULLER
Chromis xanthura

TWOSPOT DEMOISELLE
Chrysiptera biocellata

BLUELINE DEMOISELLE
Chrysiptera caeruleolineata

BLUE DEMOISELLE
Chrysiptera cyanea

BLUEHEAD DEMOISELLE
Chrysiptera rollandi

BANDED HUMBUG
Dascyllus aruanus

BLACKTAIL HUMBUG
Dascyllus melanurus

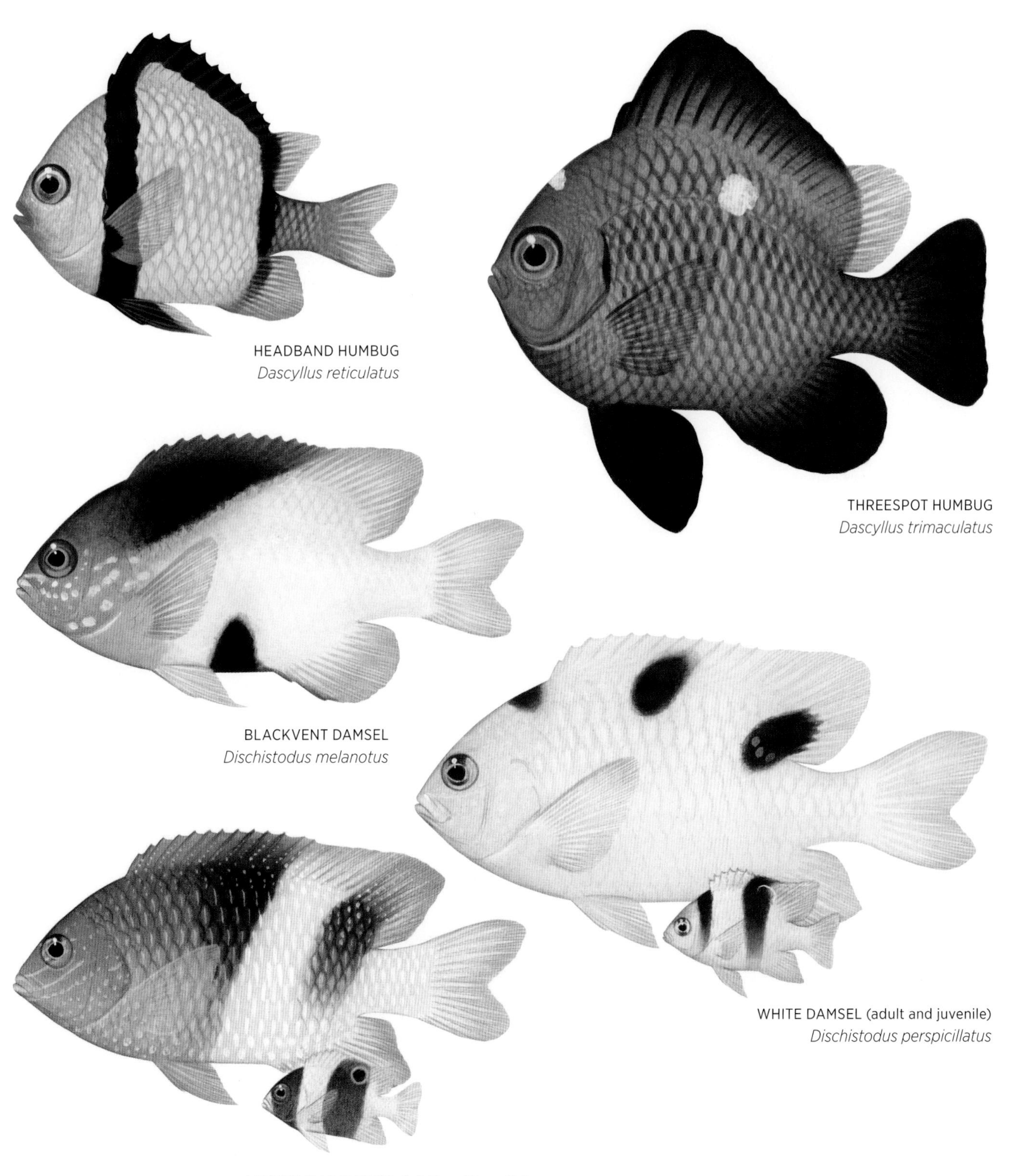

HEADBAND HUMBUG
Dascyllus reticulatus

THREESPOT HUMBUG
Dascyllus trimaculatus

BLACKVENT DAMSEL
Dischistodus melanotus

WHITE DAMSEL (adult and juvenile)
Dischistodus perspicillatus

HONEYHEAD DAMSEL (adult and juvenile)
Dischistodus prosopotaenia

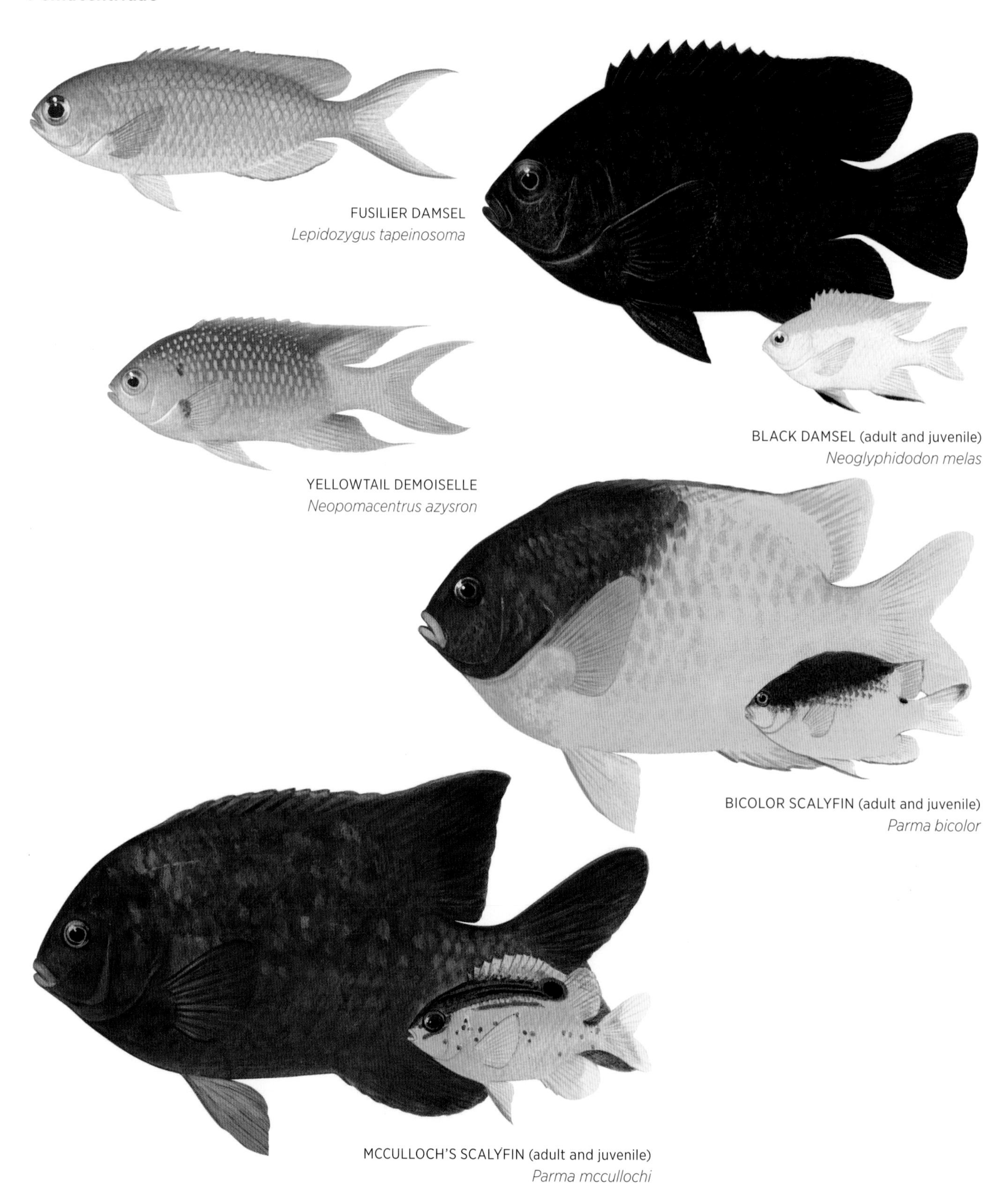

FUSILIER DAMSEL
Lepidozygus tapeinosoma

BLACK DAMSEL (adult and juvenile)
Neoglyphidodon melas

YELLOWTAIL DEMOISELLE
Neopomacentrus azysron

BICOLOR SCALYFIN (adult and juvenile)
Parma bicolor

MCCULLOCH'S SCALYFIN (adult and juvenile)
Parma mccullochi

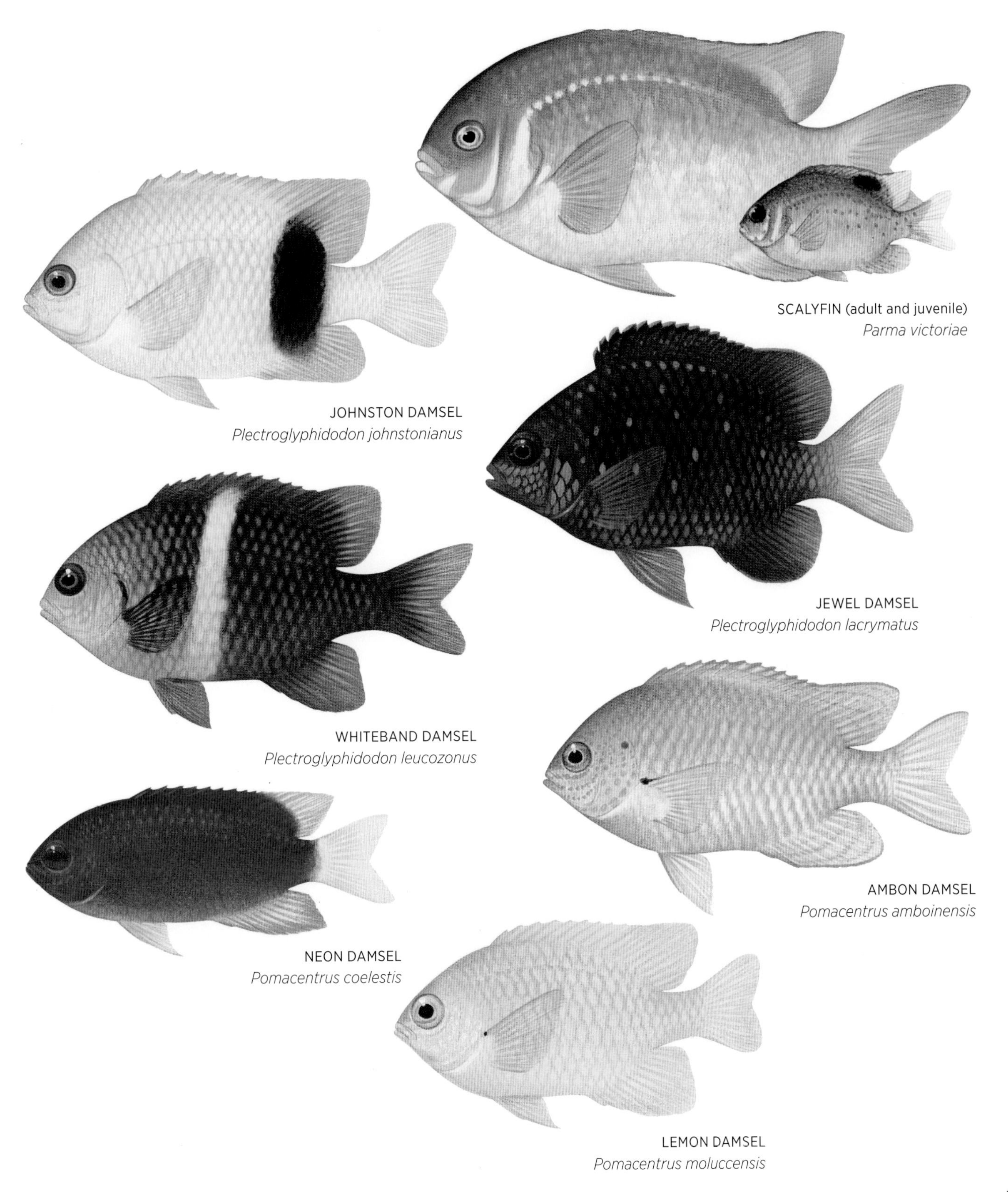

SCALYFIN (adult and juvenile)
Parma victoriae

JOHNSTON DAMSEL
Plectroglyphidodon johnstonianus

JEWEL DAMSEL
Plectroglyphidodon lacrymatus

WHITEBAND DAMSEL
Plectroglyphidodon leucozonus

AMBON DAMSEL
Pomacentrus amboinensis

NEON DAMSEL
Pomacentrus coelestis

LEMON DAMSEL
Pomacentrus moluccensis

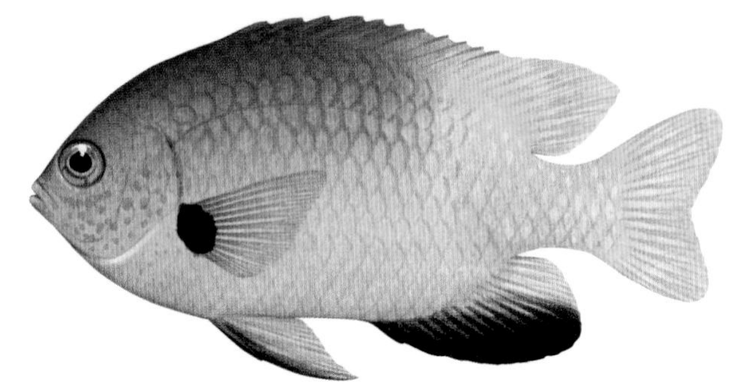
GOLDBACK DAMSEL
Pomacentrus nigromanus

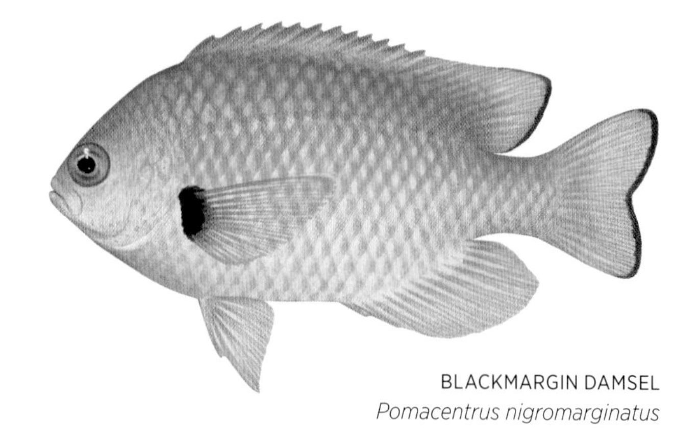
BLACKMARGIN DAMSEL
Pomacentrus nigromarginatus

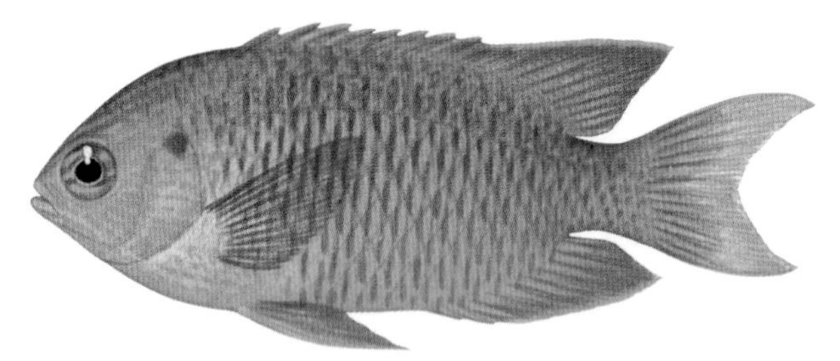
PEACOCK DAMSEL
Pomacentrus pavo

PHILIPPINE DAMSEL
Pomacentrus philippinus

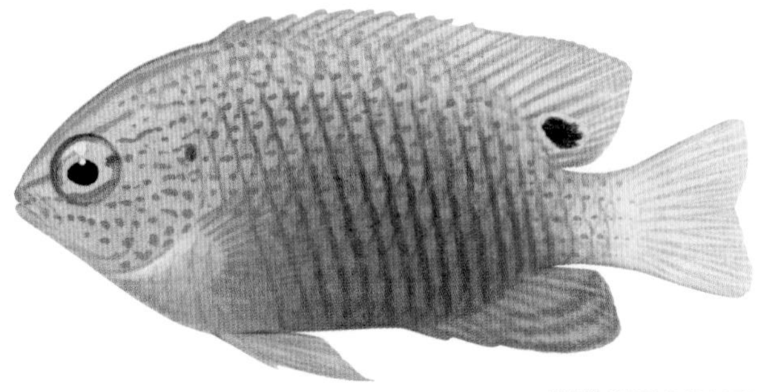
PRINCESS DAMSEL
Pomacentrus vaiuli

SPINE-CHEEK CLOWNFISH
Premnas biaculeatus

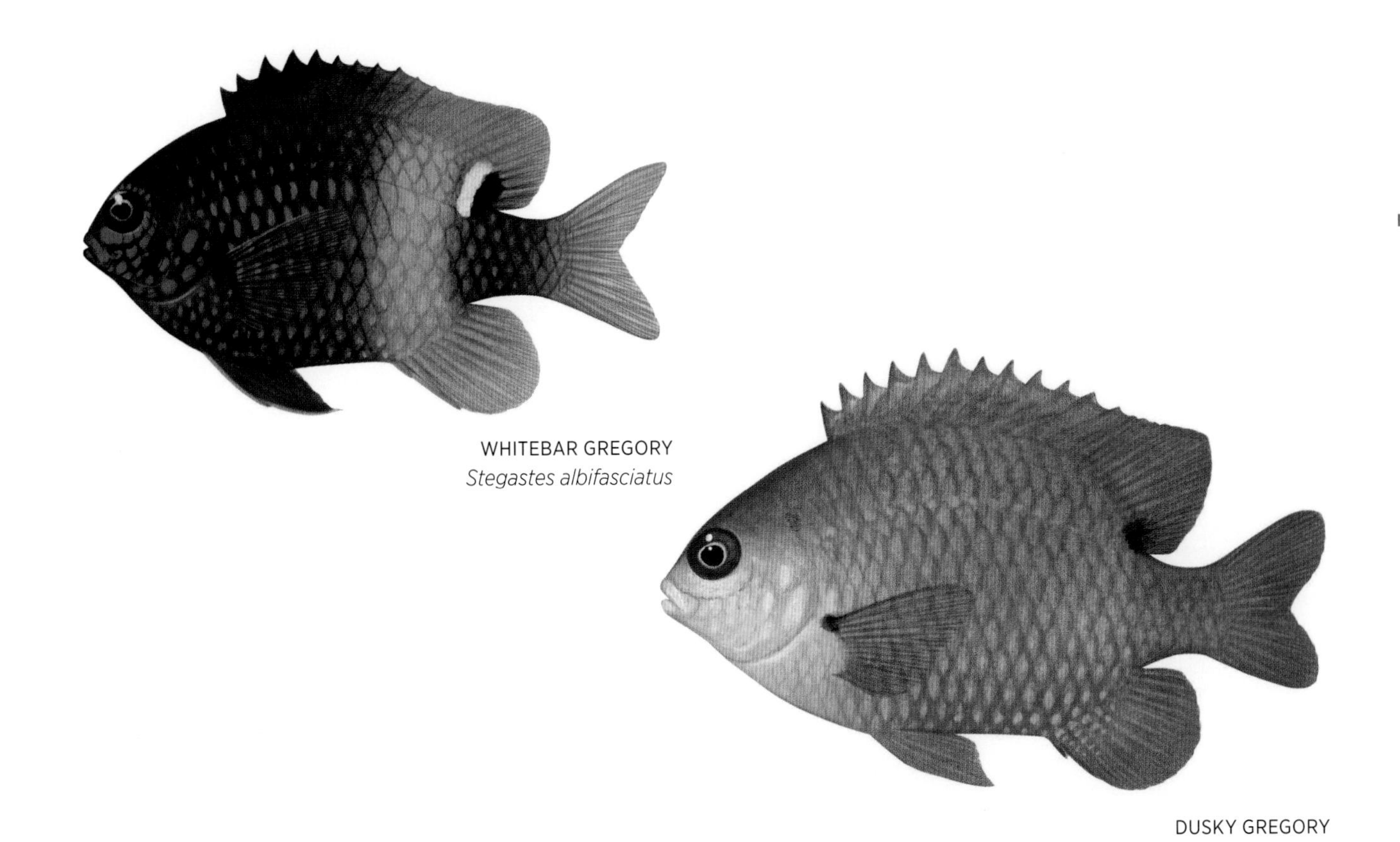

WHITEBAR GREGORY
Stegastes albifasciatus

DUSKY GREGORY
Stegastes nigricans

AUSTRALIAN POMACENTRIDAE SPECIES

Abudefduf bengalensis Bengal Sergeant
Abudefduf septemfasciatus Banded Sergeant
Abudefduf sexfasciatus Scissortail Sergeant
Abudefduf sordidus Blackspot Sergeant
Abudefduf vaigiensis Indo-Pacific Sergeant
Abudefduf whitleyi Whitley's Sergeant
Acanthochromis polyacanthus Spiny Puller
Amblyglyphidodon aureus Golden Damsel
Amblyglyphidodon batunai Batuna Damsel
Amblyglyphidodon curacao Staghorn Damsel
Amblyglyphidodon leucogaster Whitebelly Damsel
Amblypomacentrus breviceps Blackbanded Damsel
Amphiprion akindynos Barrier Reef Anemonefish
Amphiprion chrysopterus Orangefin Anemonefish
Amphiprion clarkii Clark's Anemonefish
Amphiprion latezonatus Wideband Anemonefish
Amphiprion mccullochi McCulloch's Anemonefish
Amphiprion melanopus Blackback Anemonefish
Amphiprion ocellaris Western Clown Anemonefish
Amphiprion percula Eastern Clown Anemonefish
Amphiprion perideraion Pink Anemonefish
Amphiprion polymnus Saddleback Anemonefish
Amphiprion rubrocinctus Australian Anemonefish
Amphiprion sandaracinos Orange Anemonefish
Cheiloprion labiatus Biglip Damsel
Chromis abyssicola Deepsea Puller
Chromis agilis Reef Puller
Chromis alpha Yellowspeckled Puller
Chromis amboinensis Ambon Puller
Chromis analis Yellow Puller
Chromis atripectoralis Blackaxil Puller
Chromis atripes Darkfin Puller
Chromis caudalis Dusky Puller
Chromis chrysura Stoutbody Puller
Chromis cinerascens Green Puller
Chromis delta Deep-reef Puller
Chromis elerae Twinspot Puller
Chromis flavomaculata Yellowspot Puller
Chromis fumea Smoky Puller
Chromis hypsilepis Onespot Puller

Chromis iomelas Half-and-half Puller
Chromis klunzingeri Blackhead Puller
Chromis lepidolepis Scaly Puller
Chromis lineata Lined Puller
Chromis margaritifer Whitetail Puller
Chromis mirationis Japanese Puller
Chromis nitida Yellowback Puller
Chromis retrofasciata Blackbar Puller
Chromis ternatensis Swallowtail Puller
Chromis vanderbilti Vanderbilt's Puller
Chromis viridis Blue-green Puller
Chromis weberi Weber's Puller
Chromis westaustralis West Australian Puller
Chromis xanthochira Yellow-axil Puller
Chromis xanthura Pale-tail Puller
Chrysiptera biocellata Twospot Demoiselle
Chrysiptera brownriggii Surge Demoiselle
Chrysiptera caeruleolineata Blueline Demoiselle
Chrysiptera cyanea Blue Demoiselle
Chrysiptera flavipinnis Yellowfin Demoiselle
Chrysiptera glauca Grey Demoiselle
Chrysiptera hemicyanea Azure Demoiselle
Chrysiptera notialis Southern Demoiselle
Chrysiptera rex Pink Demoiselle
Chrysiptera rollandi Bluehead Demoiselle
Chrysiptera starcki Starck's Demoiselle
Chrysiptera talboti Talbot's Demoiselle
Chrysiptera taupou South Seas Demoiselle
Chrysiptera tricincta Threeband Demoiselle
Chrysiptera unimaculata Onespot Demoiselle
Dascyllus aruanus Banded Humbug
Dascyllus melanurus Blacktail Humbug
Dascyllus reticulatus Headband Humbug
Dascyllus trimaculatus Threespot Humbug
Dischistodus chrysopoecilus Whitepatch Damsel
Dischistodus darwiniensis Banded Damsel
Dischistodus melanotus Blackvent Damsel
Dischistodus perspicillatus White Damsel
Dischistodus prosopotaenia Honeyhead Damsel
Dischistodus pseudochrysopoecilus Monarch Damsel
Hemiglyphidodon plagiometopon Lagoon Damsel
Lepidozygus tapeinosoma Fusilier Damsel
Mecaenichthys immaculatus Immaculate Damsel
Neoglyphidodon melas Black Damsel
Neoglyphidodon nigroris Scarface Damsel
Neoglyphidodon oxyodon Bluestreak Damsel
Neoglyphidodon polyacanthus Multispine Damsel
Neopomacentrus azysron Yellowtail Demoiselle
Neopomacentrus bankieri Chinese Demoiselle
Neopomacentrus cyanomos Regal Demoiselle
Neopomacentrus filamentosus Brown Demoiselle
Neopomacentrus taeniurus Freshwater Demoiselle
Parma alboscapularis Black Scalyfin
Parma bicolor Bicolor Scalyfin
Parma mccullochi McCulloch's Scalyfin
Parma microlepis White-ear
Parma occidentalis Western Scalyfin
Parma oligolepis Bigscale Scalyfin
Parma polylepis Banded Scalyfin
Parma unifasciata Girdled Scalyfin
Parma victoriae Scalyfin
Plectroglyphidodon dickii Dick's Damsel
Plectroglyphidodon imparipennis Brighteye Damsel
Plectroglyphidodon johnstonianus Johnston Damsel
Plectroglyphidodon lacrymatus Jewel Damsel
Plectroglyphidodon leucozonus Whiteband Damsel
Plectroglyphidodon phoenixensis Phoenix Damsel
Pomacentrus adelus Obscure Damsel
Pomacentrus alexanderae Alexander's Damsel
Pomacentrus amboinensis Ambon Damsel
Pomacentrus australis Australian Damsel
Pomacentrus bankanensis Speckled Damsel
Pomacentrus brachialis Charcoal Damsel
Pomacentrus chrysurus Whitetail Damsel
Pomacentrus coelestis Neon Damsel
Pomacentrus grammorhynchus Bluespot Damsel
Pomacentrus imitator Imitator Damsel
Pomacentrus lepidogenys Scaly Damsel
Pomacentrus limosus Muddy Damsel
Pomacentrus littoralis Smoky Damsel
Pomacentrus milleri Miller's Damsel
Pomacentrus moluccensis Lemon Damsel
Pomacentrus nagasakiensis Blue-scribbled Damsel
Pomacentrus nigromanus Goldback Damsel
Pomacentrus nigromarginatus Blackmargin Damsel
Pomacentrus pavo Peacock Damsel
Pomacentrus philippinus Philippine Damsel
Pomacentrus reidi Grey Damsel
Pomacentrus tripunctatus Threespot Damsel
Pomacentrus vaiuli Princess Damsel
Pomacentrus wardi Ward's Damsel
Pomachromis richardsoni Richardson's Damsel
Premnas biaculeatus Spine-cheek Clownfish
Pristotis obtusirostris Gulf Damsel
Stegastes albifasciatus Whitebar Gregory
Stegastes apicalis Yellowtip Gregory
Stegastes fasciolatus Pacific Gregory
Stegastes gascoynei Coral Sea Gregory
Stegastes lividus Bluntsnout Gregory
Stegastes nigricans Dusky Gregory
Stegastes obreptus Western Gregory

WRASSES

Labridae

WESTERN BLUE GROPER
Achoerodus gouldii

The family of Wrasses is one of the largest and most diverse of all fish families, with about 450 species found throughout the world's tropical and temperate waters. They have deep, oval to elongate, compressed to rounded bodies with large cycloid scales. The mouth usually has large fleshy lips and well-developed teeth, with separate canines often projecting from the front or sides of the mouth. Most have pharyngeal teeth in the throat, for crushing the shells of prey such as molluscs and crustaceans. Wrasses have a single continuous dorsal fin with 8–21 spines and 6–21 rays, and the anal fin usually has three spines and 7–18 rays. They range in size from tiny reef dwellers like the 9 cm long **Filamentous Flasher Wrasse**, to the enormous **Humphead Maori Wrasse**, which can reach over 2 m in length. Wrasses are strong swimmers and mainly use their well-developed pectoral fins to propel themselves, with the tail used only for sudden bursts of speed.

Wrasses are capable of sex reversal and have a complicated reproductive strategy. Juveniles develop into either females or 'initial-phase' males. A population consisting of young initial-phase males and females will be dominated by a large adult 'terminal-phase' male. If this dominant male is removed, a female is able to change sex and assume the characteristics and functions of the terminal-phase male. Initial-phase males and females will spawn in mass aggregations in the water column above the reef. Terminal-phase males usually spawn in pairs with their female of choice. Wrasses often have strikingly different colour patterns at these different life stages, which can make identification difficult. Juveniles may change colour and pattern completely as they develop into initial-phase males or females, and again if they develop into terminal-phase males.

Wrasses can be found throughout Australian coastal waters. *Notolabrus*, *Pictilabrus* and *Pseudolabrus* species inhabit rocky reefs and seaweed beds around Australia's southern half, while *Macropharyngodon*, *Cheilinus* and *Halichoeres* species are mostly found in tropical waters, favouring areas of algae-covered rubble near coral reefs. *Halichoeres* is the largest Australian genus of Wrasses, with 21 species. Most Wrasses inhabit fairly shallow water, though certain *Bodianus* species can be found at depths of more than 200 m.

Wrasses display a wide variety of lifestyles and feeding strategies. All are carnivorous, with most feeding on benthic invertebrates such as crustaceans, urchins, molluscs and worms, and some on small fishes. Most have sharp, projecting teeth and terminal mouths, and are well adapted to picking up small prey from the substrate. The vividly coloured *Cirrhilabrus* and *Paracheilinus* species, however, take zooplankton from the water column. The Cleanerfishes, which are well known to divers in tropical waters, have their own distinctive feeding strategy. They take up station on a reef, advertising their presence with a distinctive undulating swimming motion and their vividly striped colour pattern. Other fishes visit these stations and allow the Cleanerfishes to remove parasites and damaged skin from their bodies, even from inside the mouth and gill cavities. Juveniles of some other species of Wrasse also perform this service.

The brightly coloured Moon Wrasses are very active, opportunistic predators that are widespread and commonly seen on reefs. They will often accompany fishes that are feeding in sand or rubble, and dart in to seize any small prey disturbed by the activity. Large individuals of **Redblotched Wrasse** often turn over rocks or break off substantial branches of coral with their teeth, searching for crabs or molluscs sheltering there, and a Moon Wrasse will usually be nearby, waiting for an opportunity to snap up a morsel.

Some medium-sized Wrasses, such as *Hemigymnus* species, take mouthfuls of sand, filter out and swallow any tiny invertebrates, then spit the sand out again. *Iniistius* species prefer open sandy areas, where they feed on benthic invertebrates. Like many other small Wrasses, they will dive into the sand in the blink of an eye when threatened, and sleep buried under the sand at night. The curious **Slingjaw Wrasse** has a highly protrusible tube-like mouth, which it can shoot suddenly forward to suck in small invertebrates and fishes.

The Australian Blue Gropers are large residents of deeper southern coastal reefs, and the eastern and western species have slightly different forms. They are both superb food fishes and have been intensely targeted by anglers and spearfishers. Their slow growth rate means that they are easily overfished, and they have disappeared from much of their original range. Both Eastern and Western Blue Gropers are now protected in many areas.

Some other large Wrasses, such as Tuskfishes, are also highly sought-after as food fishes, particularly the **Baldchin Groper** and the **Blackspot Tuskfish**, both of which can grow to over 80 cm in length and weigh 7–8 kg. These two species, as well as the Humphead Maori Wrasse, have suffered from overfishing and the Humphead Maori Wrasse is now completely protected in Australian waters.

DIAMOND WRASSE (male and female)
Anampses caeruleopunctatus

BLUE-AND-YELLOW WRASSE (male and female)
Anampses lennardi

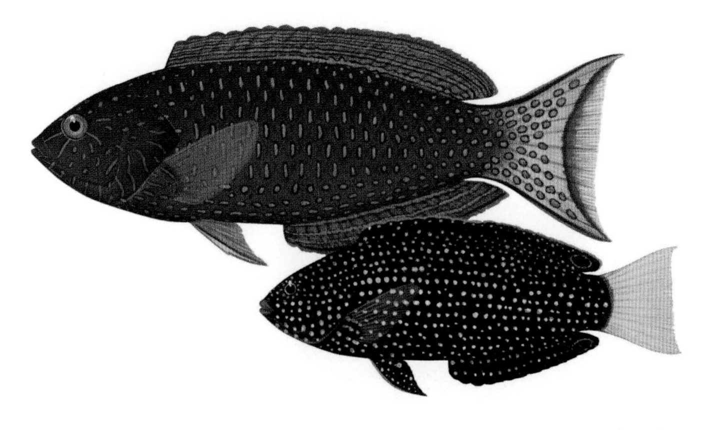

SPECKLED WRASSE (male and female)
Anampses meleagrides

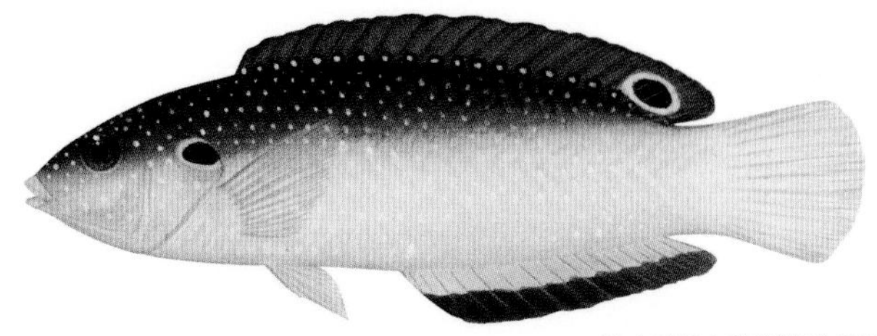

BLACKBACK WRASSE
Anampses neoguinaicus

LYRETAIL PIGFISH
Bodianus anthioides

CORAL PIGFISH
Bodianus axillaris

DIANA'S PIGFISH
Bodianus diana

FOXFISH
Bodianus frenchii

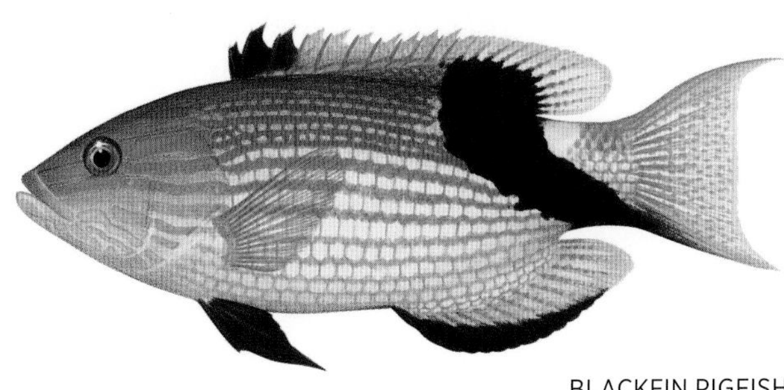

BLACKFIN PIGFISH
Bodianus loxozonus

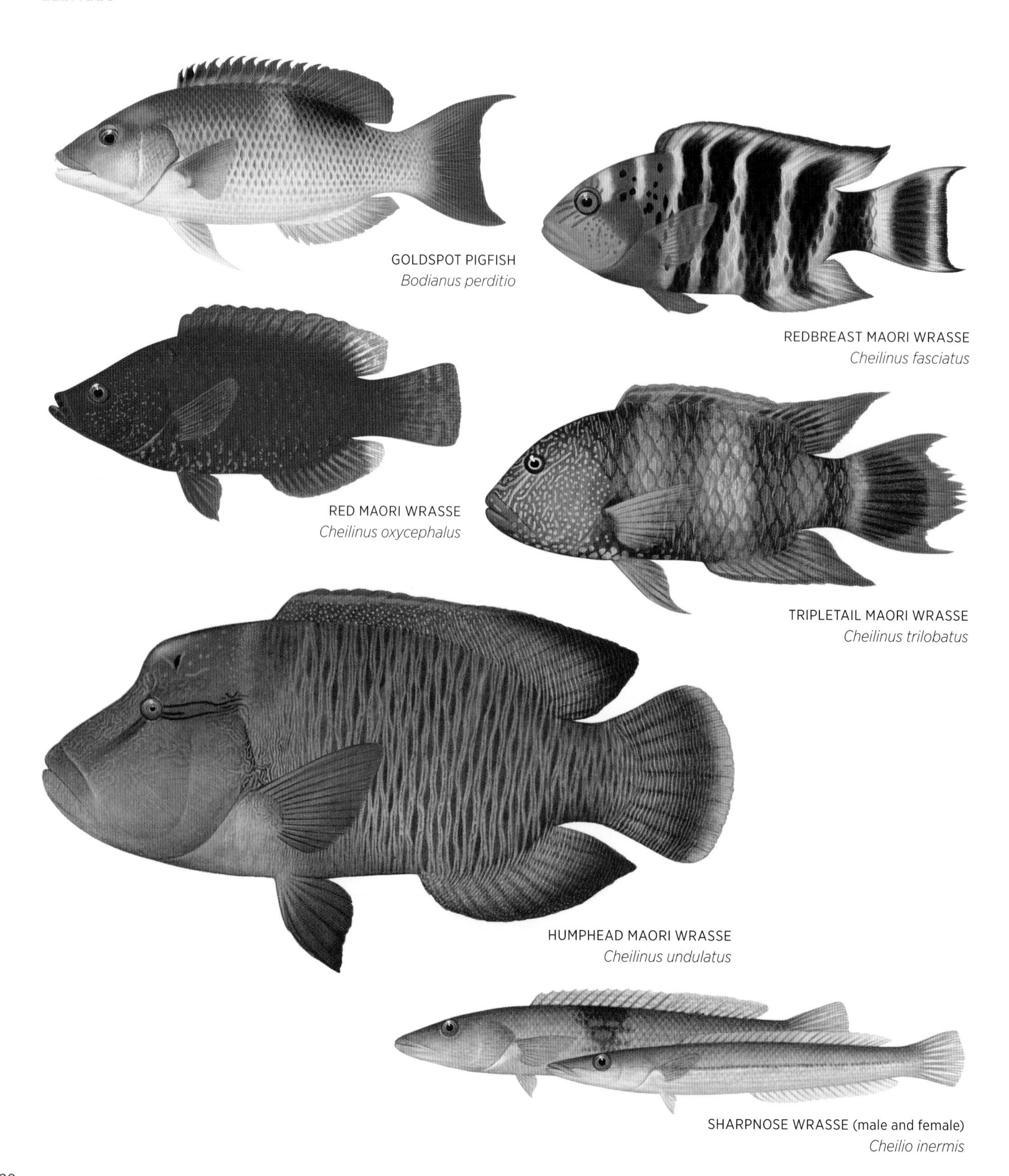

GOLDSPOT PIGFISH
Bodianus perditio

REDBREAST MAORI WRASSE
Cheilinus fasciatus

RED MAORI WRASSE
Cheilinus oxycephalus

TRIPLETAIL MAORI WRASSE
Cheilinus trilobatus

HUMPHEAD MAORI WRASSE
Cheilinus undulatus

SHARPNOSE WRASSE (male and female)
Cheilio inermis

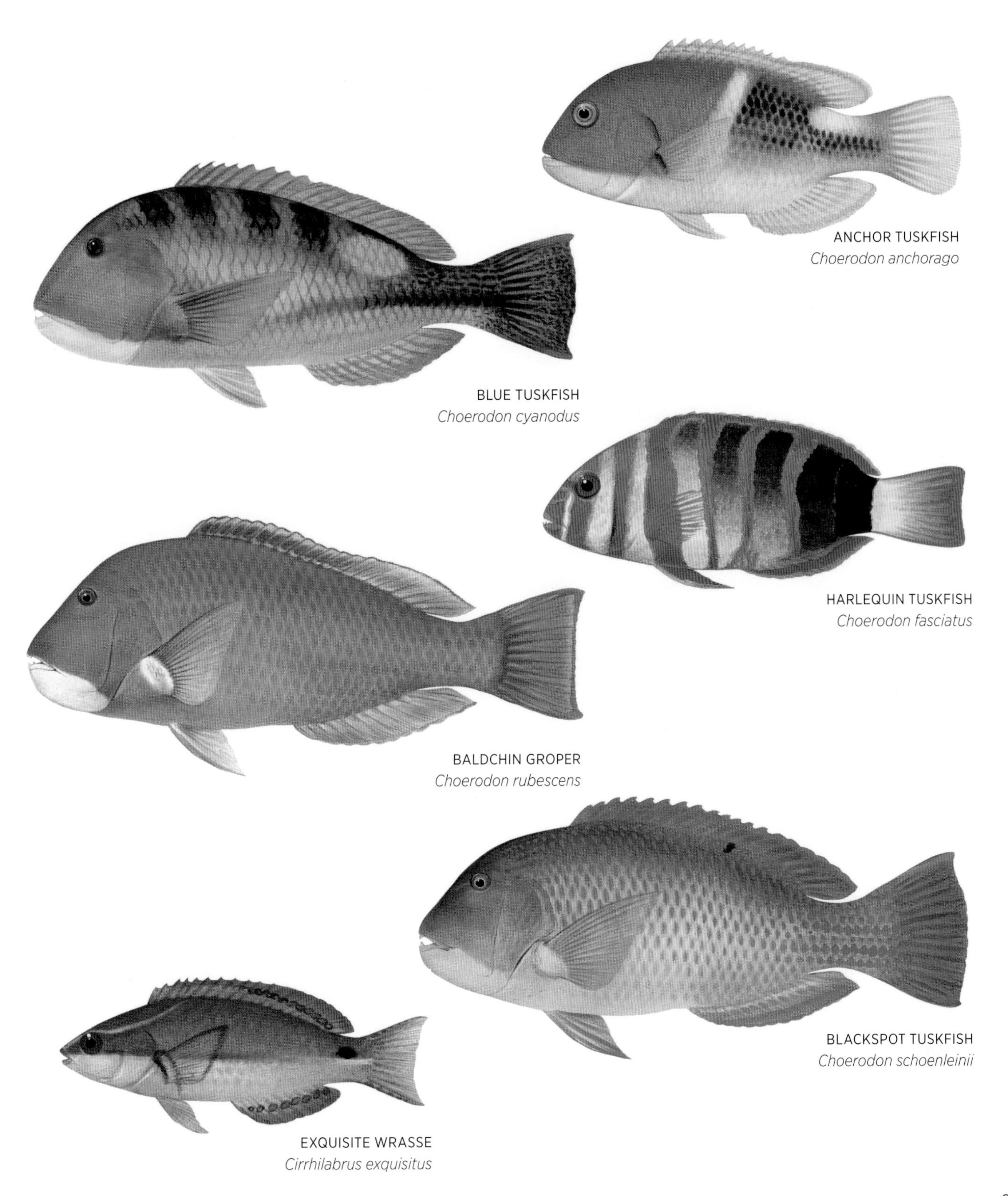

ANCHOR TUSKFISH
Choerodon anchorago

BLUE TUSKFISH
Choerodon cyanodus

HARLEQUIN TUSKFISH
Choerodon fasciatus

BALDCHIN GROPER
Choerodon rubescens

BLACKSPOT TUSKFISH
Choerodon schoenleinii

EXQUISITE WRASSE
Cirrhilabrus exquisitus

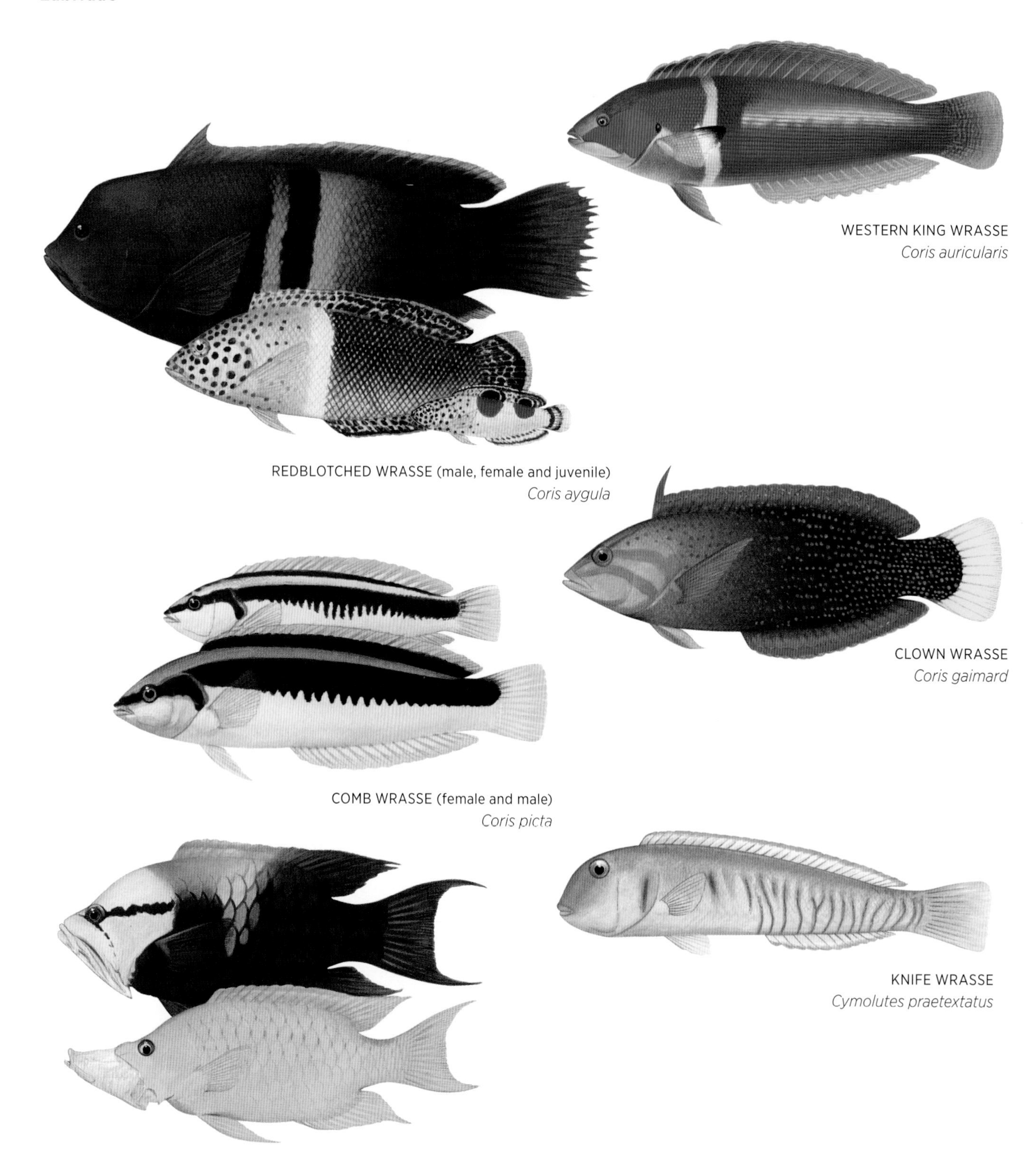

WESTERN KING WRASSE
Coris auricularis

REDBLOTCHED WRASSE (male, female and juvenile)
Coris aygula

CLOWN WRASSE
Coris gaimard

COMB WRASSE (female and male)
Coris picta

KNIFE WRASSE
Cymolutes praetextatus

SLINGJAW WRASSE (dark form and yellow form)
Epibulus insidiator

BIRDNOSE WRASSE (male and female)
Gomphosus varius

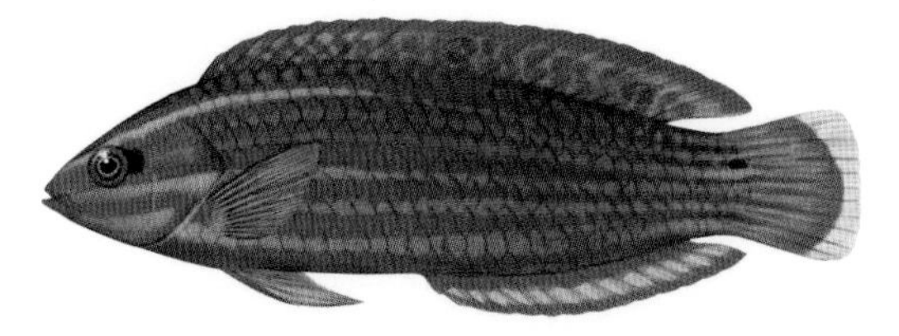

FALSE-EYE WRASSE
Halichoeres biocellatus

PASTEL-GREEN WRASSE (male and female)
Halichoeres chloropterus

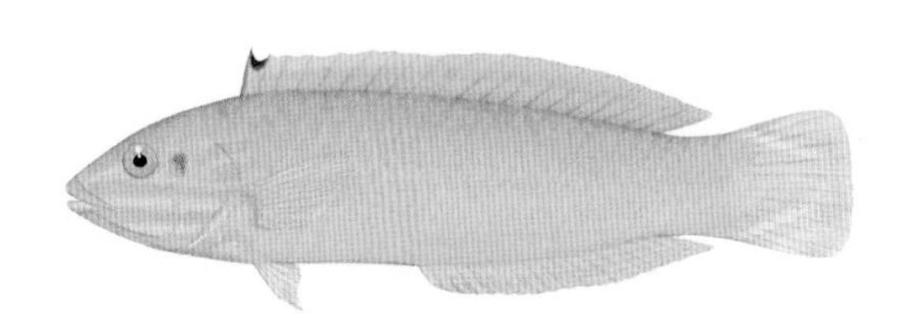

GOLDEN WRASSE
Halichoeres chrysus

CHECKERBOARD WRASSE (male and female)
Halichoeres hortulanus

DUSKY WRASSE (adult and juvenile)
Halichoeres marginatus

THREE-EYE WRASSE (male and female)
Halichoeres melanurus

EARMUFF WRASSE
Halichoeres melasmapomus

ZIGZAG WRASSE
Halichoeres scapularis

CLOUD WRASSE (male and female)
Halichoeres nebulosus

THREESPOT WRASSE
Halichoeres trimaculatus

FIVEBAND WRASSE
Hemigymnus fasciatus

THICKLIP WRASSE
Hemigymnus melapterus

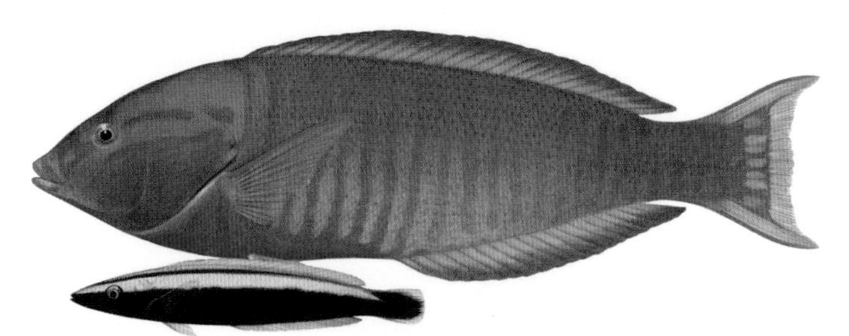

RINGED SLENDER WRASSE (adult and juvenile)
Hologymnosus annulatus

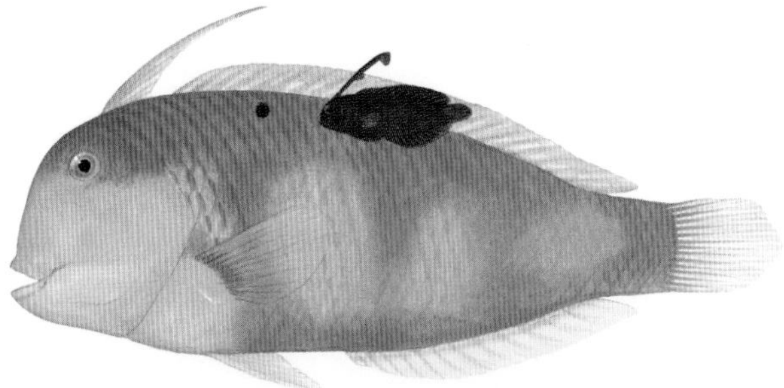

LEAF WRASSE (adult and juvenile)
Iniistius dea

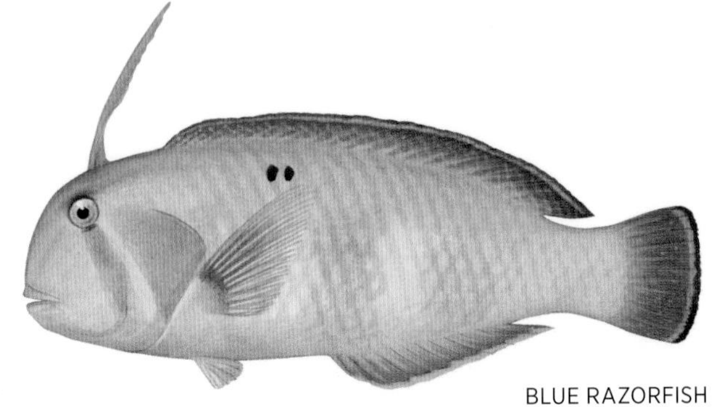

BLUE RAZORFISH
Iniistius pavo

ONELINE WRASSE
Labrichthys unilineatus

COMMON CLEANERFISH
Labroides dimidiatus

YELLOWBACK TUBELIP (male and female)
Labropsis xanthonota

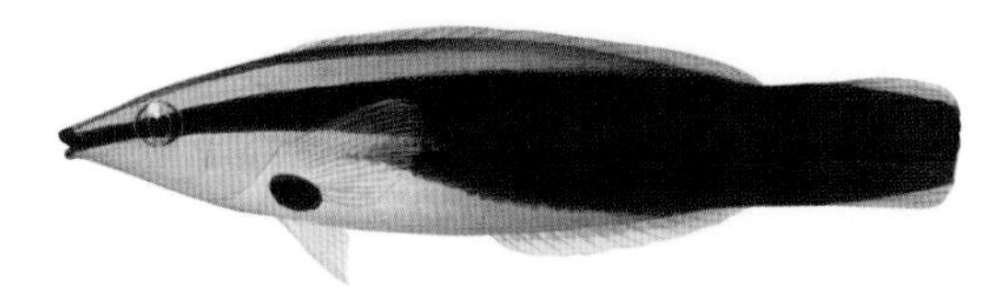

BREASTSPOT CLEANERFISH
Labroides pectoralis

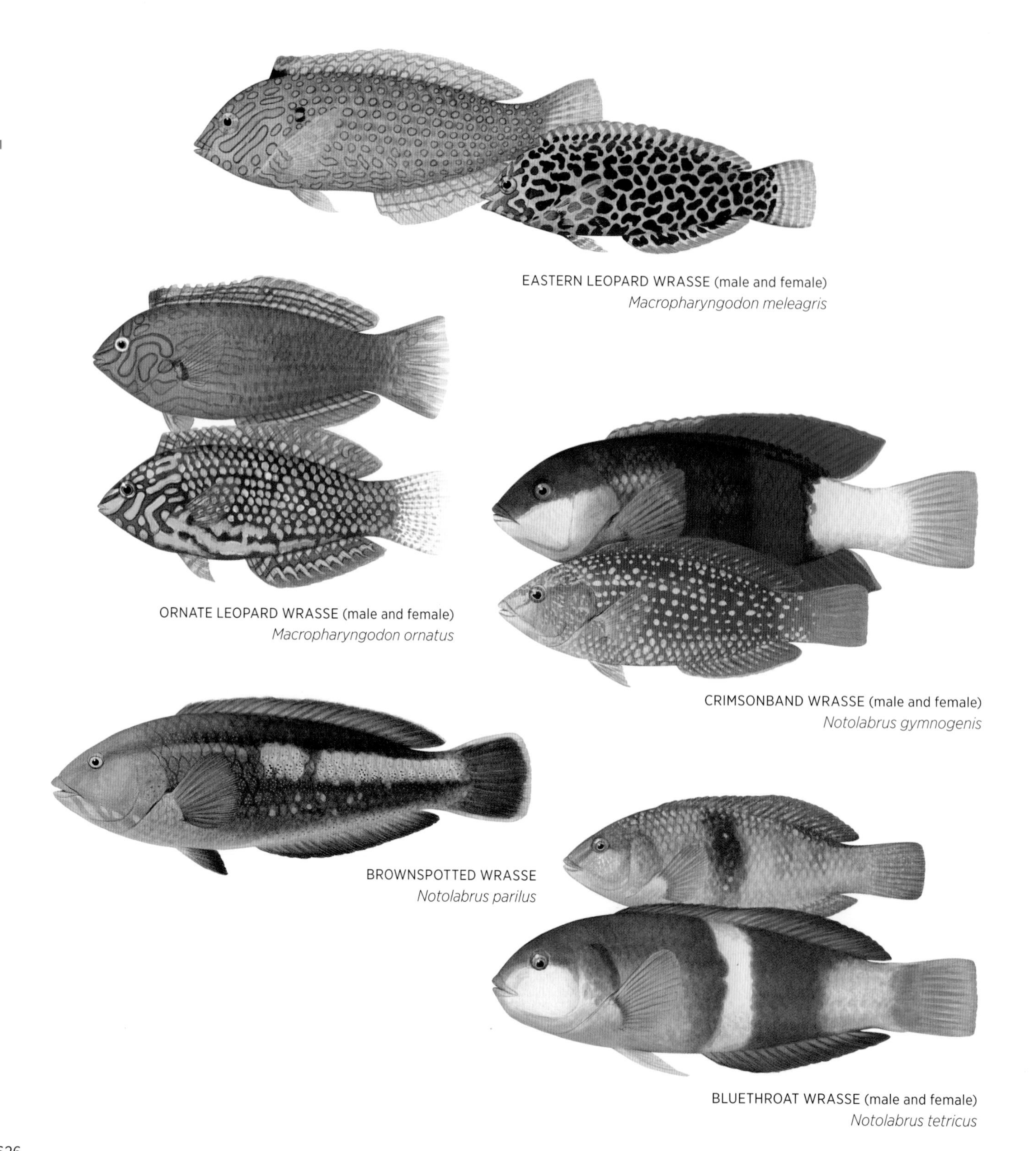

EASTERN LEOPARD WRASSE (male and female)
Macropharyngodon meleagris

ORNATE LEOPARD WRASSE (male and female)
Macropharyngodon ornatus

CRIMSONBAND WRASSE (male and female)
Notolabrus gymnogenis

BROWNSPOTTED WRASSE
Notolabrus parilus

BLUETHROAT WRASSE (male and female)
Notolabrus tetricus

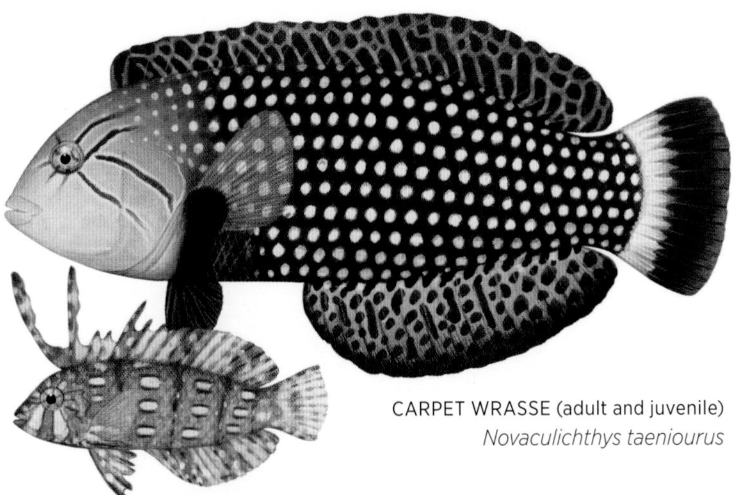

CARPET WRASSE (adult and juvenile)
Novaculichthys taeniourus

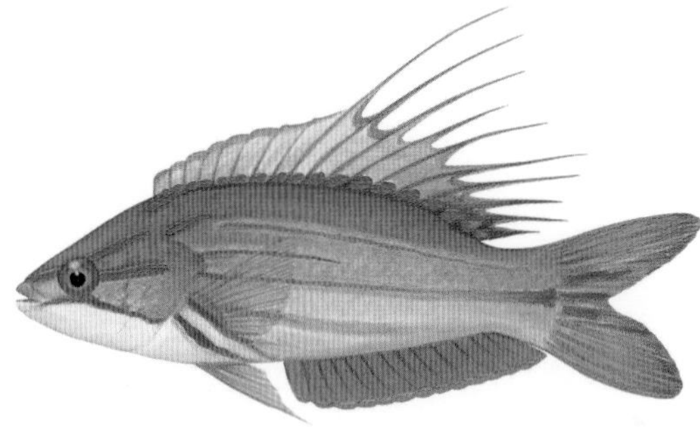

FILAMENTOUS FLASHER WRASSE
Paracheilinus filamentosus

SENATOR WRASSE (female and male)
Pictilabrus laticlavius

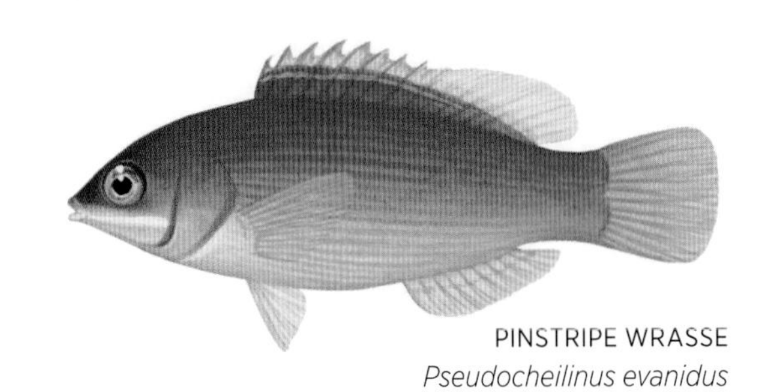

PINSTRIPE WRASSE
Pseudocheilinus evanidus

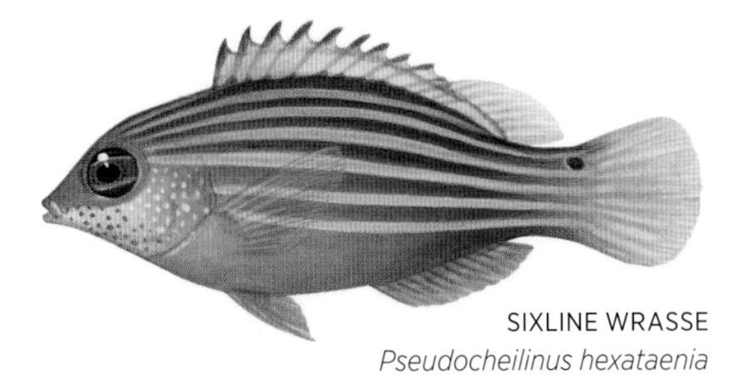

SIXLINE WRASSE
Pseudocheilinus hexataenia

JAPANESE WRASSE (male and female)
Pseudocoris yamashiroi

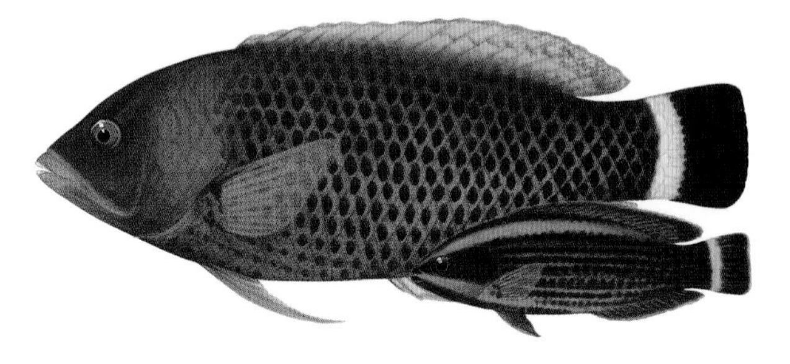

CHISELTOOTH WRASSE (adult and juvenile)
Pseudodax moluccanus

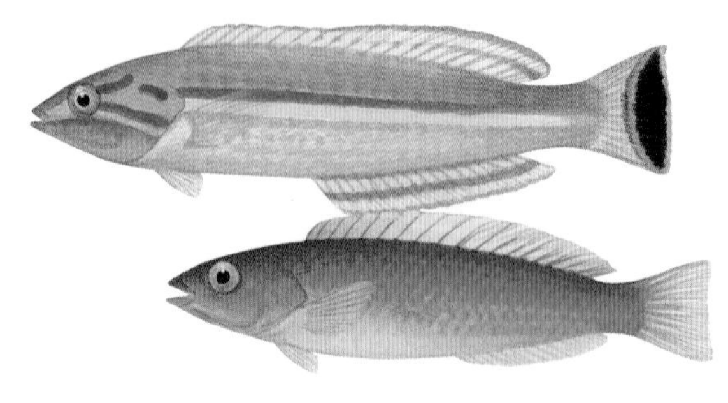

CANDY WRASSE (male and female)
Pseudojuloides cerasinus

GUNTHER'S WRASSE (male and female)
Pseudolabrus guentheri

COCKEREL WRASSE
Pteragogus enneacanthus

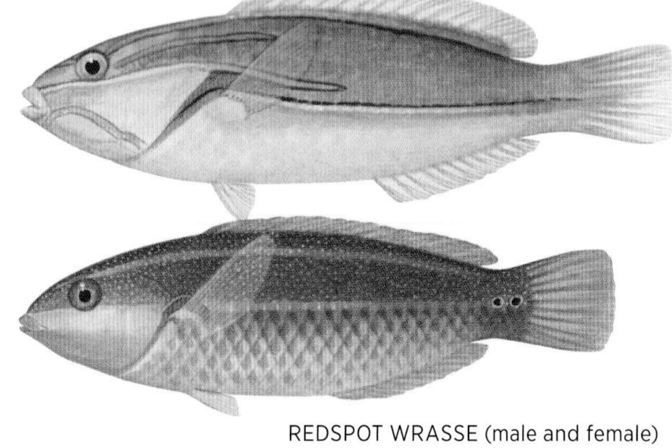

REDSPOT WRASSE (male and female)
Stethojulis bandanensis

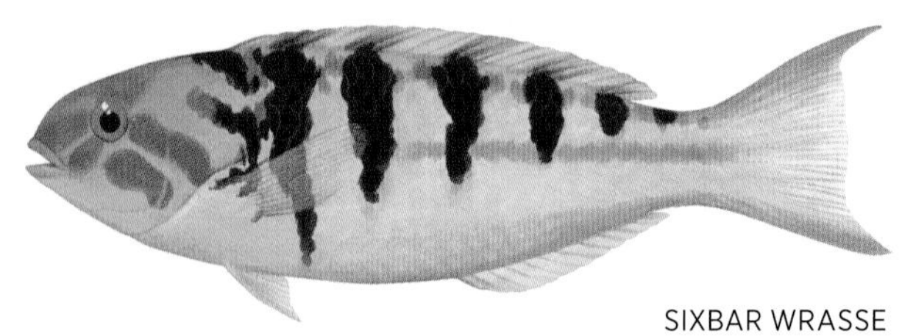

SIXBAR WRASSE
Thalassoma hardwicke

JANSEN'S WRASSE
Thalassoma jansenii

MOON WRASSE (adult and juvenile)
Thalassoma lunare

SURGE WRASSE (male and female)
Thalassoma purpureum

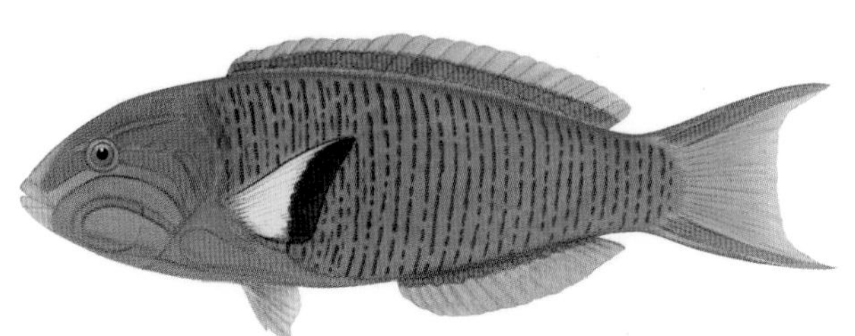

GREEN MOON WRASSE
Thalassoma lutescens

RED-RIBBON WRASSE (female and male)
Thalassoma quinquevittatum

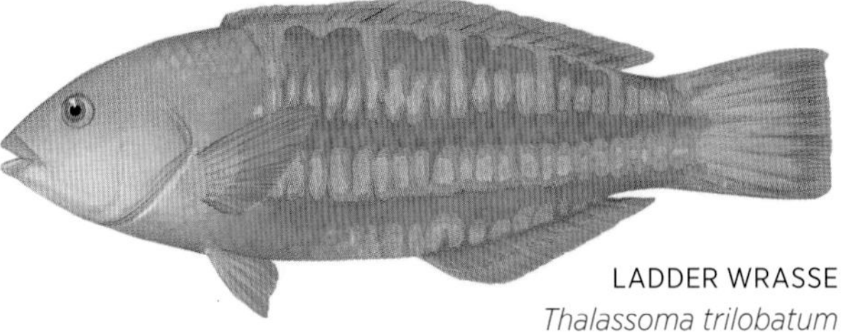

LADDER WRASSE
Thalassoma trilobatum

AUSTRALIAN LABRIDAE SPECIES

Achoerodus gouldii Western Blue Groper
Achoerodus viridis Eastern Blue Groper
Anampses caeruleopunctatus Diamond Wrasse
Anampses elegans Elegant Wrasse
Anampses femininus Bluetail Wrasse
Anampses geographicus Scribbled Wrasse
Anampses lennardi Blue-and-yellow Wrasse
Anampses melanurus Blacktail Wrasse
Anampses meleagrides Speckled Wrasse
Anampses neoguinaicus Blackback Wrasse
Anampses twistii Yellowbreast Wrasse
Austrolabrus maculates Blackspotted Wrasse
Bodianus anthioides Lyretail Pigfish
Bodianus axillaris Coral Pigfish
Bodianus bilunulatus Saddleback Pigfish
Bodianus bimaculatus Twospot Pigfish
Bodianus diana Diana's Pigfish
Bodianus flavifrons Masked Pigfish
Bodianus flavipinnis Yellowfin Pigfish
Bodianus frenchii Foxfish
Bodianus izuensis Striped Pigfish
Bodianus loxozonus Blackfin Pigfish
Bodianus mesothorax Eclipse Pigfish
Bodianus perditio Goldspot Pigfish
Bodianus unimaculatus Eastern Blackspot Pigfish
Bodianus vulpinus Western Blackspot Pigfish
Cheilinus chlorourus Floral Maori Wrasse
Cheilinus fasciatus Redbreast Maori Wrasse
Cheilinus oxycephalus Red Maori Wrasse
Cheilinus trilobatus Tripletail Maori Wrasse
Cheilinus undulatus Humphead Maori Wrasse
Cheilio inermis Sharpnose Wrasse
Choerodon anchorago Anchor Tuskfish
Choerodon cauteroma Purple Tuskfish
Choerodon cyanodus Blue Tuskfish
Choerodon fasciatus Harlequin Tuskfish
Choerodon frenatus Bridled Tuskfish
Choerodon graphicus Graphic Tuskfish
Choerodon jordani Dagger Tuskfish
Choerodon monostigma Darkspot Tuskfish
Choerodon paynei Payne's Tuskfish
Choerodon rubescens Baldchin Groper
Choerodon schoenleinii Blackspot Tuskfish
Choerodon sugillatum Wedgetail Tuskfish
Choerodon venustus Venus Tuskfish
Choerodon vitta Redstripe Tuskfish
Choerodon zamboangae Eyebrow Tuskfish
Cirrhilabrus bathyphilus Deepwater Wrasse
Cirrhilabrus condei Conde's Wrasse
Cirrhilabrus cyanopleura Blueside Wrasse
Cirrhilabrus exquisitus Exquisite Wrasse
Cirrhilabrus laboutei Laboute's Wrasse
Cirrhilabrus lineatus Lavender Wrasse
Cirrhilabrus morrisoni Morrison's Wrasse
Cirrhilabrus punctatus Finespot Wrasse
Cirrhilabrus randalli Randall's Wrasse
Cirrhilabrus scottorum Scott's Wrasse
Cirrhilabrus temminckii Peacock Wrasse
Conniella apterygia Connie's Wrasse
Coris auricularis Western King Wrasse
Coris aurilineata Goldlined Wrasse
Coris aygula Redblotched Wrasse
Coris batuensis Variegated Wrasse
Coris bulbifrons Doubleheader
Coris caudimacula Spot-tail Wrasse
Coris dorsomacula Pinklined Wrasse
Coris gaimard Clown Wrasse
Coris picta Comb Wrasse
Coris pictoides Pixy Wrasse
Coris sandeyeri Eastern King Wrasse
Cymolutes praetextatus Knife Wrasse
Cymolutes torquatus Razor Wrasse
Decodon pacificus Ten-tooth Wrasse
Diproctacanthus xanthurus Yellowtail Wrasse
Dotalabrus alleni Little Rainbow Wrasse
Dotalabrus aurantiacus Castelnau's Wrasse
Epibulus insidiator Slingjaw Wrasse
Eupetrichthys angustipes Snakeskin Wrasse
Gomphosus varius Birdnose Wrasse
Halichoeres argus Argus Wrasse
Halichoeres biocellatus False-eye Wrasse
Halichoeres brownfieldi Brownfield's Wrasse

Halichoeres chloropterus Pastel-green Wrasse
Halichoeres chrysus Golden Wrasse
Halichoeres hartzfeldii Orangeline Wrasse
Halichoeres hortulanus Checkerboard Wrasse
Halichoeres leucurus Chain-line Wrasse
Halichoeres margaritaceus Pearly Wrasse
Halichoeres marginatus Dusky Wrasse
Halichoeres melanochir Orangefin Wrasse
Halichoeres melanurus Three-eye Wrasse
Halichoeres melasmapomus Earmuff Wrasse
Halichoeres miniatus Cheek-ring Wrasse
Halichoeres nebulosus Cloud Wrasse
Halichoeres nigrescens Bubblefin Wrasse
Halichoeres ornatissimus Ornamental Wrasse
Halichoeres prosopeion Twotone Wrasse
Halichoeres scapularis Zigzag Wrasse
Halichoeres trimaculatus Threespot Wrasse
Halichoeres zeylonicus Goldstripe Wrasse
Hemigymnus fasciatus Fiveband Wrasse
Hemigymnus melapterus Thicklip Wrasse
Hologymnosus annulatus Ringed Slender Wrasse
Hologymnosus doliatus Pastel Slender Wrasse
Hologymnosus longipes Pale Slender Wrasse
Hologymnosus rhodonotus Red Slender Wrasse
Iniistius aneitensis Whiteblotch Razorfish
Iniistius dea Leaf Wrasse
Iniistius jacksonensis Keelhead Razorfish
Iniistius pavo Blue Razorfish
Iniistius pentadactylus Fivefinger Razorfish
Labrichthys unilineatus Oneline Wrasse
Labroides bicolor Bicolor Cleanerfish
Labroides dimidiatus Common Cleanerfish
Labroides pectoralis Breastspot Cleanerfish
Labropsis australis Southern Tubelip
Labropsis manabei Tailblotch Tubelip
Labropsis xanthonota Yellowback Tubelip
Leptojulis cyanopleura Shoulderspot Wrasse
Macropharyngodon choati Choat's Wrasse
Macropharyngodon kuiteri Kuiter's Wrasse
Macropharyngodon meleagris Eastern Leopard Wrasse
Macropharyngodon negrosensis Black Leopard Wrasse
Macropharyngodon ornatus Ornate Leopard Wrasse
Notolabrus fucicola Purple Wrasse
Notolabrus gymnogenis Crimsonband Wrasse
Notolabrus inscriptus Inscribed Wrasse
Notolabrus parilus Brownspotted Wrasse
Notolabrus tetricus Bluethroat Wrasse
Novaculichthys macrolepidotus Seagrass Wrasse
Novaculichthys taeniourus Carpet Wrasse
Ophthalmolepis lineolatus Southern Maori Wrasse
Oxycheilinus bimaculatus Little Maori Wrasse
Oxycheilinus celebicus Slender Maori Wrasse
Oxycheilinus digrammus Violetline Maori Wrasse
Oxycheilinus nigromarginatus Blackmargin Maori Wrasse
Oxycheilinus unifasciatus Ringtail Maori Wrasse
Paracheilinus filamentosus Filamentous Flasher Wrasse
Paracheilinus flavianalis Yellowfin Flasher Wrasse
Pictilabrus brauni Braun's Wrasse
Pictilabrus laticlavius Senator Wrasse
Pictilabrus viridis False Senator Wrasse
Pseudocheilinus evanidus Pinstripe Wrasse
Pseudocheilinus hexataenia Sixline Wrasse
Pseudocheilinus ocellatus Whitebarred Pink Wrasse
Pseudocheilinus octotaenia Eightline Wrasse
Pseudocoris heteroptera Zebra Wrasse
Pseudocoris yamashiroi Japanese Wrasse
Pseudodax moluccanus Chiseltooth Wrasse
Pseudojuloides cerasinus Candy Wrasse
Pseudojuloides elongatus Long Green Wrasse
Pseudolabrus biserialis Redband Wrasse
Pseudolabrus guentheri Gunther's Wrasse
Pseudolabrus luculentus Luculent Wrasse
Pseudolabrus mortonii Rosy Wrasse
Pteragogus cryptus Cryptic Wrasse
Pteragogus enneacanthus Cockerel Wrasse
Pteragogus flagellifer Cocktail Wrasse
Stethojulis bandanensis Redspot Wrasse
Stethojulis interrupta Brokenline Wrasse
Stethojulis strigiventer Silverstreak Wrasse
Stethojulis trilineata Three-ribbon Wrasse
Suezichthys arquatus Painted Rainbow Wrasse
Suezichthys aylingi Crimson Rainbow Wrasse
Suezichthys bifurcatus Striped Rainbow Wrasse
Suezichthys cyanolaemus Bluethroat Rainbow Wrasse
Suezichthys devisi Australian Rainbow Wrasse
Suezichthys gracilis Slender Rainbow Wrasse
Suezichthys notatus Northern Rainbow Wrasse
Suezichthys soelae Soela Wrasse
Thalassoma amblycephalum Bluehead Wrasse
Thalassoma hardwicke Sixbar Wrasse
Thalassoma jansenii Jansen's Wrasse
Thalassoma lunare Moon Wrasse
Thalassoma lutescens Green Moon Wrasse
Thalassoma nigrofasciatum Blackbarred Wrasse
Thalassoma purpureum Surge Wrasse
Thalassoma quinquevittatum Red-ribbon Wrasse
Thalassoma septemfasciatum Solarfin Wrasse
Thalassoma trilobatum Ladder Wrasse
Wetmorella albofasciata Doubleline Wrasse
Wetmorella nigropinnata Possum Wrasse
Xenojulis margaritaceus Pinkspeckled Wrasse
Xiphocheilus typus Bluetooth Tuskfish

WEED WHITINGS

Odacidae

BLUE WEED WHITING
Haletta semifasciata

Weed Whitings are found only in temperate waters of Australia and New Zealand, and there are about 12 species in the family. They have elongate to oval, rounded to compressed bodies with small to moderate cycloid scales and a small mouth. The single dorsal fin has 14–27 spines and 9–22 rays, the anal fin usually has three spines and 7–14 rays, and the caudal fin varies from lunate to diamond-shaped.

Weed Whitings are, as their common name suggests, usually found in areas of dense algal growth. *Siphonognathus* species are small, elongate fishes that reach about 10–15 cm in length. In Australia, they are found in areas of weed and seagrass, and near rocky outcrops along the south coast. They are not often seen by divers, as they are well camouflaged and rarely leave the shelter of the weed. The **Pencil Weed Whiting** forms small schools and is one of the more common species found in Australian waters. The extremely elongate **Tubemouth**, a strange fish with a fleshy barbel on the upper lip, reaches a maximum length of about 50 cm and was originally placed in its own family. The **Rainbow Cale** is a deeper-bodied, colourful fish that is reasonably common on southern Australian coastal reefs with plentiful weed, where it feeds mainly on molluscs. The teeth are fused into a parrot-like beak and for this reason it was originally classified as a Parrotfish (Scaridae, p. 634). The **Herring Cale** reaches about 40 cm in length and is herbivorous, feeding on broad-leafed algae. It is found on exposed rocky reefs along the southern Australian coastline, where it moves sinuously amongst the weed. The **Blue Weed Whiting** also reaches about 40 cm in length, and it is found over sand and weed areas in shallow coastal waters and estuaries around southern Australia.

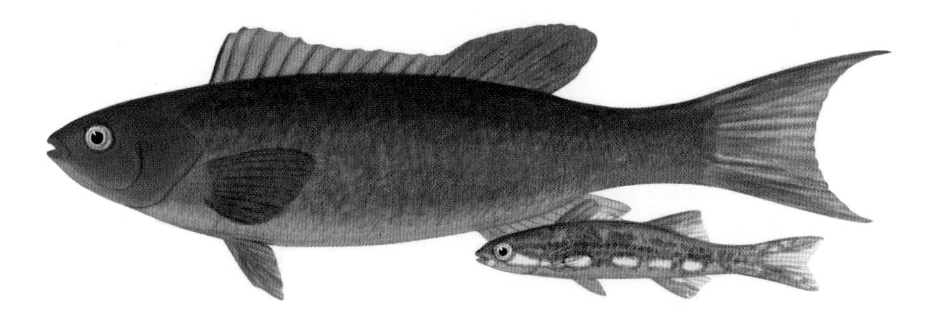

HERRING CALE (adult and juvenile)
Odax cyanomelas

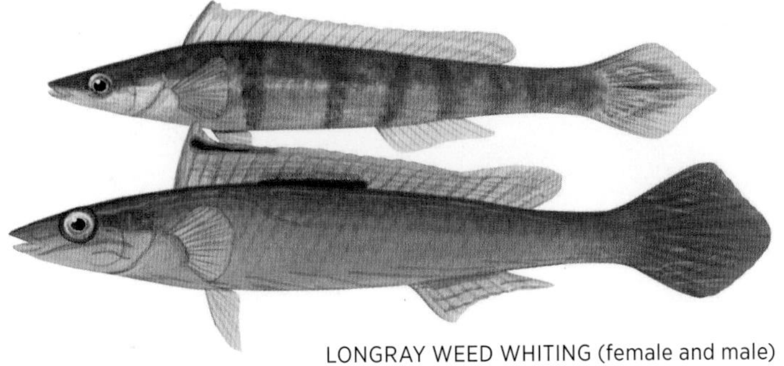

LONGRAY WEED WHITING (female and male)
Siphonognathus radiatus

PENCIL WEED WHITING (female and male)
Siphonognathus beddomei

TUBEMOUTH
Siphonognathus argyrophanes

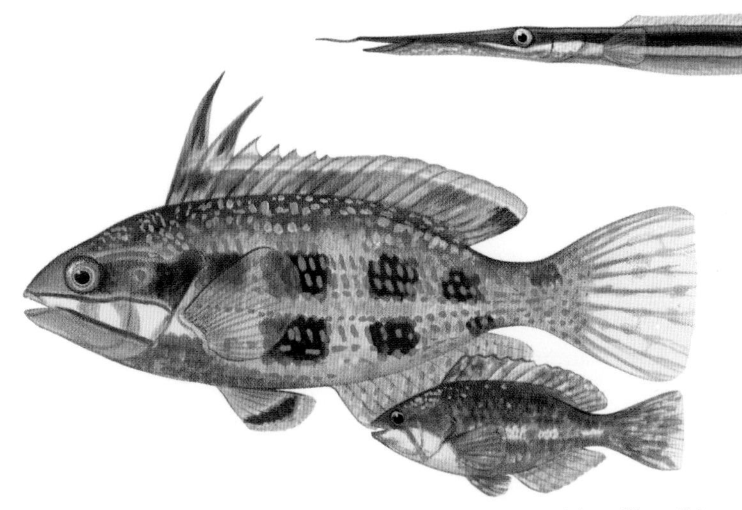

RAINBOW CALE (adult and juvenile)
Odax acroptilus

AUSTRALIAN ODACIDAE SPECIES

Haletta semifasciata Blue Weed Whiting
Neoodax balteatus Little Weed Whiting
Odax acroptilus Rainbow Cale
Odax cyanomelas Herring Cale
Siphonognathus argyrophanes Tubemouth
Siphonognathus attenuatus Slender Weed Whiting
Siphonognathus beddomei Pencil Weed Whiting
Siphonognathus caninis Sharpnose Weed Whiting
Siphonognathus radiatus Longray Weed Whiting
Siphonognathus tanyourus Longtail Weed Whiting

PARROTFISHES

Scaridae

BLACKVEIN PARROTFISH
Scarus rubroviolaceus

Parrotfishes are found in tropical waters of the Atlantic, Indian and Pacific oceans, and there are about 88 species in the family. They have oval, rounded to slightly compressed bodies with large cycloid scales. Their small mouth has teeth fused to form a parrot-like beak. The single dorsal fin has nine spines and 10 rays, the anal fin has three spines and nine rays, and the caudal fin is rounded to lunate.

Parrotfishes are colourful residents of Australian coral reefs, mostly occurring in northern waters. They use their powerful teeth to scrape algae from rock and dead coral, often taking in large amounts of the coral or rock with each bite. A few larger species also feed on live coral. Parrotfishes use pharyngeal teeth to grind the algae and coral together, and periodically excrete large amounts of the crushed coral. With their constant scraping and excreting, Parrotfishes are the primary producers of sand and sediment in many coral-reef areas. A few species, including the **Spinytooth Parrotfish**, use a different feeding strategy, inhabiting weedy areas and consuming mainly broad-leafed algae.

Parrotfishes are closely related to Wrasses (Labridae, p. 617), and employ the same method of swimming (using only the pectoral fins) and much the same sexual strategy. Parrotfishes are hermaphroditic, with a large 'terminal-phase' male maintaining a territory and a harem of females. If the terminal-phase male is removed, one of the females can change sex to replace it. Juvenile Parrotfishes often occur in large mixed-species schools, constantly moving and grazing over the reef. Parrotfishes are active by day; at night they shelter in crevices in the reef and secrete a mucus cocoon that completely envelopes their body. This cocoon is perhaps to protect them from carnivorous Moray Eels (Muraenidae, p. 158), which hunt at night and use smell to locate their prey.

Parrotfish species are remarkably consistent in physical form and are usually differentiated by their distinctive colour patterns. It can be difficult to identify a particular species under water, as juveniles, females and terminal-phase males all have different markings. The impressive **Bumphead Parrotfish** is the largest member of the family, reaching up to 1.2 m in length. It is sometimes seen in large groups on outer reefs, making a great deal of noise under water as it crunches and scrapes corals. The **Bluebarred Parrotfish** is widely distributed in Australia, ranging into more southern waters than others in the family.

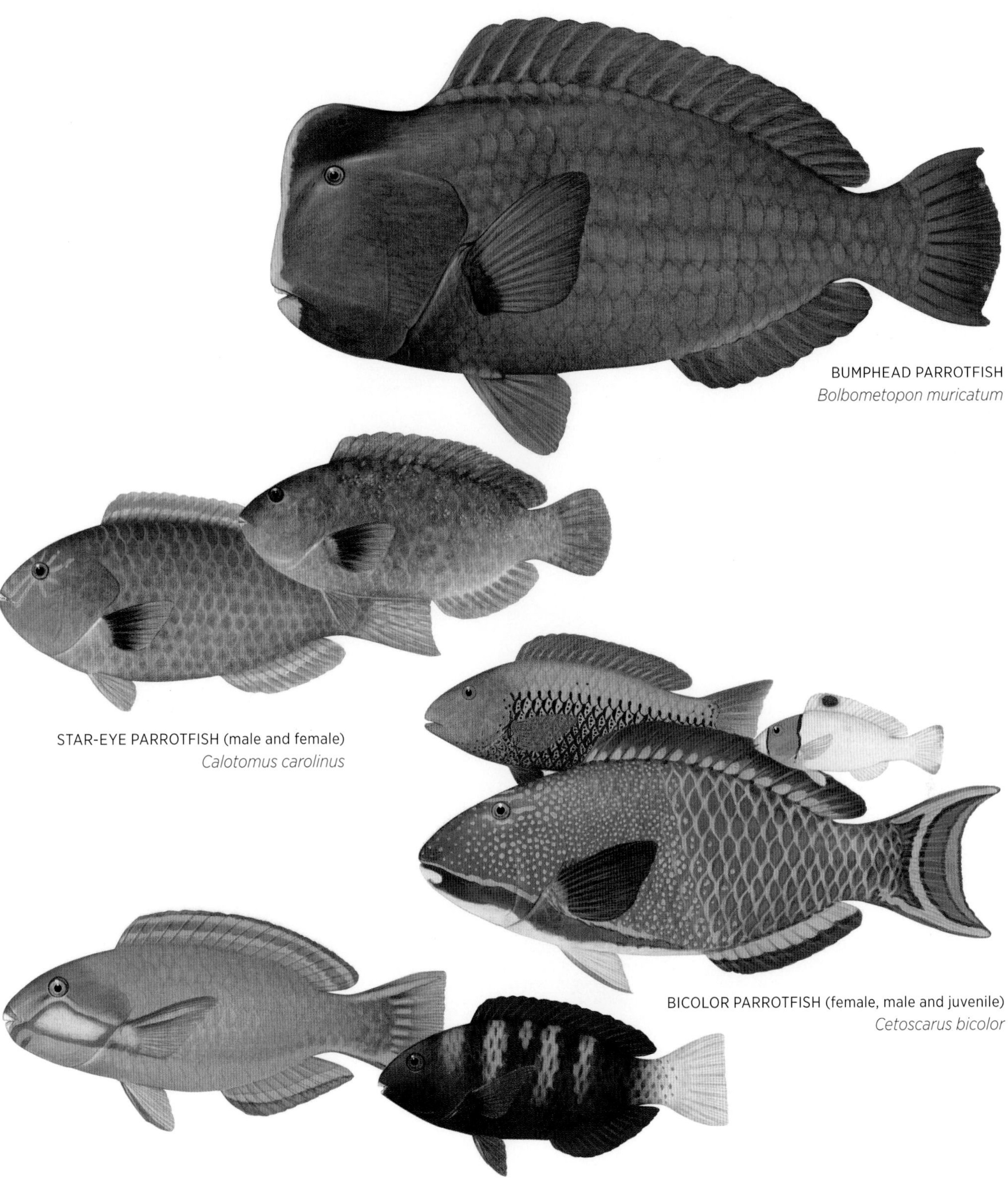

BUMPHEAD PARROTFISH
Bolbometopon muricatum

STAR-EYE PARROTFISH (male and female)
Calotomus carolinus

BICOLOR PARROTFISH (female, male and juvenile)
Cetoscarus bicolor

BLEEKER'S PARROTFISH (male and female)
Chlorurus bleekeri

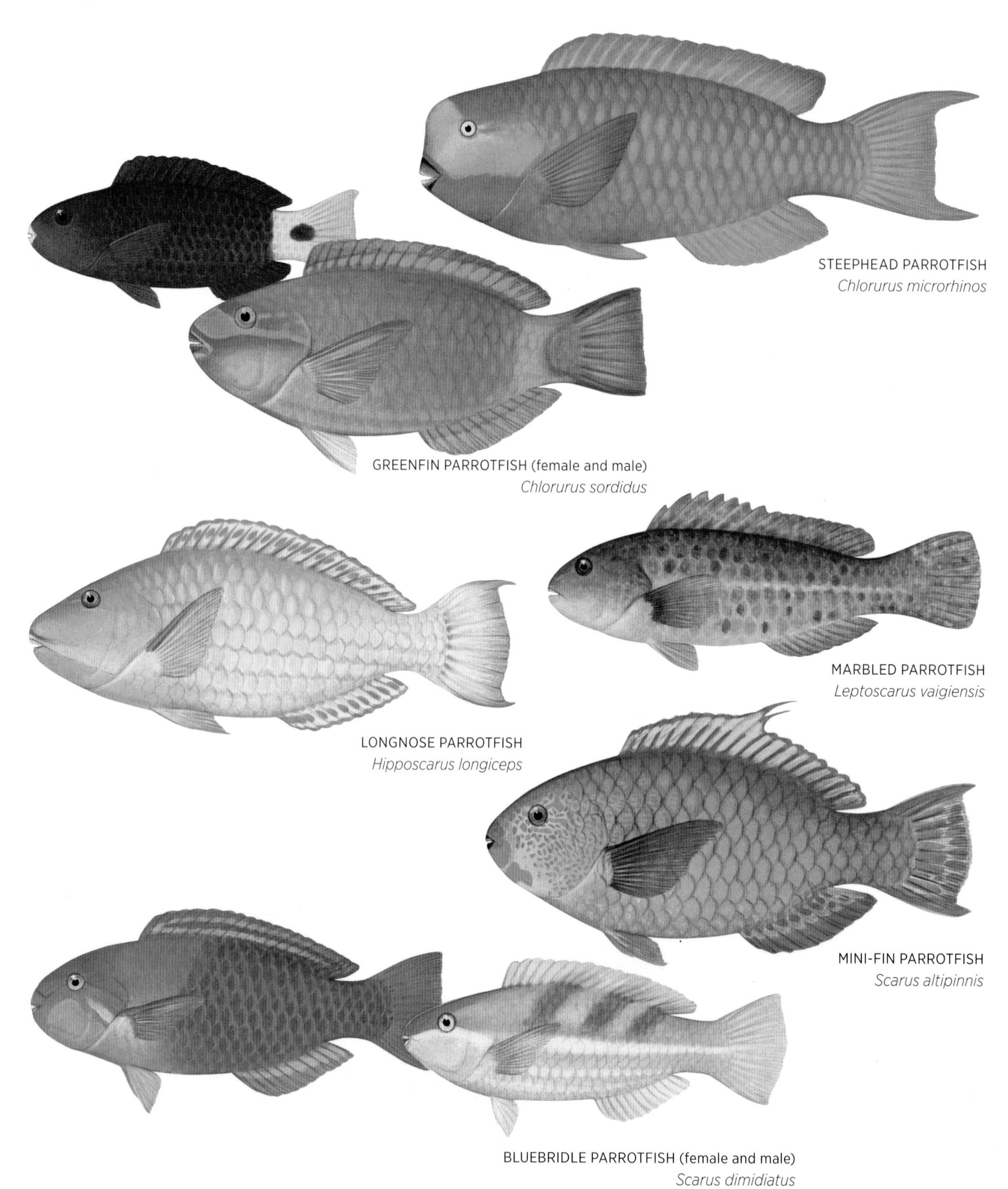

STEEPHEAD PARROTFISH
Chlorurus microrhinos

GREENFIN PARROTFISH (female and male)
Chlorurus sordidus

MARBLED PARROTFISH
Leptoscarus vaigiensis

LONGNOSE PARROTFISH
Hipposcarus longiceps

MINI-FIN PARROTFISH
Scarus altipinnis

BLUEBRIDLE PARROTFISH (female and male)
Scarus dimidiatus

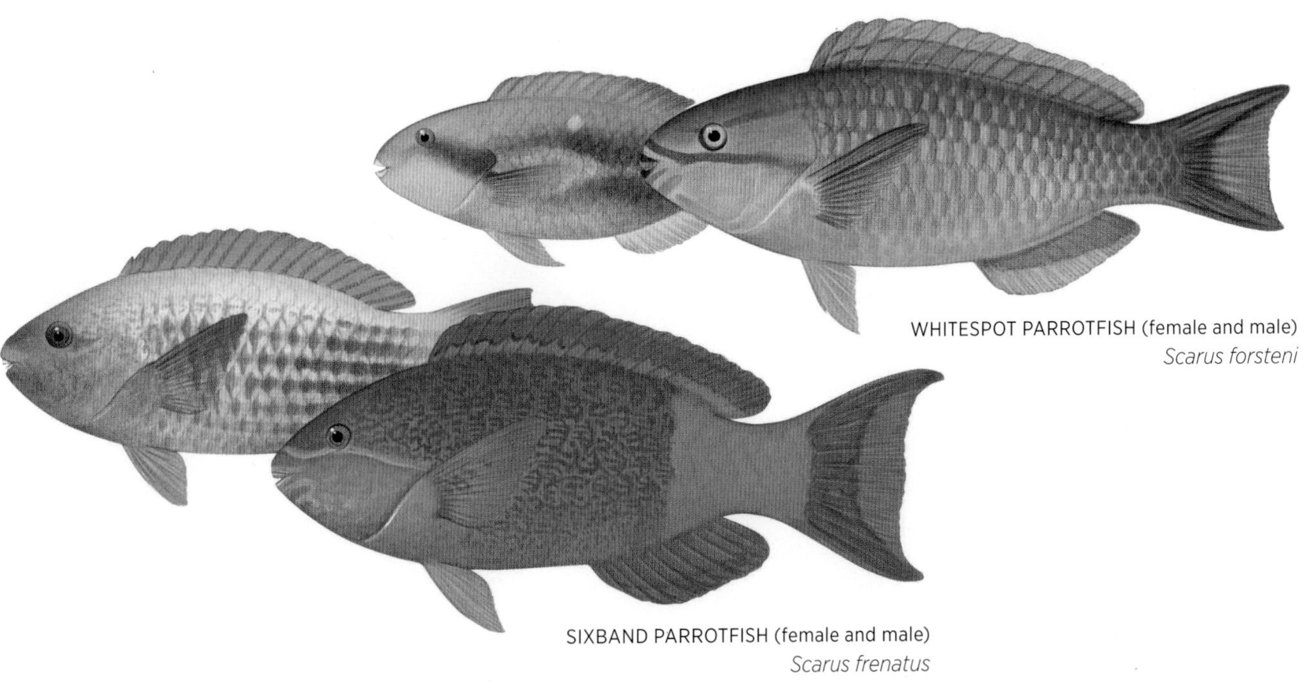

WHITESPOT PARROTFISH (female and male)
Scarus forsteni

SIXBAND PARROTFISH (female and male)
Scarus frenatus

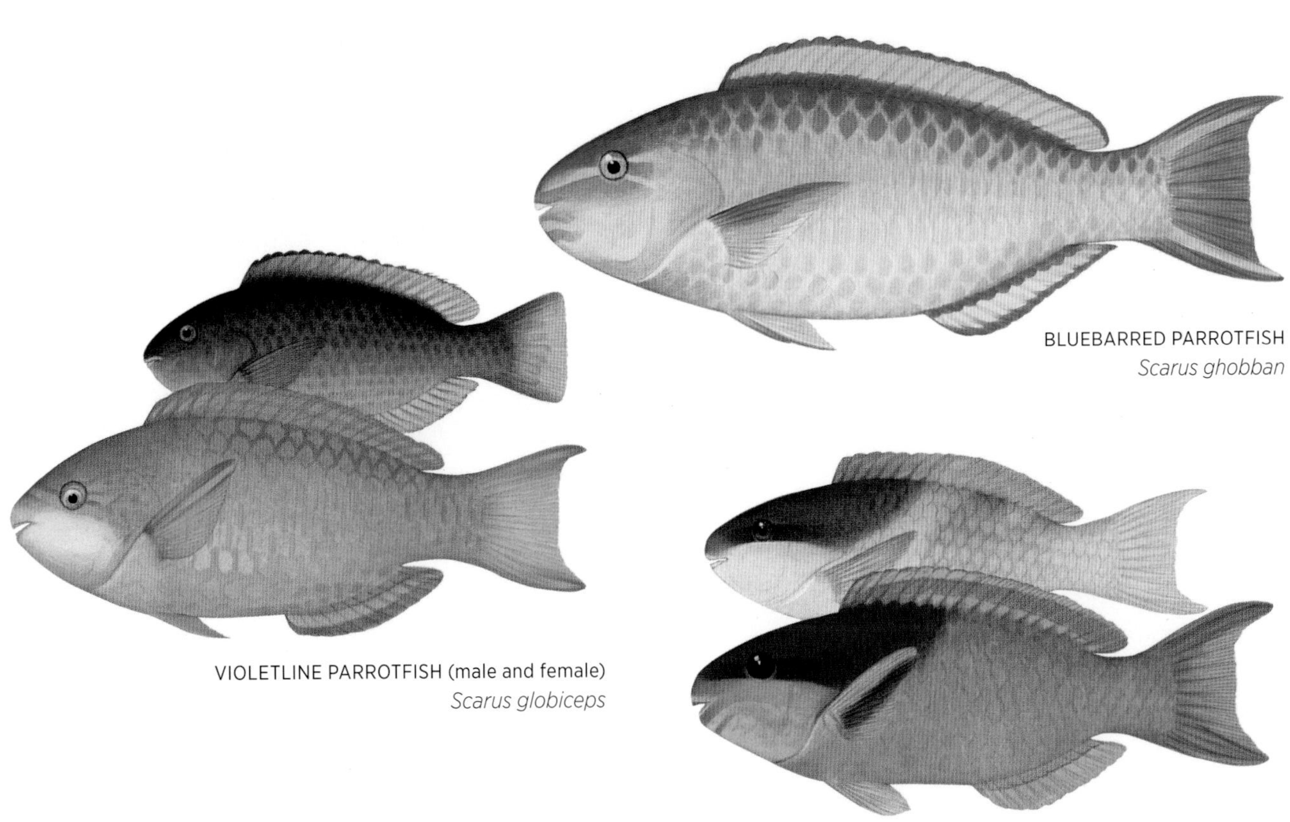

BLUEBARRED PARROTFISH
Scarus ghobban

VIOLETLINE PARROTFISH (male and female)
Scarus globiceps

DARKCAP PARROTFISH (female and male)
Scarus oviceps

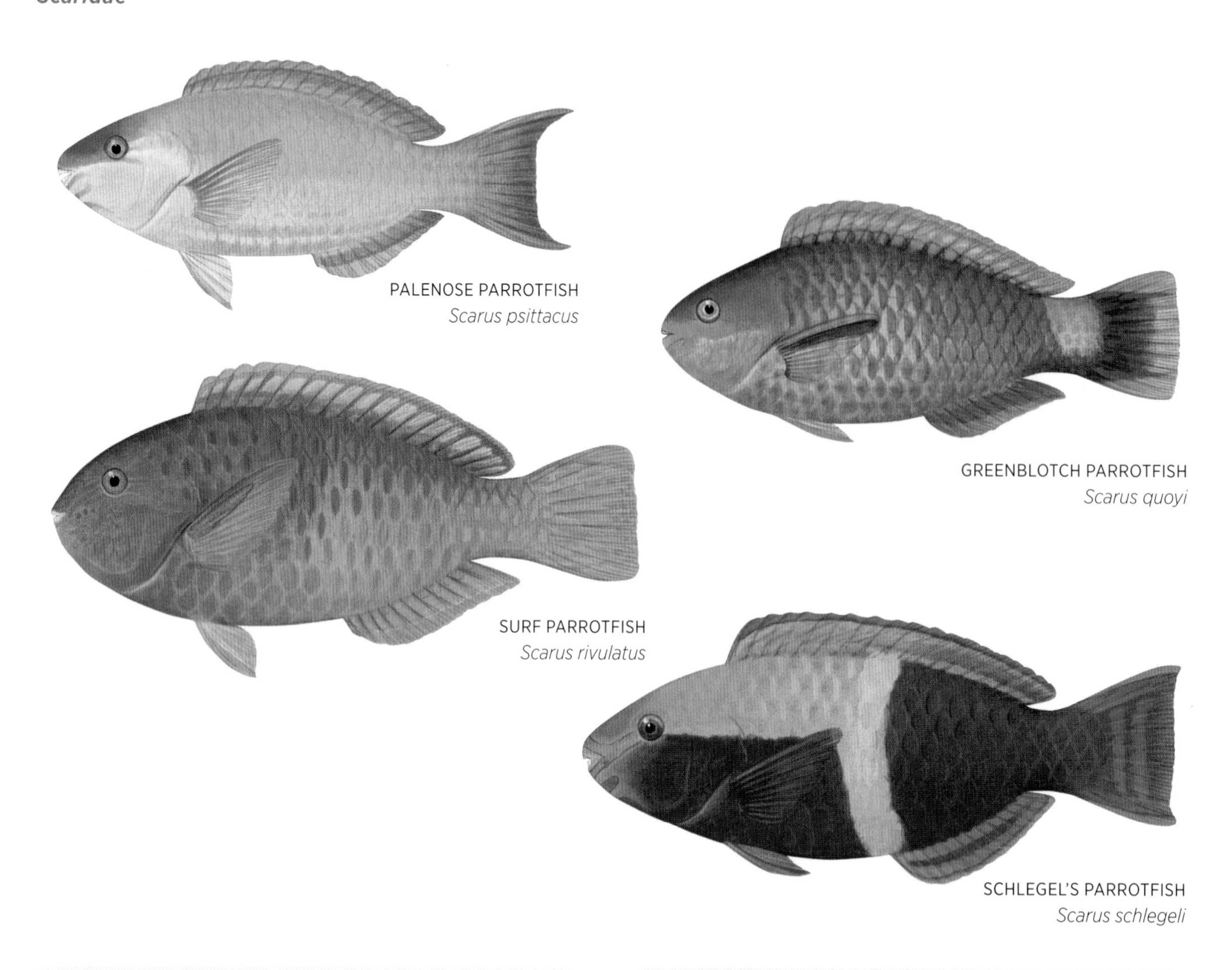

PALENOSE PARROTFISH
Scarus psittacus

GREENBLOTCH PARROTFISH
Scarus quoyi

SURF PARROTFISH
Scarus rivulatus

SCHLEGEL'S PARROTFISH
Scarus schlegeli

AUSTRALIAN SCARIDAE SPECIES

Bolbometopon muricatum Bumphead Parrotfish
Calotomus carolinus Star-eye Parrotfish
Calotomus spinidens Spinytooth Parrotfish
Cetoscarus bicolor Bicolor Parrotfish
Chlorurus bleekeri Bleeker's Parrotfish
Chlorurus frontalis Reefcrest Parrotfish
Chlorurus japanensis Redtail Parrotfish
Chlorurus microrhinos Steephead Parrotfish
Chlorurus oedema Knothead Parrotfish
Chlorurus sordidus Greenfin Parrotfish
Hipposcarus longiceps Longnose Parrotfish
Leptoscarus vaigiensis Marbled Parrotfish
Scarus altipinnis Mini-fin Parrotfish
Scarus chameleon Chameleon Parrotfish
Scarus dimidiatus Bluebridle Parrotfish
Scarus flavipectoralis Yellowfin Parrotfish
Scarus forsteni Whitespot Parrotfish
Scarus frenatus Sixband Parrotfish
Scarus ghobban Bluebarred Parrotfish
Scarus globiceps Violetline Parrotfish
Scarus longipinnis Highfin Parrotfish
Scarus niger Swarthy Parrotfish
Scarus oviceps Darkcap Parrotfish
Scarus prasiognathos Greencheek Parrotfish
Scarus psittacus Palenose Parrotfish
Scarus quoyi Greenblotch Parrotfish
Scarus rivulatus Surf Parrotfish
Scarus rubroviolaceus Blackvein Parrotfish
Scarus schlegeli Schlegel's Parrotfish
Scarus spinus Yellowhead Parrotfish
Scarus tricolor Tricolour Parrotfish
Scarus xanthopleura Red Parrotfish

EELPOUTS

Zoarcidae

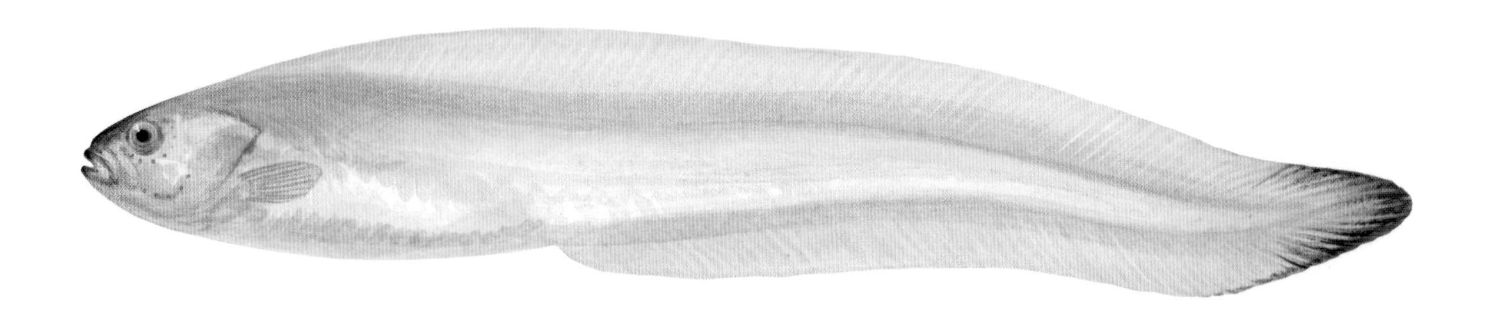

LIMP EELPOUT
Melanostigma gelatinosum

Eelpouts are found worldwide, mainly in cool temperate and polar waters. There are about 230 species in this large family and they are most diverse in the North Pacific and North Atlantic. They are elongate, eel-like fishes, and many have a gelatinous body with scales that are either very small and embedded in the skin or absent altogether. Their moderate-sized mouth often has fleshy lips, a projecting lower jaw, and small, conical teeth, and there are many sensory pores on the head. They have long dorsal and anal fins, both of which are continuous with the caudal fin, and the ventral fins are absent or rudimentary and placed well forward beneath the throat.

Eelpouts are mostly small fishes, growing to 10–20 cm in length, with the largest species reaching about 1.1 m. A few species occur in shallow tide pools and over coastal reefs, but most inhabit deep water, to 2000 m or more. The deepwater species are poorly known. The **Limp Eelpout** has no scales and the lateral line is absent. It is widely distributed in the southern hemisphere and has been recorded from about 40 m to over 2500 m.

AUSTRALIAN ZOARCIDAE SPECIES

Melanostigma gelatinosum Limp Eelpout

THORNFISHES

Bovichtidae

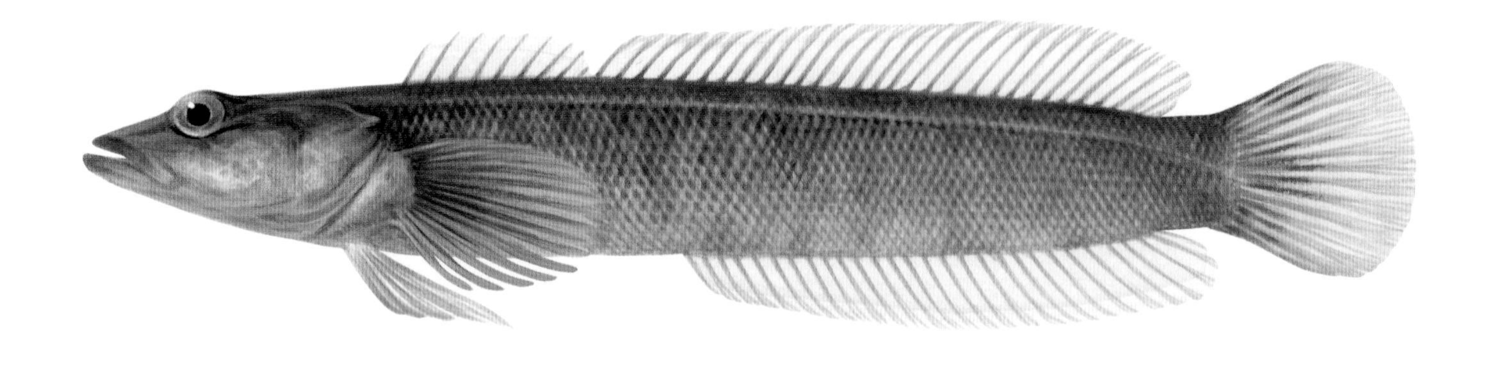

FLATHEAD CONGOLLI
Halaphritis platycephala

There are about 11 species of Thornfish and they are found in temperate waters around southern Australia, New Zealand and southern South America. Thornfishes have an elongate body with small ctenoid scales, a large, flattened head with large eyes on the dorsal surface, and a prominent spine on the operculum. They have two dorsal fins, the first with 8–9 spines and the second with 18–25 rays; the anal fin has no spines and 18–25 rays; and the caudal fin is rounded. The pectoral fins have noticeably thickened lower rays.

Thornfishes are bottom dwelling and found on reefs and rocky areas in shallow coastal waters around south-eastern Australia and Tasmania. The **Dragonet** reaches about 30 cm in length and is common in tide pools and under jetties. The **Flathead Congolli**, slightly smaller at about 19 cm, inhabits caves and deep crevices in coastal reefs, and is thus rarely seen. Thornfishes are predators of small benthic crustaceans such as prawns.

DRAGONET
Bovichtus angustifrons

AUSTRALIAN BOVICHTIDAE SPECIES

Bovichtus angustifrons Dragonet
Halaphritis platycephala Flathead Congolli

CONGOLLI

Pseudaphritidae

CONGOLLI
Pseudaphritis urvillii

The **Congolli**, the sole member of this family, occurs only in freshwater streams in south-eastern Australia. It has an elongate, rounded body with small ctenoid scales, a flattened head with a pointed snout, and large eyes. It has two dorsal fins, the first with 7–8 spines and the second with 19–22 rays. The anal fin has two spines and 21–22 rays and the caudal fin is truncate to slightly rounded.

The Congolli is found in the lower reaches of coastal streams, usually near vegetation or debris and often partially buried in the sediment. Adults move into estuaries to spawn and are remarkable for their ability to travel directly from salt water to fresh water without harm. The Congolli reaches about 36 cm in length and feeds on small invertebrates such as insects, molluscs and crustaceans, and small fishes.

AUSTRALIAN PSEUDAPHRITIDAE SPECIES
Pseudaphritis urvillii Congolli

SWALLOWERS

Chiasmodontidae

FIRELINE SWALLOWER
Pseudoscopelus scriptus

Swallowers are found in deep oceanic waters around the world, and there are about 15 species in the family. They have an elongate, compressed body with no scales, and small prickles or spinules on the skin. The head has a very lumpy dorsal surface with many sensory pores and a very large mouth, reaching to behind the eyes, with numerous long, sharp teeth. Swallowers have two dorsal fins, the first with 9–13 spines and the second with 18–29 rays; the anal fin has one spine and 17–29 rays; and the caudal fin is forked. The mouth and stomach are enormously distensible, and Swallowers are capable of consuming prey as large as themselves. The skeleton is much reduced, with modifications to allow greater expansion of the stomach. For example, the bones that support the anal fin are not attached to the muscles, as they are in other fishes, and the pelvic bones are not attached to each other and are very mobile. Swallowers are pelagic in deep mid to bottom waters down to at least 3000 m and are predators of other fishes. They are rarely encountered in deepwater trawls and very little is known of their biology.

AUSTRALIAN CHIASMODONTIDAE SPECIES

Chiasmodon niger Black Swallower
Dysalotus oligoscolus Smooth Swallower
Kali macrura Longnose Swallower
Kali normani Norman's Swallower
Pseudoscopelus scriptus Fireline Swallower

GAPERS

Champsodontidae

SNYDER'S GAPER
Champsodon snyderi

Gapers are found in tropical and temperate waters of the Indo-Pacific region and there are about 13 species in the family. They have an elongate, compressed body with small, very rough scales. There are two spines on the cheekbone that project downwards over the upper jaw, and a large spine on the angle of the preoperculum. They have dorsally placed eyes and a large, oblique mouth with outer comb-like teeth and inner, hinged, needle-like teeth. There are two dorsal fins, the first short-based with five spines and the second much longer with 17–20 rays. The anal fin is similar to the second dorsal fin, with no spines and 17–20 rays; the ventral fins are large; and the caudal fin is forked. They have a ladder-like lateral line of sensory papillae, with dorsal and ventral lines crossed at regular intervals by vertical lines.

Gapers are found in deep water from about 100 to 1000 m. They occur in large schools and make nightly migrations towards the surface to feed, preying mainly on pelagic crustaceans and other fishes. They reach a maximum length of about 15 cm.

AUSTRALIAN CHAMPSODONTIDAE SPECIES

Champsodon atridorsalis Blackfin Gaper
Champsodon guentheri Gunther's Gaper
Champsodon longipinnis Longfin Gaper
Champsodon machaeratus Knife Gaper
Champsodon nudivittis Nakedband Gaper
Champsodon pantolepis Sheath Gaper
Champsodon sagittus Arrow Gaper
Champsodon snyderi Snyder's Gaper
Champsodon vorax Greedy Gaper

GRUBFISHES

Pinguipedidae

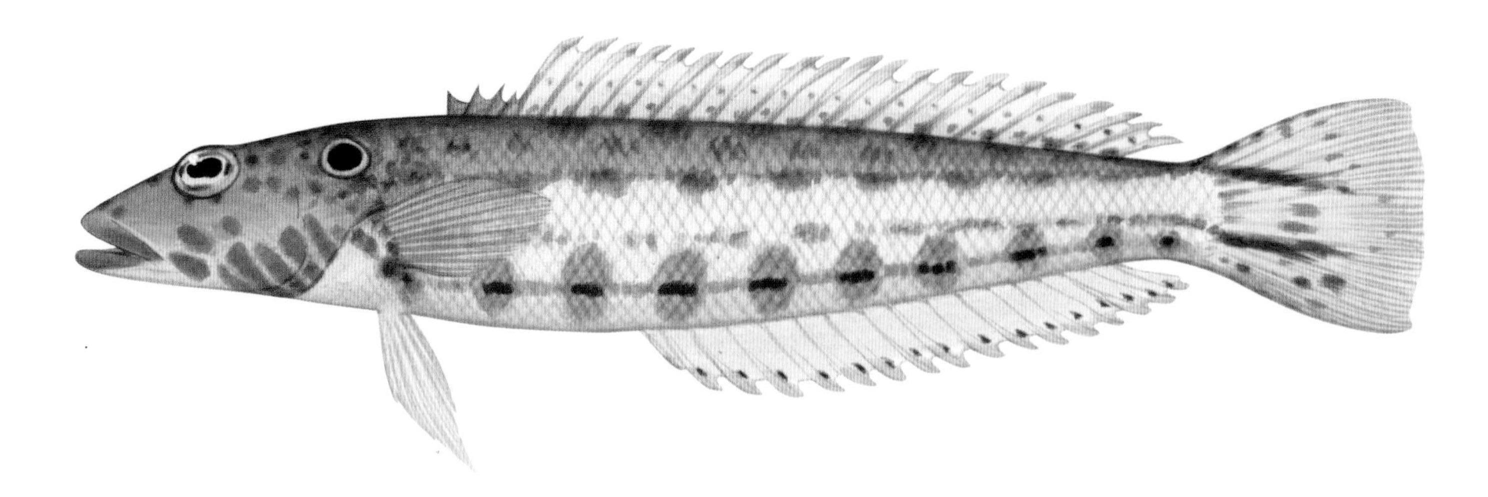

SPOTHEAD GRUBFISH
Parapercis clathrata

There are about 54 species of Grubfish and they are found in temperate coastal waters in the southern hemisphere. They have an elongate, cylindrical body with small ctenoid scales and usually cycloid scales on the cheeks. Their flattened head has prominent eyes on the dorsal surface and a large mouth with fleshy lips, inner villiform teeth, outer comb-like teeth and anterior canines. The single deeply notched dorsal fin has 4–7 spines and 19–27 rays, the anal fin has one weak spine and 16–19 rays, and the caudal fin is truncate to deeply lunate.

Grubfishes are bottom dwellers found mainly in shallow coastal waters, with some species occurring in deeper waters down to about 360 m. In Australia Grubfishes mainly inhabit tropical areas, and are common on sandy and rubble bottoms near coral reefs. They glide just above the substrate, regularly stopping and propping themselves up on their strong ventral fins and swivelling their eyes to observe their surroundings. The **Wavy Grubfish** and several other species are found around the south coast. Grubfishes are predators of small benthic crustaceans such as crabs and shrimps, and small fishes. They reach a maximum length of about 30 cm. The males are territorial and defend a harem of females from other males. Some species are hermaphroditic, with females capable of undergoing sex reversal to replace a missing dominant male.

SHARPNOSE GRUBFISH
Parapercis cylindrica

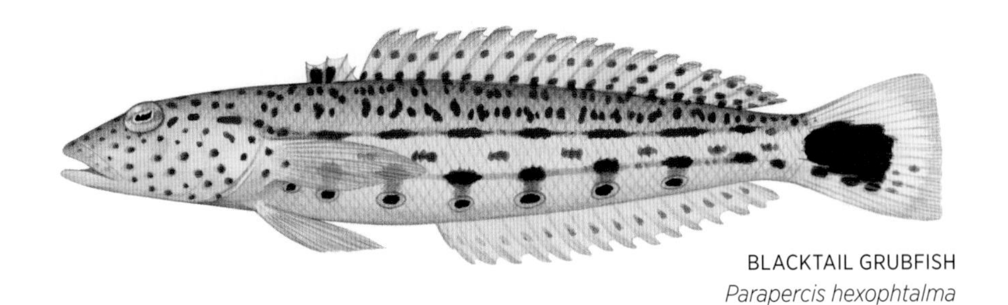

BLACKTAIL GRUBFISH
Parapercis hexophtalma

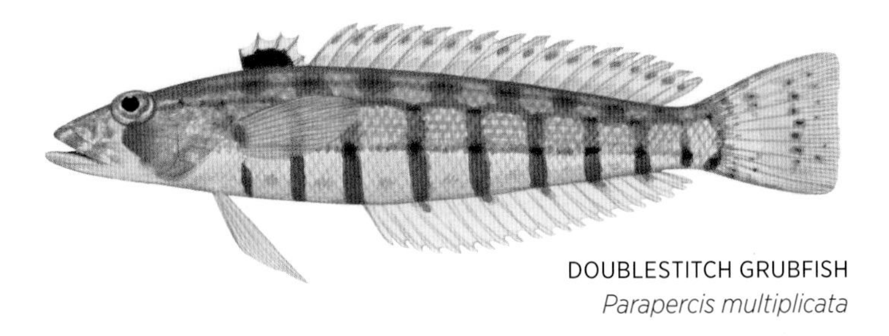

DOUBLESTITCH GRUBFISH
Parapercis multiplicata

AUSTRALIAN PINGUIPEDIDAE SPECIES

Parapercis alboguttata Bluenose Grubfish
Parapercis allporti Barred Grubfish
Parapercis australis Southern Grubfish
Parapercis binivirgata Redbanded Grubfish
Parapercis biordinis Double-row Grubfish
Parapercis clathrata Spothead Grubfish
Parapercis cylindrica Sharpnose Grubfish
Parapercis diplospilus Doublespot Grubfish
Parapercis gushikeni Rosy Grubfish
Parapercis haackei Wavy Grubfish
Parapercis hexophtalma Blacktail Grubfish
Parapercis lineopunctata Dotlined Grubfish
Parapercis macrophthalma Narrowbarred Grubfish
Parapercis millepunctata Thousand-spot Grubfish
Parapercis mimaseana Banded Grubfish
Parapercis multiplicata Doublestitch Grubfish
Parapercis naevosa Western Barred Grubfish
Parapercis nebulosa Pinkbanded Grubfish
Parapercis ramsayi Spotted Grubfish
Parapercis schauinslandii Lyretail Grubfish
Parapercis snyderi Snyder's Grubfish
Parapercis somaliensis Somali Grubfish
Parapercis stricticeps Whitestreak Grubfish
Parapercis xanthozona Peppered Grubfish

SAND DIVERS

Trichonotidae

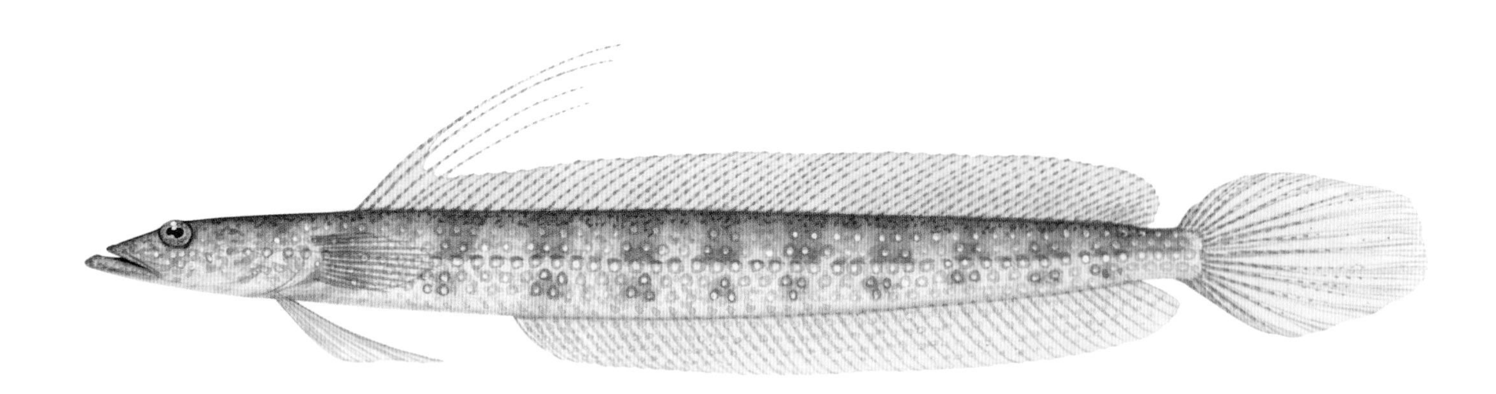

SPOTTED SAND DIVER
Trichonotus setiger

There are about eight species of Sand Diver and they are found only in the Indo-West Pacific. They have a very elongate, rounded body with large, cycloid scales, a pointed snout with a large mouth, and a protruding fleshy-tipped lower jaw. Their distinctive eyes have radiating filaments of the iris projecting into the dorsal margin of the pupil. They have a single long, sail-like dorsal fin with 3–8 spines, some of which are elongated into trailing filaments, and 39–47 rays. The anal fin is also long based, with one spine and 34–42 rays, the caudal fin is pointed, and the ventral fins are often large and fan-like.

Sand Divers are found in tropical coastal waters of Australia over sandy bottoms near reefs, from the shallows down to about 150 m. They hover in small groups, several metres above the bottom, and can quickly dive beneath the sand if disturbed. They are territorial, with males maintaining a harem of females, and the fan-like ventral fins are used in courtship and territorial displays. Sand Divers feed on small crustaceans and zooplankton, and reach a maximum length of approximately 20 cm.

AUSTRALIAN TRICHONOTIDAE SPECIES

Trichonotus blochii Bloch's Sand Diver
Trichonotus elegans Elegant Sand Diver
Trichonotus setiger Spotted Sand Diver

SANDBURROWERS

Creediidae

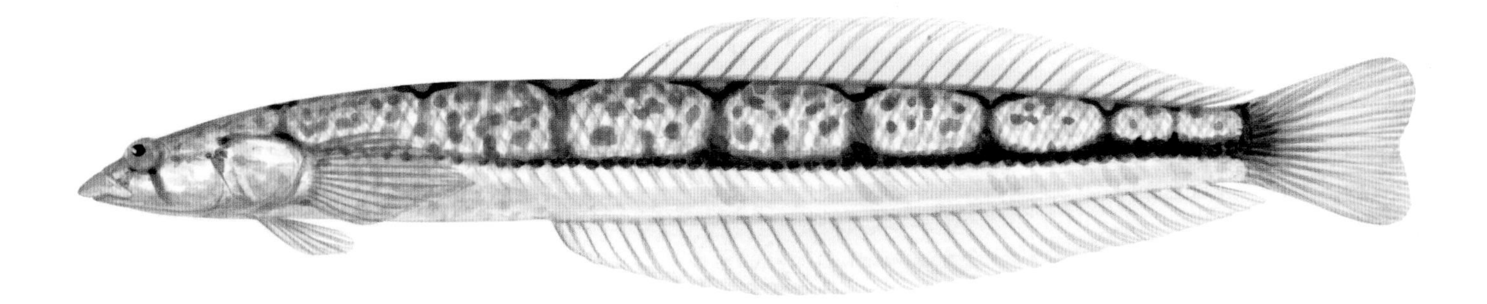

TOMMYFISH
Limnichthys fasciatus

Sandburrowers are found only in the Indo-Pacific region and there are about 14 species in the family. They have an elongate, rounded body with large cycloid scales, a sharp, pointed snout, and close-set, dorsally directed eyes. They have a single long-based dorsal fin, originating well back on the body, with no spines and 12–43 unbranched rays. The anal fin is longer than the dorsal fin, with no spines and 23–40 unbranched rays, the ventral fins are very small or absent, and the caudal fin is rounded.

Sandburrowers are found right around the coast of Australia, on sandy bottoms from the shallows down to about 350 m. They are reasonably common, but are rarely seen due to their burrowing behaviour. They rest on or just beneath the sand, with their dorsally placed eyes exposed, waiting for prey such as small crustaceans and other invertebrates to pass by. They reach a maximum length of about 10 cm. The **Tommyfish** is quite common off the east and west coasts of Australia in sandy tidal pools and shallow waters near reefs, where it lives in small colonies.

AUSTRALIAN CREEDIIDAE SPECIES

Creedia alleni Allen's Sandburrower
Creedia haswelli Slender Sandburrower
Creedia partimsquamigera Halfscale Sandburrower
Limnichthys fasciatus Tommyfish
Limnichthys nitidus Elegant Sandburrower
Schizochirus insolens Brokenfin Sandburrower

DUCKBILLS

Percophidae

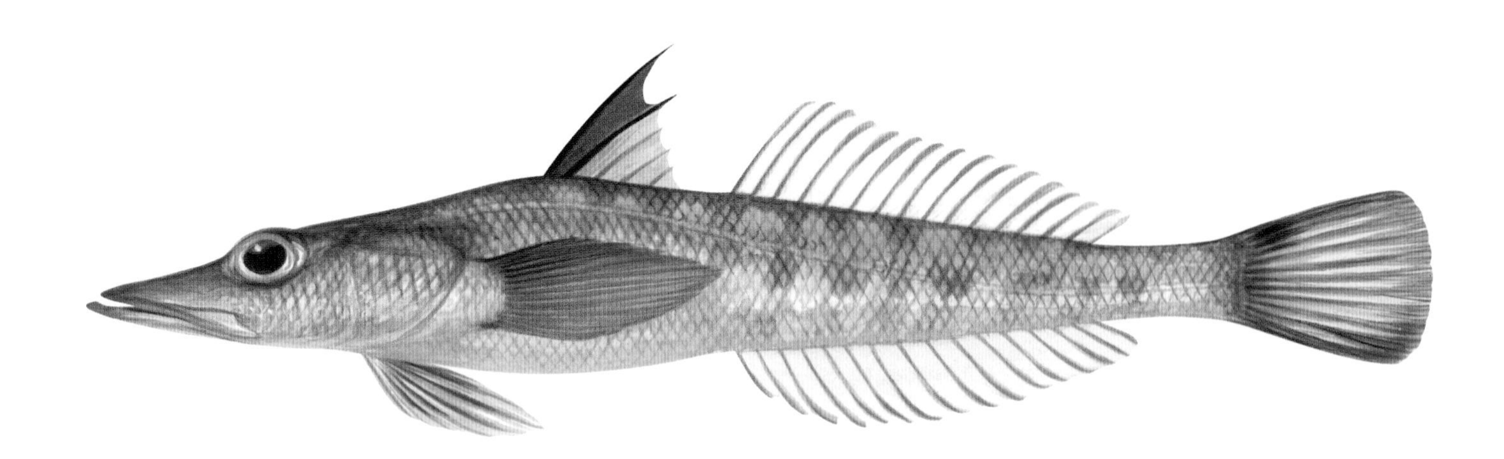

SHARPNOSE DUCKBILL
Bembrops filifera

Duckbills are found in the Atlantic, Indo-West Pacific and South-East Pacific, and there are about 44 species in the family. They have an elongate, rounded body with large to moderate ctenoid or cycloid scales, and a large, flattened head with a pointed snout. They have a large mouth with fine teeth, and close-set, dorsally directed eyes. Duckbills have two dorsal fins (except for one endemic New Zealand species, which lacks the first dorsal fin), the first fin with 2–6 spines and the second with 13–22 rays. The anal fin is similar to the soft dorsal fin, with no spines and 15–24 rays, and the caudal fin is truncate to rounded. They resemble the Flatheads (Platycephalidae, p. 430), but differ in having the ventral fins positioned in front of the pectoral fins, rather than behind.

Duckbills are predominantly deepwater, bottom-dwelling fishes, found over soft substrates from about 100 to 600 m in tropical and subtropical regions. The **Broad Duckbill**, however, also occurs in very shallow southern coastal waters of Australia. Duckbills are rarely encountered, with many species known from only a few specimens, and the family is poorly known. They grow to about 30 cm in length and are carnivorous, probably feeding mainly on small crustaceans and fishes.

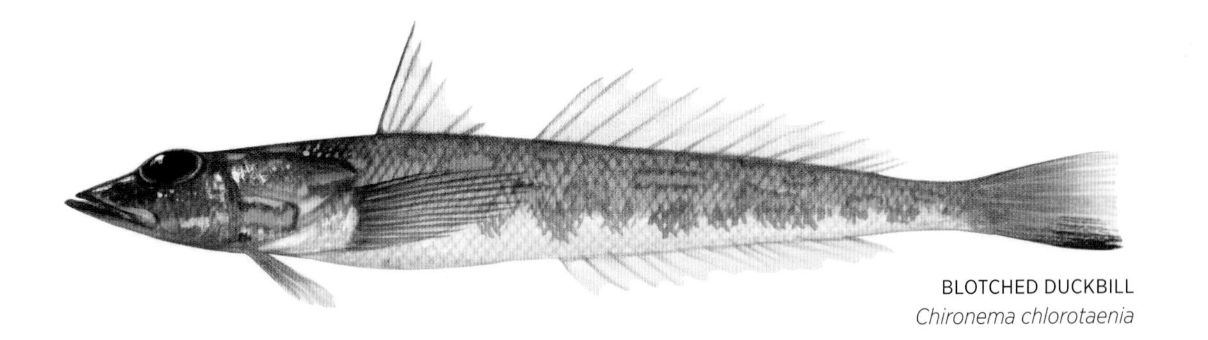

BLOTCHED DUCKBILL
Chironema chlorotaenia

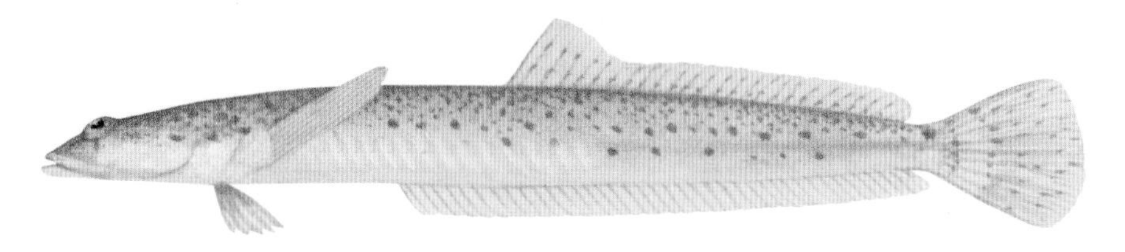

OBTUSE DUCKBILL
Squamicreedia obtusa

AUSTRALIAN PERCOPHIDAE SPECIES

Acanthaphritis barbata Barbel Duckbill
Acanthaphritis ozawai Yellowspotted Duckbill
Bembrops caudimacula Tailspot Duckbill
Bembrops curvatura Bigscale Duckbill
Bembrops filifera Sharpnose Duckbill
Bembrops morelandi Fourspot Duckbill
Bembrops nelsoni Smallscale Duckbill
Bembrops philippinus Philippine Duckbill
Bembrops platyrhynchus Spotted Duckbill
Chironema chlorotaenia Blotched Duckbill
Enigmapercis reducta Broad Duckbill
Matsubaraea fusiforme Narrow Duckbill
Squamicreedia obtusa Obtuse Duckbill

SANDFISHES

Leptoscopidae

PINK SANDFISH
Crapatalus munroi

Sandfishes are found only around southern Australia and New Zealand, and the family contains about five species. They have an elongate, tapering, rounded body with large cycloid scales and a broad, flattened head with small, dorsally directed eyes. The wide mouth is oblique to nearly vertical, with villiform or canine teeth, and the jaws and upper operculum are bordered by a fringe of small fleshy filaments. They have a long dorsal fin with no spines and 32–38 rays. The anal fin is similar to the dorsal fin, with 36–37 rays, the ventral fins are placed well forward beneath the throat, and the caudal fin is slightly rounded.

Sandfishes are found in shallow coastal waters, often along surf beaches close to the shoreline, and in areas with soft sand and sediment bottoms. They are ambush predators, similar in form and strategy to the Stargazers (Uranoscopidae, p. 652): burrowing into the sand with only the dorsal surface of the head and eyes exposed, and lunging out to snatch passing prey. The fringed filaments around the mouth and gill openings prevent sand entering when the fishes are buried. They are moderately common in southern and south-eastern Australian waters, but are not often encountered, remaining hidden in the sand unless disturbed. Sandfishes reach a maximum length of about 40 cm and feed on small crustaceans and other invertebrates, and small fishes. They are not well known and the specimens collected thus far may represent more species than has been thought.

AUSTRALIAN LEPTOSCOPIDAE SPECIES

Crapatalus munroi Pink Sandfish
Lesueurina platycephala Flathead Sandfish

SANDLANCES

Ammodytidae

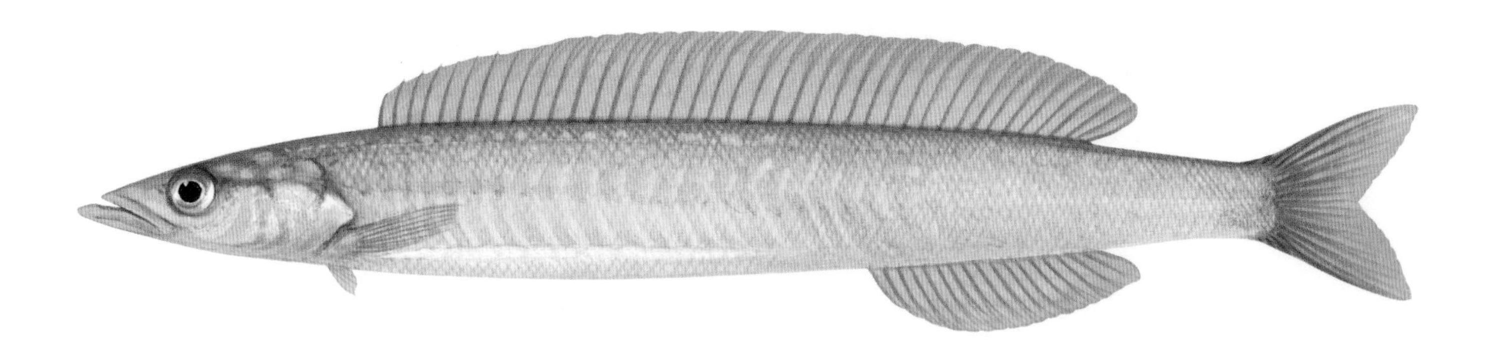

MITSUKURI'S SANDLANCE
Bleekeria mitsukurii

There are about 23 species of Sandlance and they are found in temperate and tropical waters around the world, including the cold waters of the Arctic. Sandlances have a very elongate, cyclindrical body with minute cycloid scales or no scales at all. They have a pointed snout with a large mouth, and the lower jaw projects well beyond the upper. Some species have teeth and some do not. The single long dorsal fin has no spines and 49–54 rays, the anal fin has no spines and 14–36 rays, the caudal fin is forked, and the ventral fins are usually rudimentary or absent.

Sandlances are found near the bottom over sandy substrates and are rarely encountered, being able to burrow rapidly into the sand at any sign of danger. They are small fishes, reaching about 15 cm in length. There are few records of them from Australian waters, and the two *Bleekeria* species may in fact be misidentified and represent new species.

AUSTRALIAN AMMODYTIDAE SPECIES

Ammodytoides vagus Sandlance
Bleekeria mitsukurii Mitsukuri's Sandlance
Bleekeria viridianguilla Eel Sandlance

STARGAZERS

Uranoscopidae

COMMON STARGAZER
Kathetostoma laeve

Stargazers are found in the Atlantic, Indian and Pacific oceans and there are about 50 species in the family. They have an elongate, rounded, tapering body, compressed posteriorly, with either no scales or small cycloid scales. Stargazers have a massive, dorsally flattened head covered in bony plates, with small, dorsally directed eyes. The large, vertical mouth has small, fine teeth and sometimes large canines in the lower jaw. The lips, and sometimes the upper operculum, are bordered by a fringe of fleshy filaments. The dorsal fin has 2–10 spines and 12–18 rays, with some species having a separate spinous fin and others a continuous fin. Some lack dorsal-fin spines altogether. The anal fin has no spines and 12–18 rays, the pectoral fins are large and broad, the ventral fins are placed well forward beneath the throat, and the caudal fin is truncate to rounded. There is a large, venomous spine on the shoulder above the pectoral fin.

Stargazers are found in most Australian waters, over sandy and muddy bottoms from the shallows down to about 900 m, with the majority of species found in deep waters. Stargazers are perfectly adapted ambush predators, resting buried in the sediment with head and eyes just at the surface, and lunging out to snatch passing prey. Some have a worm-like lure on the floor of the mouth, which they wriggle enticingly to attract small fishes. They are quite common, but are not often seen as they usually remain beneath the sediment. The fringed filaments on the lips and operculum prevent sand from entering the mouth and gills. Some species move around over the substrate at night to hunt, and at this time they are sometimes taken in nets. Stargazers reach a maximum length of about 75 cm and are excellent food fishes, though they should be handled with great care as the shoulder spines can cause serious wounds and are reported to be extremely venomous in some species.

FRINGE STARGAZER
Ichthyscopus barbatus

BANDED STARGAZER
Ichthyscopus fasciatus

KAI STARGAZER
Uranoscopus kaianus

AUSTRALIAN URANOSCOPIDAE SPECIES

Ichthyscopus barbatus Fringe Stargazer
Ichthyscopus fasciatus Banded Stargazer
Ichthyscopus insperatus Doubleband Stargazer
Ichthyscopus nigripinnis Blackfin Stargazer
Ichthyscopus sannio Spotted Stargazer
Ichthyscopus spinosus Spiny Stargazer
Kathetostoma canaster Speckled Stargazer
Kathetostoma laeve Common Stargazer
Kathetostoma nigrofasciatum Deepwater Stargazer
Pleuroscopus pseudodorsalis Scaled Stargazer
Uranoscopus affinis Armed Stargazer
Uranoscopus bicinctus Marbled Stargazer
Uranoscopus cognatus Yellowtail Stargazer
Uranoscopus japonicus Japanese Stargazer
Uranoscopus kaianus Kai Stargazer
Uranoscopus oligolepis Naked-nape Stargazer
Uranoscopus sulphureus Whitemargin Stargazer
Uranoscopus terraereginae Queensland Stargazer
Xenocephalus armatus Bulldog Stargazer
Xenocephalus australiensis Australian Stargazer
Xenocephalus cribratus Ringed Stargazer

CONVICT BLENNIES

Pholidichthyidae

SNAKE CONVICT BLENNY
Pholidichthys anguis

Convict Blennies are found in tropical waters from the Philippines to northern Australia and the Solomon Islands. There are two species in the family. They have an elongate, eel-like body with no scales, and a small head with a blunt snout and moderate-sized mouth. The single long dorsal fin has no spines and 72–98 rays, the anal fin has no spines and 54–81 rays, and both fins are continuous with the caudal fin. The ventral fins are very small and placed well forward, beneath the pectoral fins.

In Australia the **Snake Convict Blenny** is restricted to mud and sand bottoms in northern coastal areas. Juveniles of this species may occur in large schools, but adults are very rarely encountered – only one adult specimen, about 25 cm in length, has ever been recorded. Virtually nothing is known of its biology.

AUSTRALIAN PHOLIDICHTHYIDAE SPECIES

Pholidichthys anguis Snake Convict Blenny

THREEFINS

Tripterygiidae

WESTERN JUMPING BLENNY (male and female)
Lepidoblennius marmoratus

Threefins are found in tropical and temperate waters of the Atlantic, Indian and Pacific oceans, and are most diverse in the Indo-Pacific region. This large family contains about 150 species. Threefins have an elongate, slightly compressed body with ctenoid scales, large eyes and a moderate-sized mouth with broad bands of conical teeth. They are distinct in having three dorsal fins, the first with 3–4 spines, the second with 8–16 spines and the third with 7–17 rays. The anal fin has 0–2 spines and 14–32 rays, the ventral fins are small with one very small spine and 2–3 rays, and the caudal fin is rounded.

Threefins are found in shallow coastal waters right around Australia, often in the intertidal zone, with some species occurring in deeper waters to about 550 m. Most reach a maximum length of about 5–6 cm and are well camouflaged, and although common they are often overlooked by divers. Many have different male and female colour patterns. Most are found on shallow, algae-covered reefs, often around jetty pylons and amongst rich, sessile invertebrate growth. The Jumping Blennies are common in intertidal areas around southern Australia and are often seen out of the water on the edge of rock pools, flipping back into the water at the approach of danger. Threefins are predators of small benthic invertebrates. The family is not well known and more species probably await description.

RINGSCALE THREEFIN (female and male)
Enneapterygius atrogulare

VARIABLE THREEFIN
Forsterygion varium

BLACKTHROAT THREEFIN (female and male)
Helcogramma decurrens

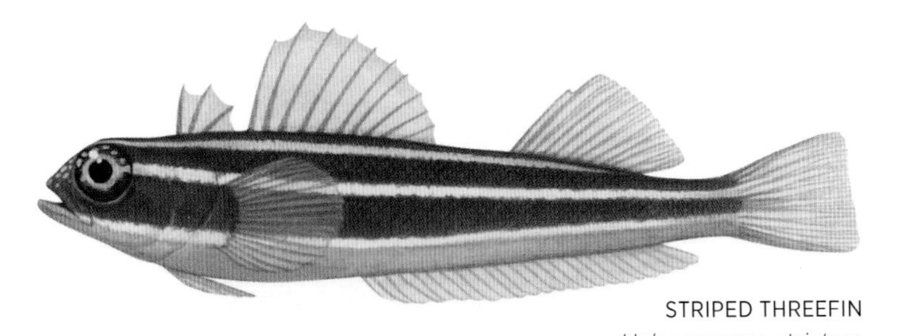

STRIPED THREEFIN
Helcogramma striatum

CLARK'S THREEFIN
Trinorfolkia clarkei

AUSTRALIAN TRIPTERYGIIDAE SPECIES

Apopterygion alta Tasselled Threefin
Brachynectes fasciatus Barred Threefin
Ceratobregma acanthops Spiny-eye Threefin
Ceratobregma helenae Orangebar Threefin
Enneapterygius atrogulare Ringscale Threefin
Enneapterygius bahasa Blacktail Threefin
Enneapterygius clea Clea's Threefin
Enneapterygius elegans Elegant Threefin
Enneapterygius flavoccipitis Two-colour Threefin
Enneapterygius gracilis Yellow-black Threefin
Enneapterygius hemimelas Halfblack Threefin
Enneapterygius howensis Lord Howe Threefin
Enneapterygius larsonae Blackhead Threefin
Enneapterygius mirabilis Miracle Threefin
Enneapterygius namarrgon Lightning-man Threefin
Enneapterygius nanus Pygmy Threefin
Enneapterygius philippinus Minute Threefin
Enneapterygius pyramis Pyramid Threefin
Enneapterygius rufopileus Blackcheek Threefin
Enneapterygius similis Masked Threefin
Enneapterygius triserialis Whitespotted Threefin
Enneapterygius tutuilae Highfin Threefin
Enneapterygius ziegleri Ziegler's Threefin
Forsterygion lapillum Common Threefin
Forsterygion varium Variable Threefin
Grahamina gymnota Estuarine Threefin
Helcogramma decurrens Blackthroat Threefin
Helcogramma gymnauchen Redfin Threefin
Helcogramma springeri Springer's Threefin
Helcogramma striatum Striped Threefin
Lepidoblennius haplodactylus Eastern Jumping Blenny
Lepidoblennius marmoratus Western Jumping Blenny
Norfolkia brachylepis Scalyfin Threefin
Norfolkia leeuwin Leeuwin Threefin
Norfolkia squamiceps Scalyhead Threefin
Norfolkia thomasi Thomas' Threefin
Springerichthys kulbickii Kulbicki's Threefin
Trianectes bucephalus Bighead Threefin
Trinorfolkia clarkei Clark's Threefin
Trinorfolkia cristata Crested Threefin
Trinorfolkia incisa Notched Threefin
Ucla xenogrammus Largemouth Threefin

BLENNIES

Blenniidae

REDSTREAKED BLENNY
Cirripectes stigmaticus

This is a very large and diverse family, with about 360 species. Blennies are found in tropical and temperate waters around the world, with a few species occurring in brackish and occasionally fresh waters. They have elongate, tapering, rounded to compressed bodies with no scales, large eyes, and dentition ranging from movable comb-like teeth to enormous curved canines. The single dorsal fin has 3–17 spines and 9–119 rays, and the anal fin has two spines, which are often very small or embedded. The ventral fins are reduced, with one very small spine (often embedded) and 1–4 rays, and the caudal fin is rounded to lunate, sometimes with elongate upper and lower lobes.

In Australia Blennies are very common residents of reefs and shallow, algae-covered rocky areas. A few species are found along the southern coast, though the vast majority occur in warmer northern waters. They have a wide variety of habits and many display complex mimicry.

Sabretooth Blennies have enormous fangs in the lower jaw, and many feed on the fins and tissue of other fishes. Some of these species dart out from sheltering outcrops to snatch a morsel, while others mimic Cleanerfishes (Labridae, p. 617), which allows them to easily approach other fishes. Some Sabretooth Blennies use their fangs only for defence. Fangblennies have large, venomous canines that discourage possible predators and thus allow them to be free-swimming, often well out in the open, feeding on zooplankton. Other Blenny species mimic the Fangblennies, some for the protection this appearance offers, others in order to approach larger fishes that would normally be indifferent to Fangblennies, so that they can feed on their fins and skin. Most Blennies, however, are cryptic and remain on the bottom near the shelter of crevices or amongst branching corals. They are mainly herbivorous, using their comb-like teeth to scrape algae from the substrate, and usually live in small holes, into which they retreat tail first at any hint of danger. Rockskippers inhabit tide pools and shallow reef areas, often skipping across rocks to move between pools.

Blennies can be difficult to differentiate under water, as many are very similar in shape, with only slight variations in colour patterns. Most are small fishes, reaching 5–10 cm in length, but the **Hairtail Blenny** can reach 50 cm or more, and the **Wavyline Rockskipper** reaches about 18 cm.

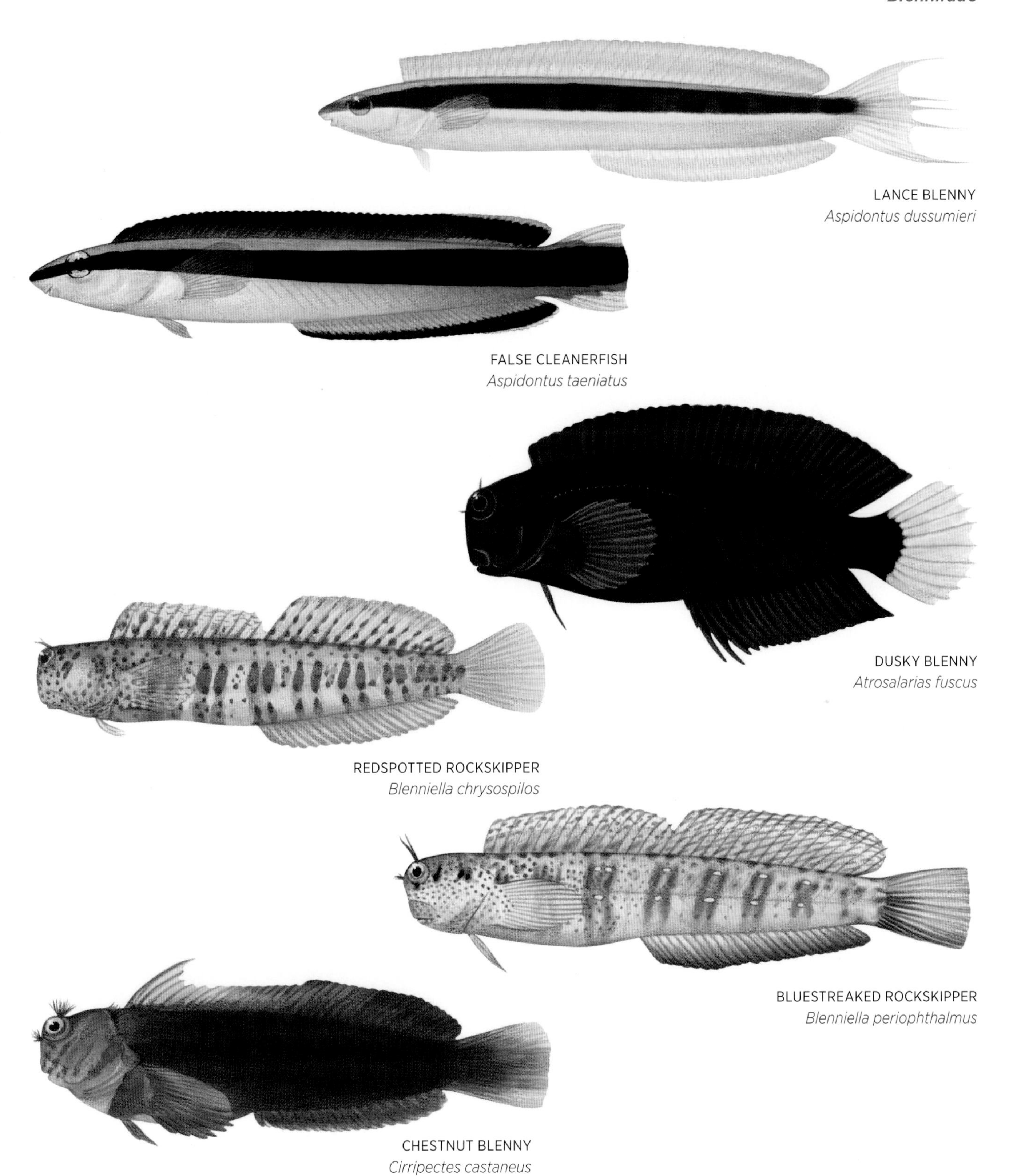

LANCE BLENNY
Aspidontus dussumieri

FALSE CLEANERFISH
Aspidontus taeniatus

DUSKY BLENNY
Atrosalarias fuscus

REDSPOTTED ROCKSKIPPER
Blenniella chrysospilos

BLUESTREAKED ROCKSKIPPER
Blenniella periophthalmus

CHESTNUT BLENNY
Cirripectes castaneus

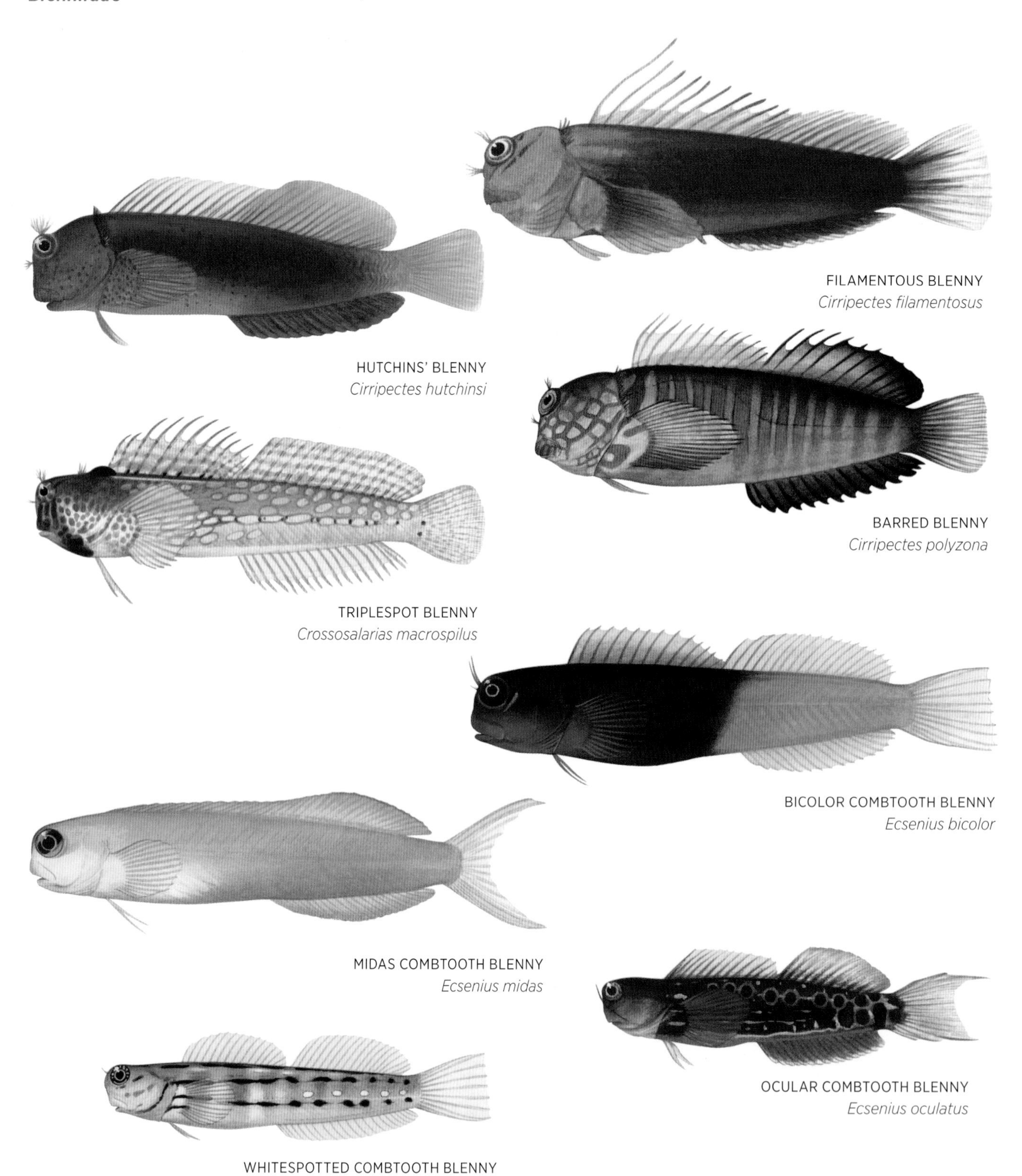

FILAMENTOUS BLENNY
Cirripectes filamentosus

HUTCHINS' BLENNY
Cirripectes hutchinsi

BARRED BLENNY
Cirripectes polyzona

TRIPLESPOT BLENNY
Crossosalarias macrospilus

BICOLOR COMBTOOTH BLENNY
Ecsenius bicolor

MIDAS COMBTOOTH BLENNY
Ecsenius midas

OCULAR COMBTOOTH BLENNY
Ecsenius oculatus

WHITESPOTTED COMBTOOTH BLENNY
Ecsenius trilineatus

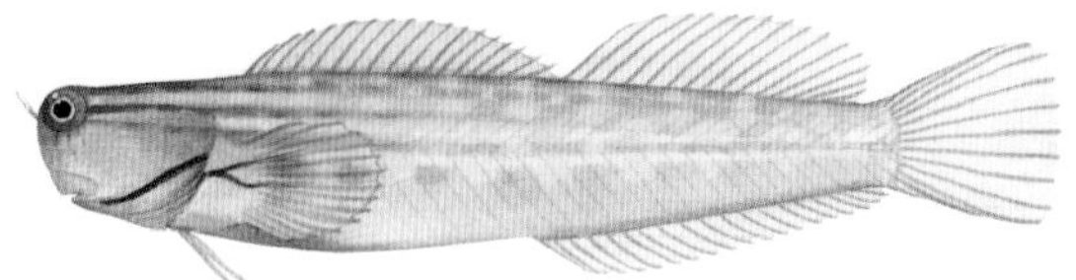

PALESPOTTED COMBTOOTH BLENNY
Ecsenius yaeyamaensis

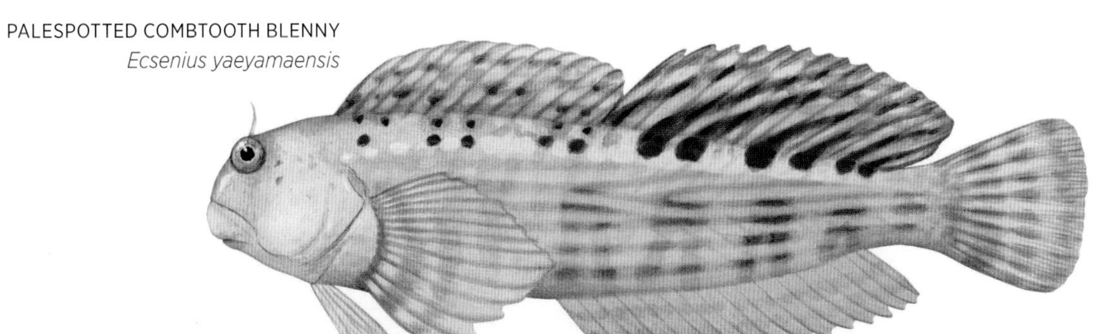

WAVYLINE ROCKSKIPPER
Entomacrodus decussatus

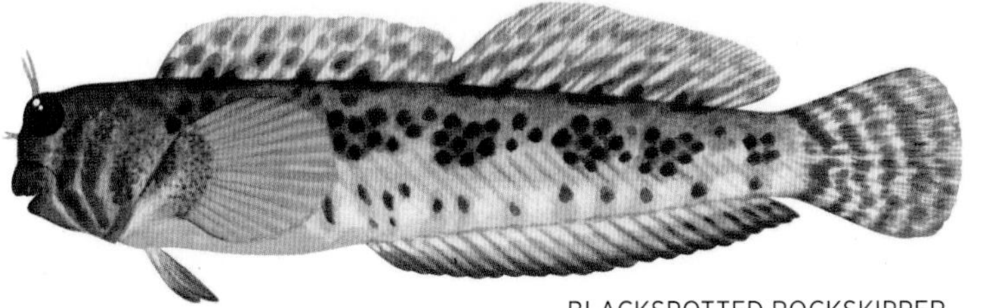

BLACKSPOTTED ROCKSKIPPER
Entomacrodus striatus

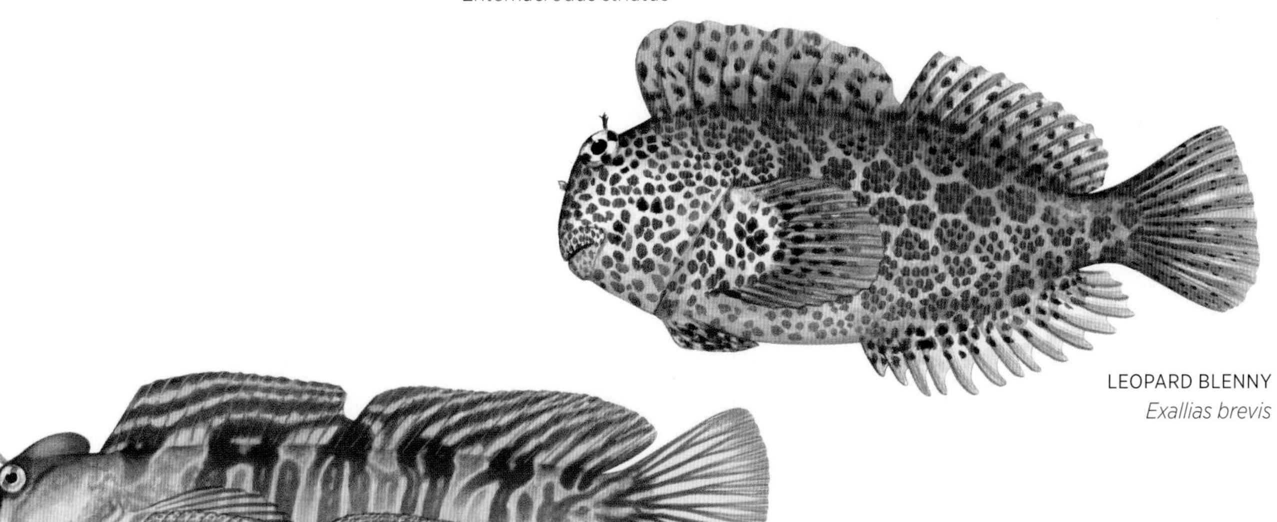

LEOPARD BLENNY
Exallias brevis

RIPPLED ROCKSKIPPER (male and female)
Istiblennius edentulus

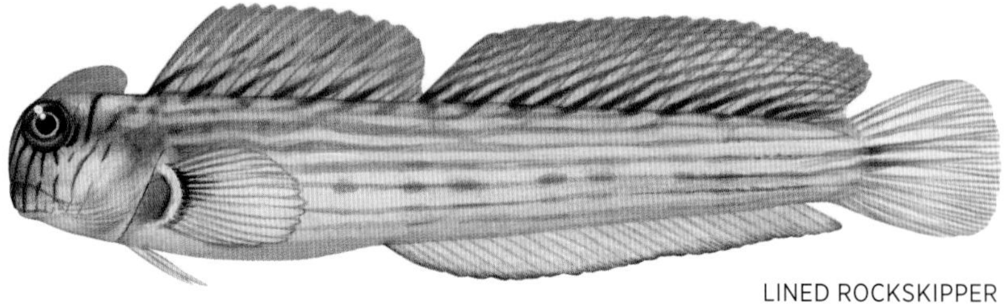

LINED ROCKSKIPPER
Istiblennius lineatus

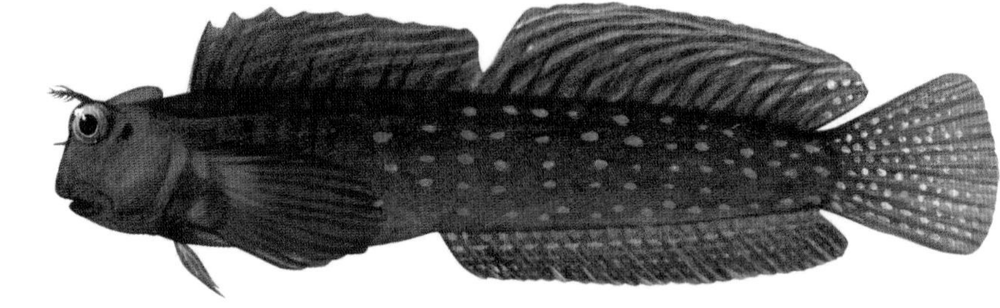

PEACOCK ROCKSKIPPER
Istiblennius meleagris

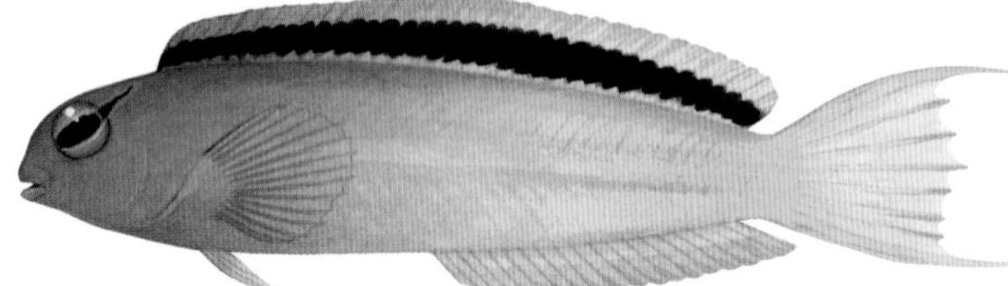

EYELASH FANGBLENNY
Meiacanthus atrodorsalis

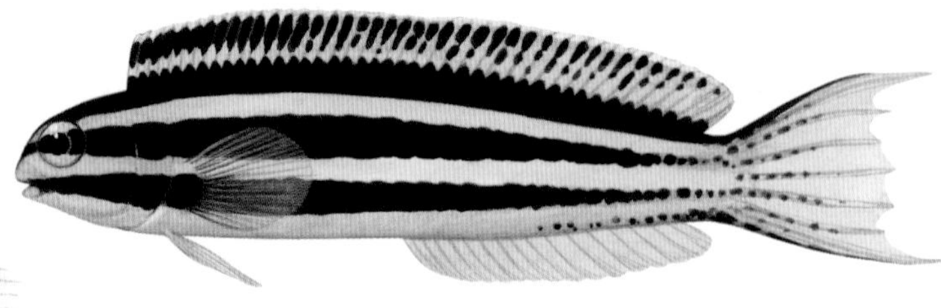

LINESPOT FANGBLENNY
Meiacanthus grammistes

GERMAIN'S BLENNY
Omobranchus germaini

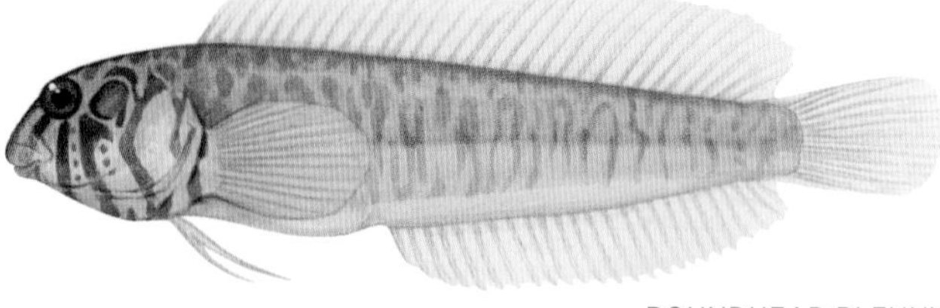

ROUNDHEAD BLENNY
Omobranchus lineolatus

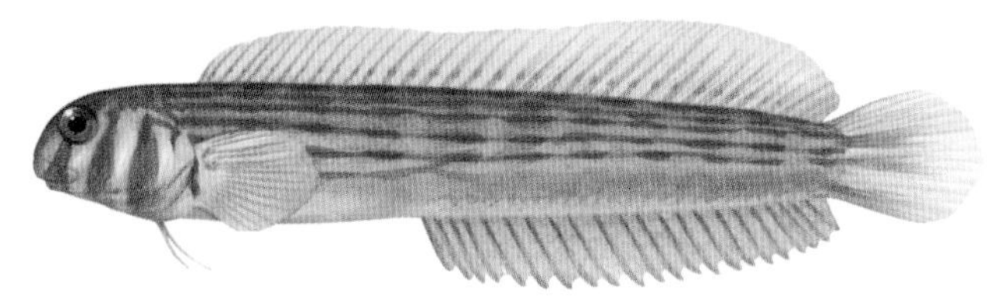

MUZZLED BLENNY
Omobranchus punctatus

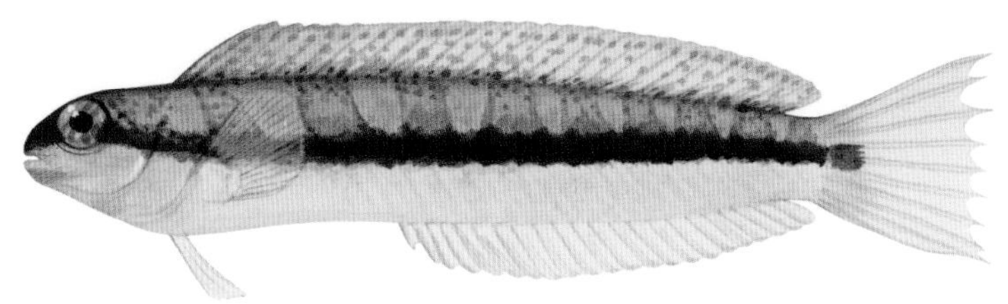

SHORTHEAD SABRETOOTH BLENNY
Petroscirtes breviceps

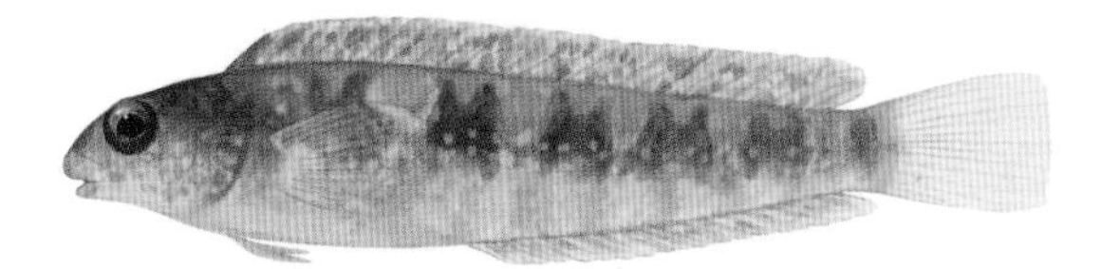

BROWN SABRETOOTH BLENNY
Petroscirtes lupus

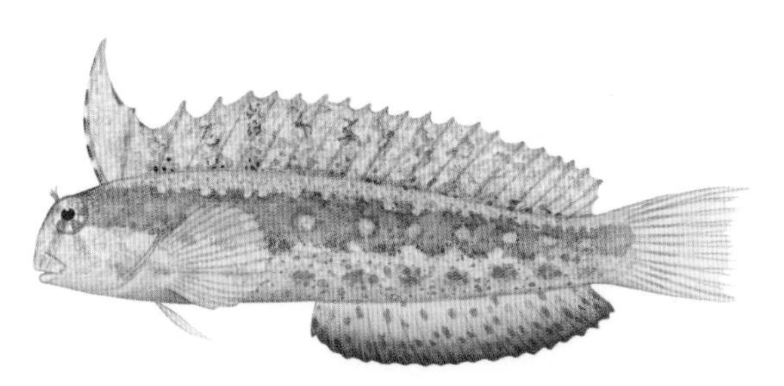

CRESTED SABRETOOTH BLENNY
Petroscirtes mitratus

BLUESTRIPED FANGBLENNY
Plagiotremus rhinorhynchos

PIANO FANGBLENNY
Plagiotremus tapeinosoma

STARRY BLENNY
Salarias ramosus

HAIRTAIL BLENNY
Xiphasia setifer

AUSTRALIAN BLENNIIDAE SPECIES

Aspidontus dussumieri Lance Blenny
Aspidontus taeniatus False Cleanerfish
Atrosalarias fuscus Dusky Blenny
Blenniella bilitonensis Biliton Rockskipper
Blenniella chrysospilos Redspotted Rockskipper
Blenniella paula Bluedash Rockskipper
Blenniella periophthalmus Bluestreaked Rockskipper
Cirripectes alboapicalis Whitedotted Blenny
Cirripectes alleni Kimberley Blenny
Cirripectes castaneus Chestnut Blenny
Cirripectes chelomatus Lady Musgrave Blenny
Cirripectes filamentosus Filamentous Blenny
Cirripectes hutchinsi Hutchins' Blenny
Cirripectes polyzona Barred Blenny
Cirripectes quagga Zebra Blenny
Cirripectes stigmaticus Redstreaked Blenny
Crossosalarias macrospilus Triplespot Blenny
Ecsenius aequalis Fourline Combtooth Blenny
Ecsenius alleni Allen's Combtooth Blenny
Ecsenius australianus Australian Combtooth Blenny
Ecsenius bicolor Bicolor Combtooth Blenny
Ecsenius fourmanoiri Blackstriped Combtooth Blenny
Ecsenius lineatus Lined Combtooth Blenny
Ecsenius lividanalis Blackass Combtooth Blenny
Ecsenius mandibularis Queensland Combtooth Blenny
Ecsenius midas Midas Combtooth Blenny
Ecsenius namiyei Namiye's Combtooth Blenny
Ecsenius oculatus Ocular Combtooth Blenny
Ecsenius schroederi Schroeder's Combtooth Blenny
Ecsenius stictus Smallspotted Combtooth Blenny
Ecsenius tigris Tiger Combtooth Blenny
Ecsenius trilineatus Whitespotted Combtooth Blenny
Ecsenius yaeyamaensis Palespotted Combtooth Blenny
Enchelyurus ater Black Blenny
Enchelyurus kraussi Krauss' Blenny
Entomacrodus decussatus Wavyline Rockskipper
Entomacrodus striatus Blackspotted Rockskipper
Entomacrodus thalassinus Twinspot Rockskipper
Exallias brevis Leopard Blenny
Glyptoparus delicatulus Delicate Blenny
Istiblennius dussumieri Dussumier's Rockskipper
Istiblennius edentulus Rippled Rockskipper
Istiblennius lineatus Lined Rockskipper
Istiblennius meleagris Peacock Rockskipper
Laiphognathus multimaculatus Manyspot Blenny
Meiacanthus atrodorsalis Eyelash Fangblenny
Meiacanthus ditrema Schooling Fangblenny
Meiacanthus grammistes Linespot Fangblenny
Meiacanthus lineatus Lined Fangblenny
Meiacanthus luteus Yellow Fangblenny
Meiacanthus naevius Birthmark Fangblenny
Meiacanthus phaeus Twilight Fangblenny
Meiacanthus reticulatus Reticulate Fangblenny
Mimoblennius atrocinctus Mimic Blenny
Nannosalarias nativitatus Throatspot Blenny
Omobranchus anolius Oyster Blenny
Omobranchus elongatus Chevron Blenny
Omobranchus ferox Gossamer Blenny
Omobranchus germaini Germain's Blenny
Omobranchus lineolatus Roundhead Blenny
Omobranchus punctatus Muzzled Blenny
Omobranchus robertsi Roberts' Blenny
Omobranchus rotundiceps Rotund Blenny
Omobranchus verticalis Vertical Blenny
Omox biporos Omox Blenny
Parablennius intermedius Horned Blenny
Parablennius laticlavius Crested Blenny
Parablennius postoculomaculatus False Tasmanian Blenny
Parablennius tasmanianus Tasmanian Blenny
Parenchelyurus hepburni Hepburn's Blenny
Parenchelyurus hyena Hyena Blenny
Petroscirtes breviceps Shorthead Sabretooth Blenny
Petroscirtes fallax Yellow Sabretooth Blenny
Petroscirtes lupus Brown Sabretooth Blenny
Petroscirtes mitratus Crested Sabretooth Blenny
Petroscirtes variabilis Variable Sabretooth Blenny
Petroscirtes xestus Smooth Sabretooth Blenny
Plagiotremus laudandus Bicolor Fangblenny
Plagiotremus rhinorhynchos Bluestriped Fangblenny
Plagiotremus tapeinosoma Piano Fangblenny
Rhabdoblennius ellipes Barchin Blenny
Salarias alboguttatus Whitespotted Blenny
Salarias ceramensis Seram Blenny
Salarias fasciatus Banded Blenny
Salarias guttatus Breastspot Blenny
Salarias ramosus Starry Blenny
Salarias sexfilum Spalding's Blenny
Salarias sinuosus Fringelip Blenny
Stanulus seychellensis Seychelles Blenny
Stanulus talboti Talbot's Blenny
Xiphasia matsubarai Matsubara's Blenny
Xiphasia setifer Hairtail Blenny

WEEDFISHES AND SNAKE BLENNIES

Clinidae

YELLOW CRESTED WEEDFISH
Cristiceps aurantiacus

There are about 80 species in this family, and they are found in temperate waters of the Atlantic, Indian and Pacific oceans, with many species endemic to southern Australia. The Australian species can be divided into two groups:

The Weedfishes have an elongate, compressed body, with small cycloid scales. They have small tentacles, or cirri, on the head above the eyes and nostrils, and sometimes on the anterior fin spines. The single dorsal fin is often deeply notched or with the first three spines completely separate. The first three spines are usually elevated, followed by 22–40 spines and 1–8 rays. The anal fin has two spines and 15–28 rays, and the ventral fins have thickened, unbranched rays and are placed well forward, beneath the throat. Weedfishes are found on rocky reefs around southern Australia, in dense weed, seagrass, kelp and other algal growths, with many species associated with a particular type of algal growth. Most species are rarely seen by divers as they are small, beautifully camouflaged and secretive in their habits. The **Johnston's Weedfish** is the largest in the family, reaching a maximum length of 40 cm, though most are much smaller at 5–15 cm.

The Snake Blennies have a very elongate, eel-like body with minute, non-overlapping cycloid scales. They have a single long, low dorsal fin with 40–84 spines and 1–4 rays, and the anal fin usually has two spines and 25–62 rays; both fins are usually connected to the caudal fin. Snake Blennies are found near seagrass and weed beds along the southern coast and are not often encountered. They are rare and cryptic – some species bury themselves in the sand, emerging only to feed – and the group is poorly known. The largest, the **Frosted Snake Blenny**, reaches a maximum length of about 17 cm.

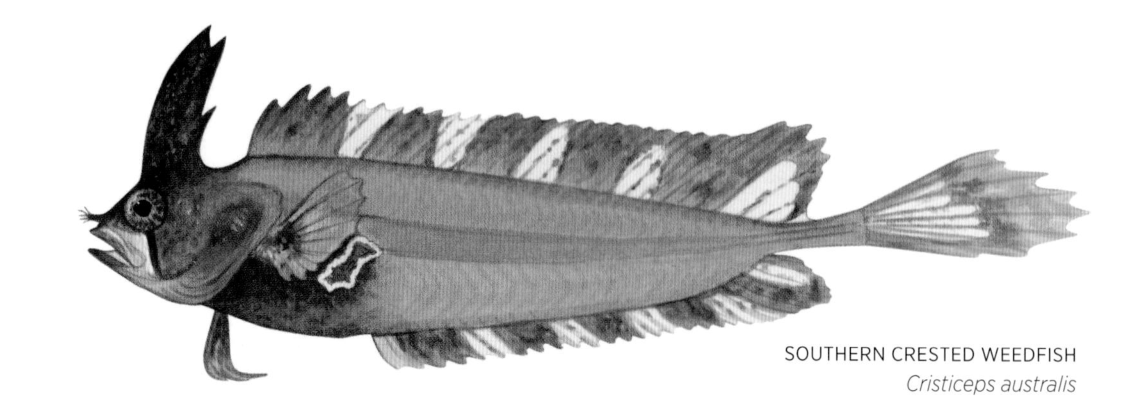

SOUTHERN CRESTED WEEDFISH
Cristiceps australis

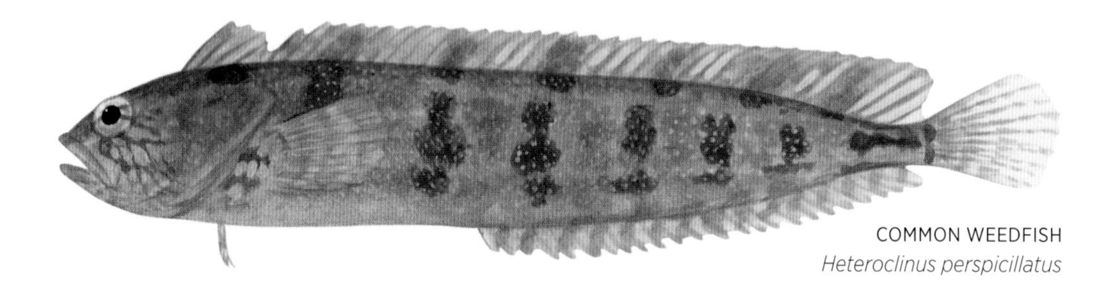

COMMON WEEDFISH
Heteroclinus perspicillatus

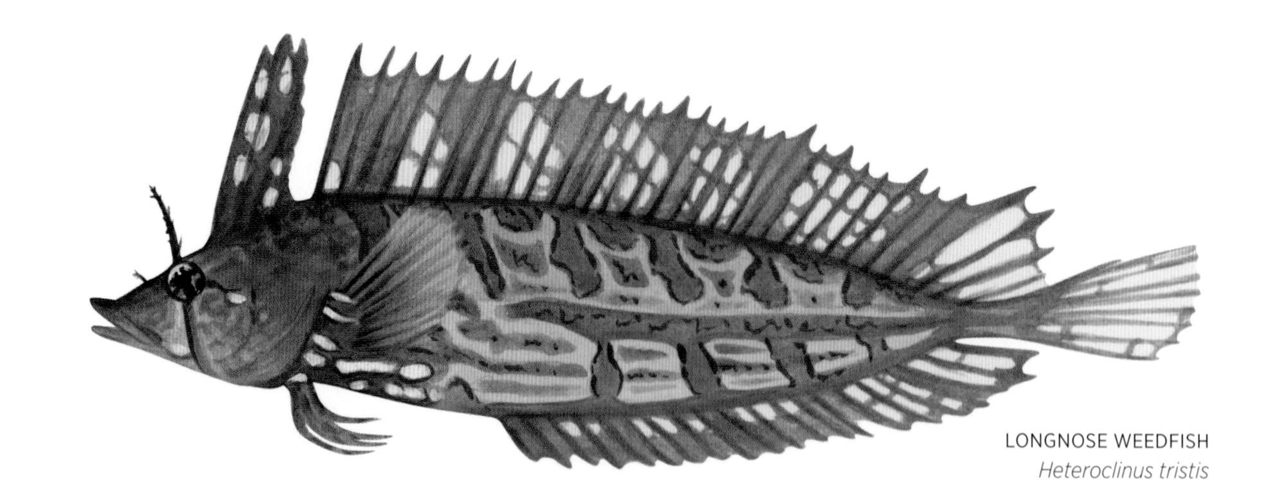

LONGNOSE WEEDFISH
Heteroclinus tristis

BANDED WEEDFISH
Heteroclinus whiteleggii

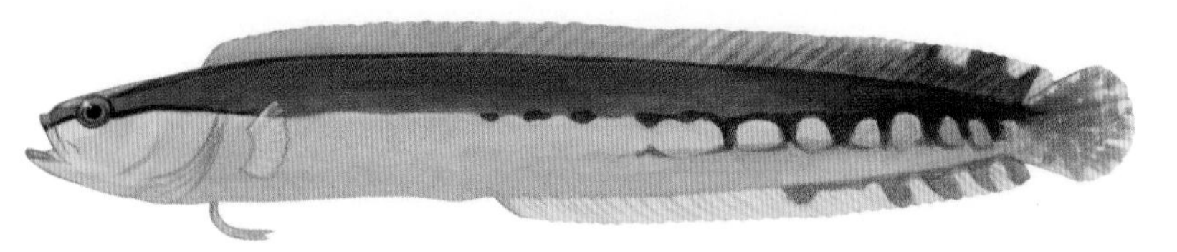

BLACKBACK SNAKE BLENNY
Ophiclinus gracilis

AUSTRALIAN CLINIDAE SPECIES

Cristiceps argyropleura Silverside Weedfish
Cristiceps aurantiacus Yellow Crested Weedfish
Cristiceps australis Southern Crested Weedfish
Heteroclinus adelaidae Adelaide Weedfish
Heteroclinus antinectes Natal Weedfish
Heteroclinus eckloniae Kelp Weedfish
Heteroclinus equiradiatus Sevenbar Weedfish
Heteroclinus heptaeolus Ogilby's Weedfish
Heteroclinus johnstoni Johnston's Weedfish
Heteroclinus klunzingeri Short-tassel Weedfish
Heteroclinus macrophthalmus Large-eye Weedfish
Heteroclinus nasutus Largenose Weedfish
Heteroclinus perspicillatus Common Weedfish
Heteroclinus puellarum Little Weedfish
Heteroclinus roseus Rosy Weedfish
Heteroclinus tristis Longnose Weedfish
Heteroclinus whiteleggii Banded Weedfish
Heteroclinus wilsoni Wilson's Weedfish
Ophiclinops hutchinsi Earspot Snake Blenny
Ophiclinops pardalis Spotted Snake Blenny
Ophiclinops varius Variegated Snake Blenny
Ophiclinus antarcticus Dusky Snake Blenny
Ophiclinus brevipinnis Shortfin Snake Blenny
Ophiclinus gabrieli Frosted Snake Blenny
Ophiclinus gracilis Blackback Snake Blenny
Ophiclinus ningulus Variable Snake Blenny
Ophiclinus pectoralis Whiteblotch Snake Blenny
Peronedys anguillaris Eel Snake Blenny
Springeratus caledonicus Caledonian Weedfish
Sticharium clarkae Clark's Snake Blenny
Sticharium dorsale Slender Snake Blenny

CLINGFISHES

Gobiesocidae

BROAD CLINGFISH (dorsal and lateral view)
Creocele cardinalis

Clingfishes are found in coastal waters of the Atlantic, Indian and Pacific oceans, with a few species occurring in deep water and some in brackish and fresh waters. There are about 140 species in the family. They have elongate, compressed to flattened bodies with no scales, and are usually coated with a thick mucus. There are large sensory pores on the body and on the broad, flattened head. The single dorsal fin has no spines and 3–11 rays, and the anal fin has no spines and 2–13 rays. The ventral fins are modified to form a sucking disc, which Clingfishes use to adhere to the substrate.

In Australia Clingfishes are found mainly around the southern coastline, in seagrass beds and on shallow wave-washed reefs amongst algae. The **Deepwater Clingfish** is found at the greatest depths, to more than 400 m. Some species live beneath rocks or in sponges and most are very small, from 3 to 5 cm in length, and cryptic, so although quite common they are rarely noticed by divers. The Cleaner Clingfishes are exceptions, being brilliantly patterned with red spots and blue lines, and they are often seen clinging to larger fishes, where they feed on parasites or the mucus of their hosts.

Shore Eels are highly modified Clingfishes, with an eel-like body and gill openings reduced to a small slit beneath the throat. The sucking disc is rudimentary or absent and they have no dorsal or anal fin rays. They are small fishes, from 5 to 12 cm in length, often with exceptionally vivid colour patterns. Shore Eels are found mainly in seagrass beds, on sandy bottoms and beneath rubble in shallow coastal areas, though the rare **Deepwater Shore Eel** occurs from about 160 to 350 m. Their small size and cryptic habits mean that most species are rarely encountered. Clingfishes and Shore Eels feed on microscopic invertebrates.

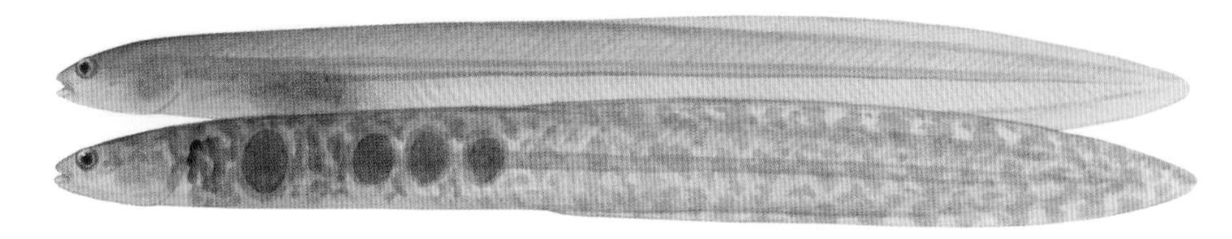

COMMON SHORE EEL (plain colour form and blotched colour form)
Alabes dorsalis

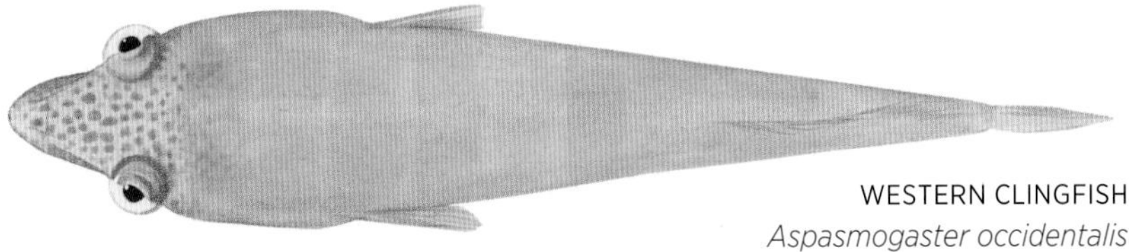

WESTERN CLINGFISH
Aspasmogaster occidentalis

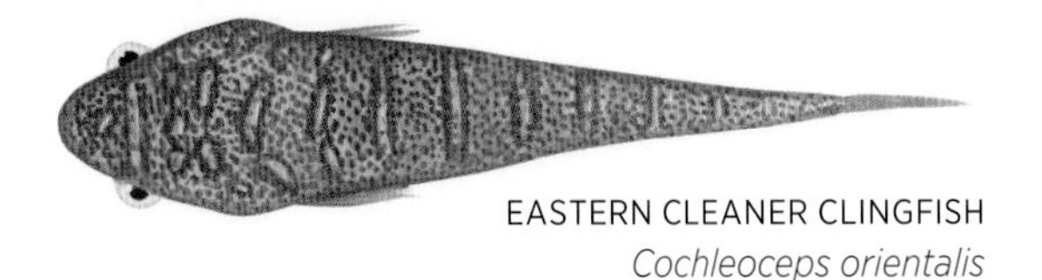

TASMANIAN CLINGFISH
Aspasmogaster tasmaniensis

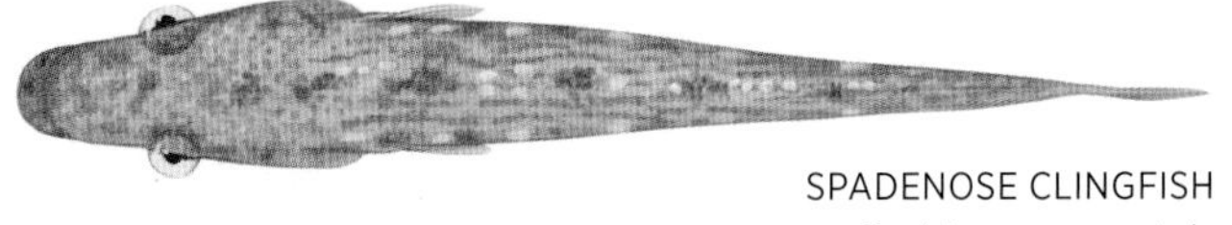

WESTERN CLEANER CLINGFISH
Cochleoceps bicolor

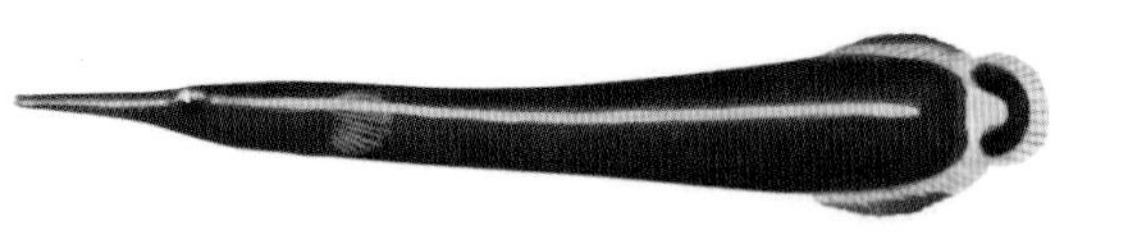

EASTERN CLEANER CLINGFISH
Cochleoceps orientalis

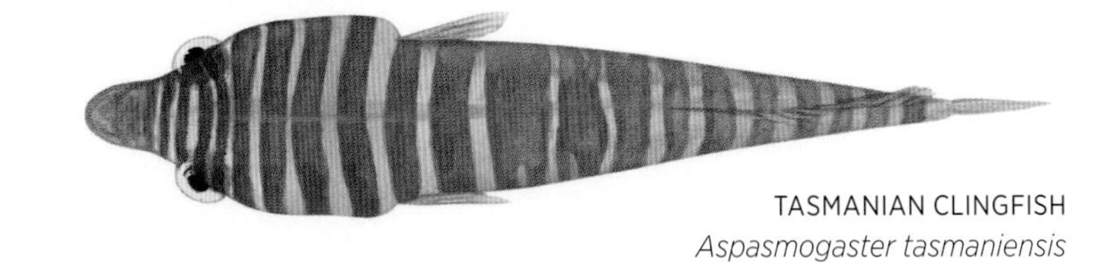

SPADENOSE CLINGFISH
Cochleoceps spatula

STRIPED CLINGFISH
Diademichthys lineatus

AUSTRALIAN GOBIESOCIDAE SPECIES

Alabes bathys Deepwater Shore Eel
Alabes brevis Short Shore Eel
Alabes dorsalis Common Shore Eel
Alabes elongata Elongate Shore Eel
Alabes gibbosa Hunchback Shore Eel
Alabes hoesei Dwarf Shore Eel
Alabes obtusirostris Pugnose Shore Eel
Alabes occidentalis Western Shore Eel
Alabes parvulus Pygmy Shore Eel
Alabes scotti Scott's Shore Eel
Aspasmogaster costata Pink Clingfish
Aspasmogaster liorhyncha Smoothsnout Clingfish
Aspasmogaster occidentalis Western Clingfish
Aspasmogaster tasmaniensis Tasmanian Clingfish
Cochleoceps bassensis Broadhead Clingfish
Cochleoceps bicolor Western Cleaner Clingfish
Cochleoceps orientalis Eastern Cleaner Clingfish
Cochleoceps spatula Spadenose Clingfish
Cochleoceps viridis Green Clingfish
Conidens samoensis Samoan Clingfish
Creocele cardinalis Broad Clingfish
Diademichthys lineatus Striped Clingfish
Discotrema crinophila Crinoid Clingfish
Kopua kuiteri Deepwater Clingfish
Lepadichthys caritus Precious Clingfish
Lepadichthys frenatus Bridled Clingfish
Lepadichthys sandaracatus Shark Bay Clingfish
Parvicrepis parvipinnis Smallfin Clingfish
Posidonichthys hutchinsi Posidonia Clingfish

DRAGONETS AND STINKFISHES

Callionymidae

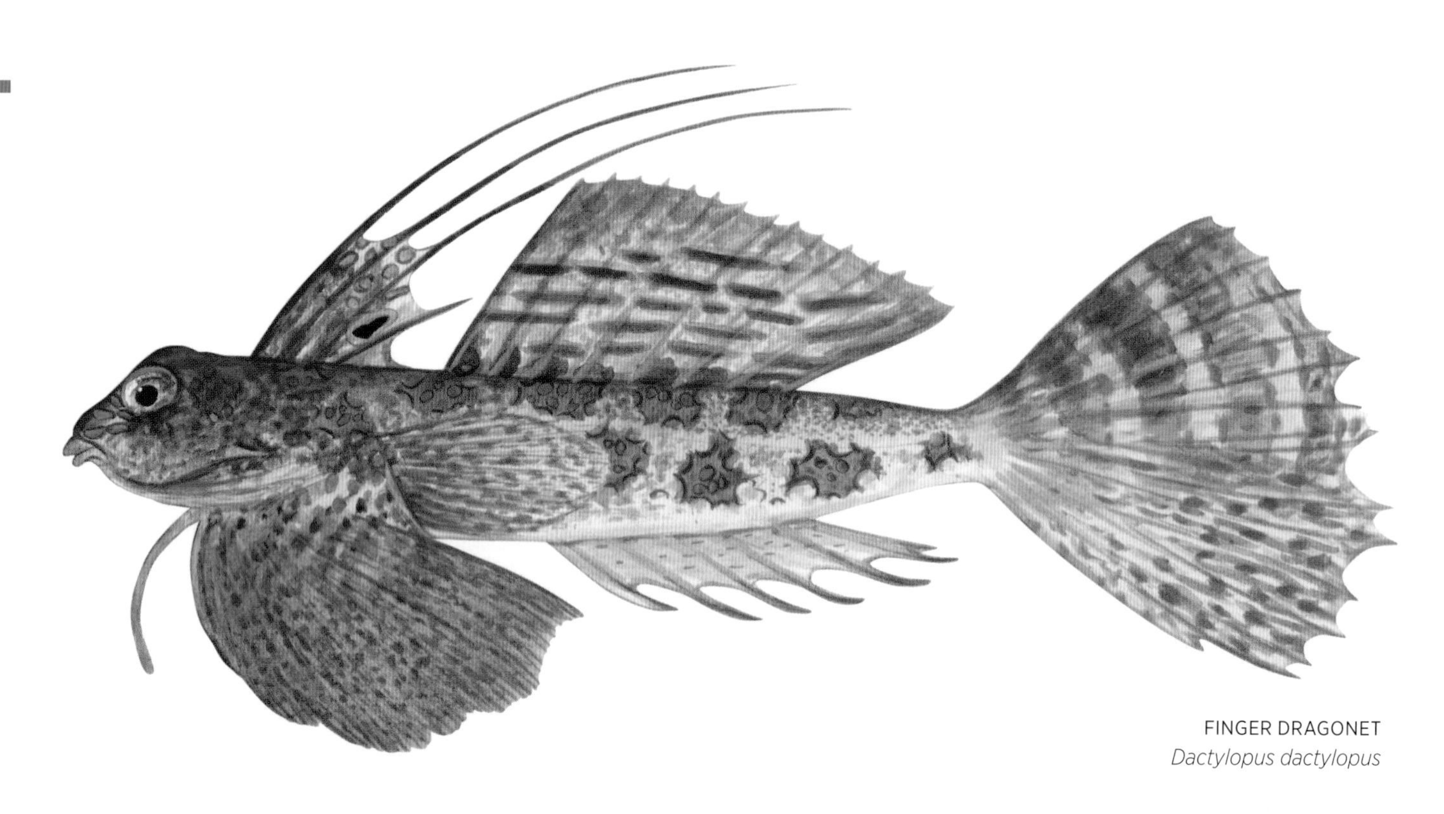

FINGER DRAGONET
Dactylopus dactylopus

There are about 180 species in this family, and they are found mainly in warm temperate and tropical waters of the Indo-Pacific. They have an elongate, rounded body with no scales, and a broad, flattened head with large eyes on the dorsal surface. Dragonets and Stinkfishes have a prominent, erectable, barbed spine on the preoperculum, and a small, downwardly directed, protrusible mouth. The gill openings are reduced to a small hole. They have two dorsal fins, the first usually with four spines that are often extremely elongate, and the second with 7–10 rays. The anal fin has no spines and 6–10 rays, and the large and often colourful ventral fins are placed well forward, beneath the throat. The caudal fin is rounded, and elongate in many species.

Dragonets and Stinkfishes are mainly found in shallow coastal waters, with a few species occurring in deeper waters to about 500 m, usually over sandy or muddy bottoms. Males and females often have quite different colouring, males usually having ornate, sail-like dorsal fins and brighter colours than females. Stinkfishes take their name from the strong odour they emit when captured, and they should not be eaten as the bitter-tasting flesh causes nausea. The **Common Stinkfish** is one of the largest in the family, reaching about 35 cm in length. The **Finger Dragonet** has a separate, finger-like ray before the fan-like ventral fins, and ornate, sail-like dorsal and caudal fins. It grows to about 22 cm and is common in shallow bays along the west coast of Australia. Dragonets and Stinkfishes feed on small invertebrates, which they pick from the substrate with their protrusible mouth. They are often encountered in prawn trawl nets in northern Australian waters and occasionally by divers over open sandy bottoms.

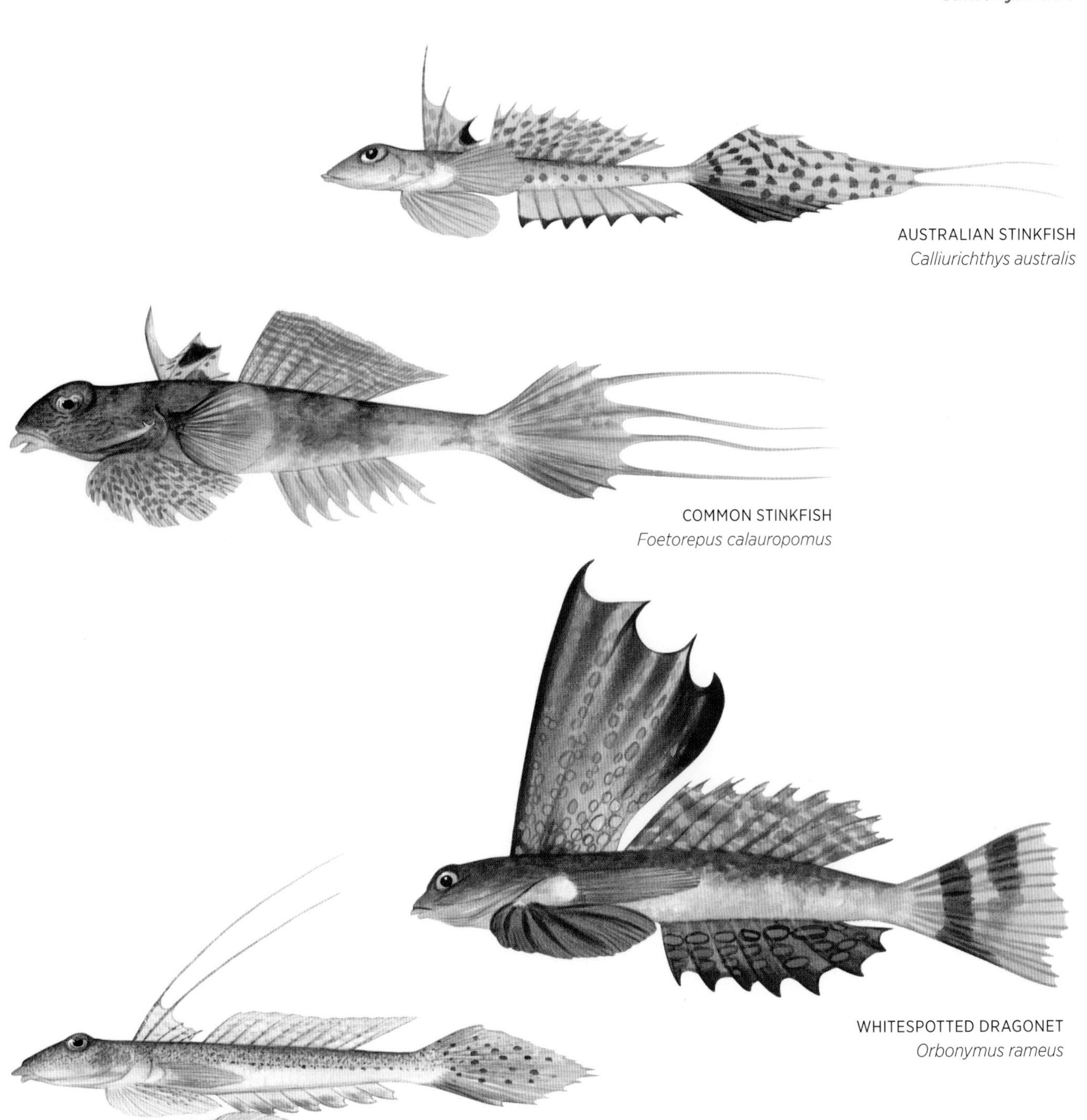

AUSTRALIAN STINKFISH
Calliurichthys australis

COMMON STINKFISH
Foetorepus calauropomus

WHITESPOTTED DRAGONET
Orbonymus rameus

LONGSPINE DRAGONET (lateral and dorsal view)
Pseudocalliurichthys goodladi

Callionymidae

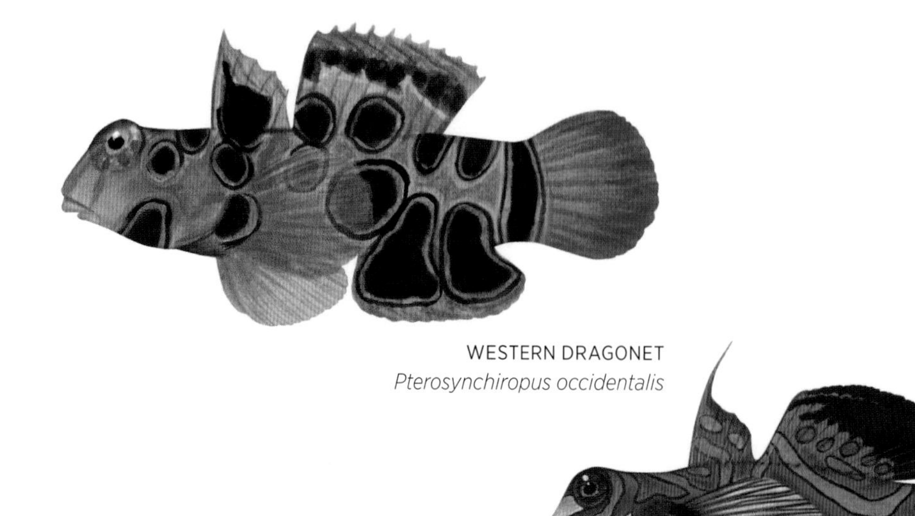

WESTERN DRAGONET
Pterosynchiropus occidentalis

MANDARINFISH
Pterosynchiropus splendidus

AUSTRALIAN CALLIONYMIDAE SPECIES

Anaora tentaculata Weedy Dragonet
Bathycallionymus bifilum Western Ocellate Dragonet
Bathycallionymus kailolae Kailola's Dragonet
Bathycallionymus moretonensis Ocellate Dragonet
Calliurichthys afilum Lowfin Stinkfish
Calliurichthys australis Australian Stinkfish
Calliurichthys grossi Longnose Stinkfish
Calliurichthys ogilbyi Ogilby's Stinkfish
Calliurichthys scaber Japanese Stinkfish
Dactylopus dactylopus Finger Dragonet
Diplogrammus goramensis Goram Dragonet
Diplogrammus xenicus Northern Dragonet
Eocallionymus papilio Painted Stinkfish
Foetorepus apricus Exposed Stinkfish
Foetorepus australis Eastern Deepwater Stinkfish
Foetorepus calauropomus Common Stinkfish
Foetorepus delandi Deland's Stinkfish
Foetorepus grandoculis Bigeye Stinkfish
Foetorepus paxtoni Paxton's Stinkfish
Foetorepus phasis Bight Stinkfish
Minysynchiropus claudiae Claudia's Dragonet
Neosynchiropus morrisoni Morrison's Dragonet
Neosynchiropus ocellatus Marble Dragonet
Neosynchiropus stellatus Starry Dragonet
Orbonymus rameus Whitespotted Dragonet
Paradiplogrammus corallinus Coral Dragonet
Paradiplogrammus enneactis Mangrove Dragonet
Pseudocalliurichthys brevianalis Shortfin Dragonet
Pseudocalliurichthys delicatulus Delicate Dragonet
Pseudocalliurichthys goodladi Longspine Dragonet
Pseudocalliurichthys pleurostictus Sidespotted Dragonet
Pseudocalliurichthys simplicicornis Simplespine Dragonet
Pterosynchiropus occidentalis Western Dragonet
Pterosynchiropus splendidus Mandarinfish
Repomucenus annulatus Ringed Dragonet
Repomucenus belcheri Flathead Dragonet
Repomucenus calcaratus Spotted Dragonet
Repomucenus filamentosus Filamentous Dragonet
Repomucenus keeleyi Keeley's Dragonet
Repomucenus leucobranchialis Whitegill Dragonet
Repomucenus limiceps Rough-head Dragonet
Repomucenus macdonaldi Greyspotted Dragonet
Repomucenus octostigmatus Eightspot Dragonet
Repomucenus russelli Russell's Dragonet
Repomucenus sphinx Sphinx Dragonet
Repomucenus sublaevis Multifilament Dragonet
Spinicapitichthys draconis Barb Dragonet

DEEPSEA DRAGONETS

Draconettidae

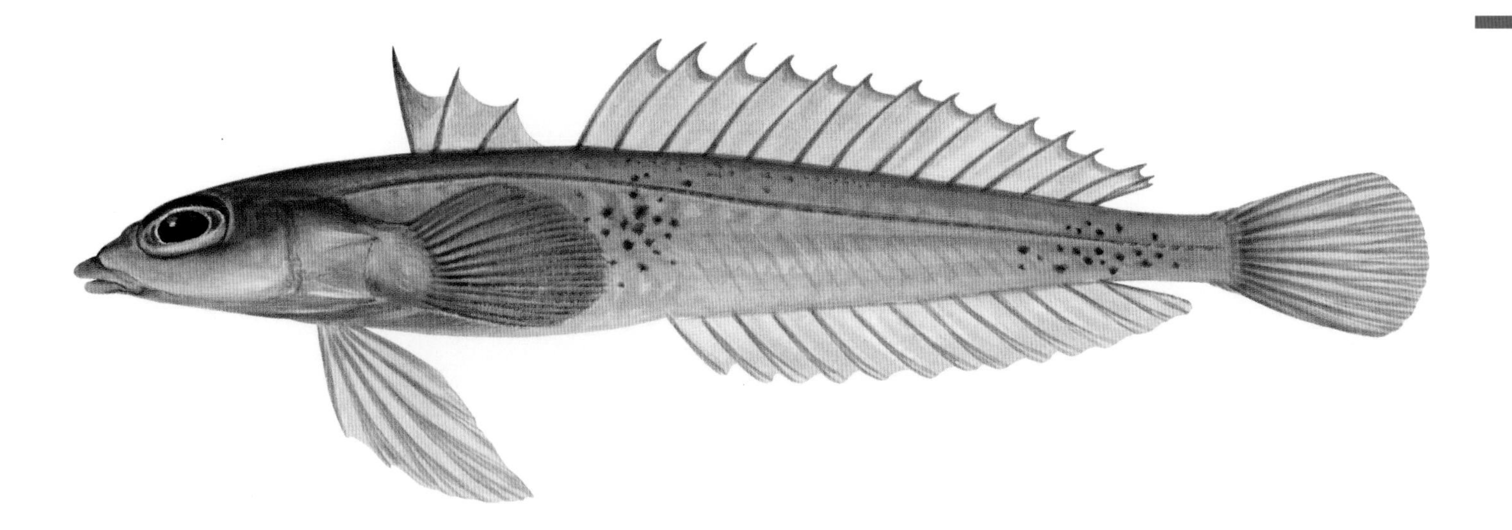

UNCOMMON DRAGONET
Centrodraco insolitus

Deepsea Dragonets are found in tropical and warm temperate waters of the western Pacific, Indian and Atlantic oceans. There are about 12 species in the family. They have an elongate, rounded body with no scales. The flattened head has two spines on the operculum, a pointed snout with protrusible mouth, and large eyes. There are two dorsal fins, the first with three spines and the second with 12–15 rays, and the anal fin has no spines and 12–15 rays. All rays in the dorsal fin, except the last, are unbranched. The ventral fins are placed well forward, beneath the throat.

Deepsea Dragonets are rare, living in waters from about 130 to 600 m in depth, over sandy or muddy bottoms. They are very similar to the Dragonets (Callionymidae, p. 670), but differ in having spines on the operculum and not on the preoperculum. Very little is known of their biology, but they probably have similar dietary habits to the Dragonets, which feed on small benthic invertebrates.

AUSTRALIAN DRACONETTIDAE SPECIES

Centrodraco insolitus Uncommon Dragonet

GUDGEONS

Eleotridae

EMPIRE GUDGEON
Hypseleotris compressa

Gudgeons are found in temperate to tropical marine, brackish and fresh waters around the world. There are about 155 species in the family. They have an elongate, rounded body with small to moderate-sized cycloid or ctenoid scales, no lateral line, and a large head with large, oblique mouth. There are two dorsal fins, the first with 2–8 spines and the second with 8–17 rays; the anal fin has one spine and 7–13 rays; and the pectoral and caudal fins are broad and rounded. The ventral fins are separate, differentiating them from the very similar Gobies (Gobiidae, p. 678), most of which have the ventral fins joined to form a disc.

The great majority of Australian Gudgeons are found in fresh water, with only a few, such as the **Cryptic Sea Gudgeon**, occurring in shallow coastal environments. They are found in most freshwater habitats, from shallow desert pools to swift-flowing streams. Many have a very limited distribution, with some found only in a single stream. The **Cave Gudgeon**, a unique species that has no skin pigmentation and lacks eyes, is found only in a subterranean aquifer in the Cape Range near Exmouth in Western Australia. The **Southern Purple-spotted Gudgeon** is a colourful species, common and widely distributed in east-coast drainages. The **Sleepy Cod** is one of the largest Gudgeons, growing to around 45 cm in length and weighing about 3 kg, and it is found in many northern and north-eastern drainages. Most Gudgeons are much smaller, from about 4 to 10 cm.

Male and female Gudgeons often have different colour patterns, particularly in the breeding season, when the males' colours become more intense. Female Gudgeons lay adhesive eggs on rocks or other hard substrates and the male guards the eggs until they hatch, fanning them with his fins to ensure a flow of oxygenated water. The juveniles of many stream-dwelling species are swept downstream into estuaries, then migrate back upstream as they mature. Gudgeons are carnivorous, feeding on small invertebrates and insects, with the larger species also preying on crustaceans and small fishes. Several of the larger species are popular food fishes in South-East Asia and are farmed extensively across the region.

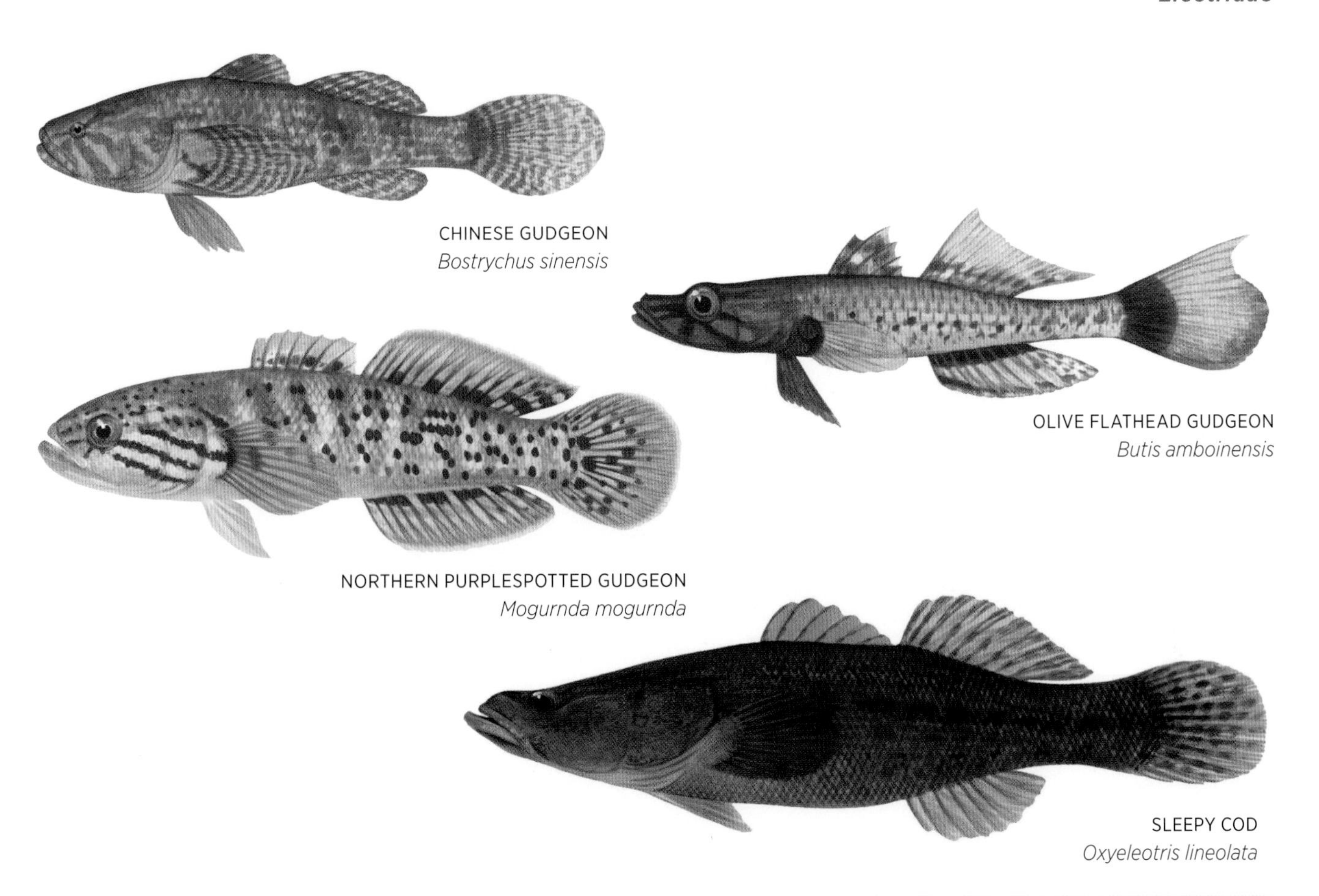

CHINESE GUDGEON
Bostrychus sinensis

OLIVE FLATHEAD GUDGEON
Butis amboinensis

NORTHERN PURPLESPOTTED GUDGEON
Mogurnda mogurnda

SLEEPY COD
Oxyeleotris lineolata

AUSTRALIAN ELEOTRIDAE SPECIES

Bostrychus sinensis Chinese Gudgeon
Bostrychus zonatus Sunset Gudgeon
Bunaka gyrinoides Greenback Gudgeon
Butis amboinensis Olive Flathead Gudgeon
Butis butis Crimsontip Gudgeon
Butis koilomatodon Crested Gudgeon
Calumia godeffroyi Tail-face Calumia
Calumia profunda Longjaw Calumia
Eleotris acanthopoma Spine-cheek Gudgeon
Eleotris fusca Brown Spine-cheek Gudgeon
Eleotris melanosoma Black Spine-cheek Gudgeon
Gobiomorphus australis Striped Gudgeon
Gobiomorphus coxii Cox Gudgeon
Hypseleotris aurea Golden Carp Gudgeon
Hypseleotris compressa Empire Gudgeon
Hypseleotris ejuncida Slender Carp Gudgeon
Hypseleotris galii Firetail Gudgeon
Hypseleotris kimberleyensis Barnett River Gudgeon
Hypseleotris klunzingeri Western Carp Gudgeon
Hypseleotris regalis Regent Carp Gudgeon
Incara multisquamatus Finescale Gudgeon
Kimberleyeleotris hutchinsi Mitchell Gudgeon
Kimberleyeleotris notata Drysdale Gudgeon
Milyeringa veritas Cave Gudgeon
Mogurnda adspersa Southern Purplespotted Gudgeon
Mogurnda clivicola Flinders Ranges Mogurnda
Mogurnda larapintae Desert Mogurnda
Mogurnda mogurnda Northern Purplespotted Gudgeon
Mogurnda oligolepis Kimberley Mogurnda
Mogurnda thermophila Dalhousie Mogurnda
Odonteleotris macrodon Sinuous Gudgeon
Ophieleotris margaritacea Snakehead Gudgeon
Ophiocara porocephala Spangled Gudgeon
Oxyeleotris aruensis Aru Gudgeon
Oxyeleotris fimbriata Fimbriate Gudgeon
Oxyeleotris lineolata Sleepy Cod
Oxyeleotris nullipora Poreless Gudgeon
Oxyeleotris selheimi Blackbanded Gudgeon
Parascyllium collare Collar Carpetshark
Philypnodon grandiceps Flathead Gudgeon
Prionobutis microps Smalleye Gudgeon
Thalasseleotris adela Cryptic Sea Gudgeon

WRIGGLERS

Xenisthmidae

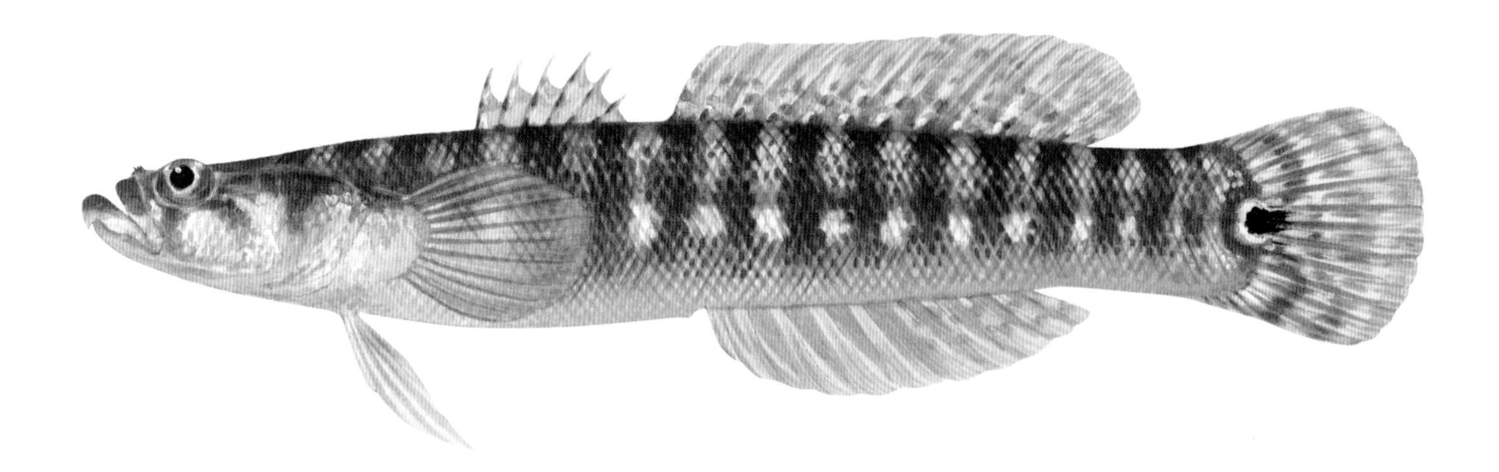

BULLSEYE WRIGGLER
Xenisthmus polyzonatus

Wrigglers are found only in the Indo-Pacific region, and the family contains about 12 species. They have an elongate, rounded body, either with small cycloid scales on the head and body, or scales on the head but not the body. They have a slightly flattened head with a large mouth and a projecting, thickened lower jaw with the lower margin of the lip unattached. Wrigglers usually have two dorsal fins, the first with 2–6 spines and the second with 0–1 spine and 8–33 rays, and an anal fin with 0–1 spine and 8–26 rays. The ventral fins are separate, not united to form a disc as they generally are in the very similar Gobies (Gobiidae, p. 678). The pectoral fins are broadly rounded and the caudal fin is rounded to slightly forked.

Wrigglers are found on coral reefs, over sand and rubble bottoms in tropical waters. Some species are able to dive beneath the sand to escape predators. They are uncommon and the family is poorly known. Some species are known from only one or two specimens and more species probably await description. Wrigglers are carnivorous and feed on small crustaceans and fishes.

AUSTRALIAN XENISTHMIDAE SPECIES

Allomicrodesmus dorotheae Toothpick Wriggler
Tyson belos Tyson's Arrow Goby
Xenisthmus chi Chi Wriggler
Xenisthmus clarus Clear Wriggler
Xenisthmus eirospilus Spotted Wriggler
Xenisthmus polyzonatus Bullseye Wriggler
Xenisthmus semicinctus Halfbelt Wriggler

SAND GOBIES

Kraemeriidae

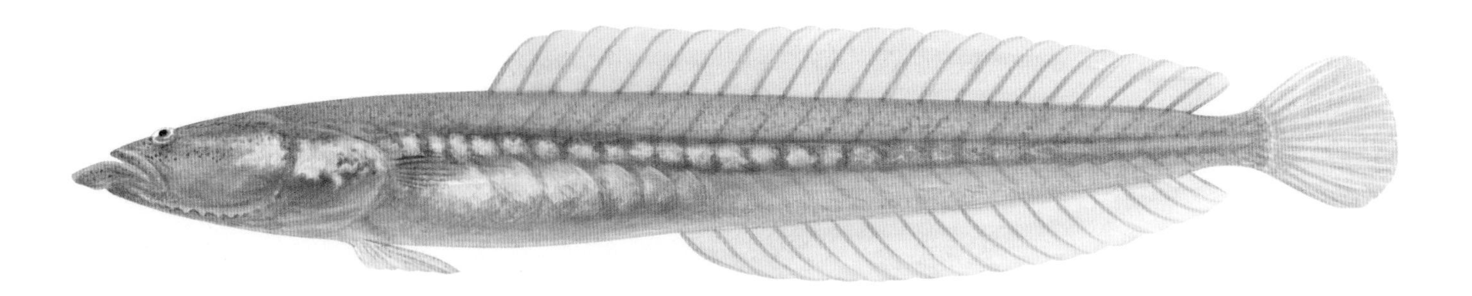

MER SAND DART
Kraemeria merensis

There are about eight species of Sand Goby and they are found only in tropical waters of the Indo-Pacific region. They have a very elongate, rounded body with no scales, and a small head with very small, close-set, dorsally oriented eyes. The large mouth has a distinctive, projecting lower jaw and an unusual bilobed tongue. There are small flaps of skin on the lower edges of the peroperculum and operculum. The single dorsal fin has 4–5 spines and 13–17 unbranched rays, and the anal fin has one spine and 11–14 unbranched rays. The pectoral fins are small and the ventral fins are separate or joined (but do not form a disc).

Sand Gobies burrow into the sand or mud and in Australia are found near coral reefs and in shallow coastal and estuarine environments in northern waters. Because of their burrowing habit they are rarely encountered and very little is known of their biology.

AUSTRALIAN KRAEMERIIDAE SPECIES

Kraemeria merensis Mer Sand Dart

GOBIES

Gobiidae

DESERT GOBY
Chlamydogobius eremius

This is the largest of all marine fish families and the second-largest of all fish families (second only to the Carps – Cyprinidae, p. 192). Gobies are small fishes found in virtually all marine and estuarine environments, and most freshwater habitats, around the world. There are at least 1500 species, possibly more than 2000 (experts disagree and the exact number will probably never be known). New species are frequently discovered and many more await description.

Gobies have an elongate, rounded body, usually with large cycloid or ctenoid scales (though some species are without scales), and they have no lateral line. Most have a bluntly rounded head and a small mouth with rows of small conical teeth. Sensory papillae and pores on the head can be important characteristics for differentiating the various species. Gobies have two dorsal fins, the first with 5–10 spines, the second with one spine and 5–37 rays, and an anal fin with one spine and 5–36 rays. The pectoral fins are broad and the ventral fins are joined together to form a sucking disc, except in some coral-reef species, where they are separate.

The habits of Gobies are extremely diverse. Mudskippers, found in mangroves and on mudflats across northern Australia, are capable of spending considerable time out of water and can move with great agility on dry land, even climbing up mangrove roots and branches. They are able to absorb oxygen through their skin, which is well supplied with blood vessels. The various species of Shrimpgoby live commensally in burrows with small alpheid shrimps. The shrimps excavate and maintain the burrow, and the Gobies, which have far more acute vision, serve as lookouts. The shrimps emerge from the burrow cautiously, always keeping one feeler touching the Goby for any sign of alarm.

Gobies are very common on Australian coral reefs and in nearby sand and rubble areas. Some, such as the Coral Gobies, live within colonies of *Acropora* and other branching corals, never leaving the shelter they provide. The Eviotas are also found amongst corals, both encrusting and branching forms. Others, such as Whipgobies, live on gorgonians and seawhips, lying with their small, semi-transparent bodies positioned lengthwise along the axis of the branches. Many other species live on sandy areas near reefs, often in burrows. The large

Glider Gobies, which reach about 15–18 cm in length, are often found in pairs. It is common to see them carrying quite large pieces of rubble in their mouth as they excavate and maintain their burrow. Pygmy Gobies reach a maximum length of about 2–3 cm, and are often found hovering near the shelter of coral or in caves on outer-reef slopes. Atom Gobies are even smaller and one member of this group, *Trimmatom nanus*, from the Chagos Archipelago, previously held the record for the world's smallest vertebrate, with mature females growing to about 8 mm in length. (The record is now held by an Australian Infantfish – Schindleriidae, p. 691.)

Sandgobies of the genus *Nesogobius* are endemic to the southern and south-eastern coastal waters of Australia, where they are found on sandy bottoms in shallow bays and estuaries, often near seagrass. There are quite a few undescribed species in this genus. There are also many species of Goby found in fresh water. Desert Gobies (genus *Chlamydogobius*) are found in springs and pools in central Australia and are capable of gulping air at the surface to survive in oxygen-poor waters. Several have very restricted distribution, with some species found only in a single spring or pool system, and are at risk from reduced water flow caused by over-use of artesian water. There are also Gobies that inhabit fast-flowing clear streams in tropical Queensland, such as the rare **Allen's Stiphodon**, which is known from a single specimen. Many species inhabit estuarine environments and these, along with some species from the lower reaches of freshwater systems, appear to have marine larval stages.

Most Gobies are carnivorous, feeding on very small invertebrates such as crustaceans, molluscs and worms, the eggs of other fishes, and zooplankton. Some species also consume foraminiferans and sponges. Female Gobies lay adhesive eggs, usually on the undersurface of rocks or dead shells, which are guarded by the males.

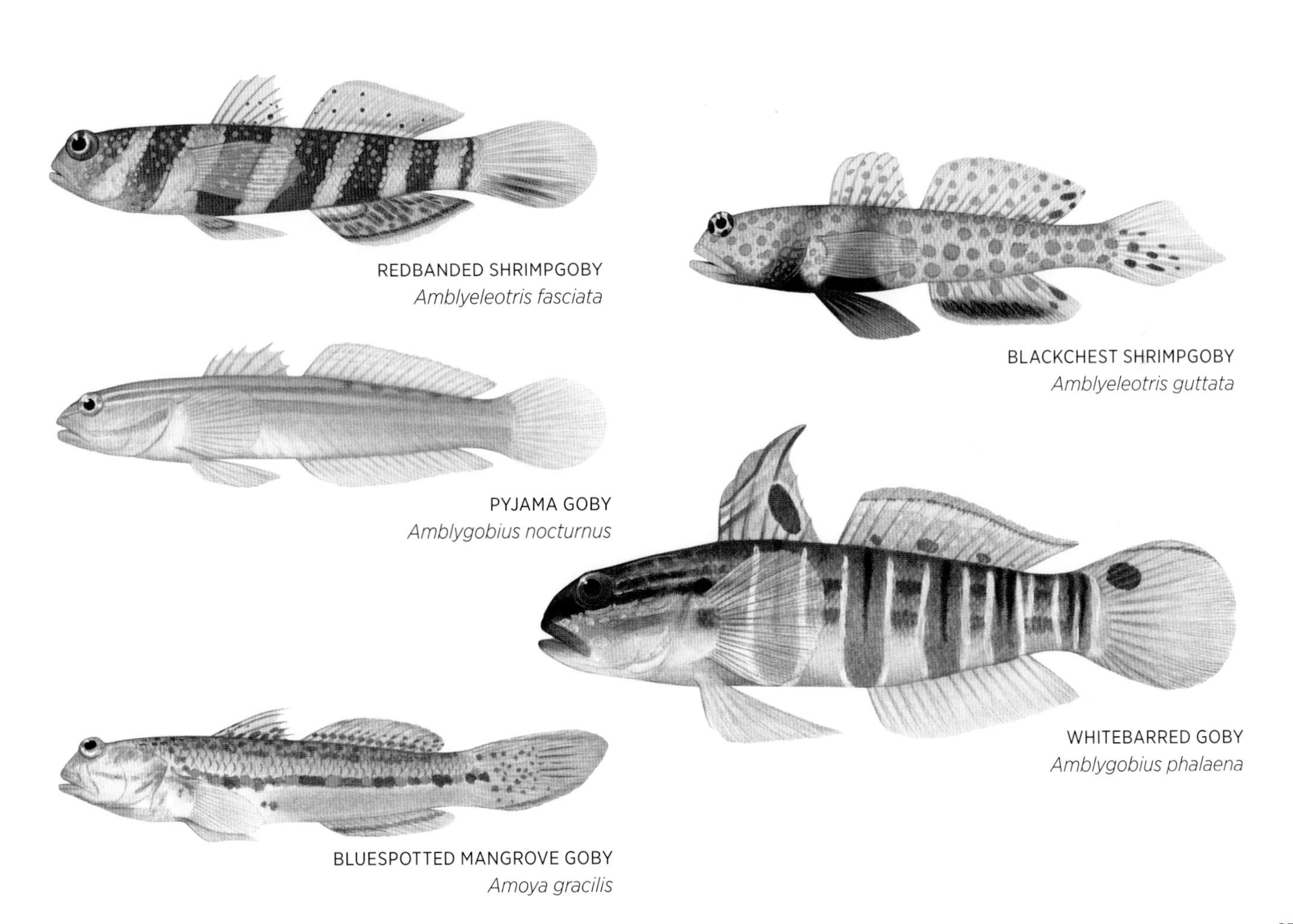

REDBANDED SHRIMPGOBY
Amblyeleotris fasciata

BLACKCHEST SHRIMPGOBY
Amblyeleotris guttata

PYJAMA GOBY
Amblygobius nocturnus

WHITEBARRED GOBY
Amblygobius phalaena

BLUESPOTTED MANGROVE GOBY
Amoya gracilis

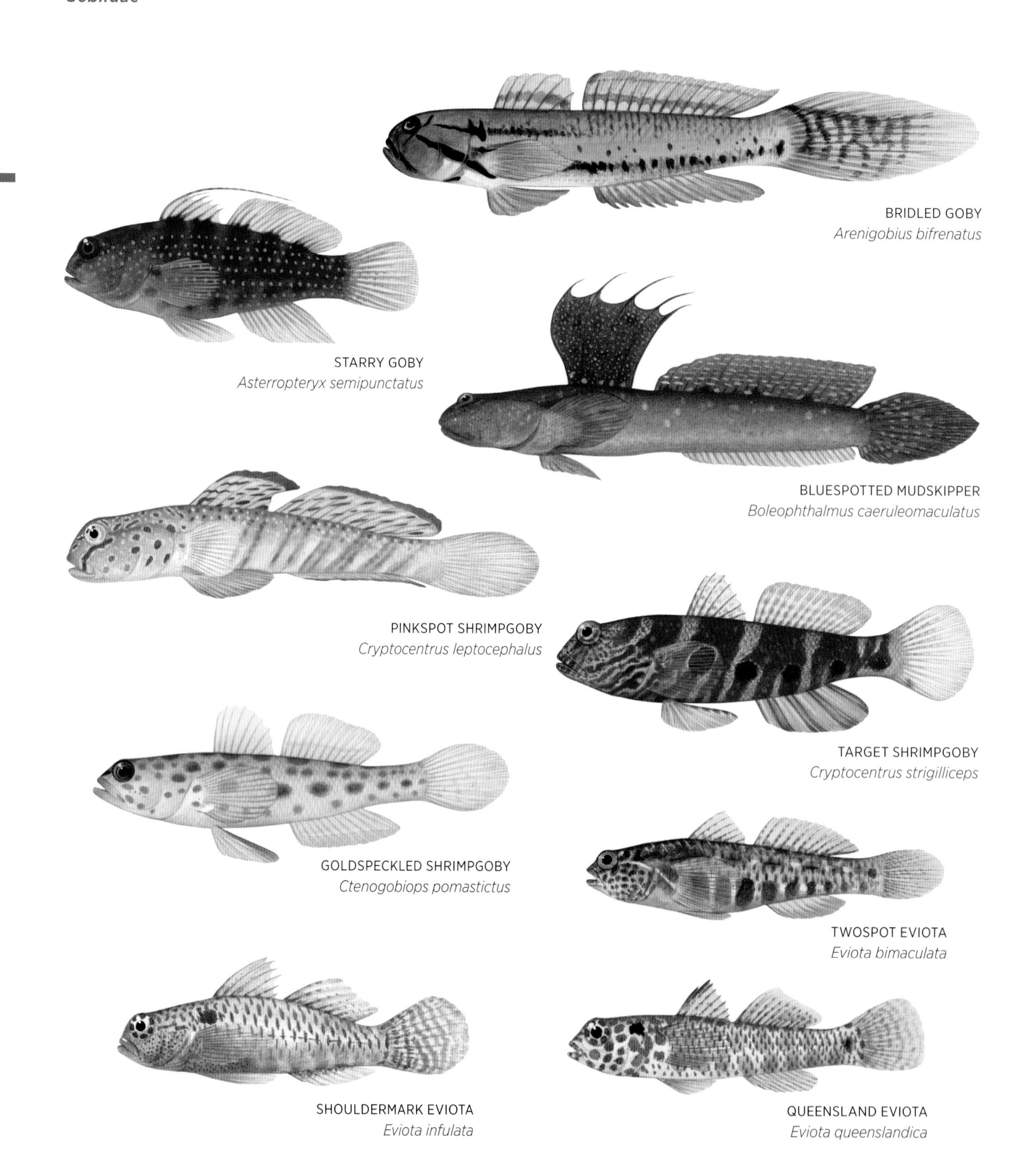

BRIDLED GOBY
Arenigobius bifrenatus

STARRY GOBY
Asterropteryx semipunctatus

BLUESPOTTED MUDSKIPPER
Boleophthalmus caeruleomaculatus

PINKSPOT SHRIMPGOBY
Cryptocentrus leptocephalus

TARGET SHRIMPGOBY
Cryptocentrus strigilliceps

GOLDSPECKLED SHRIMPGOBY
Ctenogobiops pomastictus

TWOSPOT EVIOTA
Eviota bimaculata

SHOULDERMARK EVIOTA
Eviota infulata

QUEENSLAND EVIOTA
Eviota queenslandica

MUD-REEF GOBY
Exyrias belissimus

TWOSPOT SANDGOBY
Fusigobius duospilus

NEOPHYTE SANDGOBY
Fusigobius neophytus

MANGROVE FLATHEAD GOBY
Glossogobius circumspectus

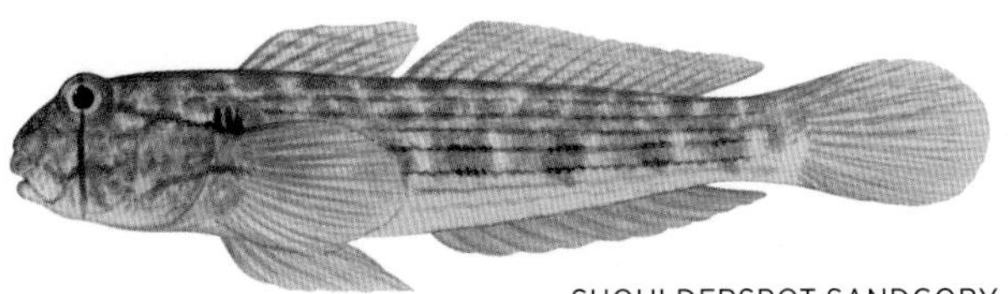

SHOULDERSPOT SANDGOBY
Gnatholepis anjerensis

MAORI CORAL GOBY
Gobiodon histrio

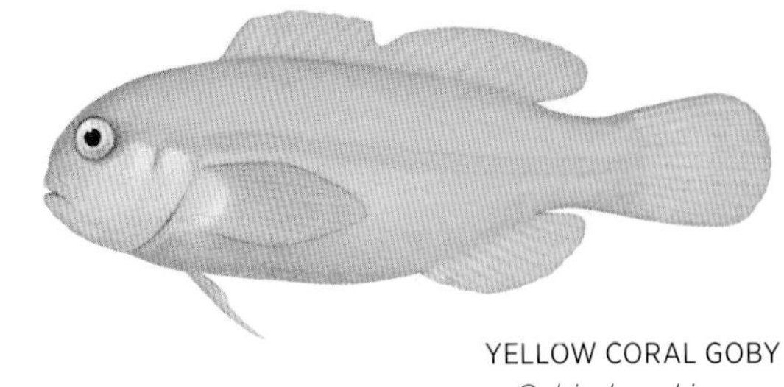

YELLOW CORAL GOBY
Gobiodon okinawae

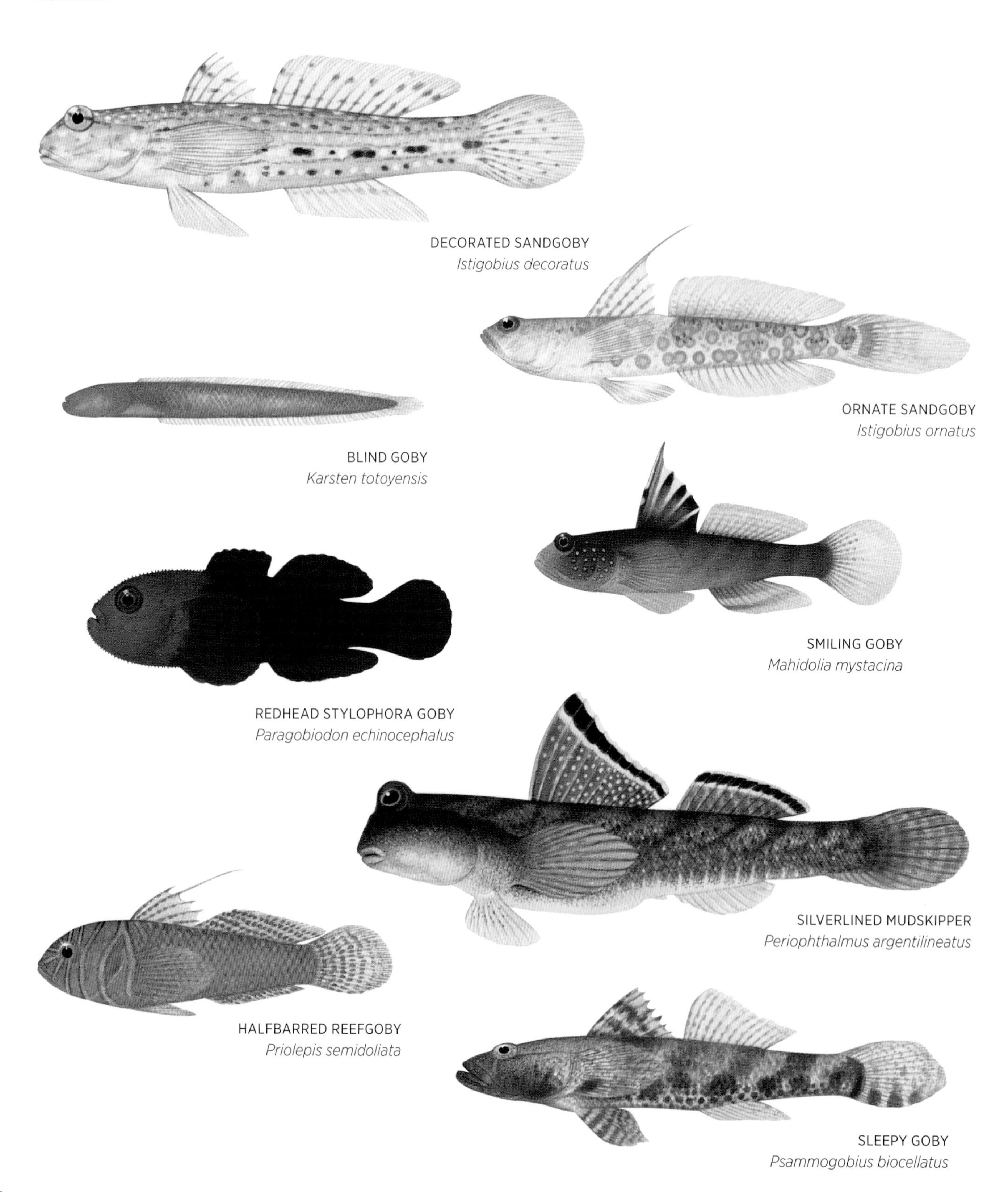

DECORATED SANDGOBY
Istigobius decoratus

ORNATE SANDGOBY
Istigobius ornatus

BLIND GOBY
Karsten totoyensis

SMILING GOBY
Mahidolia mystacina

REDHEAD STYLOPHORA GOBY
Paragobiodon echinocephalus

SILVERLINED MUDSKIPPER
Periophthalmus argentilineatus

HALFBARRED REEFGOBY
Priolepis semidoliata

SLEEPY GOBY
Psammogobius biocellatus

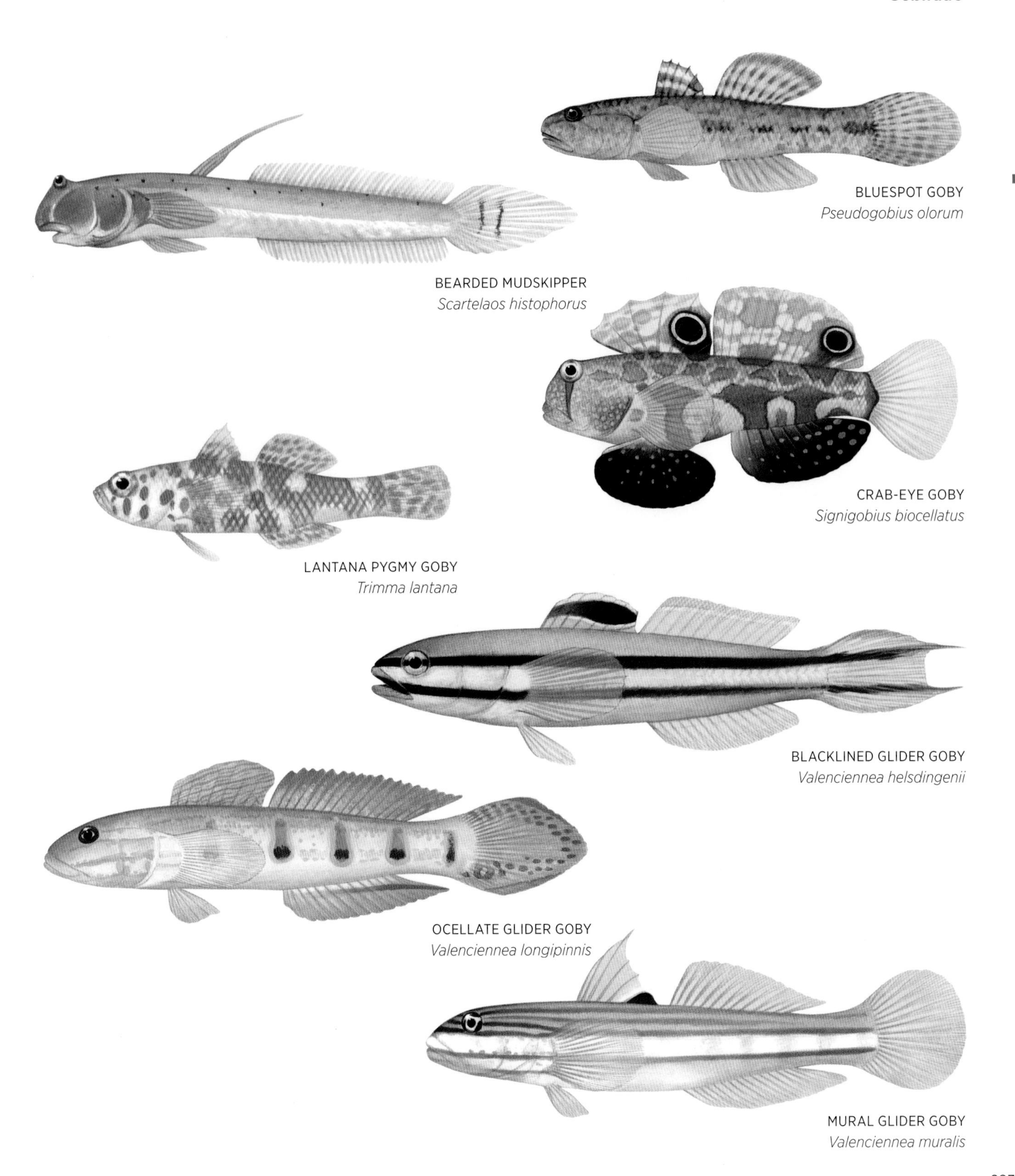
BLUESPOT GOBY
Pseudogobius olorum
BEARDED MUDSKIPPER
Scartelaos histophorus
CRAB-EYE GOBY
Signigobius biocellatus
LANTANA PYGMY GOBY
Trimma lantana
BLACKLINED GLIDER GOBY
Valenciennea helsdingenii
OCELLATE GLIDER GOBY
Valenciennea longipinnis
MURAL GLIDER GOBY
Valenciennea muralis

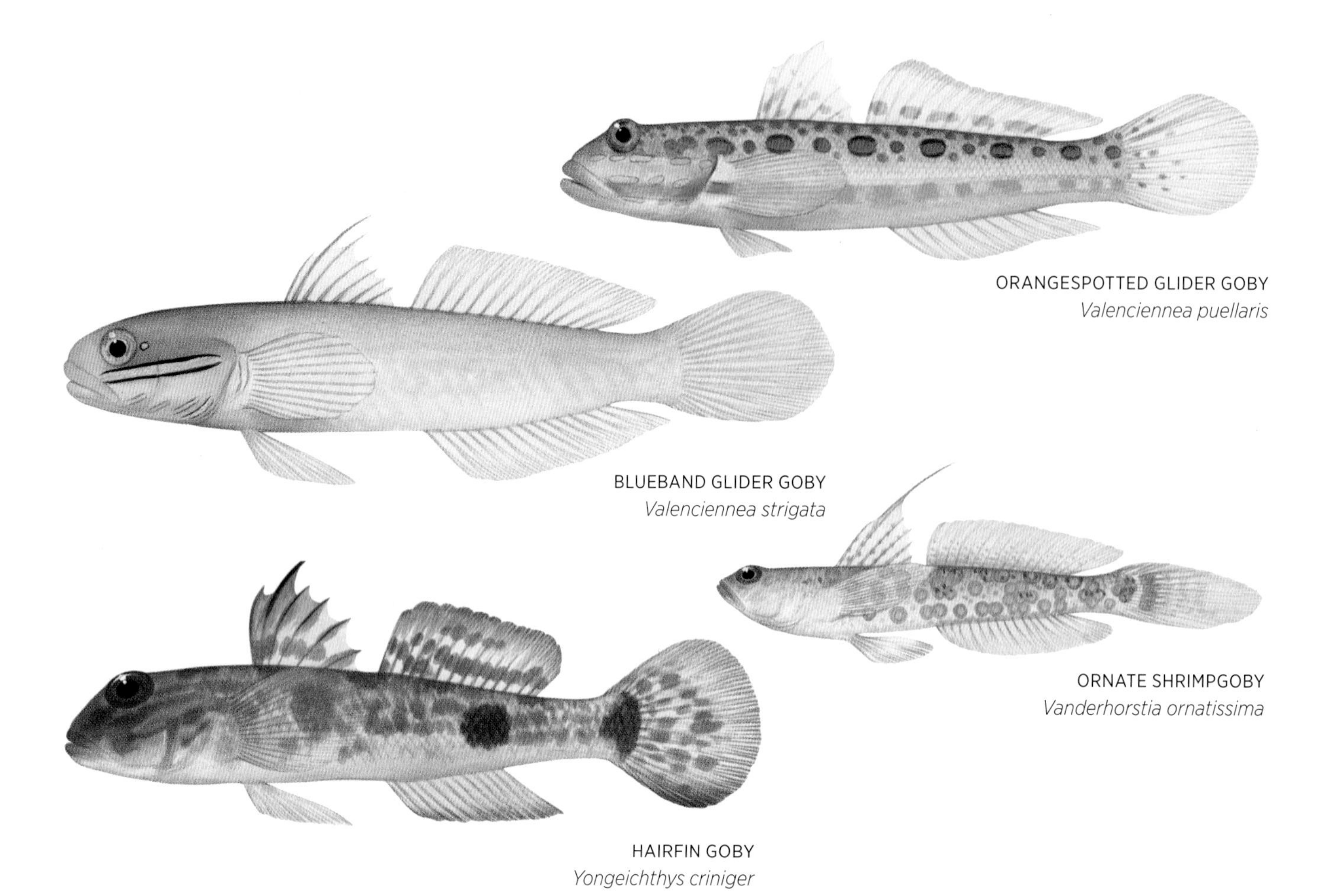

ORANGESPOTTED GLIDER GOBY
Valenciennea puellaris

BLUEBAND GLIDER GOBY
Valenciennea strigata

ORNATE SHRIMPGOBY
Vanderhorstia ornatissima

HAIRFIN GOBY
Yongeichthys criniger

AUSTRALIAN GOBIIDAE SPECIES

Acanthogobius flavimanus Yellowfin Goby
Acentrogobius caninus Green-shoulder Goby
Acentrogobius janthinopterus Robust Mangrove Goby
Acentrogobius pflaumii Striped Sandgoby
Acentrogobius viridipunctatus Greenspotted Goby
Afurcagobius suppositus Southwestern Goby
Afurcagobius tamarensis Tamar Goby
Amblyeleotris callopareia Beautiful-cheek Shrimpgoby
Amblyeleotris diagonalis Diagonal Shrimpgoby
Amblyeleotris fasciata Redbanded Shrimpgoby
Amblyeleotris fontanesii Giant Shrimpgoby
Amblyeleotris guttata Blackchest Shrimpgoby
Amblyeleotris gymnocephala Mask Shrimpgoby
Amblyeleotris macronema Longspine Shrimpgoby
Amblyeleotris ogasawarensis Redspotted Shrimpgoby
Amblyeleotris periophthalma Broadbanded Shrimpgoby
Amblyeleotris randalli Sailfin Shrimpgoby
Amblyeleotris rhyax Volcano Shrimpgoby
Amblyeleotris rubrimarginata Redmargin Shrimpgoby
Amblyeleotris steinitzi Steinitz's Shrimpgoby
Amblyeleotris wheeleri Burgundy Shrimpgoby
Amblygobius bynoensis Bynoe Goby
Amblygobius decussatus Crosshatch Goby
Amblygobius nocturnus Pyjama Goby
Amblygobius phalaena Whitebarred Goby
Amblygobius rainfordi Old Glory Goby
Amblygobius sphynx Sphynx Goby
Amblyotrypauchen artocephalus Armour Eel Goby
Amoya gracilis Bluespotted Mangrove Goby
Amoya madraspatensis Manyband Goby
Amoya moloanus Barcheek Amoya
Apocryptodon madurensis Peppered Mudskipper
Arcygobius baliurus Isthmus Goby
Arenigobius bifrenatus Bridled Goby
Arenigobius frenatus Halfbridled Goby
Arenigobius leftwichi Oyster Goby
Asterropteryx ensiferus Cheekspine Goby
Asterropteryx semipunctatus Starry Goby

Asterropteryx spinosus Eyebar Spinygoby
Austrolethops wardi Nudey Goby
Awaous acritosus Roman-nose Goby
Barbuligobius boehlkei Cryptic Bearded Goby
Bathygobius coalitus Whitespotted Frillgoby
Bathygobius cocosensis Cocos Frillgoby
Bathygobius cotticeps Cheekscaled Frillgoby
Bathygobius cyclopterus Cyclops Frillgoby
Bathygobius fuscus Dusky Frillgoby
Bathygobius krefftii Krefft's Frillgoby
Bathygobius laddi Ladd's Frillgoby
Bathygobius meggitti Meggitt's Frillgoby
Boleophthalmus birdsongi Birdsong's Mudskipper
Boleophthalmus caeruleomaculatus Bluespotted Mudskipper
Bryaninops amplus Large Whipgoby
Bryaninops erythrops Fire-coral Goby
Bryaninops isis Isis Goby
Bryaninops loki Loki Whipgoby
Bryaninops natans Purple-eye Goby
Bryaninops nexus Upside-down Goby
Bryaninops ridens Porites Goby
Bryaninops tigris Antipatharia Goby
Bryaninops yongei Seawhip Goby
Cabillus lacertops Lizard Cabillus
Cabillus macrophthalmus Bigeye Cabillus
Cabillus tongarevae Tail-bar Cabillus
Calamiana mindora Stripe-face Calamiana
Callogobius clitellus Saddled Flaphead Goby
Callogobius depressus Flathead Goby
Callogobius flavobrunneus Slimy Goby
Callogobius hasseltii Hasselt's Flaphead Goby
Callogobius maculipinnis Ostrich Goby
Callogobius mucosus Sculptured Goby
Callogobius okinawae Tailspot Flaphead Goby
Callogobius sclateri Tripleband Goby
Caragobius rubristriatus Red Eel Goby
Caragobius urolepis Scaleless Wormgoby
Chlamydogobius eremius Desert Goby
Chlamydogobius gloveri Dalhousie Goby
Chlamydogobius japalpa Finke Goby
Chlamydogobius micropterus Elizabeth Springs Goby
Chlamydogobius ranunculus Tadpole Goby
Chlamydogobius squamigenus Edgbaston Goby
Cristatogobius rubripectoralis Redfin Crested Goby
Cryptocentroides gobioides Crested Oyster Goby
Cryptocentroides insignis Insignia Goby
Cryptocentrus bulbiceps Bluelined Shrimpgoby
Cryptocentrus caeruleomaculatus Bluespotted Shrimpgoby
Cryptocentrus cebuanus Cebu Shrimpgoby
Cryptocentrus cinctus Yellow Shrimpgoby
Cryptocentrus fasciatus Y-bar Shrimpgoby
Cryptocentrus inexplicatus Inexplicable Shrimpgoby
Cryptocentrus insignitus Signal Goby
Cryptocentrus leptocephalus Pinkspot Shrimpgoby
Cryptocentrus leucostictus Saddled Shrimpgoby
Cryptocentrus maudae Maude's Shrimpgoby
Cryptocentrus strigilliceps Target Shrimpgoby
Cryptocentrus tentaculatus Tentacle Shrimpgoby
Ctenogobiops aurocingulus Goldstreaked Shrimpgoby
Ctenogobiops feroculus Fierce Shrimpgoby
Ctenogobiops maculosus Silverspot Shrimpgoby
Ctenogobiops pomastictus Goldspeckled Shrimpgoby
Ctenogobiops tangaroai Tangaroa Shrimpgoby
Discordipinna griessingeri Spikefin Goby
Drombus dentifer Yellow Drombus
Drombus globiceps Kranji Drombus
Drombus halei Hale's Drombus
Drombus lepidothorax White-edge Drombus
Drombus ocyurus Bluemarked Drombus
Drombus simulus Pinafore Goby
Drombus triangularis Brown Drombus
Echinogobius hayashii Cheekstreak Goby
Egglestonichthys bombylios Egglestone's Bumblebee Goby
Eviota afelei Afele's Eviota
Eviota albolineata Whitelined Eviota
Eviota bifasciata Twostripe Eviota
Eviota bimaculata Twospot Eviota
Eviota cometa Comet Eviota
Eviota distigma Distigma Eviota
Eviota fasciola Barred Eviota
Eviota herrei Herre's Eviota
Eviota hoesei Doug's Eviota
Eviota infulata Shouldermark Eviota
Eviota inutilis Chestspot Eviota
Eviota lachdeberei Redlight Eviota
Eviota melasma Headspot Eviota
Eviota monostigma Singlespot Eviota
Eviota nebulosa Palespot Eviota
Eviota nigriventris Red-and-black Eviota
Eviota pellucida Neon Eviota
Eviota prasina Rubble Eviota
Eviota prasites Hairfin Eviota
Eviota punctulata Finespot Eviota
Eviota queenslandica Queensland Eviota
Eviota readerae Sally's Eviota
Eviota sebreei Striped Eviota
Eviota sigillata Sign Eviota
Eviota sparsa Sparse Eviota
Eviota spilota Shoulderspot Eviota
Eviota storthynx Rosy Eviota
Eviota variola Finspot Eviota
Eviota zebrina Zebra Eviota

Eviota zonura Greenbanded Eviota
Exyrias belissimus Mud-reef Goby
Exyrias puntang Puntang Goby
Favonigobius exquisitus Exquisite Sandgoby
Favonigobius lateralis Southern Longfin Goby
Favonigobius lentiginosus Eastern Longfin Goby
Favonigobius melanobranchus Blackthroat Goby
Favonigobius punctatus Yellowspotted Sandgoby
Feia nota Palemarked Feia Goby
Feia nympha Feia Goby
Fusigobius aureus Golden Sandgoby
Fusigobius duospilus Twospot Sandgoby
Fusigobius gracilis Slender Sandgoby
Fusigobius humeralis Sidespot Sandgoby
Fusigobius inframaculatus Innerspot Sandgoby
Fusigobius melacron Blacktip Sandgoby
Fusigobius neophytus Neophyte Sandgoby
Fusigobius signipinnis Flasher Sandgoby
Glossogobius aureus Golden Flathead Goby
Glossogobius bicirrhosus Bearded Flathead Goby
Glossogobius celebius Celebes Flathead Goby
Glossogobius circumspectus Mangrove Flathead Goby
Glossogobius concavifrons Concave Flathead Goby
Glossogobius giuris Tank Goby
Gnatholepis anjerensis Shoulderspot Sandgoby
Gnatholepis gymnocara Nakedcheek Sandgoby
Gobiodon acicularis Needlespine Coral Goby
Gobiodon axillaris Red-striped Coral Goby
Gobiodon brochus Rasp Coral Goby
Gobiodon ceramensis Ceram Coral Goby
Gobiodon citrinus Lemon Coral Goby
Gobiodon heterospilos Peppered Coral Goby
Gobiodon histrio Maori Coral Goby
Gobiodon oculolineatus Eyeline Coral Goby
Gobiodon okinawae Yellow Coral Goby
Gobiodon quinquestrigatus Fiveline Coral Goby
Gobiodon rivulatus Rippled Coral Goby
Gobiodon spilophthalmus Whitelined Coral Goby
Gobiodon unicolor Plain Coral Goby
Gobiopsis angustifrons Narrow Barbel Goby
Gobiopsis aporia Poreless Barbel Goby
Gobiopsis arenaria Patchwork Goby
Gobiopsis macrostoma Longjaw Goby
Gobiopsis malekulae Striped Barbel Goby
Gobiopterus mindanensis Mindanao Glassgoby
Gobiopterus semivestitus Glassgoby
Hazeus elati Eilat Sandgoby
Hemigobius hoevenii Banded Mullet Goby
Istigobius decoratus Decorated Sandgoby
Istigobius diadema Spectacled Goby
Istigobius goldmanni Goldmann's Sandgoby
Istigobius hoesei Hoese's Sandgoby
Istigobius nigroocellatus Blackspotted Sandgoby
Istigobius ornatus Ornate Sandgoby
Istigobius rigilius Orangespotted Sandgoby
Istigobius spence Reticulate Sandgoby
Karsten totoyensis Blind Goby
Larsonella pumilus Dwarf Slippery Goby
Lobulogobius morrigu Eyebar Coral Goby
Lobulogobius omanensis Giant Lobe Goby
Lophogobius bleekeri Dark Mangrove Goby
Lotilia graciliosa Whitecap Shrimpgoby
Lubricogobius ornatus Ornate Slippery Goby
Luposicya lupus Cup-sponge Goby
Macrodontogobius wilburi Wilbur's Goby
Mahidolia mystacina Smiling Goby
Minysicya caudimaculata Blackspot Minigoby
Mugilogobius filifer Threadfin Mangrove Goby
Mugilogobius littoralis Beachrock Mangrove Goby
Mugilogobius mertoni Chequered Mangrove Goby
Mugilogobius notospilus Freshwater Mangrove Goby
Mugilogobius platynotus Flatback Mangrove Goby
Mugilogobius platystomus Island Mangrove Goby
Mugilogobius rivulus Drain Mangrove Goby
Mugilogobius stigmaticus Blackspot Mangrove Goby
Mugilogobius wilsoni Wilson's Mangrove Goby
Myersina macrostoma Flagfin Goby
Myersina nigrivirgata Blackline Shrimpgoby
Nesogobius hinsbyi Hinsby's Goby
Nesogobius pulchellus Sailfin Goby
Oplopomus caninoides Shy Lagoon Goby
Oplopomus oplopomus Pretty Lagoon Goby
Opua diacanthus Twospine Sandgoby
Oxuderces wirzi Peacock Mudskipper
Oxyurichthys auchenolepis Scaly-nape Tentacle Goby
Oxyurichthys cornutus Horned Tentacle Goby
Oxyurichthys microlepis Maned Tentacle Goby
Oxyurichthys ophthalmonema Eyebrow Goby
Oxyurichthys uronema Longtail Tentacle Goby
Paedogobius kimurai Babyface Goby
Pandaka lidwilli Lidwill's Dwarf Goby
Pandaka rouxi Roux's Dwarf Goby
Parachaeturichthys polynema Taileye Goby
Paragobiodon echinocephalus Redhead Stylophora Goby
Paragobiodon lacunicolus Blackfin Coral Goby
Paragobiodon melanosoma Black Coral Goby
Paragobiodon modestus Redhead Pocillopora Goby
Paragobiodon xanthosoma Emerald Coral Goby
Parkraemeria ornata Ornate Sand-diving Goby
Periophthalmodon freycineti Giant Mudskipper
Periophthalmus argentilineatus Silverlined Mudskipper
Periophthalmus darwini Darwin's Mudskipper

Periophthalmus gracilis Slender Mudskipper
Periophthalmus minutus Minute Mudskipper
Periophthalmus murdyi Ed's Mudskipper
Periophthalmus novaeguineaensis New Guinea Mudskipper
Periophthalmus weberi Weber's Mudskipper
Phyllogobius platycephalops Flathead Sponge Goby
Pleurosicya annandalei Solenocaulon Ghost Goby
Pleurosicya bilobatus Seagrass Ghost Goby
Pleurosicya boldinghi Softcoral Ghost Goby
Pleurosicya coerulea Bluecoral Ghost Goby
Pleurosicya elongata Slender Sponge Goby
Pleurosicya fringilla Staghorn Thicket Goby
Pleurosicya labiata Barrel-sponge Goby
Pleurosicya micheli Stony Coral Ghost Goby
Pleurosicya mossambica Many-host Ghost Goby
Pleurosicya muscarum Sinularia Ghost Goby
Pleurosicya plicata Lobed Ghost Goby
Pleurosicya prognatha Beaky Acropora Goby
Priolepis cincta Girdled Reefgoby
Priolepis compita Crossroads Reefgoby
Priolepis fallacincta False-girdled Reefgoby
Priolepis inhaca Network Reefgoby
Priolepis kappa Kappa Reefgoby
Priolepis nuchifasciata Threadfin Reefgoby
Priolepis pallidicincta Palebarred Reefgoby
Priolepis profunda Orange Convict Reefgoby
Priolepis psygmophilia Southern Reefgoby
Priolepis semidoliata Halfbarred Reefgoby
Psammogobius biocellatus Sleepy Goby
Pseudogobius olorum Bluespot Goby
Pseudogobius poicilosoma Northern Fatnose Goby
Psilogobius prolatus Orangespotted Shrimpgoby
Redigobius balteatus Rhinohorn Goby
Redigobius bikolanus Speckled Goby
Redigobius chrysosoma Spotfin Goby
Redigobius macrostoma Largemouth Goby
Scartelaos histophorus Bearded Mudskipper
Schismatogobius insignum Scaleless Goby
Sicyopterus lagocephalus Blue Stream Goby
Signigobius biocellatus Crab-eye Goby
Silhouettea evanida Vanishing Silhouette Goby
Silhouettea hoesei Hoese's Silhouette Goby
Silhouettea insinuans Phantom Silhouette Goby
Stiphodon alleni Allen's Stiphodon
Stonogobiops larsonae Larson's Shrimpgoby
Stonogobiops xanthorhinica Yellowface Shrimpgoby
Sueviota atrinasa Blacknose Sueviota
Sueviota lachneri Ernie's Sueviota
Sueviota larsonae Larson's Sueviota
Taenioides cirratus Bearded Wormgoby
Taenioides gracilis Slender Eel Goby
Taenioides limicola Bearded Eel Goby
Taenioides mordax Eastern Eel Goby
Taenioides purpurascens Purple Eel Goby
Tasmanogobius gloveri Glover's Tasman Goby
Tasmanogobius lasti Scary's Tasman Goby
Tasmanogobius lordi Lord's Tasman Goby
Tomiyamichthys latruncularia Fan Shrimpgoby
Tridentiger trigonocephalus Trident Goby
Trimma anaima Sharpeye Pygmy Goby
Trimma benjamini Ringeye Pygmy Goby
Trimma caesiura Caesiura Pygmy Goby
Trimma emeryi Emery's Pygmy Goby
Trimma hoesei Forktail Pygmy Goby
Trimma lantana Lantana Pygmy Goby
Trimma macrophthalma Large-eye Pygmy Goby
Trimma milta Moorea Pygmy Goby
Trimma nasa Nasal Pygmy Goby
Trimma necopina Australian Pygmy Goby
Trimma okinawae Orange-red Pygmy Goby
Trimma stobbsi Stobbs' Pygmy Goby
Trimma striata Redlined Pygmy Goby
Trimma taylori Yellow Cave Goby
Trimma tevegae Bluestripe Pygmy Goby
Trimma unisquamis Blackmargin Pygmy Goby
Trimmatom eviotops Redbarred Atom Goby
Trimmatom macropodus Bigfoot Atom Goby
Trimmatom nanus Dwarf Atom Goby
Trimmatom pharus Lighthouse Atom Goby
Trimmatom zapotes Hard-drinking Atom Goby
Trypauchen microcephalus Comb Goby
Trypauchenichthys sumatrensis Indonesian Eel Goby
Trypauchenichthys typus Typical Eel Goby
Valenciennea alleni Allen's Glider Goby
Valenciennea decora Decorated Glider Goby
Valenciennea helsdingenii Blacklined Glider Goby
Valenciennea immaculata Immaculate Glider Goby
Valenciennea longipinnis Ocellate Glider Goby
Valenciennea muralis Mural Glider Goby
Valenciennea parva Little Glider Goby
Valenciennea puellaris Orangespotted Glider Goby
Valenciennea randalli Greenband Glider Goby
Valenciennea sexguttata Sixspot Glider Goby
Valenciennea strigata Blueband Glider Goby
Valenciennea wardii Broadbarred Glider Goby
Vanderhorstia ambanoro Twinspot Shrimpgoby
Vanderhorstia lanceolata Lanceolate Shrimpgoby
Vanderhorstia longimana Longfin Shrimpgoby
Vanderhorstia mertensi Slender Shrimpgoby
Vanderhorstia ornatissima Ornate Shrimpgoby
Yoga pyrops Fire-eye Goby
Yongeichthys criniger Hairfin Goby

WORMFISHES

Microdesmidae

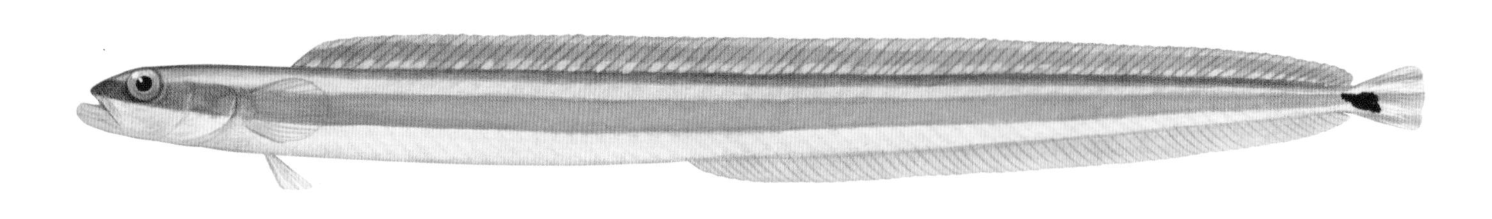

CURIOUS WORMFISH
Gunnellichthys curiosus

Wormfishes are found in tropical and subtropical waters around the world, and there are about 30 species in the family. They have a very elongate, eel-like, compressed body with small cycloid scales embedded in the skin, and a small head with an oblique mouth and a robust, projecting lower jaw. They have a single, very long dorsal fin with 10–28 spines and 28–66 rays. The anal fin has no spines and 23–61 rays, and the caudal fin is rounded.

Wormfishes are found in northern Australian waters over sand and mud bottoms, into which they burrow, in a range of shallow coastal environments from estuaries to coral reefs. They are most often seen hovering near the bottom above sandy areas adjacent to reefs, feeding on zooplankton and small benthic crustaceans. At the approach of danger they rapidly disappear into their burrows. Most Wormfishes live in small colonies and reach a maximum length of about 12 cm.

AUSTRALIAN MICRODESMIDAE SPECIES

Gunnellichthys copleyi Copley's Wormfish
Gunnellichthys curiosus Curious Wormfish
Gunnellichthys monostigma Onespot Wormfish
Gunnellichthys pleurotaenia Blacklined Wormfish
Gunnellichthys viridescens Yellowstripe Wormfish
Paragunnellichthys seychellensis Seychelles Wormfish

DARTGOBIES

Ptereleotridae

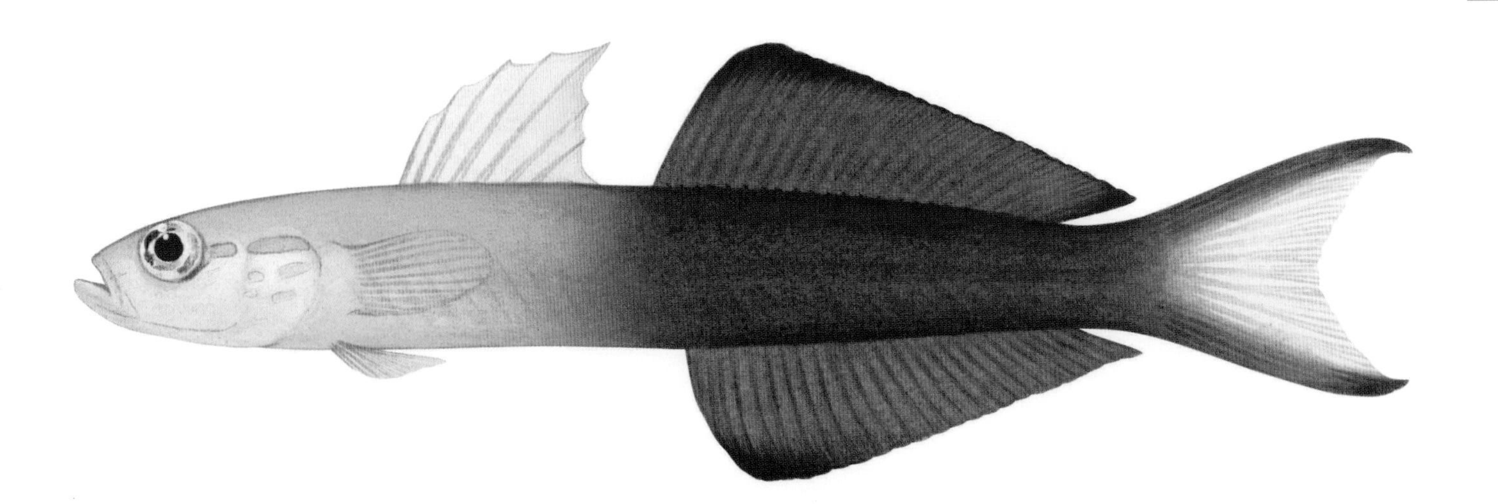

ARROW DARTGOBY
Ptereleotris evides

Dartgobies are found in tropical and subtropical waters around the world, and there are about 36 species in the family. They have elongate, rounded to compressed bodies with small, embedded scales, a small head with an oblique mouth, and the lower jaw is robust and projecting in many species. They have two dorsal fins, the first with six spines and the second with one spine and 9–37 rays. The anal fin has one spine and 9–36 rays, and the caudal fin is rounded to forked, sometimes with long, trailing filaments on the upper and lower lobes.

Dartgobies are found over sand and rubble areas near coral reefs in northern Australian waters. They typically hover above the substrate, sometimes well above the bottom and often in small groups, feeding on zooplankton. The magnificent Firegobies, which reach about 6–7 cm in length, are found in pairs or small groups in deeper waters at the base of reefs, down to about 60 m. They hover near the bottom and disappear into burrows at any hint of danger. *Ptereleotris* species grow to about 12–15 cm and usually occur in small groups as juveniles, often with several individuals retreating into the same burrow, and in pairs as adults. They tend to hover higher in the water column than other members of the family. The Dartfishes, which reach a maximum length of about 3–4 cm, are found over shallow coastal reefs and in mangrove areas, where they form schools. They do not live in burrows, preferring the shelter of mangroves or branching corals.

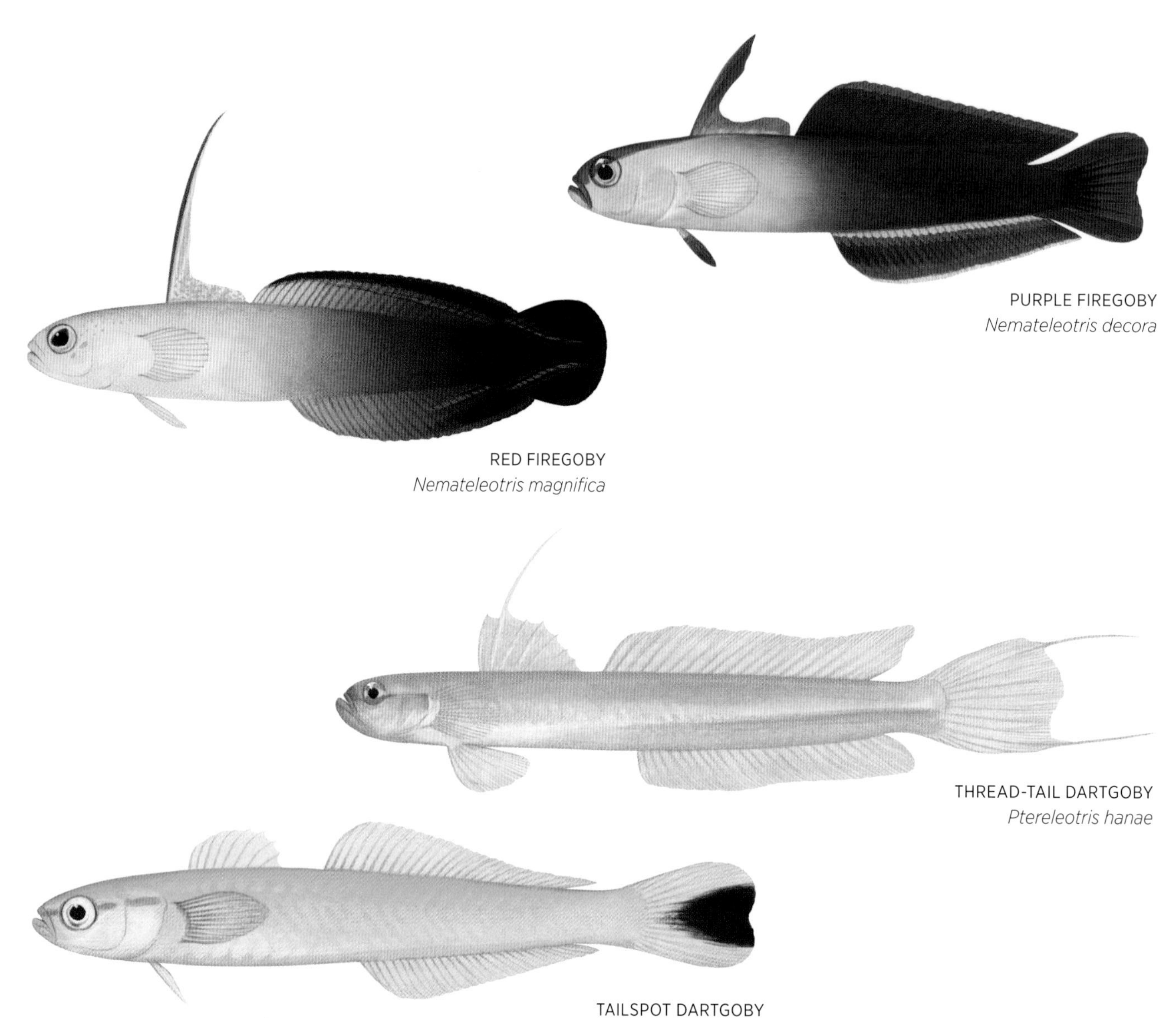

PURPLE FIREGOBY
Nemateleotris decora

RED FIREGOBY
Nemateleotris magnifica

THREAD-TAIL DARTGOBY
Ptereleotris hanae

TAILSPOT DARTGOBY
Ptereleotris heteroptera

AUSTRALIAN PTERELEOTRIDAE SPECIES

Aioliops novaeguineaensis New Guinea Mini Dartgoby
Aioliops tetrophthalmus Four-eye Mini Dartgoby
Nemateleotris decora Purple Firegoby
Nemateleotris magnifica Red Firegoby
Oxymetopon typus Sailfin Ribbongoby
Parioglossus formosus Yellowstriped Dartfish
Parioglossus marginalis Blackmargin Dartfish
Parioglossus palustris Tailspot Dartfish
Parioglossus philippinus Philippine Dartfish
Parioglossus rainfordi Rainford's Dartfish
Parioglossus raoi Rao's Dartfish
Ptereleotris evides Arrow Dartgoby
Ptereleotris grammica Lined Dartgoby
Ptereleotris hanae Thread-tail Dartgoby
Ptereleotris heteroptera Tailspot Dartgoby
Ptereleotris microlepis Greeneye Dartgoby
Ptereleotris monoptera Lyretail Dartgoby
Ptereleotris uroditaenia Flagtail Dartgoby
Ptereleotris zebra Zebra Dartgoby

INFANTFISHES

Schindleriidae

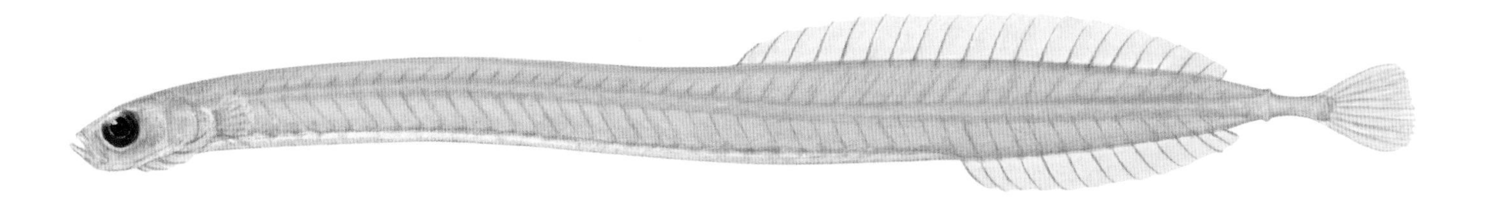

STOUT FLOATER
Schindleria brevipinguis

The three known species in this family are found in oceanic waters around the world, often near coral reefs. Infantfishes have a very small, elongate, transparent body with no scales, large eyes, and a small, oblique mouth. The single dorsal fin has no spines and 15–22 rays, the anal fin has no spines and 10–18 rays, and the ventral fins are absent. The caudal fin is slightly forked and placed on the end of a distinctive elongate bone that forms the end of the vertebral column.

Infantfishes are so named because adults display many of the characteristics of fish larvae, with which they are often confused. They have a poorly ossified skeleton, with many elements of cartilage and bone not developed at all. They reach a maximum length of about 2.2 cm and are sometimes found in large schools near coral reefs. The **Stout Floater**, found in waters off the Great Barrier Reef, holds the record for the world's smallest fish: males are sexually mature at only 6.5 mm, females at 7–8 mm, and the largest specimen recorded was 8.5 mm in length. Very little is known of the biology of Infantfishes, and more species have recently been discovered but are as yet undescribed.

AUSTRALIAN SCHINDLERIIDAE SPECIES

Schindleria brevipinguis Stout Floater
Schindleria pietschmanni Pietschmann's Floater
Schindleria praematura Premature Floater

NURSERYFISHES

Kurtidae

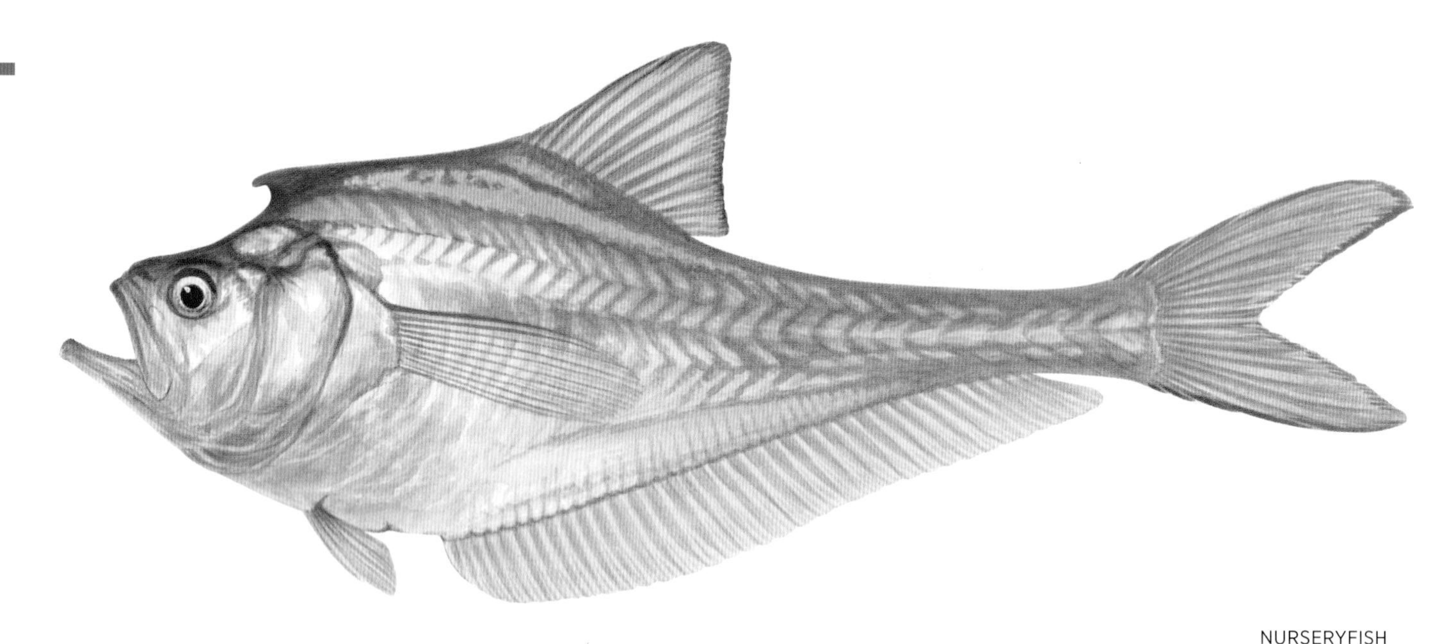

NURSERYFISH
Kurtus gulliveri

Nurseryfishes are found in coastal marine, brackish and fresh waters of South-East Asia and northern Australia, and there are two known species. They have a deep, compressed, tapering body with a distinctive humped back, small cycloid scales and a large, oblique mouth with villiform teeth. The single tall dorsal fin has 5–7 very small spines, two larger spines and 12–14 rays; the long-based anal fin has three spines and 31–47 rays; and the caudal fin is deeply forked.

Male Nurseryfishes develop a hook-shaped projection on the forehead that is used for carrying eggs. The females lay a mass of eggs held together by a twisted cord of membranes, and the males carry them with the hook, like two bunches of grapes, one on each side of the head. It is thought that the strange method of carrying the eggs is an adaptation to ensure the embryos have constant access to well-oxygenated water.

The single Australian species occurs in small schools, mainly in estuaries and the lower reaches of slow-moving, muddy northern rivers. It feeds on small fishes and crustaceans, and grows to a maximum length of about 60 cm. It is reported to be an excellent food fish and occasionally appears in the nets of coastal prawn trawlers.

AUSTRALIAN KURTIDAE SPECIES
Kurtus gulliveri Nurseryfish

BATFISHES

Ephippidae

ROUNDFACE BATFISH (adult and juvenile)
Platax teira

Ephippidae

Batfishes are found in tropical and temperate waters of the Atlantic, Indian and Pacific oceans, and there are about 16 species in the family. They have deep, oval to circular bodies with small ctenoid scales that extend onto the base of the fins, and a small mouth with bands of brush-like teeth. They have a single, sometimes deeply notched dorsal fin with 5–9 spines and 19–38 rays. The anal fin has three spines and 15–27 rays, and the caudal fin is truncate to wedge-shaped.

Batfishes are found in northern Australia, in coastal waters, estuaries and over coral reefs. Juveniles have greatly elongated dorsal and anal fins, and are often brightly coloured. Some juveniles imitate floating leaves in very shallow, inshore tropical waters, moving slowly and perfectly mimicking the colouring of a large dead leaf. Others mimic colourful, toxic flatworms. Adults of the various species are very similar to each other, being overall silvery with dark bars on the head, and are generally found in deeper waters, over dropoffs and outer reef slopes. The **Shortfin Batfish**, however, prefers inshore waters over soft bottoms. The **Threadfin Scat** is found in coastal waters and estuaries; it has a distinctive angular dorsal profile, and elongate dorsal and ventral fin spines. The **Roundface Batfish** is one of the largest members of the family, reaching about 60 cm in length. It can become quite accustomed to divers and is often easily approached. Batfishes feed mainly on benthic invertebrates such as crustaceans, as well as zooplankton and small fishes.

ROUND BATFISH (adult and juvenile)
Platax orbicularis

LONGFIN BATFISH (adult and juvenile)
Platax pinnatus

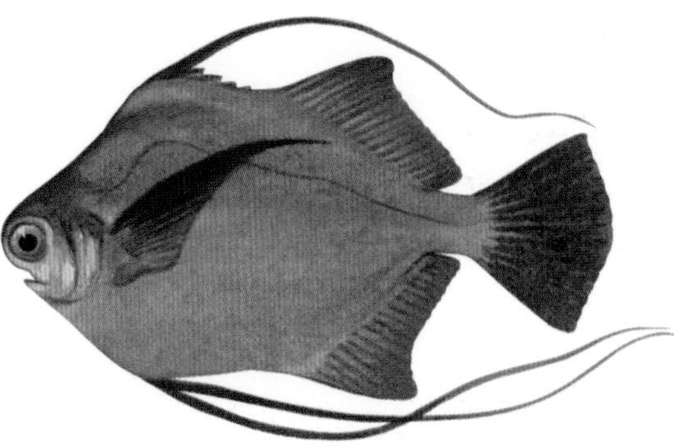

THREADFIN SCAT
Rhinoprenes pentanemus

SHORTFIN BATFISH
Zabidius novemaculeatus

AUSTRALIAN EPHIPPIDAE SPECIES

Platax batavianus Humphead Batfish
Platax orbicularis Round Batfish
Platax pinnatus Longfin Batfish
Platax teira Roundface Batfish
Rhinoprenes pentanemus Threadfin Scat
Zabidius novemaculeatus Shortfin Batfish

SCATS

Scatophagidae

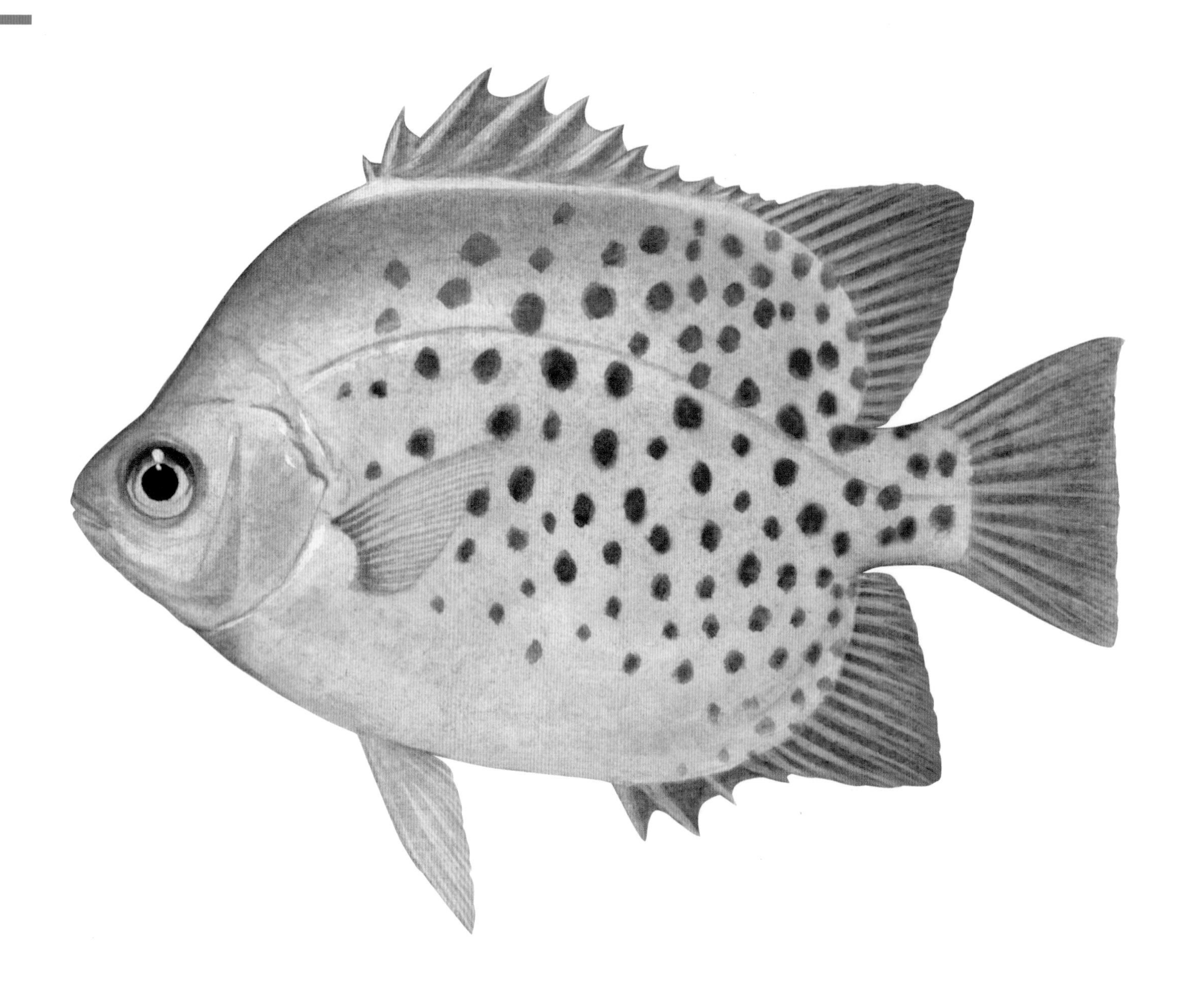

SPOTTED SCAT
Scatophagus argus

Scats are found in coastal waters of the Indo-West Pacific, and the family contains about four species. They have a deep, compressed body with very small ctenoid scales that extend onto the fins, a small head with a steep forehead, and a small mouth with bands of minute, movable, tricuspid teeth. The single, deeply notched dorsal fin has 11–12 spines and 15–18 rays, the anal fin has four spines and 14–17 rays, and the caudal fin is truncate.

Scats occur in northern Australia, in sheltered coastal waters such as harbours and mangroves, and sometimes over coral reefs. Juveniles are found in estuaries and the lower reaches of northern rivers. Scats feed on organic detritus, small invertebrates and algae, and reach a maximum length of about 33 cm. They are popular aquarium fishes, though the Australian species have venomous fin spines and care should be taken when handling them.

STRIPED SCAT
Selenotoca multifasciata

AUSTRALIAN SCATOPHAGIDAE SPECIES

Scatophagus argus Spotted Scat

Selenotoca multifasciata Striped Scat

RABBITFISHES

Siganidae

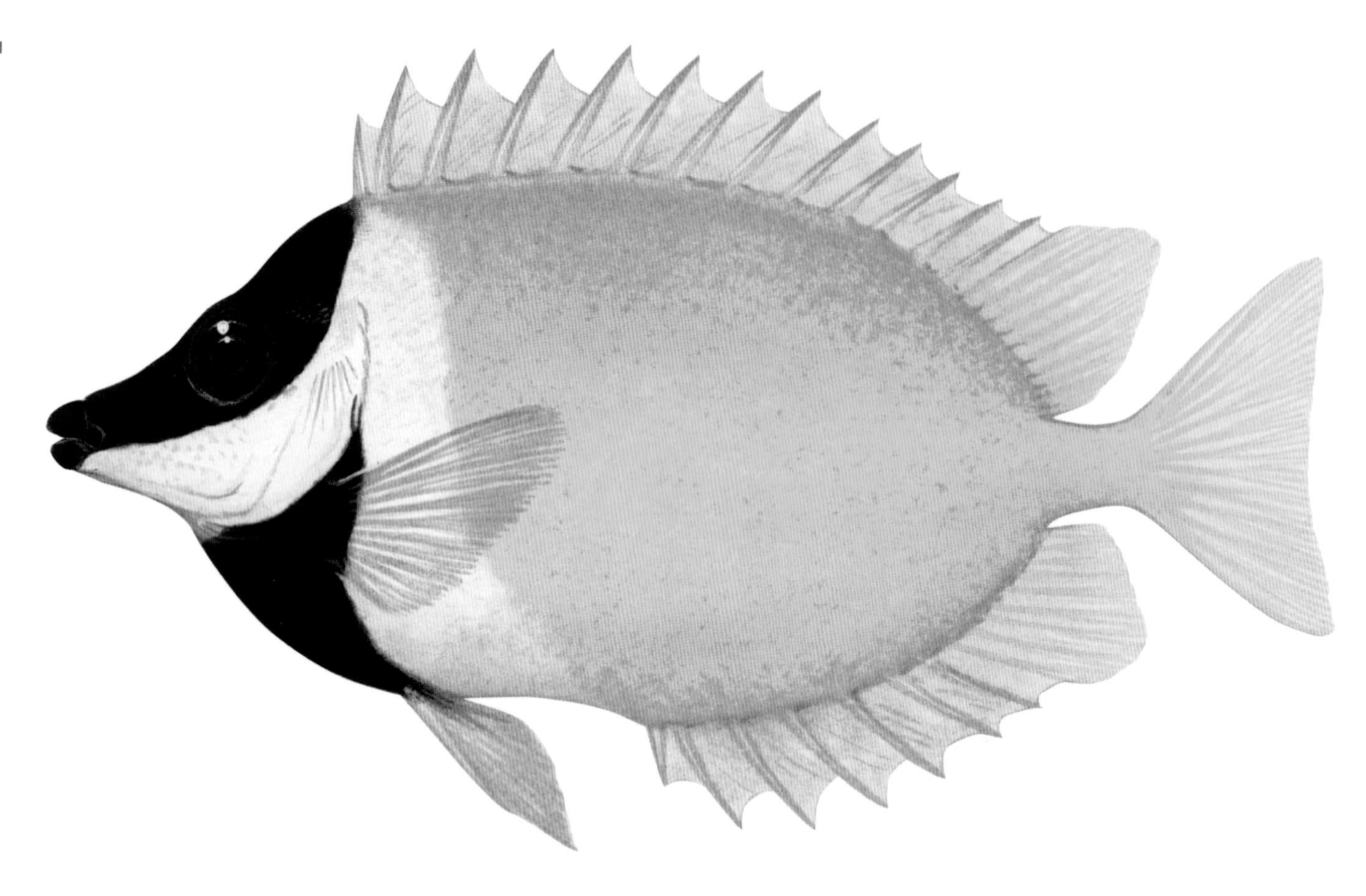

FOXFACE
Siganus vulpinus

Rabbitfishes are found in tropical waters of the eastern Mediterranean and the Indo-West Pacific. There are about 27 species in the family. They have an oval, compressed body with minute scales, a small mouth with large fleshy lips, and single rows of bicuspid teeth in the jaws. The single dorsal fin has 13 strong spines, the first embedded and forward-pointing, and 10 rays. The anal fin has seven spines and nine rays; the ventral fins have one spine, then three rays and another spine; and the caudal fin is slightly forked. The fin spines are venomous and can inflict very painful stings.

Rabbitfishes occur in northern Australian coastal waters and occasionally in estuaries. Some colourful species are found over coral reefs and these often occur in pairs. Other species form small to large schools and are found over shallow rocky reefs and weed beds, sometimes in great abundance. Rabbitfishes are primarily herbivorous, feeding on filamentous algae, other seaweeds and seagrass, but opportunistically taking small invertebrates. Some species feed on sponges and sessile colonial tunicates. Rabbitfishes reach a maximum length of about 40 cm and are utilised as food fishes, though they often have a somewhat weedy taste.

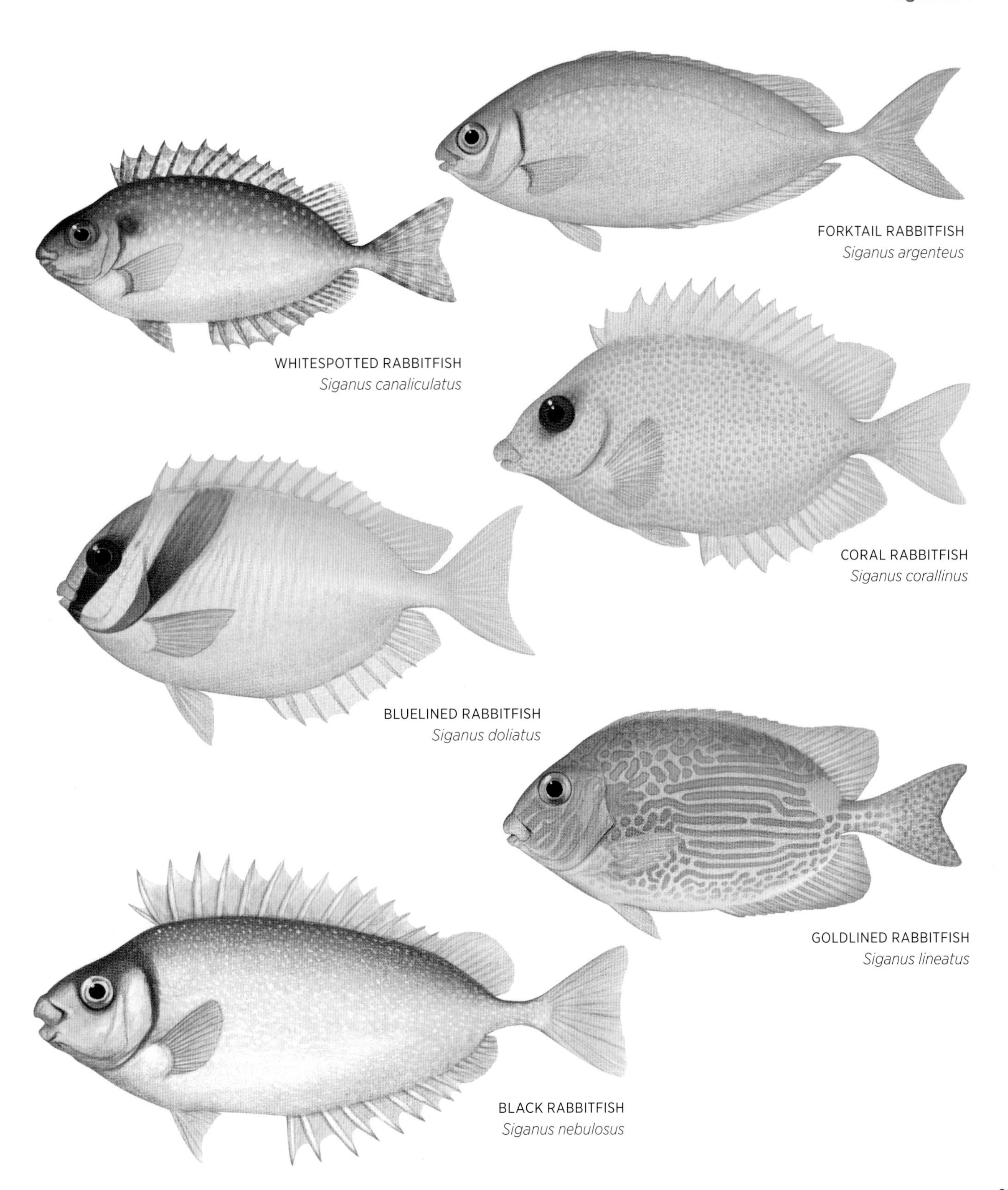

FORKTAIL RABBITFISH
Siganus argenteus

WHITESPOTTED RABBITFISH
Siganus canaliculatus

CORAL RABBITFISH
Siganus corallinus

BLUELINED RABBITFISH
Siganus doliatus

GOLDLINED RABBITFISH
Siganus lineatus

BLACK RABBITFISH
Siganus nebulosus

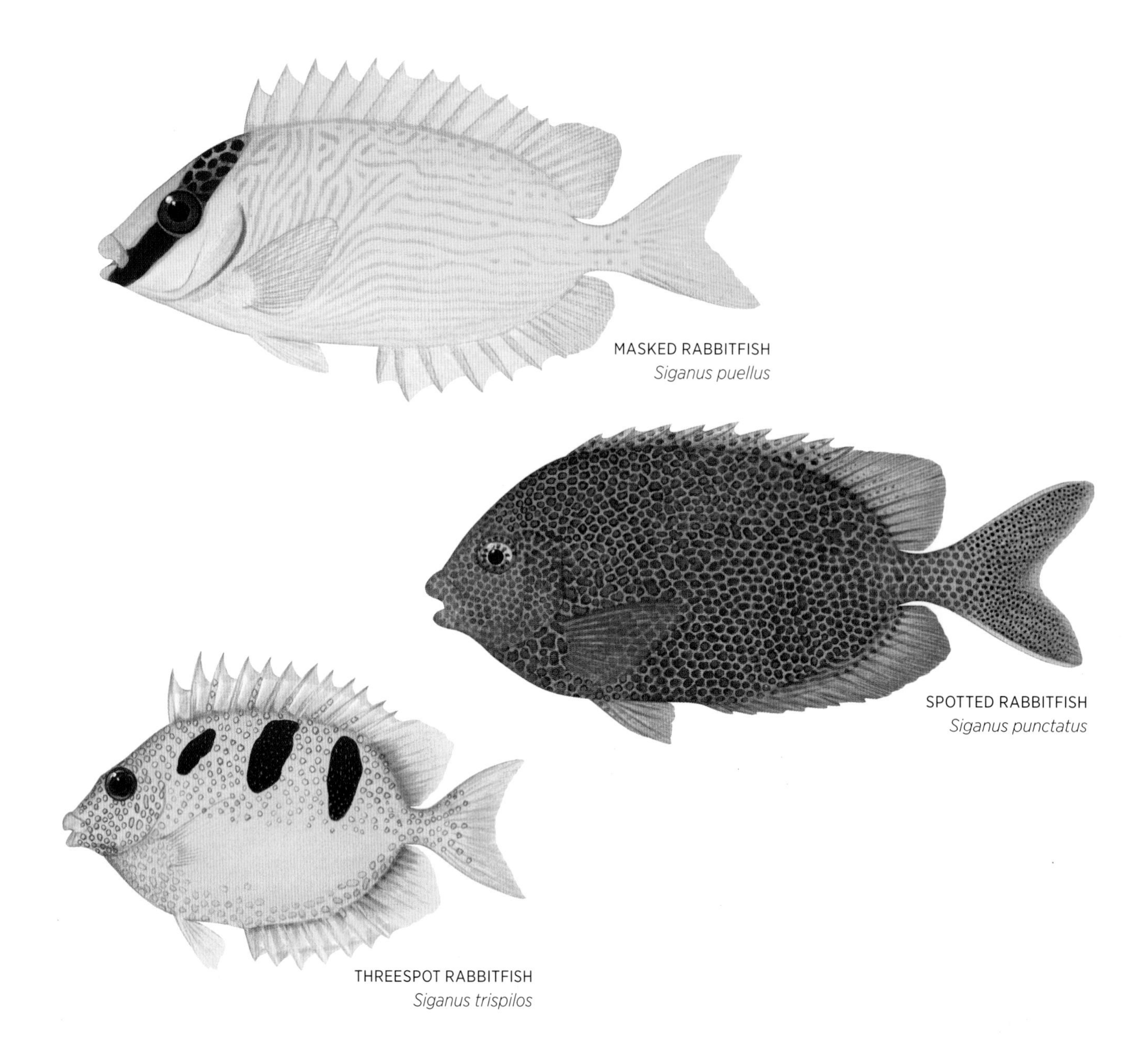

MASKED RABBITFISH
Siganus puellus

SPOTTED RABBITFISH
Siganus punctatus

THREESPOT RABBITFISH
Siganus trispilos

AUSTRALIAN SIGANIDAE SPECIES

Siganus argenteus Forktail Rabbitfish
Siganus canaliculatus Whitespotted Rabbitfish
Siganus corallinus Coral Rabbitfish
Siganus doliatus Bluelined Rabbitfish
Siganus javus Java Rabbitfish
Siganus lineatus Goldlined Rabbitfish
Siganus nebulosus Black Rabbitfish
Siganus puellus Masked Rabbitfish
Siganus punctatissimus Finespotted Rabbitfish
Siganus punctatus Spotted Rabbitfish
Siganus spinus Scribbled Rabbitfish
Siganus trispilos Threespot Rabbitfish
Siganus unimaculatus Blackblotch Foxface
Siganus vermiculatus Maze Rabbitfish
Siganus virgatus Doublebar Rabbitfish
Siganus vulpinus Foxface

LOUVAR

Luvaridae

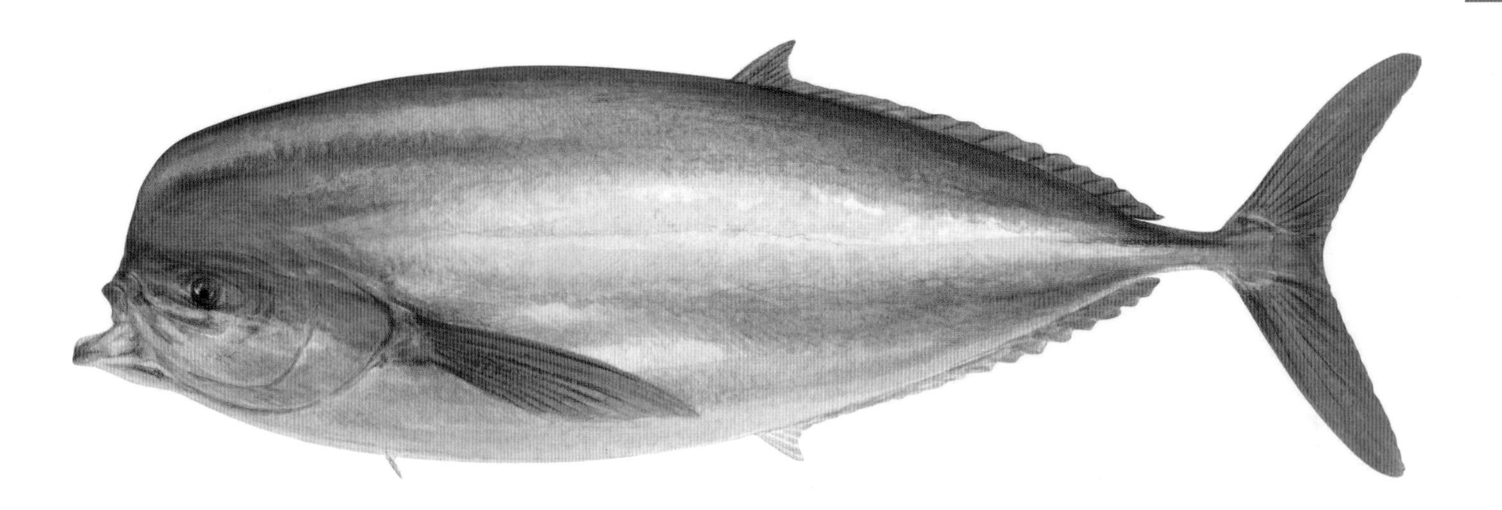

LOUVAR
Luvarus imperialis

The single species in this family is found worldwide in oceanic waters. The **Louvar** has an oval, slightly compressed body with minute granular scales, a rounded head profile with a small mouth, and a fleshy keel on the caudal peduncle. The single dorsal fin has two spines and 20–22 rays, though the anterior 9–10 rays are lost by maturity. The anal fin has no spines and 18 rays (the first four are lost with growth), the ventral fins are rudimentary or absent, and the caudal fin is lunate.

The Louvar is a large, solitary, pelagic fish that roams the vast expanses of the open ocean. It can reach over 2 m in length. The species produces perhaps the highest number of eggs of all fishes – a large female was found to contain an estimated 47 million eggs. This profusion of eggs is perhaps to compensate for the rarity of mating encounters. The Louvar feeds on jellyfishes, salps, ctenophores and other planktonic organisms, and is only very occasionally encountered in temperate waters; most records come from strandings after storms.

AUSTRALIAN LUVARIDAE SPECIES
Luvarus imperialis Louvar

MOORISH IDOL

Zanclidae

MOORISH IDOL
Zanclus cornutus

The **Moorish Idol** is the only species in this family and it is found in tropical waters in the Indo-Pacific region. It has a very deep, compressed body with minute scales, a projecting snout with a small mouth, and long, bristle-like teeth. There is a small forward-pointing bony projection above each eye in adult fish. The single dorsal fin has 6–7 spines, the third spine elongated into a long trailing filament, and 39–42 rays. The anal fin has three spines and 31–37 rays, and the caudal fin is truncate.

The Moorish Idol is quite common in northern Australian waters and is usually found in small groups over coral and rocky reefs, from shallow water down to about 180 m. It reaches a maximum length of about 25 cm and feeds on a variety of benthic organisms, including sponges, algae and invertebrates.

AUSTRALIAN ZANCLIDAE SPECIES

Zanclus cornutus Moorish Idol

SURGEONFISHES

Acanthuridae

VELVET SURGEONFISH
Acanthurus nigricans

There are about 80 species of Surgeonfish, and they are found in tropical and subtropical waters around the world. They have oval to elongate, compressed bodies with minute scales and thick skin. Their small mouth has thick lips, and either flattened teeth with serrated edges, or numerous comb-like teeth. The single long dorsal fin has 4–9 spines and 19–31 rays, the anal fin has 2–3 spines and 19–36 rays, and the caudal fin is truncate to slightly forked.

Surgeonfishes are named for the razor-sharp spines they possess on the caudal peduncle, which are used for defence. These spines are often surrounded by an area of skin in a bright, contrasting colour to advertise their presence. Surgeonfishes can be divided into two subfamilies:

ACANTHURINAE Members of this group usually have a single scalpel-like spine on the caudal peduncle, which folds forward into a groove, and three anal-fin spines. (The spine on the caudal peduncle is remarkably sharp, and the lightest touch can cause a serious gash.) The Sawtails, which are sometimes placed in a separate subfamily, have 2–6 spines on bony plates on the caudal peduncle. Acanthurinae species are found in northern Australian waters, usually over rocky or coral reefs. A few species feed on zooplankton but most are herbivorous, grazing on algae. They are found individually, in small groups, or in large schools and are common in most of their range.

NASINAE Unicornfishes have 1–2 fixed, keel-like spines on bony plates on the caudal peduncle, and two anal-fin spines.

Many species develop a forward-projecting hump or long, finger-like protuberance on the forehead, which gives them their common name. They are found individually or in schools, over coral reefs, outer reef slopes and dropoffs, and are generally shy and difficult to approach. The **Ringtail Unicornfish** is one of the largest members of the group, reaching up to 1 m in length. The spectacular **Bignose Unicornfish** has sail-like fins and long, trailing filaments on the caudal fin. Unicornfishes feed on either algae or zooplankton, depending on the species.

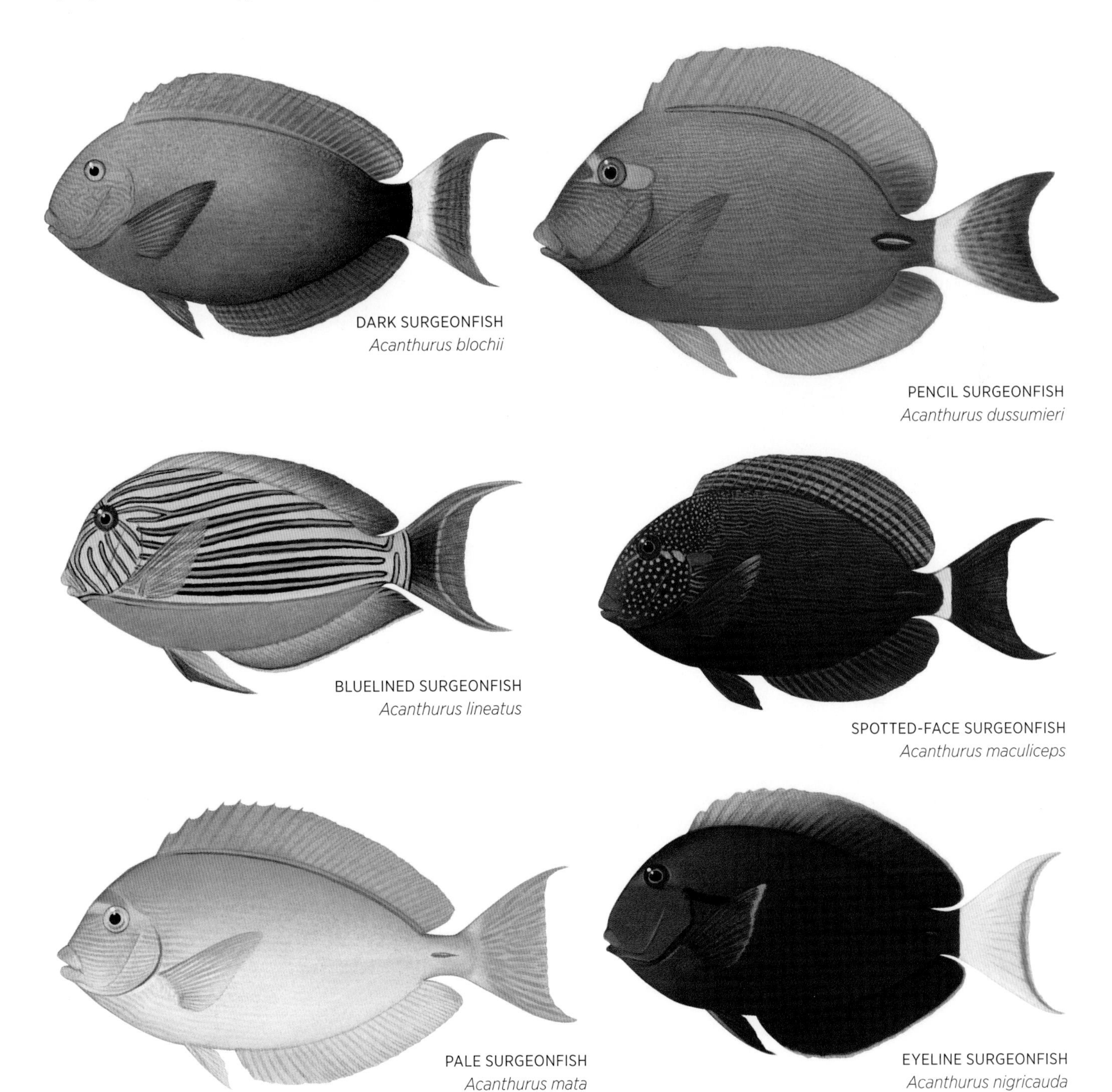

DARK SURGEONFISH
Acanthurus blochii

PENCIL SURGEONFISH
Acanthurus dussumieri

BLUELINED SURGEONFISH
Acanthurus lineatus

SPOTTED-FACE SURGEONFISH
Acanthurus maculiceps

PALE SURGEONFISH
Acanthurus mata

EYELINE SURGEONFISH
Acanthurus nigricauda

DUSKY SURGEONFISH
Acanthurus nigrofuscus

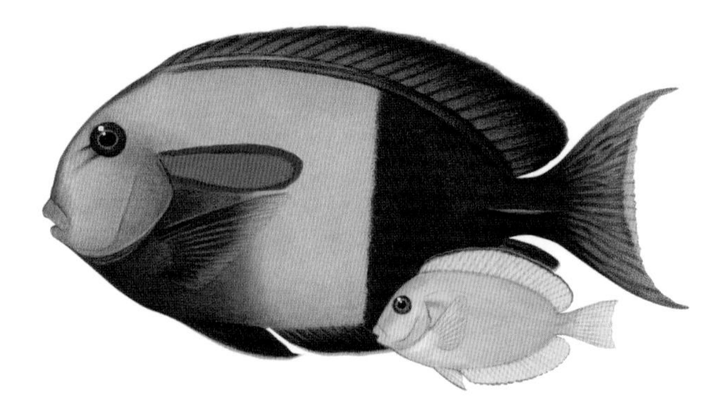

ORANGEBLOTCH SURGEONFISH (adult and juvenile)
Acanthurus olivaceus

MIMIC SURGEONFISH (adult and juvenile)
Acanthurus pyroferus

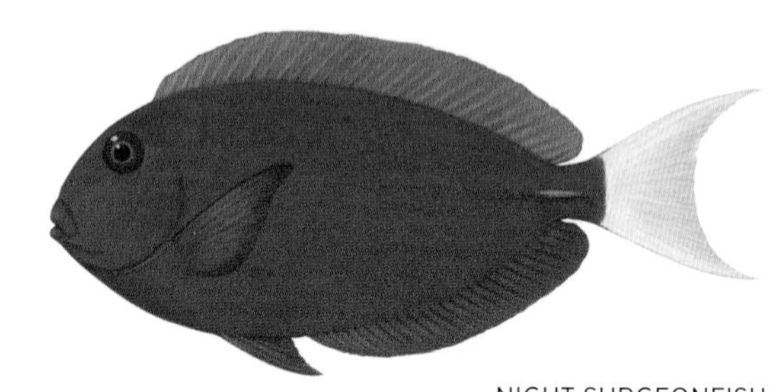

NIGHT SURGEONFISH
Acanthurus thompsoni

CONVICT SURGEONFISH
Acanthurus triostegus

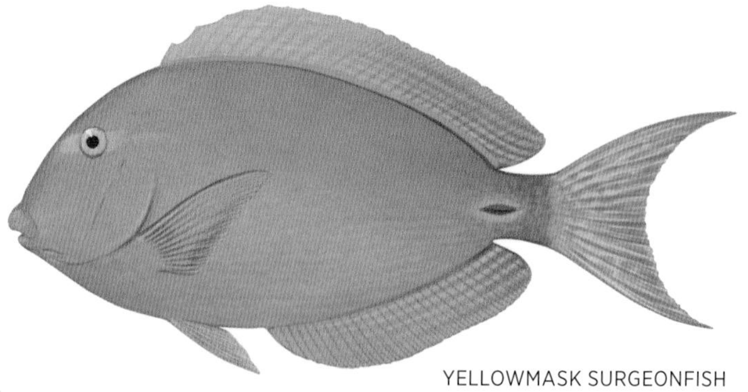

YELLOWMASK SURGEONFISH
Acanthurus xanthopterus

TWOSPOT BRISTLETOOTH
Ctenochaetus binotatus

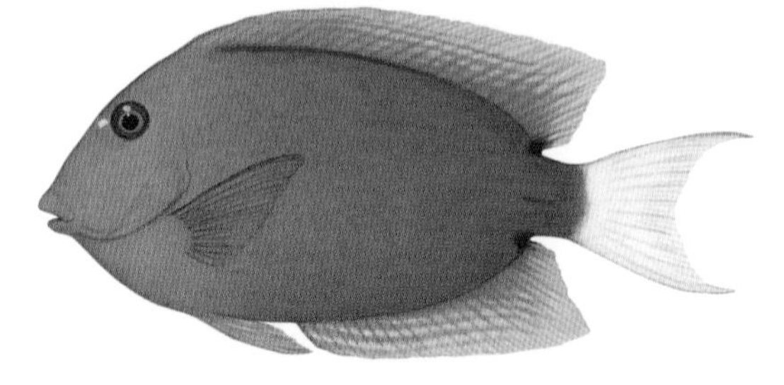

ORANGE-TIP BRISTLETOOTH
Ctenochaetus tominiensis

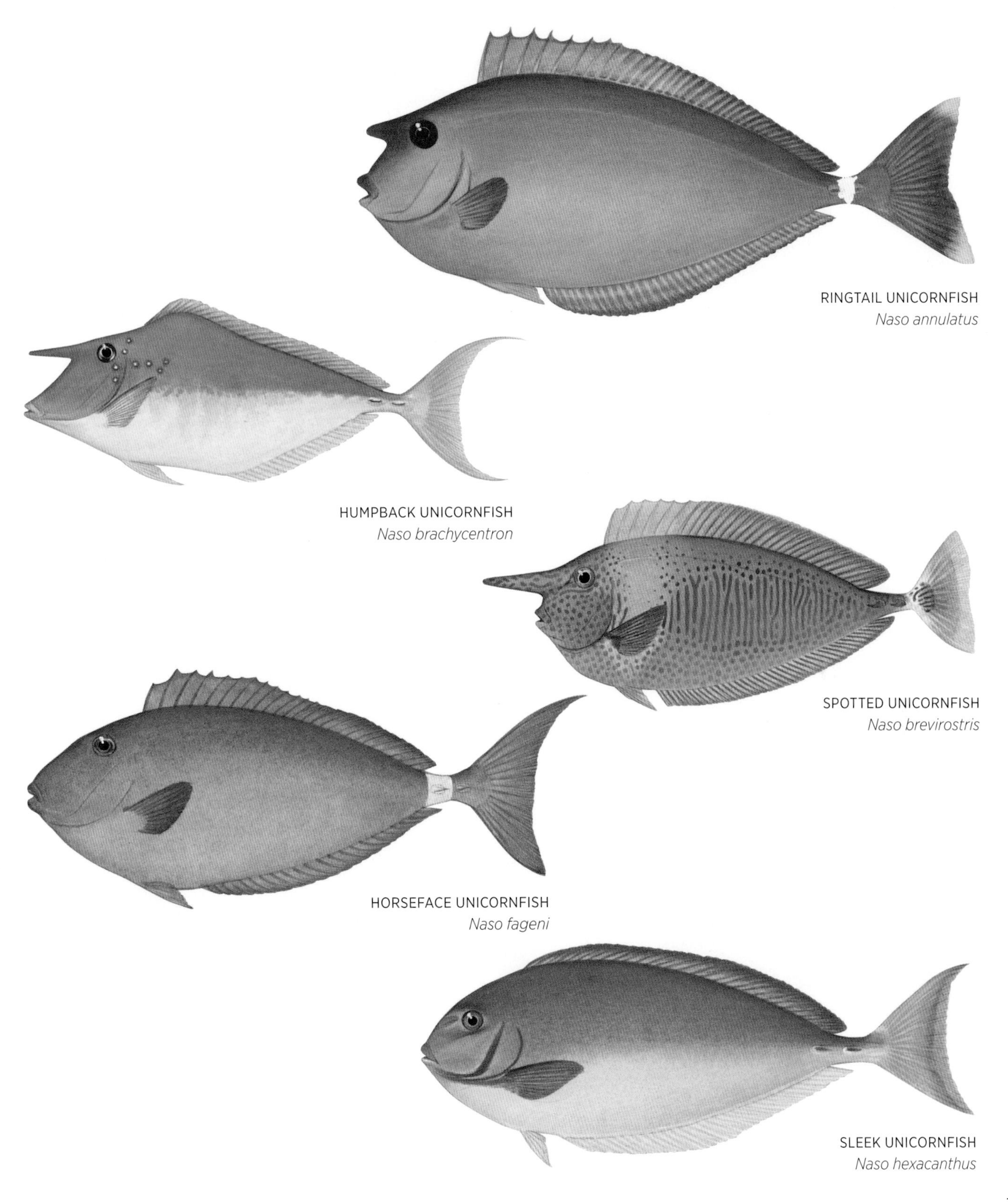

RINGTAIL UNICORNFISH
Naso annulatus

HUMPBACK UNICORNFISH
Naso brachycentron

SPOTTED UNICORNFISH
Naso brevirostris

HORSEFACE UNICORNFISH
Naso fageni

SLEEK UNICORNFISH
Naso hexacanthus

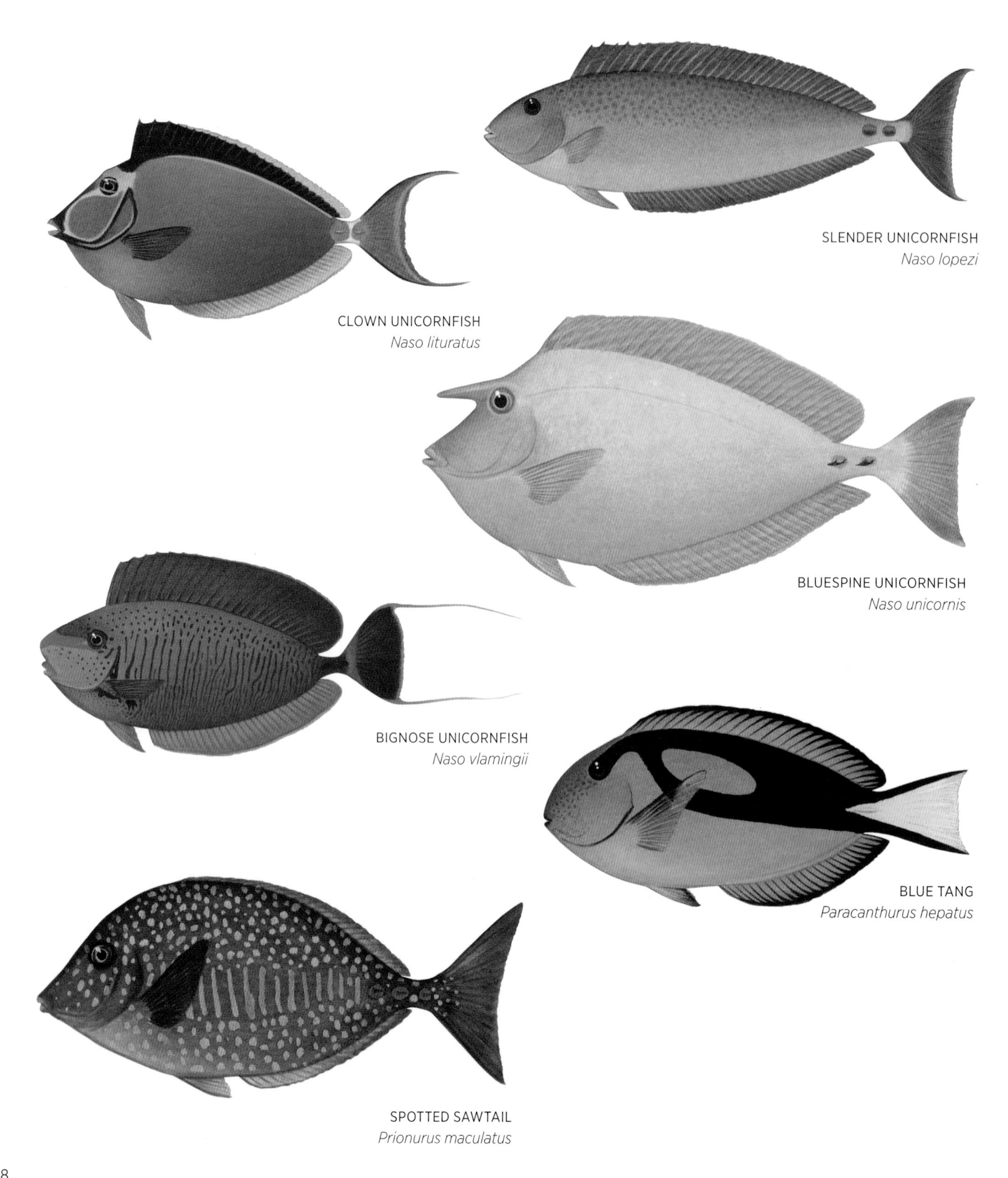

SLENDER UNICORNFISH
Naso lopezi

CLOWN UNICORNFISH
Naso lituratus

BLUESPINE UNICORNFISH
Naso unicornis

BIGNOSE UNICORNFISH
Naso vlamingii

BLUE TANG
Paracanthurus hepatus

SPOTTED SAWTAIL
Prionurus maculatus

BROWN TANG
Zebrasoma scopas

SAILFIN TANG (adult and juvenile)
Zebrasoma veliferum

AUSTRALIAN ACANTHURIDAE SPECIES

Acanthurus albipectoralis Whitefin Surgeonfish
Acanthurus auranticavus Ringtail Surgeonfish
Acanthurus bariene Eyespot Surgeonfish
Acanthurus blochii Dark Surgeonfish
Acanthurus dussumieri Pencil Surgeonfish
Acanthurus fowleri Horseshoe Surgeonfish
Acanthurus grammoptilus Inshore Surgeonfish
Acanthurus guttatus Whitespotted Surgeonfish
Acanthurus lineatus Bluelined Surgeonfish
Acanthurus maculiceps Spotted-face Surgeonfish
Acanthurus mata Pale Surgeonfish
Acanthurus nigricans Velvet Surgeonfish
Acanthurus nigricauda Eyeline Surgeonfish
Acanthurus nigrofuscus Dusky Surgeonfish
Acanthurus nigroris Greyhead Surgeonfish
Acanthurus olivaceus Orangeblotch Surgeonfish
Acanthurus pyroferus Mimic Surgeonfish
Acanthurus thompsoni Night Surgeonfish
Acanthurus triostegus Convict Surgeonfish
Acanthurus xanthopterus Yellowmask Surgeonfish
Ctenochaetus binotatus Twospot Bristletooth
Ctenochaetus cyanocheilus Yelloweye Bristletooth
Ctenochaetus striatus Lined Bristletooth
Ctenochaetus tominiensis Orange-tip Bristletooth
Naso annulatus Ringtail Unicornfish
Naso brachycentron Humpback Unicornfish
Naso brevirostris Spotted Unicornfish
Naso caeruleacauda Blue Unicornfish
Naso caesius Silverblotched Unicornfish
Naso fageni Horseface Unicornfish
Naso hexacanthus Sleek Unicornfish
Naso lituratus Clown Unicornfish
Naso lopezi Slender Unicornfish
Naso maculatus Scribbled Unicornfish
Naso mcdadei Squarenose Unicornfish
Naso minor Blackspine Unicornfish
Naso thorpei Thorpe's Unicornfish
Naso thynnoides Onespine Unicornfish
Naso tonganus Humpnose Unicornfish
Naso unicornis Bluespine Unicornfish
Naso vlamingii Bignose Unicornfish
Paracanthurus hepatus Blue Tang
Prionurus maculatus Spotted Sawtail
Prionurus microlepidotus Australian Sawtail
Zebrasoma scopas Brown Tang
Zebrasoma veliferum Sailfin Tang

LONGFIN ESCOLAR

Scombrolabracidae

LONGFIN ESCOLAR
Scombrolabrax heterolepis

The single species in this family is found in deep waters of the Atlantic, Indian and Pacific oceans. It has an elongate, compressed body with irregular, fragile scales and large eyes. The large, slightly protrusible mouth has long, fang-like teeth in the anterior of the jaws. There are two dorsal fins, the first long-based with 12 spines and the second short-based with 15–16 rays. The anal fin has 2–3 spines and 16–18 rays, the pectoral fins are very long, and the caudal fin is forked. The **Longfin Escolar** reaches about 30 cm in length and is pelagic in deep water from about 150 to 900 m. It is only occasionally encountered in deepwater trawls and nothing is known of its biology, though its dentition suggests it is a predator of other fishes.

AUSTRALIAN SCOMBROLABRACIDAE SPECIES
Scombrolabrax heterolepis Longfin Escolar

BARRACUDAS

Sphyraenidae

GREAT BARRACUDA
Sphyraena barracuda

About 21 species of Barracuda are found in tropical and temperate waters around the world. They have an elongate, rounded body with small cycloid scales, a large mouth with sharp teeth, and fang-like canines in the anterior of the jaws. They have two widely separated dorsal fins, the first with four strong spines and the second with one spine and nine rays. The anal fin has two spines and 7–9 rays, and the caudal fin is slightly to deeply forked.

Barracudas are fierce predators of fishes and squid. There are reports of attacks on humans, but these are unsubstantiated and could simply be conjecture based on the fishes' fearsome dentition. Barracudas are found around Australia, juveniles often occurring in estuaries and adults in coastal waters over reefs and well offshore. The **Snook**, which reaches about 1 m in length, is found in southern Australian waters over seagrass beds and rocky reefs. Most of the Australian species are found in northern waters, often in large schools. Some species, such as the **Pickhandle Barracuda**, **Military Barracuda** and **Heller's Barracuda**, are largely nocturnal. They hover or circle slowly over reef pinnacles during the day (they can often be quite closely approached by divers in these periods of inactivity) then disperse to hunt over the reef at night. Other species, such as the **Yellowtail Barracuda**, occur in small, fast-moving schools near reefs and hunt during the day. Juvenile **Great Barracudas** are found in small schools, while large adults, which can reach up to 2 m in length, are almost always solitary and occur over outer-reef slopes and bommies. The Great Barracuda develops large black blotches on the flanks as it matures.

Barracudas are often implicated in cases of ciguatera poisoning. Ciguatera is caused by ciguatoxin, which is derived from certain species of algae. The toxin gradually builds up in the flesh of large, long-lived fishes that prey on algae-eating fishes. Large Barracudas should therefore not be eaten. The various species of Barracuda are difficult to tell apart , differing mainly in the degree of development of the gill-rakers.

PICKHANDLE BARRACUDA
Sphyraena jello

SNOOK
Sphyraena novaehollandiae

STRIPED BARRACUDA
Sphyraena obtusata

MILITARY BARRACUDA
Sphyraena putnamae

BLACKFIN BARRACUDA
Sphyraena qenie

AUSTRALIAN SPHYRAENIDAE SPECIES

Sphyraena acutipinnis Sharpfin Barracuda
Sphyraena barracuda Great Barracuda
Sphyraena flavicauda Yellowtail Barracuda
Sphyraena forsteri Blackspot Barracuda
Sphyraena helleri Heller's Barracuda
Sphyraena jello Pickhandle Barracuda
Sphyraena novaehollandiae Snook
Sphyraena obtusata Striped Barracuda
Sphyraena putnamae Military Barracuda
Sphyraena qenie Blackfin Barracuda

SNAKE MACKERELS

Gempylidae

ESCOLAR
Lepidocybium flavobrunneum

There are about 24 species of Snake Mackerel, and they are found in tropical and temperate waters around the world. They have elongate, rounded to compressed bodies with minute, cycloid scales. The large mouth has a protruding lower jaw, rows of sharp teeth, and large canines in the anterior of the jaws. There are two dorsal fins, the first long-based with many long spines and the second short-based and followed by several detached finlets. The anal fin has 1–3 spines and 8–35 rays, and is also followed by several detached finlets, and the caudal fin is forked. The **Escolar** has fleshy keels on the caudal peduncle, but these are not present in other members of the family.

Snake Mackerels are pelagic from the surface to near the bottom at depths of 1000 m or more. Many make vertical migrations towards the surface at night to feed. They are voracious predators of other fishes, pelagic crustaceans and squid. Some species of Snake Mackerel occur in large schools and several of these, including the Gemfishes and **Barracouta**, are important commercially. Others, such as the more solitary Escolar and **Oilfish**, are sometimes taken by longline fisheries. Snake Mackerels are generally excellent food fishes, though the Oilfish, which can reach up to 3 m in length, has flesh containing an oil that can cause diarrhoea if consumed. Most Snake Mackerels grow to about 1 m in length. However, a few species, such as the **Sackfish**, **Small Gemfish** and **Black Snake Mackerel**, which are occasionally found in deepwater trawls, reach a maximum length of only about 30 cm.

Gempylidae

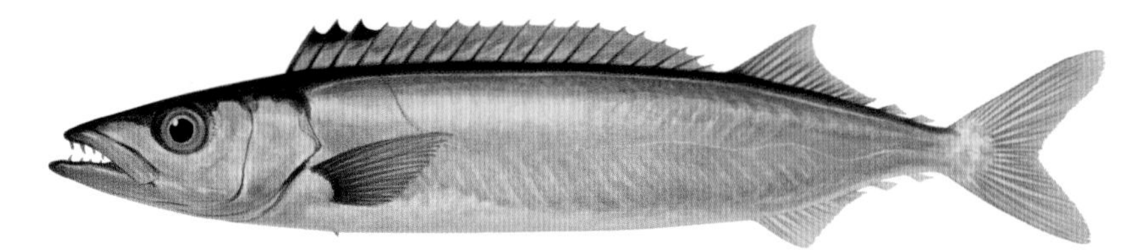

GEMFISH
Rexea solandri

OILFISH
Ruvettus pretiosus

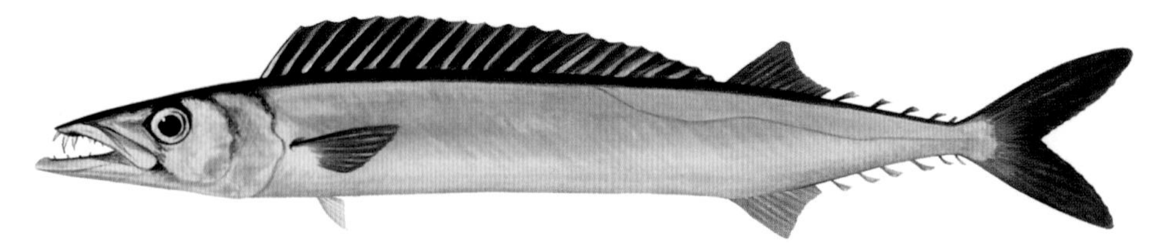

BARRACOUTA
Thyrsites atun

AUSTRALIAN GEMPYLIDAE SPECIES

Diplospinus multistriatus Striped Escolar
Gempylus serpens Snake Mackerel
Lepidocybium flavobrunneum Escolar
Nealotus tripes Black Snake Mackerel
Neoepinnula orientalis Sackfish
Nesiarchus nasutus Black Gemfish
Paradiplospinus gracilis Slender Escolar
Promethichthys prometheus Singleline Gemfish
Rexea antefurcata Longfin Gemfish
Rexea bengalensis Small Gemfish
Rexea prometheoides Royal Gemfish
Rexea solandri Gemfish
Rexichthys johnpaxtoni Paxton's Gemfish
Ruvettus pretiosus Oilfish
Thyrsites atun Barracouta
Thyrsitoides marleyi Black Snoek
Tongaichthys robustus Tonga Escolar

CUTLASSFISHES AND HAIRTAILS

Trichiuridae

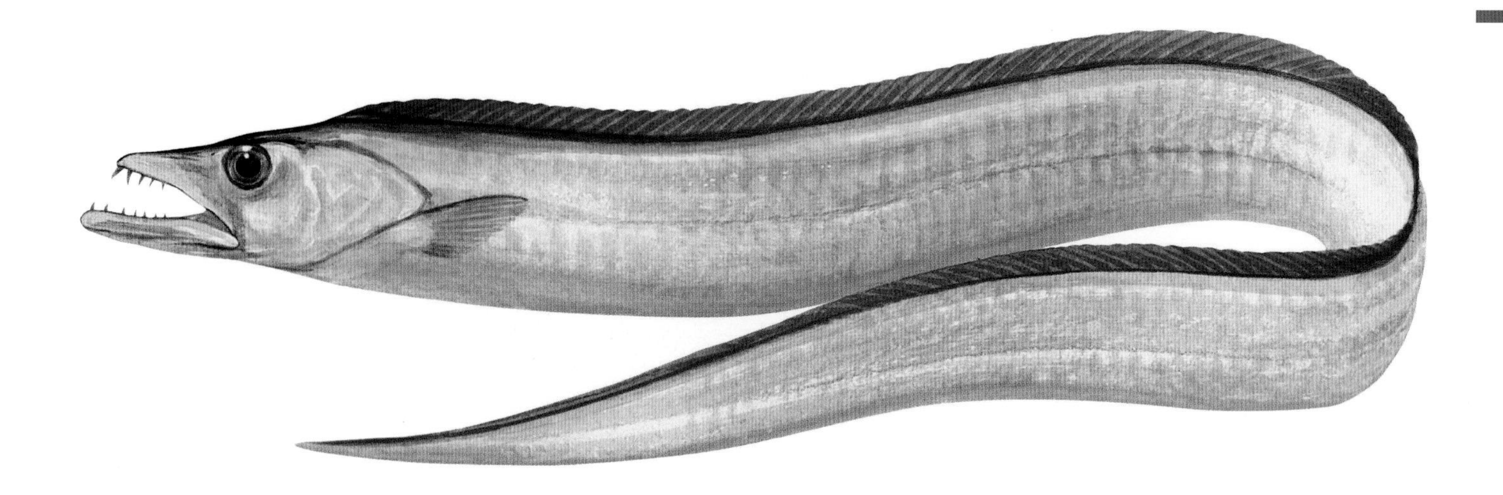

LARGEHEAD HAIRTAIL
Trichiurus lepturus

Cutlassfishes and Hairtails are found in tropical and temperate waters around the world. There are about 39 species in the family. They have a very elongate to ribbon-like, highly compressed, silvery body with no scales. The jaws are large and pointed, with a protruding lower jaw, and there are large canine teeth and fangs in the anterior of both jaws. They have a single, very long dorsal fin, with the anterior, spinous part shorter than the rayed part, and sometimes a slight notch between the two parts. The anal fin usually has two spines and 56–121 rays, but in Hairtails the anal fin is buried in the skin. The ventral fins are reduced or absent and the caudal fin is either forked or tapers to a thread-like filament, giving the Hairtails their common name.

Cutlassfishes and Hairtails are voracious predators of other fishes, crustaceans and squid. They are generally found near the bottom in deep water, down to about 2000 m, but are occasionally encountered at the surface. The **Largehead Hairtail**, which migrates towards the surface to feed during the day and descends to the bottom at night, is sometimes taken by anglers in inshore waters and estuaries. It is caught commercially in many areas of the world and is an excellent food fish. The **Frostfish** can reach 2 m in length and is an important commercial species in some areas of the world, but is taken only incidentally in Australia, where it occurs in southern waters. Most other species grow to between 20 cm and 1 m in length. They are occasionally encountered in deepwater trawls in northern Australian and South-East Asian waters, but are too small and scarce to be important commercially.

FROSTFISH
Lepidopus caudatus

AUSTRALIAN TRICHIURIDAE SPECIES

Aphanopus beckeri Becker's Scabbardfish
Aphanopus capricornis Capricorn Scabbardfish
Aphanopus mikhailini Mikhailin's Scabbardfish
Assurger anzac Razorback Scabbardfish
Benthodesmus elongatus Slender Frostfish
Benthodesmus macrophthalmus Bigeye Frostfish
Benthodesmus tuckeri Tucker's Frostfish
Benthodesmus vityazi Vityaz Frostfish
Lepidopus caudatus Frostfish
Lepturacanthus savala Spiny Hairtail
Tentoriceps cristatus Crested Hairtail
Trichiurus auriga Pearly Hairtail
Trichiurus australis Australian Hairtail
Trichiurus lepturus Largehead Hairtail
Trichiurus nickolensis Short-tail Hairtail

MACKERELS AND TUNAS

Scombridae

YELLOWFIN TUNA
Thunnus albacares

Mackerels and Tunas are found in tropical and temperate waters throughout the world. There are about 51 species in the family. They have elongate, cylindrical to compressed, tapering bodies, with minute cycloid scales (except for the **Butterfly Mackerel**, which has large scales). The large mouth has a single row of small, sharp teeth in each jaw. They have two dorsal fins: the first has spines and can be depressed into a groove in the dorsal surface, and the second has 9–27 rays and is followed by a series of detached finlets. The anal fin is similar to the second dorsal fin and is also followed by a series of detached finlets. The caudal fin is deeply forked, and the narrow caudal peduncle usually has one or more fleshy keels.

The strange Butterfly Mackerel forms its own subfamily, the Gasterochismatinae. The species is named for the fan-like ventral fin of the juvenile form, which reduces in size as the fish matures. It is found occasionally in southern Australian waters and is sometimes taken as bycatch by longline fisheries targeting other Tuna species. It grows to about 1.6 m in length. The remaining Mackerels and Tunas belong to the subfamily Scombrinae, and can be divided into four groups:

SCOMBRINI This group includes the **Mouth Mackerel** and **Blue Mackerel**, both of which have an adipose eyelid, form large schools in coastal waters and feed on zooplankton. The Mouth Mackerel is encountered in dense, fast-swimming schools in northern Australian waters. It swims with its mouth wide open, filtering plankton from the water. The small Blue Mackerel reaches about 40 cm in length and occurs in large schools in coastal waters and offshore, most commonly around southern Australia. It is a favourite with young shore-based anglers, as it is often abundant in harbours and around jetties and is easily caught with a handline.

SCOMBEROMORINI Members of this group include the **Spanish Mackerel**, **Shark Mackerel**, **Scad Mackerel**, **School Mackerel**, **Grey Mackerel**, **Spotted Mackerel** and the **Wahoo**. All are predators of other fishes, squid and crustaceans, and most are superb food fishes. The Spanish Mackerel makes seasonal migrations along the east and west coasts of Australia to and from spawning grounds in the north. Reaching up to 2.2 m in length, it is highly prized as a food fish and is an important recreational and commercial species. The other Mackerels are also sought-after by anglers, but are smaller in size. The Wahoo is a wide-ranging oceanic species and is reputed to be one of the fastest-swimming fishes in the sea. It reaches over 2 m in length and is encountered offshore in northern Australian waters.

SARDINI This group includes the Bonitos and the **Dogtooth Tuna**: species which are intermediate between Mackerels and Tunas. The Dogtooth Tuna is a large predator, reaching a maximum length of nearly 2.5 m and weighing up to 130 kg. It is found in northern Australian waters, individually or in small groups, over deep dropoffs and outer reef slopes. The Bonitos are schooling fishes and are pelagic in coastal waters. They are small, fast-swimming predators of other fishes and reach a maximum length of about 1 m. In Australia the **Leaping Bonito** is found mainly in northern waters, the **Australian Bonito** in south-eastern waters and the **Oriental Bonito** in south-western waters. They are excellent food fishes, though often much maligned – which is probably due to insufficient care being taken of caught fish.

THUNNINI Including the Tunas, **Frigate Mackerel** and **Albacore**, this group is unique in that its members are able to maintain a body temperature higher than that of the surrounding water. Blood is pumped through specialised vascular heat exchangers, allowing a higher metabolic rate and greater muscle activity. Because of their high body temperature, these fishes must be cooled immediately once caught or the flesh will rapidly spoil.

Tunas are amongst the fastest-swimming of all fishes and are voracious predators of other small schooling fishes, crustaceans and squid. They are wide ranging in oceanic waters and some reach very large sizes. The **Northern Bluefin Tuna** can reach over 3 m in length and 670 kg in weight, while the **Yellowfin Tuna**, **Bigeye Tuna** and **Southern Bluefin Tuna** grow to over 2 m in length and weigh up to about 190 kg. These Tunas are some of the most important of all commercial fishes. They are intensively fished around the world by longline, driftnet and purse seine. Smaller species are also very important commercially: the **Skipjack Tuna** makes up about 40 per cent of the total world Tuna catch. Tunas have been severely overfished in many parts of the world, often in international waters where fishing is difficult to police. The Southern Bluefin Tuna has been particularly affected. There is an urgent need for more responsible fishing practices worldwide to prevent the total collapse of Tuna populations.

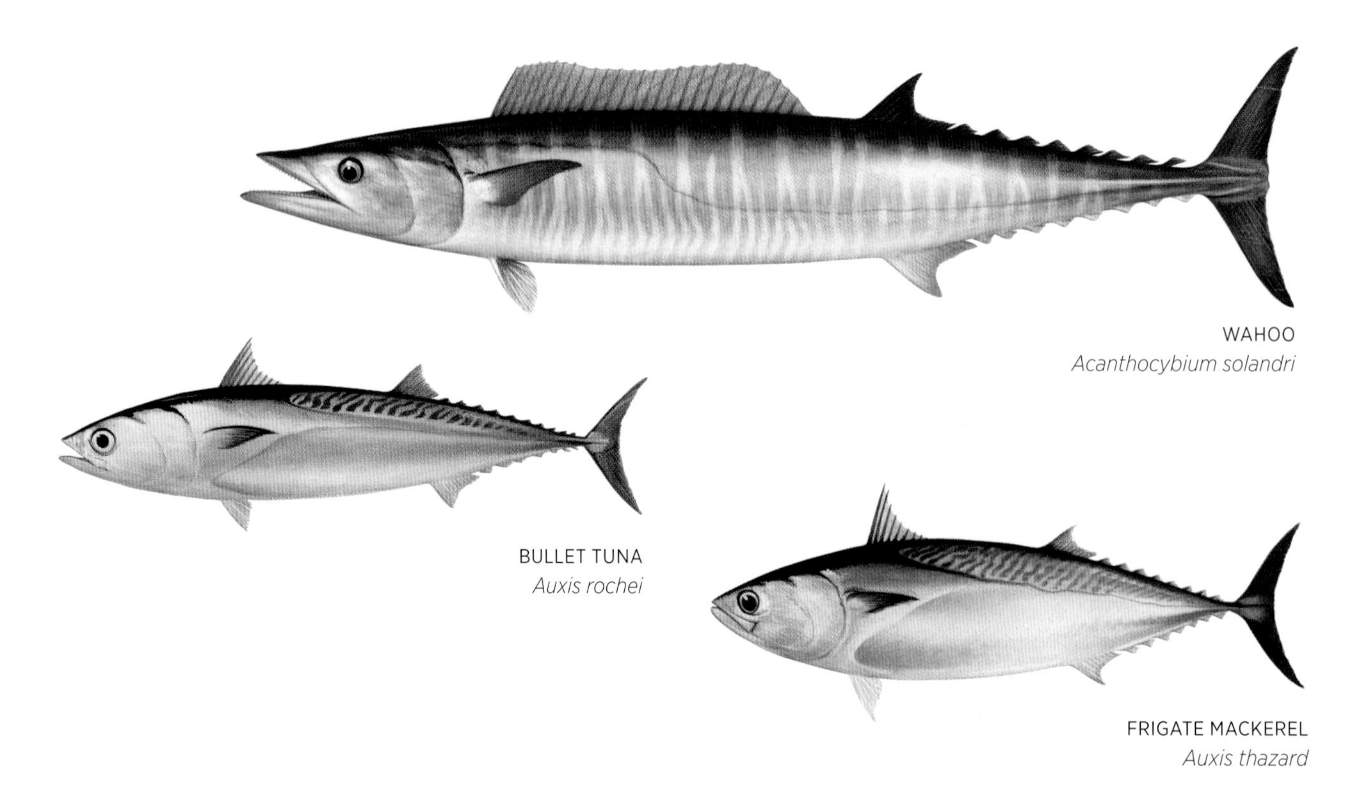

WAHOO
Acanthocybium solandri

BULLET TUNA
Auxis rochei

FRIGATE MACKEREL
Auxis thazard

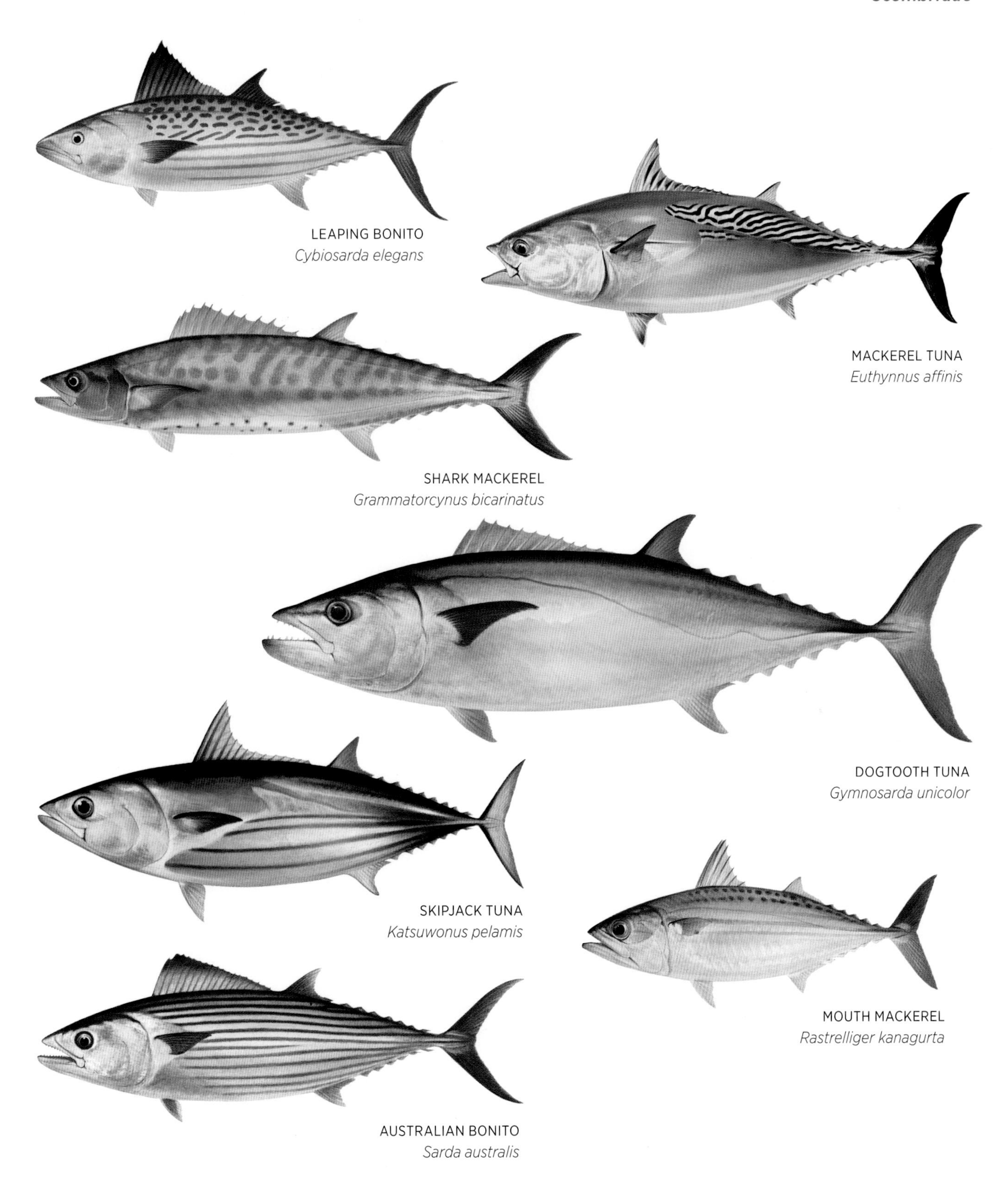

LEAPING BONITO
Cybiosarda elegans

MACKEREL TUNA
Euthynnus affinis

SHARK MACKEREL
Grammatorcynus bicarinatus

DOGTOOTH TUNA
Gymnosarda unicolor

SKIPJACK TUNA
Katsuwonus pelamis

MOUTH MACKEREL
Rastrelliger kanagurta

AUSTRALIAN BONITO
Sarda australis

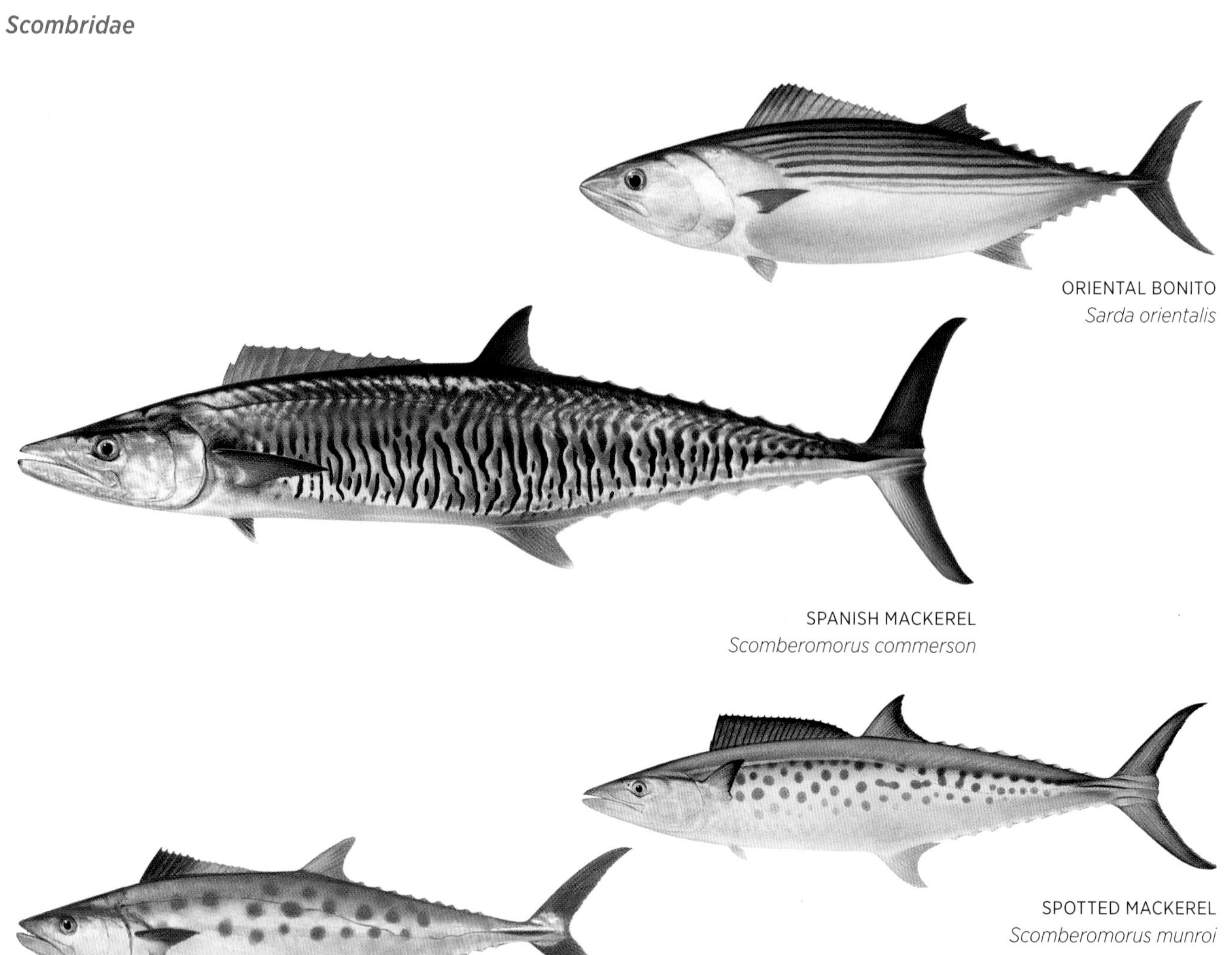

ORIENTAL BONITO
Sarda orientalis

SPANISH MACKEREL
Scomberomorus commerson

SPOTTED MACKEREL
Scomberomorus munroi

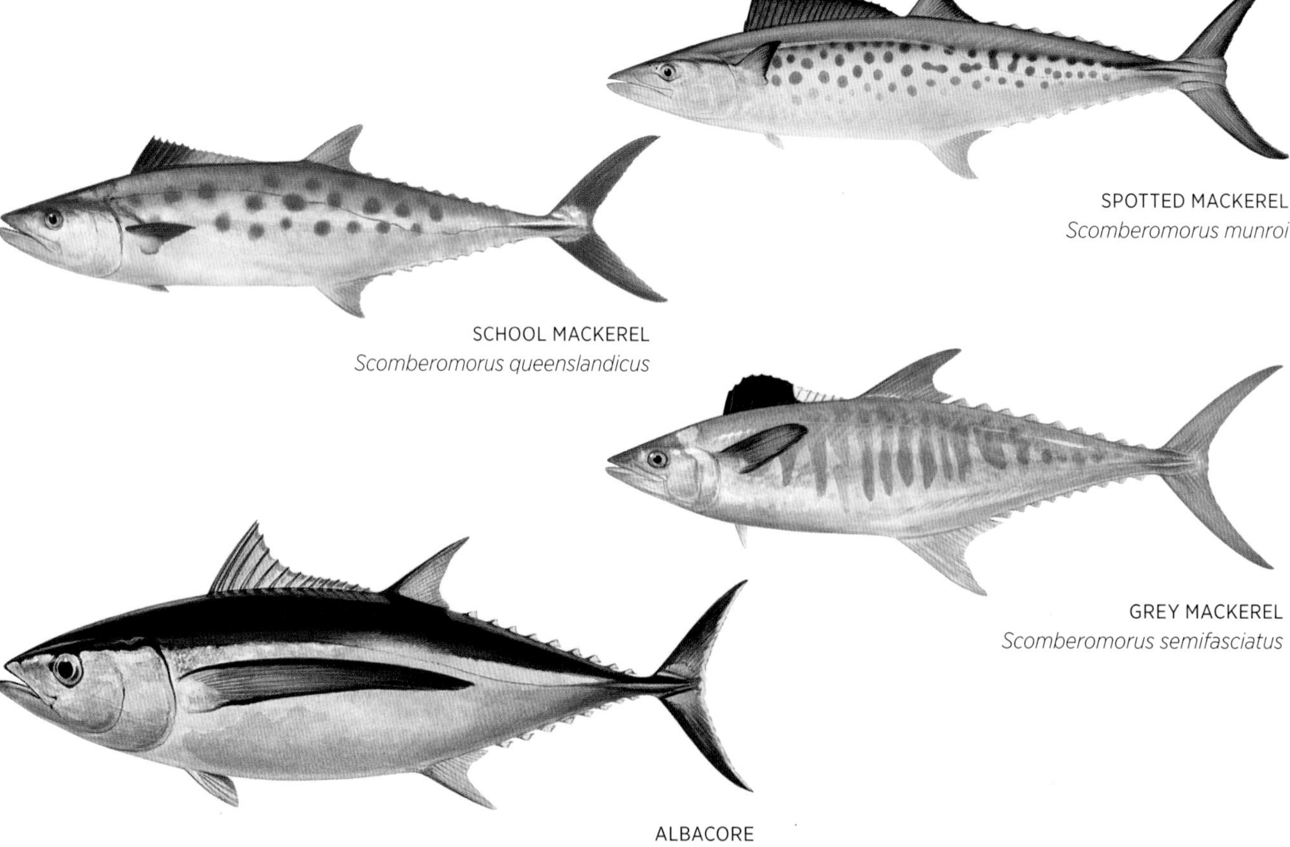

SCHOOL MACKEREL
Scomberomorus queenslandicus

GREY MACKEREL
Scomberomorus semifasciatus

ALBACORE
Thunnus alalunga

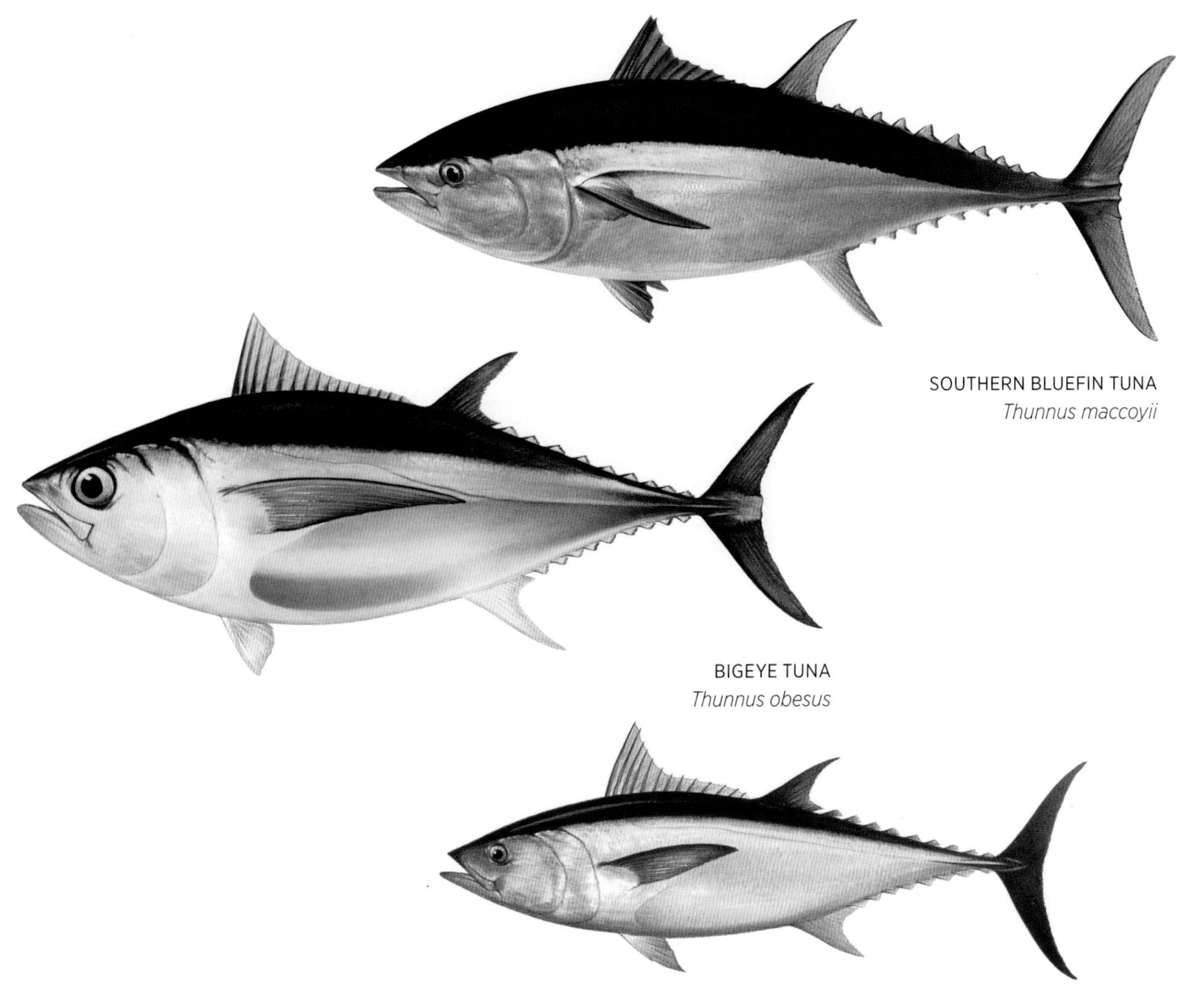

SOUTHERN BLUEFIN TUNA
Thunnus maccoyii

BIGEYE TUNA
Thunnus obesus

LONGTAIL TUNA
Thunnus tonggol

AUSTRALIAN SCOMBRIDAE SPECIES

Acanthocybium solandri Wahoo
Allothunnus fallai Slender Tuna
Auxis rochei Bullet Tuna
Auxis thazard Frigate Mackerel
Cybiosarda elegans Leaping Bonito
Euthynnus affinis Mackerel Tuna
Gasterochisma melampus Butterfly Mackerel
Grammatorcynus bicarinatus Shark Mackerel
Grammatorcynus bilineatus Scad Mackerel
Gymnosarda unicolor Dogtooth Tuna
Katsuwonus pelamis Skipjack Tuna
Rastrelliger kanagurta Mouth Mackerel
Sarda australis Australian Bonito
Sarda orientalis Oriental Bonito
Scomber australasicus Blue Mackerel
Scomberomorus commerson Spanish Mackerel
Scomberomorus munroi Spotted Mackerel
Scomberomorus queenslandicus School Mackerel
Scomberomorus semifasciatus Grey Mackerel
Thunnus alalunga Albacore
Thunnus albacares Yellowfin Tuna
Thunnus maccoyii Southern Bluefin Tuna
Thunnus obesus Bigeye Tuna
Thunnus orientalis Northern Bluefin Tuna
Thunnus tonggol Longtail Tuna

SWORDFISH

Xiphiidae

SWORDFISH
Xiphias gladius

The only species in this family, the **Swordfish**, is found in deep waters around the world. It has an elongate, cylindrical body with no scales and large eyes. The snout is formed into a very long 'spear', which is flattened in cross-section. The Swordfish has two dorsal fins: the first is very tall and stiff, placed well forward on the body, and has 38–45 rays; the second is very small, placed just before the caudal fin, and has 4–5 rays. The anal fin is also tall, placed at about mid-body, with 12–16 rays, and is followed by a second small fin that is similar to and directly opposite the second dorsal fin. The long, curved pectoral fins are placed low on the body and the ventral fins are absent. The caudal fin is deeply forked, with notches before it on the dorsal and ventral surface of the caudal peduncle, which bears a single fleshy keel on each side. The Swordfish is superficially very similar to the Billfishes (Istiophoridae, p.723) and was formerly included in the same family. However, it differs in lacking ventral fins and in having a flattened spear (the Billfishes have prominent ventral fins and a rounded spear).

The Swordfish is an extremely large and powerful fish, reaching up to 4.5 m in length and weighing up to 540 kg. It is a predator of other fishes and squid, and is usually found in deep water down to about 600 m. At times it rises to the surface and occasionally it is encountered in inshore waters. The Swordfish is an important commercial species, caught by longline in many parts of the world, and a superb food fish. It has been overfished in many areas, and the average size of caught fish has declined as the large, prime breeding fish have been removed; populations in some areas are at risk of collapse.

AUSTRALIAN XIPHIIDAE SPECIES
Xiphias gladius Swordfish

BILLFISHES

Istiophoridae

SAILFISH
Istiophorus platypterus

There are about 11 species of Billfish, and they are found worldwide in tropical and temperate waters. They have an elongate, slightly compressed, tapering body with narrow, pointed scales embedded in the skin. The snout is formed into a long spear, which is rounded in cross-section. Billfishes have two dorsal fins: the first is long-based and able to be depressed into a groove in the back, and the second is much smaller and situated just behind the first. There are two anal fins: the first is tall, is placed at mid-body and can be depressed into a groove, and the second is similar to and opposite the second dorsal fin. The pectoral fins are long and curved, the ventral fins are long and thin, and the caudal fin is large and deeply forked. There are two fleshy keels on each side of the caudal peduncle, and notches on the dorsal and ventral surfaces just before the caudal fin.

Billfishes are fast-swimming predators of other fishes and squid. Around Australia they are found near the surface in oceanic and coastal waters. They possess a vascular heat exchanger similar to Tunas (Scombridae, p. 717), which warms only the blood supply to the brain and eyes.

The **Black Marlin** is one of the largest of Billfishes, reaching a maximum length of about 4.6 m and weighing up to 700 kg. It is distinguished from all other Billfishes by its rigid pectoral fins, which are held at right angles to the body and cannot be folded flat. The **Blue Marlin** is even larger, reaching up to 5 m in length and 820 kg in weight. In both species the larger specimens are always females.

Billfishes are pelagic in open oceanic waters, but occasionally approach the coast, particularly in areas where there are deep dropoffs. The **Sailfish** is found in northern Australian waters and has a tall, sail-like dorsal fin and an elongate, compressed body. Like most Billfishes it is highly sought-after by gamefishers, as it fights hard and makes spectacular leaps. The **Shortbill Spearfish** is a smaller species, reaching about 2.3 m in length, and is occasionally taken by longliners targeting Tuna. The **Striped Marlin** grows to about 3.5 m in length and in Australia it is found mainly in warmer northern waters. Billfishes are caught commercially in many areas of the world and are fine food fishes. In Australia they are taken incidentally and usually exported. Billfishes are mainly solitary, though the Sailfish occurs in small groups, and in some species males form groups around a large female during mating.

BLACK MARLIN
Makaira indica

BLUE MARLIN
Makaira nigricans

SHORTBILL SPEARFISH
Tetrapturus angustirostris

STRIPED MARLIN
Tetrapturus audax

AUSTRALIAN ISTIOPHORIDAE SPECIES

Istiophorus platypterus Sailfish
Makaira indica Black Marlin
Makaira nigricans Blue Marlin
Tetrapturus angustirostris Shortbill Spearfish
Tetrapturus audax Striped Marlin

AMARSIPA

Amarsipidae

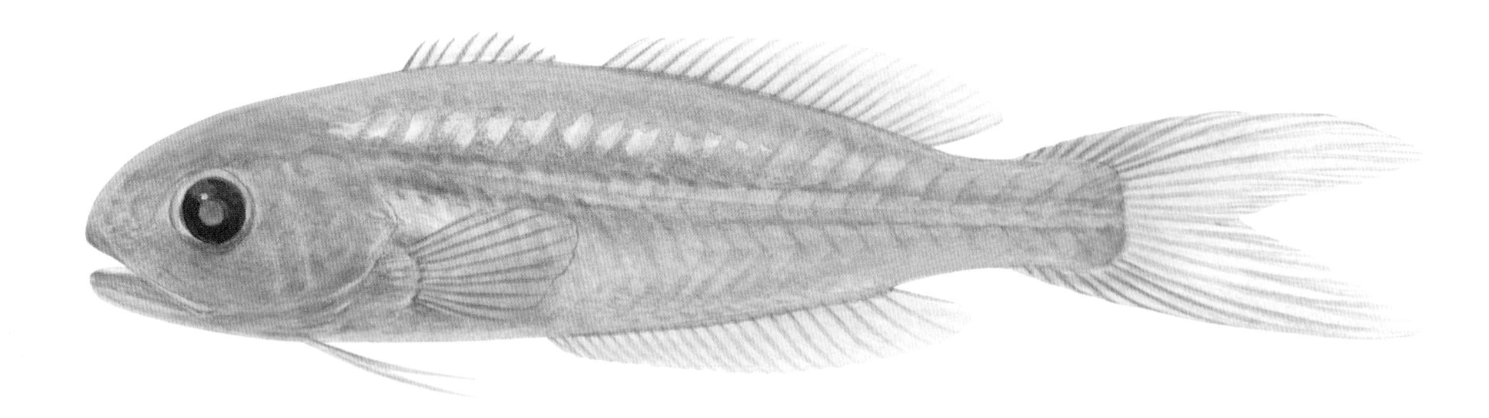

AMARSIPA (juvenile)
Amarsipus carlsbergi

The **Amarsipa** is the only species in this family, and it is found in tropical oceanic waters of the Indian and Pacific oceans. It has an elongate, compressed body with small, fragile cycloid scales, and a small head with large eyes. The moderate-sized mouth bears small, sharp teeth, and the upper jaw is covered by the suborbital bones when the mouth is closed. The Amarsipa has two dorsal fins, the first with 9–12 weak spines and the second with 22–27 rays. The anal fin has 27–32 rays and no spines, the ventral fins are placed well forward beneath the throat, and the caudal fin is forked. The Amarsipa reaches about 22 cm in length. It was only recently discovered, with specimens taken by midwater trawls. Almost all records of this species are of juveniles, which are found near the surface, while the very few adult specimens have been taken at depths to about 130 m at night. Almost nothing is known of the species' biology and feeding habits.

AUSTRALIAN AMARSIPIDAE SPECIES

Amarsipus carlsbergi Amarsipa

TREVALLAS, WAREHOUS, RUDDERFISHES AND RUFFES

Centrolophidae

BLUE-EYE TREVALLA
Hyperoglyphe antarctica

There are about 28 species in this family, and they are found in tropical and temperate waters around the world. They have oval to elongate, compressed to rounded bodies with small, fragile cycloid scales that extend onto the fins. They have a bluntly rounded snout, many sensory pores on the head, quite large eyes surrounded by adipose tissue, and a small, slightly inferior mouth. The upper jaw is covered by the suborbital bones when the mouth is closed. The dorsal fin is either continuous, with 0–5 graduated spines attached to the rayed part, or notched, with 5–9 spines detached from the rayed part – the spines and rays are often very similar and difficult to distinguish from each other. The anal fin has 0–3 spines and 15–41 rays, and the caudal fin is slightly to deeply forked.

In Australia Trevallas, Warehous, Rudderfishes and Ruffes are generally found in deep waters off the southern coast, down to about 900 m. They feed on pelagic crustaceans, salps, jellyfishes and small fishes. Several species are important commercially. The large **Blue-eye Trevalla**, which can reach up to 1.4 m in length, is considered one of the finest food fishes in Australian waters. It is taken by longline and trawl over rocky bottoms in southern waters. Warehous are also excellent food fishes and are caught commercially off south-eastern Australia. The **Blue Warehou** and **Silver Warehou** are often found in shallower waters than other members of the family, occasionally occurring inshore and in estuaries. Ruffes and Rudderfishes are found in the same areas but are much rarer, only occasionally encountered by deepwater trawlers. Butterfishes and Driftfishes, which reach about 20–30 cm in length, are found in deeper offshore waters around northern Australia. Juveniles of this family are usually encountered in surface waters and are often associated with jellyfishes, floating debris and weed.

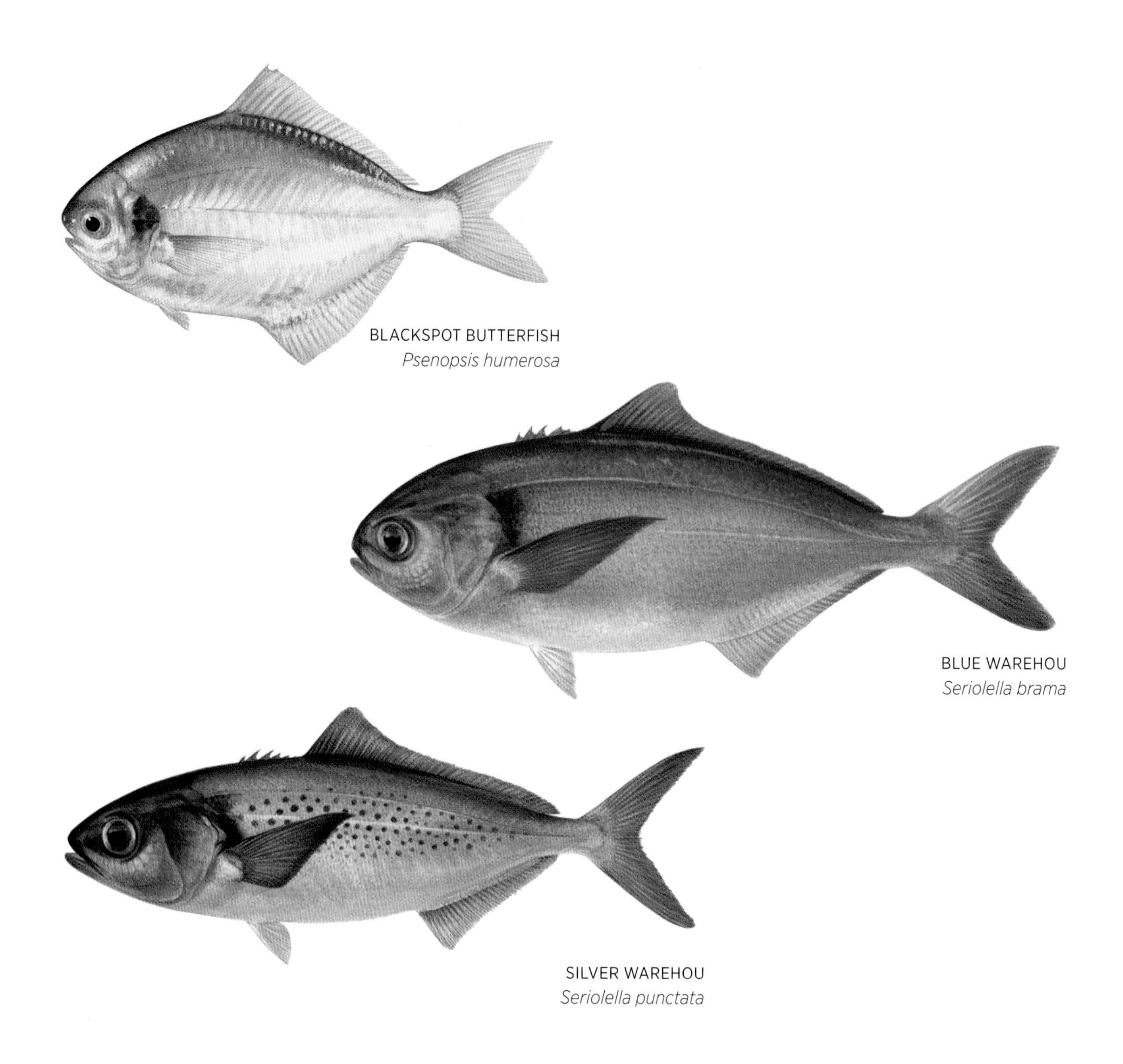

BLACKSPOT BUTTERFISH
Psenopsis humerosa

BLUE WAREHOU
Seriolella brama

SILVER WAREHOU
Seriolella punctata

AUSTRALIAN CENTROLOPHIDAE SPECIES

Centrolophus niger Rudderfish
Hyperoglyphe antarctica Blue-eye Trevalla
Icichthys australis Southern Ruffe
Psenopsis anomala White Butterfish
Psenopsis humerosa Blackspot Butterfish
Psenopsis obscura Obscure Driftfish
Schedophilus huttoni New Zealand Ruffe
Schedophilus maculatus Raft-fish
Schedophilus velaini Ocean Blue-eye Trevalla
Seriolella brama Blue Warehou
Seriolella caerulea White Warehou
Seriolella punctata Silver Warehou
Tubbia tasmanica Tasmanian Rudderfish

DRIFTFISHES

Nomeidae

COASTAL CUBEHEAD
Cubiceps whiteleggii

Driftfishes are found in tropical and temperate waters around the world, and there are about 16 species in the family. They have oval to elongate bodies with thin, fragile scales that extend onto the base of the fins. They have a bluntly rounded snout, sensory pores on the head and back, large eyes and a small mouth. The upper jaw is covered by the expanded suborbital bones when the mouth is closed. They have two dorsal fins: the first with 9–12 spines, which can be depressed into a groove on the dorsal surface; and the second with 0–3 spines and 15–32 rays. The anal fin has 1–3 spines and 14–30 rays, and the caudal fin is forked.

Driftfishes are named for the habits of the juveniles, which 'drift' in oceanic waters and are usually associated with large jellyfishes, other cnidarians and ctenophores, floating debris or weed. The juvenile **Bluebottle-fish** lives commensally with the Bluebottle (Portuguese Man o'War), immune to the hydrozoan's stinging tentacles, and feeds on small pelagic invertebrates. The adults, which were unknown until recently, have been taken by deepwater bottom trawls in waters off Japan. Cubeheads are small fishes, reaching about 20–40 cm in length. They are found around Australia, usually in deeper offshore waters down to about 750 m, with some making vertical migrations towards the surface at night. The **Blue Cubehead** is found in surface waters offshore and the **Coastal Cubehead** is found near the bottom at depths of 180–550 m over the continental shelf.

Psenes species reach about 20 cm in length and in Australia are widespread in warm temperate and tropical oceanic waters. They occur mainly near the surface and adults are often associated with drifting debris. The **Blackrag** reaches about 80 cm in length and is found closer to the bottom, where it feeds on zooplankton and small fishes.

BLUEBOTTLE-FISH
Nomeus gronovii

AUSTRALIAN NOMEIDAE SPECIES

Cubiceps baxteri Black Cubehead
Cubiceps caeruleus Blue Cubehead
Cubiceps capensis Cape Cubehead
Cubiceps kotlyari Kotlyar's Cubehead
Cubiceps pauciradiatus Longfin Cubehead
Cubiceps whiteleggii Coastal Cubehead
Nomeus gronovii Bluebottle-fish
Psenes arafurensis Dusky Driftfish
Psenes cyanophrys Freckled Driftfish
Psenes hillii Hill's Driftfish
Psenes pellucidus Blackrag

DEEPSEA DRIFTFISHES

Ariommatidae

INDIAN DRIFTFISH
Ariomma indicum

Deepsea Driftfishes are found in deep tropical and subtropical waters of the Atlantic, Indian and Pacific oceans. There are about seven species in the family. They have an oval, compressed body with large, fragile cycloid scales. They have a rounded snout, large eyes surrounded by adipose tissue, and a small mouth, with the upper jaw covered by the suborbital bones when the mouth is closed. There are two dorsal fins: the first with 10–13 spines, which can be depressed into a groove on the dorsal surface; and the second with one spine and 13–16 rays. The anal fin has 1–3 spines and 13–18 rays, and the caudal fin is deeply forked. There are two low, fleshy keels on each side of the caudal peduncle.

Deepsea Driftfishes are found in large schools near the bottom over muddy substrates, down to about 750 m. They feed on zooplankton and small pelagic invertebrates such as jellyfishes. The **Indian Driftfish** and **Slope Driftfish** grow to about 25 cm in length and are found off the north and north-eastern coasts of Australia, from depths of about 50 to 450 m. They appear in deepwater trawl catches, sometimes in large numbers, and are excellent food fishes.

AUSTRALIAN ARIOMMATIDAE SPECIES
Ariomma indicum Indian Driftfish
Ariomma luridum Slope Driftfish

SQUARETAILS

Tetragonuridae

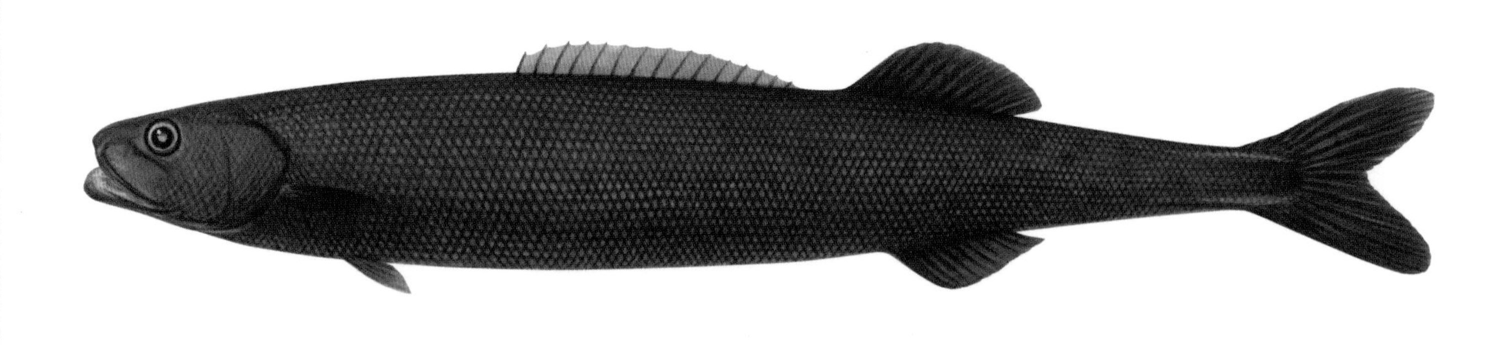

SMALLEYE SQUARETAIL
Tetragonurus cuvieri

The three species of Squaretail are widely distributed in tropical and temperate waters around the world. They have an elongate, cylindrical body with small, ridged scales in distinctive spiralling rows, which extend onto the base of the fins. They have a bluntly rounded snout and a large mouth, a robust lower jaw with a row of blade-like teeth, and small pointed teeth in the upper jaws. The upper jaw is concealed by the suborbital bones when the mouth is closed, and the lower jaw fits almost completely into a recess in the upper jaw. Squaretails have two dorsal fins: the first with 10–20 short spines, which can be depressed into a groove in the dorsal surface; and the second with 10–17 rays. The anal fin has one spine and 10–16 rays, and the caudal fin is deeply forked. The caudal peduncle is long, almost square in cross-section, and bears a pair of prominent scaly keels on each side.

Squaretails are pelagic in oceanic mid to surface waters and are fast-swimming predators, feeding mainly on jellyfishes and other ctenophores. They are rarely encountered, though occasionally taken by longline, and are reported to be poisonous to eat. They reach a maximum length of about 50 cm.

AUSTRALIAN TETRAGONURIDAE SPECIES
Tetragonurus cuvieri Smalleye Squaretail
Tetragonurus pacificus Pacific Squaretail

DEEPSEA BOARFISHES

Caproidae

ROBUST DEEPSEA BOARFISH
Antigonia capros

Caproidae

Deepsea Boarfishes are found in deep waters of the Atlantic, Indian and Pacific oceans, and the family contains about 11 species. They have a very deep, compressed body (sometimes deeper than it is long), with small, ridged, ctenoid scales, large eyes and a small, protrusible mouth. The single dorsal fin has 8–9 spines and 25–30 rays, the anal fin has three spines and 25–27 rays, and the caudal fin is truncate.

Deepsea Boarfishes occur around most of Australia, except along the south coast, near the bottom from about 50 to 600 m. They grow to about 20 cm in length and occasionally appear in deepwater trawls as bycatch. They feed on small crustaceans and molluscs.

AUSTRALIAN CAPROIDAE SPECIES

Antigonia capros Robust Deepsea Boarfish
Antigonia malayana Malayan Deepsea Boarfish
Antigonia rhomboidea Rhomboid Deepsea Boarfish
Antigonia rubescens Sharpsnout Deepsea Boarfish
Antigonia rubicunda Rosy Deepsea Boarfish

PLEURONECTIFORMES

Pleuronectiforms have an extremely compressed body, with both eyes on the same side of the head. Eye position is normal in the free-swimming larval form, but during development one eye migrates over the dorsal profile to the other side of the head. Adults spend their lives on the substrate, with their blind side (which is usually flatter than the eyed side and much paler in colour) facing the bottom. Individuals of a given species usually have the eyes always on the same side, though some species have both left- and right-eyed individuals. The various families are differentiated by the position of the fin origins, shape of the ventral fins and placement of the eyes. Many species are able to rapidly change their colour pattern to precisely imitate that of the substrate. There are 14 families in the order, 10 of which occur in Australian waters.

SPINY TURBOTS

Psettodidae

AUSTRALIAN HALIBUT
Psettodes erumei

This family is found in waters around West Africa and in the Indo-West Pacific, and contains only three species. They have an oval, thick body with small, finely ctenoid scales, and a large mouth with large canine teeth. The eyes may be on the right or left side of the head, with the upper eye located on the dorsal margin of the head. The dorsal fin has spinous anterior rays and originates posterior to the upper eye. The anal fin also has spinous anterior rays. The dorsal and anal fins are not attached to the caudal fin, and the ventral fins are symmetrical, with one spine and five rays.

The **Australian Halibut** is found in northern Australian waters over sand and mud bottoms, from the shallows down to about 100 m. Although it is a typical flatfish in form, it is unusual in that it sometimes swims in an upright position. It is a predator of other fishes, including pelagic species, indicating that it often leaves the substrate to feed. The Australian Halibut grows to about 64 cm in length. It is an excellent food fish and is taken commercially in waters to the north of Australia.

AUSTRALIAN PSETTODIDAE SPECIES
Psettodes erumei Australian Halibut

LARGESCALE FLOUNDERS

Citharidae

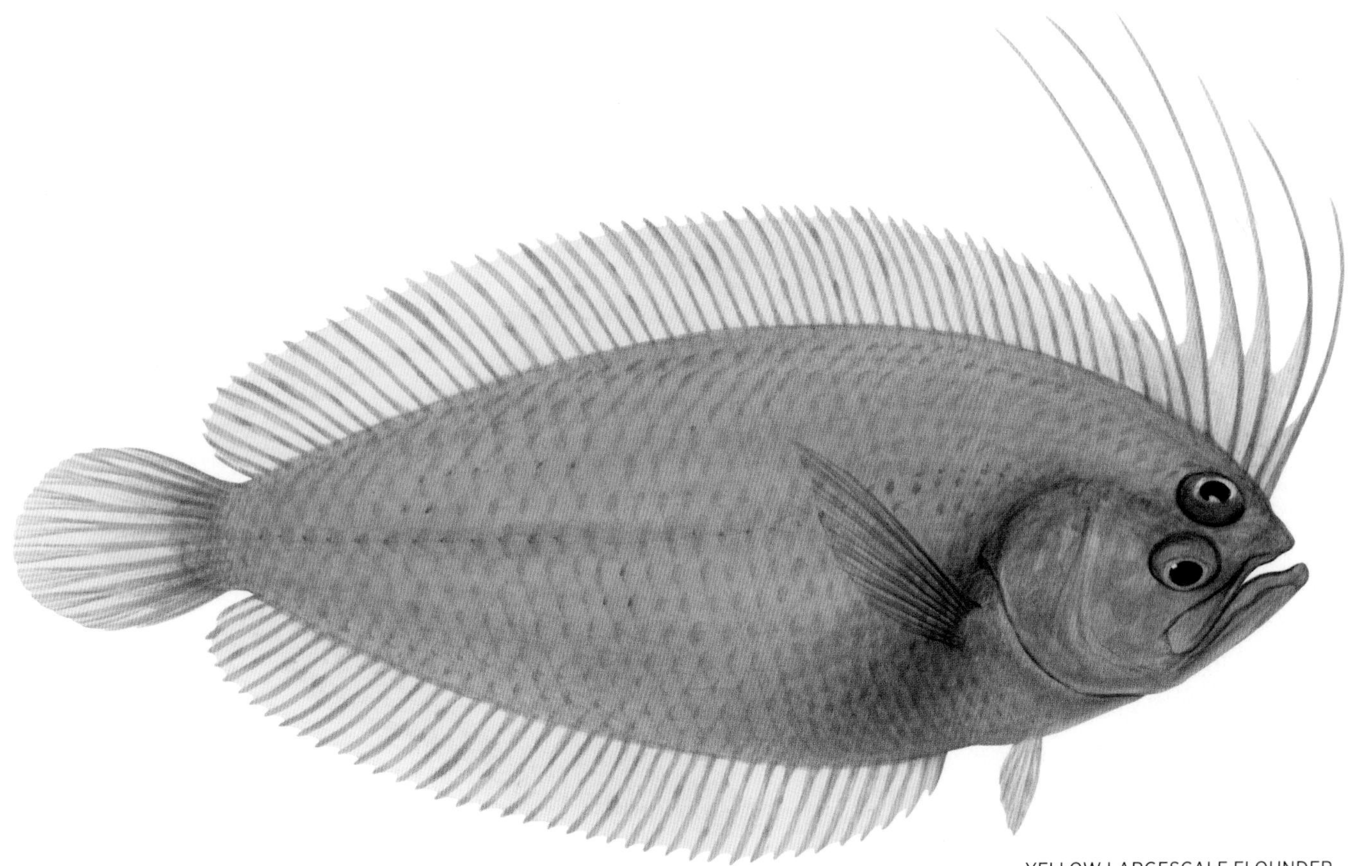

YELLOW LARGESCALE FLOUNDER
Brachypleura novaezeelandiae

Largescale Flounders are found in the Mediterranean and Indo-West Pacific, and there are about six species in the family. They have an elongated, oval, moderately thin body, with ctenoid scales on the eyed side and weakly ctenoid or cycloid scales on the blind side. The large mouth reaches to below the eyes and, depending on the species, the eyes are either on the left or right side of the head. The origin of the dorsal fin is above or before the anterior margin of the upper eye, and both the dorsal fin and anal fin have no spines and are not attached to the caudal fin. The ventral fins are short-based and symmetrical, with one spine and five rays, and the caudal fin is rounded. Male Large-scale Flounders have elongated anterior dorsal-fin rays.

The **Scale-eye Flounder** is the largest of the Australian species, reaching a maximum length of about 36 cm. It is occasionally encountered in deepwater trawls in northern waters. The **Yellow Largescale Flounder** is also found off northern Australia, over sand and mud bottoms from depths of about 20 to 70 m, and it grows to about 14 cm in length. Largescale Flounders feed on small benthic crustaceans and invertebrates.

AUSTRALIAN CITHARIDAE SPECIES

Brachypleura novaezeelandiae Yellow Largescale Flounder
Citharoides macrolepidotus Branchray Flounder
Citharoides macrolepis Common Largescale Flounder
Citharoides orbitalis Orbital Flounder
Lepidoblepharon ophthalmolepis Scale-eye Flounder

SAND FLOUNDERS

Paralichthyidae

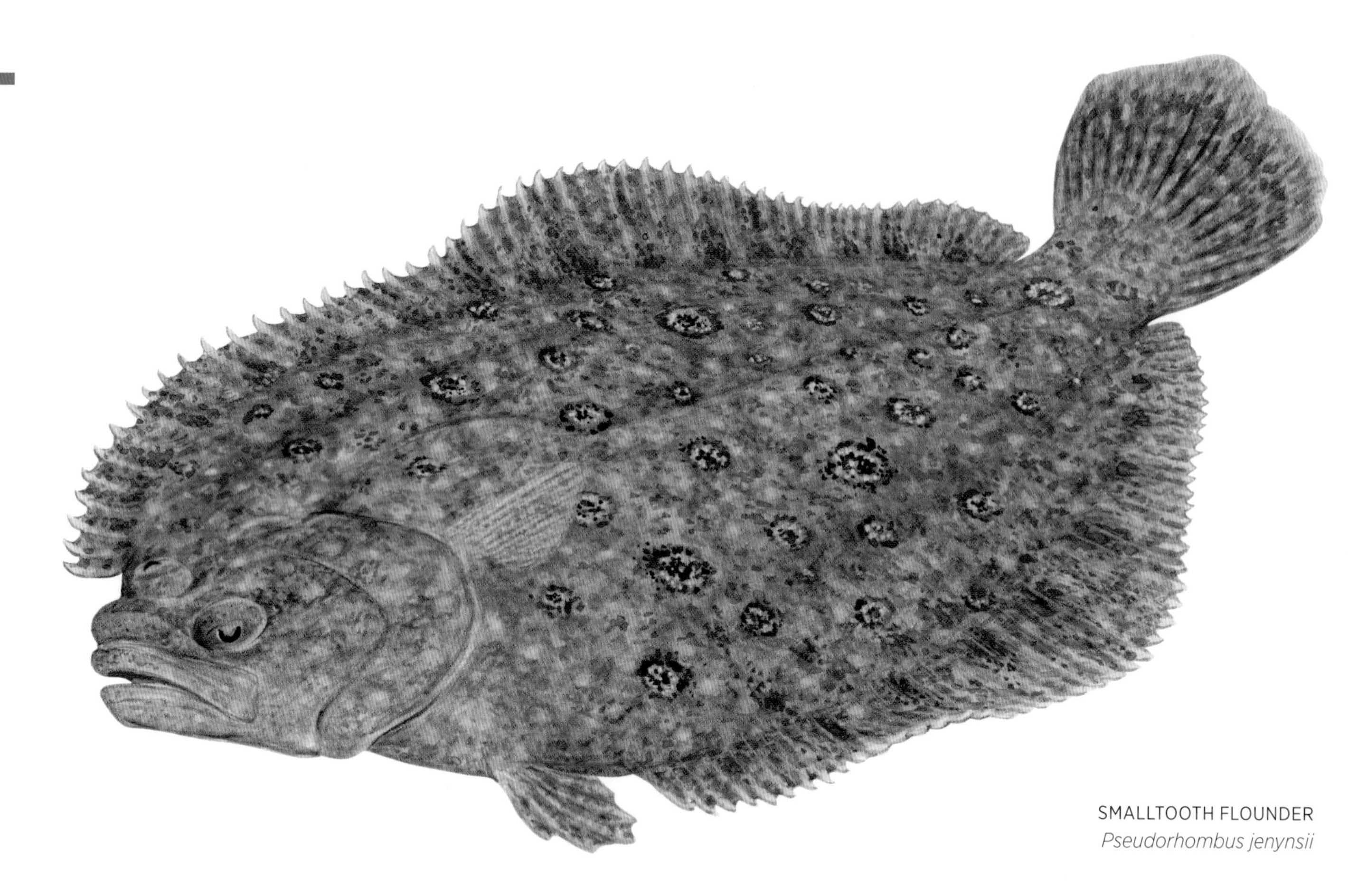

SMALLTOOTH FLOUNDER
Pseudorhombus jenynsii

The Sand Flounders are found in tropical and temperate waters of the Atlantic, Indian and Pacific oceans. There are about 105 species in the family. They have an oval body with small cycloid or ctenoid scales, and a large mouth with a single row of teeth in the jaws. The eyes are usually on the left side of the head, set close together and separated by a bony ridge. The origin of the dorsal fin is anterior to the upper eye, and the ventral fins are symmetrical, usually with six rays. The dorsal and anal fins are not attached to the wedge-shaped caudal fin. There is a distinct curve in the lateral line over the pectoral fins.

Sand Flounders are found on sand and mud bottoms, from the shallows down to about 150 m, and feed on benthic invertebrates such as crustaceans, and small fishes. The **Smalltooth Flounder** and **Largetooth Flounder** are found right around the coast of Australia, though they are less common along the south coast. The Largetooth Flounder is one of the largest members of the family, growing to about 45 cm in length, and it is often taken by anglers. Various other species of Sand Flounder inhabit shallow northern waters, most reaching 20–30 cm in length, and are often taken as bycatch by prawn trawlers. Sand Flounders are all excellent food fishes.

LARGETOOTH FLOUNDER
Pseudorhombus arsius

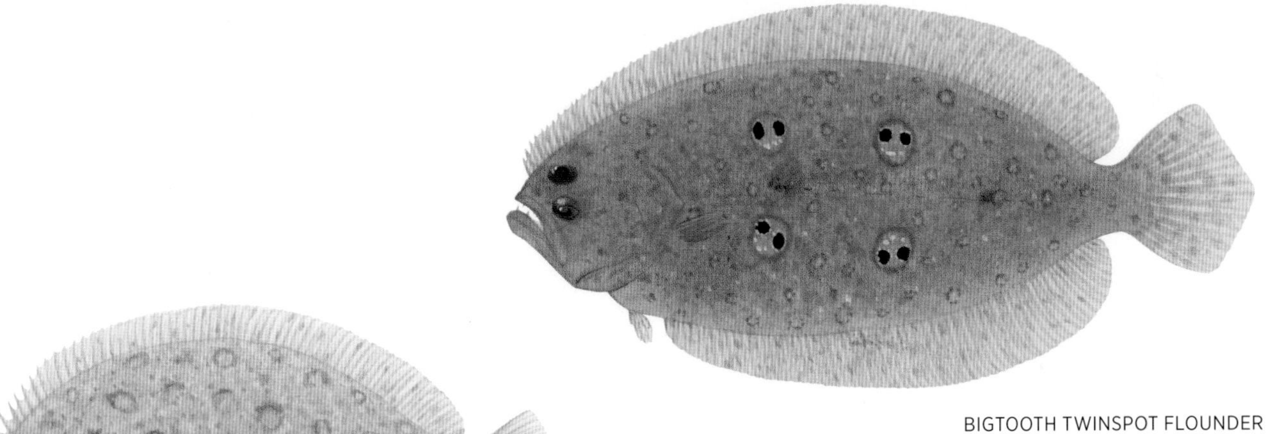

BIGTOOTH TWINSPOT FLOUNDER
Pseudorhombus diplospilus

DEEP FLOUNDER
Pseudorhombus elevatus

AUSTRALIAN PARALICHTHYIDAE SPECIES

Cephalopsetta ventrocellata Bighead Flounder
Pseudorhombus argus Peacock Flounder
Pseudorhombus arsius Largetooth Flounder
Pseudorhombus diplospilus Bigtooth Twinspot Flounder
Pseudorhombus dupliciocellatus Three Twinspot Flounder
Pseudorhombus elevatus Deep Flounder
Pseudorhombus jenynsii Smalltooth Flounder
Pseudorhombus megalops Bigeye Flounder
Pseudorhombus neglectus Neglected Flounder
Pseudorhombus quinquocellatus Five-eye Flounder
Pseudorhombus spinosus Spiny Flounder
Pseudorhombus tenuirastrum Slender Flounder
Pseudorhombus triocellatus Three-ring Flounder

LEFTEYE FLOUNDERS

Bothidae

OVAL FLOUNDER
Bothus myriaster

Lefteye Flounders are found in the Atlantic, Indian and Pacific oceans, and the family contains about 140 species. They have elongate to deep, oval bodies with small cycloid or ctenoid scales, and a moderate-sized, arched mouth. The eyes are usually on the left side of the head. The origin of the dorsal fin is above or before the anterior margin of the upper eye. The ventral fins are not symmetrical and each has 6–7 rays; the ventral fin on the blind side is placed dorsally to the ventral midline and is posterior to the ventral fin on the eyed side. The dorsal and anal fins are not attached to the caudal fin, which is rounded to wedge-shaped. The pectoral fin on the eyed side is longer than the pectoral fin on the blind side and is sometimes greatly elongated.

In some *Asterorhombus* species the first dorsal-fin ray is modified to form a fishing apparatus similar to that of the Anglerfishes (Antennaridae, p. 294). They use this ray in a similar fashion to the Anglerfishes, waving the small fleshy lure at the tip of the 'fishing rod' to attract prey within striking distance. Many Lefteye Flounders are sexually dimorphic, with males differing from females in having some elongated fin rays, darker patterns on the (normally pale) blind side, and different spacing between the eyes.

Lefteye Flounders are found right around the Australian coastline, occurring on sand and mud bottoms, sometimes near coral reefs, from shallow inshore waters down to at least 300 m. Most are small, from about 5 to 40 cm in length, and feed on small benthic crustaceans and other invertebrates.

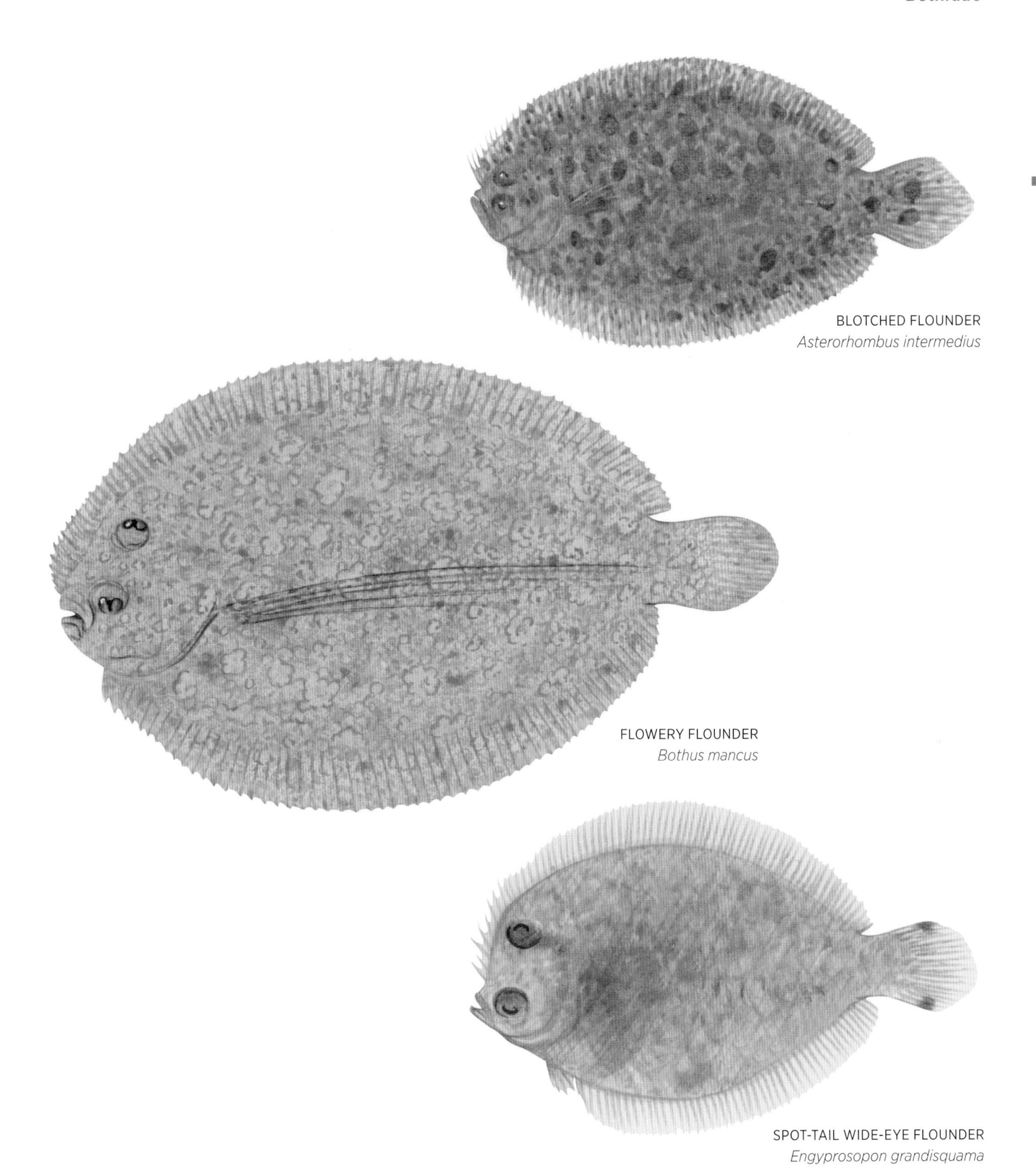

BLOTCHED FLOUNDER
Asterorhombus intermedius

FLOWERY FLOUNDER
Bothus mancus

SPOT-TAIL WIDE-EYE FLOUNDER
Engyprosopon grandisquama

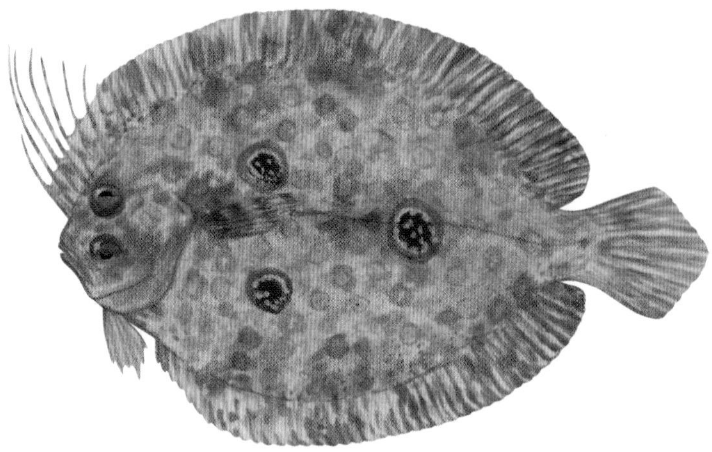

THREESPOT FLOUNDER
Grammatobothus polyophthalmus

AUSTRALIAN BOTHIDAE SPECIES

Arnoglossus andrewsi Andrew's Flounder
Arnoglossus armstrongi Armstrong's Flounder
Arnoglossus aspilos Spotless Lefteye Flounder
Arnoglossus bassensis Bass Strait Flounder
Arnoglossus elongatus Long Lefteye Flounder
Arnoglossus fisoni Fison's Flounder
Arnoglossus japonicus Japanese Lefteye Flounder
Arnoglossus macrolophus Largecrest Lefteye Flounder
Arnoglossus micrommatus Smalleye Flounder
Arnoglossus muelleri Mueller's Flounder
Arnoglossus nigrifrons Manyspot Lefteye Flounder
Arnoglossus tenuis Dwarf Lefteye Flounder
Arnoglossus waitei Waite's Flounder
Asterorhombus bleekeri Bleeker's Flounder
Asterorhombus fijiensis Angler Flatfish
Asterorhombus filifer Longfin Flounder
Asterorhombus intermedius Blotched Flounder
Asterorhombus osculus Bumphead Flounder
Bothus mancus Flowery Flounder
Bothus myriaster Oval Flounder
Bothus pantherinus Leopard Flounder
Bothus swio Swio Flounder
Chascanopsetta lugubris Pelican Flounder
Crossorhombus azureus Chain-mail Flounder
Crossorhombus howensis Lord Howe Flounder
Crossorhombus valderostratus Broadbrow Flounder
Engyprosopon grandisquama Spot-tail Wide-eye Flounder
Engyprosopon hureaui Hureau's Wide-eye Flounder
Engyprosopon latifrons Regan's Wide-eye Flounder
Engyprosopon longipterum Longfin Wide-eye Flounder
Engyprosopon maldivensis Olive Wide-eye Flounder
Grammatobothus pennatus Pennant Flounder
Grammatobothus polyophthalmus Threespot Flounder
Japonolaeops dentatus Eyetooth Flounder
Kamoharaia megastoma Widemouth Flounder
Laeops kitaharae Kitahara's Flounder
Laeops parviceps Smallhead Flounder
Lophonectes gallus Crested Flounder
Neolaeops microphthalmus Crosseye Flounder
Parabothus coarctatus Smallscale Flounder
Parabothus kiensis Yellowspotted Flounder
Parabothus polylepis Manyscale Flounder
Parabothus taiwanensis Taiwanese Flounder
Psettina gigantea Combscale Flounder
Psettina iijimae Leaf Flounder
Psettina senta Rough Flounder
Psettina tosana Ragged Flounder
Psettina variegata Variegated Flounder
Taeniopsetta ocellata Ocellate Flounder
Tosarhombus longimanus Longarm Flounder

RIGHTEYE FLOUNDERS

Poecilopsettidae

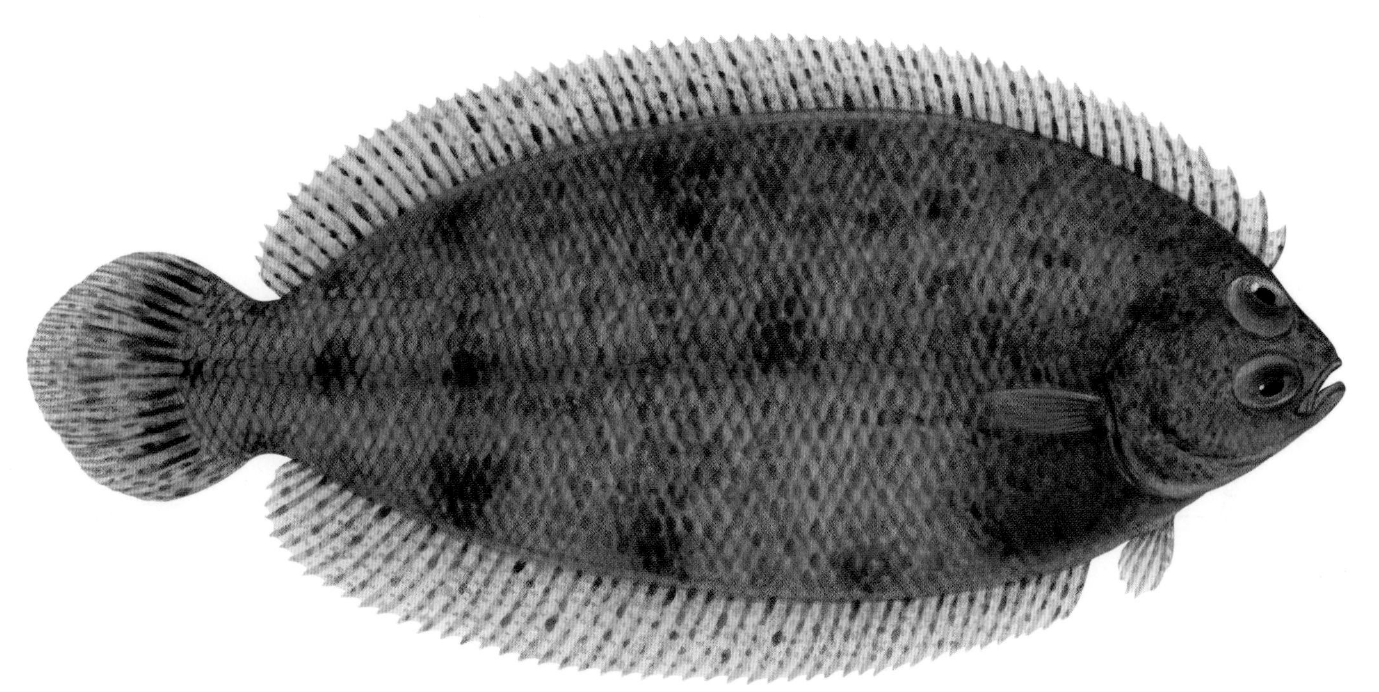

TILED RIGHTEYE FLOUNDER
Poecilopsetta plinthus

Righteye Flounders are found in the Atlantic, Indian and Pacific oceans, mainly in deep waters, and there are about 20 species in the family. They have an elongate, oval body with cycloid or ctenoid scales, and a small mouth. The eyes are on the right side of the head, and the origin of the dorsal fin is above the upper eye. The ventral fins are symmetrical, and the dorsal and anal fins are not attached to the caudal fin. Righteye Flounders are found in northern Australian waters over mud and sand bottoms down to about 450 m, and feed on small benthic invertebrates. They are small fishes, reaching a maximum size of about 10–20 cm, and are rarely encountered.

AUSTRALIAN POECILOPSETTIDAE SPECIES

Nematops macrochirus Longfin Righteye Flounder
Nematops microstoma Smallmouth Righteye Flounder
Poecilopsetta colorata Coloured Righteye Flounder
Poecilopsetta macrocephala Bighead Righteye Flounder
Poecilopsetta natalensis African Righteye Flounder
Poecilopsetta plinthus Tiled Righteye Flounder
Poecilopsetta praelonga Narrowbody Righteye Flounder

AUSTRAL RIGHTEYE FLOUNDERS

Rhombosoleidae

GREENBACK FLOUNDER
Rhombosolea tapirina

There are about 19 species of Austral Righteye Flounder, and they are found in waters around southern Australia and New Zealand, with one species occurring in the Atlantic Ocean. They have elongate to deep, oval bodies with ctenoid or cycloid scales, and a small mouth. In some the snout is formed into a hook shape anterior to the mouth. The large eyes are set close together on the right side of the head and the origin of the dorsal fin is before the upper eye, sometimes on the tip of the snout. The ventral fins are not symmetrical: the ventral fin on the eyed side has a longer base with more rays and is often attached to the anal fin. The dorsal and anal fins are not attached to the caudal fin, which is rounded.

Ammotretis species resemble the Soles (Soleidae, p.748) in having a hook-shaped snout before the mouth. They are found in the shallow coastal waters of southern Australia and most are quite small, from 10 to 30 cm in length. The family contains an important commercial species, the **Greenback Flounder**, which grows to about 45 cm in length. It is found along the southern coast of Australia from the shallows down to about 100 m, over sand, mud or shell bottoms. Austral Righteye Flounders are excellent food fishes and quite common in trawl catches, though most species are not large enough to be considered worth processing.

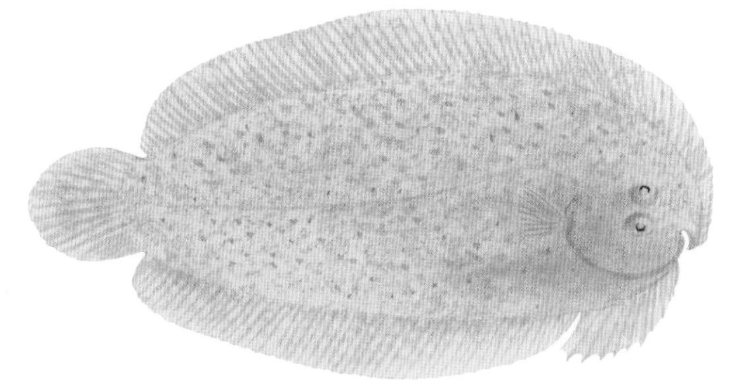

ELONGATE FLOUNDER
Ammotretis elongates

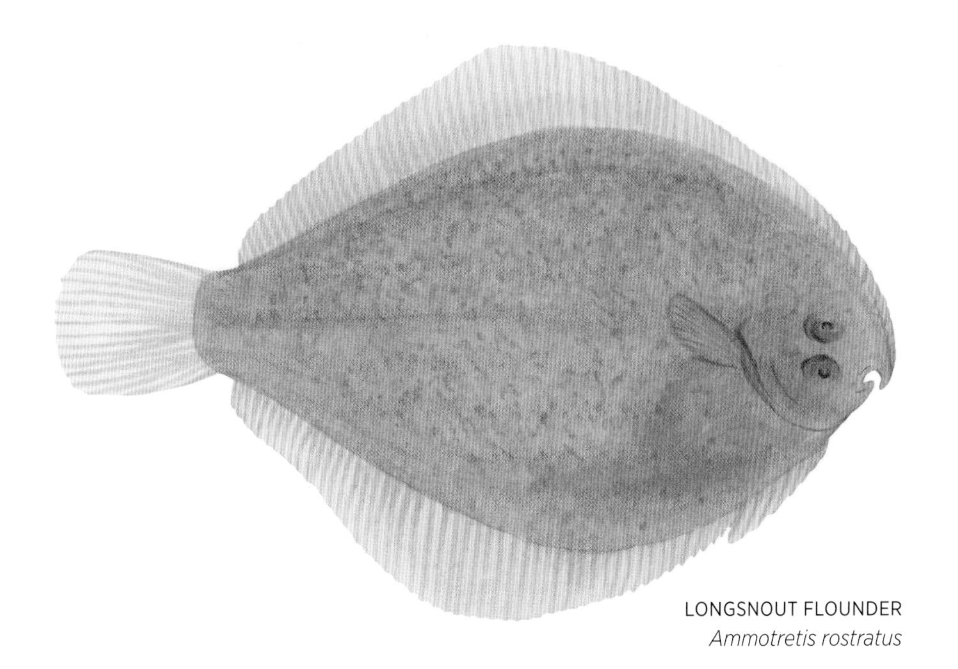

LONGSNOUT FLOUNDER
Ammotretis rostratus

AUSTRALIAN RHOMBOSOLEIDAE SPECIES

Ammotretis brevipinnis Shortfin Flounder
Ammotretis elongatus Elongate Flounder
Ammotretis lituratus Spotted Flounder
Ammotretis macrolepis Largescale Flounder
Ammotretis rostratus Longsnout Flounder
Azygopus pinnifasciatus Banded-fin Flounder
Psammodiscus ocellatus Freckled Righteye Flounder
Rhombosolea tapirina Greenback Flounder
Taratretis derwentensis Derwent Flounder

DEEPSEA FLOUNDERS

Achiropsettidae

SPOTTED DEEPSEA FLOUNDER
Mancopsetta maculata

Deepsea Flounders are found in deep, cold waters of the southern hemisphere, and the family contains about six species. They have an elongate, oval, very thin body, with a large mouth that reaches to below the lower eye. The dorsal profile has a deep indentation anterior to the eyes, which are partially covered with small scales. The eyes are on the left side of the head and the origin of the dorsal fin is above the anterior edge of the upper eye. The pectoral fins are absent or rudimentary.

Deepsea Flounders are large fishes, with the **Armless Deepsea Flounder** reaching up to 57 cm in length. In Australia Deepsea Flounders are found in cold southern waters from depths of about 100 to 1000 m. The biology of Deepsea Flounders is not well known and the relationships between this family and others in the order are uncertain.

AUSTRALIAN ACHIROPSETTIDAE SPECIES
Mancopsetta maculata Spotted Deepsea Flounder
Mancopsetta milfordi Armless Deepsea Flounder

CRESTED FLOUNDERS

Samaridae

COCKATOO FLOUNDER
Samaris cristatus

There are about 20 species of Crested Flounder, and they are found mainly in deep tropical and subtropical waters of the Indo-Pacific. They have an elongate, oval body with ctenoid or cycloid scales, and a small mouth. The eyes are on the right side of the head, and the origin of the dorsal fin is anterior to the upper eye. In some species the dorsal fin has greatly elongated anterior rays. The ventral fins have five rays and are symmetrical, and the dorsal and anal fins are not attached to the caudal fin.

In Australia Crested Flounders are found mainly in deep northern waters, from about 20 to 300 m, though the **Threespot Righteye Flounder** occurs in shallow sandy areas near coral reefs. Most species in the family are small, measuring about 10–20 cm in length, and rarely encountered. They feed on small benthic invertebrates.

AUSTRALIAN SAMARIDAE SPECIES

Plagiopsetta glossa Tongue Flatfish
Samaris cristatus Cockatoo Flounder
Samaris macrolepis Bigfin Righteye Flounder
Samariscus huysmani Huysman's Flounder
Samariscus sunieri Sunier's Flounder
Samariscus triocellatus Threespot Righteye Flounder

SOLES

Soleidae

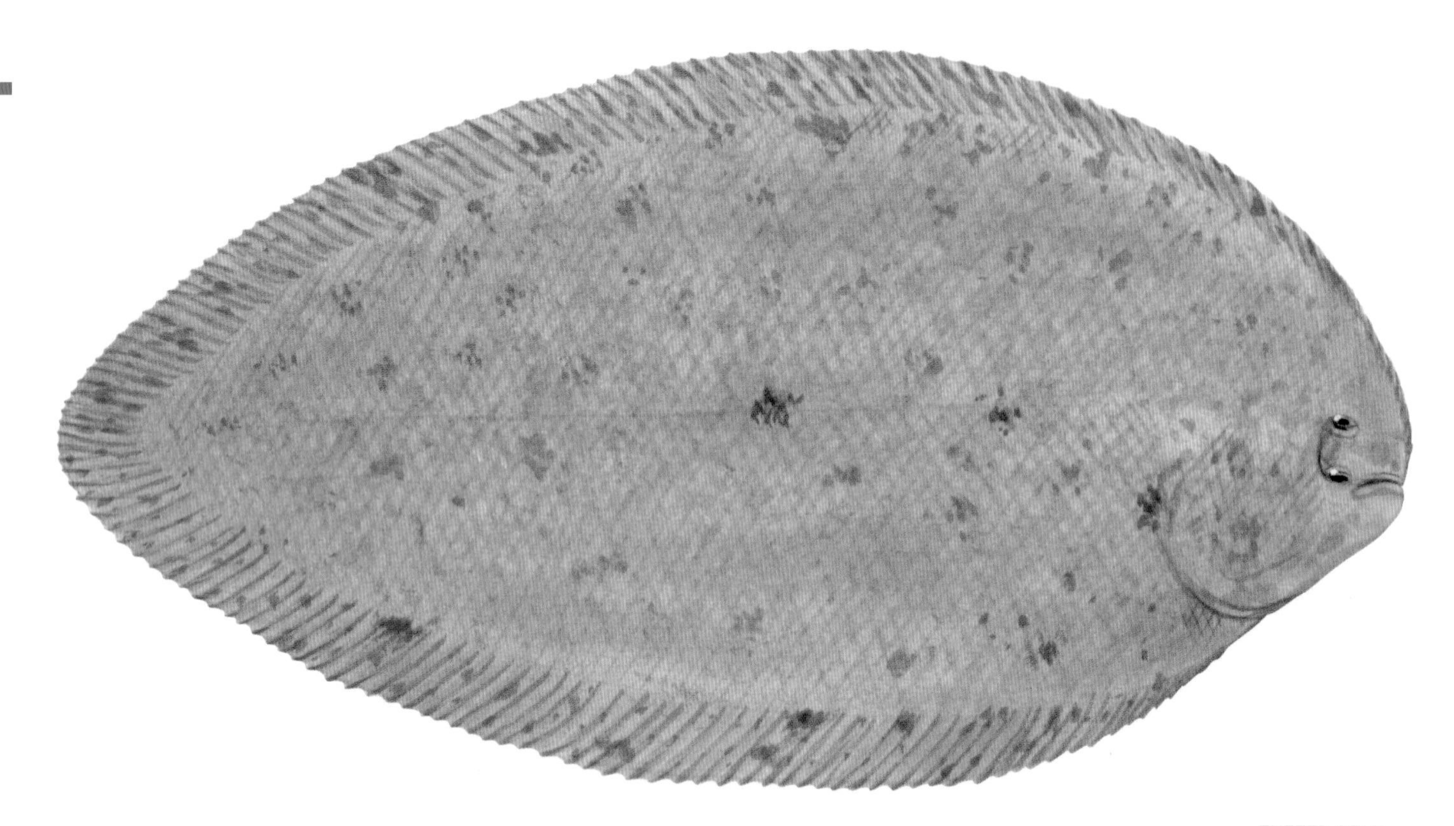

TUFTED SOLE
Dexillus muelleri

The large family of Soles contains about 130 species. They are found in tropical and temperate waters from Europe and around Africa to Australia and Japan, with a few species found in fresh water. Soles have an elongate body with moderately large cycloid or ctenoid scales, and a small mouth, often with a hook-shaped snout and curved lower jaw. The small eyes are on the right side of the head, and the margin of the preoperculum is hidden by skin and scales. The origin of the dorsal fin is anterior to the upper eye, the pectoral fin is sometimes absent, and the ventral fins may be symmetrical or asymmetrical and are attached to the anal fin. The dorsal and anal fins are attached to the caudal fin in some species. Many Soles have striking patterns of bands or spots.

Soles are found right around the coast of Australia, occurring on sand and mud bottoms mainly in coastal waters, shallow bays and estuaries, down to about 60 m. In other areas of the world Soles are very important commercially, but in Australia they are usually taken only as bycatch by bottom trawlers and many Australian species are too small to be processed. One species that appears regularly in south-eastern Australian trawl catches is the **Black Sole**, which reaches about 35 cm in length and is highly esteemed as a food fish. The **Peacock Sole** grows to about 22 cm and is found in northern Australian waters. It produces a toxin in its skin that repels predators, including sharks, and may also help it to capture the small benthic invertebrates on which it feeds. Three species of Sole are found in fresh water in far northern Australia: the **Saltpan Sole** and **Freshwater Sole** spend their entire lives in freshwater rivers and pools, while the **Kimberley Sole** is less well known and may only move into the lower reaches of freshwater rivers to breed.

THICKRAY SOLE
Aesopia cornuta

DARKSPOTTED SOLE
Aseraggodes melanospilos

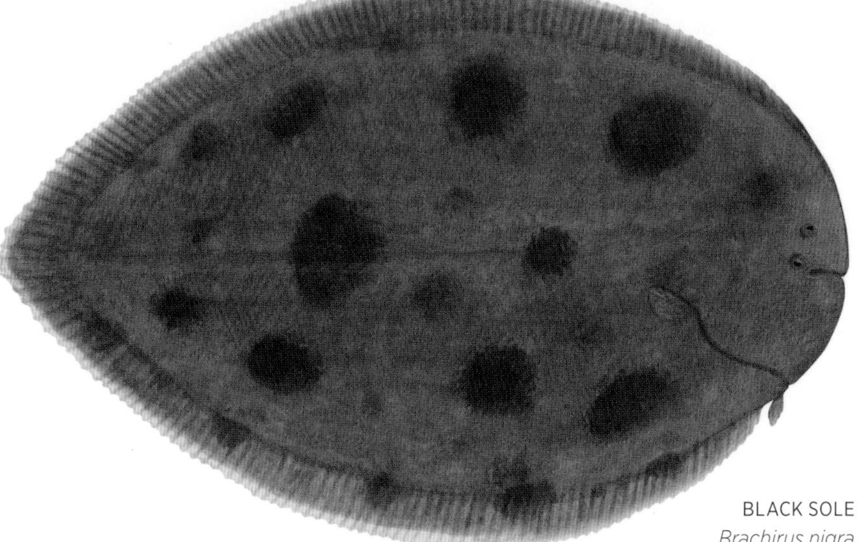

BLACK SOLE
Brachirus nigra

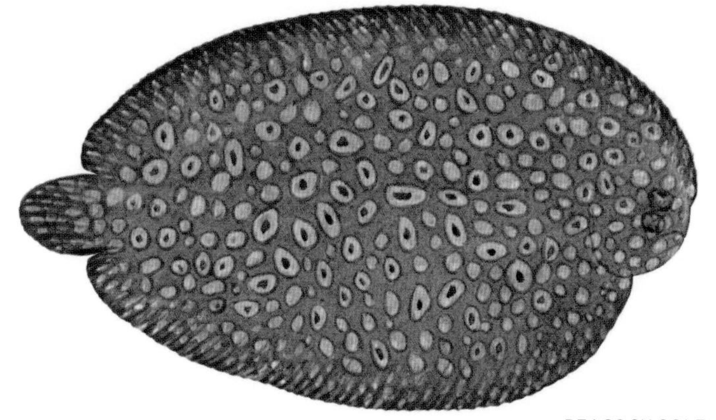

PEACOCK SOLE
Pardachirus pavoninus

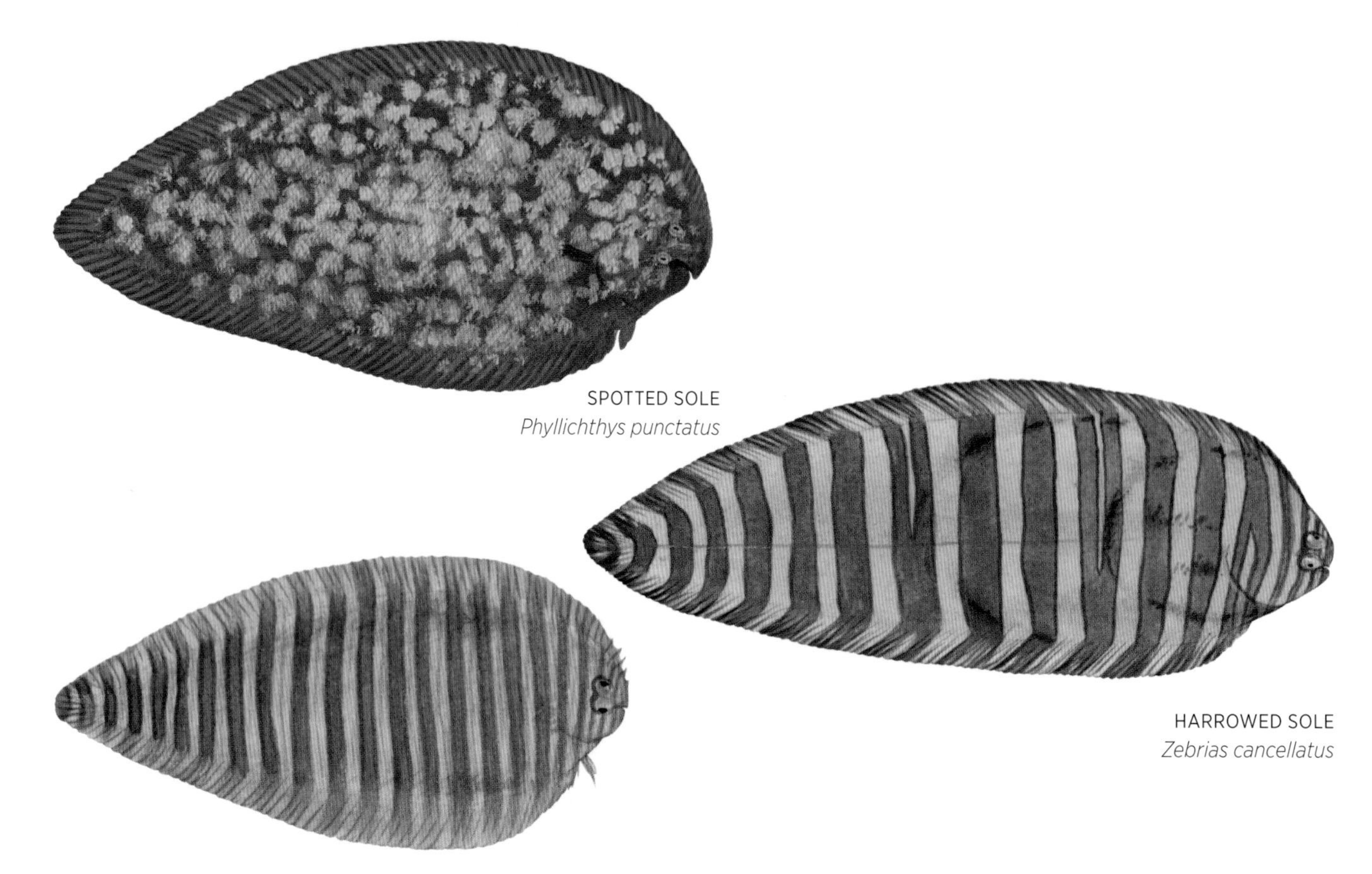

SPOTTED SOLE
Phyllichthys punctatus

HARROWED SOLE
Zebrias cancellatus

WICKER-WORK SOLE
Zebrias craticulus

AUSTRALIAN SOLEIDAE SPECIES

Aesopia cornuta Thickray Sole
Aseraggodes bahamondei Island Sole
Aseraggodes haackeanus Southern Sole
Aseraggodes klunzingeri Kimberley Sole
Aseraggodes macleayanus Narrowbanded Sole
Aseraggodes melanospilos Darkspotted Sole
Aseraggodes melanostictus Dappled Sole
Aseraggodes normani Norman's Sole
Aseraggodes ramsaii Ramsay's Flounder
Aseraggodes whitleyi Bearded Sole
Brachirus annularis Annulate Sole
Brachirus aspilos Dusky Sole
Brachirus breviceps Shorthead Sole
Brachirus fitzroiensis Fitzroy Sole
Brachirus nigra Black Sole
Brachirus orientalis Oriental Sole
Brachirus salinarum Saltpan Sole
Brachirus selheimi Freshwater Sole
Dexillus muelleri Tufted Sole
Heteromycteris hartzfeldi Hooknose Sole
Paradicula setifer Paradice's Sole
Pardachirus hedleyi Southern Peacock Sole
Pardachirus pavoninus Peacock Sole
Pardachirus poropterus Mottled Sole
Phyllichthys punctatus Spotted Sole
Phyllichthys sclerolepis Hardscale Sole
Phyllichthys sejunctus Checker Sole
Pseudaesopia japonica Japanese Sole
Rendahlia jaubertensis Jaubert Sole
Soleichthys heterorhinos Tiger Sole
Soleichthys maculosus Whiteblotched Sole
Soleichthys microcephalus Smallhead Sole
Soleichthys oculofasciatus Banded-eye Sole
Soleichthys serpenpellis Snakeskin Sole
Zebrias cancellatus Harrowed Sole
Zebrias craticulus Wicker-work Sole
Zebrias munroi Munro's Sole
Zebrias penescalaris Duskybanded Sole
Zebrias quagga Zebra Sole
Zebrias scalaris Manyband Sole

TONGUE SOLES

Cynoglossidae

LEMON TONGUE SOLE
Paraplagusia bilineata

Tongue Soles are found in tropical and temperate waters around the world. There are about 127 species in the family, a few of which occur in fresh water. They have an elongate, compressed body that tapers to a point posteriorly, with small cycloid or ctenoid scales, a small, curved mouth, and the snout formed into a hook shape before the mouth. The small eyes are set close together and are usually on the left side of the head. The margin of the preoperculum is obscured by skin and scales. The origin of the dorsal fin is well before the eyes, and both the dorsal fin and anal fin are attached to the caudal fin. The pectoral fins are absent and usually only the right ventral fin is present.

Tongue Soles are widely distributed in Australian waters and are found from the intertidal zone to depths of more than 1500 m, on mud and sand bottoms. Most are small, from 20 to 30 cm in length. They are excellent food fishes and are important commercially in other areas of the world. In Australia they occasionally appear as bycatch in bottom trawls, but most Australian species are too small to be utilised as food fishes. The **Fourline Tongue Sole** is an exception: it reaches about 40 cm in length in northern waters and is often taken commercially in small quantities. One species in the family, the **Freshwater Tongue Sole**, occurs in the East Alligator River in the Northern Territory and is known only from a few specimens. Tongue Soles feed on small benthic invertebrates.

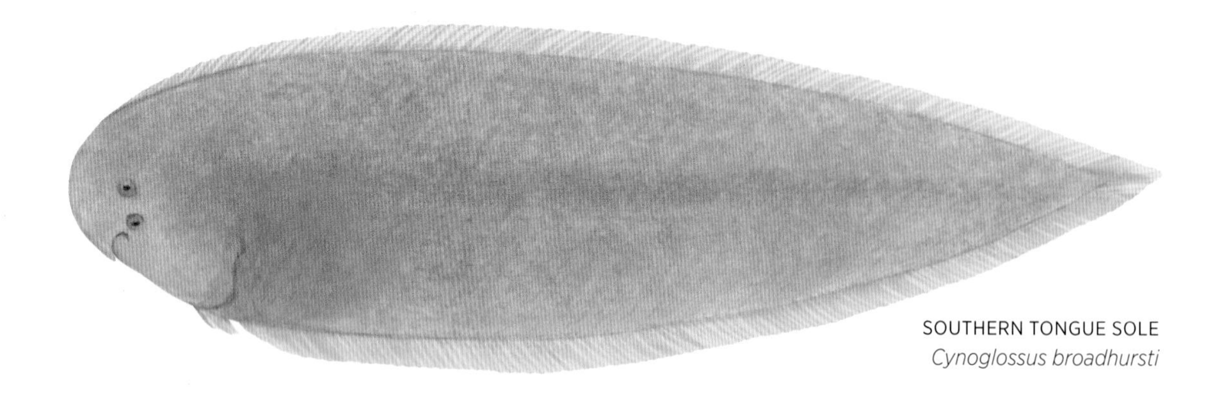

SOUTHERN TONGUE SOLE
Cynoglossus broadhursti

AUSTRALIAN CYNOGLOSSIDAE SPECIES

Cynoglossus bilineatus Fourline Tongue Sole
Cynoglossus broadhursti Southern Tongue Sole
Cynoglossus heterolepis Freshwater Tongue Sole
Cynoglossus kopsii Kops' Tongue Sole
Cynoglossus maccullochi McCulloch's Tongue Sole
Cynoglossus macrophthalmus Longnose Tongue Sole
Cynoglossus maculipinnis Spotfin Tongue Sole
Cynoglossus ogilbyi Ogilby's Tongue Sole
Cynoglossus puncticeps Spotted Tongue Sole
Cynoglossus suyeni Suyen's Tongue Sole
Paraplagusia bilineata Lemon Tongue Sole
Paraplagusia blochii Bloch's Tongue Sole
Paraplagusia longirostris Pinocchio Tongue Sole
Paraplagusia sinerama Dusky Tongue Sole
Symphurus australis Deepwater Tongue Sole
Symphurus microrhynchus Small-lip Tongue Sole

TETRAODONTIFORMES

Tetraodontiforms are the mostly highly evolved of the bony fishes. Most have very few teeth in the jaws (usually four), and many bones of the skeleton are fused together or absent. The ribs and ventral fins are absent in many species and the number of vertebrae is much reduced. Most have highly modified scales formed into spines, prickles or plates. There are nine families in the order, all of which occur in Australian waters.

SPIKEFISHES

Triacanthodidae

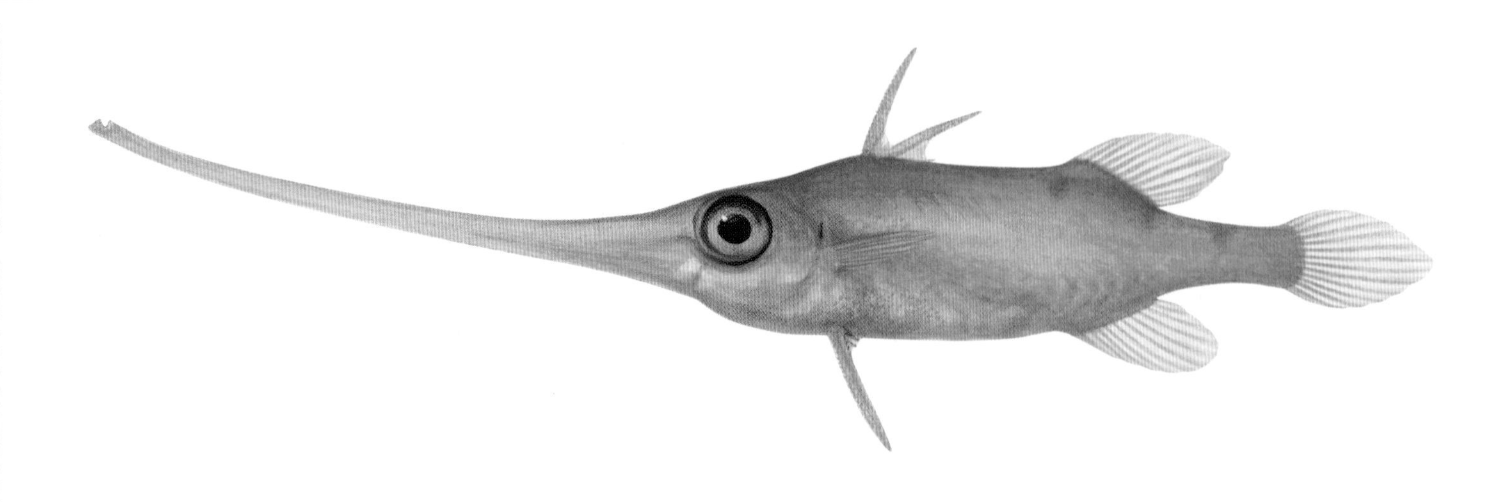

ALCOCK'S SPIKEFISH
Halimochirurgus alcocki

Spikefishes are found in tropical and subtropical waters of the Western Atlantic and Indo-Pacific, and there are about 21 species in the family. They have deep to elongate, slightly compressed bodies with thick skin that is covered with minute, spiny scales, giving it a sandpapery feel. They have large eyes and a small mouth, and in some species the snout is formed into a tube. The gill openings are reduced to a small slit before the pectoral fin. They have a single, deeply notched dorsal fin with six spines and 12–18 rays: the first spine is the largest, with the others decreasing in size and sometimes very small and concealed. The anal fin has 11–16 rays, the ventral fins have one large, strong spine and several very small rays, and the caudal fin is rounded to truncate.

Spikefishes are bottom dwellers, mainly occurring in deep water from about 35 to 900 m. In Trumpetsnouts, the snout is formed into a very long tube, with the small mouth at its tip oriented vertically as opposed to horizontally. The **Alcock's Spikefish** and **Longsnout Spikefish** also have a long, tube-like snout. Spikefishes are sometimes taken in large numbers by bottom trawlers and are considered a nuisance, as their strong, serrated spines easily become entangled in nets and are difficult to remove. Spikefishes reach a maximum length of 10–20 cm and feed on small benthic invertebrates.

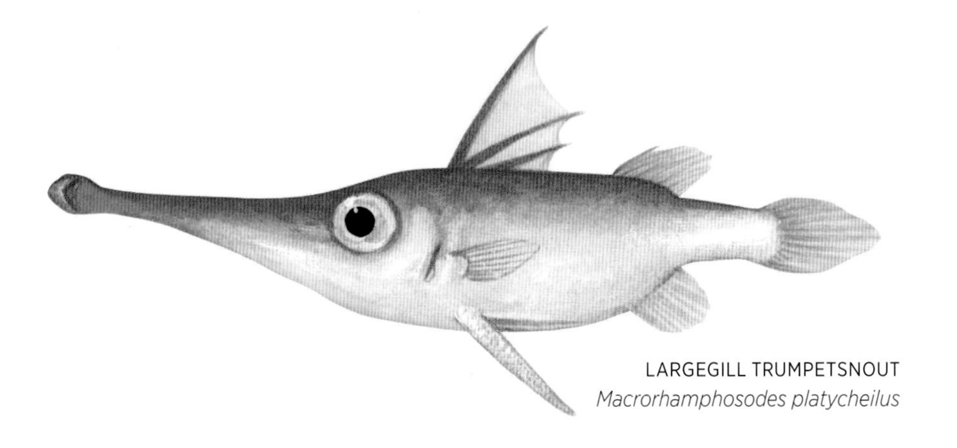

LARGEGILL TRUMPETSNOUT
Macrorhamphosodes platycheilus

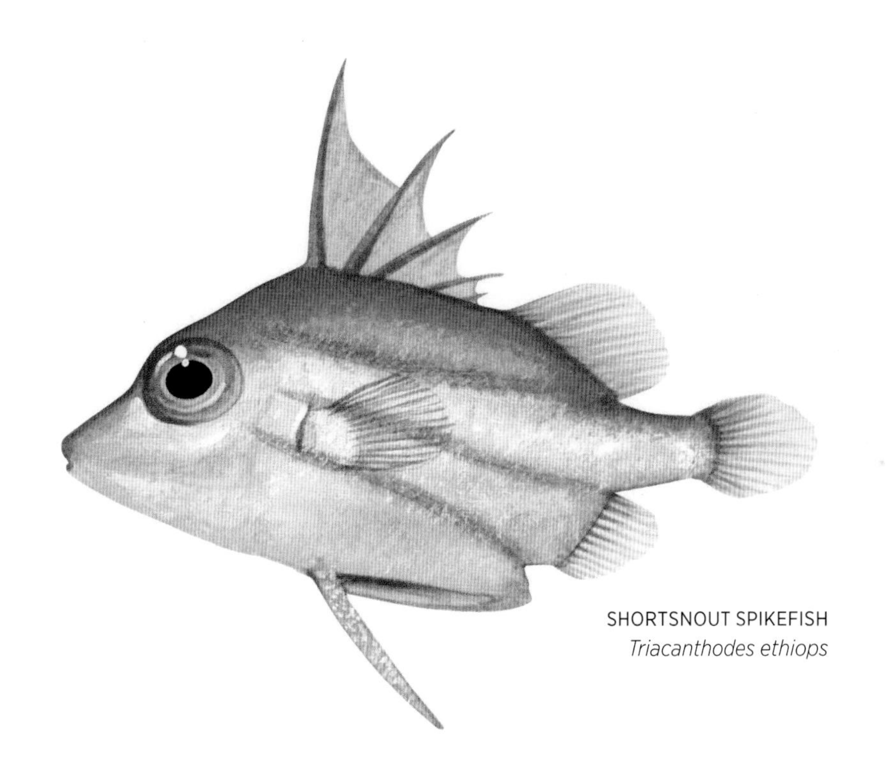

SHORTSNOUT SPIKEFISH
Triacanthodes ethiops

AUSTRALIAN TRIACANTHODIDAE SPECIES

Bathyphylax bombifrons Boomer Spikefish
Bathyphylax omen Signal Spikefish
Halimochirurgus alcocki Alcock's Spikefish
Halimochirurgus centriscoides Longsnout Spikefish
Macrorhamphosodes platycheilus Largegill Trumpetsnout
Macrorhamphosodes uradoi Common Trumpetsnout
Paratriacanthodes herrei Thickspine Spikefish
Paratriacanthodes retrospinis Sawspine Spikefish
Triacanthodes ethiops Shortsnout Spikefish
Tydemania navigatoris Fleshylip Spikefish

TRIPODFISHES

Triacanthidae

SILVER TRIPODFISH
Triacanthus nieuhofi

Tripodfishes are found in tropical waters of the Indo-Pacific, and there are about seven species in the family. They have a deep, slightly compressed, tapering body with a long caudal peduncle. Their thick skin is covered in small spiny scales, giving it a sandpapery feel. The gill openings are reduced to a small slit before the pectoral fin, and the mouth is small and bears large incisor and molariform teeth. They have two dorsal fins: the first with five spines (the anterior spine longest, serrated and very strong), and the second with 20–26 rays. The anal fin has 13–22 rays, the ventral fins have one very large, strong, serrated spine and no visible rays, and the caudal fin is deeply forked.

Tripodfishes are found in shallow northern Australian coastal waters over sand and weed bottoms, and sometimes in estuaries. They are frequently taken by prawn trawlers and are disliked for their tendency to become entangled in the nets – the tripod of strong, serrated, sharp dorsal and ventral spines makes them very difficult to remove. They reach a maximum length of about 30 cm and feed on small benthic invertebrates.

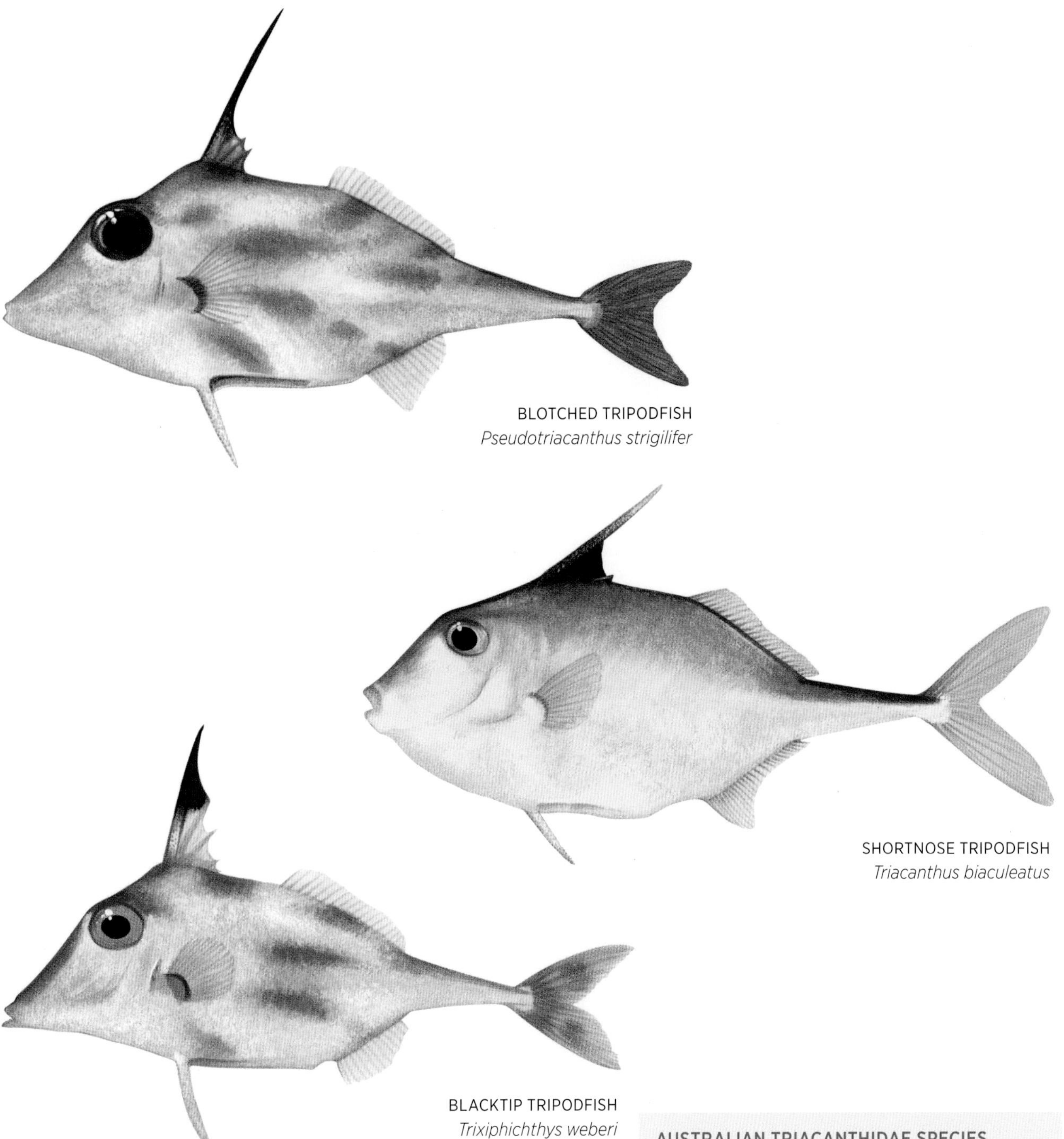

BLOTCHED TRIPODFISH
Pseudotriacanthus strigilifer

SHORTNOSE TRIPODFISH
Triacanthus biaculeatus

BLACKTIP TRIPODFISH
Trixiphichthys weberi

AUSTRALIAN TRIACANTHIDAE SPECIES

Pseudotriacanthus strigilifer Blotched Tripodfish
Triacanthus biaculeatus Shortnose Tripodfish
Triacanthus nieuhofi Silver Tripodfish
Tripodichthys angustifrons Yellowfin Tripodfish
Tripodichthys blochii Longtail Tripodfish
Trixiphichthys weberi Blacktip Tripodfish

TRIGGERFISHES

Balistidae

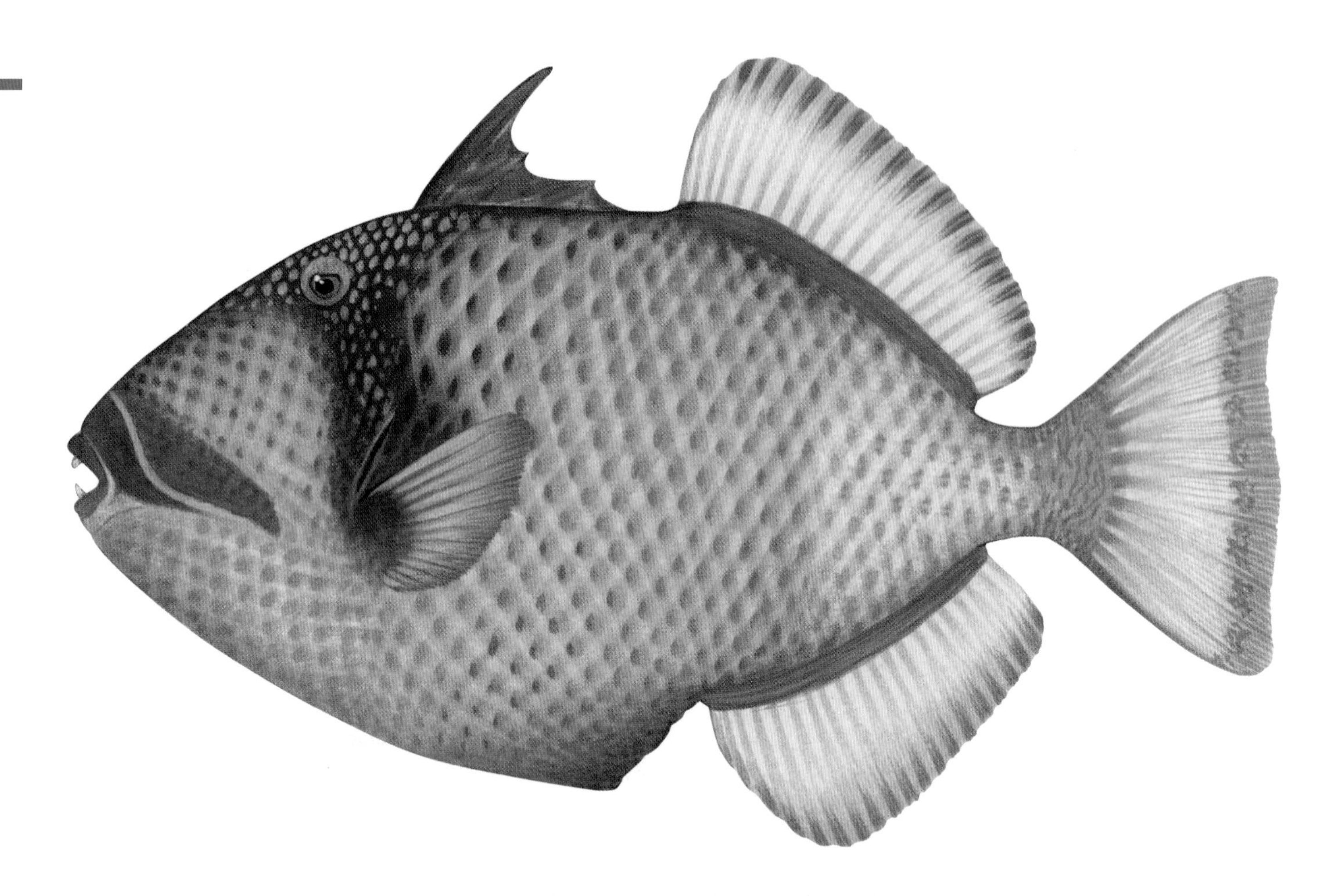

TITAN TRIGGERFISH
Balistoides viridescens

Triggerfishes are found in tropical and warm temperate waters of the Atlantic, Indian and Pacific oceans, and there are about 40 species in the family. They have a deep, moderately compressed body with very tough skin and large, thick, plate-like scales. The small mouth has large, strong teeth; the eyes are placed very close to the dorsal profile; and the gill openings are reduced to a slit before the pectoral fin. They have two dorsal fins: the first with three spines, the second of which forms a locking mechanism that holds the larger first spine in a vertical position; and the second fin with 23–36 rays. The ventral fins are rudimentary, appearing on the end of the long pelvic bone in the form of four pairs of enlarged scales. The anal fin has no spines and 21–31 rays, and the caudal fin is rounded to lunate. Triggerfishes swim by undulating the dorsal and anal fins, with the caudal fin used only for sudden bursts of speed.

In Australia Triggerfishes mainly inhabit coral and rocky reefs in northern waters, and feed on a wide variety of benthic invertebrates such as crustaceans, echinoderms, molluscs and corals, and small fishes. Some species, such as the **Redtooth Triggerfish**, feed on zooplankton and occur in large schools in midwater. The **Whitespotted Triggerfish** is pelagic in oceanic waters and is occasionally encountered over outer reefs. Most species, however, are closely associated with the substrate and take refuge in holes in the reef, wedging themselves firmly in place with their robust, lockable dorsal spine. The **Titan Triggerfish** is a large, aggressive species reaching over 80 cm in length. The female builds a nest of coral rubble and defends it against any threat, having been known to vigorously attack and bite divers. *Rhinecanthus* species are superbly patterned with contrasting stripes, and occur over sand and rubble areas near coral reefs. The **Clown Triggerfish** inhabits similar areas and is equally stunning, with blue and white spots and bright-yellow lips. The **Black Triggerfish** is commonly encountered on coral reefs in northern Australian waters and feeds mainly on algae.

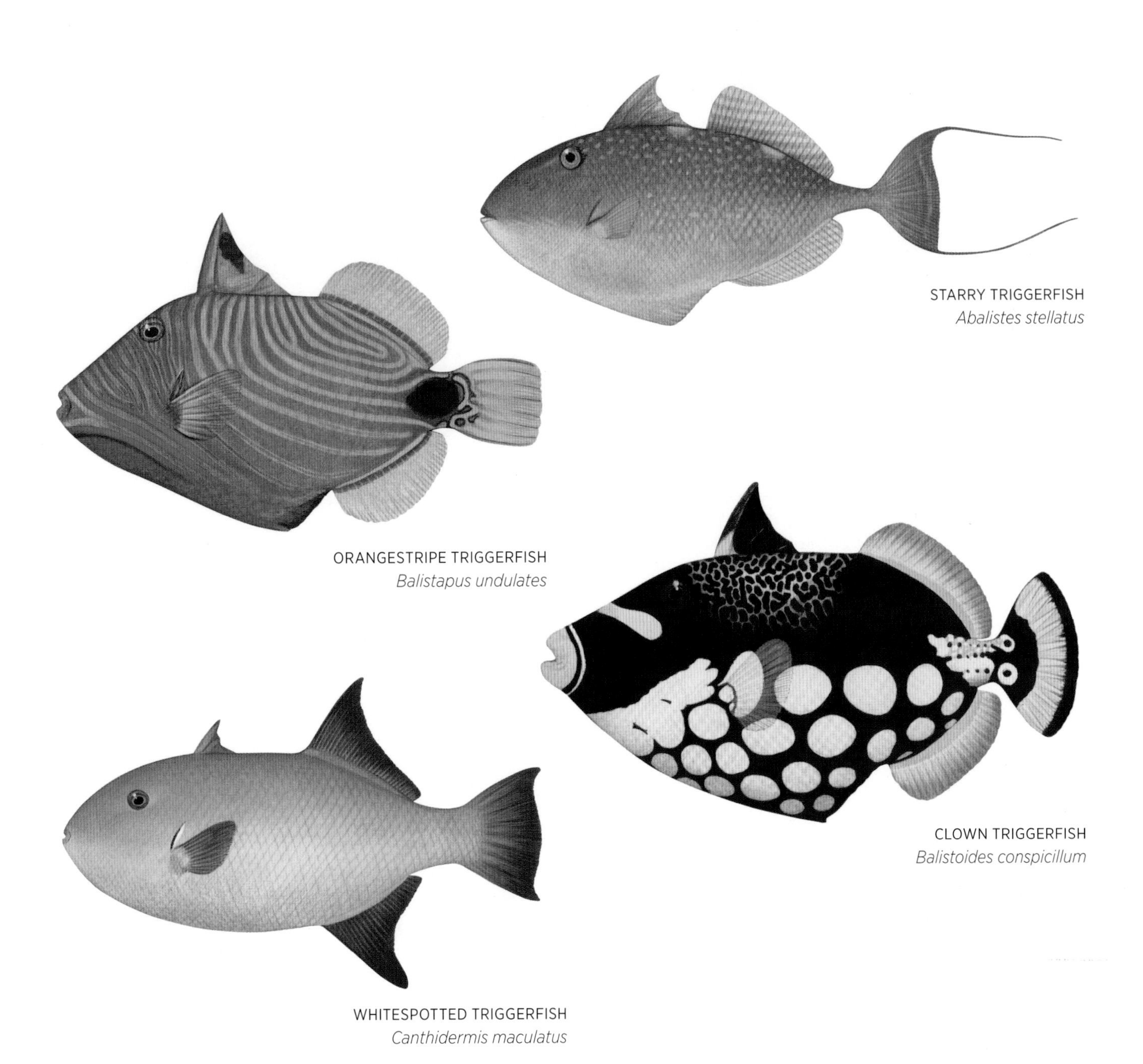

STARRY TRIGGERFISH
Abalistes stellatus

ORANGESTRIPE TRIGGERFISH
Balistapus undulates

CLOWN TRIGGERFISH
Balistoides conspicillum

WHITESPOTTED TRIGGERFISH
Canthidermis maculatus

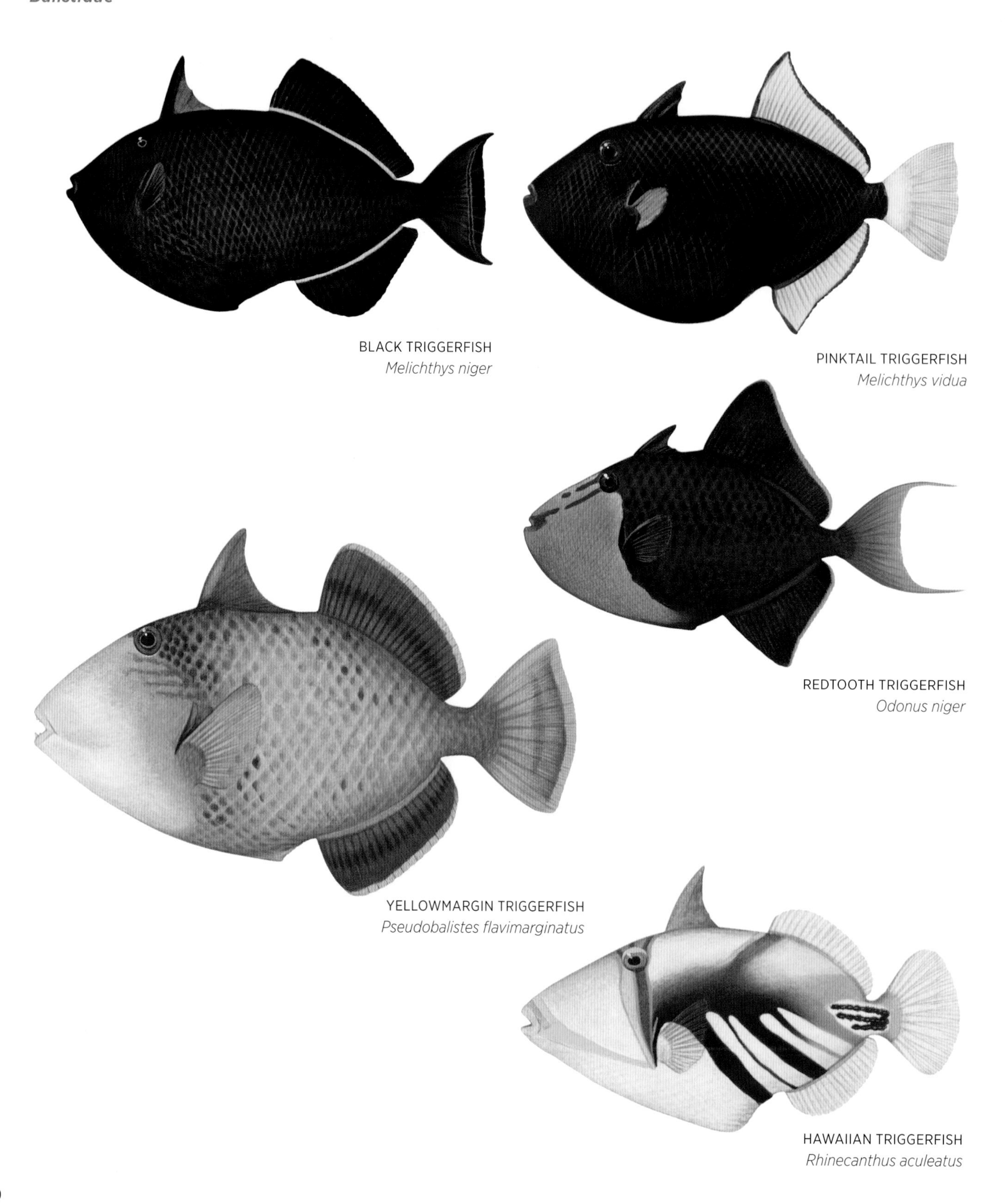

BLACK TRIGGERFISH
Melichthys niger

PINKTAIL TRIGGERFISH
Melichthys vidua

REDTOOTH TRIGGERFISH
Odonus niger

YELLOWMARGIN TRIGGERFISH
Pseudobalistes flavimarginatus

HAWAIIAN TRIGGERFISH
Rhinecanthus aculeatus

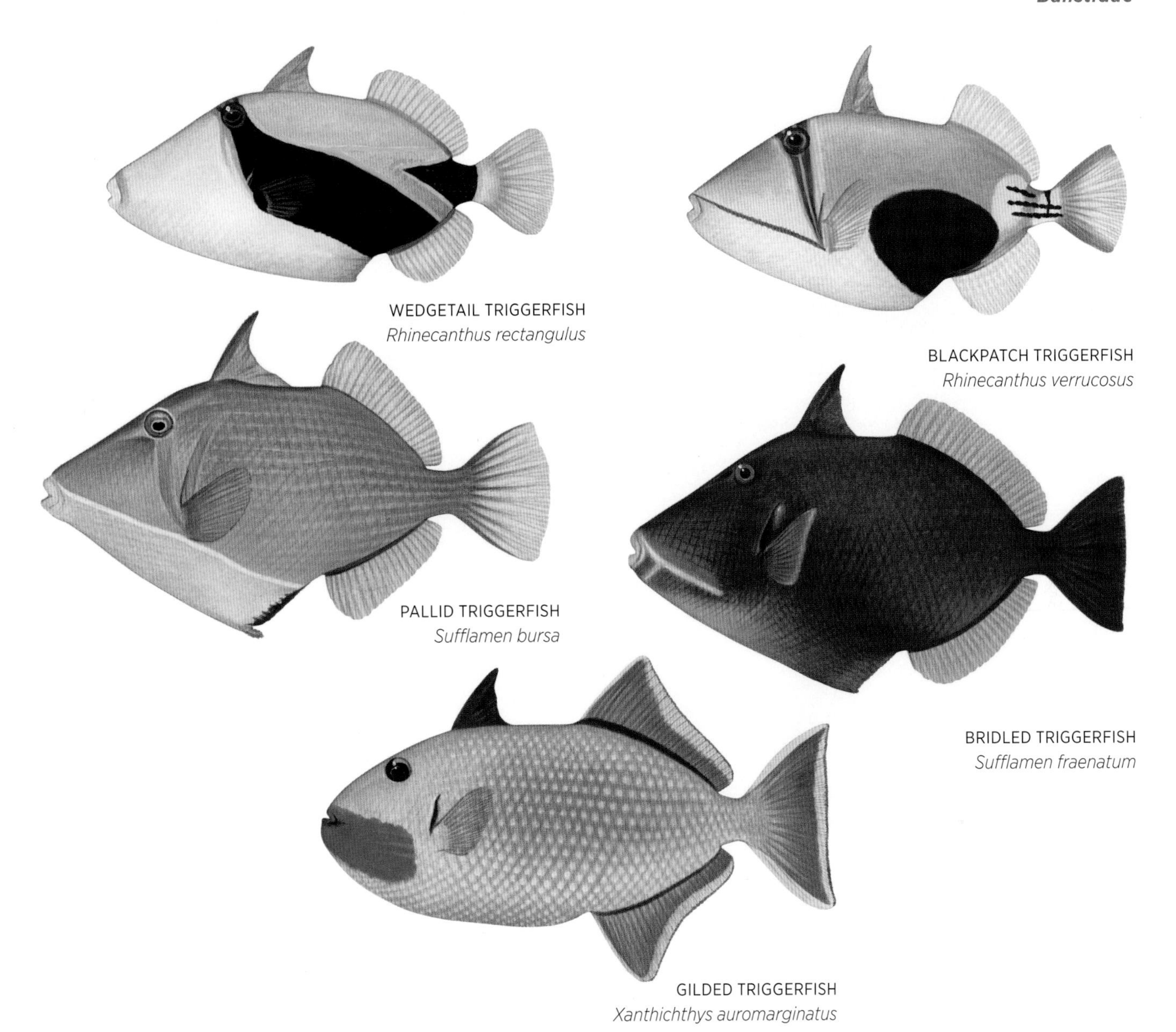

WEDGETAIL TRIGGERFISH
Rhinecanthus rectangulus

BLACKPATCH TRIGGERFISH
Rhinecanthus verrucosus

PALLID TRIGGERFISH
Sufflamen bursa

BRIDLED TRIGGERFISH
Sufflamen fraenatum

GILDED TRIGGERFISH
Xanthichthys auromarginatus

AUSTRALIAN BALISTIDAE SPECIES

Abalistes filamentosus Hairfin Triggerfish
Abalistes stellatus Starry Triggerfish
Balistapus undulatus Orangestripe Triggerfish
Balistoides conspicillum Clown Triggerfish
Balistoides viridescens Titan Triggerfish
Canthidermis maculatus Whitespotted Triggerfish
Melichthys niger Black Triggerfish
Melichthys vidua Pinktail Triggerfish
Odonus niger Redtooth Triggerfish
Pseudobalistes flavimarginatus Yellowmargin Triggerfish
Pseudobalistes fuscus Yellowspotted Triggerfish
Rhinecanthus aculeatus Hawaiian Triggerfish
Rhinecanthus lunula Halfmoon Triggerfish
Rhinecanthus rectangulus Wedgetail Triggerfish
Rhinecanthus verrucosus Blackpatch Triggerfish
Sufflamen bursa Pallid Triggerfish
Sufflamen chrysopterum Eye-stripe Triggerfish
Sufflamen fraenatum Bridled Triggerfish
Xanthichthys auromarginatus Gilded Triggerfish
Xanthichthys caeruleolineatus Blueline Triggerfish
Xanthichthys lineopunctatus Lined Triggerfish
Xenobalistes tumidipectoris Wingkeel Triggerfish

LEATHERJACKETS

Monocanthidae

OCEAN JACKET
Nelusetta ayraudi

Leatherjackets are found in temperate and tropical waters around the world, and are most diverse in Australian waters. There are about 102 species in the family. They have deep, almost circular to elongate, compressed bodies with tough, leathery skin and small, prickly scales that give them a sand-papery feel. The eyes are placed very close to the dorsal profile. They have a small mouth with large, pointed teeth, and in some species the anterior teeth have flattened cutting edges. The gill openings are reduced to a small slit before the pectoral fin. Leatherjackets have two dorsal fins: the first with two spines, the first of which can be locked in a vertical position by the much smaller second spine; and the second fin long-based with 22–52 unbranched rays. The anal fin is similar to the second dorsal fin, with 20–62 unbranched rays. The ventral fins are reduced to a spiny knob on the end of a long pelvic bone, which can be raised or lowered in some species to contract or expand the ventral profile. The caudal fin varies in shape depending on the species, from truncate to rounded, lunate to pointed. Leatherjackets swim by undulating the dorsal and anal fins, with the caudal fin used only for sudden bursts of speed.

Leatherjackets are widely distributed in Australian waters, particularly around the southern coast, where they are common in weed beds and over rocky reefs. They occur from the shallows down to 200 m or more. Many species are sexually dimorphic, with males usually more colourful and larger than females. Males of some species have patches of elongated bristles or sharp spines on the caudal peduncle.

Leatherjackets range in size from the 8 cm **Southern Pygmy Leatherjacket** to the coral-reef dwelling **Scrawled Leatherjacket**, which can reach over 1 m. Many are excellent food fishes, particularly the larger southern Australian species. The **Ocean Jacket**, which is pelagic in offshore waters, is fished commercially in large quantities off the south coast. The **Tasselled Leatherjacket** is found in coastal Australian waters, mainly in the north, and its body is covered with camouflaging weed-like dermal filaments.

The small, brilliantly coloured **Harlequin Filefish** is found on northern reefs, usually in pairs, sheltering in staghorn corals. The **Bearded Leatherjacket** has a very elongate body and a long barbel on the lower jaw. It is found in weed areas around northern Australia. The **Blacksaddle Filefish**, which is found on coral reefs, mimics a toxic Puffer (Tetraodontidae, p. 773) to deter predators. Leatherjackets are opportunistic feeders, grazing on algae and epiphytes growing on seagrass, and preying on benthic invertebrates and coral polyps.

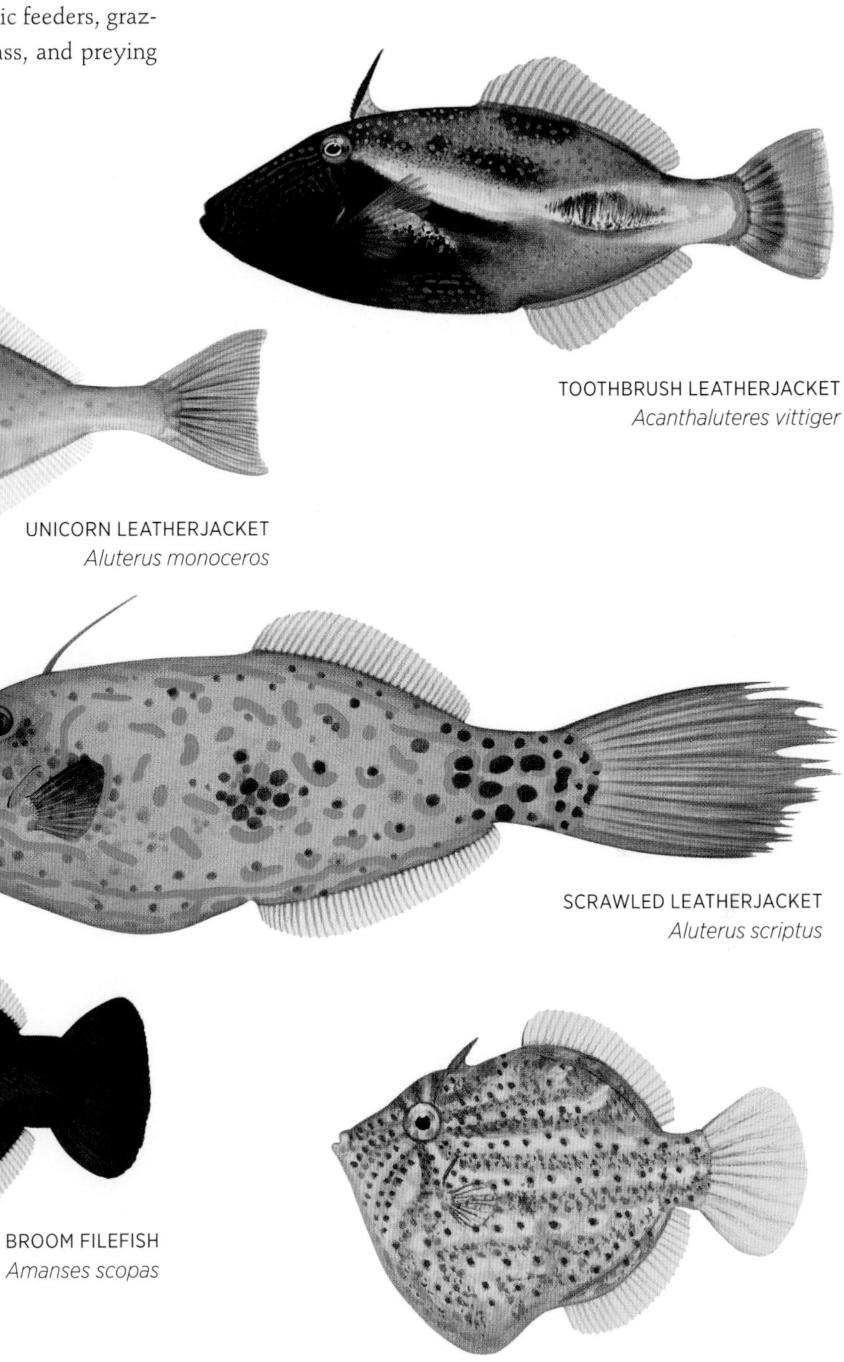

TOOTHBRUSH LEATHERJACKET
Acanthaluteres vittiger

UNICORN LEATHERJACKET
Aluterus monoceros

SCRAWLED LEATHERJACKET
Aluterus scriptus

BROOM FILEFISH
Amanses scopas

SOUTHERN PYGMY LEATHERJACKET
Brachaluteres jacksonianus

BEARDED LEATHERJACKET
Anacanthus barbatus

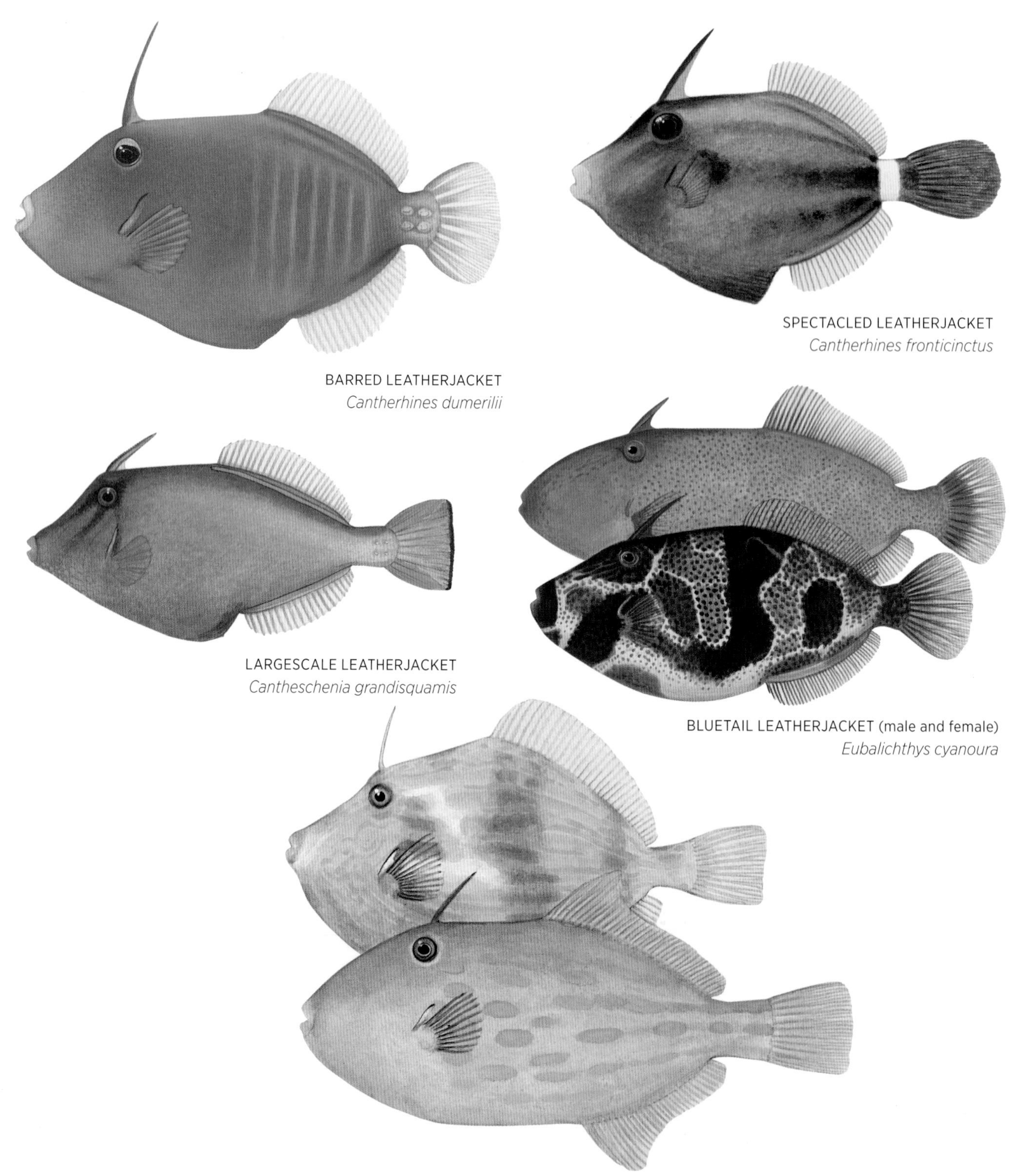

BARRED LEATHERJACKET
Cantherhines dumerilii

SPECTACLED LEATHERJACKET
Cantherhines fronticinctus

LARGESCALE LEATHERJACKET
Cantheschenia grandisquamis

BLUETAIL LEATHERJACKET (male and female)
Eubalichthys cyanoura

MOSAIC LEATHERJACKET (male and female)
Eubalichthys mosaicus

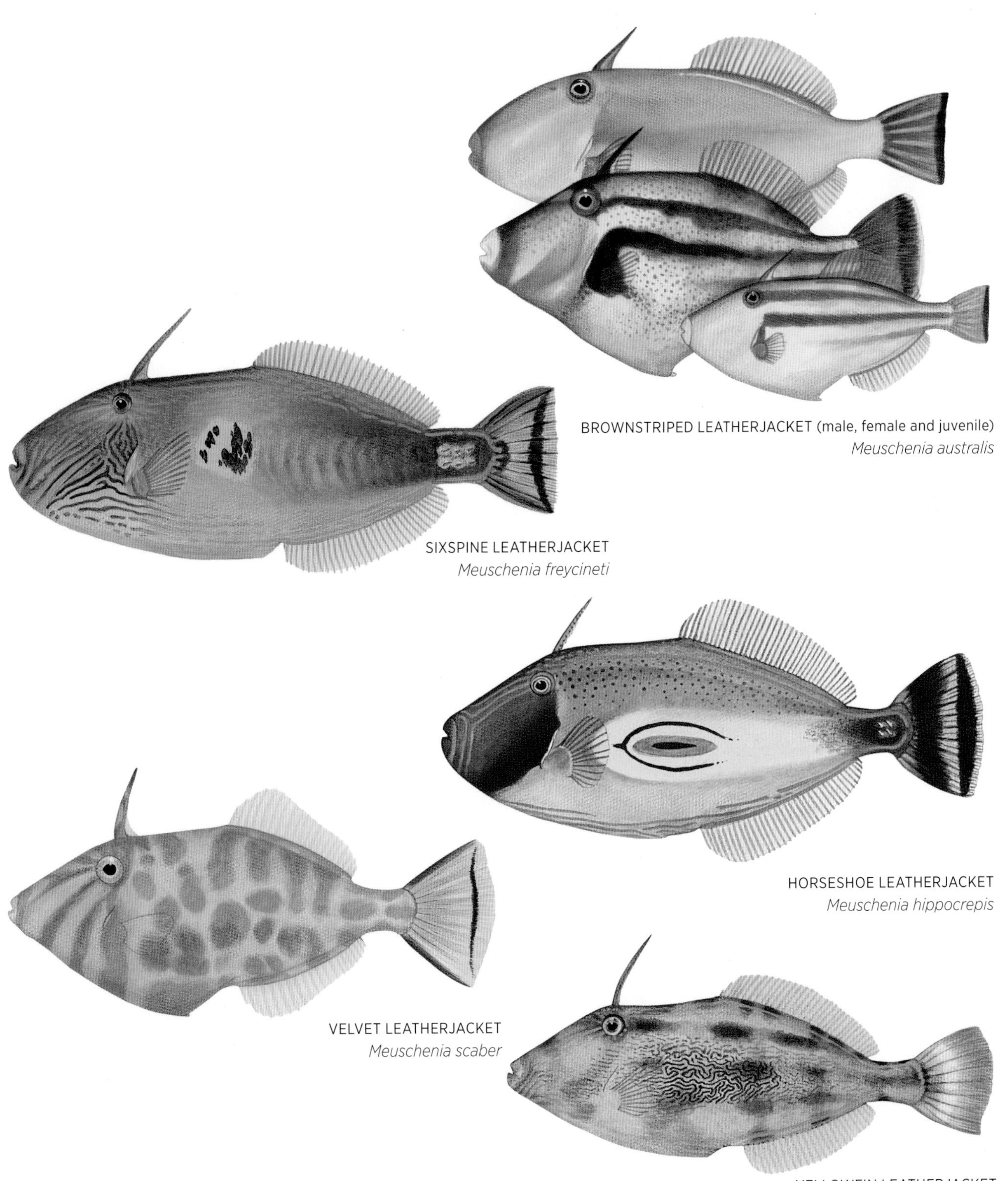

BROWNSTRIPED LEATHERJACKET (male, female and juvenile)
Meuschenia australis

SIXSPINE LEATHERJACKET
Meuschenia freycineti

HORSESHOE LEATHERJACKET
Meuschenia hippocrepis

VELVET LEATHERJACKET
Meuschenia scaber

YELLOWFIN LEATHERJACKET
Meuschenia trachylepis

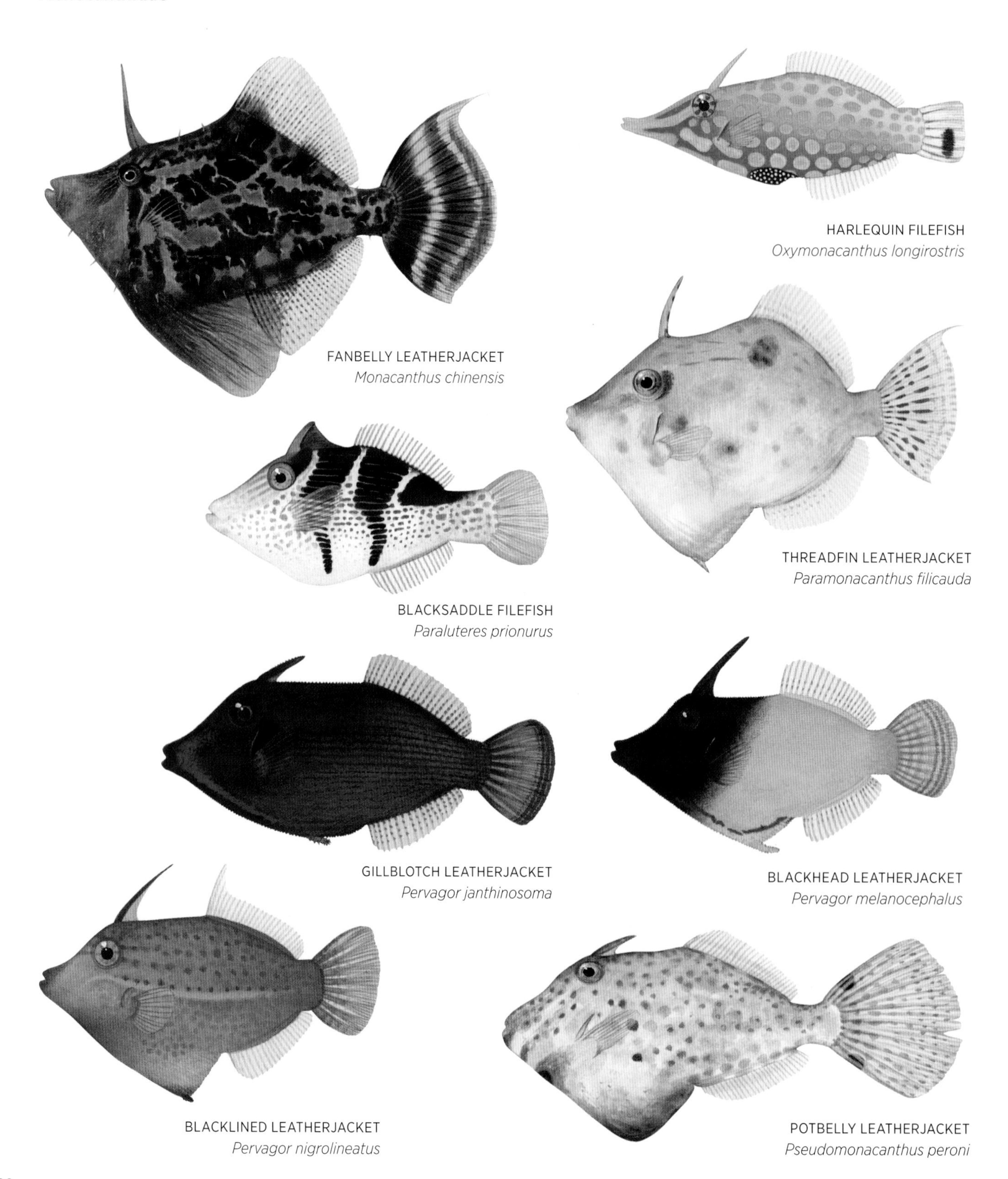

HARLEQUIN FILEFISH
Oxymonacanthus longirostris

FANBELLY LEATHERJACKET
Monacanthus chinensis

THREADFIN LEATHERJACKET
Paramonacanthus filicauda

BLACKSADDLE FILEFISH
Paraluteres prionurus

GILLBLOTCH LEATHERJACKET
Pervagor janthinosoma

BLACKHEAD LEATHERJACKET
Pervagor melanocephalus

BLACKLINED LEATHERJACKET
Pervagor nigrolineatus

POTBELLY LEATHERJACKET
Pseudomonacanthus peroni

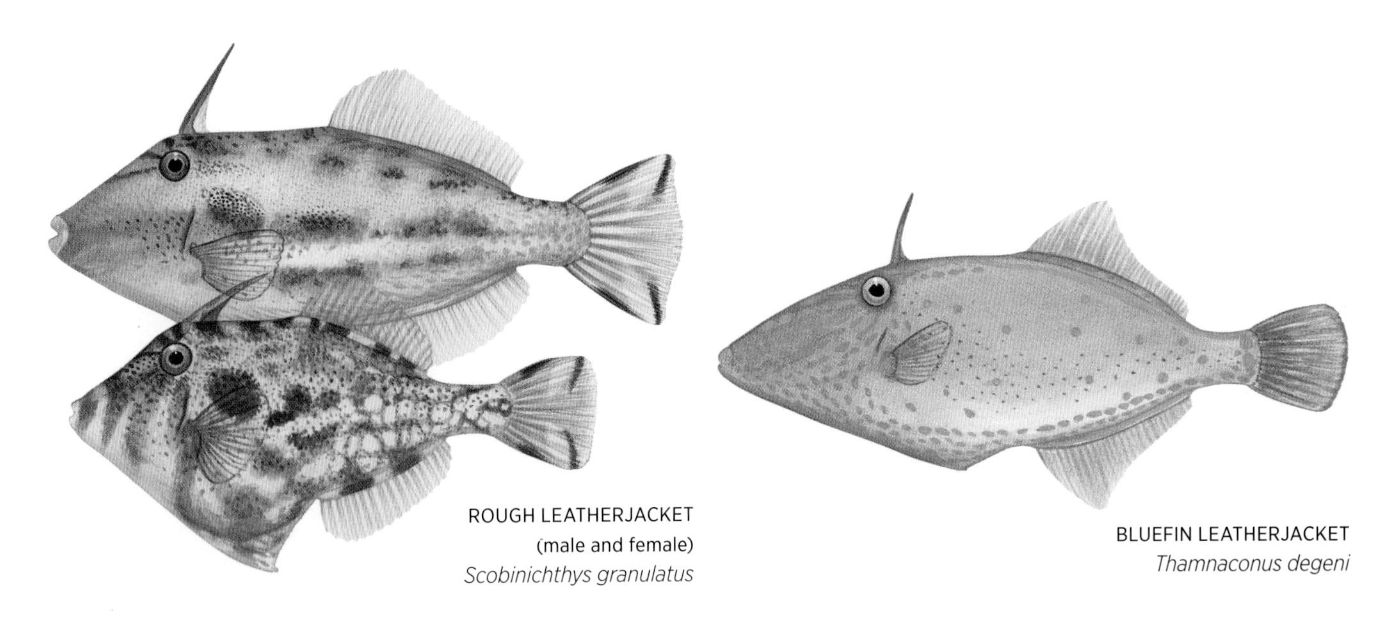

ROUGH LEATHERJACKET
(male and female)
Scobinichthys granulatus

BLUEFIN LEATHERJACKET
Thamnaconus degeni

AUSTRALIAN MONOCANTHIDAE SPECIES

Acanthaluteres brownii Spinytail Leatherjacket
Acanthaluteres spilomelanurus Bridled Leatherjacket
Acanthaluteres vittiger Toothbrush Leatherjacket
Acreichthys radiatus Radial Leatherjacket
Acreichthys tomentosus Bristle-tail Leatherjacket
Aluterus monoceros Unicorn Leatherjacket
Aluterus scriptus Scrawled Leatherjacket
Amanses scopas Broom Filefish
Anacanthus barbatus Bearded Leatherjacket
Brachaluteres jacksonianus Southern Pygmy Leatherjacket
Brachaluteres taylori Taylor's Pygmy Leatherjacket
Cantherhines dumerilii Barred Leatherjacket
Cantherhines fronticinctus Spectacled Leatherjacket
Cantherhines pardalis Honeycomb Leatherjacket
Cantheschenia grandisquamis Largescale Leatherjacket
Cantheschenia longipinnis Smoothspine Leatherjacket
Chaetodermis penicilligera Tasselled Leatherjacket
Colurodontis paxmani Paxman's Leatherjacket
Eubalichthys bucephalus Black Reef Leatherjacket
Eubalichthys caeruleoguttatus Bluespotted Leatherjacket
Eubalichthys cyanoura Bluetail Leatherjacket
Eubalichthys gunnii Gunn's Leatherjacket
Eubalichthys mosaicus Mosaic Leatherjacket
Eubalichthys quadrispinis Fourspine Leatherjacket
Meuschenia australis Brownstriped Leatherjacket
Meuschenia flavolineata Yellowstriped Leatherjacket
Meuschenia freycineti Sixspine Leatherjacket
Meuschenia galii Bluelined Leatherjacket
Meuschenia hippocrepis Horseshoe Leatherjacket
Meuschenia scaber Velvet Leatherjacket
Meuschenia trachylepis Yellowfin Leatherjacket
Meuschenia venusta Stars-and-stripes Leatherjacket
Monacanthus chinensis Fanbelly Leatherjacket
Nelusetta ayraudi Ocean Jacket
Oxymonacanthus longirostris Harlequin Filefish
Paraluteres prionurus Blacksaddle Filefish
Paramonacanthus choirocephalus Pigface Leatherjacket
Paramonacanthus filicauda Threadfin Leatherjacket
Paramonacanthus japonicus Japanese Leatherjacket
Paramonacanthus lowei Lowe's Leatherjacket
Paramonacanthus otisensis Dusky Leatherjacket
Paramonacanthus pusillus Sinhalese Leatherjacket
Pervagor alternans Yelloweye Leatherjacket
Pervagor aspricaudus Orangetail Leatherjacket
Pervagor janthinosoma Gillblotch Leatherjacket
Pervagor melanocephalus Blackhead Leatherjacket
Pervagor nigrolineatus Blacklined Leatherjacket
Pseudalutarius nasicornis Rhinoceros Leatherjacket
Pseudomonacanthus elongatus Fourband Leatherjacket
Pseudomonacanthus peroni Potbelly Leatherjacket
Rudarius excelsus Diamond Leatherjacket
Rudarius minutus Minute Leatherjacket
Scobinichthys granulatus Rough Leatherjacket
Thamnaconus analis Darkvent Leatherjacket
Thamnaconus degeni Bluefin Leatherjacket
Thamnaconus hypargyreus Yellowspotted Leatherjacket
Thamnaconus modestoides Modest Leatherjacket
Thamnaconus striatus Manyline Leatherjacket
Thamnaconus tessellatus Manyspot Leatherjacket

BOXFISHES AND COWFISHES

Ostraciidae

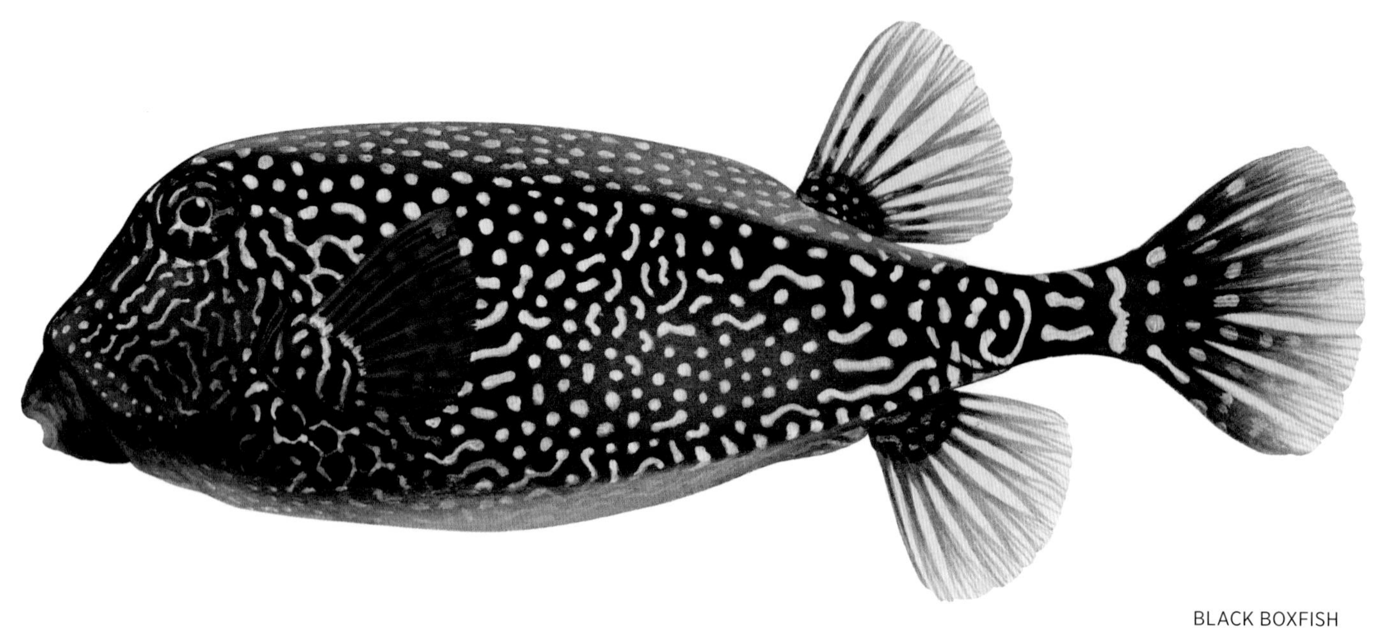

BLACK BOXFISH
Ostracion meleagris

There are about 33 species in this family, and they are found in tropical and temperate waters of the Atlantic, Indian and Pacific oceans. Their body is encased in a rigid armour of interlocking bony plates, often with spiny projections. There are openings between the plates for the fins, eyes, mouth and gill openings. They have a small mouth with fleshy lips and small, conical teeth, and the gill openings are reduced to a small slit before the pectoral fin. The single dorsal fin has no spines and 9–13 rays; the anal fin is very similar to the dorsal fin, also with 9–13 rays; the ventral fins are absent; and the caudal fin is rounded to truncate.

Boxfishes and Cowfishes swim by undulating the dorsal and anal fins, with the caudal fin used only for sudden bursts of speed. They are often sexually dimorphic, with the males and females displaying quite different shapes and colours.

In Australia members of this family are usually found close to the bottom, over rocky or coral reefs, around jetties and in sponge gardens, down to about 100 m. The beautifully patterned Cowfishes are commonly encountered over rocky reefs and seagrass beds in southern waters. The **Yellow Boxfish** and **Black Boxfish** are residents of coral reefs in northern waters. The small **Rigid Boxfish** and **Spiny Boxfish** are found in deep water, down to about 200 m, off the south coast, and are occasionally washed ashore after storms. Some species of Boxfish are known to secrete a toxic substance that deters predators and can kill other fishes in confined spaces such as aquariums. Boxfishes and Cowfishes feed on small benthic invertebrates and reach a maximum length of about 40 cm (the **Western Smooth Boxfish**). They feed mainly on small benthic invertebrates.

EASTERN SMOOTH BOXFISH
Anoplocapros inermis

WHITEBARRED BOXFISH
Anoplocapros lenticularis

SHAW'S COWFISH (female and male)
Aracana aurita

ORNATE COWFISH
Aracana ornate

SPINY BOXFISH
Capropygia unistriata

LONGHORN COWFISH
Lactoria cornuta

ROUNDBELLY COWFISH (juvenile)
Lactoria diaphana

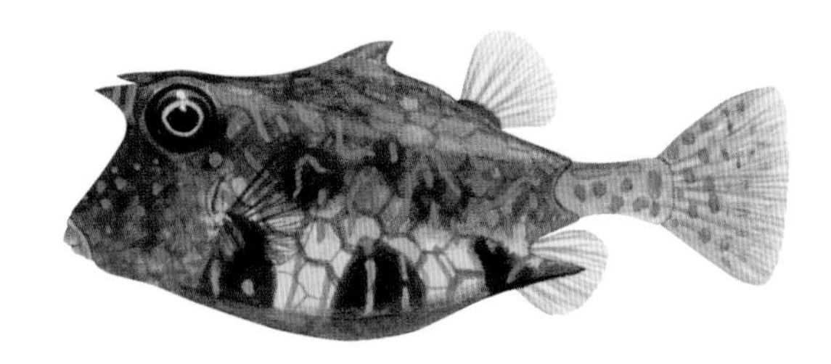

THORNBACK COWFISH (juvenile)
Lactoria fornasini

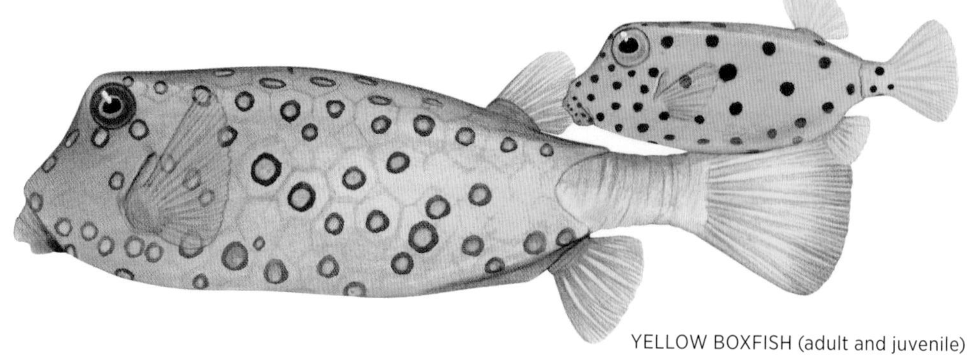

YELLOW BOXFISH (adult and juvenile)
Ostracion cubicus

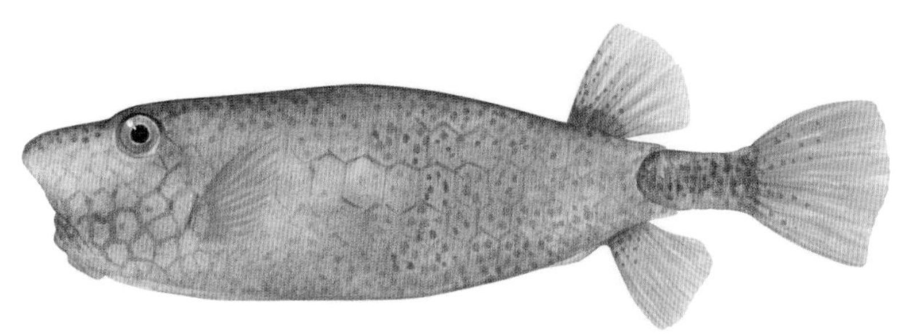

HORN-NOSE BOXFISH
Rhynchostracion rhinorhynchos

HUMPBACK TURRETFISH
Tetrosomus gibbosus

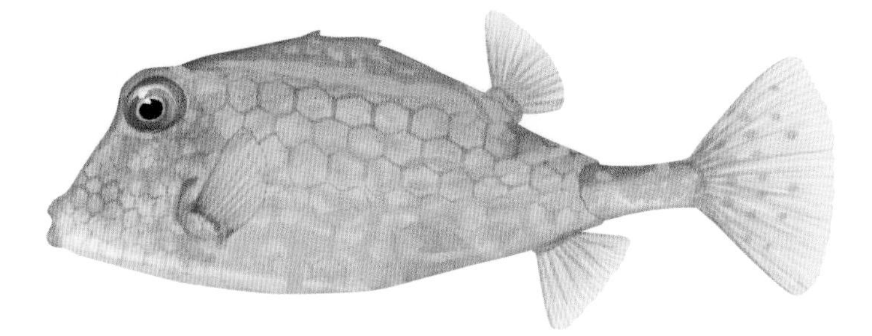

SMALLSPINE TURRETFISH (adult)
Tetrosomus concatenatus

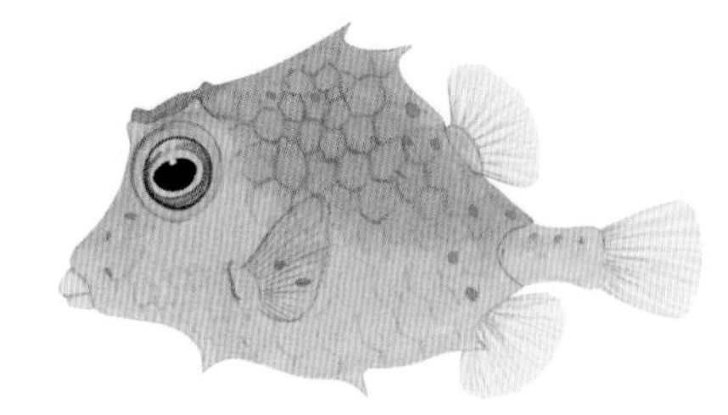

SMALLSPINE TURRETFISH (juvenile)
Tetrosomus concatenatus

AUSTRALIAN OSTRACIIDAE SPECIES

Acanthostracion bucephalus Bighead Boxfish
Anoplocapros amygdaloides Western Smooth Boxfish
Anoplocapros inermis Eastern Smooth Boxfish
Anoplocapros lenticularis Whitebarred Boxfish
Aracana aurita Shaw's Cowfish
Aracana ornata Ornate Cowfish
Caprichthys gymnura Rigid Boxfish
Capropygia unistriata Spiny Boxfish
Kentrocapros flavofasciatus Yellowstriped Boxfish
Lactoria cornuta Longhorn Cowfish
Lactoria diaphana Roundbelly Cowfish
Lactoria fornasini Thornback Cowfish
Ostracion cubicus Yellow Boxfish
Ostracion meleagris Black Boxfish
Ostracion solorensis Striped Boxfish
Polyplacapros tyleri Tyler's Boxfish
Rhynchostracion nasus Shortnose Boxfish
Rhynchostracion rhinorhynchos Horn-nose Boxfish
Tetrosomus gibbosus Humpback Turretfish
Tetrosomus concatenatus Smallspine Turretfish

THREETOOTH PUFFER

Triodontidae

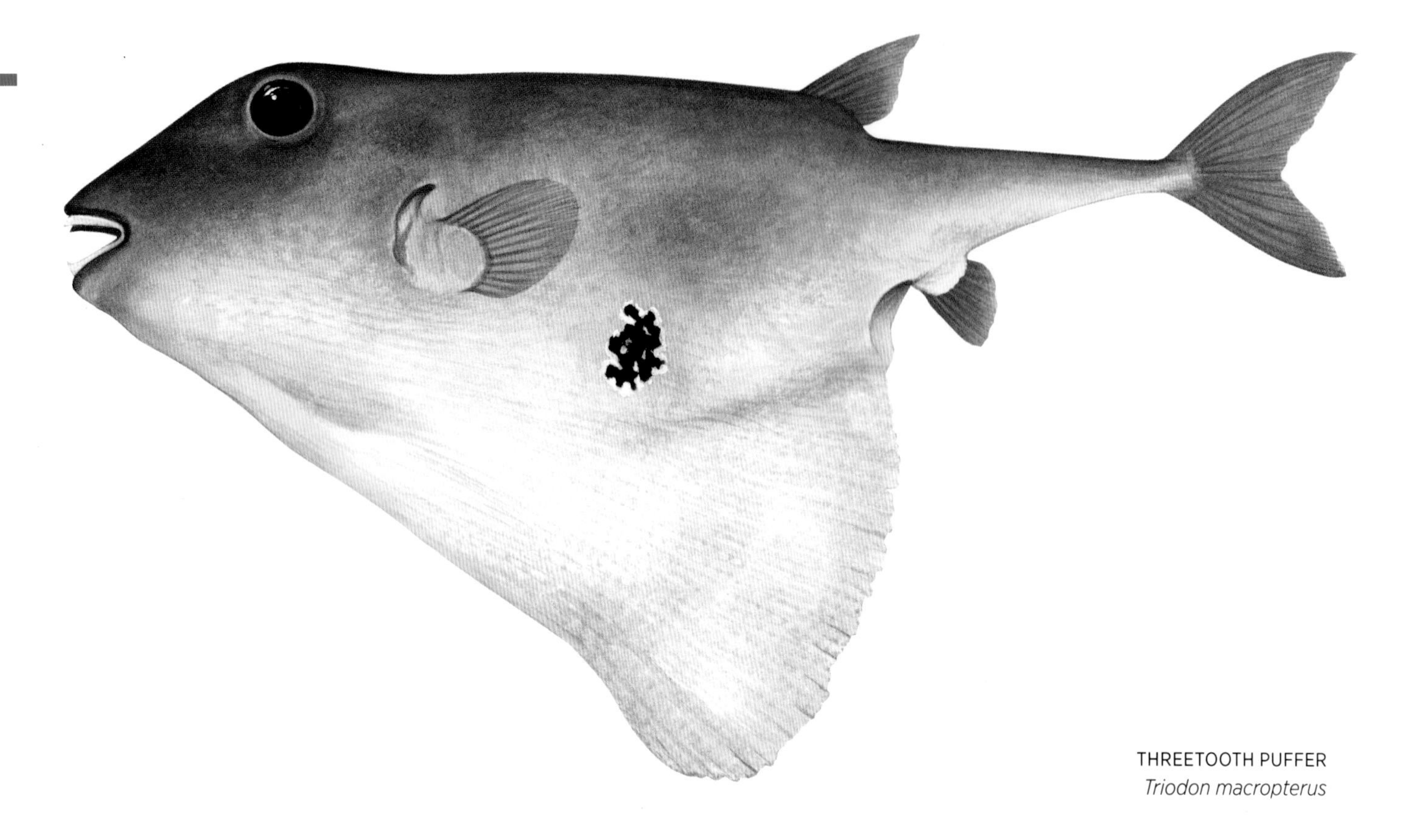

THREETOOTH PUFFER
Triodon macropterus

The single species in this family, the **Threetooth Puffer**, is found in deep waters of the Indo-West Pacific. It has an elongate, moderately compressed body with leathery skin. The eyes are placed very close to the dorsal profile and the gill openings are reduced to a small slit before the pectoral fin. The mouth is small and the teeth are fused into sharp, biting plates, two in the upper jaw and one in the lower jaw. There is a single small dorsal fin, usually with 11 rays and sometimes preceded by two rudimentary dorsal spines. The anal fin is similar to the dorsal fin and also has 11 rays. The ventral fins are absent and the belly is expanded into a broad flap that is supported by the very long pelvic bone.

This distinctive fish is found in deep water down to about 300 m, and reaches a maximum length of about 60 cm. It is rarely encountered in Australia, but is occasionally taken by deepwater trawlers. It is used as a food fish in Japan.

AUSTRALIAN TRIODONTIDAE SPECIES

Triodon macropterus Threetooth Puffer

PUFFERS

Tetraodontidae

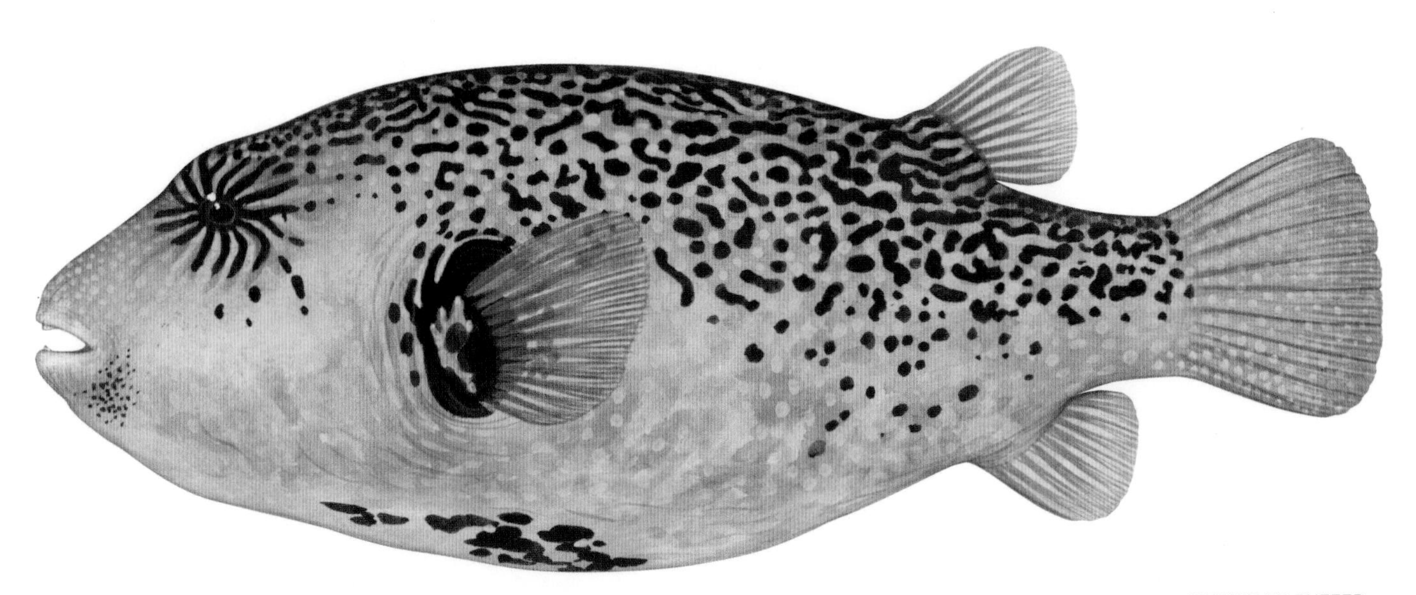

SCRIBBLED PUFFER
Arothron mappa

Puffers are found worldwide in tropical and temperate waters, with a few species entering fresh water. There are about 130 species in this large family. Puffers have elongate, rounded to cylindrical, inflatable bodies, with no scales and often small prickles on the skin. The eyes are placed very close to the dorsal profile and the gill openings are reduced to a small slit before the pectoral fin. They have a small mouth and the teeth are fused into four plates – two in each jaw, with a suture on the midline, forming a sharp, powerful beak. Puffers have a single dorsal fin with 7–18 rays; the anal fin is similar to the dorsal fin with 7–18 rays; the ventral fins are absent; and the caudal fin is rounded to truncate. They swim by undulating the dorsal and anal fins, with the caudal fin used only for sudden bursts of speed.

As a defence mechanism, Puffers can inflate their body by swallowing either water or air into their distensible stomach, making it difficult for a predator to swallow them. Many species are also extremely toxic, with the poison concentrated in the skin, ovaries, blood and other visceral tissues. Most Puffers, though not all, have no toxin in the muscles, which means they can be eaten if prepared correctly. In Japan they are considered a delicacy and are prepared by specially trained chefs; however, many people die from poisioning each year after eating flesh that has been improperly prepared by untrained amateurs. No member of this family should ever be eaten unless prepared by someone fully qualified in the process.

Some species, such as the **Starry Puffer**, can grow to 1 m in length, but most are much smaller at around 10–20 cm. The **Weeping Toadfish**, or Blowie, is well known to southern Australian anglers. It occurs in large schools, sometimes in near-plague proportions, devouring any bait presented for other fishes. It has been known to attack divers. *Lagocephalus* species, which can reach up to 1 m in length, are found in open coastal waters in the north, often in schools, and they too have been known to attack swimmers. Their powerful beaks can chop out a neat V-shaped piece of flesh, and they often bite caught fish struggling on a line.

The Tobies are small fishes, reaching about 12 cm in length, with an elongate snout. They are found near coral reefs in northern Australian waters and are often brightly coloured with spots and saddles. The **Ringed Toadfish** and **Bluespotted Toadfish** are commonly encountered by divers in south-west Australian coastal waters, often resting on the bottom, and are easily approached. Puffers feed on a wide variety of benthic invertebrates and small fishes.

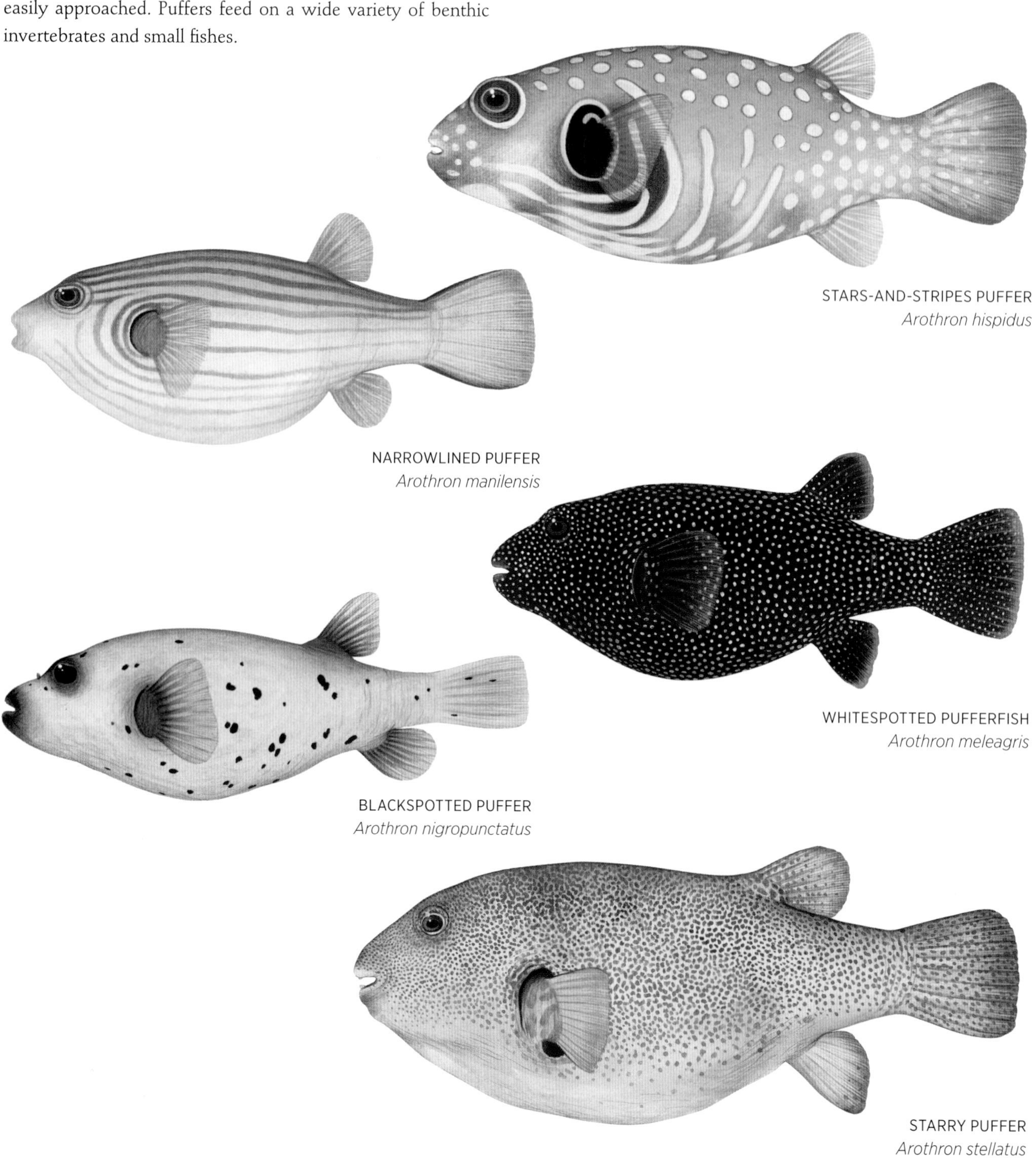

STARS-AND-STRIPES PUFFER
Arothron hispidus

NARROWLINED PUFFER
Arothron manilensis

WHITESPOTTED PUFFERFISH
Arothron meleagris

BLACKSPOTTED PUFFER
Arothron nigropunctatus

STARRY PUFFER
Arothron stellatus

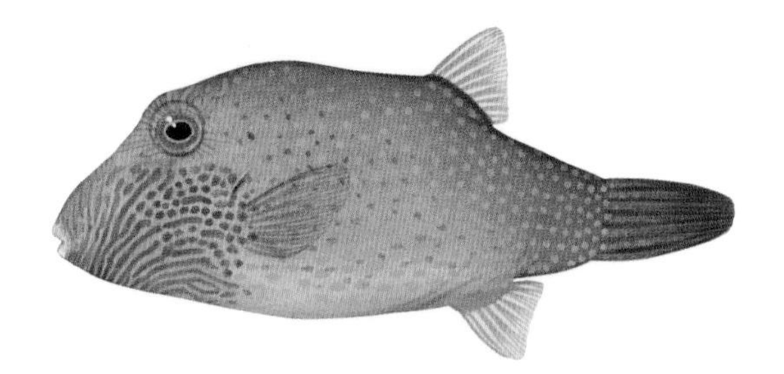

AMBON TOBY
Canthigaster amboinensis

CROWNED TOBY
Canthigaster axiologus

BLACKSPOT TOBY
Canthigaster bennetti

COMPRESSED TOBY
Canthigaster compressa

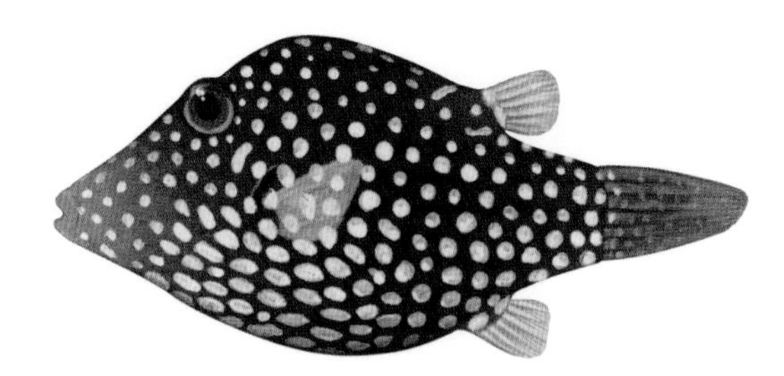

SPOTTED TOBY
Canthigaster janthinoptera

BLACKSADDLE TOBY
Canthigaster valentini

PRICKLY TOADFISH
Contusus brevicaudus

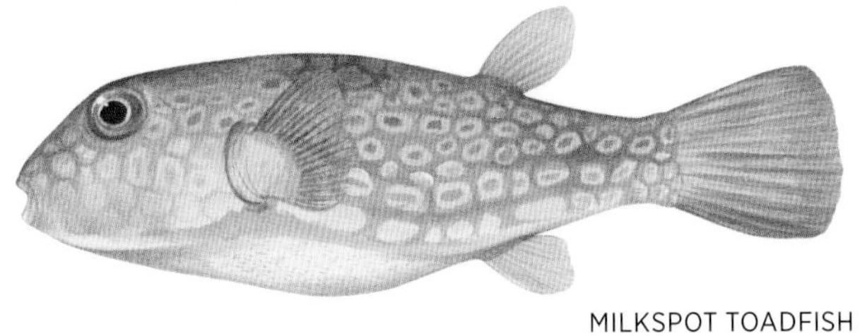

MILKSPOT TOADFISH
Chelonodon patoca

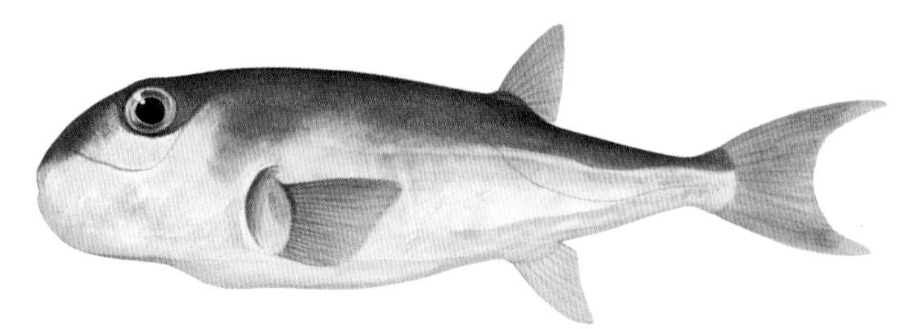

ROUGH GOLDEN TOADFISH
Lagocephalus lunaris

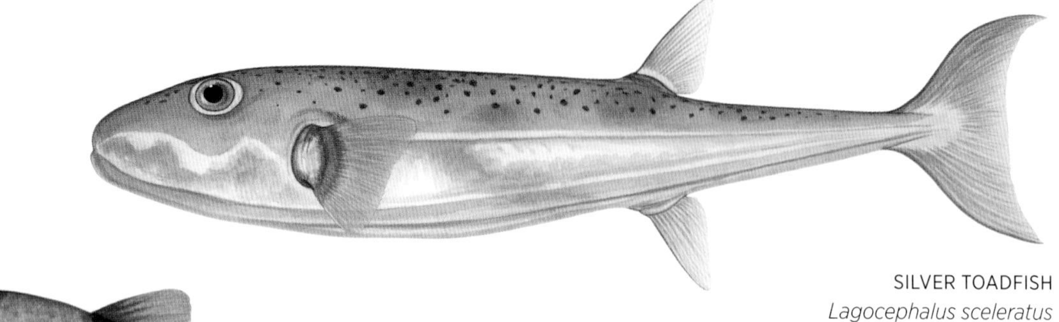

SILVER TOADFISH
Lagocephalus sceleratus

DARWIN TOADFISH
Marilyna darwinii

BLUESPOTTED TOADFISH
Omegophora cyanopunctata

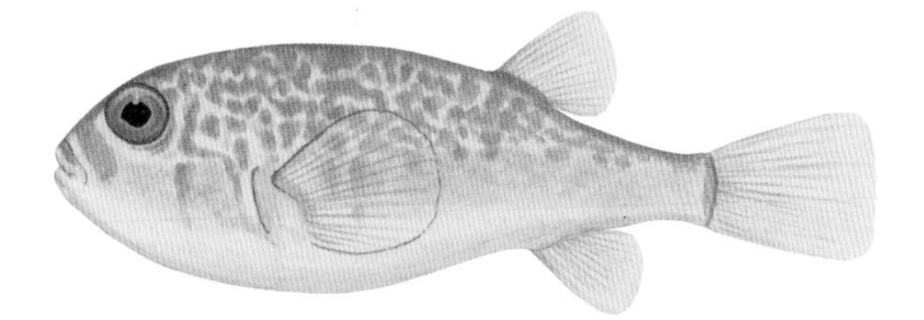

HALSTEAD'S TOADFISH
Reicheltia halsteadi

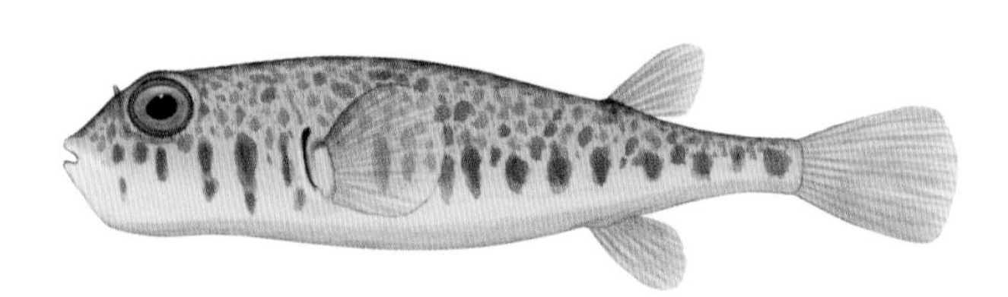

COMMON TOADFISH
Tetractenos hamiltoni

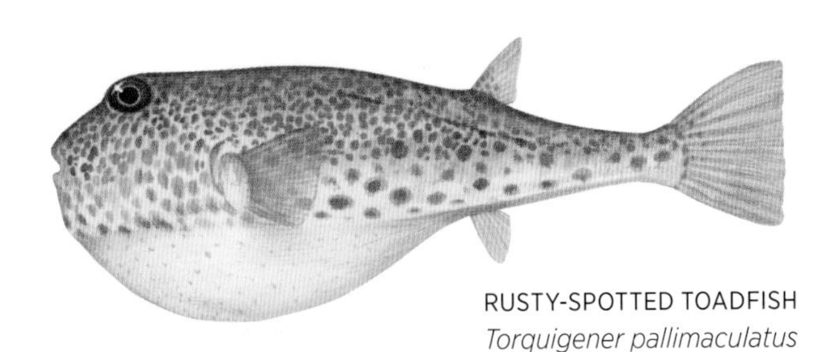

RUSTY-SPOTTED TOADFISH
Torquigener pallimaculatus

WEEPING TOADFISH
Torquigener pleurogramma

AUSTRALIAN TETRAODONTIDAE SPECIES

Arothron caeruleopunctatus Bluespotted Puffer
Arothron firmamentum Starry Toadfish
Arothron hispidus Stars-and-stripes Puffer
Arothron immaculatus Yellow-eye Puffer
Arothron manilensis Narrowlined Puffer
Arothron mappa Scribbled Puffer
Arothron meleagris Whitespotted Pufferfish
Arothron nigropunctatus Blackspotted Puffer
Arothron reticularis Reticulate Toadfish
Arothron stellatus Starry Puffer
Canthigaster amboinensis Ambon Toby
Canthigaster axiologus Crowned Toby
Canthigaster bennetti Blackspot Toby
Canthigaster callisterna Clown Toby
Canthigaster compressa Compressed Toby
Canthigaster epilampra Lantern Toby
Canthigaster janthinoptera Spotted Toby
Canthigaster ocellicincta Shy Toby
Canthigaster papua Netted Toby
Canthigaster rivulata Ocellate Toby
Canthigaster valentini Blacksaddle Toby
Chelonodon dapsilis Plentiful Toby
Chelonodon patoca Milkspot Toadfish
Contusus brevicaudus Prickly Toadfish
Contusus richei Barred Toadfish
Feroxodon multistriatus Ferocious Puffer
Lagocephalus cheesemanii Cheeseman's Puffer
Lagocephalus inermis Smooth Golden Toadfish
Lagocephalus lagocephalus Ocean Puffer
Lagocephalus lunaris Rough Golden Toadfish
Lagocephalus sceleratus Silver Toadfish
Lagocephalus spadiceus Brownback Toadfish
Marilyna darwinii Darwin Toadfish
Marilyna meraukensis Merauke Toadfish
Marilyna pleurosticta Banded Toadfish
Omegophora armilla Ringed Toadfish
Omegophora cyanopunctata Bluespotted Toadfish
Polyspina piosae Orangebarred Puffer
Reicheltia halsteadi Halstead's Toadfish
Sphoeroides pachygaster Balloonfish
Tetractenos glaber Smooth Toadfish
Tetractenos hamiltoni Common Toadfish
Tetraodon erythrotaenia Redstripe Toadfish
Torquigener altipinnis Highfin Toadfish
Torquigener andersonae Anderson's Toadfish
Torquigener hicksi Hicks' Toadfish
Torquigener hypselogeneion Northern Toadfish
Torquigener pallimaculatus Rusty-spotted Toadfish
Torquigener parcuspinus Yelloweye Toadfish
Torquigener paxtoni Paxton's Toadfish
Torquigener perlevis Spineless Toadfish
Torquigener pleurogramma Weeping Toadfish
Torquigener squamicauda Scalytail Toadfish
Torquigener tuberculiferus Fringe-gill Toadfish
Torquigener vicinus Orangespotted Puffer
Torquigener whitleyi Whitley's Toadfish
Tylerius spinosissimus Finespine Pufferfish

PORCUPINEFISHES

Diodontidae

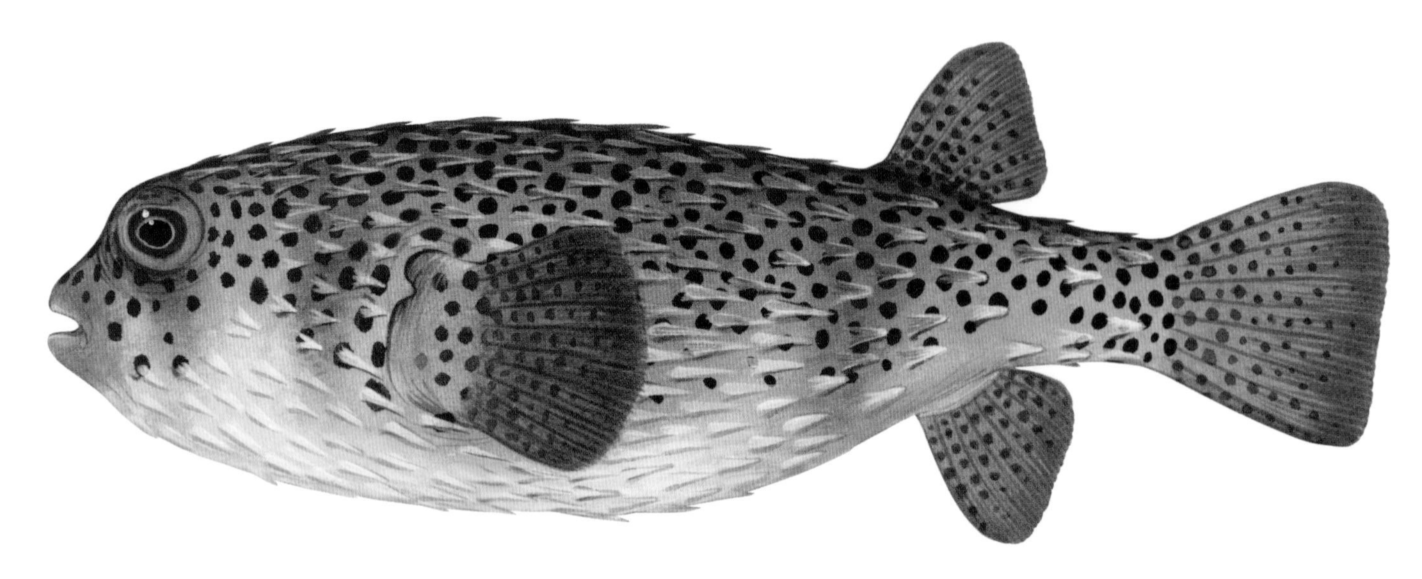

SPOTTED PORCUPINEFISH
Diodon hystrix

Porcupinefishes are found in tropical and temperate waters of the Atlantic, Indian and Pacific oceans, and there about 19 species in the family. They have an elongate, cylindrical body with no scales, and numerous movable or fixed spines on the head and body. Their small, broad mouth has teeth fused into single plates, one in each jaw, without a suture along the midline. They have large eyes placed very close to the dorsal profile, and the gill openings are reduced to a slit before the pectoral fin. The single dorsal fin has 11–18 rays, the anal fin is similar to the dorsal fin, the ventral fins are absent, and the caudal fin is rounded. Puffers swim by undulating the dorsal and anal fins, with the caudal fin used only for sudden bursts of speed.

Like the Puffers (Tetraodontidae, p.773), Porcupinefishes can inflate their bodies by taking water or air into the stomach. Many species can also erect long spines on the surface of the body, making it extremely difficult for any predator to swallow them. In addition, some species are reported to be toxic, and no member of this family should be eaten.

Porcupinefishes are found in coastal waters over rocky and coral reefs. Most are nocturnal, sheltering in crevices and caves by day, and feeding at night on hard-shelled benthic invertebrates such as molluscs, crabs and echinoderms. Most Australian species are found in tropical waters, although the **Globefish** is often encountered by divers on rocky reefs and around jetties in southern Australian waters. The largest Australian species is the **Spotfin Porcupinefish**, which reaches a maximum length of about 55 cm and has short, fixed spines on the head and body.

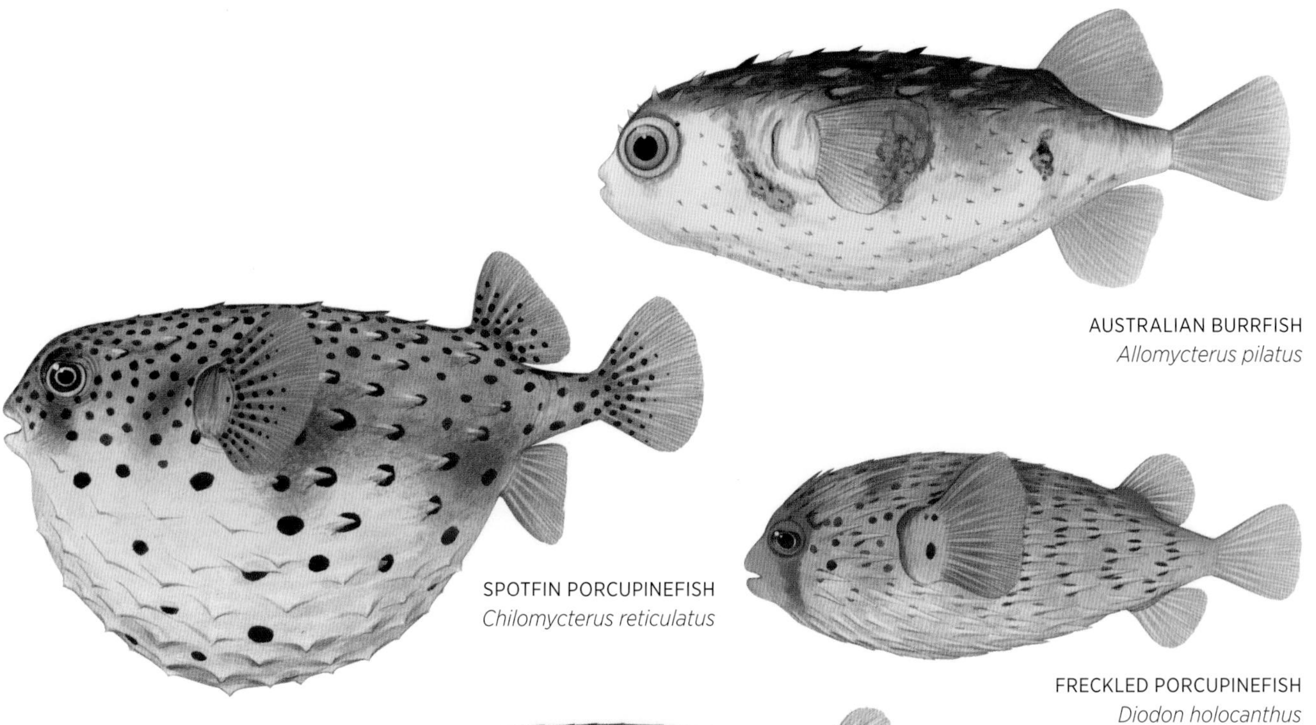

AUSTRALIAN BURRFISH
Allomycterus pilatus

SPOTFIN PORCUPINEFISH
Chilomycterus reticulatus

FRECKLED PORCUPINEFISH
Diodon holocanthus

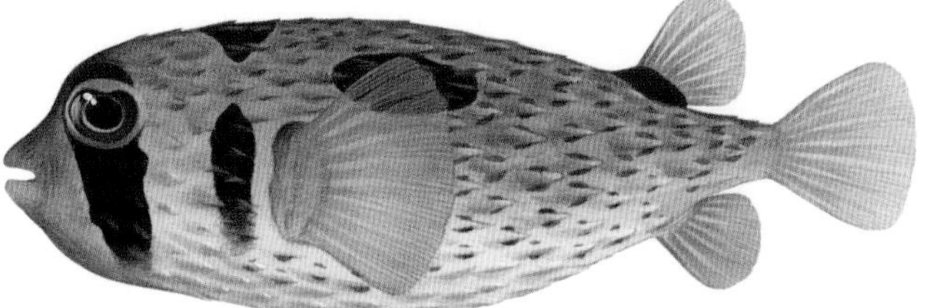

BLACKBLOTCHED PORCUPINEFISH
Diodon liturosus

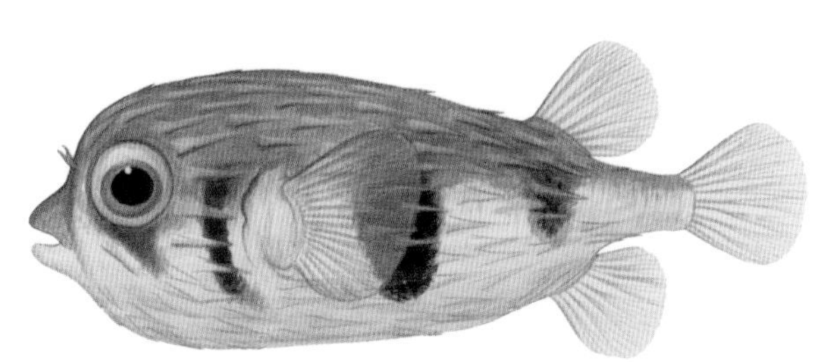

GLOBEFISH
Diodon nicthemerus

LONGSPINE PORCUPINEFISH
Tragulichthys jaculiferus

AUSTRALIAN DIODONTIDAE SPECIES

Allomycterus pilatus Australian Burrfish
Chilomycterus reticulatus Spotfin Porcupinefish
Cyclichthys hardenbergi Plain Porcupinefish
Cyclichthys orbicularis Shortspine Porcupinefish
Cyclichthys spilostylus Spotbase Burrfish
Dicotylichthys punctulatus Threebar Porcupinefish
Diodon holocanthus Freckled Porcupinefish
Diodon hystrix Spotted Porcupinefish
Diodon liturosus Blackblotched Porcupinefish
Diodon nicthemerus Globefish
Lophodiodon calori Fourbar Porcupinefish
Tragulichthys jaculiferus Longspine Porcupinefish

SUNFISHES

Molidae

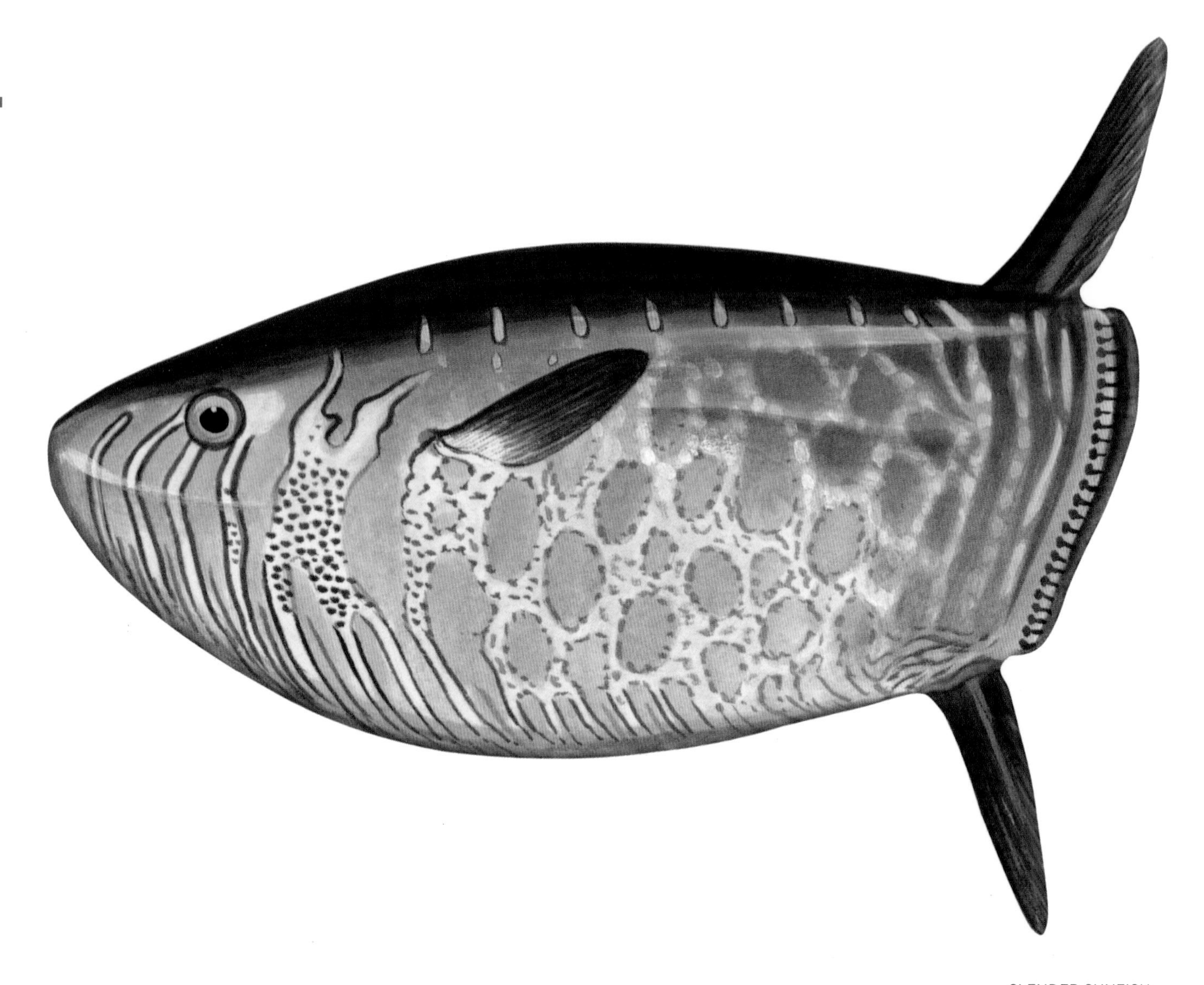

SLENDER SUNFISH
Ranzania laevis

Sunfishes are found around the world in tropical and temperate waters, and the family contains four species. They are very distinctive fishes, with a disc-like, compressed body and minute scales. The gill openings are reduced to a small hole before the pectoral fin and they have a small mouth with teeth fused into two plates, one in each jaw. They have a high, short-based dorsal fin with 15–21 rays, an anal fin very similar to the dorsal fin with 15–19 rays, and no ventral fins. The caudal fin has also been lost and the rear of the body is formed into a leathery flap that is supported by modified dorsal and anal fin rays. They swim using only their powerful dorsal and anal fins.

Sunfishes are usually encountered in oceanic surface waters, where they are often seen lying on their side basking in the sun, but they are also known to dive to great depths. They feed mainly on jellyfishes, salps and other soft-bodied pelagic animals, though bottom-dwelling invertebrates have also been found in their stomach contents. Sunfishes occasionally come into inshore waters and in some areas of the world form a large part of the bycatch from driftnet fisheries. The **Ocean Sunfish** is one of the largest of all bony fishes, reaching up to 4 m in length and weighing up to 2 tonnes. Large females have been found to contain an estimated 300 million eggs. The **Sharptail Sunfish** and **Short Sunfish** both reach about 3 m in length and are rarely encountered in Australian waters, though specimens are occasionally found stranded after storms. The **Slender Sunfish** is a fast-swimming species that often occurs in schools. It reaches about 90 cm in length and is also seldom encountered, usually living well offshore.

SHORT SUNFISH
Mola ramsayi

AUSTRALIAN MOLIDAE SPECIES

Masturus lanceolatus Sharptail Sunfish
Mola mola Ocean Sunfish
Mola ramsayi Short Sunfish
Ranzania laevis Slender Sunfish

SARCOPTERYGII

LUNGFISHES

CERATODONTIFORMES

This is an extremely ancient group of fishes, with fossil records going back about 250 million years. Ceratodontiforms have functional lungs as well as gills, and inhabit fresh waters. The dorsal and anal fins are continuous with the tapering caudal fin, and the ventral fins are placed well towards the rear of the body. There are three families in the order, one of which occurs in Australia.

AUSTRALIAN LUNGFISH

Ceratodontidae

AUSTRALIAN LUNGFISH
Neoceratodus forsteri

The **Australian Lungfish**, the sole member of this family, is found only in the river systems of south-eastern Queensland. Lungfishes are amongst the most ancient of all living vertebrates and have remained almost unchanged for over 100 million years. Once thought to be the link between fishes and reptiles, it is now clear that both arose from a common ancestor, with the reptiles moving onto land and Lungfishes remaining aquatic. They are the only fishes that possess functional lungs as well as gills. This enables them to live in very muddy or stagnant waters, coming to the surface to gulp air if the gills cannot provide enough oxygen.

The Australian Lungfish generally lies immobile on the bottom during the day and feeds nocturnally on frogs, small fishes, worms, snails, crustaceans and some plant material, including fruit and flowers that fall into the water. The teeth are fused into plates, with projections for cutting and grinding. Females lay pea-sized, jelly-like eggs which adhere to aquatic vegetation.

Although once widely distributed in Australia, when discovered by Europeans the species was restricted to the Mary and Burnett rivers in Queensland, and was threatened with extinction. Translocation by conservation groups has somewhat lessened the pressure on numbers, though loss of suitable breeding habitats has meant that very few juveniles are being added to the population. This slow-moving fish grows to 2 m in length and can weigh up to 45 kg, and can live to be at least 100 years old.

AUSTRALIAN CERATODONTIDAE SPECIES

Neoceratodus forsteri Australian Lungfish

GLOSSARY

abyssal plain The seafloor of the ocean basins, usually below 4000 m in depth. (See Fig. 3, p. 7.)

adipose eyelid A thick transparent layer of tissue that partially or wholly covers the eye of some fishes.

adipose fin A small fleshy fin without supporting rays.

Agnathans The jawless fishes, including the Hagfishes (Myxini, p. 25) and Lampreys (Petromyzontida, p. 31).

anterior Relating to the front of the body.

axillary Relating to the angle formed by the point of attachment of a ventral or pectoral fin to the body.

axillary process A scale-like or fleshy projection in the angle formed by the attachment of the ventral or pectoral fin to the body.

barbel A fleshy tentacle found in some fishes, usually located near the mouth.

benthic Relating to animals living on the substrate or sea floor.

bicuspid teeth Teeth with two points or cusps.

bilobed Having two lobes or projections.

bioluminescence The production of light by chemical means from living tissue. In fishes this light is either created through symbiotic bacteria or within the tissues of the fish itself.

biomass The total amount, by weight, of a living organism in a given area.

branchial Relating to the gills.

branchiostegal rays Thin bony rays supporting the gill membrane beneath the head. (See Fig. 5, p. 18.)

bucklers Modified scales that form flattened bony plates.

cartilage Tough, flexible connective tissue, forming the skeleton of sharks and rays.

cartilaginous Made of cartilage.

caudal Relating to the hind part of the body, or to the tail.

caudal fin The tail fin. (See Fig. 5, p. 18.)

caudal peduncle The part of the body between the rear of the anal-fin base and the origin of the caudal fin. (See Fig. 5, p. 18.)

cephalofoil The flattened, laterally expanded head of Hammerhead Sharks.

ceratioid Anglerfishes Deepwater Anglerfishes with dwarf males that attach to the females (and in some cases fuse permanently with and parasitise them).

Chondrichthyans The cartilaginous fishes, comprising the sharks, rays and chimaeras (Chondrichthyes, p. 37).

ciguatera An illness caused by ingesting poison present in the tissues of some larger predatory reef fishes, derived from algae that is consumed by herbivores and passed up the food chain.

claspers The sexual organs of male sharks and rays, for internal fertilisation of females.

commensal Relating to a relationship between two species or animals in which one member benefits from the relationship while the other is unaffected.

compressed body A laterally flattened body.

conical Having the shape of a cone.

continental shelf Where the sea floor slopes gradually downward from the shore to about 200 m in depth. (See Fig. 3, p. 7.)

continental slope Where the sea floor slopes steeply from the edge of the continental shelf down to about 2000 m. (See Fig. 3, p. 7.)

continental rise The gently sloping area between the abyssal plain and the steeper continental slope. (See Fig. 3, p. 7.)

copepods Minute swimming crustaceans that often occur in enormous numbers in zooplankton and are an important food source for many planktivorous fishes.

crumenal organ A small pouch in the upper rear of the throat in some fishes, formed of interlocking gillrakers, that helps to break down food.

cryptic Relating to creatures that remain hidden or are well camouflaged.

ctenoid scales Scales with fine teeth on the exposed edge. (See Fig. 6, p. 19.)

cycloid scales Scales that are smooth on the exposed edge. (See Fig. 6, p. 19.)

deep bodied With a body that is laterally flattened and dorso-ventrally expanded.

denticles Tooth-like scales with an outer layer of enamel-like material, found on the skin of sharks and rays.

denticular teeth Denticles modified to function as teeth, as found in males of some Anglerfish species (Lophiiformes, p. 291).

dermal papillae Small finger-like outgrowths of the skin.

dermal spinules Small spines formed from the skin.

dimorphism *see* sexual dimorphism

disc The flattened shape formed by the head, body and pectoral fins of Skates (Rajidae, p. 121) and rays.

distensible Able to be expanded or inflated.

dorsal Relating to the upper surface of the body.

electrogenic organs Specialised tissues that are able to generate an electrical charge.

electroreceptors Specialised cells that are sensitive to the minute electrical fields produced by all living organisms.

endemic Naturally occurring only within a certain area.

esca The fleshy lure, often containing bioluminescent bacteria, found in many Anglerfishes (Lophiiformes, p. 291).

eutrophication A depletion of dissolved oxygen in waters. Usually caused by abnormally high nutrient levels promoting excessive growth of some organisms in an ecosystem, thus using up available oxygen and causing the subsequent death of many other organisms in the ecosystem.

external fertilisation The reproductive strategy used by the majority of fishes, whereby eggs and sperm are both released into the water and fertilisation occurs outside the body of the female.

filamentous Long and thin; thread-like.

filter feeding The method of feeding by straining small organisms from the water, usually using well-developed gillrakers.

finlet A small fin following the second dorsal fin, as found in Tunas and Mackerels (Scombridae, p. 717).

foraminiferans A group of amoeboid protists (simple micro-organisms).

fusiform Having a shape that is wide in the middle and tapers at each end.

gestation The period of development of the embryo, from fertilisation of the egg until birth.

gillrakers Bony finger-like projections on the anterior edges of the gill arches, often used to filter small organisms from water passing through the gills.

gills The organs via which fishes absorb oxygen from the water.

Gondwana The large southern landmass that existed more than 150 million years ago, before movements of the Earth's crust broke it apart. (See Evolution and Geological History, p. 4.)

gonopodium A secondary sexual organ, derived from the modified anal fin, as in Mosquitofishes (Poeciliidae, p. 343).

herbivorous Relating to fishes that feed on plant material.

hermaphroditic Having both male and female reproductive tissue.

illicium A modified dorsal-fin spine, usually with a fleshy lure (esca) at its tip, as in Anglerfishes (Lophiiformes, p. 291).

incisor teeth Teeth with flattened cutting edges, common in herbivorous fishes.

inferior mouth A mouth located below the overhanging snout, as found in bottom-feeding fishes.

infraorbital Located below the eye.

internal fertilisation The reproductive strategy used by sharks, rays and certain other fishes, whereby the male introduces the sperm into the body of the female.

keel A fleshy or bony ridge.

lamellae Tissue in the form of thin sheets or plates.

lateral Relating to the side of the body.

lateral line A line of sensory cells running along the lateral surface of the body of fishes, often extending onto the head, and composed of neuromasts. The neuromasts can be free-standing, in simple pores or canals, or beneath pored scales. (See Fig. 5, p. 18.)

leptocephalus Transparent elongate or leaf-like pelagic larvae, found in Anguilliforms (p. 153) and related groups.

lunate Crescent shaped.

melanistic form A dark colour form of an animal, caused by increased amounts of the pigment melanin.

metamorphose To change form, as from larvae to adult, or from male to female.

molariform teeth Flattened crushing teeth.

nasal barbels Small fleshy tentacles near the nostril, as seen in many sharks.

nasal organ The organ responsible for the sense of smell. (See Senses, p. 22.)

neurocranium The part of the skull encasing the brain.

neuromasts Sensory cells composed of hair-like projections that are sensitive to vibration or movement in the surrounding water. They can be either free-standing (as in some deepwater fishes) or enclosed in a gelatinous dome.

nictitating membrane A tough protective eyelid that can be moved over the eye, found in some sharks.

nocturnal Relating to animals that are active at night.

notch An indentation or sharp dip in the outline.

omnivorous Relating to fishes that take advantage of a great variety of food sources.

opercular bones The series of flattened bones that together make up the operculum.

opercular spine A spine situated on the operculum. (See Fig. 5, p. 18.)

operculum The bony plate covering the gill arches and forming the cheek of the fish. (See Fig. 5, p. 18.)

opportunistic Relating to fishes that take any food item that becomes available.

oral disc The circular mouth of Lampreys (Petromyzontida, p. 31), which contains concentric rows of horny teeth.

ossified skeleton A skeleton consisting of embryonic cartilage that has developed into bone.

Osteichthyans The bony fishes (Actinopterygii, p. 141 and Sarcopterygii, p 783).

otolith A small bone found in the inner ear of fishes.

oviparous The reproductive strategy whereby females produce eggs that hatch outside the body.

ovoviviparous The reproductive strategy whereby females produce eggs with a developed shell, which hatch inside the body.

palate The roof of the mouth.

pelagic In open water; not associated with the substrate.

pelagic spawning The release of eggs and sperm into the water column for external fertilisation.

pelvic fins *see* ventral fins.

pharyngeal teeth Grinding teeth located in the throat or pharynx region, used for breaking up hard-shelled prey items.

pheromone A chemical compound broadcast by an animal to attract others; usually used in mating to attract the opposite sex.

photophore A small light-producing organ that produces light using symbiotic bacteria or bioluminescent tissue.

phytoplankton Microscopic green algae. (See Pelagic Fishes, p. 7.)

placenta A mass of tissue produced in the uterus that functions to sustain developing embryos.

placental yolk sac A sac attached to the embryo, containing nutrients for the embryo's development.

planktonic Relating to micro-organisms that drift freely with ocean currents, such as phytoplankton or zooplankton.

preoperculum The bony plate on the cheek of fishes, between the eye and the operculum. (See Fig. 5, p. 18.)

protrusible Able to be protruded or extended outwards, as with the mouths and jaws of many fishes.

pterygiophore A small bone in the dorsal or ventral musculature, supporting the external fin spines and rays. (See Fig. 7, p. 20.)

rostral cartilage Cartilage supporting the nose or snout, as in the Skates (Rajidae, p. 121).

rostrum A projecting nose or snout.

sagitta One of the small bones (otoliths) in the inner ear of fishes.

scavengers Fishes that feeds on dead animals.

sclerotic bone The thin plate of bone within the eye, developed from cartilage in the tough outer covering of the eye.

scutes Sharp bony scales, such as those along the caudal peduncle of some Trevallies (Carangidae, p. 497).

sedentary Relating to animals that move or swim very little.

sensory canals Shallow grooves or troughs that form part of the sensory system.

sensory pores Small openings leading to sensory cells such as neuromasts or electroreceptors.

sequential hermaphrodites Individuals that possess both female and male reproductive tissue, developing and functioning as one sex before changing to the other, as in many Wrasses (Labridae, p. 617).

sessile Relating to animals that are permanently attached to the substrate.

sexual dimorphism In sexually dimorphic species the males and females take different forms, shapes or colours.

sister group A group of fishes that has evolved from the same ancestral form as another group.

spinule A small spine.

spiracle An opening from the exterior to the pharynx, allowing passage of water for respiration, as in sharks and rays.

squalene oil A low-density oil found in the liver of sharks, used to maintain buoyancy. (See Digestion and Excretion, p. 21.)

suborbital bones The bones below the eye.

symbiotic Relating to the relationship between two species or animals that live together for mutual benefit, often one residing within the tissues of another.

tectonic movement The movement of the plates that form the Earth's crust.

Teleosts The spiny-rayed fishes, which are members of Division Teleostei in the class Actinopterygii (p. 141).

terminal mouth A forwardly directed mouth positioned at the front of the head.

torpor A state of dormancy or a greatly lowered rate of activity in an animal.

transverse Relating to something positioned crossways on the body.

tricuspid teeth Teeth with three points or cusps.

tubercle A small, hard lump on the surface of the body.

upwelling The movement of masses of cold, deep water towards the surface.

vascular heat exchanger A system of interwoven capillaries that serves to retain heat generated by muscular activity. (See Pelagic Fishes, p. 7.)

ventral Relating to the lower surface or underneath of the body.

ventral (pelvic) fins Paired fins on the ventral surface of fishes that correspond to the hind legs of terrestrial vertebrates. (See Fig. 5, p. 18.)

ventral profile The outline of the lower surface of the body.

villiform In the form of numerous small, thin projections, as in the teeth of many fishes.

viviparous The reproductive strategy whereby females give birth to live young that have developed from embryos within the body of the female.

zooplanktivorous Relating to fishes that feed on zooplankton.

zooplankton Aggregations of a variety of small pelagic animals and larvae, found in the upper layers of the ocean. (See Pelagic Fishes, p. 7.)

BIBLIOGRAPHY

Aizawa, M, Mastsuura, K & Fujii, E 1990, *Fishes Trawled off Suriname and French Guiana*, Japan Marine Fishery Resource Research Center, Tokyo.

Allen, GR 1975, *Anemonefishes*, T.F.H. Publications, Neptune, New Jersey.

——1981, *Butterfly and Angelfishes of the* World, Mergus Publishers, Melle, Germany.

——1982, *A Field Guide to Inland Fishes of Western Australia*, Western Australian Museum.

——1985, *Snappers of the World*, FAO Species Catalogue, vol. 6, Food and Agriculture Organisation of the United Nations, Rome.

——1991, *Damselfishes of the World*, Mergus Publishers, Melle, Germany

——1991, *Field Guide to the Freshwater Fishes of New Guinea*, Christensen Research Institute, Madang.

——& Cross NJ 1982, *Rainbowfishes of Australia and Papua New Guinea*, Angus and Robertson Publishers, Sydney.

——& Steene, RC 1988, *Fishes of Christmas Island; Indian Ocean*, Christmas Island Natural History Association.

——Steene, R, Humann, P & DeLoach, N 2003, *Reef Fish Identification: Tropical Pacific*, New World Publications, California.

——Midgley, SH & Allen, M 2002, *Field Guide to the Freshwater Fishes of Australia*, Western Australian Museum.

——& Robertson, DR 1994, *Fishes of the Tropical Eastern Pacific*, University of Hawaii Press, Honolulu.

——& Swainston, R 1988, *The Marine Fishes of North-Western Australia*, Western Australian Museum.

——& Swainston, R 1993, *Reef Fishes of New Guinea*, Christensen Research Institute, Madang.

——& Talbot, FH 1985, 'Review of the Snappers of the genus Lutjanus from the Indo-Pacific, with the description of a new species', *Indo-Pacific Fishes*, no. 11, Bishop Museum Press, Honolulu.

Amaoka, K, Matsuura, K, Inada, T, Takeda, M, Hatanaka, H & Okada, K 1990, *Fishes Collected by the R/V Shinkai Maru around New Zealand*, Japan Marine Fishery Resource Research Center, Tokyo.

Anderson, T (ed.) 1987, *CSIRO Research for Australia*, vol. 14: Oceanography, CSIRO, Canberra.

Australian Bureau of Agricultural and Resource Economics (ABARE) 2008, *Australian Fisheries Statistics 2007*, comps R Wood, L Hohnen & A Peat, Fisheries Research and Development Corporation, Canberra.

Berra, TM 2007, *Freshwater Fish Distribution*, The University of Chicago Press.

Boltovsky, D (ed.) 1999, *South Atlantic Zooplankton*, vol. 1, Backhuys Publishers, Leiden.

Borradaile, LA & Potts, FA, Eastham, LES, Saudners, JT 1951, *The Invertebrata*, Cambridge University Press.

Burton, R & Burton, M 1975, *Encyclopedia of Fish*, Octopus Books, London.

Cadwallader, PL & Backhouse, GN 1983, *A Guide to the Freshwater Fish of Victoria*, Victorian Government Printing Office, Melbourne.

Calabi, S 1990, *Trout and Salmon of the World*, The Wellfleet Press, Edison, New Jersey.

Carpenter, KE 1988, *Fusilier Fishes of the World*, FAO Species Catalogue, vol. 8, Food and Agriculture Organisation of the United Nations, Rome.

——& Allen GR 1989, *Emperor Fishes and Large-eye Breams of the World*, FAO Species Catalogue, vol. 9, Food and Agriculture Organisation of the United Nations, Rome.

Carpenter, KE (ed.) 2002, *The Living Marine Resources of the Western Central Atlantic*, FAO Species Identification Guide for Fishery Purposes, vol. 1–3, Food and Agriculture Organisation of the United Nations, Rome.

——& Niem, VH (eds) 2001, *The Living Marine Resources of the Western Central Pacific*, FAO Species Identification Guide For Fishery Purposes, vol. 1–6, Food and Agriculture Organisation of the United Nations, Rome.

Cohen, DM, Inada, T, Iwamoto, T & Scialabba, N 1990, *Gadiform Fishes of the World*, FAO Species Catalogue, vol. 10, Food and Agriculture Organisation of the United Nations, Rome.

Coleman, N 1978, *Australian Fisherman's Fish Guide*, Bay Books, Sydney.

Collette, BB & Nauen, CE 1983, *Scombrids of the World*, FAO Species Catalogue, vol. 2, Food and Agriculture Organisation of the United Nations, Rome.

Compagno, LJV 1984, *Sharks of the World*, FAO Species Catalogue, vol. 4, parts 1 and 2, Food and Agriculture Organisation of the United Nations, Rome.

——Ebert, DA & Smale, MJ 1989, *Guide to the Sharks and Rays of Southern Africa*, New Holland, London.

Cousteau, JY & Dumas, F 1953, *The Silent World*, Hamish Hamilton, London.

Cushing, DH & Walsh, JJ (eds) 1976, *The Ecology of the Seas*, Blackwell Scientific Publications, Oxford.

Dakin, WJ 1980, *Australian Seashores*, rev. edn, Angus and Robertson Publishers, Sydney.

Daley, R K, Stevens, JD, Last, PR & Yearsley, GK 2002, *Field Guide to Australian Sharks and Rays*, CSIRO Publishing, Collingwood, Victoria.

Dawes, J 1991, *Livebearing Fishes*, Blandford Press, London.

——2002, *The Concise Encyclopedia of Popular Freshwater Tropical Fish*, Parragon Books Ltd, Bath.

Day, F 1888, *The Fishes of India*, vol. 1, London.

Debelius, H 1998, *Guide de Poissons: Mediterranee et Atlantique*, PLB Editions, Frankfurt.

Eckert, R & Randall, D 1978, *Animal Physiology*, W.H. Freeman and Company, San Francisco.

Edgar, GJ 1997, *Australian Marine Life: The Plants and Animals of Temperate Waters*, Reed Books, Kew, Victoria.

——2001, *Australian Marine Habitats in Temperate Waters*, New Holland Publishers, Sydney.

——Last, PR & Wells, MW 1982, *Coastal Fishes of Tasmania and Bass Strait*, Cat and Fiddle Press, Hobart, Tasmania.

Bibliography

Edmonds, C 1975, *Dangerous Marine Animals of the Indo-Pacific Region*, Wedneil Publications, Newport, Victoria.

Ellis, R 1983, *The Book of Sharks*, Alfred A. Knopf, New York.

——& McCosker, JE 1991, *Great White Shark*, HarperCollins Publishers, New York.

Endean, R 1982, *Australia's Great Barrier Reef*, University of Queensland Press.

Eschmeyer, WNE & Herald, ES 1983, *A Field Guide to Pacific Coast Fishes of North America*, Houghton Mifflin, Boston.

Fautin, DG & Allen, GR 1992, *Field Guide to Anemone Fishes and their Host Sea Anemones*, Western Australian Museum.

Filleul, A, 2001 *Poissons de Mer: Guide Scientifique a l'usage des Pêcheurs de France et d'Ailleurs*, Mame, Tours.

Fletcher, WJ & Santoro, K (eds) 2008, *State of the Fisheries Report 2007/08*, Department of Fisheries, Western Australia.

Frank, S 1969, *The Pictorial Encyclopedia of Fishes*, The Hamlyn Publishing Group, London.

George, D & George, J 1979, *Marine Life: An Illustrated Encyclopedia of Invertebrates in the Sea*, Rigby Limited, Adelaide, South Australia.

Gloerfelt-Tarp, T & Kailola, J 1984, *Trawled Fishes of Southern Indonesia and Northwestern Australia*, The Australian Development Assistance Bureau, Canberra, Directorate General of Fisheries, Jakarta & German Agency for Technical Cooperation, Eschborn.

Gomon, MD Bray, D & Kuiter, R (eds) 2008, *Fishes of Australia's Southern Coast*, New Holland Publishers, Sydney.

Grant, EM 1982, *Guide to Fishes*, The Department of Harbours and Marine, Brisbane.

——1987, *Fishes of Australia*, E.M. Grant Publishers, Queensland.

Hall, H 1990, *Sharks: The Perfect Predators*, Blake Publishing, California.

Halstead, B 2000, *Coral Sea Reef Guide*, Sea Challengers, Monterey, California.

Harvey, N & Caton, B 2003, *Coastal Management in Australia*, Oxford University Press, Melbourne.

Heemstra, PC & Randall, JE 1993, *Groupers of the World*, FAO Species Catalogue, vol. 16, Food and Agriculture Organisation of the United Nations, Rome.

Helfman, GS, Collette, BB & Facey, DE 1997, *The Diversity of Fishes*, Blackwell Science, Malden, Massachusetts.

Herbert, B & Peeters, J 1995, *Freshwater Fishes of Far North Queensland*, Queensland Department of Primary Industries, Brisbane.

Herring, PJ & Clarke, MR 1971, *Deep Oceans*, Arthur Barker Ltd, London.

Hoese, DF, Bray, DJ, Allen GR, Paxton, JR, Wells, A & Beesley, PL 2006, *Zoological Catalogue of Australia*, vol. 35, parts 1–3, CSIRO Publishing & Australian Biological Resources Study.

Humann, P 1996, *Reef Fish Identification: Florida, Caribbean, Bahamas*, Paramount Miller Graphics Inc., Florida.

Hungerford, R 1972, *Anglers Omnibus*, Pollard Publishing, Wollstonecraft.

Hutchings, P & Saenger, P 1987, *Ecology of Mangroves*, University of Queensland Press.

Hutchins, JB 1979, *The Fishes of Rottnest Island*, Creative Research, Perth.

——1980, *Dangerous Fishes of Western Australia*, Western Australian Museum.

——1991, 'Dispersal of tropical fishes to temperate seas in the southern hemisphere', *Journal of the Royal Society of Western Australia*, vol. 74.

——& Swainston, R 1986, *Sea Fishes of Southern Australia*, Swainston Publishing, Perth.

Imberger, J (ed.) 1983, 'Physical oceanography in Australia', reprinted from the *Australian Journal of Marine and Freshwater Research*, vol. 34.

Jubb, RA 1967, *Freshwater Fishes of Southern Africa*, Gothic Printing Company, Cape Town.

Kailola, PJ, Williams, MJ, Stewart, PC, Reichelt, RE, McNee, A & Grieve C (eds) 1993, *Fisheries Resources*, Bureau of Resource Sciences, Department of Primary Industries and Energy & Fisheries Research and Development Corporation, Canberra.

Kluge, AG & Waterman, AJ, 1971, *Chordate Structure and Function*, Macmillan, New York.

Knox, GA 1994, *The Biology of the Southern Ocean*, Cambridge University Press.

Kuiter, RH 1993, *Coastal Fishes of South-Eastern Australia*, University of Hawaii Press, Honolulu.

——1996, *Guide to the Sea Fishes of Australia: A Comprehensive Reference for Divers and Fishermen*, New Holland Publishers, Sydney.

—— 2002, *Fairy and Rainbow Wrasses and their Relatives*, TMC Publishing, Chorleywood.

—— 2007, *World Atlas of Marine Fishes*, IKAN Unterwasserarchiv, Frankfurt.

—— & Debelius, H 2001, *Surgeonfishes, Rabbitfishes and their Relatives: A Comprehensive Guide to Acanthuroidei*, TMC Publishing, Chorleywood.

Kyushin, K, Amaoka, K, Nakaya, K & Ida, S 1977, *Fishes of Indian Ocean*, Hiroshige Ehara, Tokyo.

Larcombe, J & Begg, G (eds) 2007, *Fishery Status Reports, Status of Fish Stocks Managed by the Australian Government*, Bureau of Rural Sciences.

Last, PR & Stevens, JD 1994, *Sharks and Rays of Australia*, CSIRO Publishing, Collingwood, Victoria.

—— 2009, *Sharks and Rays of Australia*, 2nd edn, CSIRO Publishing, Collingwood, Victoria.

—— Scott, EOG & Talbot, FH 1983, *Fishes of Tasmania*, Tasmanian Fisheries Development Authority, Hobart.

Levine, JS 1985, *Undersea Life*, Stewart, Tabori and Chang, New York.

——1993, *The Coral Reef at Night*, Harry N. Abrams Inc., New York.

Lloris, D, Matallanas, J & Oliver, P 2005, *Hakes of the World*, FAO Species Catalogue for Fisheries Purposes No. 2, Food and Agriculture Organisation of the United Nations, Rome.

Long, AJ 1995, *The Rise of Fishes*, University of New South Wales Press.

Louisy, P 2002, *Poissons Marins, Europe et Méditerranée*, Ulmer, Paris.

Lythgoe, J & Lythgoe, G 1991, *Fishes of the Sea*, Blandford Press, London.

Maitre-Allain, T & Louisy, P 1990, *Poissons de Mer*, Arthaud, Paris.

Marshall, AJ & Williams, WD 1972, *Textbook of Zoology: Invertebrates*, The Macmillan Press, London.

Marshall, TC 1982, *Tropical Fishes of the Great Barrier Reef*, Angus and Robertson Publishers, Sydney.

May, JL, Garrey, J & Maxwell, H 1986, *Field Guide to Trawl Fish from Temperate Waters of Australia*, CSIRO Division of Fisheries Research, Hobart.

Masuda, H 1984, *Field Guide to Sea Fishes*, Tokai University Press.

—— 1987, *Sea Fishes of the World*, Yama-Kei Publishers, Tokyo.

—— Amaoka, K, Araga, C, Uyeno, T & Yoshino, T (eds) 1984, *The Fishes of the Japanese Archipelago*, Tokai University Press.

Matthews, D 1996, *Sharks: The Mysterious Killers*, Discovery Channel Books, Avenel, New Jersey.

McDiarmid, M 1996, *Shark Attack*, Parragon Books Ltd, Bath.

McDowell, R 1996, *Freshwater Fishes of South-Eastern Australia*, Reed Books, Kew, Victoria.

McKay, RJ 1992, *Sillaginid Fishes of the World*, FAO Species Catalogue, vol. 14, Food and Agriculture Organisation of the United Nations, Rome.

Moreland, JJ 1967, *Marine Fishes of New Zealand*, A.H. and A.W. Reed, Wellington.

Morgan, GJ & Wells, FE 1991, 'Zoogeographic provinces of the Humboldt, Benguela and Leewuin Current systems', *Journal of the Royal Society of Western Australia*, vol. 74.

Motomura, H 2004, *Threadfins of the World*, FAO Species Catalogue for Fisheries Purposes No. 3, Food and Agriculture Organisation of the United Nations, Rome.

Moyle, PB & Cech, JJ 1982, *Fishes: An Introduction to Ichthyology*, Prentice-Hall Inc., Englewood Cliffs, New Jersey.

Munro, ISR 1967, *The Fishes of New Guinea*, Department of Agriculture, Stock and Fisheries, Port Moresby.

Muus, BJ, Nielson, JG, Dahlstrom, P, Olesen Nystrom, B 1998, *Guide des Poissons de Mer et Pêche*, Delachaux et Niestle, Paris.

Myers, RF 1989, *Micronesian Reef Fishes*, Coral Graphics, Barrigada, Guam.

Nakamura, I 1986, *Important Fishes Trawled off Patagonia*, Japan Marine Fishery Resource Research Center, Toyko.

Nakamura, I 1985, *Billfishes of the World*, FAO Species Catalogue, vol. 5, Food and Agriculture Organisation of the United Nations, Rome.

—— & Parin, NV 1993, *Snake Mackerels and Cutlassfishes of the World*, FAO Species Catalogue, vol. 15, Food and Agriculture Organisation of the United Nations, Rome.

Neira, FJ, Miskiewicz, AG & Trnski, T 1998, *Larvae of Temperate Australian Fishes*, University of Western Australia Press.

Nelson, JS 2006, *Fishes of the World*, 4th edn, John Wiley and Sons Inc., New York.

Newton, G & Boshier, J (eds) 2001, *Coasts and Oceans*, Australian State of the Environment Committee & CSIRO Publishing, Collingwood, Victoria.

Nouvian, C 2007, *The Deep: Extraordinary Creatures of the Abyss*, University of Chicago Press, London.

Paulin, C 1998, *Common New Zealand Marine Fishes*, Canterbury University Press.

—— Stewart, A, Roberts, C & McMillian, P 2001, *New Zealand Fishes: A Complete Guide*, Te Papa Press, Wellington.

Phillips, R, & Rix, M 1985, *A Guide to the Freshwater Fish of Britain, Ireland and Europe*, Pan Books, London.

Pietsch, TW 2009, *Oceanic Anglerfishes: Extraordinary Diversity in the Deep Sea*, University of California Press.

Pownall, P 1977, *Commercial Fish of Australia*, Australian Government Publishing Service, Canberra.

Quero, JC, Porche, P, & Vayne, JJ 2003, *Guide des Poissons de l'Atlantique Europeen*, Delachaux et Niestle, Switzerland.

Randall, DJ & Farrell, AP 1997, *Deep-Sea Fishes*, Academic Press, San Diego.

Randall, JE 1992, *Red Sea Reef Fishes*, Immel Publishing, London.

——Allen, GR & Steene, RC 1990, *Fishes of the Great Barrier Reef and Coral Sea*, University of Hawaii Press, Honolulu.

Ritz, D, Swadling, K, Hosie, G & Cazassus, F 2003, *Guide to the Zooplankton of South-Eastern Australia*, Fauna of Tasmania Handbook No. 10, University of Tasmania.

Russell, BC 1990, *Nemipterid Fishes of the World*, FAO Species Catalogue, vol. 12, Food and Agriculture Organisation of the United Nations, Rome.

Sainsbury, KJ, Kailola, PJ & Leyland, GG 1985, *Continental Shelf Fishes of Northern and North-Western Australia*, CSIRO Division of Fisheries Research, Hobart.

Sakurai, A, Sakamoto, Y, & Mori, F 1993, *Aquarium Fish of the World*, Chronicle Books, San Francisco.

Sammon, R 1996, *Rhythm of the Reef: A Day In The Life Of The Coral Reef*, Swan Hill Press, Shrewsbury.

Scott, TD, Glover, CJM & Southcott, RV 1980, *The Marine and Freshwater Fishes of South Australia*, D.J. Woolman, Adelaide.

Serventy, V & Raymond, R 1980, *Lakes and Rivers of Australia*, Summit Books, Sydney.

Shepherd, SA & Thomas, IM (eds) 1982, *Marine Invertebrates of Southern Australia: Part 1*, South Australian Museum.

Smith, MM & Heemstra, PC 1986, *Smiths' Sea Fishes*, Macmillan, Johannesburg.

Spoczynska, JOI & Spoczynski, M (ill.) 1976, *An Age of Fishes*, David and Charles, London.

Stafford-Deitsch, J 1987, *Shark: A Photographer's Story*, Headline Book Publishing, London.

——1989, *L'Homme-Requin*, Headline Book Publishing, London.

Steel, R & Harvey, AP 1979, *The Encyclopaedia of Prehistoric Life*, Mitchell Beazley Publishers, London.

Steene, RC 1978, *Butterfly and Angelfishes of the World*, A.H. and A.W. Reed, Wellington.

Stevens, JD 1987, *Sharks*, Golden Press, Sydney.

Taylor, G 1994, *Whale Sharks*, Angus and Robertson Publishers, Sydney.

Taylor, L 2000, *Sharks and other Sea Creatures*, Reader's Digest, New York.

Taylor, R & Taylor, V 1998, *Sharks: Silent Hunters of the Deep*, Reader's Digest, Sydney.

Tee-Van, J (ed.) 1948, *Fishes of Western North Atlantic*, vol.1–5, Sears Foundation for Marine Research, Yale University.

Thain, M & Hickman, M 2004, *Penguin Dictionary of Biology*, 11th edn, Penguin Books, London.

Thomas, S 2002, *The Encounter, 1802: Art of the Flinders and Baudin Voyages*, Art Gallery of South Australia.

Thomson, JM 1978, *A Field Guide to the Common Sea and Estuary Fishes of Non-Tropical Australia*, William Collins Publishers, Sydney.

Thresher, RE,1980, *Reef Fish*, The Palmetto Publishing Company, St Petersburg, Florida.

Tricas, TC, Deacon, K, Last, P, McCosker, JE, Walker, TI & Taylor, L 1997, *Sharks and Rays*, Time-Life Book Series, Weldon Owen, San Francisco.

Van der Elst, R 1992, *Everyone's Guide to Sea Fishes of Southern Africa*, Struik Publishers, Cape Town.

Vivien, MH & Harmelin, JG 1991, *Guide des Poissons de la Mediterranee*, Delachaux et Niestle, Neuchâtel.

Wagner, R & Unmack, PJ 2000, *Fishes of the Lake Eyre Catchment of Central Australia*, Queensland Department of Primary Industries, Brisbane.

Wallace, J 1995, *Coral Reefs: Exploring the World Below*, Friedman/Fairfax Publishers, New York.

Watson, W & Walker jr, HJ 2004, *The World's Smallest Vertebrate,* Schindleria brevipinguis: *A New Paedomorphic Species in the Family Schindleriidae (Perciformes: Gobioidei)*, Records of the Australian Museum, vol. 56, pp. 139–142.

Wells S & Hanna, N 1992, *The Greenpeace Book of Coral Reefs*, Sterling Publishing Co. Inc., New York.

Whitehead, PJP 1985, *Clupeoid Fishes of the World*, FAO Species Catalogue, vol. 7, Parts 1 and 2, Food and Agriculture Organisation of the United Nations, Rome.

Whitley, GP 1980, *G.P. Whitley's Handbook of Australian Fishes*, Jack Pollard Publishing, North Sydney.

——1983, *Australian Sharks*, Lloyd O'Neil, Melbourne.

Wu, N 1998, *Splendors of the Seas*, Könemann Verlagsgesellschaft, Cologne.

Yearsley, GK, Last, PR & Hoese, DF (eds) 2006, *Standard Names of Australian Fishes*, paper 9, CSIRO Marine and Atmospheric Research, Hobart.

Zeitschel, B & Gerlach, SA (eds) 1973, *The Biology of the Indian Ocean*, Chapman and Hall, London.

Online References

Aidan Martin, R, *Biology of Sharks and Rays*, ReefQuest Centre for Shark Research, <www.elasmo-research.org/index.html>.

Australian Museum, *Fishes*, <http://australianmuseum.net.au/Fishes>.

Beeton, RJS, Buckley, KI, Jones, GJ, Morgan, D, Reichelt, RE & Trewin, D 2006, *Australia State of the Environment 2006: Independent Report to the Australian Government Minister for the Environment and Heritage*, <www.environment.gov.au/soe/2006/publications/report/transmittal.html>.

Department of the Environment, Water, Heritage and the Arts, 2009, *Australian Faunal Directory*, Australian Biological Resources Study, Canberra, <www.environment.gov.au/biodiversity/abrs/online-resources/fauna/afd/index.html>.

Eschmeyer, WN & Fong, JD 2008, *Species of Fishes by Family/Subfamily*, California Academy of Sciences, <research.calacademy.org/redirect?url=http://researcharchive.calacademy.org/research/Ichthyology/catalog/fishcatmain.asp>.

Froese, R & Pauly, D (eds) 2009, *FishBase*, <www.fishbase.org>.

Kochi University, *Catalogue of Fishes of Kochi Prefecture*, <www.kochi-u.ac.jp/w3museum/Fish_Labo/FishCatalog/Orders.html>.

Palaeos, *The Vertebrates: Glossary*, <www.palaeos.com/Vertebrates/Lists/Glossary/Glossary.html>.

Shao, KT 2005, *The Fish Database of Taiwan*, Biodiversity Research Center, Academia Sinica, Taiwan, <fishdb.sinica.edu.tw>.

Unmack, PJ, *Australian Desert Fishes Descriptions*, <www.desertfishes.org/australia/fish/index.html>.

ACKNOWLEDGEMENTS

Over the years people from all walks of life have shared their knowledge and enthusiasm for fishes and brought me material to work from. Many fishers, both professional and amateur, have kept unusual fish aside for me and allowed me to fossick through their catches, taking photos and specimens – they are far too numerous to mention individually but I thank them all and hope to meet again. Thanks to Pedro for many years of boat trips, fish and fish talk.

Special thanks to Dr Gerry Allen, formerly of the Western Australian Museum, for his encouragement, generosity and assistance in the preparation of numerous illustrations of tropical species. Many of these appeared in the museum's publication *The Marine Fishes of North-western Australia* and a variety of other guide books. Special thanks also to Dr Barry Hutchins, formerly of the WA Museum, for sharing his vast knowledge of Australia's temperate fishes and providing material for illustrations that appeared in our self-published book *Sea Fishes of Southern Australia*. Both these eminent scientists have been major contributors to my ongoing education on the fishes of Australia and have provided me with some fantastic opportunities to see fishes in their natural habitat.

Dr Peter Last and Dr John Stevens, both of the CSIRO, provided source material and expertise for many paintings of sharks and rays, which appeared in the definitive CSIRO publication *Sharks and Rays of Australia*, a superb volume that covers in great detail every shark and ray known from Australian waters.

Around the world there is a small army of dedicated scientists such as these, working to understand the taxonomy, distribution and biology of fishes. Some spend many years working on a few poorly known species, others dedicate a lifetime's work to studying a single order. Still others undertake the complex synthesis of the vast amount of information thus generated, to present overall guides to major groups of fishes, or to the fauna of certain regions. The information in this book is drawn from such works, as well as communications from fisheries biologists and commercial fishing operators, and a variety of other sources. Though it is impossible to thank them individually, all of these people have contributed in some way, no matter how small or large, to the preparation of this book.

Special thanks to Jo Turner from Penguin Books for her enthusiasm and encouragement from concept to publication, to Jess Redman for her painstaking and rigorous editing, and to Claire Tice for the superb layout work. Thanks also to Jolly Read and Prof. Jessica Meeuwig for comments and suggestions on the introductory text.

Much of the scanning and preparation of illustrations for publication was carried out by Catherine Swainston, and the bibliography was prepared by Zoe Swainston.

To my parents Tony and Bella Swainston: my thanks for your tremendous support over many years. To my children Matt, Ben and Zoe: thanks for putting up with my numerous long absences in the field, and for countless cups of tea delivered to me while working on this book.

INDEX

Page numbers in bold indicate illustrations. Species that are not illustrated or mentioned in the family text (i.e. only appear in a species list) are not included in the index.

Index

D

E

G

Index

N

U

X

Y

Z

VIKING

UK | USA | Canada | Ireland | Australia
India | New Zealand | South Africa | China

Viking is part of the Penguin Random House group of companies
whose addresses can be found at global.penguinrandomhouse.com.

First published by Viking, an imprint of Penguin Random House Australia Pty Ltd 2010
This edition published by Viking 2020

Maps by Damien Demaj and Ed Merritt (concepts by Roger Swainston)

Cover and text design by Claire Tice © Penguin Random House Australia Pty Ltd
Typeset in Stempel Schneidler and Gotham Narrow by Post Pre-Press Group, Brisbane, Australia
Scanning by Anima, Western Australia
Printed and bound in China

ISBN 978 1 76104 054 2

penguin.com.au